TODAY'S TECHNICIAN™

CLASSROOM MANUAL

For Basic Automotive Service & Systems

SIXTH EDITION

Chris Hadfield

John Witthauer

Cengage

Australia • Brazil • Canada • Mexico • Singapore • United Kingdom • United States

Cengage

Today's Technician: Classroom Manual for Basic Automotive Service & Systems,
Sixth Edition
Chris Hadfield
John Witthauer

SVP, Higher Education & Skills Product: Erin Joyner

Product Director: Matthew Seeley

Senior Product Manager: Katie McGuire

Product Assistant: Kimberly Klotz

Director, Learning Design: Rebecca von Gillern

Senior Manager, Learning Design: Leigh Hefferon

Learning Designer: Mary Clyne

Marketing Director: Sean Chamberland

Marketing Manager: Andrew Ouimet

Director, Content Creation: Juliet Steiner

Content Creation Manager: Alexis Ferraro

Senior Content Manager: Meaghan Tomaso

Digital Delivery Lead: Amanda Ryan

Art Designer: Erin Griffin

Text Designer: Angela Sheehan

Cover Designer: Angela Sheehan

Cover image(s): Umberto Shtanzman/Shutterstock.com

Library of Congress Control Number: 2020901507

Book Only ISBN: 978-1-337-79568-5
Package ISBN: 978-1-337-79566-1

Cengage
5191 Natorp Boulevard
Mason, OH 45040
USA

Notice to the Reader

Publisher does not warrant or guarantee any of the products described herein or perform any independent analysis in connection with any of the product information contained herein. Publisher does not assume, and expressly disclaims, any obligation to obtain and include information other than that provided to it by the manufacturer. The reader is expressly warned to consider and adopt all safety precautions that might be indicated by the activities described herein and to avoid all potential hazards. By following the instructions contained herein, the reader willingly assumes all risks in connection with such instructions. The publisher makes no representations or warranties of any kind, including but not limited to, the warranties of fitness for particular purpose or merchantability, nor are any such representations implied with respect to the material set forth herein, and the publisher takes no responsibility with respect to such material. The publisher shall not be liable for any special, consequential, or exemplary damages resulting, in whole or part, from the readers' use of, or reliance upon, this material.

Printed at CLDPC, USA, 06-24

TODAY'S TECHNICIAN™

CLASSROOM MANUAL

For Basic Automotive Service & Systems

SIXTH EDITION

CONTENTS

PREFACE

Thanks to the support the *Today's Technician* series has received from those who teach automotive technology, Cengage, the leader in automotive-related textbooks, is able to live up to its promise to regularly provide new editions of texts of this series. We have listened and responded to our critics and our fans and present this new updated and revised sixth edition. By revising this series on a regular basis, we can respond to changes in the industry, changes in technology, changes in the accreditation process, and to the ever-changing needs of those who teach automotive technology.

The *Today's Technician* series features textbooks that cover all mechanical and electrical systems of automobiles and light trucks. Principally, the individual titles correspond to the areas for National Institute for Automotive Service Excellence (ASE) certification.

All titles in the *Today's Technician* series include remedial skills and theories common to all of the certification areas and advanced or specific subject areas that reflect the latest technological trends.

Today's Technician: Basic Automotive Service & System, 6e is designed to give students a chance to develop the same skills and gain the same knowledge that today's successful technician has. This edition also reflects the changes in the guidelines established by the ASE Education Foundation in 2017.

The purpose of the ASE Education Foundation is to evaluate technician training programs against standards developed by the automotive industry and recommend qualifying programs for accreditation by ASE. Programs can earn ASE accreditation upon the recommendation of the ASE Education Foundation. ASE Education Foundation's national standards reflect the skills that students must master. ASE Education Foundation accreditation ensures that the training programs meet or exceed industry-recognized, uniform standards of excellence.

The technician of today and for the future must know the underlying theory of all automotive systems and be able to service and maintain those systems. Dividing the material into two volumes, a Classroom Manual and a Shop Manual, provides the reader with the information needed to begin a successful career as an automotive technician without interrupting the learning process by mixing cognitive and performance learning objectives into one volume.

The design of Cengage Learning's *Today's Technician* series was based on features known to promote improved student learning. The design was further enhanced by a careful study of survey results, in which the respondents were asked to value particular features. Some of these features can be found in other textbooks, while others are unique to this series.

Each Classroom Manual contains the principles of operation for each system and subsystem. The Classroom Manual also discusses design variations in key components used by the different vehicle manufacturers and considers emerging technologies that will be standard or optional features in the near future. This volume is organized to build upon basic facts and theories. Its primary objective is to help the reader gain an understanding of how each system and subsystem operates. This understanding is necessary to diagnose the complex automobiles of today and tomorrow. Although the basics contained in the Classroom Manual provide the knowledge needed for diagnostics, diagnostic procedures appear only in the Shop Manual. An understanding of the underlying theories is also a requirement for competence in the skill areas covered in the Shop Manual.

A spiral–bound Shop Manual delivers hands-on learning experiences with step-by-step instructions for diagnostic and repair procedures. Photo Sequences are used to illustrate some of the common service procedures. Other common procedures are listed and accompanied with fine-line drawings and photos that let the reader visualize and conceptualize the finest details of the procedure. This volume explains the reasons for performing the procedures, as well as the circumstances when each particular service is appropriate.

The two volumes are designed to be used together and are arranged in corresponding chapters. Not only are the chapters in the volumes linked together, but the contents of the chapters are also linked. The linked content is indicated by marginal callouts that refer the reader to the chapter and page where the same topic is addressed in the companion volume. This valuable feature saves users the time and trouble of searching the index or table of contents to locate supporting information in the other volume. Instructors will find this feature especially helpful when planning the presentation of material and when making reading assignments.

Both volumes contain clear and thoughtfully selected illustrations, many of which are original drawings or photos specially prepared for inclusion in this series. This means that the art is a vital part of each textbook and not merely inserted to increase the number of illustrations.

The page design of this series uses available margin space to deliver helpful information efficiently without interrupting the pedagogical lesson material. This information includes examples of concepts just introduced in the text, explanations or definitions of terms that are not defined in the text, examples of common trade jargon used to describe a part or operation, and unique applications of the system or service described in the text. Many textbooks also include this information but insert it in the main body of text; this tends to interrupt the reader's thought process. By placing this information to the side of the main text, students can read through the text uninterrupted and refer to the additional information when it is best for them.

Jack Erjavec
Series Editor

HIGHLIGHTS OF THIS EDITION—CLASSROOM MANUAL

The new sixth edition has been fully updated to align with the new ASE Education Foundation accreditation model introduced in 2017.

This edition's full-color design with new photographs and updated line art helps readers better visualize processes and procedures. All chapters have been updated and arranged for easier learning progression. This edition is designed to build skills and knowledge as the student advances through the chapters. The entire text was updated with information about new tools, equipment, and vehicle systems in use today and expected in the near future. Coverage of hybrid vehicles and emerging technologies, along with revised and updated coverage of electronic engine controls, prepares students for today's workplace.

What's New to This Edition

- All chapters have been updated and arranged for easier learning progression. This edition is designed to build skills and knowledge as the student advances through the chapters.
- New and updated photos to show modern industry.

- New Chapter 1, "Introduction to the Automotive Industry." This chapter highlights the different types of businesses, careers, training, and ethics needed to operate a successful operation.
- New Chapter 2, "Introduction to the automobile." This chapter explains the complex systems and the purpose for today's automobiles.
- New Chapter 4, "Fasteners and measurements." This chapter combines the 5th edition Chapters 4 and 5 to consolidate and prioritize basic fasteners and measurement tools for the entry-level technician.
- New Chapter 5, "Electrical and electronic systems." This chapter corresponds with the ASE Maintenance and Light repair requirements needed for the importance of basic electrical fundamentals.
- New Chapter 6, "Accessory Lighting and Electrical Systems." This chapter provides a deeper dive onto electrical systems.
- New Chapter 7, "Engine Systems." This chapter provides a better and basic understanding for engine operation.
- New Chapter 11, "Heating and Air Conditioning Systems." This chapter focuses on operation and theory of the HVAC systems working together.
- New Chapter 12, "Engine Performance Theories." This chapter is designed to align with ASE and MLR required material.

HIGHLIGHTS OF THIS EDITION—SHOP MANUAL

Along with the Classroom Manual, the Shop Manual was revised and updated in both text and figures to emphasize the changing technology in tools, equipment, and procedures. The chapters parallel with and complement the Classroom Manual's layout and closely match a typical automotive school's sequence of instruction. The sixth edition includes even more all-new Job Sheets focused on skill building.

New content cross-references current industry practices to help the student understand and practice real-world workplace competencies, and employer-focused soft skills are integrated into each topic. The revised edition features a new chatper on Safety Inspection and Preventive Maintenance which focuses on the importance on complete safety inspections and proper documentation.

What's New to This Edition

- All chapters have been updated and arranged for easier learning progression. This edition is designed to build skills and knowledge as the student advances through the chapters.
- All chapters include new task sheets that include "Required Supplemental Tasks" (RST) that align with the ASE standards required for Entry-Level Technician Certification.
- Task sheets also have new grading rubric with explanation of the rubric terms.
- Task sheets also added "instructor check points" these are designed to stop the students and let the instructor quiz and or inspect the progress of each student or groups of students.
- All chapters align and support the Classroom manual and all changes are followed in the Shop manuals.
- New Chapter 12, Safety Inspection and Preventive maintenance." This chapter focuses on the importance of a complete safety inspection to provide the customer an overview of the current condition of their vehicle.
- This chapter also stresses the importance of proper documentation of the safety inspections finding and suggesting options for preventive maintenance.

CLASSROOM MANUAL

Features of the Classroom Manual include the following:

Cognitive Objectives

These objectives outline the chapter's contents and identify what students should know and be able to do upon completion of the chapter. Each topic is divided into small units to promote easier understanding and learning.

Terms to Know List

A list of key terms appears immediately after the Objectives. Students will see these terms discussed in the chapter. Definitions can also be found in the Glossary at the end of the manual.

Cross-References to the Shop Manual

References to the appropriate page in the Shop Manual appear whenever necessary. Although the chapters of the two manuals are synchronized, material covered in other chapters of the Shop Manual may be fundamental to the topic discussed in the Classroom Manual.

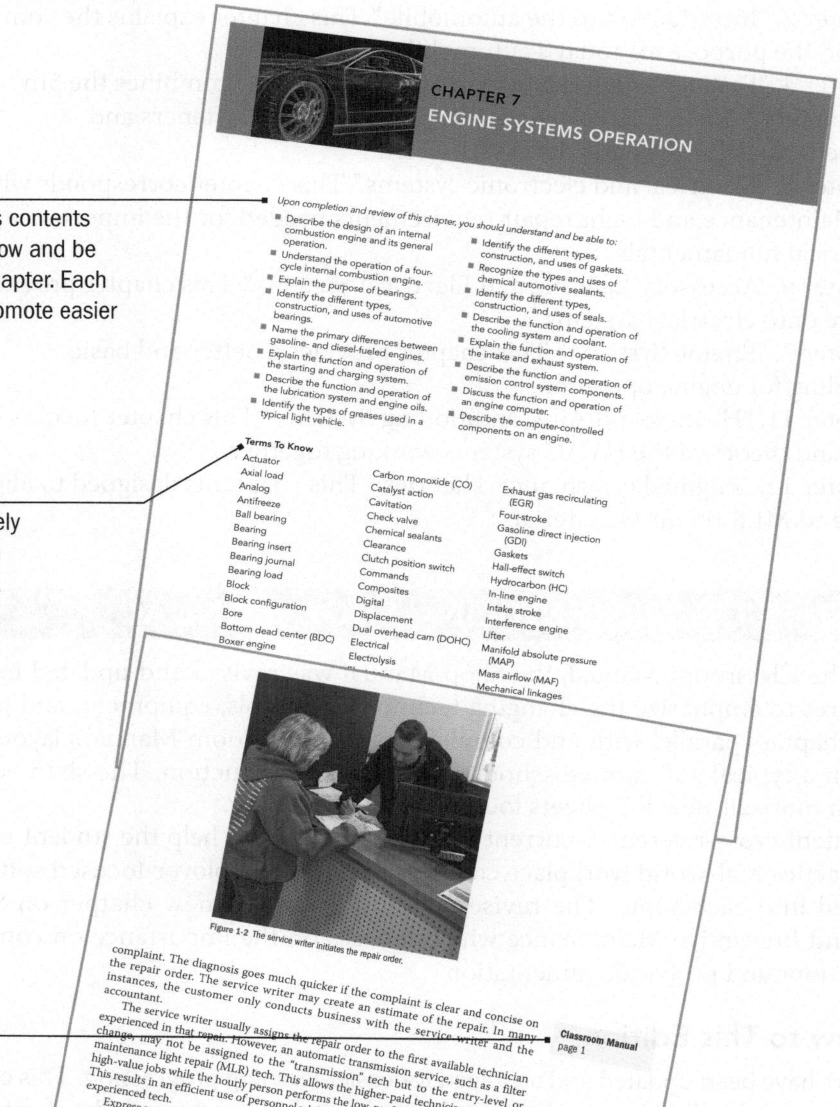

CHAPTER 7
ENGINE SYSTEMS OPERATION

Upon completion and review of this chapter, you should understand and be able to:

- Describe the design of an internal combustion engine and its general operation.
- Understand the operation of a four-cycle internal combustion engine.
- Explain the purpose of bearings.
- Identify the different types, construction, and uses of automotive bearings.
- Name the primary differences between gasoline- and diesel-fueled engines.
- Explain the function and operation of the starting and charging system.
- Describe the function and operation of the lubrication system and engine oils.
- Identify the types of greases used in a typical light vehicle.
- Identify the different types, construction, and uses of gaskets.
- Recognize the types and uses of chemical automotive sealants.
- Identify the different types, construction, and uses of seals.
- Describe the function and operation of the cooling system and coolant.
- Explain the function and operation of the intake and exhaust system.
- Describe the function and operation of emission control system components.
- Discuss the function and operation of an engine computer.
- Describe the computer-controlled components on an engine.

Terms To Know

Actuator
Axial load
Analog
Antifreeze
Ball bearing
Bearing
Bearing insert
Bearing journal
Bearing load
Block
Block configuration
Bore
Bottom dead center (BDC)
Boxer engine
Carbon monoxide (CO)
Catalyst action
Cavitation
Check valve
Chemical sealants
Clearance
Clutch position switch
Commands
Composites
Digital
Displacement
Dual overhead cam (DOHC)
Electrical
Electrolysis
Exhaust gas recirculating (EGR)
Four-stroke
Gasoline direct injection (GDI)
Gaskets
Hall-effect switch
Hydrocarbon (HC)
In-line engine
Intake stroke
Interference engine
Lifter
Manifold absolute pressure (MAP)
Mass airflow (MAF)
Mechanical linkages

Figure 1-2 The service writer initiates the repair order.

complaint. The diagnosis goes much quicker if the complaint is clear and concise on the repair order. The service writer may create an estimate of the repair. In many instances, the customer only conducts business with the service writer and the accountant.

The service writer usually assigns the repair order to the first available technician experienced in that repair. However, an automatic transmission service, such as a filter change, may not be assigned to the "transmission" tech but to the entry-level or maintenance light repair (MLR) tech. This allows the higher-paid technician to perform high-value jobs while the hourly person performs the low-profit, quick-turnaround jobs. This results in an efficient use of personnel while providing excellent training for the less experienced tech.

Express service is a term used to emphasize the speed of basic and regularly schedule maintenance. As mentioned in the Classroom Manual, express service gives the customer a quick turnaround and a detailed inspection of the vehicle. To speed things up more, some dealerships have service attendants. These are people who assist the service writer. The attendant has several duties. One is greeting the customer if the writer is busy. One protocol is to never let the customer stand around waiting to be noticed. A quick greeting and "How may we help you?" immediately upon the customer's arrival indicate that person is recognized and welcomed. A second duty consists of placing disposable floor mats and seat covers in the car and retrieving the odometer mileage and the **vehicle identification number (VIN)**. The third duty, but not necessarily the last, is to begin the vehicle inspection. Dealership service attendants check all lighting, tire wear and air pressure, parking brakes, horn, and warning lights, and perform a walk-around visual check for damage to the vehicle body. The attendant starts the inspection sheet and marks the result of this first phase. This is done while the service order is being prepared and will save time in the express service bay. In addition, any defects found can now be shown immediately to the customer by the writer. This express service job may become a bigger repair job for items such as tires or a lighting problem. Part of this third duty is to remove the ignition key from the customer's key ring and return all other keys to the customers. Explain, if questioned, that the shop can replace a lost ignition key easily, but

Classroom Manual
page 1

The vehicle identification number (VIN) is a number assigned to a specific vehicle. Information on the engine size and type, transmission, year of production, and other vehicle data is encoded in the VIN.

Margin Notes

The most important terms to know are highlighted and defined in the margin. Common trade jargon also appears in the margin and gives some of the common terms used for components. This helps students understand and speak the language of the trade, especially when conversing with an experienced technician.

A Bit of History

This feature gives the student a sense of the evolution of the automobile. This feature not only contains nice-to-know information but also should spark some interest in the subject matter.

Author's Note

This feature includes simple explanations, stories, or examples of complex topics. These are included to help students understand difficult concepts.

Summary

Each chapter concludes with summary statements that contain the important topics of the chapter. These are designed to help the reader review the contents.

Review Questions

Short-answer essay, fill in the blank, and multiple-choice questions follow each chapter. These questions are designed to accurately assess the student's competence in the stated objectives at the beginning of the chapter.

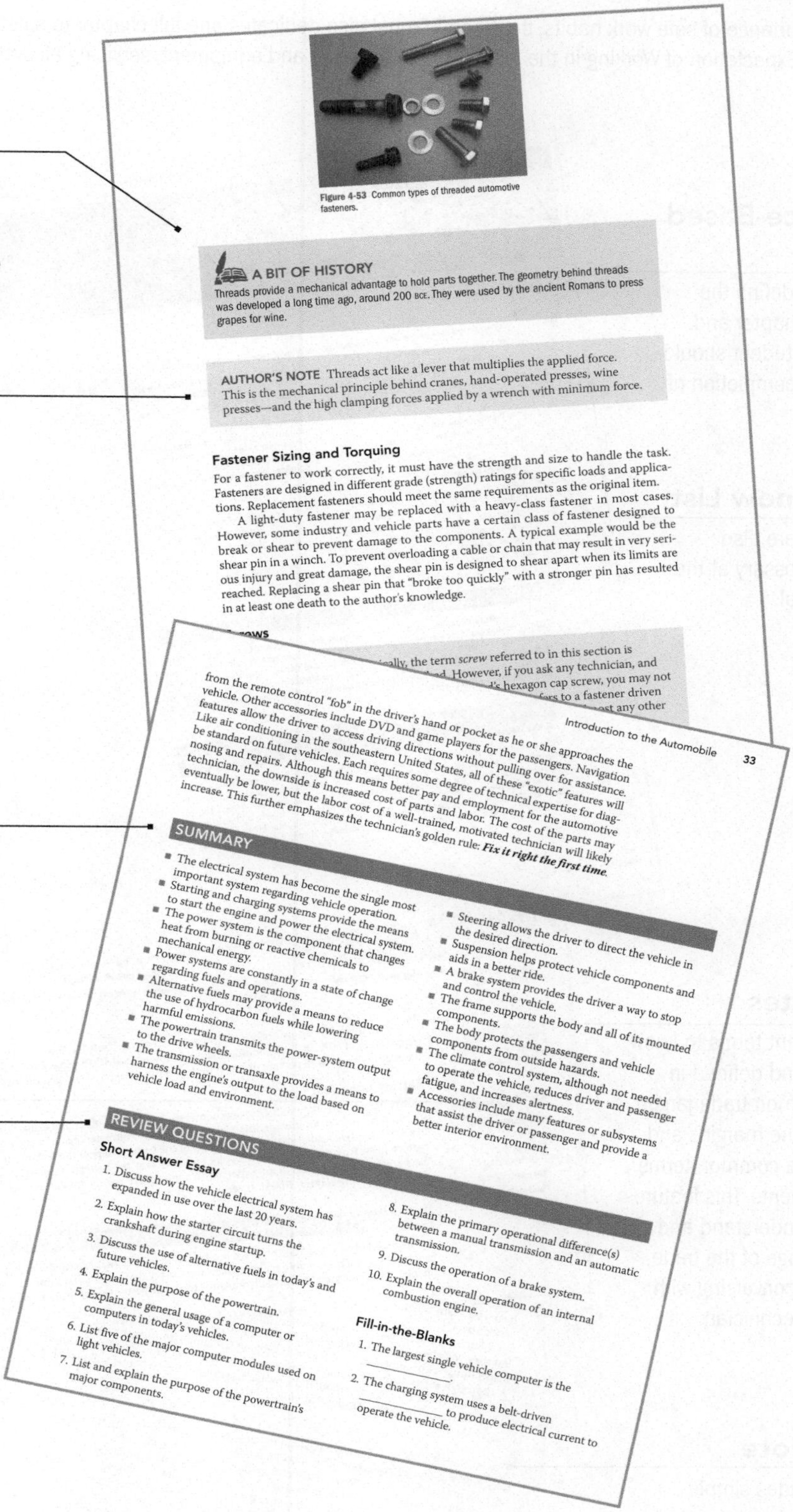

Figure 4-53 Common types of threaded automotive fasteners.

A BIT OF HISTORY

Threads provide a mechanical advantage to hold parts together. The geometry behind threads was developed a long time ago, around 200 BCE. They were used by the ancient Romans to press grapes for wine.

AUTHOR'S NOTE Threads act like a lever that multiplies the applied force. This is the mechanical principle behind cranes, hand-operated presses, wine presses—and the high clamping forces applied by a wrench with minimum force.

Fastener Sizing and Torquing

For a fastener to work correctly, it must have the strength and size to handle the task. Fasteners are designed in different grade (strength) ratings for specific loads and applications. Replacement fasteners should meet the same requirements as the original item.

A light-duty fastener may be replaced with a heavy-class fastener in most cases. However, some industry and vehicle parts have a certain class of fastener designed to break or shear to prevent damage to the components. A typical example would be the shear pin in a winch. To prevent overloading a cable or chain that may result in very serious injury and great damage, the shear pin is designed to shear apart when its limits are reached. Replacing a shear pin that "broke too quickly" with a stronger pin has resulted in at least one death to the author's knowledge.

Introduction to the Automobile 33

from the remote control "fob" in the driver's hand or pocket as he or she approaches the vehicle. Other accessories include DVD and game players for the passengers. Navigation features allow the driver to access driving directions without pulling over for assistance. Like air conditioning in the southeastern United States, all of these "exotic" features will be standard on future vehicles. Each requires some degree of technical expertise for diagnosing and repairs. Although this means better pay and employment for the automotive technician, the downside is increased cost of parts and labor. The cost of the parts may eventually be lower, but the labor cost of a well-trained, motivated technician will likely increase. This further emphasizes the technician's golden rule: ***Fix it right the first time.***

SUMMARY

- The electrical system has become the single most important system regarding vehicle operation.
- Starting and charging systems provide the means to start the engine and power the electrical system.
- The power system is the component that changes heat from burning or reactive chemicals to mechanical energy.
- Power systems are constantly in a state of change regarding fuels and operations.
- Alternative fuels may provide a means to reduce the use of hydrocarbon fuels while lowering harmful emissions.
- The powertrain transmits the power-system output to the drive wheels.
- The transmission or transaxle provides a means to harness the engine's output to the load based on vehicle load and environment.
- Steering allows the driver to direct the vehicle in the desired direction.
- Suspension helps protect vehicle components and aids in a better ride.
- A brake system provides the driver a way to stop and control the vehicle.
- The frame supports the body and all of its mounted components.
- The body protects the passengers and vehicle components from outside hazards.
- The climate control system, although not needed to operate the vehicle, reduces driver and passenger fatigue, and increases alertness.
- Accessories include many features or subsystems that assist the driver or passenger and provide a better interior environment.

REVIEW QUESTIONS

Short Answer Essay

1. Discuss how the vehicle electrical system has expanded in use over the last 20 years.
2. Explain how the starter circuit turns the crankshaft during engine startup.
3. Discuss the use of alternative fuels in today's and future vehicles.
4. Explain the purpose of the powertrain.
5. Explain the general usage of a computer or computers in today's vehicles.
6. List five of the major computer modules used on light vehicles.
7. List and explain the purpose of the powertrain's major components.
8. Explain the primary operational difference(s) between a manual transmission and an automatic transmission.
9. Discuss the operation of a brake system.
10. Explain the overall operation of an internal combustion engine.

Fill-in-the-Blanks

1. The largest single vehicle computer is the ______.
2. The charging system uses a belt-driven ______ to produce electrical current to operate the vehicle.

SHOP MANUAL

To stress the importance of safe work habits, the Shop Manual also dedicates one full chapter to safety. Other important features of this manual include: Expectation of Working in the industry, knowing tools and equipment, servicing all systems covered and vehicle inspections.

Performance-Based Objectives

These objectives define the contents of the chapter and define what the student should have learned on completion of the chapter.

Terms to Know List

Terms in this list are also defined in the Glossary at the end of the manual.

Margin Notes

The most important terms to know are highlighted and defined in the margin. Common trade jargon also appears in the margins and gives some of the common terms used for components. This feature helps students understand and speak the language of the trade, especially when conversing with an experienced technician.

Author's Note

This feature includes simple explanations, stories, or examples of complex topics. These are included to help students understand difficult concepts.

Basic Tools Lists

Each chapter begins with a list of the basic tools needed to perform the tasks included in the chapter.

Special Tools Lists

Whenever a special tool is required to complete a task, it is listed in the margin next to the procedure.

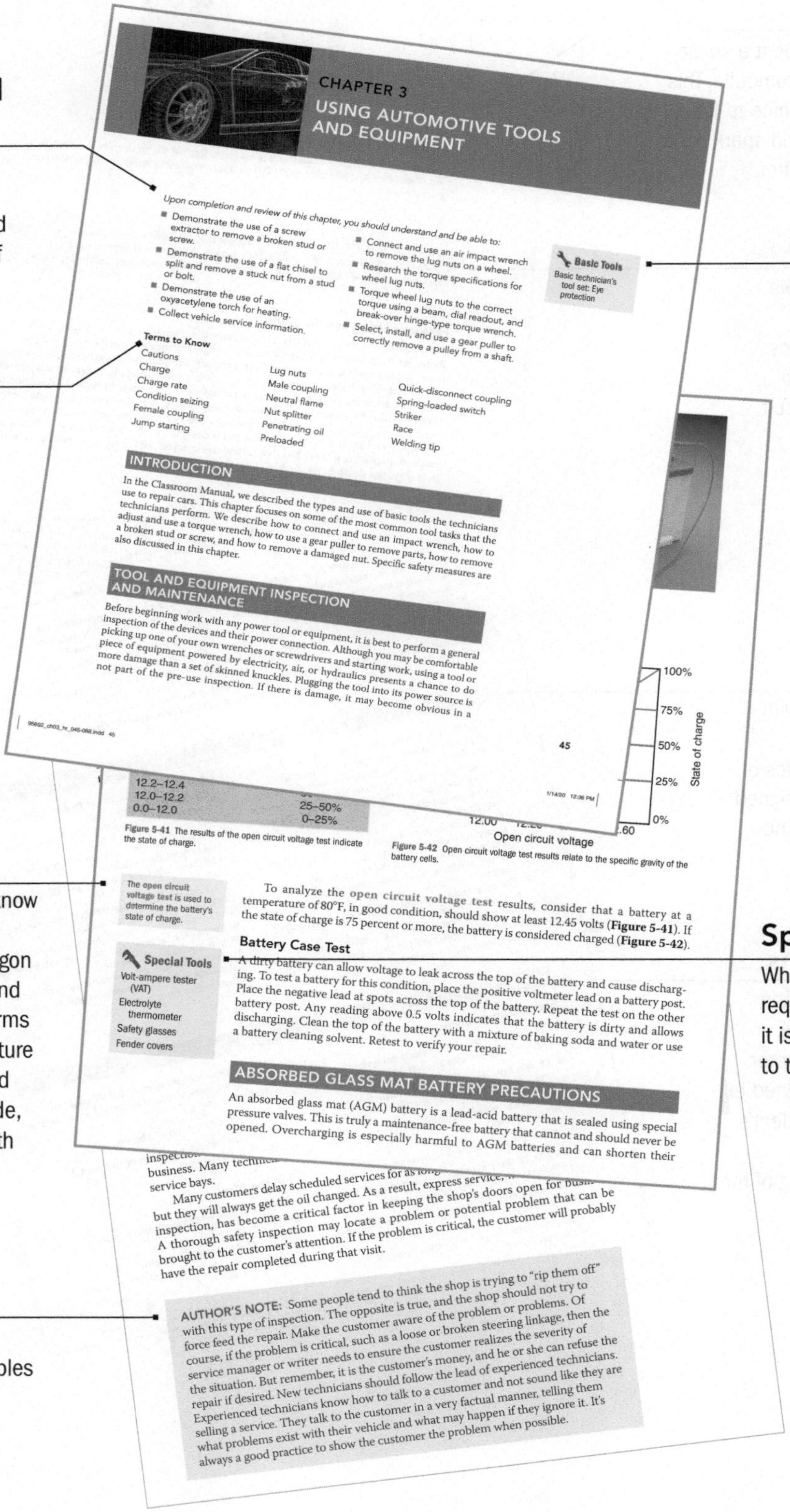

CHAPTER 3
USING AUTOMOTIVE TOOLS AND EQUIPMENT

Upon completion and review of this chapter, you should understand and be able to:

- Demonstrate the use of a screw extractor to remove a broken stud or screw.
- Demonstrate the use of a flat chisel to split and remove a stuck nut from a stud or bolt.
- Demonstrate the use of an oxyacetylene torch for heating.
- Collect vehicle service information.
- Connect and use an air impact wrench to remove the lug nuts on a wheel.
- Research the torque specifications for wheel lug nuts.
- Torque wheel lug nuts to the correct torque using a beam, dial readout, and break-over hinge-type torque wrench.
- Select, install, and use a gear puller to correctly remove a pulley from a shaft.

Basic Tools
Basic technician's tool set: Eye protection

Terms to Know

Cautions	Lug nuts	Quick-disconnect coupling
Charge	Male coupling	Spring-loaded switch
Charge rate	Neutral flame	Striker
Condition seizing	Nut splitter	Race
Female coupling	Penetrating oil	Welding tip
Jump starting	Preloaded	

INTRODUCTION

In the Classroom Manual, we described the types and use of basic tools the technicians use to repair cars. This chapter focuses on some of the most common tool tasks that the technicians perform. We describe how to connect and use an impact wrench, how to adjust and use a torque wrench, how to use a gear puller to remove parts, how to remove a broken stud or screw, and how to remove a damaged nut. Specific safety measures are also discussed in this chapter.

TOOL AND EQUIPMENT INSPECTION AND MAINTENANCE

Before beginning work with any power tool or equipment, it is best to perform a general inspection of the devices and their power connection. Although you may be comfortable picking up one of your own wrenches or screwdrivers and starting work, using a tool or piece of equipment powered by electricity, air, or hydraulics presents a chance to do more damage than a set of skinned knuckles. Plugging the tool into its power source is not part of the pre-use inspection. If there is damage, it may become obvious in a

45

12.2–12.4	
12.0–12.2	25–50%
0.0–12.0	0–25%

Figure 5-41 The results of the open circuit voltage test indicate the state of charge.

Figure 5-42 Open circuit voltage test results relate to the specific gravity of the battery cells.

The open circuit voltage test is used to determine the battery's state of charge.

To analyze the open circuit voltage test results, consider that a battery at a temperature of 80°F, in good condition, should show at least 12.45 volts (Figure 5-41). If the state of charge is 75 percent or more, the battery is considered charged (Figure 5-42).

Battery Case Test

Special Tools
Volt-ampere tester (VAT)
Electrolyte thermometer
Safety glasses
Fender covers

A dirty battery can allow voltage to leak across the top of the battery and cause discharging. To test a battery for this condition, place the positive voltmeter lead on a battery post. Place the negative lead at spots across the top of the battery. Repeat the test on the other battery post. Any reading above 0.5 volts indicates that the battery is dirty and allows discharging. Clean the top of the battery with a mixture of baking soda and water or use a battery cleaning solvent. Retest to verify your repair.

ABSORBED GLASS MAT BATTERY PRECAUTIONS

An absorbed glass mat (AGM) battery is a lead-acid battery that is sealed using special pressure valves. This is truly a maintenance-free battery that cannot and should never be opened. Overcharging is especially harmful to AGM batteries and can shorten their

Many customers delay scheduled services for as long... but they will always get the oil changed. As a result, express service, ... inspection, has become a critical factor in keeping the shop's doors open for business. A thorough safety inspection may locate a problem or potential problem that can be brought to the customer's attention. If the problem is critical, the customer will probably have the repair completed during that visit.

AUTHOR'S NOTE: Some people tend to think the shop is trying to "rip them off" with this type of inspection. The opposite is true, and the shop should not try to force feed the repair. Make the customer aware of the problem or problems. Of course, if the problem is critical, such as a loose or broken steering linkage, then the service manager or writer needs to ensure the customer realizes the severity of the situation. But remember, it is the customer's money, and he or she can refuse the repair if desired. New technicians should follow the lead of experienced technicians. Experienced technicians know how to talk to a customer and not sound like they are selling a service. They talk to the customer in a very factual manner, telling them what problems exist with their vehicle and what may happen if they ignore it. It's always a good practice to show the customer the problem when possible.

Photo Sequences

Many procedures are illustrated in detailed Photo Sequences. These photographs show the students what to expect when they perform particular procedures. They also familiarize students with a system or type of equipment that the school might not have.

References to the Classroom Manual

References to the appropriate page in the Classroom Manual appear whenever necessary. Although the chapters of the two manuals are synchronized, material covered in other chapters of the Classroom Manual may be fundamental to the topic discussed in the Shop Manual.

Service Tips

Whenever a shortcut or special procedure is appropriate, it is described in the text. Generally, these tips describe common procedures used by experienced technicians.

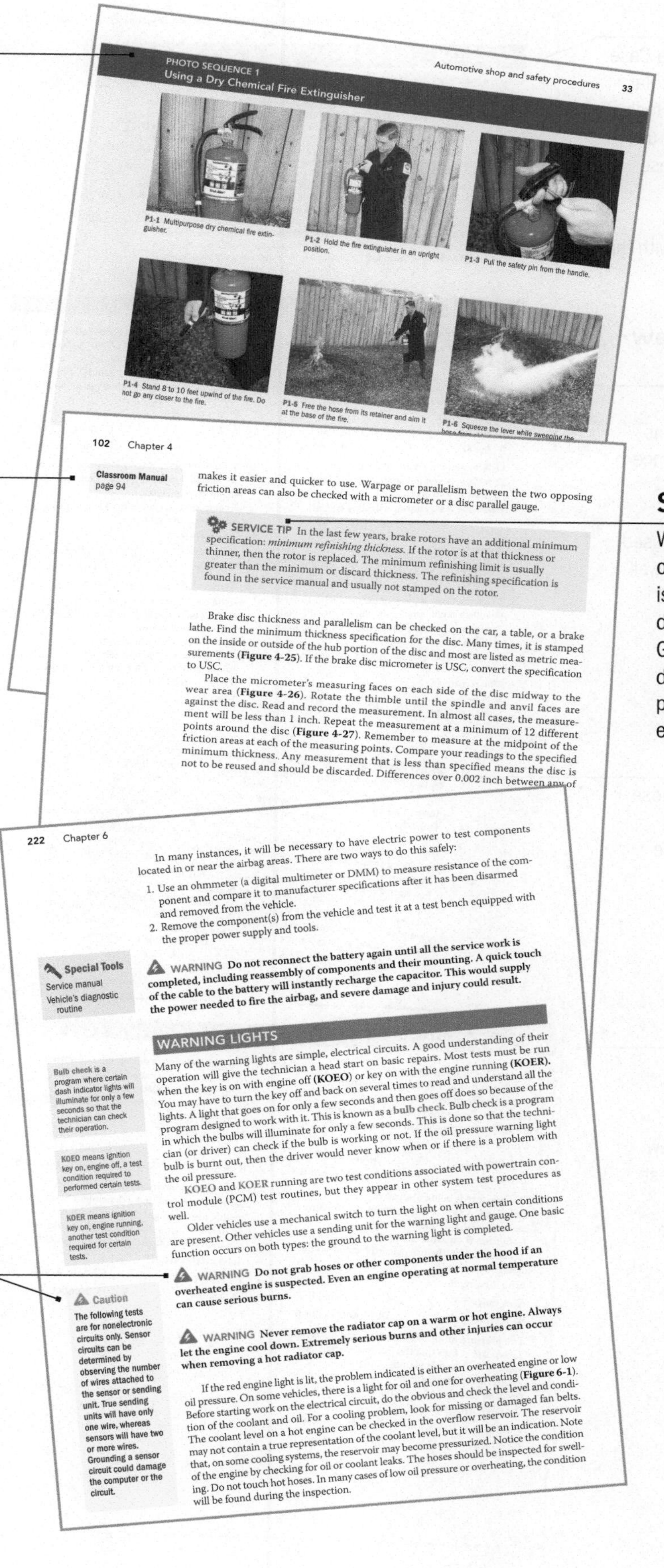
PHOTO SEQUENCE 1
Using a Dry Chemical Fire Extinguisher
Automotive shop and safety procedures 33
P1-1 Multipurpose dry chemical fire extinguisher.
P1-2 Hold the fire extinguisher in an upright position.
P1-3 Pull the safety pin from the handle.
P1-4 Stand 8 to 10 feet upwind of the fire. Do not go any closer to the fire.
P1-5 Free the hose from its retainer and aim it at the base of the fire.
P1-6 Squeeze the lever while sweeping the hose...

102 Chapter 4

Classroom Manual page 94

makes it easier and quicker to use. Warpage or parallelism between the two opposing friction areas can also be checked with a micrometer or a disc parallel gauge.

SERVICE TIP In the last few years, brake rotors have an additional minimum specification: *minimum refinishing thickness*. If the rotor is at that thickness or thinner, then the rotor is replaced. The minimum refinishing limit is usually greater than the minimum or discard thickness. The refinishing specification is found in the service manual and usually not stamped on the rotor.

Brake disc thickness and parallelism can be checked on the car, a table, or a brake lathe. Find the minimum thickness specification for the disc. Many times, it is stamped on the inside or outside of the hub portion of the disc and most are listed as metric measurements (**Figure 4-25**). If the brake disc micrometer is USC, convert the specification to USC.

Place the micrometer's measuring faces on each side of the disc midway to the wear area (**Figure 4-26**). Rotate the thimble until the spindle and anvil faces are against the disc. Read and record the measurement. In almost all cases, the measurement will be less than 1 inch. Repeat the measurement at a minimum of 12 different points around the disc (**Figure 4-27**). Remember to measure at the midpoint of the friction areas at each of the measuring points. Compare your readings to the specified minimum thickness. Any measurement that is less than specified means the disc is not to be reused and should be discarded. Differences over 0.002 inch between any of

222 Chapter 6

In many instances, it will be necessary to have electric power to test components located in or near the airbag areas. There are two ways to do this safely:

1. Use an ohmmeter (a digital multimeter or DMM) to measure resistance of the component and compare it to manufacturer specifications after it has been disarmed and removed from the vehicle.
2. Remove the component(s) from the vehicle and test it at a test bench equipped with the proper power supply and tools.

Special Tools
Service manual
Vehicle's diagnostic routine

WARNING **Do not reconnect the battery again until all the service work is completed, including reassembly of components and their mounting. A quick touch of the cable to the battery will instantly recharge the capacitor. This would supply the power needed to fire the airbag, and severe damage and injury could result.**

WARNING LIGHTS

Bulb check is a program where certain dash indicator lights will illuminate for only a few seconds so that the technician can check their operation.

KOEO means ignition key on, engine off, a test condition required to performed certain tests.

KOER means ignition key on, engine running, another test condition required for certain tests.

Many of the warning lights are simple, electrical circuits. A good understanding of their operation will give the technician a head start on basic repairs. Most tests must be run when the key is on with engine off (**KOEO**) or key on with the engine running (**KOER**). You may have to turn the key off and back on several times to read and understand all the lights. A light that goes on for only a few seconds and then goes off does so because of the program designed to work with it. This is known as a bulb check. Bulb check is a program in which the bulbs will illuminate for only a few seconds. This is done so that the technician (or driver) can check if the bulb is working or not. If the oil pressure warning light bulb is burnt out, then the driver would never know when or if there is a problem with the oil pressure.

KOEO and KOER running are two test conditions associated with powertrain control module (PCM) test routines, but they appear in other system test procedures as well.

Older vehicles use a mechanical switch to turn the light on when certain conditions are present. Other vehicles use a sending unit for the warning light and gauge. One basic function occurs on both types: the ground to the warning light is completed.

WARNING **Do not grab hoses or other components under the hood if an overheated engine is suspected. Even an engine operating at normal temperature can cause serious burns.**

Caution
The following tests are for nonelectronic circuits only. Sensor circuits can be determined by observing the number of wires attached to the sensor or sending unit. True sending units will have only one wire, whereas sensors will have two or more wires. Grounding a sensor circuit could damage the computer or the circuit.

WARNING **Never remove the radiator cap on a warm or hot engine. Always let the engine cool down. Extremely serious burns and other injuries can occur when removing a hot radiator cap.**

If the red engine light is lit, the problem indicated is either an overheated engine or low oil pressure. On some vehicles, there is a light for oil and one for overheating (**Figure 6-1**). Before starting work on the electrical circuit, do the obvious and check the level and condition of the coolant and oil. For a cooling problem, look for missing or damaged fan belts. The coolant level on a hot engine can be checked in the overflow reservoir. The reservoir may not contain a true representation of the coolant level, but it will be an indication. Note that, on some cooling systems, the reservoir may become pressurized. Notice the condition of the engine by checking for oil or coolant leaks. The hoses should be inspected for swelling. Do not touch hot hoses. In many cases of low oil pressure or overheating, the condition will be found during the inspection.

Cautions and Warnings

Cautions appear throughout the text to alert the reader to potentially hazardous materials or unsafe conditions. Warnings advise the student of things that can go wrong if instructions are not followed or if an incorrect part or tool is used.

Case Studies

Each chapter ends with a Case Study describing a particular vehicle problem and the logical steps a technician might use to solve the problem. These studies focus on system diagnosis skills and help student gain familiarity with the process.

CASE STUDY

This is a problem that was national in scope, but it only concerned one brand of vehicle and only for a model year or two. The customers complained that the vent air was not cool when the air conditioning was on in MAX or NORMAL. The heater worked fine. This system used stepper motors on the blend doors. Routine electrical tests did not locate the problem. The doors could be heard moving back and forth as the controls were moved. Studying the service manual's system operation section revealed a possible cause. When the ignition key was moved to ON, the A/C computer ran a test on its circuits. Part of this test was the search for a 35-mA load on each of the motors as they moved the doors through their entire range of movement. A DMM set to amps and connected between the heat blend door motor and the computer showed 35 mA at full cold and at full hot. Supposedly the system was functioning correctly. Further physical investigation found the problem on top of the air box. The arm from the motor was held in place to the door lever with a small plastic clip. The clip had broken and was hanging down. As the motor moved the door toward the cold position, the clip would hook a ridge of the air box. The motor would try to push the door further, but the 35-mA load was reached, and the computer assumed this was the limit of door travel and stopped the motor. The end result was the door never reached its cold position even though the ampere reading was correct. Replacing the clip with a better connection solved the problem.

ASE–Style Review Questions

Each chapter contains ASE-style review questions that reflect the performance objectives listed at the beginning of the chapter. These questions can be used to review the chapter as well as to prepare for the ASE certification exam.

ASE-STYLE REVIEW QUESTIONS

1. Keyless entry system is being tested.
   Technician A says that all doors and windows should always be closed when test the keyless remote.
   Technician B says the driver's window should be down when testing the keyless remote.
   Who is correct?
   A. A only C. Both A and B
   B. B only D. Neither A nor B
2. Lighting systems are being discussed.
   Technician A says inoperative marking lights may be caused by an open circuit breaker.
   Technician B says testing a lighting system without relays can normally be done with a test light.
   Who is correct?
   A. A only C. Both A and B
   B. B only D. Neither A nor B
3. A customer complains that the headlights turn on and off, generally when traveling on a rough road.
   Technician A says the circuit breaker may be weak or failing.
   Technician B says a loose ground on the relay may be the fault.
   Who is correct?
   A. A only C. Both A and B
4. A power window is stuck about halfway up and will not move up or down. There is a "clunk" each time the switch is operated.
   Technician A says the drive gear is stripped.
   Technician B says the window motor switch is shorting to ground.
   Who is correct?
   A. A only C. Both A and B
   B. B only D. Neither A nor B
5. Instrument gauge diagnosis is being discussed.
   Technician A says a test light can be used to test digital gauge systems.
   Technician B says supplying 12 VDC to a gauge will cause it to read full.
   Who is correct?
   A. A only C. Both A and B
   B. B only D. Neither A nor B
6. Warning light systems are being discussed.
   Technician A says grounding an oil pressure warning light wire at the sending unit will cause the gauge to read low.
   Technician B says grounding the engine temperature sending unit will cause the warning light to come on and the gauge to read hot.
   Who is correct?
   A. A only C. Both A and B

dipstick hole. You will be...

Customer Care

This feature highlights those little things a technician can do or say to enhance customer relations.

CUSTOMER CARE Although a cylinder leakage test can identify the primary cause of low compression readings, remember to keep your mind and eyes open to other problems when repairing the engine. Do not automatically inform a customer that an exhaust valve is leaking and a valve job will cure the problem. Consider the engine mileage, oil condition, pressure and usage, and degree of leakage past the rings. If the engine has high miles, it is very likely that the rings are significantly worn. By replacing the faulty valve(s) and restoring the others to like-new condition, you will increase the compression and combustion pressures. The change may be significant enough that the old rings will not seal as well as they had been before the valve job. In this case, the customer would be back complaining about blue smoke from oil consumption or about the engine still lacking adequate power. This may very well be an example of when a replacement engine would be the most cost-effective repair for the customer. Evaluate the whole situation, and make your recommendation to the customer. If you offer the customer repair options, be very clear about explaining the possible consequences.

ASE Challenge Questions

Each technical chapter ends with five ASE challenge questions. These are not more review questions; rather, they test the students' ability to apply general knowledge to the contents of the chapter.

ASE CHALLENGE QUESTIONS

1. Tires are being discussed.
   Technician A says that the contact patch will change due to different tire inflation.
   Technician B says the contact patch is where the tire supports the weight of the vehicle.
   Who is correct?
   A. A only C. Both A and B
   B. B only D. Neither A nor B
2. Tire pressure is being discussed.
   Technician A says that the proper tire pressure for each tire is on the side of the tire.
   Technician B says that the proper tire pressure can be found on the tire placard located on the drivers door frame.
   Who is correct?
   A. A only C. Both A and B
   B. B only D. Neither A nor B
3. Tire pressure monitor system is being discussed.
   Technician A says that the indirect system can identify the tire that is under inflated.
   Technician B says the direct system needs to be relearned after a tire rotation.
   Who is correct?
   A. A only C. Both A and B
   B. B only D. Neither A nor B
4. Two technicians are discussing coil springs.
   Technician A says that when the spring compresses it is called a jounce.
   Technician B says that when the spring extends it is called a rebound.
   Who is correct?
   A. A only C. Both A and B
   B. B only D. Neither A nor B
5. Suspension systems are being discussed.
   Technician A says that a dependent suspension is mainly used for light duty application.
   Technician B says that independent suspension can mainly be found on heavy duty trucks.
   Who is correct?
   A. A only C. Both A and B
   B. B only D. Neither A nor B

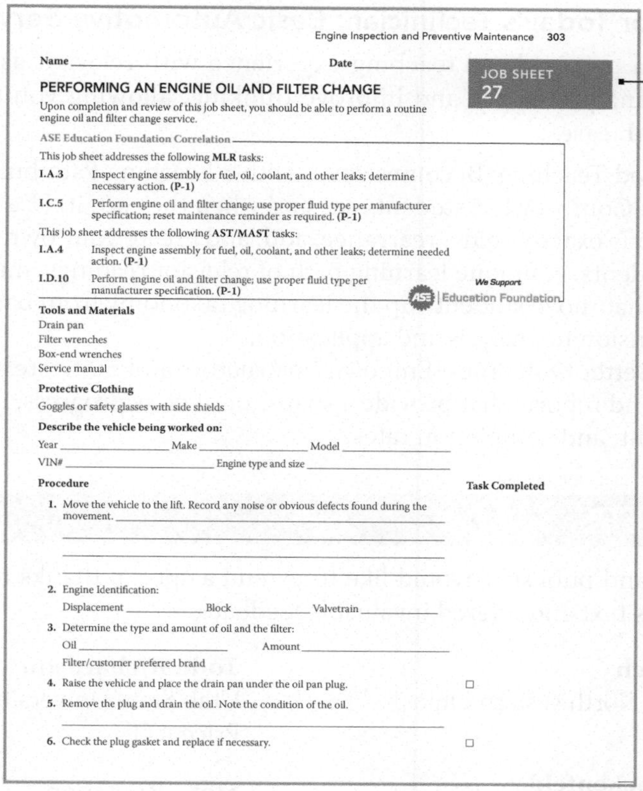
Engine Inspection and Preventive Maintenance 303

Name ______________ Date ______________

JOB SHEET 27

PERFORMING AN ENGINE OIL AND FILTER CHANGE

Upon completion and review of this job sheet, you should be able to perform a routine engine oil and filter change service.

ASE Education Foundation Correlation

This job sheet addresses the following **MLR** tasks:

I.A.3 Inspect engine assembly for fuel, oil, coolant, and other leaks; determine necessary action. (P-1)

I.C.5 Perform engine oil and filter change; use proper fluid type per manufacturer specification; reset maintenance reminder as required. (P-1)

This job sheet addresses the following **AST/MAST** tasks:

I.A.4 Inspect engine assembly for fuel, oil, coolant, and other leaks; determine needed action. (P-1)

I.D.10 Perform engine oil and filter change; use proper fluid type per manufacturer specification. (P-1)

We Support ASE | Education Foundation

Tools and Materials

Drain pan

Filter wrenches

Box-end wrenches

Service manual

Protective Clothing

Goggles or safety glasses with side shields

Describe the vehicle being worked on:

Year ______ Make ______ Model ______

VIN# ______ Engine type and size ______

Procedure **Task Completed**

1. Move the vehicle to the lift. Record any noise or obvious defects found during the movement. ______

2. Engine Identification:
   Displacement ______ Block ______ Valvetrain ______

3. Determine the type and amount of oil and the filter:
   Oil ______ Amount ______
   Filter/customer preferred brand

4. Raise the vehicle and place the drain pan under the oil pan plug. □

5. Remove the plug and drain the oil. Note the condition of the oil. ______

6. Check the plug gasket and replace if necessary. □

Job Sheets

Located at the end of each chapter, the Job Sheets provide a format for students to perform procedures covered in the chapter. A reference to the ASE Education Foundation Task addressed by the procedure is referenced on the Job Sheet.

SUPPLEMENTS

Instructor Resources

The *Today's Technician* series offers a robust set of instructor resources, available online at Cengage's Instructor Resource Center (http://login.cengage.com) and on DVD. The following tools have been provided to meet any instructor's classroom preparation needs:

- Lesson Plans provide lecture outlines, teaching tips, and complete answers to end-of-chapter questions.
- Power Point presentations include images, videos, and animations that coincide with each chapter's content coverage.
- Cengage Learning Testing Powered by Cognero® delivers hundreds of test questions in a flexible, online system. You can choose to author, edit, and manage test bank content from multiple Cengage Learning solutions and deliver tests from your LMS, or you can simply download editable Word documents from the Instructor Resource Center.
- An Image Gallery includes photos and illustrations from the text.
- The Job Sheets from the Shop Manual are provided in Word format.
- End-of-Chapter Review Questions are also provided in Word format, with a separate set of text rejoinders available for instructors' reference.
- To complete this powerful suite of planning tools, a pair of correlation guides map this edition's content to the ASE Education Foundation tasks and to the previous edition.

MindTap for Today's Technician: Basic Automotive Service & Systems, 6e

MindTap is a personalized teaching experience with relevant assignments that guide students to analyze, apply, and improve thinking, allowing you to measure skills and outcomes with ease.

- Personalized Teaching: Becomes yours with a Learning Path built with key student objectives. Control what students see and when they see it. Use it as-is or match to your syllabus exactly–hide, rearrange, add, and create your own content.
- Guide Students: A unique learning path of relevant readings, multimedia, and activities that move students up the learning taxonomy from basic knowledge and comprehension to analysis and application.
- Promote Better Outcomes: Empower instructors and motivate students with analytics and reports that provide a snapshot of class progress, time in course, engagement, and completion rates.

REVIEWERS

The author and publisher would like to extend a special thanks to the instructors who reviewed this text and offered invaluable feedback:

Rodney Batch
University of Northwestern Ohio
Lima, OH

Matthew Bockenfeld
Vatterott College, Joplin Campus
Joplin, MO

Shih-Wei "Jack"
Chen Utah State University Eastern
Price, UT

Cory Peck
Vatterott College
St. Louis, MO

Todd Richardson
Utah State University Eastern
Price, UT

Steve Roessner
University of Northwestern Ohio
Lima, OH

Joshua Summey
Richmond High School / Ivy Tech
Richmond, IN

CHAPTER 1

INTRODUCTION TO THE AUTOMOTIVE INDUSTRY

Upon completion and review of this chapter, you should understand and be able to:

- Describe the different types of businesses supporting the automotive industry.
- Describe the main differences between dealerships, independents, franchises, and service stations.
- Explain how departments can increase business.
- Describe the general duties and responsibilities of the service manager.
- Explain the legal and ethical responsibilities of a business and its employees.
- Describe the role of accrediting agencies and other industry associations within the automotive industry.
- Understand the several types of training for automotive technicians.

Terms To Know

ASE Education Foundation
Business ethics
Ethical word
Express service
Franchise
General manager
Investment
Lifelong learning
National Institute for Automotive Service Excellence (ASE)
Vendors

INTRODUCTION

Employees must work effectively within the organization so they can show a profitable return for their **investment** within the company. This includes knowing the company's organization and the general duties of each department, working with the various departments, and protecting the investment through safe, profitable, and **ethical work** practices.

Investment is the amount of time, money, and other resources used to establish a business.

Ethical work Moral principles a technician uses in their job, which encompasses traits like: Reliability,dependability, dedication,productivity,cooperation,character,integrity,responsibility,quality,teamwork and professionalism.

GLOBAL AND U.S. AUTOMOTIVE BUSINESS

Automotive service centers used to be small operations. Usually, there was a person who owned and managed the business while performing repairs and customer service at the same time. Even manufacturer repair service centers were on a small scale compared to today's multimillion-dollar shops. Small operations still exist today, but most automotive repair businesses must meet the demand for highly trained technicians, expensive diagnostic equipment, customer waiting rooms, extensive parts departments, and the facilities and staff to support them. The various levels of business interlock throughout the industry (**Figure 1-1**).

Shop Manual page 1

A modern-day automotive repair business includes every type of employee position from custodian to general manager. There are accountants, clerks, salespeople, service writers, technicians, department and service managers, vehicle and service preps, lot technicians, parts managers, service directors, and a general manager to oversee the entire

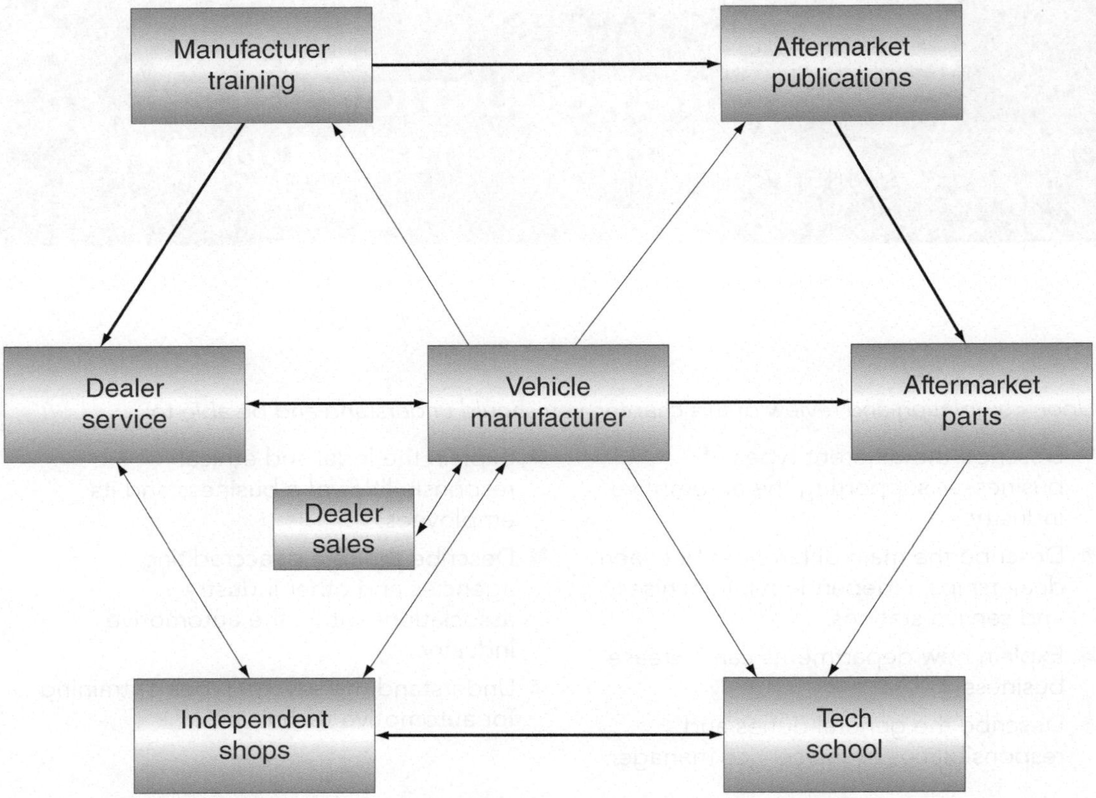

Figure 1-1 Relationship between automotive business and technician education and training.

operation. All these positions are important for the business to thrive. In any case, the person who performs the work that directly impacts profit-making is the technician.

LOCAL AUTOMOTIVE REPAIR BUSINESS

Local car repair shops are a great community asset offering great employment opportunities. They are probably one of the largest single-type employers in the area. Most people refer to the place they bring their vehicle as having "their technician" or "their mechanic" working on their vehicle. Individual manufacturing plants may employ several hundred people, but all those employees and their families have vehicles that must be maintained and repaired. That requires many repair shops, sales, and parts vendors with their numerous trained technicians and supporting staff, ranging from the cleanup crews to accountants to attorneys and the waste and recycling companies. Many people who work in the automotive business do not realize their importance to the community at large. Not only do they support the automotive side, they pay taxes, property and personal, that help support community activities. Many employers and businesses participate in local activities and fundraisers. The automotive employer and its employees provide a great amount of money and labor to support local charities, churches, and schools—not to mention the money they spend locally for their personal or professional needs. The individual technician plays just as important a role in his or her community as within the automotive industry itself.

Shop Manual
page 1

Dealership Operations

Normally, dealerships are automotive businesses operating under independent ownership in conjunction with a vehicle manufacturer. Operations of this type usually sell and service

Figure 1-2 Modern-day dealerships' repair focus is manufacturer specific.

only certain makes of vehicles. The owners conduct day-to-day business, either directly or through a **general manager.** The business must conform to certain standards set forth by the vehicle manufacturer. These standards may include policies on new car sales, accounting and financing procedures, service operations, and the layout of the facilities. In some instances, the manufacturer may withdraw the **franchise** license if standards and profit margins are not met. Some dealerships are large, and some are small. Some dealerships combine different manufacturers. **Figure 1-2** shows an example of a modern and technically advanced dealership.

A **general manager** is the person in charge of that business. In the automotive repair business, the manager supervises all aspects of the business directly or through department managers.

Franchise are businesses locally owned and operated under the policies of a larger company. For example, McDonald's is a franchise operation. A person (or group) decides to build a store and purchases the name recognition rights. They must follow rules and can buy and sell certain items directly from the manufacturer.

Sales and service departments working together can create a customer relationship that draws repeat business and favorable word-of-mouth advertising. The technician is required to have extensive knowledge of the makes and models sold at the dealership and is required to perform warranty repairs. Technicians at a dealership often have extensive training and updating on new models.

Shop Manual
page 2

AUTHOR'S NOTE In early 2008, Ford, General Motors, and Chrysler announced that they as individual corporations were combining various dealerships to increase profit and reduce costs. The current urban-area dealer separation of 5 to 10 miles presents too much competition among dealers selling the same brand and results in lower sales for all. Most rural areas, where the dealerships can be 20 to 40 miles apart, should not be affected. This decision by the corporations could mean the end of many locally owned, urban-based dealerships. As a side note, Daimler sold Chrysler to investors in 2007, and in 2009 Fiat acquired a controlling interest in Chrysler.

Independent Repair Facilities

Independent repair shops usually concentrate only on vehicle repairs and service (**Figure 1-3**). They may stock some of the most common repair parts, but most of the needed parts come from aftermarket **vendors** or the local dealership. Independents

Vendors are businesses that sell parts or service to other businesses and the public. For example, NAPA is an automotive vendor.

Figure 1-3 Independent shops repair almost any type of make and model of vehicle.

Figure 1-4 Franchise businesses usually specialize in one system of the vehicle but may perform related system repairs.

usually take in any vehicle that needs repair, although some may specialize in certain types of vehicles or types of repairs. This places a heavy technical load on the technicians. They must know how to repair almost any vehicle and all automotive systems. Independent shop owners usually pay for their technicians' update training in addition to salaries and travel expenses. Most independents make use of classes offered by local parts vendors or the local technical schools.

One advantage of working for an independent is that they are often small enough that the vehicle owner can have a relationship with the technician. Often a healthy relationship can lead to easier repair sales, more maintenance, less major repairs, and a long-term repeat customer.

Other Automotive Repair Businesses

Franchise automotive businesses are independently owned and operated. They are licensed under the business umbrella of a national or regional chain (**Figure 1-4**). The owner must follow the policies of the company issuing the franchise license. Precision Tune and Quik Lube are two examples. These types of businesses are set up to perform only certain types of repairs or maintenance. Businesses such as Precision Tune make engine-performance repairs or other vehicle maintenance, whereas businesses such as Quik Lube usually perform oil and lubrication work. Since their inception, each business has changed its operations to include work not directly related to its original base business.

Local service stations are another part of the repair business. Usually, they handle brakes, tires, minor tune-ups, and other light repairs. Other local businesses repair only certain systems of the automobiles. Generally, they specialize in brakes, tires, exhaust, and related repairs. However, most specialized shops, including service stations, are realizing that they cannot survive today without accepting repair work outside of their original business intentions, and it is common for smaller shops to convert to a franchise where they will have more business.

All these automotive businesses make up the automotive repair facilities in this country. To support them, there are numerous parts makers and suppliers, hazardous waste disposal companies, and many other local, national, and international firms that contribute to keeping a vehicle on the road.

The trend in dealership operation is the megadealership. As mentioned earlier, dealerships must meet certain corporate standards set by the vehicle manufacturer but are locally owned. In recent years, Roger Penske (see Figure 1-4) and others have formed corporations that own multiple dealerships throughout the world under one company umbrella. Some of these corporations are publicly traded, meaning an individual can buy shares in the business through the New York Stock Exchange or, as it is commonly known, Wall Street. However, the general operation of the local dealership is the same as before; only the owners and some business policies have changed.

Express Services

Express services are used by many dealerships and some independent shops to expedite quick services such as oil changes and create additional business for the shop through vehicle inspections. The idea originated with franchises like Jiffy Lube and Precision Tune in the late 1980s, and now almost all dealerships have an express service. A well-conducted inspection can also bring in additional services such as tires, brakes, and tune-ups for the business. Many technicians start their career, or have long-lived careers, in the express service bays.

Express services refer to quick routine preventive maintenance, primarily oil changes, tire rotation, and vehicle safety inspections.

Shop Manual
page 5

Many customers delay scheduled services for as long as possible when money is short, but they will always get the oil changed. As a result, express service, with a vehicle safety inspection, has become a critical factor in keeping the shop's doors open for business. A thorough safety inspection may locate a problem or potential problem that can be brought to the customer's attention. If the problem is critical, the customer will probably have the repair completed during that visit.

AUTHOR'S NOTE Some people tend to think the shop is trying to "rip them off" with this type of inspection. The opposite is true, and the shop should not try to force feed the repair. Make the customer aware of the problem or problems. Of course, if the problem is critical, such as a loose or broken steering linkage, then the service manager or writer needs to ensure the customer realizes the severity of the situation. But remember, it is the customer's money, and he or she can refuse the repair if desired. New technicians should follow the lead of experienced technicians. Experienced technicians know how to talk to a customer and not sound like they are selling a service. They talk to the customer in a very factual manner, telling them what problems exist with their vehicle and what may happen if they ignore it. It's always a good practice to show the customer the problem when possible.

Typically, new technicians are hired as Entry-Level or Maintenance and Light Repair (MLR) Technicians. Almost all new technicians work in the express service section and then advance from that position to full technicians. Generally speaking, the job title of most students completing an automotive education or training program will be related to MLR, also referred to as general service technician. The tasks outlined in this book are related to the job of that technician. After some time, your employer will want to have you do more complicated repairs and diagnostics. The service manager or writer is another position in the automotive industry. This person is generally referred to as the supervisor for an MLR technician. This person will work very closely with a new employee, like yourself. This position is responsible for managing workflow and relationships with employees and customers, and sometimes must perform customer service and sales. This person gets to take a good look at the new employee to see if this person fits into the overall enterprise as he or she performs the important tasks needed to make the business presentable and profitable. Most service managers were once technicians. Many of them started their career reading a textbook and attending a class just like you are. Service managers must show a responsible and progressive history of good customer service and high skill level and technical expertise.

Over the next several chapters, in both the Classroom and Shop Manuals, you'll be introduced to the various technical entry-level tasks an MLR technician must achieve.

TECHNICIAN TRAINING AND CERTIFICATION

If there is one thing about the automotive industry that will remain true forever, it is that this is an industry where technology will always change. It is for this reason that a student in an automotive education or training program must realize that, to keep up with the industry, one must commit oneself to always learning new things. We call this **Lifelong learning**. Successful technicians all have one thing in common. They periodically go through update training to keep their skills sharp and knowledge level high.

Lifelong learning Continuing to learn and acquire the latest skills required by industry advances in technology.

The technician's training usually involves some type of certification. Certification is usually done through technical training offered by some type of educational agency or certification by an outside source, such as those discussed next. But technical training is not all that is required. The technician not only has technical responsibilities but also legal and ethical responsibilities he or she must be aware of and must practice.

Legal Responsibilities

Not only are the technicians expected to be technically proficient, they must understand the legal side of automotive repair. The customer is paying for professional service and has a right to expect it. Poor repair work, damage to the vehicle, and a poor attitude toward the customer may lead to legal proceedings against the technician and the business. Keep in mind that a vehicle must operate properly to keep the passengers inside it safe. It is part of a technician's job to maintain and inspect a vehicle so that it is kept this way.

In addition to legal responsibilities to the customer, federal, state, and local laws and regulations require business owners and employees to help protect the community and environment from harm. Automotive technicians share this legal responsibility. Proper handling of hazardous materials and waste is of prime concern to the automotive industry. Not only can a poorly tuned engine emit hazardous materials and gases, the by-products of poor repair work also may damage the environment. The business owner is ultimately responsible for the actions of the firm's employees, but the technician plays a large part in fulfilling the legal responsibilities of the company.

Business Ethics

Most people hold the automotive industry in high esteem. Two segments of the industry, sales and service, are the exceptions. Much of this distrust stems from the high cost of vehicles and their repairs. Further supporting this distrust is the fact that most people do not understand the workings of an automobile, and they have no reference with which to compare the costs of buying and maintaining a car to that of other large purchases. Educating customers is one way to resolve the problem. Very few automobile owners understand that their vehicle's computer is as powerful as the personal computer they have at home or that the technicians are constantly attending some type of training. A business will survive much better in today's highly competitive automotive arena if it can educate its customers to some extent. A customer may pay well over $1,000 for a personal computer (PC) and hesitate at paying $600 for a vehicle computer. It is easy to see the benefits of purchasing a new PC because it is colorful, right in front of your face, and provides happiness. It is difficult to feel the same about a vehicle's computer because it doesn't provide the same feeling. A technician or service writer can win over the customer by showing the capabilities of that $600 computer versus the PC. Consumers have different feelings about vehicle repair, especially when it can be their main mode of transportation needed to go to work to make the money it takes to live life.

All repairs to a vehicle are difficult for a customer to deal with. Most consumers want to pay an upfront price for the vehicle and expect it to last for years with only minimal maintenance. A vehicle is an investment in a consumer's life. In many cases, it can be the most expensive investment for a person.

Automotive businesses expect to be treated fairly by their suppliers and must treat their own customers in the same manner—this is the essence of **business ethics.** By treating each customer as a valued individual, charging a fair price for service, and admitting and correcting mistakes, the shop and all its employees can expect to be in business for a long time.

Business ethics are hard to define. Because one person's idea of ethics may be different from another's, the managers set the rules. Treating customers, managers, and employees fairly is the basic rule.

NATIONAL INSTITUTE FOR AUTOMOTIVE SERVICE EXCELLENCE (ASE)

The **National Institute for Automotive Service Excellence (ASE)** tests technicians for their technical knowledge. Technicians are authorized to wear the shoulder patch only if they have successfully completed a written test and have 2 years' working experience. They must recertify every 5 years to maintain their standing. The Master Automotive Technician patch is given only if a technician passes all eight automotive tests and meets the experience requirement (**Figure 1-5**). Certifications are also offered in heavy truck, alternative fuels,

National Institute for Automotive Service Excellence (ASE) is the acronym for the National Institute for Automotive Service Excellence.

Figure 1-5 ASE Certification Levels.

parts, machining, service writing, and school buses. Automotive students that have completed a certain amount of education can certify through ASE as a student. This certification is called ASE Student Certification. Students can get certified in all eight areas (A1–A8) plus two new certifications, MLR and Automobile Service Technology. Students are not required to have any work experience for these certifications. Student certifications will give you the edge you need when applying for your first automotive job. Student certifications are valid for 2 years. The regular ASE automotive certification tests are given throughout the year at a testing facility and are solely computer based. Technicians with a minimum of 1 year of experience can take a new MLR certification (G1). This is a regular ASE certification that is valid for 2 years. The eight main areas of regular ASE testing for technicians are Engine Repair (A1), Automatic Transmissions (A2), Manual Drivetrain and Axles (A3), Brakes (A4), Steering and Suspension (A5), Electrical/Electronics (A6), Heating and Air Conditioning (A7), and Engine Performance (A8). Upon earning all eight certifications, you will be considered an ASE Master Technician. All regular ASE certifications are valid for 5 years. In 2009, a new test was added called Light Duty Diesel (A9) in response to the increase in the use of light duty diesels in cars and light-duty trucks. ASE also offers an Advanced Engine Performance Specialist (L1). This certification tests technicians' knowledge of diagnosis and repair of computer-controlled engine systems.

Those who choose to be ASE-certified technicians are showing their customers and their employer that they are part of an elite crowd and will hold up to the highest standards of automotive excellence.

A repair facility can display the blue ASE sign and use the ASE symbol in its advertising if it has automotive technicians who are ASE certified and if it adheres to the ASE standards of excellence. The ASE sign and shoulder patch can be a valid advertising tool. ASE also publicizes the technician's role within the automotive business and to the public at large. Learn more at ASE's Web site: www.ase.com. You should also visit the entry-level certification Web site at www.ase.com/entr-level/.

ASE EDUCATION FOUNDATION

ASE Education Foundation is a nonprofit organization that evaluates and accredits entry-level automotive technology education programs.

The **ASE Education Foundation** certifies automotive training programs. The programs may be offered at public and private high schools or postsecondary schools and manufacturer schools. Students who graduate from an ASE Education Foundation-certified program have an advantage over other persons applying for an automotive technician position.

To meet ASE Education Foundation standards for certification, the training program must be capable of teaching standard tasks published by the ASE Education Foundation. The program must have all the tools required by the ASE Education Foundation and have ASE-certified instructors in each area requesting certification. The program is certified for 5 years with an update at the halfway point. Once certified, the school can use the ASE Education Foundation logo in its advertising and catalog (**Figure 1-6**). Generally, high schools will be certified in MLR. Students who attend and graduate from an ASE Education Foundation-certified high school program can likely get hired as an entry-level technician. But moving beyond that position will most likely require some postsecondary training at an ASE Education Foundation school and ASE certification. Learn more at: https://www.aseeducationfoundation.org/

Figure 1-6 The right to use the ASE Education Foundation logo in advertising is awarded after successfully completing the ASE Education Foundation training program procedures.

AUTOMOTIVE YOUTH EDUCATIONAL SYSTEMS (AYES)

The **Automotive Youth Education Systems (AYES)** is a part of the ASE family. AYES is an education-to-work program directed toward high school students. AYES focus is the student's success by helping create those partnerships between business and education. It is supported by many of the vehicle manufacturers and other automotive-based organizations. The student receives excellent education and training in basic automotive repair. This training will aid the graduate in gaining employment and entering postsecondary or manufacturer programs for further training.

AYES: Effective January 1, 2018 NATEF and AYES have merged to be come ASE education Foundation

AUTOMATIC TRANSMISSION REBUILDERS ASSOCIATION (ATRA)

The **Automatic Transmission Rebuilders Association (ATRA)** is an association of transmission rebuilders. It offers certification to any technician who desires it, but it is directed toward technicians who only do automatic transmission repairs. The test is given annually at selected sites and is brand and model specific in its questions. The test is designed for technicians who have experience in transmission diagnosis and repairs. ATRA publishes *Gears* magazine, which may be found at its Web site: http://www.gears magazine.com.

ATRA: Worlds largest trade organization of transmission re-builders with over 2000 members throughout the United States and Canada.

MOBILE AIR CONDITIONING SOCIETY (MACS)

The **Mobile Air Conditioning Society (MACS)** is similar to ATRA in its testing. It is designed for automotive air-conditioning technicians but, like ATRA, is offered to any technician. An EPA requirement for all air-conditioning technicians is to be certified by an approved agency. This test will be discussed further in Chapter 12. Visit the MACS Web site for further information: http://www.macsw.org.

MACS: Empowers members to grow their businesses and delivers tangible members benefits.

AUTOMOTIVE ENGINE REBUILDERS ASSOCIATION (AERA)

The **Automotive Engine Rebuilders Association (AERA)** primarily provides technical, product, and education support to technicians and shops specializing in the rebuilding of automotive engines. It also offers financial grants to assist in education expenses. It charges a minimum fee for membership, as do the other organizations of this type. Learn more at AERA's Web site: http://www.aera.org.

ATRA and MACS are two examples of automotive certification organizations. There are others, but the most nationally recognized ones are ASE and the ASE Education Foundation.

AERA: Specialized network of professional engine builders, re-builders production re-manufacturers and installers with the expertise and connections to get you the information you need.

SOCIETY OF AUTOMOTIVE ENGINEERS (SAE)

The **Society of Automotive Engineers (SAE)** provides much technical support to governments, manufacturers, all levels of repair shops, and individual technicians. SAE, like other organizations, charges an annual fee for this expertise. At the technician level, information is available on repairs, future products and systems, and the overall goals for emission control, fuel economy, vehicle performance, and the technical aspects of the automotive industry. Further information on SAE can be found at its Web site: http://www.sae.org.

SAE: provides technical support it is comprised of engineers, executives, educators and students.

MANUFACTURER AND VENDOR TRAINING

Manufacturer training is conducted by schools set up and supported by a vehicle manufacturer. They range from entry-level training, usually conducted at a postsecondary school, to update training and certification of dealership technicians conducted through online classes, computer-based training (CBT), or at a manufacturer-funded site. Other technicians may attend manufacturer update training after paying a fee and if space is available. The manufacturer may elect to certify dealer technicians in various areas of the vehicle(s) and, in most cases, supports or accepts certification from the independent agencies just discussed.

Training conducted or sponsored by the local parts vendor is open to all technicians. Classes may require a fee payment. Vendor training is conducted for two reasons. First, updates on diagnostics and repairs are presented. This information is usually very pertinent, timely, and presented in a professional manner. Of course, they are also there to sell their products. The class presentation includes methods of using their product to perform quick and accurate diagnosis and repair. Although the underlying cause is to sell the vendor's products, most of these classes provide much technical information helpful to the technician. Most technicians understand this and do not mind listening to a little sales pitch if the overall information increases their income. In fact, technicians may attend to see the new diagnostic tools that will help solve problems that might be present in new-model vehicles. The Specialty Equipment Market Association (SEMA) is one such aftermarket vendor offering automotive products, training, and possible financial aid assistance to technicians. Visit its Web site at http://www.sema.org. Other organizations, such as ATRA and AERA, offer training in their respective areas. Their Web sites, respectively, are http://www.atra.com and http://www.aera.org. Many other groups also offer online information and training. The site http://www.howstuffworks.com provides much basic information with links to other automotive sites as well.

SUMMARY

- Management is responsible for all business operations.
- All employees share the legal responsibilities of the business.
- Education of the customer is another way to gain customer trust.
- ASE provides a nationwide certification program for technicians.
- ASE Education Foundation certifies automotive training programs.
- ATRA provides a means for automatic transmission technicians to gain specific skill recognition.
- MACS is designed for air-conditioning technicians.
- AERA and SAE provide technical and educational support for various establishments and technicians.
- Manufacturer and vendor training are essential for technicians wishing to keep up to date on various aspects of vehicle repair.

REVIEW QUESTIONS

Short Answer Essays

1. Describe the role of the independent shop within the automotive repair industry.
2. Explain how sales and service work together to gain customers.
3. Explain how certifying agencies can assist in promoting customer trust and validating a technician's technical knowledge.
4. Describe the purpose of the ASE Education Foundation.
5. Discuss how customer education can help an automotive repair business.
6. List the eight automotive areas tested semiannually by ASE.
7. List the steps needed for a technician to be certified as a Master Automotive Technician.

8. Discuss *your* definition of business ethics.
9. Explain why employees share in the legal responsibilities of the employer.
10. List the level of management typically found in an automotive shop.

Fill-in-the-Blanks

1. A dealership is an automotive business operating under ______________ ______________ in conjunction with an automotive manufacturer.
2. If standards or profits are not met, the manufacturer may withdraw the ______________ ______________ of the dealership.
3. The ASE Education Foundation certifies automotive ______________ ______________.
4. Once the school is certified, they can use the ______________ ______________ logo in its advertising and catalog.
5. ASE-certified technicians can wear a(n) ______________ ______________ if they have successfully completed the written exam and have 2 years' experience.
6. Vehicle manufacturer training programs are conducted by schools set up and supported by a ______________ ______________.
7. The EPA requires all air-conditioning technicians to be certified by an ______________ ______________.
8. ATRA offers certification to ______________ ______________ technicians.
9. AYSE is a part of the ______________ family.
10. ______________ automotive businesses are independently owned and operated.

Multiple Choice

1. Business operations are being discussed.
   Technician A says an independent repair facility stocks many of the commonly used repair parts used at the shop.
   Technician B says most franchise shops are usually owned by a national business chain and operated by local managers.
   Who is correct?
   A. A only
   B. B only
   C. Both A and B
   D. Neither A nor B
2. Business ethics are being discussed.
   Technician A says good business should be applied to all customers.
   Technician B says educating the customer could be a part of the shop's business ethics.
   Who is correct?
   A. A only
   B. B only
   C. Both A and B
   D. Neither A nor B
3. Legal responsibilities of the business are being discussed.
   Technician A says each employee shares in legal responsibilities.
   Technician B says the owner bears ultimate responsibility.
   Who is correct?
   A. A only
   B. B only
   C. Both A and B
   D. Neither A nor B
4. Program certification is being discussed.
   Technician A says ATRA certifies air-conditioning programs.
   Technician B says MACS certifies air-conditioning technicians.
   Who is correct?
   A. A only
   B. B only
   C. Both A and B
   D. Neither A nor B
5. *Technician A* says that, to achieve master automotive certification, nine tests must be successfully completed, including the Advanced Engine Performance Test.
   Technician B says only eight tests must be taken and only 2 years of experience are needed for master certification.
   Who is correct?
   A. A only
   B. B only
   C. Both A and B
   D. Neither A nor B
6. *Technician A* says business ethics means treating all customers honestly and fairly.
   Technician B says business ethics means treating vendors correctly.
   Who is correct?
   A. A only
   B. B only
   C. Both A and B
   D. Neither A nor B

7. The service manager's duties and responsibilities are being discussed.
   Technician A says the training of a technician is his or her responsibility.
   Technician B says he or she usually has authority over the parts manager.
   Who is correct?
   A. A only C. Both A and B
   B. B only D. Neither A nor B

8. Independent repair shops are being discussed.
   Technician A says this type of shop is needed to help reduce customer costs.
   Technician B says independents may send work to the local dealership.
   Who is correct?
   A. A only C. Both A and B
   B. B only D. Neither A nor B

9. Technician certification and training is being discussed.
   Technician A says the ASE Education Foundation ensures certain tasks are taught.
   Technician B says the ASE Education Foundation ensures that the technician is certified.
   Who is correct?
   A. A only C. Both A and B
   B. B only D. Neither A nor B

10. Air-conditioning certification is being discussed.
    Technician A says an air-conditioning technician can be EPA certified by ASE only.
    Technician B says ATRA deals with air-conditioning technicians.
    Who is correct?
    A. A only C. Both A and B
    B. B only D. Neither A nor B

CHAPTER 2
INTRODUCTION TO THE AUTOMOBILE

Upon completion and review of this chapter, you should understand and be able to:

- Describe the major systems of a vehicle and their purpose.
- Understand the major subsystems of the electrical system.
- Explain the need for and use of computerized systems on a vehicle.
- Describe the different types of power systems for a vehicle and their purpose.
- Explain the purpose of the powertrain and its major components.
- Explain the purpose of the major components of the steering and suspension systems.
- Understand the major components of the brake system and their general operation within the system.
- Explain the purpose of the frame and body.
- Understand the general operation of climate control systems and their major components.

Terms To Know

Automatic transmission
Biodiesel
Combustion chamber
Corporate Average Fuel Economy (CAFE)
Ethanol
Flex fuel
Fuel cell
Hybrid power
Manual transmission
Powertrain Control Module (PCM)
Refrigerant
Sprung
Thermodynamics
Transmission Control Module (TCM)
Vehicle stability

INTRODUCTION

The systems discussed in this chapter are the basic operating systems of a vehicle. It is necessary to understand them because they are the fundamental components of vehicle operation. They enable the operator to regulate the power produced by the engine, select a driving gear, control all of the modern accessories, and steer and brake the vehicle while experiencing a smooth, comfortable ride. The basic functions of the electrical system on an automobile are to provide power to start and run the engine and to operate some of the other basic features such as headlights and radio. A computer is not mandatory to the function of a vehicle but is needed to increase the fuel efficiency and lower the emissions output. Today's modern society and culture also demand vehicles that have similar computer operations to those found in daily life. This means that a modern vehicle will have a lot of electronic and computer features. The climate control system is also not needed for vehicle operation, but it is included as standard equipment on many vehicles sold today.

ELECTRICAL SYSTEMS

Electricity was not a feature found on some of the first automobiles. The first automobiles' engines were started by hand, and they had kerosene-fueled lamps. As vehicles progressed during the turn of the century, the only purpose of the electrical system was to start and operate the engine. Soon after the advent of electric starters and generators, interior and exterior lights were added to the vehicle. As vehicles became more complex and continued to add more electric accessories, computers became necessary to control and manage the systems. Every system on today's vehicles involves the use of electricity and electronics, ranging from measuring real-time tire pressure, to adjustment of the steering column and foot pedals, to electronic computer control of the vehicle's speed and direction. Several general subsystems are found within the overall electrical system.

Starting and Charging

Prior to the use of electric starter motors, the operator had to stand in front of the vehicle or wherever the engine was mounted, and hand crank the engine. Assuming the hand crank did not kick backward causing injury, this was a reliable but difficult method to turn the crankshaft. It was used on farm tractors as late as the mid-1950s. However, if the spark or fuel was not correct, the hand crank could kick backward injuring the operator, or, even after many attempts, the engine could still refuse to start. Modern engines that do not use an electric starter use a recoil cord to start the engine. This is common only in small engines. In the late 1920s, electric starters began to appear on vehicles. Electric-current-producing generators were added at the same time as the direct-current (DC) storage battery.

The starter system is composed of two subsystems: the starter and the control circuit. The starter motor relay or solenoid is wired directly to the battery's positive terminal. The drive end of the starter motor has a gear that can engage another gear to turn the crankshaft. The control circuit generally consists of the battery, ignition switch (or button), and the relay/solenoid (**Figure 2-1**). When power is supplied through the ignition switch to the relay/solenoid, large contacts within the relay connect the starter motor to the battery. At the same time, magnetic forces within the starter motor wiring or windings force the motor's drive gear into contact with the gear that drives the crankshaft (**Figure 2-2**). Therefore, the crankshaft moves the pistons through their cycles, drawing air and fuel into the cylinders, igniting the mixture, and causing the engine to run under its own power. By the mid-1970s, a switch called the park/neutral or clutch switch was installed into the starter motor control circuit to prevent engine startup if the transmission was in any drive gear or if the clutch was engaged (pedal up).

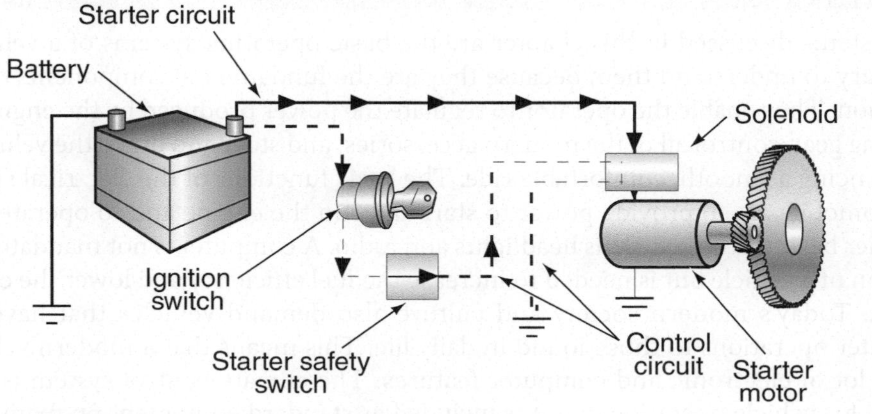

Figure 2-1 Major components of the starting system.

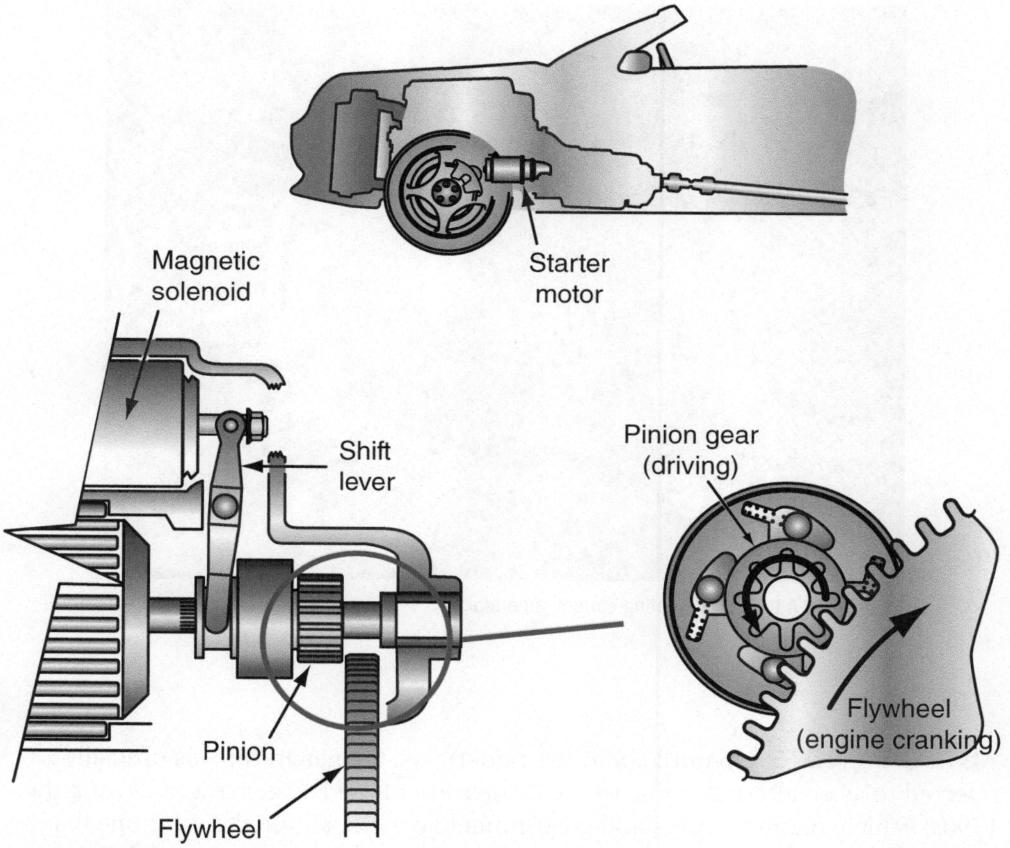

Figure 2-2 Typical location of a starter motor.

AUTHOR'S NOTE Since the early 1980s, manufacturers have inserted additional "fail-safe" devices within the starter, ignition, and fuel control circuits. Most use some type of oil pressure or crankshaft sensor to show when the engine quits running. Ford used an inertia switch that tripped open due to a roll-over accident or heavy shock. All were designed to shut down the fuel pump and reduce the chance of fire after a vehicle crash. Some, like the General Motors system, would also prevent ignition if the oil pressure was below about 20 pounds per square inch. The newest systems use the same idea, but the signals are sent to the powertrain control module (PCM), which now controls starting, ignition, and fuel control circuits. Other fail-safe devices include security/antitheft systems that disable the starter circuit and sometimes the entire electrical system.

The charging system uses a belt-driven alternating-current (AC) generator to produce the electric current to operate the vehicle and all its electrical devices (**Figure 2-3**). A rectifier with one-way electrical devices located near the generator output converts AC to DC. The battery is shared between the starting and charging systems and is used to store DC power for use when the generator is not operating, such as during engine startup. DC power is required because alternating current cannot be stored. All vehicle accessories are designed to operate on DC power, such as power windows, the heater fan, and lights.

Figure 2-3 A typical alternating-current generator.

Powertrain Control Module (PCM) A control unit used on motor vehicles. It is generally a combined control unit, consisting of the engine control unit and the transmission control unit.

Transmission Control Module (TCM) It is an electronic module that controls the operation of the transmission.

Corporate Average Fuel Economy (CAFE) is a federal requirement concerning fuel mileage or usage. It is an average calculated from a manufacturer's vehicles using the worst (lowest mileage) versus the best (highest) mileage. A company's CAFE average also considers the total number of vehicles and vehicle type sold. The most current CAFE requirement is 27.5 miles per gallon (mpg) on cars and 22.2 mpg on light trucks. In 2016, the CAFE average for cars will go up to 38.5 mpg, and in 2025, the CAFE requirement will be 54.5 mpg. This will only be possible with advanced computers and other advances to vehicle structure.

AUTHOR'S NOTE Up until about the mid-1990s, the generator was officially referred to as an alternator due to the former use of the DC generator. During the 1990s, vehicle manufacturers and government agencies adopted definitions that clarified many automotive terms. At that time, the term *alternator* was defined as the AC generator and was quickly shortened to *generator.* However, the term *alternator* is still widely accepted in shops, by customers, and by parts vendors.

Lighting, engine controls, and other subsystems receive their electrical power from the battery. The battery is then recharged by the generator based on electrical demand. In other words, if the lights are off, then power is not used but is available. These electrical subsystems are included within other vehicle systems such as the engine. They will be discussed in later chapters that pertain to accessory electrical systems.

Computer Systems

The computer system on a vehicle can be thought of as the vehicle's "intranet." Up to 80 different computer modules may be wired into one multiplexed communications bus. Data is sent and received by each computer module and acted upon per its individual programming (**Figure 2-4**). Computer control of a component is based on inputs, processing, and outputs. The largest single computer module is the **Powertrain Control Module (PCM)**. This module not only controls traffic on the vehicle's intranet but also controls the operation of various electrical actuators on the engine, sends commands to other modules, requests and processes data from sensors on the engine and other components, and makes some embedded (programmed) decisions based on that data. The PCM may control the transmission directly or provide data or commands to the **Transmission Control Module (TCM)**, which controls shift solenoids and processes information (signals) from transmission sensors.

One of the many reasons that vehicles have on-board computers is to assist the manufacturers' ability to achieve federal **Corporate Average Fuel Economy (CAFE)** and

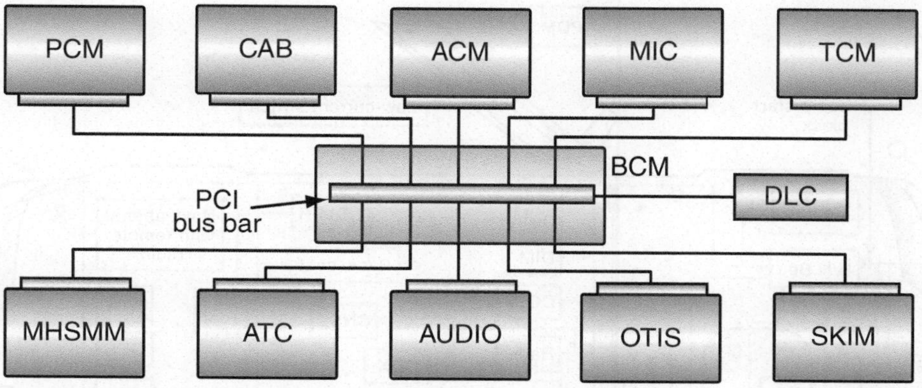

Figure 2-4 Automotive computers are networked together through multiplexing.

federal exhaust emission requirements. This just was not possible with strictly mechanical means like the carburetor and electrical/mechanical ignition systems. Electronics provided the manufacturers with a means to measure exhaust emissions, make necessary and immediate changes in the air–fuel mixture, and fire the spark plugs quickly and more accurately.

As the computer systems improved, it became more efficient to either measure or control other mechanical systems. The next two systems to be computerized, after the engine, were the automatic transmission and the Antilock Braking System (ABS). Since the mid-1990s, almost every mechanical device on the vehicle is measured, monitored, or controlled by a computer. This has greatly increased fuel mileage, reduced harmful emissions, and provided a smoother ride and happier passengers.

The major modules in the vehicle are the PCM, TCM, Body Control Module (BCM), airbag control module, instrument panel control module, and climate control module. Other modules may control antitheft devices, door locks, power seats and windows, suspension and steering, and many passenger comfort features such as radios, DVD players, and on-board navigation.

The newest major computer-controlled system is **vehicle stability**. This system incorporates data and actuators within the ABS, TCM, PCM, and automatic-ride control subsystems to reduce the loss of vehicle control. Many vehicles already have the system as standard equipment, whereas most of the others offer it as an option. On some vehicles, the computer can override the driver to prevent accidents. Other advanced computers can adapt the cruise control to the vehicle it is following and notify drivers when they are inadvertently driving into another lane. No matter what, when you start working on vehicles, there is one thing you must understand: Modern automobiles are electronically complex, and you must be able to understand the fundamentals to be successful at diagnosis and repair (**Figure 2-5**).

Vehicle stability A system that controls and processes data in critical situations to keep the vehicle safely on course by controlling each wheel separately.

In-Vehicle Communications

Recent digital communication improvements have resulted in much better data transfer within a vehicle. The most recent is known as a Controlled-Area Network (CAN). Simply put, it is an intranet system within the vehicle. The PCM acts as the controller (server), and each module (terminal) has an identification code. In action, the PCM asks each module to identify itself and transmit current data. Each individual module in turn monitors all or most of the digital messages and pulls needed information from the data stream that module requires for its function. The sequence of response is dictated by the programming in the system. **Intranet** is defined as a private working space, whereas the **"internet"** is a public working space.

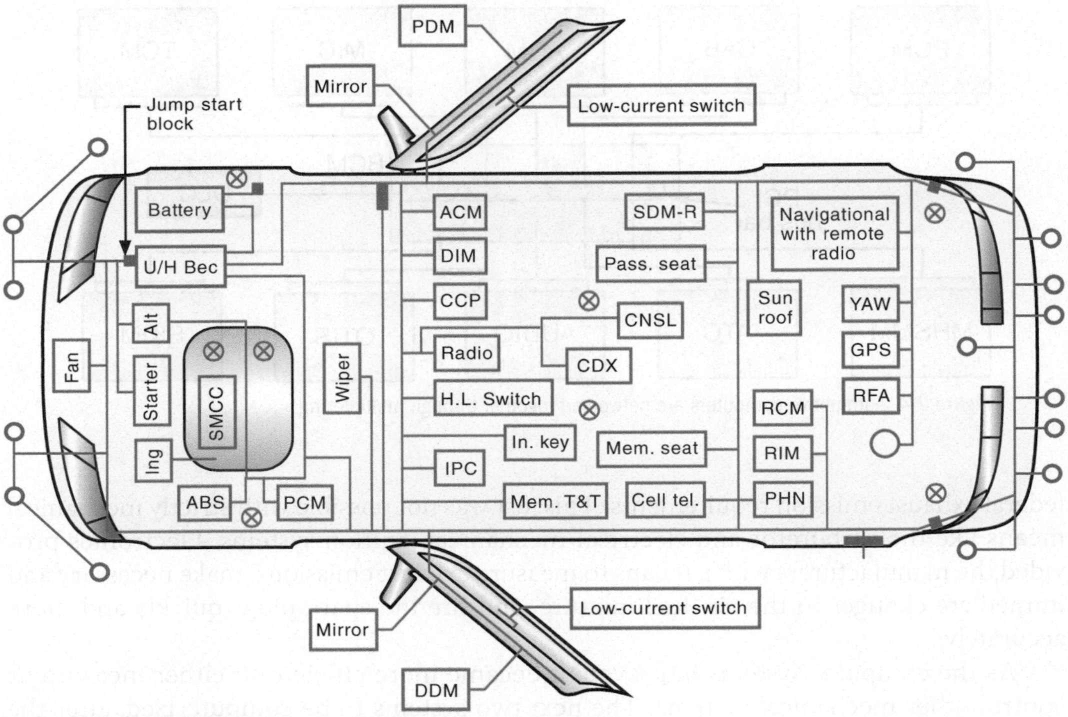

Figure 2-5 The electrical systems of today's vehicle can be complicated.

The data lines use a high and low circuit to carry the data. If one circuit fails, the same data is available on the other. The data transfers are much faster compared to the previous communication systems and can be monitored with newer scan tools. This high speed data collection and transfer allows the control system(s) to more accurately adjust vehicle component operations. This results in less fuel usage, cleaner exhaust emissions, and generally a more comfortable ride for the driver and passengers.

POWERTRAIN SYSTEMS

Modern vehicles use a variety of power sources. Most vehicles still use an internal combustion, reciprocating, piston-driven engine fueled by gasoline or diesel. A few vehicles use an electric motor with a large battery pack, and an ever-increasing number of vehicles use a combination of an electric motor and internal combustion engine (ICE). These power systems are known as Hybrid Electric Vehicles (HEVs).

Thermodynamics Physical science that deals with the relationship between heat and other forms of energy.

The **combustion chamber** is the area of the cylinder where the air-fuel mixture is compressed and ignited.

Internal Combustion Engines

The **Internal Combustion Engine (ICE)** uses expanding gases encased in a sealed chamber to apply force to a piston. This is **thermodynamics** (see Chapter 7), the use of heat to produce mechanical motion or energy (**Figure 2-6**). In a gasoline-fueled engine, air and fuel are mixed prior to entering the **combustion chamber**, either at the intake valve or in the throttle body. The air–fuel mixture is compressed by a piston to about one-eighth or less of its original volume and is ignited by a timed spark (**Figure 2-7**). In a diesel-fueled engine, first the air is compressed and then the diesel fuel is injected into the chamber under extremely high pressure. The diesel fuel is ignited by the superheated air.

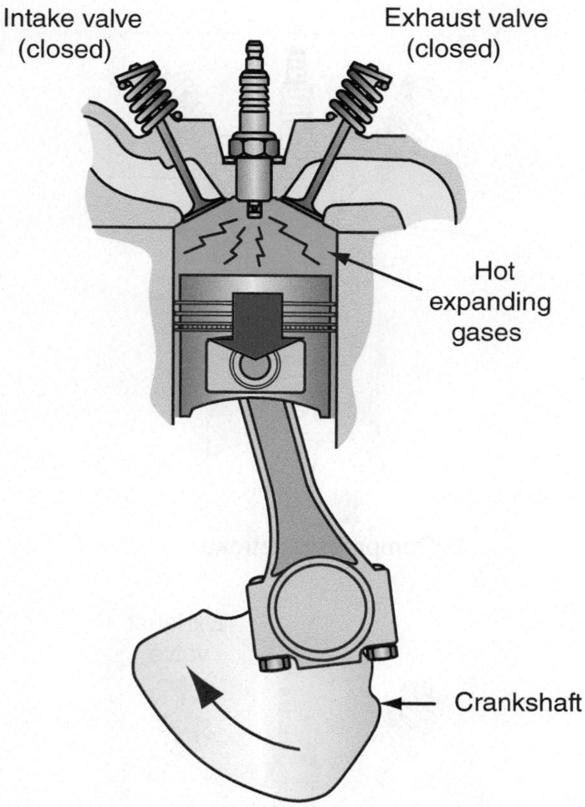

Figure 2-6 The expanding gases caused by combustion force the piston downward. Note that both valves are closed.

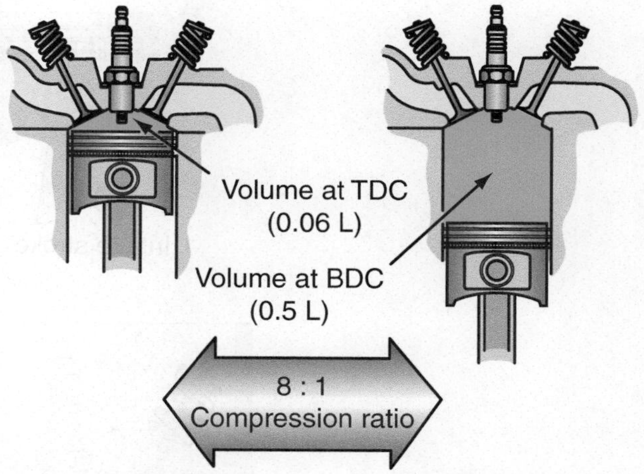

Figure 2-7 Compression ratio: One-half liter of air–fuel mixture is compressed into a 0.06-liter combustion chamber, giving a compression ratio of approximately 8:1.

AUTHOR'S NOTE When gas (vapor) is compressed, its temperature rises drastically. On a shop's air compressor, the steel line from the compressor to the storage tank reaches temperatures high enough to sear the skin if touched. On the working outlet of the air hose from the tank, the air is cool and could carry a lot of moisture. This is caused by the sudden drop in pressure between the tank and outlet end of the air hose. Moisture in the air condenses under these conditions. See Chapter 7 for more information on pressure and temperature.

Engine Operation

The standard ICE is known as a four-stroke engine. Detailed information on this type of engine is found in Chapter?, but the four "strokes" are intake, compression, power, and exhaust (**Figure 2-8**). During the intake stroke, the piston moves downward with the *intake* valve open. This allows air and fuel to fill the area vacated by the piston. The intake valve is closed, and the piston moves upward, *compressing* the air–fuel mixture to about one-eighth of its original volume. The spark plug fires near the top of piston travel, and the expanding gases *power* the piston downward turning the crankshaft. As the piston moves upward again, the *exhaust* valve opens, and the piston forces the spent gases from the bore and combustion chamber. The exhaust valve closes, the intake opens, and the cycle is repeated. It should be noted that a single piston produces a power stroke only once every two crankshaft rotations.

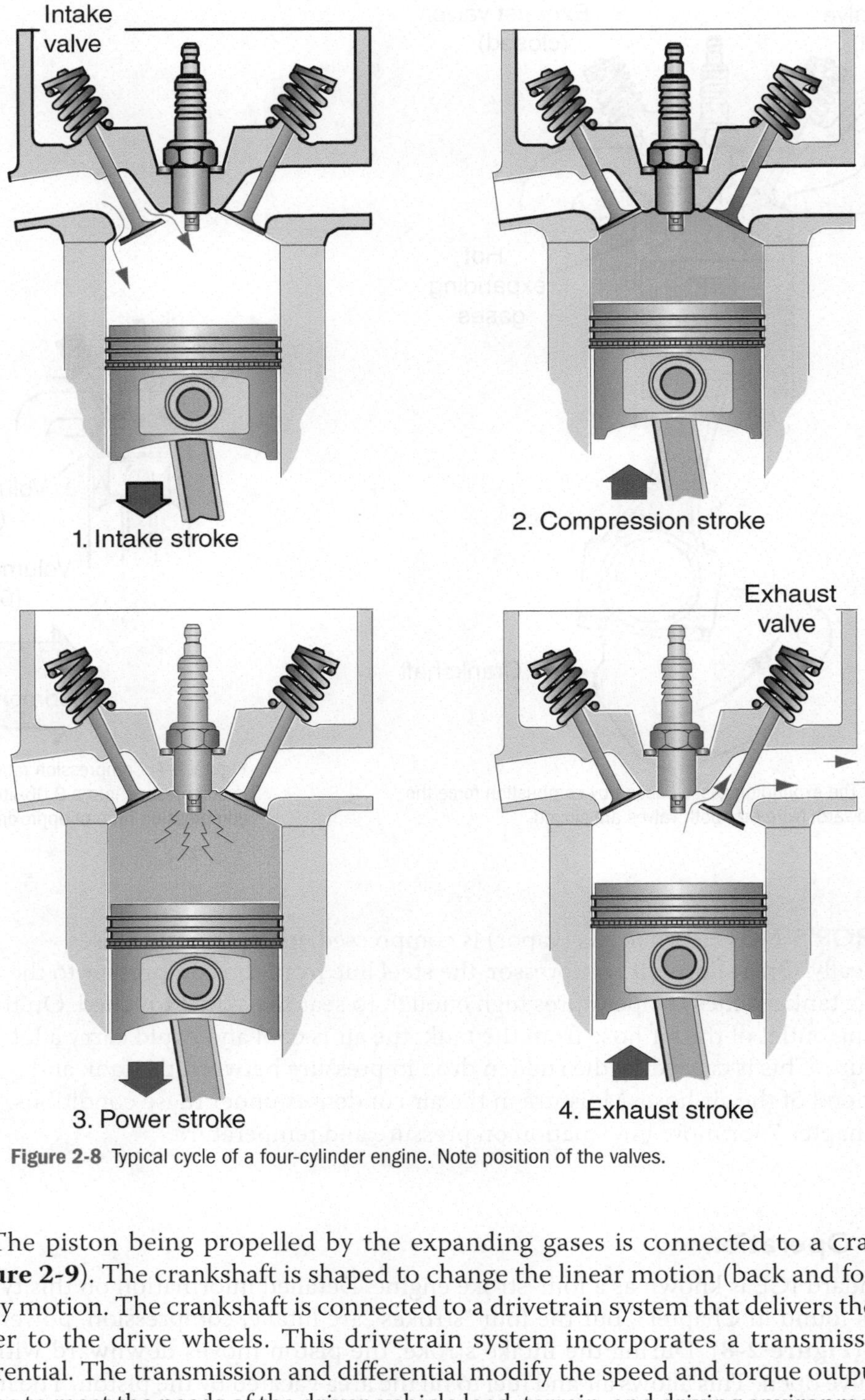

Figure 2-8 Typical cycle of a four-cylinder engine. Note position of the valves.

The piston being propelled by the expanding gases is connected to a crankshaft (**Figure 2-9**). The crankshaft is shaped to change the linear motion (back and forth) to a rotary motion. The crankshaft is connected to a drivetrain system that delivers the engine power to the drive wheels. This drivetrain system incorporates a transmission and differential. The transmission and differential modify the speed and torque output of the engine to meet the needs of the driver, vehicle load, terrain, and driving environment. This seems to be a straightforward operation, but to be effective, certain issues must be addressed throughout the entire operation.

First, the air and fuel must be in the right proportion to each other and must be delivered during particular cycles of piston travel. The best all-around air–fuel mixture is 14.7 parts air to 1 part fuel (14.7 to 1). This is accomplished by nature in a carburetor system, which mixes the fuel and air based on atmospheric pressure and engine vacuum. Atmospheric pressure is usually considered to be 14.7 pounds per square inch, and vacuum is any pressure less than atmospheric. This was a fair system for many years; however,

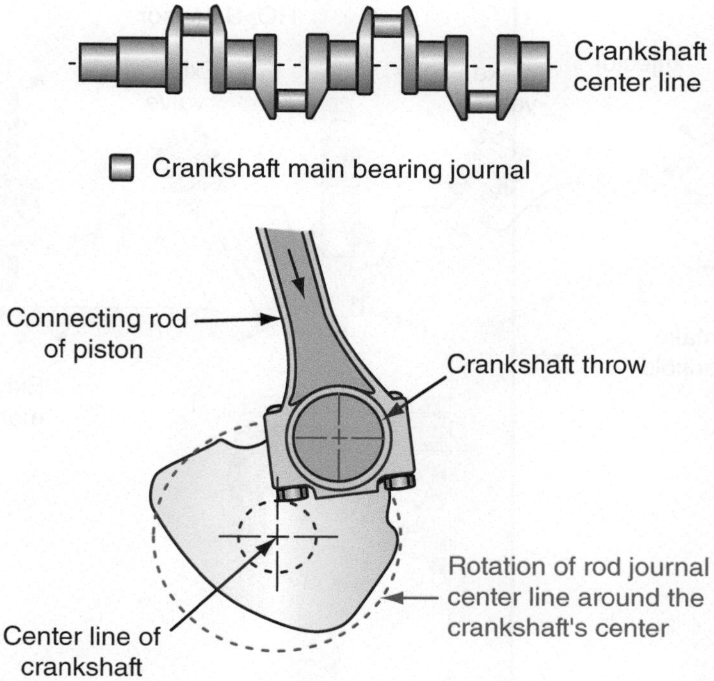

Figure 2-9 As the crankshaft throw rotates, it turns the crankshaft on its main bearings. This changes the linear motion of the piston into the rotary motion of the crankshaft. Shown is a piston one-quarter of the way through its downward movement.

it left a lot to be desired when considering the accuracy and adaptability of the air–fuel mixture for all driving conditions. Electronically controlled fuel injection is now the norm for delivering fuel to the combustion chamber (**Figure 2-10**) and is much more effective in controlling the air–fuel mixture.

The entry of the air–fuel mixture into the combustion chamber is controlled by timed mechanical valves (**Figure 2-11**). The exhaust of the spent combustion gases is similarly controlled by mechanical valves. New engine designs may use electrically controlled and actuated valves thus eliminating the need for many mechanical parts and providing better engine efficiency. Chapter 7 features a detailed explanation of the configurations of engines and their operations.

Electric-Powered Vehicles

During the late 20th century and into the 21st, millions of dollars were spent on developing vehicles that used only electric power for propulsion. Vehicle manufacturers, governments, electric utilities, and many other interested parties tried to find a zero-emission vehicle power source. The original intent was a vehicle operating on a battery (power pack), each recharge lasting about 300 miles, and capable of being quickly recharged from 80 percent discharge 1,000 times. The idea was to develop a 300,000-mile vehicle with many recyclable components. Many facets of electric power and its incorporated controls were researched, developed, and proven and now are in use on various vehicle applications today. Initially, the single biggest problem was the batteries comprising the power pack. Throughout the last few years, batteries and battery control have seen significant advances, thus allowing them to achieve enough power, sustainability, and low cost that consumers have been purchasing them in surprising numbers.

As of the beginning of 2013, of the 250 million registered vehicles in the United States, 1.6 million are HEVs and 75,000 are Electric Vehicles (EV).

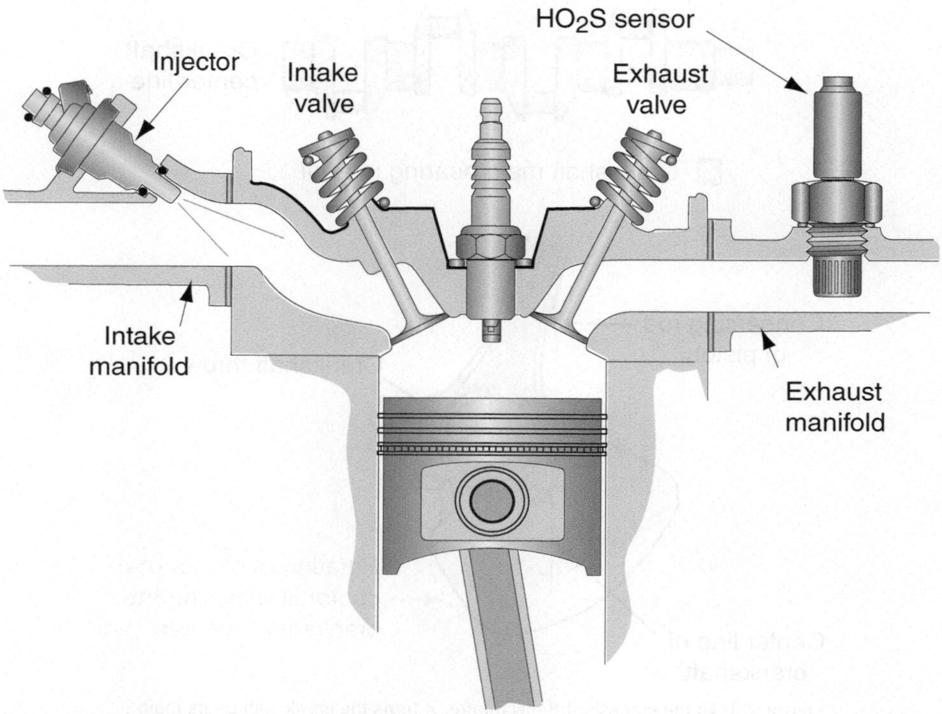

Figure 2-10 The injector opens for a specific period of time measured in milliseconds. Pressurized fuel is allowed into the air stream and into the cylinder.

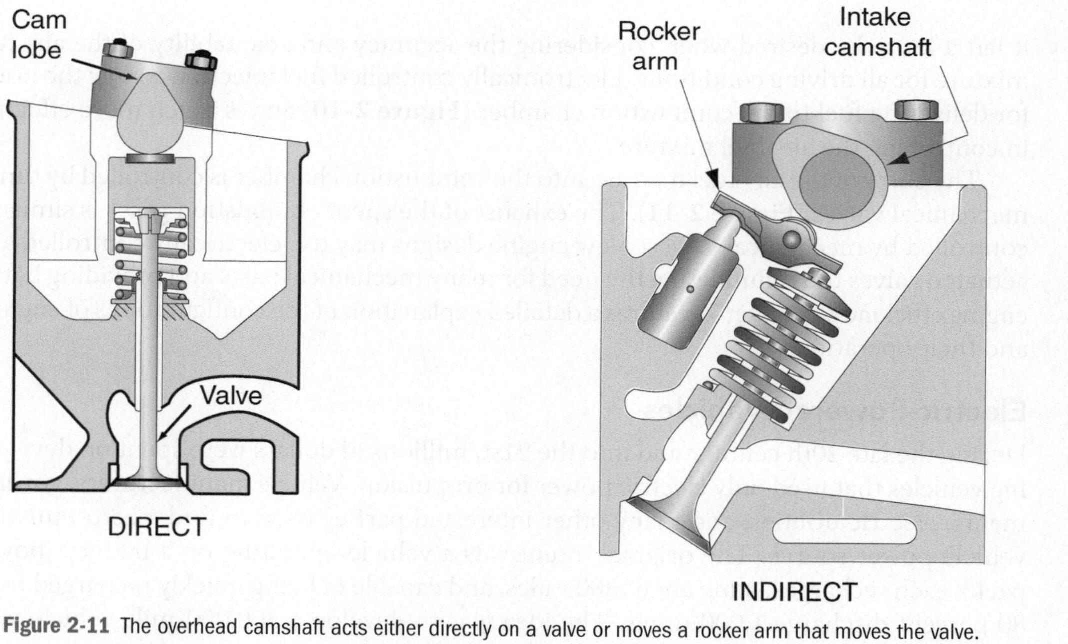

Figure 2-11 The overhead camshaft acts either directly on a valve or moves a rocker arm that moves the valve.

Hybrid power A vehicle that uses more than one means of propulsion. Hybrid vehicles use a combination of diesel or gasoline engine with an electric motor.

Hybrid Electric Vehicles

Hybrid power, in the case of motor vehicles, is a combination of an ICE coupled with a large electric-current generator and an electric power pack (**Figure 2-12**). The ICE is used to drive the generator, which recharges the power pack and, when needed, can assist in powering the vehicle.

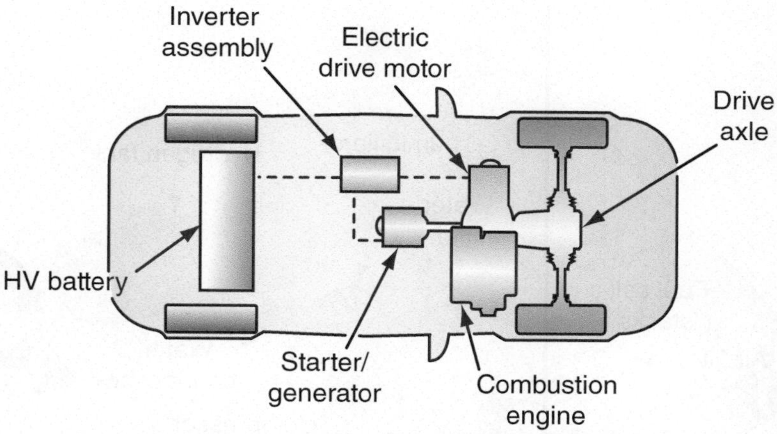

Figure 2-12 Simplified representation of a hybrid electric vehicle powertrain.

Figure 2-13 Toyota Prius was the world's first mass-produced HEV. It remains popular today but does not enjoy being the only choice for an HEV as it once did.

The power pack delivers electric current to one or more electric drive motors, either attached to a power splitter or directly to the drive wheels. On some HEVs, the engine runs at a constant speed, whereas on others the engine only operates when it is needed to recharge or provide drive power. Almost every manufacturer sells a hybrid vehicle or a hybrid version of a popular vehicle they already produce. Perhaps the most popular HEV is the Toyota Prius, with almost 3 million cars sold (**Figure 2-13**). Although hybrid-powered vehicles will continue to grow in popularity, new vehicle power systems are being developed. Much of the technology that was initially incorporated on a HEV has now made its way to other vehicles, improving their fuel economy.

Fuel Cell Power

Fuel cell technology is the utilization of electricity created by hydrogen (H) that has been run through a catalyst that strips away the hydrogen electron (**Figure 2-14**). The electron is directed to an electric device such as a motor and then routed to a second catalyst. The hydrogen ion formed after the loss of its electron goes through the second catalyst where it is combined with an oxygen molecule (O_2) and the freed hydrogen electron to form water (H_2O). This is the only emission by-product of a fuel cell. Although this technology

Fuel cell in vehicles generate electricity to power the motor, generally using oxygen and compressed hydrogen. Most fuel cell vehicles are considered zero-emission vehicles.

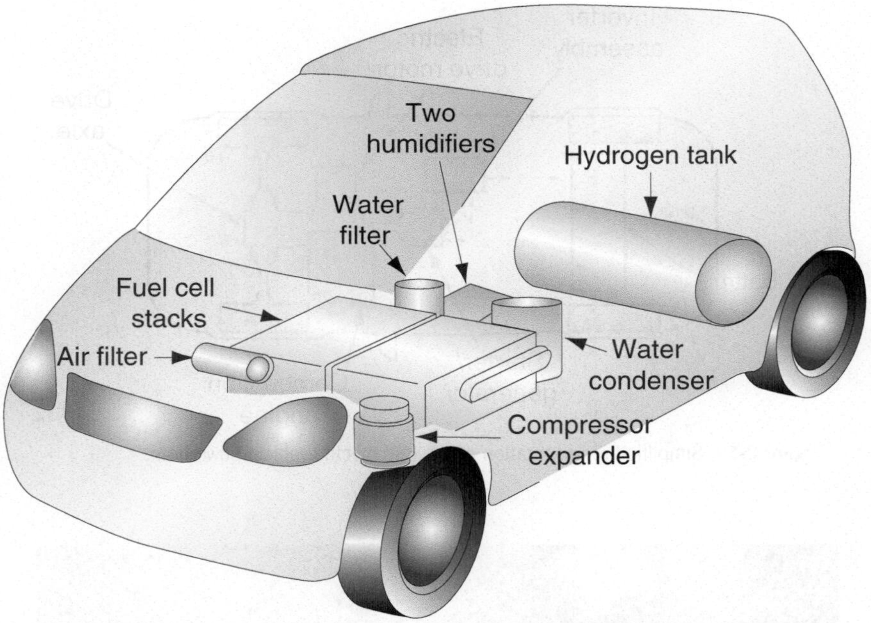

Figure 2-14 Fuel cell vehicle components. Technology has allowed engineers to design fuel cell vehicles without the loss of passenger and cargo space.

is in the first stages of research and development, it seems to work well. The system, when fully developed, will deliver sufficient power to satisfy the vehicle owner and passenger while reducing harmful emissions to near zero. A full market offering of fuel cell vehicles will be well into the future. Honda has a small fleet of fuel cell vehicles on lease to private citizens, and actual sales will follow in the next few years.

Alternative Fuels

Every type of engine requires some type of fuel. Today, hydrocarbon fuels are being blended in different chemical compounds to lower harmful exhaust emissions. A component of retailed hydrocarbon fuels—gasoline and diesel—is sulfur. Current fuel blends have reduced much of the sulfur and other compounds that create harmful emission gases during combustion. Alternative-fueled vehicles can be purchased almost anywhere. These vehicles look similar to regular vehicles because they are normal vehicles with small modifications to allow them to operate on different fuel choices.

Biodiesel A renewable fuel manufactured domestically from vegetable oils, animals fats, or recycled restaurant grease.

A new type of diesel is becoming popular in some areas of the country. **Biodiesel** is the use of modified vegetable oils. This can come from soybeans grown in a field, grease left over after cooking some foods, or algae (pond scum). These fuels are readily available and only require minor adjustments to the engine's fuel-delivery system. Most newer engines are already set up to handle anything from regular diesel to 20 percent blended diesel (known as B20). The oils are mixed with diesel to form biodiesel fuel.

Ethanol Ethyl alcohol same as the alcohol found in alcoholic beverages. mainly used as a biofuel additive for gasoline.

Starting with model year (MY) 2009, one of the main alternative fuels offered is **ethanol**, which is derived from corn, wheat, barley, sugar cane, and similar plants to produce a fuel that is environmentally friendly and has a performance like gasoline. Brazil is almost gasoline-free because of its foresight in developing an alternative to fossil fuels. However, Brazilian ethanol is made from sugar cane, of which it has plenty. Ethanol in the United States is generally made from corn or other grains and presents some problems not found in Brazil. Corn and other grains are a mainstay in the worldwide consumption of food. Moreover, U.S. corn is exported worldwide as a basic food source. More corn for ethanol means less corn for food. Governments, fuel companies, farmers, and consumers have some critical decisions to make concerning ethanol from corn and corn as a major food source.

E85 fuel is a mixture of 85 percent ethanol and 15 percent gasoline. It is cleaner burning, so harmful emissions are lowered somewhat and total gasoline usage is decreased. But sometimes information can be a little misleading. According to **Environmental Protection Agency (EPA)** testing, a 2008 Chevrolet Impala gets about 21 mpg in the city and 31 mpg on the highway using straight gasoline. The same vehicle on E85 gets 16 mpg city and 23 mpg highway. The same mileage loss applies to most of the other **flex fuel** (designed to run on E85 or other alternate fuel) vehicles. Ethanol generally sells for less than gasoline, so the loss of fuel economy is offset. Almost all "gasoline" fuels sold in 2009 had some percentage of ethanol (typically 10 percent or less). This is because blending ethanol into gasoline increases the oxidation of fuel and significantly reduces the harmful emissions output of an engine. Ethanol also burns cleaner inside the engine thus causing less long-term maintenance-related issues. Ethanol has a much higher octane rating than gasoline does, so the engine and the computer control must be modified. Some racing engines use ethanol because of this. In 2011, NASCAR decided that the fuel race cars would use would be ethanol.

Flex fuel A flexible-fuel vehicle (FFV) is an alternative fuel for vehicles. The engine is designed to run on more than one fuel, usually gasoline blended with ethanol or methanol fuel.

The consumer, with influence from the government, will eventually decide which fuel and engine/vehicle design are best suited for consumers and the environment.

Hydrogen

Some vehicle manufacturers are developing ICEs that use hydrogen for fuel. Hydrogen can be manufactured by removing it from other natural elements, and, when ignited, it burns cleanly with very few harmful emissions. BMW has produced a race vehicle fueled by hydrogen in a slightly modified ICE that produces excellent speed and great torque. The engine required some modifications to a standard gasoline-fueled engine but not enough to drive the cost out of range.

Other possible clean-burning fuels include methane retrieved from the earth, produced by rotting garbage, or made from some plants. Some of these fuels require only a source and retrofitting the engine before driving the vehicle. Vehicles with an ICE can use natural gas or propane with minimum changes to the engine or the vehicle. All the fuels and power systems discussed may be possibilities in the near future. The automotive industry, like all others, changes constantly, and vehicle power systems are changing as this text is published and read.

AUTHOR'S NOTE One of today's main drawbacks to vehicles using fuel cells or alternative fuels is the social infrastructure. Where do we refill the hydrogen tank? What are the safety concerns during fueling? Who is going to produce large quantities of biodiesel? When do we start technician training? What happens to all the people and businesses involved in the manufacturing, transportation, and sale of gasoline and diesel? Many communities, various levels of government, and fuel suppliers and retailers are eliminating this challenge at an increasing rate of growth. However, if only one-tenth of the total vehicles sold in the United States and Canada today were required to have absolutely zero harmful emissions, most of them would stop operating after the first tank of fuel because of a lack of supporting infrastructure.

Powertrains

The driveline delivers the engine's torque output to the wheels. There are several major components in this system. In a rear-wheel-drive (RWD) or four-wheel-drive (4WD) vehicle, they include a transmission, differential, axles, drive shafts, and all the mountings

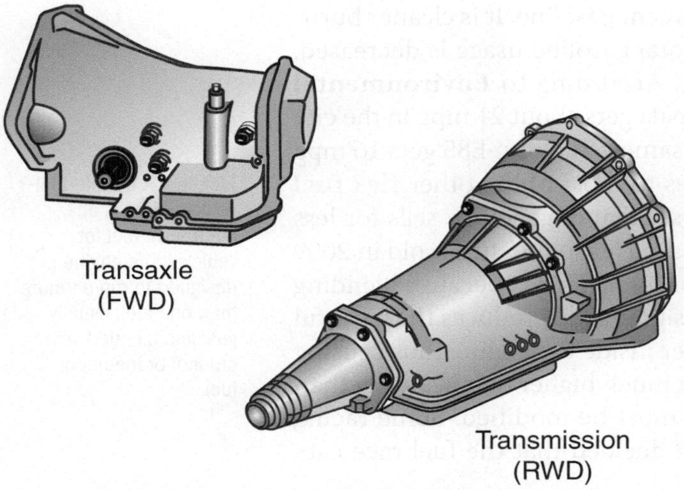

Figure 2-15 Typical transaxle (FWD) at the top and a transmission (RWD) the bottom.

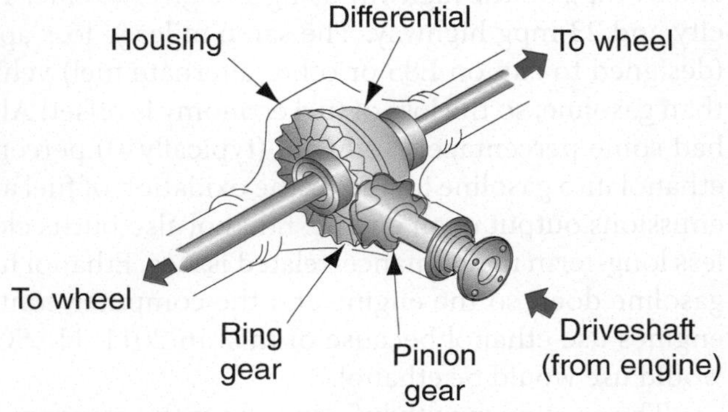

Figure 2-16 Power flow through a RWD differential.

and housings. In the front-wheel drive (FWD), in the same order, the components are the transaxle, final drive, drive axles, mountings, and housing. There is no drive shaft in a FWD vehicle. Most of the driveline is in the engine compartment or under the vehicle. These systems are covered in detail in Chapter 8.

An **automatic transmission** changes forward gears by using valves and hydraulic pressure. A **manual transmission** requires the driver to select a gear using a shift lever.

The transmission/transaxle provides a means of selecting a gear size that smoothly marries engine output to vehicle load (**Figure 2-15**). It may be an **automatic** or a **manual transmission**, where the driver selects forward gears through a lever and linkages. The differential or final drive allows the drive wheels to turn at different speeds when the vehicle turns a corner or through a curve (**Figure 2-16**). Drive shafts provide the link between the RWD transmission and differential. Drive axles deliver the power from the transaxle final drive to the wheels on a FWD vehicle.

Some vehicles have four- or all-wheel drive. Four-wheel drives allow the driver to use the front wheels to pull the vehicle when the rear wheels lose traction. All-wheel systems drive all four wheels at all times. Both systems are electronically controlled on modern vehicles.

One type of automatic transmission that is offered is the Continuously Variable Transmission (CVT). It was first introduced by SAAB in production vehicles of the late 1990s. Today, it is more common and appears on several models. Many HEVs use this type of transmission. It has no gears per se like most of the current transmissions. Instead it has two opposing pulleys whose sides can be pushed together or moved apart. As the sides of one spread open, the other pulley's sides are pushed together. This, in effect, causes the two pulleys to change their diameters. The drive belt between the two pulleys moves up and down within the pulleys as their sides move. In low-range gears or reverse, the driven pulley is closed, making a large pulley. The drive pulley is fully open, creating a small pulley. Torque is now sufficient to accelerate the vehicle. As vehicle speed increases, the driven pulley sides open (diameter gets smaller) and the drive pulley sides close (diameter gets larger), creating more speed and less torque. The applied drive ratio can be any ratio between the two extremes. See Chapter 8 for more information on this and other powertrain components.

Steering and Suspension Systems

Steering, as the word implies, allows the driver to steer the vehicle in the direction desired. The steering system provides a system of linkages that allows the driver to turn the front wheels (**Figure 2-17**). The steering wheel is connected through a shaft to the steering gear unit. The gear multiplies the input force to make steering easier.

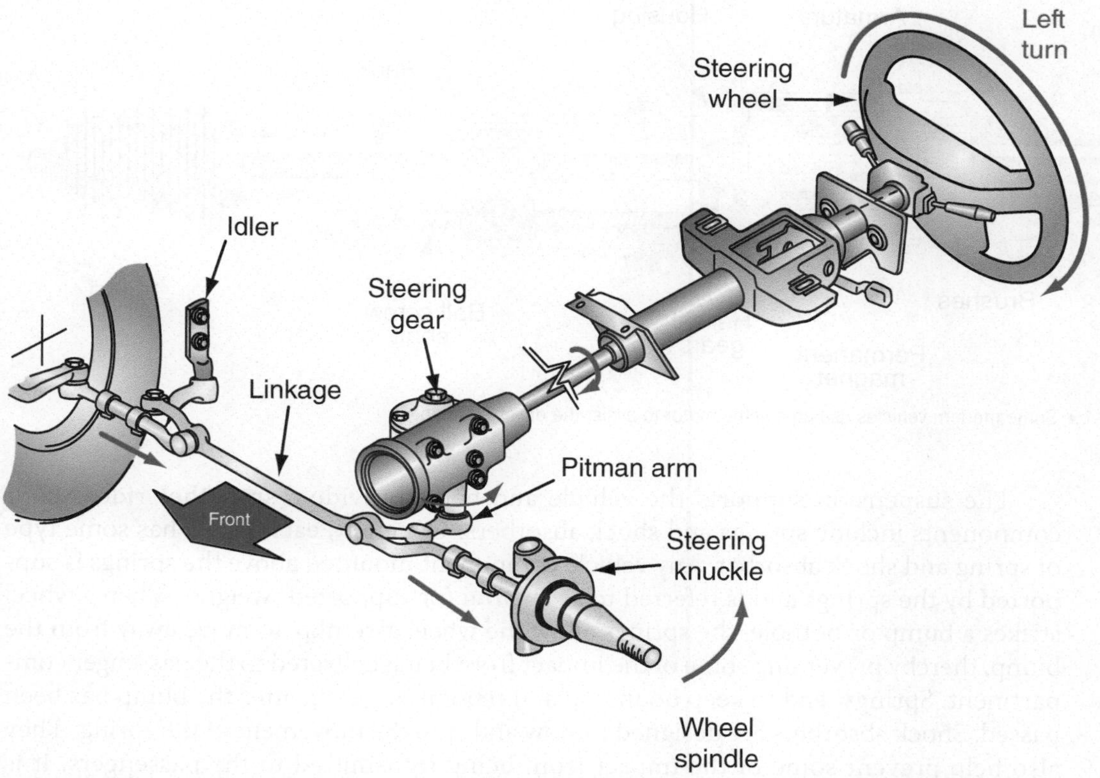

Figure 2-17 A traditional type of steering system that uses links and arms to change direction. Another name for this type of steering is the parallelogram style.

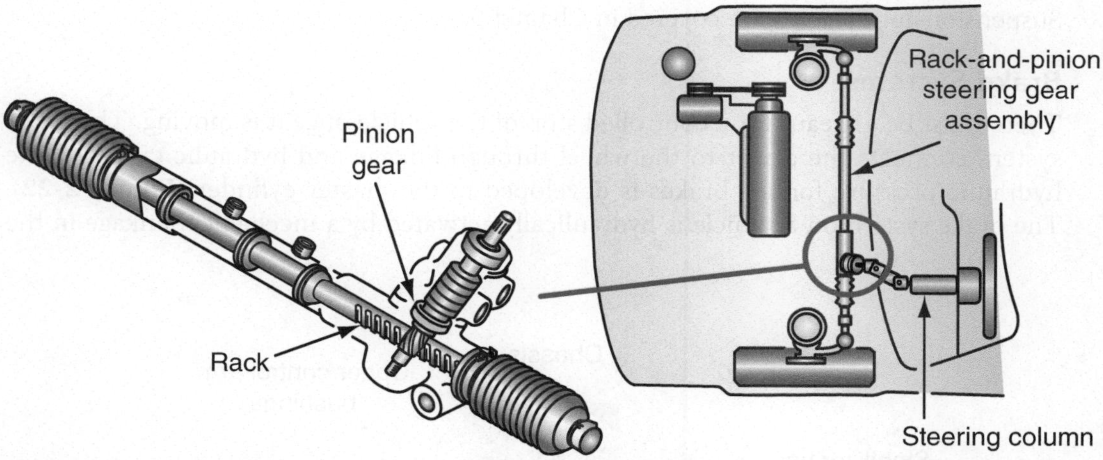

Figure 2-18 A rack-and-pinion assembly provides a direct link between the steering wheel and steering links through two gears: a pinion gear and a rack gear.

Most systems have power-assisted steering. This can be done hydraulically or electrically. In a hydraulic unit, valves in the gear unit direct pressurized hydraulic fluid against a piston. This reduces driver effort. The major components are the steering column with the steering wheel, linkage, steering knuckle, and gearing. Newer vehicles have a rack-and-pinion steering systems where the gear unit has been replaced (**Figure 2-18**). The rack system uses less room and connects the steering gear assembly directly to the wheels. Many vehicles today, and more in the future, will be using an electric motor to assist the driver instead of a hydraulic system. This system uses less room and does not need an engine-driven belt—a belt takes power to turn, and this wastes engine power and fuel (**Figure 2-19**).

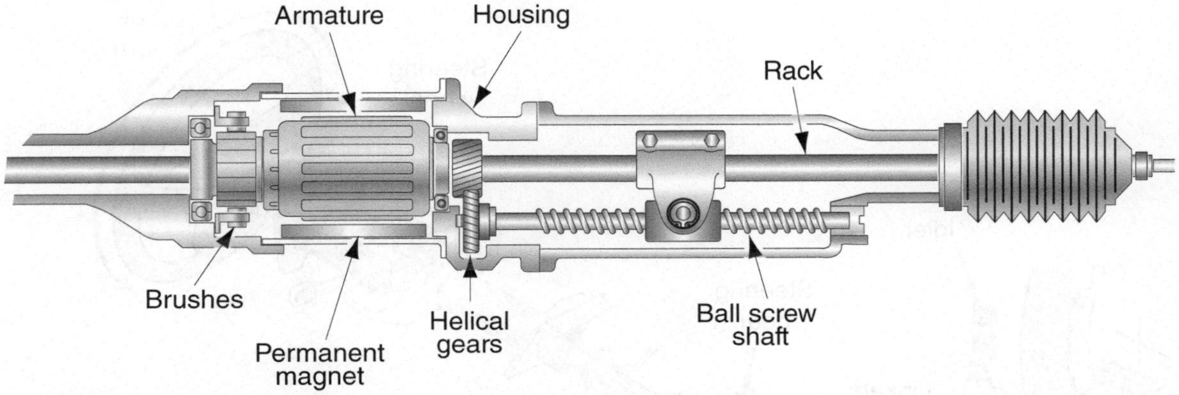

Figure 2-19 Some modern vehicles use an electric motor to assist the driver in turning.

Sprung The portion of the vehicle's total mass that is supported by the suspension. Load carrying.

The suspension supports the vehicle and helps provide a smoother ride. Major components include springs and shock absorbers. Generally, each wheel has some type of spring and shock absorber. Any vehicle component mounted above the springs is supported by the springs and is referred to as **sprung** (or supported) weight. When a wheel strikes a bump or pothole, the springs allow the wheel assembly to move away from the bump, thereby preventing some of the impact from being delivered to the passenger compartment. Springs tend to keep bounding and rebounding even after the bump has been passed. Shock absorbers are designed to slow and stop the movement of the spring. They also help prevent some of the impact from being transmitted to the passengers. It is important to remember that the shock absorbers do not support the weight of the vehicle; this is the job of the springs. There are several types and designs of suspension systems. **Figures 2-20** and **2-21** show the differences between two of the most common types. Suspension and steering are covered in Chapter 9.

Brake Systems

There must be a means for a controlled stop of the vehicle once it is moving. The brake system connects the driver to the wheel through linkage and hydraulic circuits. The hydraulic pressure for the brakes is developed in the master cylinder (**Figure 2-22**). The brake system on a vehicle is hydraulically activated by a mechanical linkage in the

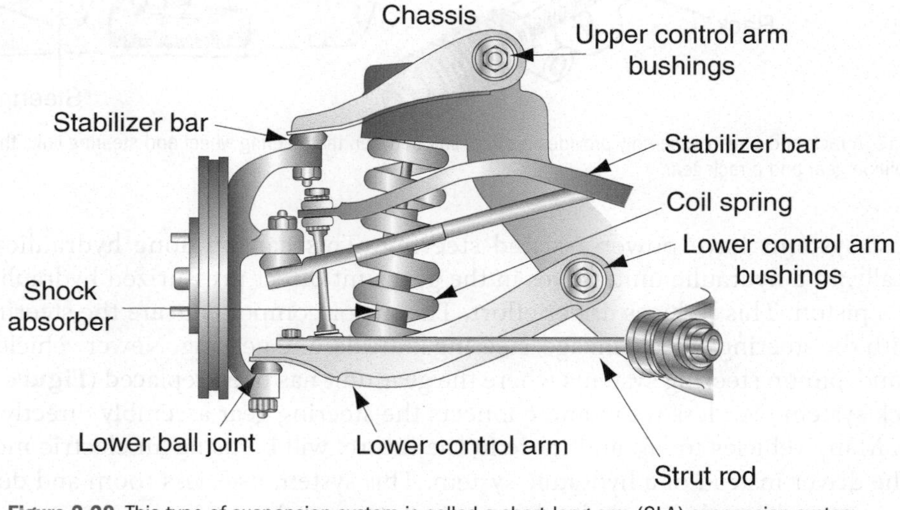

Figure 2-20 This type of suspension system is called a short-long arm (SLA) suspension system.

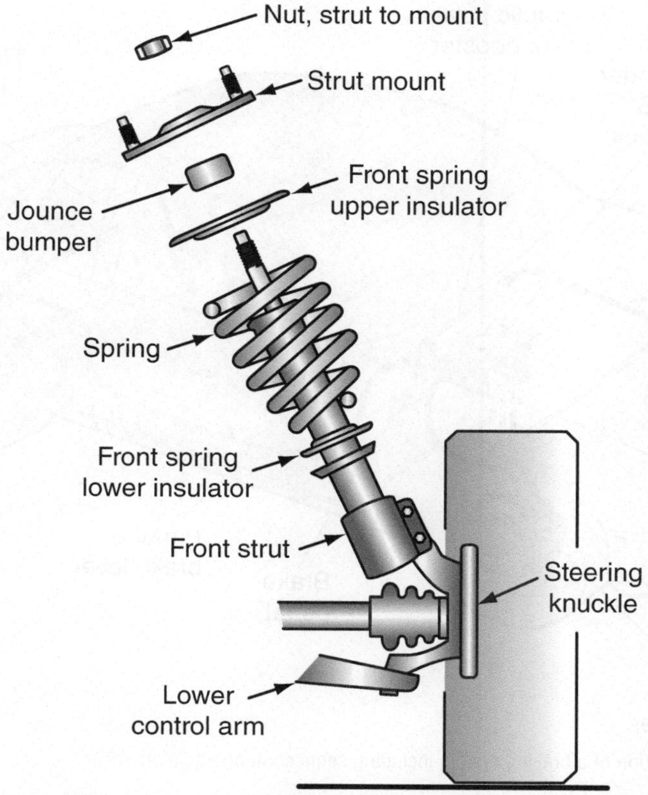

Figure 2-21 This type of suspension system is called a MacPherson strut.

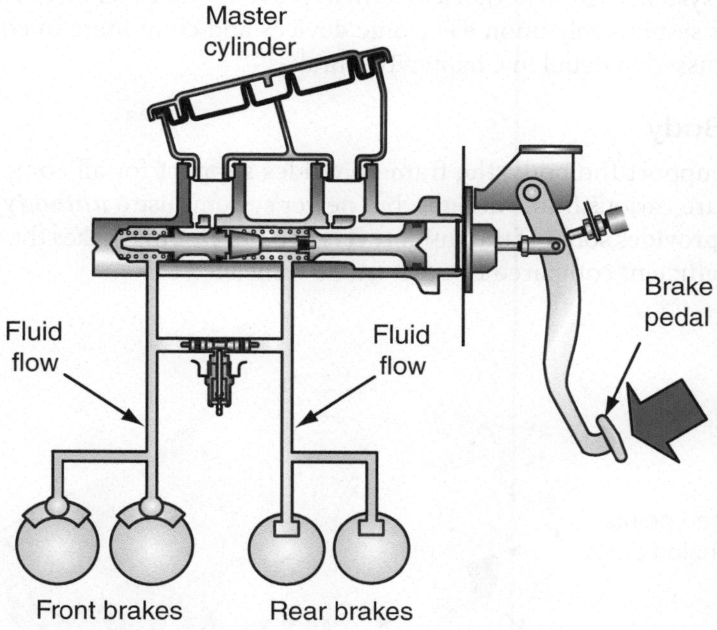

Figure 2-22 Force on the brake pedal moves the master cylinder pistons forward, forcing pressurized fluid to generate braking force at the wheels.

passenger compartment. The applied force is transmitted via hydraulics to the brakes at the front and rear axles. Mechanical devices apply the force to the braking components at the wheels. These mechanical devices include calipers, wheel cylinders, rotors, drums, pads, and shoes (**Figure 2-23**).

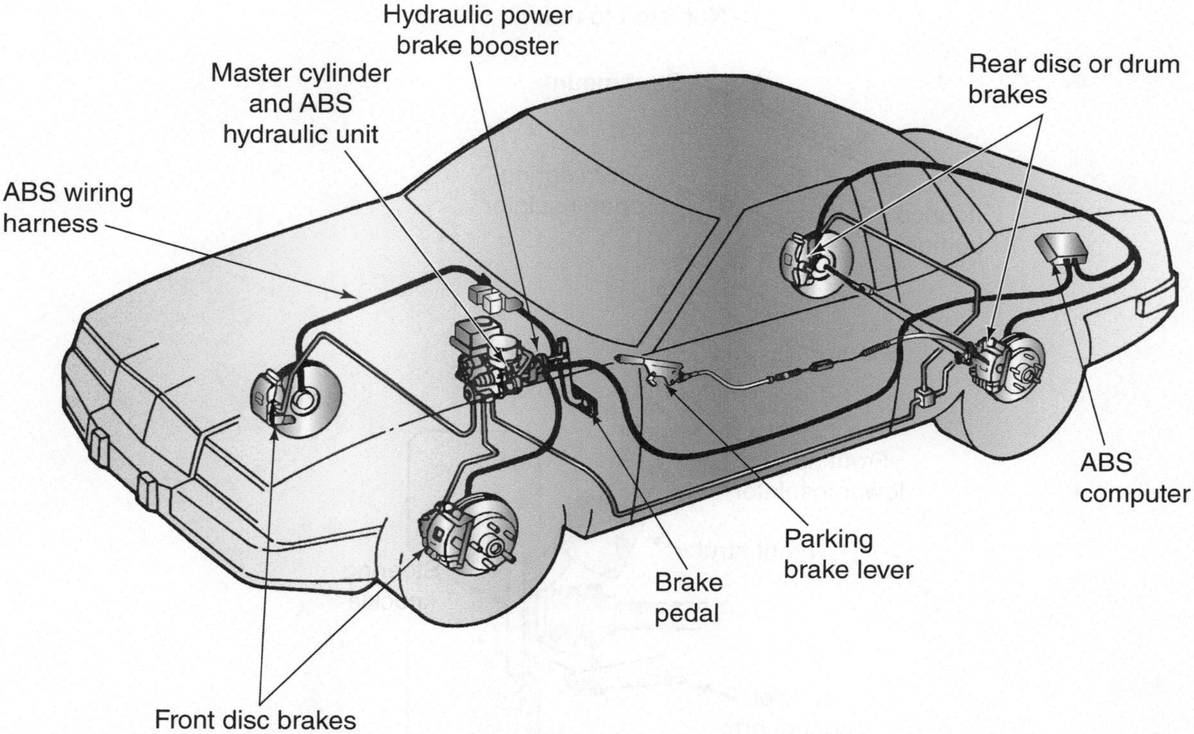

Figure 2-23 A typical layout and location of a braking system including some components of an ABS.

Many new systems use an antilock system to prevent brake lockup during emergency stops. Antilock systems rely upon electronic devices and computers to control braking. Brakes are discussed in detail in Chapter 10, "Brakes."

Frame and Body

Just as bones support the body, the frame provides support for all components of the vehicle. There are various frame designs, but newer systems use a ***unibody design*** where the body itself provides some of the support (**Figure 2-24**). This makes the vehicle lighter and more fuel efficient compared to a full-frame vehicle.

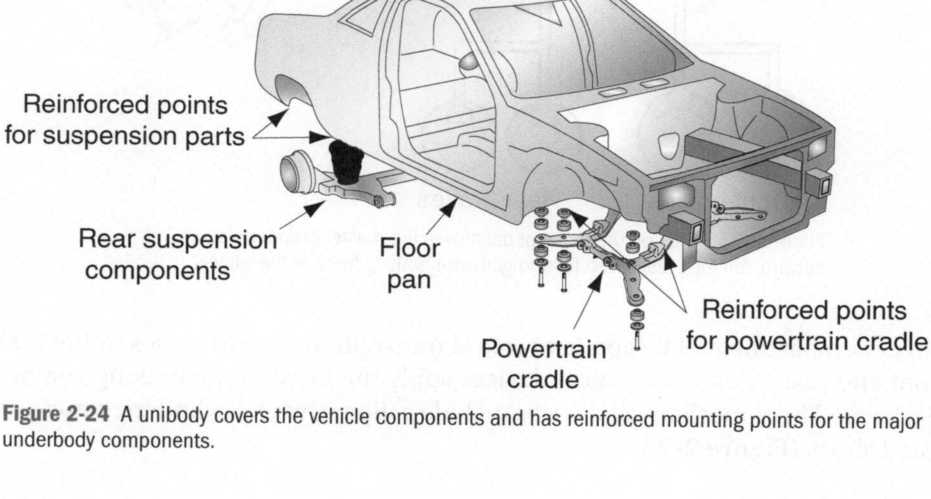

Figure 2-24 A unibody covers the vehicle components and has reinforced mounting points for the major underbody components.

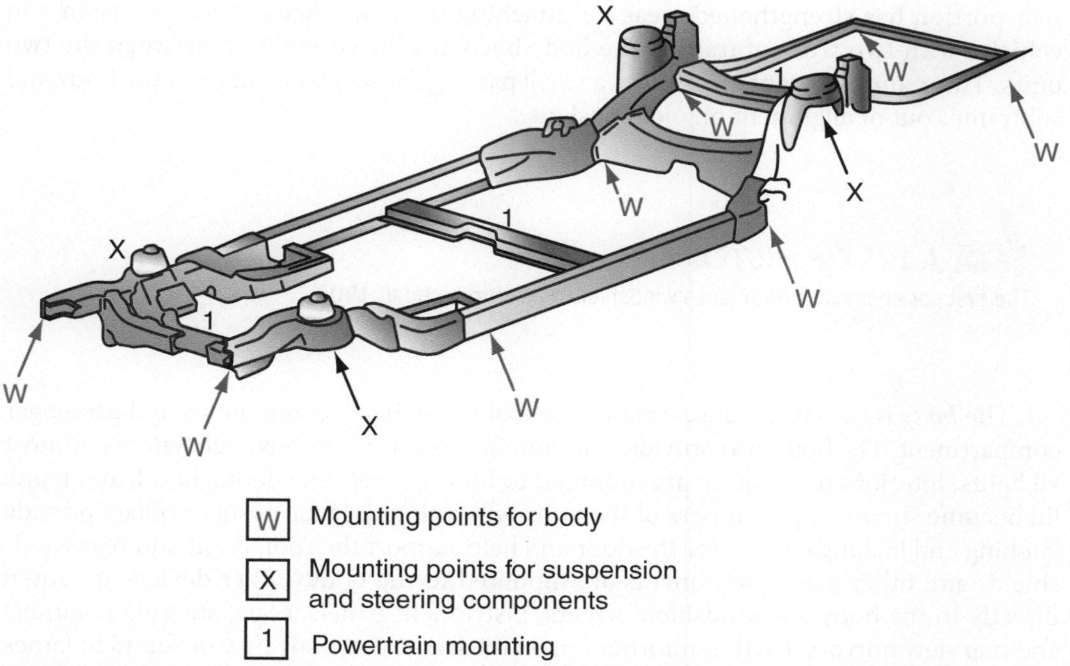

Figure 2-25 A full frame (also known as a ladder design) provides mounting points for the major underbody components.

The frame has attachment points for suspension and steering components, bumpers, and the body. A ***full frame*** has rigid parallel channels with cross-members, which help keep the parallel channels in place and provide mounting points for items such as the engine and transmission (**Figure 2-25**). Full frames are also called ladder frames because the frame looks like a ladder. Full frames are heavy but can support heavy loads. Most passenger cars now use the unibody, whereas trucks and larger sport utility vehicles still use the full frame. Older RWD cars used a version of the full frame, but modern RWD cars use a unibody design.

Unibody construction relies on the body to support the passengers and interior components. However, the unibody by itself cannot support the torque actions and movement of the powertrain. The steering and suspension systems also stress a unibody. To counter this problem, subframes are used to support the powertrain and at least some portion of the steering and suspension systems (**Figure 2-26**). The subframes are constructed to support various loads without stressing the rest of the vehicle. The subframes are also referred to as engine and transmission ***cradles***. Some vehicles only have a subframe at the front of the vehicle supporting the engine, transaxle, steering, and front suspension. The

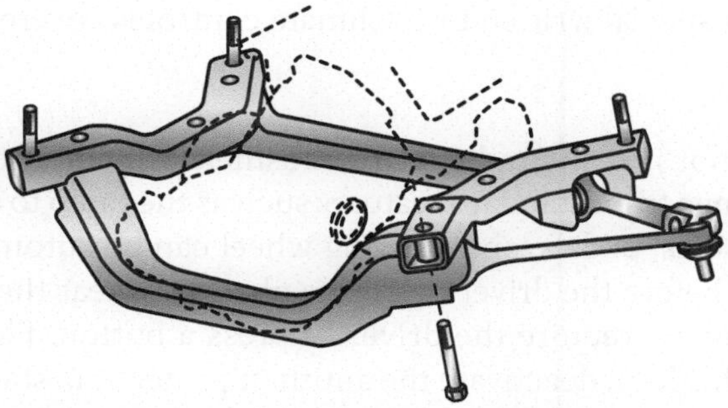

Figure 2-26 A typical subframe for supporting the engine, transaxle, and some suspension and steering components.

rear portion has strengthened areas for attaching the rear wheels and suspension. On vehicles with two true subframes, the body becomes the connection between the two units. This reduces weight greatly, but a small parking lot accident can twist the body and subframes out of alignment on older designs.

A BIT OF HISTORY

The first car enclosed with a glass windshield was introduced in 1903.

The *body* is the shell that surrounds most of the vehicle's components and passenger compartment. The body also provides a mount for several automotive subsystems. Almost all lights, interior and exterior, are mounted to body panels. The doors, hood, and trunk lid become supporting members of the body when closed. Door posts or pillars provide latching and locking devices for the door and help support the roof. Front and rear windshields and other glass components are mounted to the body. Other devices mounted directly to the body are windshield wipers, instrument panels, seats, steering columns, and rearview mirrors. Further information on the frame and body may be found in James E. Duffy, *Auto Body Repair Technology,* Fifth Edition, Cengage Learning.

The body forms the structure that moves the air aside as the vehicle moves. Because of CAFE and emissions concerns, it became necessary to make that action as smooth as possible. The body is shaped and molded to be aerodynamically fluid. Less air resistance means less fuel used with a corresponding lowering of harmful emissions. Old-time fans of NASCAR tend to be disgruntled with the racing vehicles of today because "all the cars look alike." The problem is that, from the manufacturer's viewpoint, there are only a few shapes or forms that lower air resistance on a moving vehicle. Thus, most of today's vehicles bear at least a surface resemblance to each other.

Climate Control Systems

Climate control includes the heating and air-conditioning systems. In many models, the two systems have been combined into a single system that may be controlled manually or automatically. Either system can keep the passengers comfortable by using liquids to transfer heat.

The heater system blows air over hot engine coolant flowing through a small radiator-type device called a heater core. This warms the air and the passenger compartment. Air conditioning works in a similar manner but uses a special liquid referred to as **refrigerant** and varying pressures within the system. Both systems can affect the operation of the engine, and both should be serviced or supervised by an experienced technician. In fact, federal, state, and local laws require an air-conditioning technician to be certified through a special written test. Climate control is covered in Chapter 11.

Refrigerant is a chemical that cools hot air by changing its state from liquid to vapor to liquid.

Accessories

Accessories include not just the radio and windshield wipers of the old days. Today's vehicle accessories range from standard features such as the radio to a **global positioning system (GPS).** The seats, pedals, and steering wheel can be automatically adjusted just by touching a button before the driver even gets physically near the vehicle. Some of the newer systems don't even require the driver to press a button. Electronic sensors and controls will unlock the door, deactivate the antitheft protection, start the engine, turn on the heat or air conditioning, move the steering wheel and seat to allow easy entry, and then adjust all those devices to a particular driver. This is all done based on a radio signal from the remote control "fob" in the driver's hand or pocket as he or she approaches the vehicle. Other accessories include DVD and game players for the passengers. Navigation features allow the driver to access driving directions without pulling over for assistance. Like air conditioning in the southeastern United States, all of these "exotic" features will be standard on future vehicles. Each requires some degree of technical expertise for diagnosing and repairs. Although this means better pay and employment for the automotive technician, the downside is increased cost of parts and labor. The cost of the parts may eventually be lower, but the labor cost of a well-trained, motivated technician will likely increase. This further emphasizes the technician's golden rule: ***Fix it right the first time.***

SUMMARY

- The electrical system has become the single most important system regarding vehicle operation.
- Starting and charging systems provide the means to start the engine and power the electrical system.
- The power system is the component that changes heat from burning or reactive chemicals to mechanical energy.
- Power systems are constantly in a state of change regarding fuels and operations.
- Alternative fuels may provide a means to reduce the use of hydrocarbon fuels while lowering harmful emissions.
- The powertrain transmits the power-system output to the drive wheels.
- The transmission or transaxle provides a means to harness the engine's output to the load based on vehicle load and environment.
- Steering allows the driver to direct the vehicle in the desired direction.
- Suspension helps protect vehicle components and aids in a better ride.
- A brake system provides the driver a way to stop and control the vehicle.
- The frame supports the body and all of its mounted components.
- The body protects the passengers and vehicle components from outside hazards.
- The climate control system, although not needed to operate the vehicle, reduces driver and passenger fatigue, and increases alertness.
- Accessories include many features or subsystems that assist the driver or passenger and provide a better interior environment.

REVIEW QUESTIONS

Short Answer Essays

1. Discuss how the vehicle electrical system has expanded in use over the last 20 years.
2. Explain how the starter circuit turns the crankshaft during engine startup.
3. Discuss the use of alternative fuels in today's and future vehicles.
4. Explain the purpose of the powertrain.
5. Explain the general usage of a computer or computers in today's vehicles.
6. List five of the major computer modules used on light vehicles.
7. List and explain the purpose of the powertrain's major components.
8. Explain the primary operational difference(s) between a manual transmission and an automatic transmission.
9. Discuss the operation of a brake system.
10. Explain the overall operation of an internal combustion engine.

Fill-in-the-Blanks

1. The largest single vehicle computer is the ________________.
2. The charging system uses a belt-driven ________________ to produce electrical current to operate the vehicle.

3. CAFE is a federal requirement concerning ______________ or usage.
4. ______________ is the use of heat to produce mechanical motion or energy.
5. The ______________ is shaped to change linear motion into rotary motion.
6. The entry of the air–fuel mixture is controlled by timed, mechanical ______________.
7. Vegetable oils can be mixed with diesel to form ______________.
8. Ethanol in Brazil is produced from ______________.
9. Newer vehicles have a ______________ steering system.
10. Many new antilock brake systems are used to ______________ brake lockup during emergency stops.

Multiple Choice

1. Mechanical devices apply the force to the brake components by which of the following?
   A. Calipers
   B. Wheel cylinders
   C. Pads and shoes
   D. All of the above
2. The vehicle body
   A. protects the vehicle components.
   B. provides protection for the vehicle components.
   C. provides mounting points for many passengers.
   D. All of the above.
3. CAFE helped cause the development of the
   A. unibody.
   B. full-frame support.
   C. aerodynamically shaped body.
   D. None of the above.
4. The vehicle heater
   A. blows air over the radiator for heating.
   B. may be paired with the air-conditioning system.
   C. is always a separate system.
   D. draws electric power from the battery during engine operation.
5. The following subsystems are part of the electric system, EXCEPT:
   A. ABS.
   B. ignition.
   C. starter.
   D. manual transmission.
6. When the ignition key is in the "start" position, control circuit current is routed through all of the following EXCEPT:
   A. clutch switch.
   B. P/N switch.
   C. starter motor.
   D. solenoid.
7. The data on the vehicle's communication bus is controlled by the
   A. PCM.
   B. BCM.
   C. intranet server.
   D. Both A and B.
8. Spent air–fuel mixture is routed through and around the
   A. intake valve.
   B. exhaust valve.
   C. compression valve.
   D. power valve.
9. The best air–fuel mixture ratio for reducing emissions and improving performance and fuel mileage is
   A. 14 to 7.1.
   B. 1 to 14.7.
   C. 14.7 to 1.
   D. 8 to 1.
10. A typical hybrid-powered vehicle operates with a(n)
    A. internal combustion engine.
    B. power splitter.
    C. power pack.
    D. All of the above.

CHAPTER 3

AUTOMOTIVE TOOLS AND EQUIPMENT

Upon completion and review of this chapter, you should understand and be able to:

- Describe the safe use of the common wrenches used in the automotive shop.
- Describe the correct use of the power wrenches used in the automotive shop.
- Explain the safe use of the metalworking tools commonly used in the automotive shop.
- Explain the safe use of the common threading tools used in the automotive shop.
- Understand the importance of storing, cleaning, and maintaining tools.
- Identify the major types of automotive lifts.
- Identify the types and uses of automotive test equipment.
- Describe the purpose of computerized test equipment.

Terms To Know

Adjustable wrench
Air-operated ratchet driver
Allen wrench
Ball-peen hammer
Box-end wrench
Brass hammer
Center punch
Channel-lock pliers
Chisel
Combination wrench
Computerized data
Cotter pin puller
Creeper
Diagonal cutting pliers
Diagnostic trouble codes (DTCs)
Die
Digital storage oscilloscope (DSO)
Double-post lift
Drive-on lift
File
Floor jack
Frequency
Gear puller
Hacksaw
Helicoil
High impedance
Impact wrench
Needle nose pliers
Offset screwdriver
Oxyacetylene welding and cutting outfit
Pin punch
Plastic-tip hammer
Reamer
Recessed-head screwdrivers
Roll pin punch
Rubber mallets
Scissor lift
Screwdriver
Screw extractor
Single-post lift
Socket wrench
Snapring pliers
Starter punch
Tap
Torque wrench
Thread-restoring files
Transmission jack
Twist drill
Vise grip
Wheel blocks

INTRODUCTION

Knowing how to use and care for tools properly is one of the most important parts of being an automotive technician. There is a correct tool for each and every automotive repair job. It is also important to use each tool in the correct way. Technicians use a number of these tools on every repair job. This chapter discusses the most common tools. Remember, a successful technician must know the name of each tool, what it does, and what tool works best for each automotive repair job.

Shop Manual
page 45

Most of the tools used by a technician are called hand tools. They get their name from the fact that they are operated by the technician's own muscle power. Power tools, on the other hand, get their power from electricity, compressed air, or hydraulics.

COMMON HAND TOOLS

Wrenches

Many automotive parts and components are held together or fastened with bolts and nuts. Wrenches, which come in many varied sizes, are designed to tighten or loosen these bolts and nuts. Bolts and nuts also come in varied sizes.

The size of a wrench is determined by the size of the nut or bolt head it fits on. The wrench shown in **Figure 3-1** has the number 14 stamped on it. This means the opening of the wrench measures 14 millimeters across the opening and it will fit on a bolt head or nut that measures 14 millimeters across.

Wrench sizes are given either in metric or U.S. (English) system units (see Chapter 4, "Fasteners and Measurements"). Metric wrench sizes are given in millimeters, for example, 7–13 millimeters. Sizes for U.S. (English) system wrenches are given in fractions of an inch, for example, 5/16, 3/8, 7/16, 1/2, and 9/16 inches. Because both U.S. (English) and metric system bolts and nuts are common, the technician will need both metric and U.S. (English) wrench sets.

Sometimes it is difficult to determine the size or system needed to remove a fastener. With time and experience, you will be able to "just know" which size to grab from your toolbox because you will remember the last time and will have the ability to look at a fastener and estimate the right size the first time.

Most modern vehicles use metric fasteners. But older domestic vehicles often have U.S. (English) fasteners. Sometimes a metric clamping bolt is replaced with an aftermarket non-factory U.S. (English) bolt. This can cause confusion, but again, after some experience, you will be able to determine which size you need very quickly.

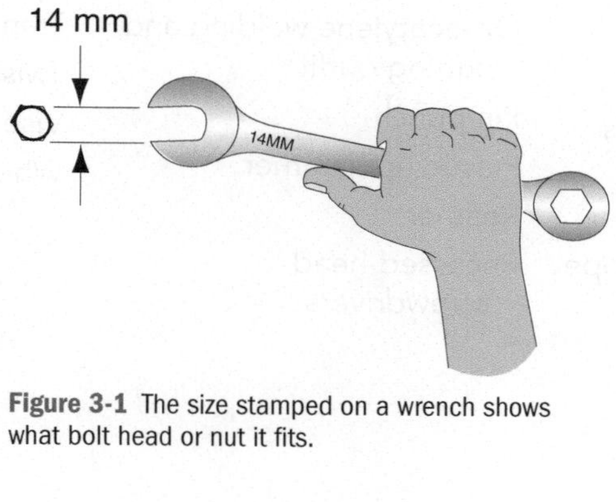

Figure 3-1 The size stamped on a wrench shows what bolt head or nut it fits.

AUTHOR'S NOTE Don't confuse the size of the wrench with the size of the fastener. A 14-millimeter wrench does not mean the fastener is 14 millimeters. The 14 millimeters on the wrench only refers to the size of the fastener's head. A 14-millimeter fastener will normally use a 17-millimeter or larger wrench to turn it.

Open-End Wrenches

An open-end wrench is one common type of wrench (**Figure 3-2**). These wrenches have an opening at the end that is placed on the bolt or nut. The opening is usually at an angle to the handle. Often one of the gripping surfaces or jaws is thicker than the other. This makes turning in a tight space easier. A hexagonal nut can be turned continuously just by turning the wrench over between each swing (**Figure 3-3**). Open-end wrenches are made in many varied sizes and shapes. Most open-end wrenches have two open ends of varied sizes.

When using open-end wrenches, always try to pull the load as shown in **Figure 3-4**. When a wrench is pulled, it is usually away from some obstruction. When a wrench is pushed, it is toward an obstruction. A hand injury can result if the wrench slips off the nut or bolt head. If an open-end wrench must be pushed toward an obstruction, do not wrap fingers around the handle. Use the heel of the hand as shown in **Figure 3-5**. Be prepared with good footing and a clean floor in case something breaks.

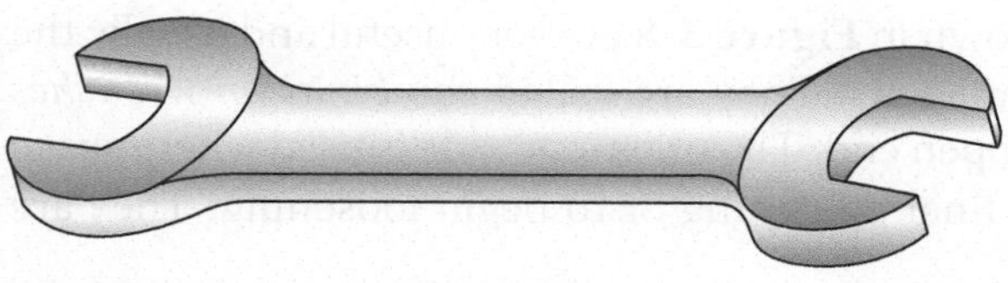

Figure 3-2 A double open-ended wrench. Open-end wrenches have only two contact points.

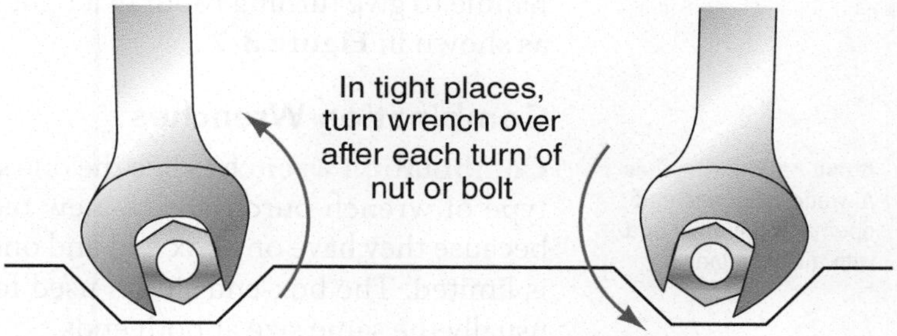

Figure 3-3 An open-end wrench is offset to allow it to fit in tight places.

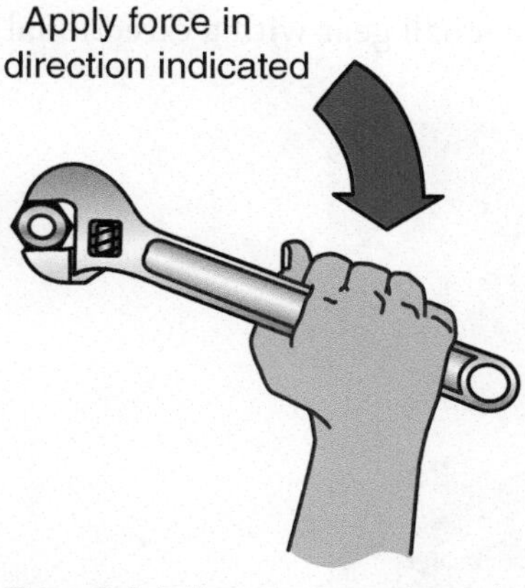

Figure 3-4 Pull on an open-end wrench when tightening in the same way as you would with an adjustable wrench.

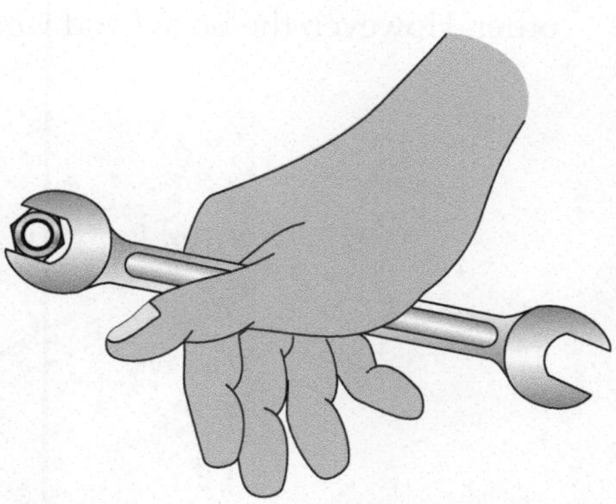

Figure 3-5 Correct way to push on an open-end wrench. Pushing a wrench to tighten or loosen a fastener should only be done when space is restricted.

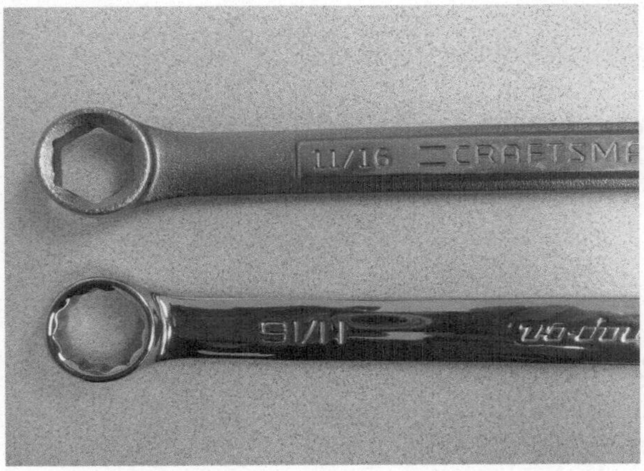

Figure 3-6 A 12-point box-end wrench (bottom) and a 6-point box-end wrench (top). A 6-point box-end wrench usually provides a better fit and reduces chances for fastener damage and incorrect torque due to slipping. However, a 12-point may be needed to remove fasteners that are already worn down on the edges.

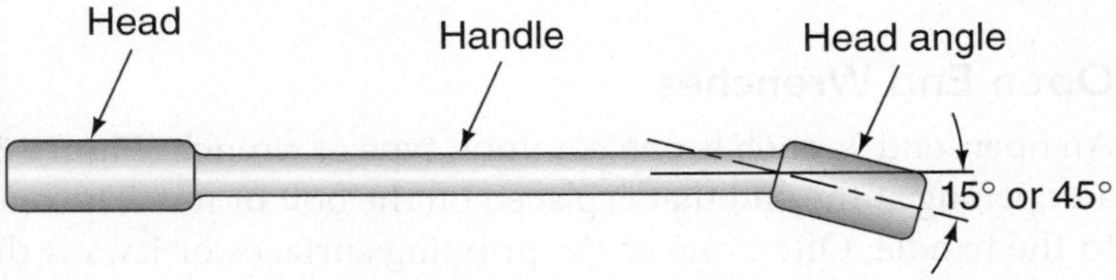

Figure 3-7 The box end of the wrench may be offset to the handle 15 or 45 degrees.

Box-End Wrenches

Box-end wrenches A wrench with a closed loop that fits over a nut or bolt head.

Box-end wrenches like the one shown in **Figure 3-6** are designed to fit around a bolt or nut. They usually have 12 or 6 points or corners to grip the nut or bolt head. They allow the technician to apply a lot of force with less chance of the wrench slipping off the nut or bolt. Like other wrenches, they come in many sizes. The angle of the head is offset to the handle to give turning room in a tight space. Two common offsets are 15 and 45 degrees, as shown in **Figure 3-7**.

Combination Wrenches

Combination wrenches A wrench with one end open and the other end with a closed loop.

Combination wrenches like the one shown in **Figure 3-8** are very useful and usually the type of wrench purchased by new technicians. They are called *combination wrenches* because they have one box end and one open end. The open-end side is used where space is limited. The box-end side is used for final tightening or to begin loosening. They are usually the same size at both ends.

Gear Wrenches

Gear wrenches are a version of a combination wrench, open at one end and boxed at the other. However, the boxed end works like a ratchet. It has a small gear with a directional

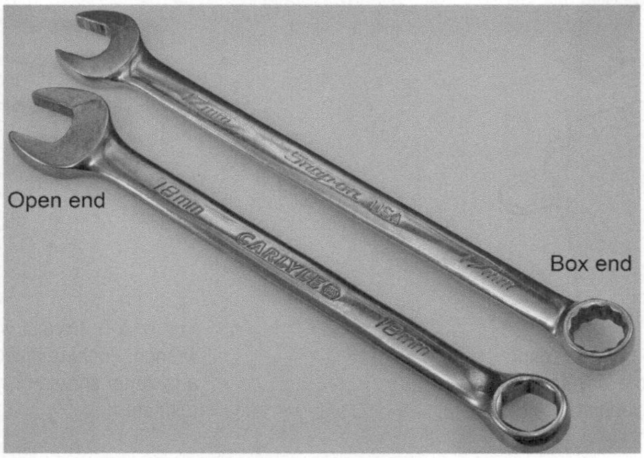

Figure 3-8 Combination wrenches.

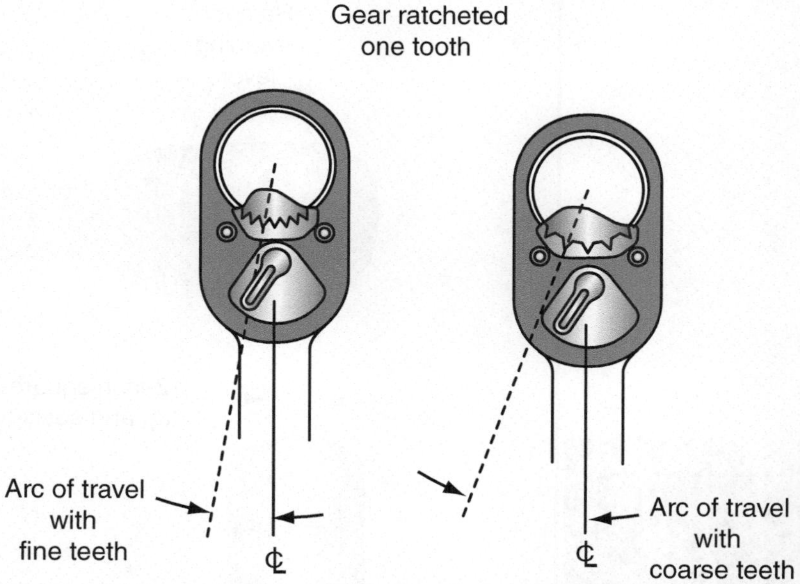

Figure 3-9 Gear wrenches with fine ratchet teeth allow for shorter wrench movement and thus can be used in more constricted spaces.

lock in the body surrounding the point section that fits over the bolt head. The body is relatively thin, like a typical combination wrench, and will fit into close areas. Once fitted over the bolt and the ratchet in the correct position, the gear wrench works like a ratchet driving a socket wrench. This allows quick tightening or loosening of fasteners in close areas. The technician must be aware of and consider two potential problems with a gear wrench. Because of the small ratcheting gear, it may not be possible to apply great torque to loosen or tighten a fastener without damaging the wrench. Usually using the open end or a regular box-end wrench to break the fastener loose or apply the final torque takes care of this problem. A second possibility also directly relates to the price and design of the wrench. In extremely close areas, the ratchet in the wrench may have too much spin or movement to help very much. The number of degrees the wrench must move before the lock "ratchets" is different on different brands and even on different models. For instance, if the outer end of the wrench must move 2 or 3 inches (about 8 to 10 degrees of rotation) before the lock shifts, then the wrench will only be useful when there is at least 2 to 3 inches of clearance (**Figure 3-9**). If not, then the wrench becomes an overpriced box-end wrench. Some of the better brands offer gear wrenches that will ratchet every 3 degrees. On a wrench 8 inches long, this is about 1 inch of travel. In other words, the wrench can fit into really confined areas. But this also means the gear teeth and ratchet will be very fine, which could mean a weaker gear, which leads back to the first problem. Tool brand should be considered closely when purchasing gear wrenches. In each case, the better, and usually more expensive, brand will be the better buy.

AUTHOR'S NOTE GearWrench™ is a brand name that has been adopted as a common name for a particular type of wrench. The actual name for a gear wrench is a *ratcheting box end*.

Socket Wrenches

Socket wrenches, or sockets, have the wrench part and the driver part made in two different pieces. The wrench part fits all the way around the bolt head or nut with little danger of slipping off. The wrench part can be removed from the driver. Sockets of many

Socket wrenches and attachments are often grouped together and called sockets. They fit all the way around a bolt or nut and can be detached from a driver.

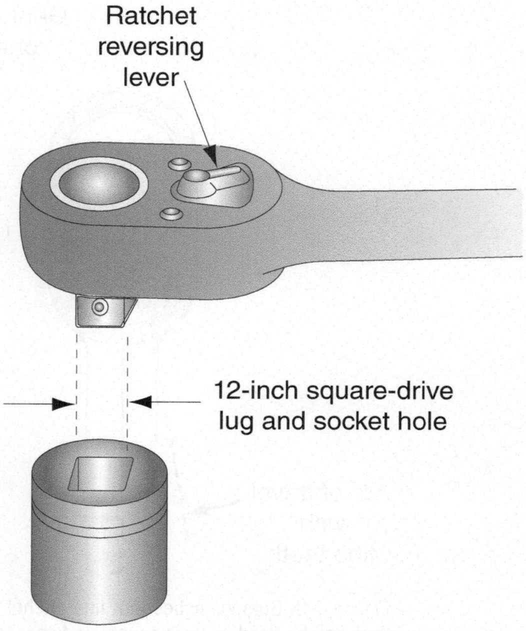

Figure 3-11 The square drive lug on the ratchet handle fits the square hole in the socket.

Figure 3-10 Typical deep-well socket set.

different sizes can be used with one handle. Sockets usually come in sets and are made in all the common English and metric wrench sizes (**Figure 3-10**).

Sockets are attached to a square drive lug on the handle by a square hole at one end of the socket (**Figure 3-11**). These drive holes and lugs are also made in different sizes. For small bolts and nuts, such as automotive trim parts, socket sets with a 1/4-inch square drive are useful. For general purpose work, a 3/8-inch drive set is popular. Heavier work requires a 1/2-inch drive socket set. Even larger sizes, 3/4 inch and 1 inch, are made for driving very large socket wrenches.

Socket wrenches come in two basic lengths: the standard or common length, and long or deep sockets. The latter is longer so that it fits over a long bolt or stud. Standard and deep sockets are shown in **Figure 3-12**. Deep sockets come in all the same common wrench sizes as the standard size sockets.

Sockets are also available with different numbers of corners, or points, inside the socket to grip the bolt or nut. The three common numbers of points are 6, 8, and 12 (**Figure 3-13**). The 12-point socket is easiest to slip over a bolt or nut because of its many

Figure 3-12 A chromed deep-well socket and an impact socket.

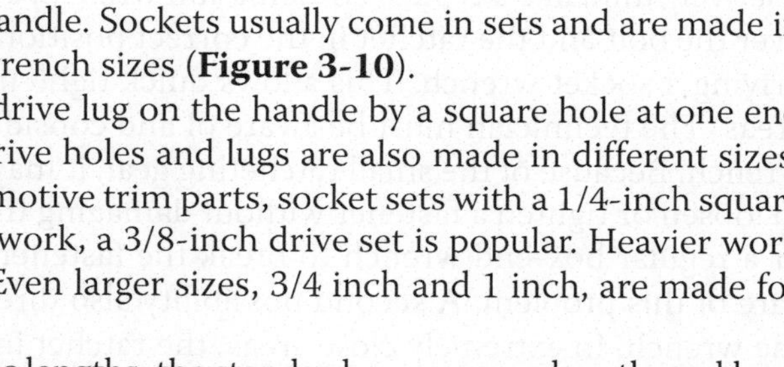

Figure 3-13 Sockets are made with different numbers of contact points. The eight-point socket (center) is often called a "double square."

corners. The six-point socket is the hardest to slip over the bolt or nut. The fewer the points or corners, the stronger the socket. The 6-point socket is stronger than the 12-point version and less likely to slip and round the corners of a fastener. It is the preferred design for most vehicle usage.

WARNING **A 12-point wrench can be used on a 6-point fastener and vice versa. However, using the wrong wrench can round off the fastener's head corners, making it difficult to properly fit a wrench tight enough to grip the fastener. Use the proper wrench, in size and points, to avoid damage to the wrench and fastener.**

Socket Drivers, Handles, and Attachments

Socket wrenches require handles or drivers to be used. Many handles and attachments are available to drive socket wrenches. The most commonly used type of driver is called a *ratchet handle* (**Figure 3-14**). It has a square drive that fits into the square hole in the socket wrench. The socket is then placed over a bolt or nut. The bolt or nut is tightened or loosened by rotating the socket handle. A freewheeling or ratchet mechanism inside the ratchet handle allows it to drive the nut in one direction and to move freely in the other direction without driving the nut. This permits fast work in a small space because the socket does not have to be removed from the nut each time it is turned. A lever on the ratchet handle allows the mechanic to choose which direction the ratchet will drive and which direction it will turn free (**Figure 3-15**).

The *speed handle* (**Figure 3-16**) is another popular socket driver. The socket wrench is installed on the end of the speed handle. The technician pushes on the end of the handle to hold the socket firmly on the nut or bolt. At the same time, the technician turns the crank-shaped handle in a direction to tighten or loosen. The combination of a swivel handle and crank allows very quick driving of a socket. The speed handle is used when many bolts or nuts must be removed or replaced. Technicians often use this handle when removing an automatic transmission fluid pan. A *breaker bar* (**Figure 3-17**) is another common socket driving tool. The breaker bar is used to break loose bolts or nuts that are very tight. The long handle allows the technician to get a lot of force into the turning. Its drive end has a hinge that will permit driving at different angles.

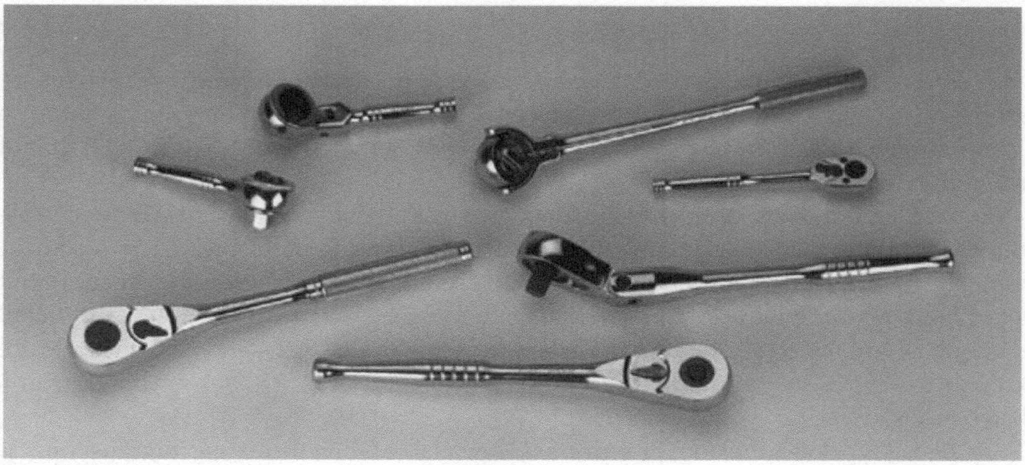

Figure 3-14 There are many types of ratchets available.

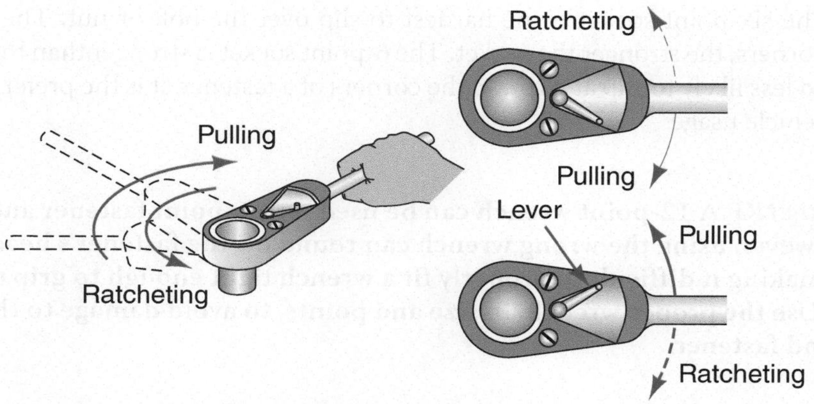

Figure 3-15 Switching the lever allows the ratchet to pull in either direction.

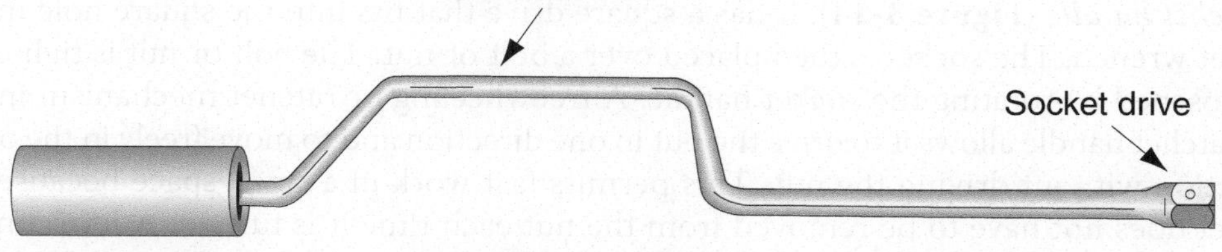

Figure 3-16 A speed handle allows for faster movement of a fastener but reduces the amount of torque that can be delivered.

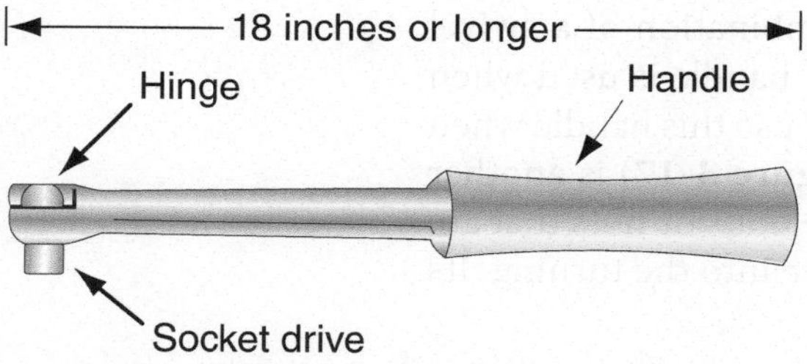

Figure 3-17 A breaker bar helps increase the amount of torque that can be delivered but is slow at turning a fastener.

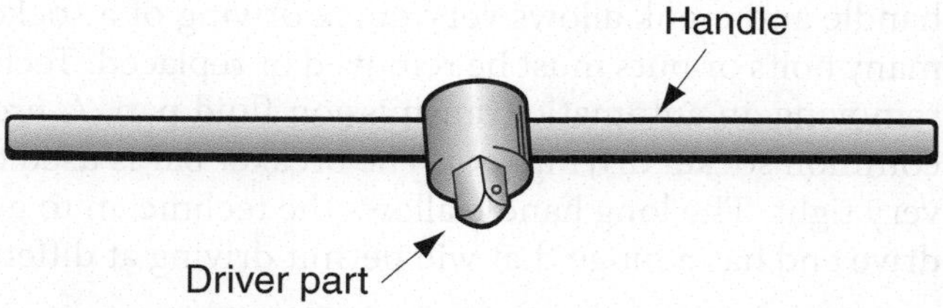

Figure 3-18 A sliding T-handle can be used in very tight spaces but usually cannot provide much torque and is very slow.

A *sliding T-handle* is also used to drive sockets. This tool has a handle shaped like the letter T. The handle slides through the driver part. The handle can be used on center or it can be moved outward to provide a longer lever for more torque. A sliding T-handle driver is shown in **Figure 3-18**.

Socket wrenches are often attached to the drivers through *extensions* (**Figure 3-19**). One end of the extension is connected to a handle or driver and the other to a socket wrench. The extension allows the socket to be used in an area where an ordinary handle would not have enough room to turn. Extensions come in a variety of lengths for the different jobs they have to do. The common lengths are 3, 6, 12, 18, 24, and 36 inches. Flexible extensions are also available that allow the technician to bend them around obstructions.

A special type of extension, or *universal,* is available when the extension must operate at a sharp angle (**Figure 3-20**). This is desirable when parts get in the way of driving a socket. It is made in two parts, which allow it to bend at near 90 degrees. Some universals

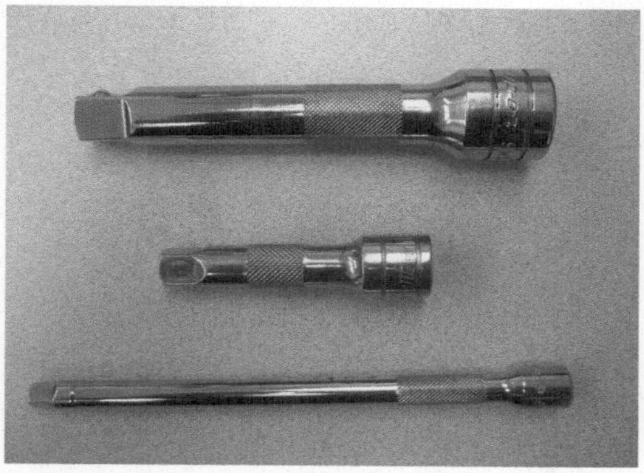

Figure 3-19 Extensions allow the wrench to reach "buried" fasteners and can be used with any matching driver. However, the longer the extension, the more the flex, which may reduce the torque being delivered.

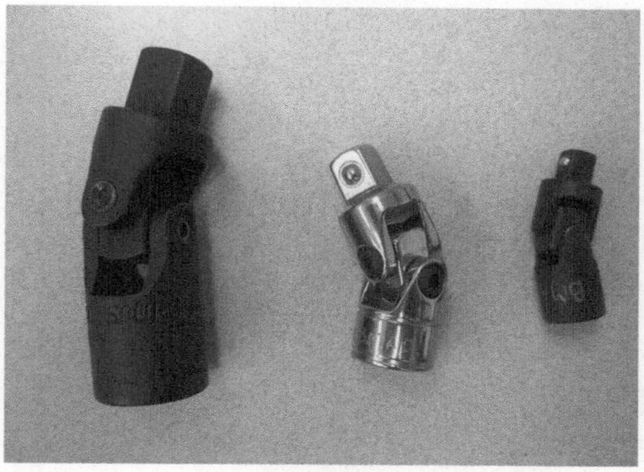

Figure 3-20 A universal allows for torque to be delivered at an angle to the socket and fastener. Torque delivered becomes less as the operating angle of the universal increases.

are part of the socket wrench itself. They are usually known as universal sockets. The primary advantage of a universal socket is the shorter overall length that allows the socket to fit into closed areas.

A socket, extension, and handle can be combined into one tool called a *nut driver* (**Figure 3-21**). The handle is often made like a screwdriver handle. Nut drivers are usually made in small sizes for very small nut and bolt heads.

Adjustable Wrenches

The wrench shown in **Figure 3-22** is called an **adjustable wrench** because it adjusts to fit bolts and nuts of varied sizes. There are adjustable wrenches from about 4 inches to about 20 inches long. The longer the wrench, the larger the opening will adjust. For example, a 6-inch adjustable wrench opens 3/4-inch wide and the 12-inch opens 15/16 inches.

Many people call the **adjustable wrench** a "crescent wrench." This is due to the fact that the Crescent Tool Company made this type of wrench for many years.

The adjustable wrench is not as strong as a box-end or open-end wrench. The adjustable jaw can break if overloaded. The technician must be careful to adjust the jaws to fit snugly against the flats of the nut or bolt head. When tightening, the wrench must be placed on the bolt or nut such that stress falls on the stationary jaw. If the wrench is used incorrectly, the adjustable jaw can break and cause injury. The right way and the wrong way to use an adjustable end wrench are shown in **Figure 3-23**.

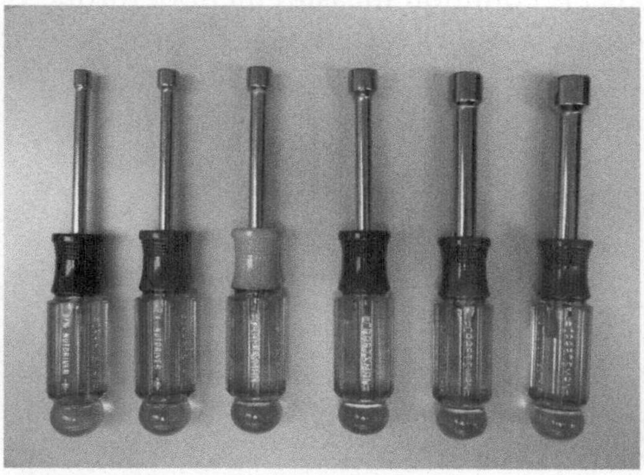

Figure 3-21 Nut drivers are usually used at low torque and on smaller fasteners. They are used for dash and instrument panel work.

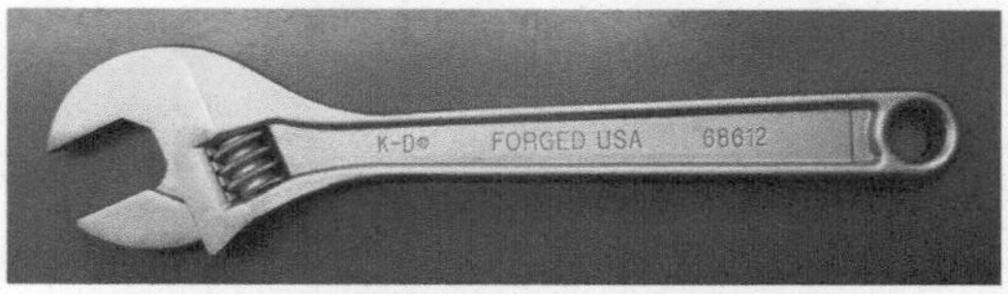

Figure 3-22 The "fit everything" wrench. The adjustable wrench provides a range of openings to fit many fasteners. It is not as strong as fixed-size wrenches.

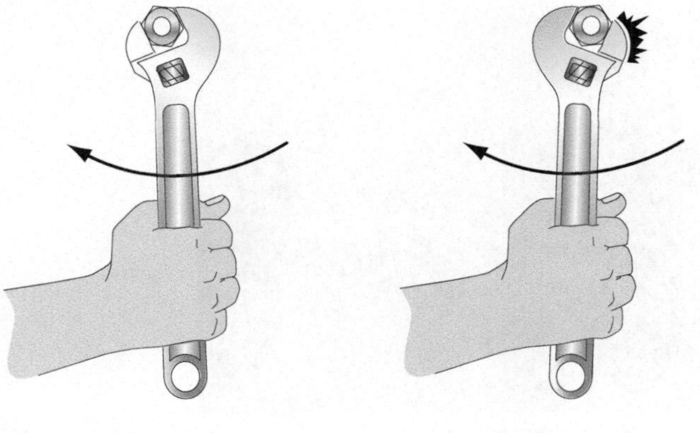

Figure 3-23 Using the adjustable wrench incorrectly can result in injury to the technician, damage to the wrench, and possible damage to the fastener or component.

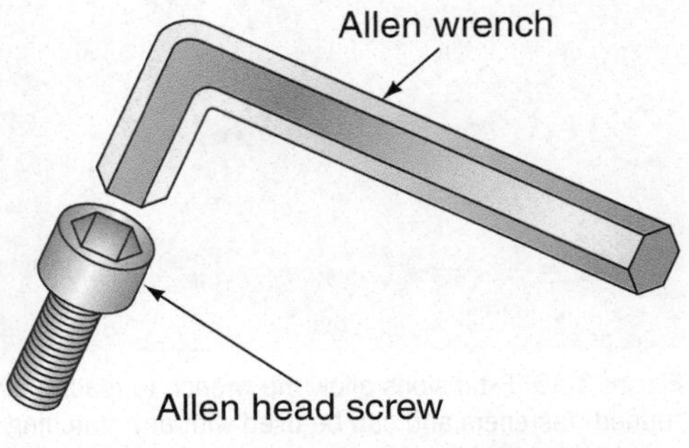

Figure 3-24 An Allen head screw and Allen head wrench set.

Allen Wrenches

An **Allen wrench** is used to tighten or loosen Allen head screws.

Some automotive components are fastened with hollow head Allen screws. These screws require special **Allen wrenches** (**Figure 3-24**). The hexagonal, or six-sided, Allen wrench fits in the hexagonal head of the Allen head screw. The fit is tight and prevents the wrench from slipping out of the screw head. Allen wrenches are available in sets. They are sized according to the size of the Allen screw in which they fit. They are made in U.S. (English) system sizes such as 3/32 and 1/8 inch. They are also made in metric sizes such as 6, 8, and 10 millimeters. Some Allen wrenches are made to fit on socket drivers, as shown in **Figure 3-25**. It is easy to lose or misplace Allen wrenches. Make sure that you take care of them. Many technicians have two or three of the commonly used Allen wrenches in their toolbox because they frequently need them and don't want to be without them, even for 1 minute, when one breaks.

Torque Wrenches

When some automotive parts are reassembled after repair, the bolts and nuts must be tightened a certain amount. Never assume the amount of tightness a ratchet or wrench has put on a fastener; always set it with a torque wrench. After a few years of experience, you will develop a feel for using ratchets and tightening down fasteners, but it is always good practice to use a torque wrench. Your reputation will depend on quality repairs.

A **torque wrench** is designed to tighten bolts or nuts to a certain tightness or torque.

A special socket handle called a **torque wrench,** which measures torque, is used for this purpose. Torque is turning or twisting force. A common measurement of torque is foot-pounds (ft.-lb.). The metric system measurement of torque is a newton-meter (N·m). This means a force, measured in pounds (or newtons), is acting through a distance

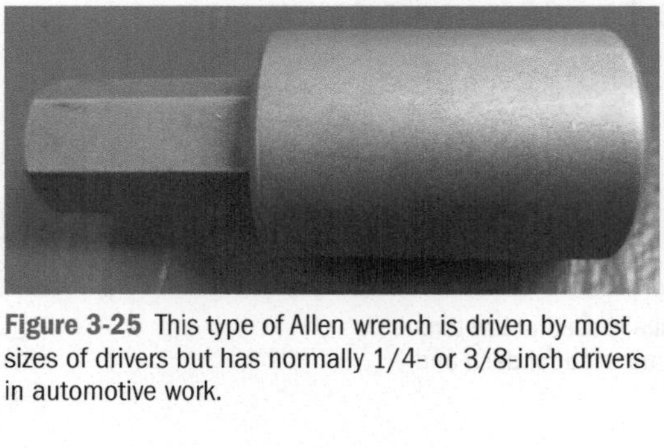

Figure 3-25 This type of Allen wrench is driven by most sizes of drivers but has normally 1/4- or 3/8-inch drivers in automotive work.

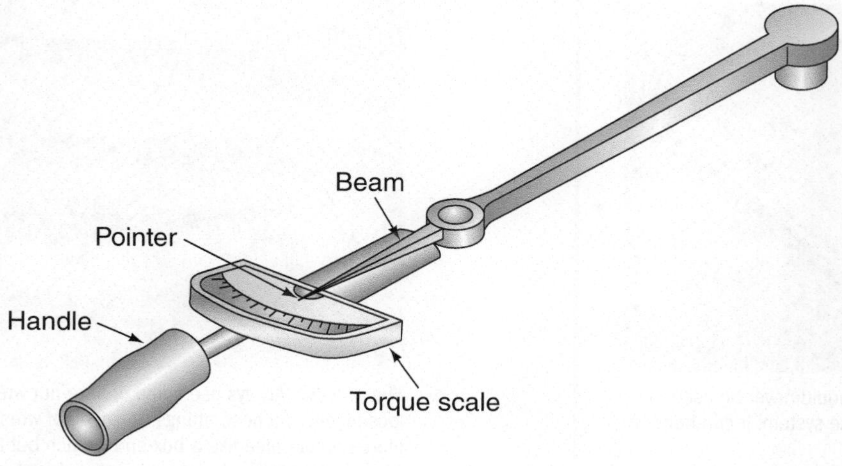

Figure 3-26 A beam-type torque wrench.

of 1 foot (or 1 meter). For example, imagine a wrench that is 1 foot long. Then push on the end of the wrench with 50 pounds of force. A force of 50 ft.-lb. (or 68 N·m) has now been applied. Torque can also be measured in inch-pounds (in.-lb.). This is common in smaller fasteners. To convert from inch-pound to foot-pound, simply multiply by 12.

There are many types of torque wrenches. One type (**Figure 3-26**) uses a beam and pointer assembly. During tightening, the beam on the wrench bends as the resistance to turning increases. The torque is shown on a scale near the handle. A popular type of torque wrench (**Figure 3-27**) has a ratchet drive head and a break-over hinge. Its adjustable handle and scale allow the mechanic to adjust the wrench to a certain torque setting. When the torque setting is reached, the break-over hinge swivels to signal the technician to stop turning. Another type has a dial on the handle (**Figure 3-28**). This dial has a needle that points to the amount of torque the wrench is delivering. Many tool suppliers offer electronic torque wrenches with a digital display.

AUTHOR'S NOTE This type of torque wrench usually produces a noticeable "click" when it breaks over at the set torque. When you use smaller torque wrenches, the click is hard to hear, but you can feel it in the handle if you are paying attention.

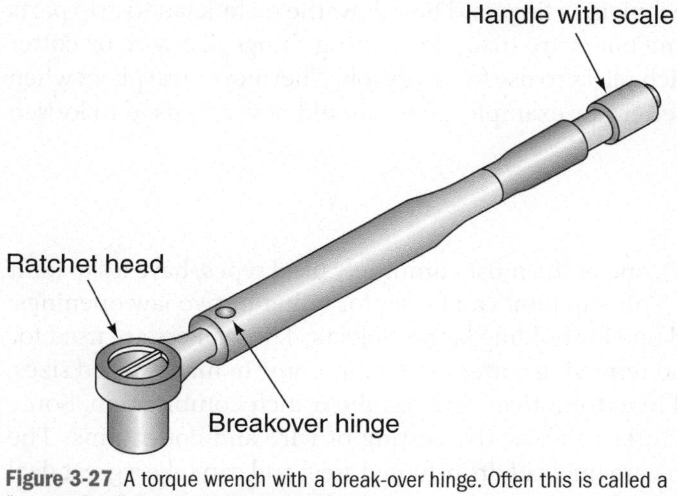

Figure 3-27 A torque wrench with a break-over hinge. Often this is called a "sensory" or "clicker-style" torque wrench.

Figure 3-28 A dial readout-type torque wrench.

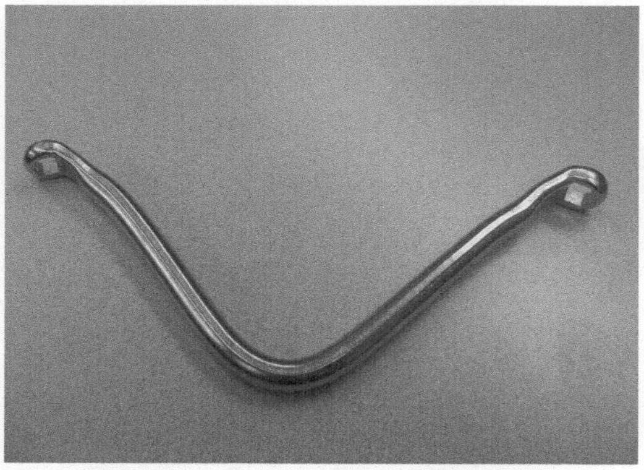

Figure 3-29 A bleeder wrench should never be used for anything except bleeding the brake system. It can bend or break under heavy torque.

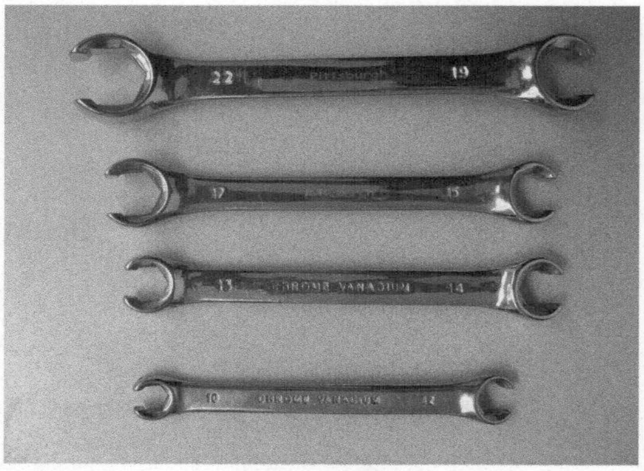

Figure 3-30 Always use a line or flare-nut wrench to loosen line and hose fittings. This type of wrench has more contact area like a box-end wrench but acts like an open-end wrench.

Several different torque measurement systems are in use. Specifications are given in metric or U.S. (English) units. Typical U.S. (English) units are inch-pounds and foot-pounds. There are 12 in.-lb. in 1 ft.-lb. Metric system torque specifications are most often given in newton-meters (N·m).

Special-Purpose Wrenches

Most automotive service jobs are done with regular open-end, box-end, combination, and socket wrenches. Some jobs, however, require special wrenches. These are usually variations of the common wrenches that have special shapes to get around obstructions. For example, a *brake bleeder wrench* (**Figure 3-29**) is a box-end wrench that is bent to fit on the small bleeder valves on brake wheel cylinders. Another special wrench is the *flare nut* or *brake line fitting wrench* (**Figure 3-30**). This one is a combination of an open-end and a box-end wrench. It is used on brake lines and older fuel line fittings. An open-end wrench might round off the soft metal fittings. A box-end wrench would be impossible to slip over a fuel line. There are, of course, many other special-purpose wrenches. Most technicians collect these wrenches for special jobs.

Pliers

Pliers act as an extension of the technician's fingers. They allow the technician to grip parts with great force. In addition, some pliers are made for cutting things like wire or cotter keys. Good technicians know which pliers to use for every job. They never use pliers when some other tool will do the job better; for example, pliers should never be used to loosen or tighten a nut.

Combination Pliers

Combination pliers (**Figure 3-31**), one of the most commonly used types, have a slip joint where the two jaws are attached. This slip joint can be set for either of two jaw openings: one for holding small objects and one for holding larger objects. These pliers are used for pulling out pins, bending wire, and removing cotter pins. They come in many varied sizes. The pliers most commonly used in automotive work are the 6-inch combination. Some combination pliers have a side cutter to allow the cutting of wire and cotter pins. The better grades of combination pliers are made of drop-forged steel and can take a great deal of hard usage.

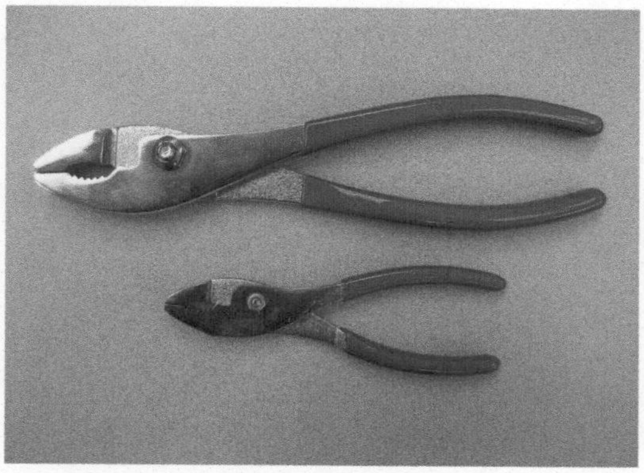

Figure 3-31 The rule for pliers use: They are to hold, not turn, fasteners or components.

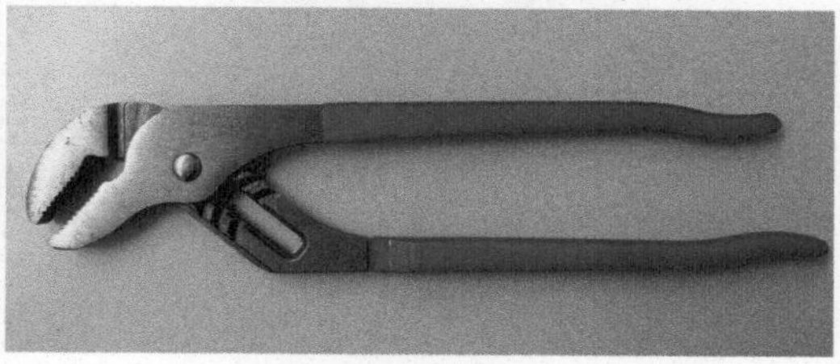

Figure 3-32 Channel-lock pliers.

Channel-Lock Pliers

Channel-lock pliers (**Figure 3-32**) are used to grip large objects. They get their name from the ribs or channels that allow the jaws to be set at many different openings. The design of the channels permits quick, nonslip adjustments with practically parallel jaws. These pliers are slim enough, with long enough handles, so that the 45-degree jaws will reach and firmly grip objects that would be out of reach of ordinary pliers. The jaw teeth are sharp and deep to take a firm grip on pipes and hoses. The interlocking design of the rib joints works well to prevent slipping under load.

Channel-lock pliers is a trade name that has been accepted as a type of pliers. These types of pliers are actually rib joint pliers.

A BIT OF HISTORY

Channel-lock pliers are often called "water pump pliers." This is the only tool with which a technician can remove the large nut from the water pump in cars made from the 1920s to the 1940s.

Diagonal Cutting Pliers

Diagonal cutting pliers (**Figure 3-33**) are used to cut electrical wire and cotter pins. Because they are made for cutting wire, their jaws have hardened cutting edges. Like other pliers, they are made in many varied sizes and are grouped by their overall length.

Diagonal cutting pliers have cutting edges on the jaw for cutting pins or wire. These pliers are often called "dikes," which is short for "diagonals."

Needle Nose Pliers

Needle nose pliers (**Figure 3-34**) have long, slender, tapering jaws that are useful in gripping small objects. They are sometimes called "long-nose pliers." Some needle nose pliers are equipped with side cutters; others are bent at right angles for hard-to-get-at places (**Figure 3-35**).

Needle nose pliers A pliers with a long nose end useful for reaching into small areas.

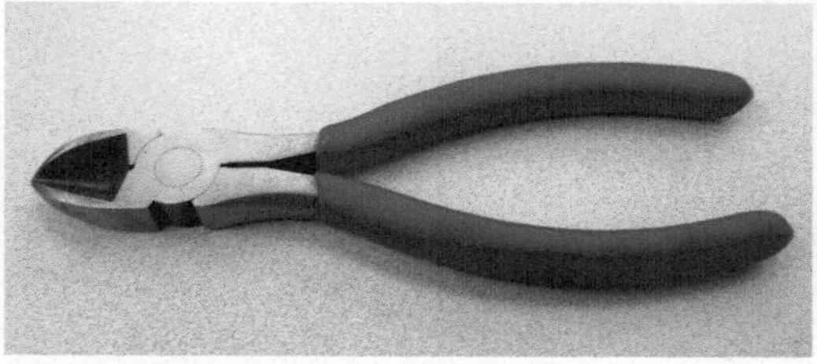

Figure 3-33 Diagonal side cutters or dykes are used to cut wires, not cable or metal.

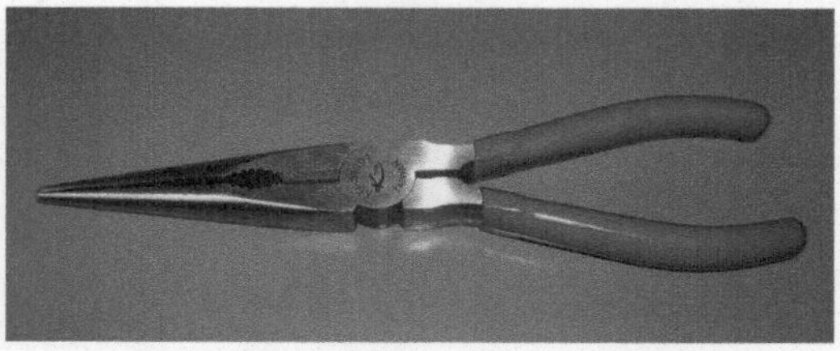

Figure 3-34 Needle nose pliers are good at picking up or holding when the component cannot be reached by other means. The longer the needle length, the less torque can be applied.Excessive torque will warp the ends.

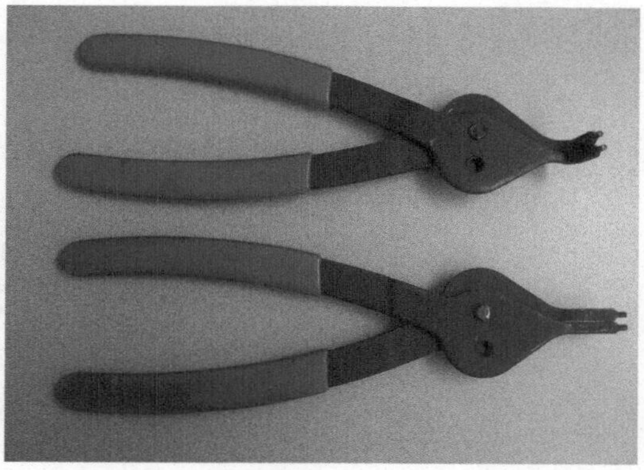

Figure 3-35 Snapring pliers are used to remove or install snaprings. They are commonly used within transmissions and some air-conditioning compressors.

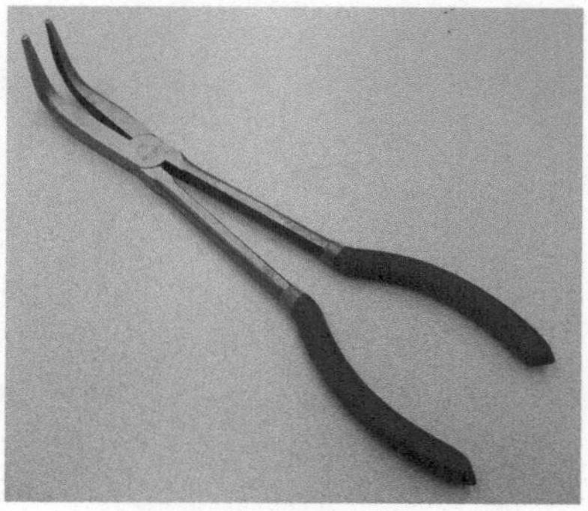

Figure 3-36 Bent needle nose pliers.

Snapring Pliers

Snapring pliers A tool designed to install and remove snap rings, which are small metal rings that hold in a bearing or other assemblies.

Many automotive components are held together with rings that snap into place under tension. Special pliers are required to remove and replace these rings. There are two major types of **snapring pliers** as shown in **Figure 3-36**. *Inside snapring pliers* have jaws that close and grip when the handles are closed like an ordinary pair of pliers. When the handles are drawn together on *outside pliers,* the jaws are opened. The jaws of each are designed to safely engage the powerful snaprings to be removed or installed.

Vise Grips

Vise grips A pliers that can be locked into position, using an over center toggle action.

Vise grips (**Figure 3-37**) are compound-lever pliers that can be locked to the part, so the technician can remove his/her hand. The adjustable lower jaw is retracted by an overcenter spring when the locking handle is released. Adjustment of the jaw opening size is made by turning a screw at the end of the primary handle. Vise grips do not take the place of wrenches. They are, however, often used to hold nuts that have had their corners rounded off by the misuse of wrenches.

AUTHOR'S NOTE "Vise Grips," like "Channel-Lock," started out as the brand name of a certain type of pliers. The brand name stuck and is generally recognized as a specific type of pliers. Vise grips pliers are actually locking pliers. In instances like this and others, the brand name has become more familiar to people than the proper technical name.

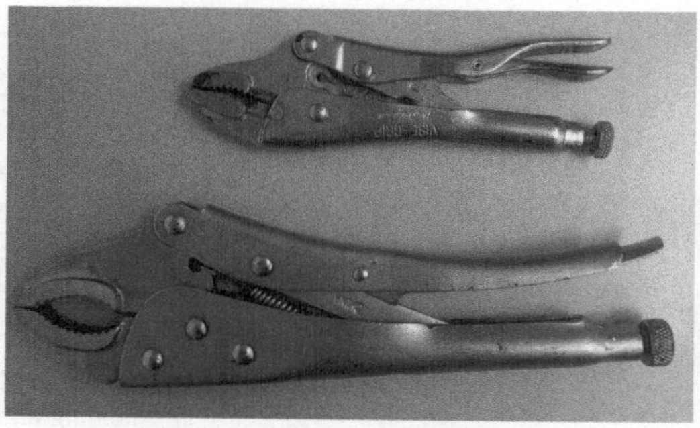

Figure 3-37 Like all pliers, vise grips are used to hold. They can provide a tight and locked fit for better holding.

Figure 3-38 This type of pliers is restricted in its use. Note the grooves used to capture and hold the ends of a hose clamp.

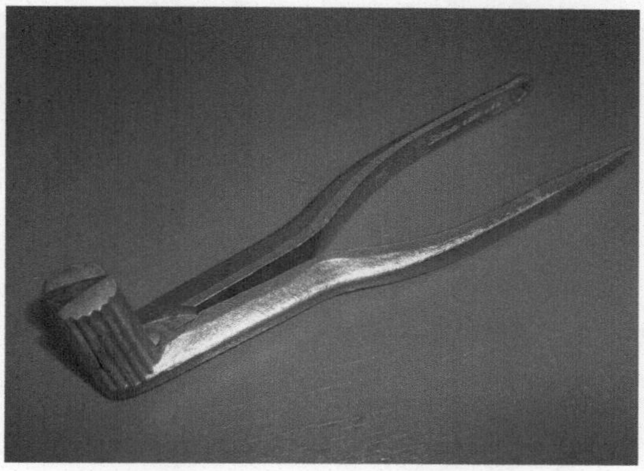

Figure 3-39 Battery terminal pliers.

Special-Purpose Pliers

There are many special-purpose pliers designed for certain repair jobs. The pliers shown in **Figure 3-38** are used to remove and replace the spring tension clamps on some coolant hoses. The special groove in the jaws grips the ends of the hose clamps.

The pliers shown in **Figure 3-39** are *battery pliers.* The jaws on these pliers are made to grip terminal nuts on battery connections. The tool shown in **Figure 3-40** is used to remove hubcaps and grease caps. The round pad on the one jaw is used as a hammer to install grease caps.

The *crimping tool* shown in **Figure 3-41** is used for electrical work. It has cutting edges to cut wire. There are special cutting edges for stripping the insulation off of wire. The jaws of the tool are designed to squeeze or crimp solderless terminals and connectors on the ends of wire.

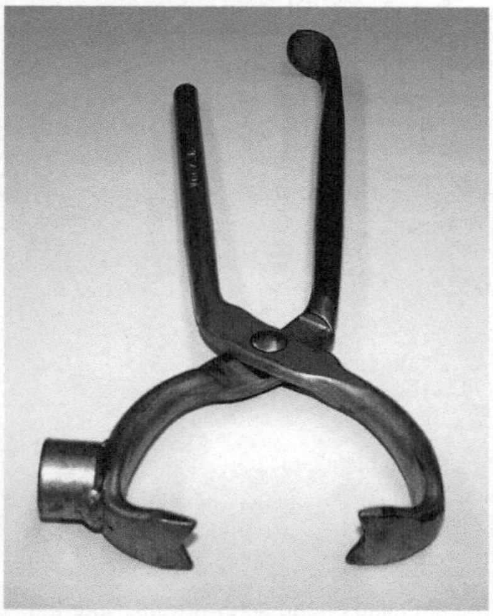

Figure 3-40 Grease cap pliers.

Figure 3-41 Crimpers are used to crimp electrical terminals onto an electrical conductor or wire. The top one can be used to crimp spark plug connectors onto spark plug wires on older vehicles.

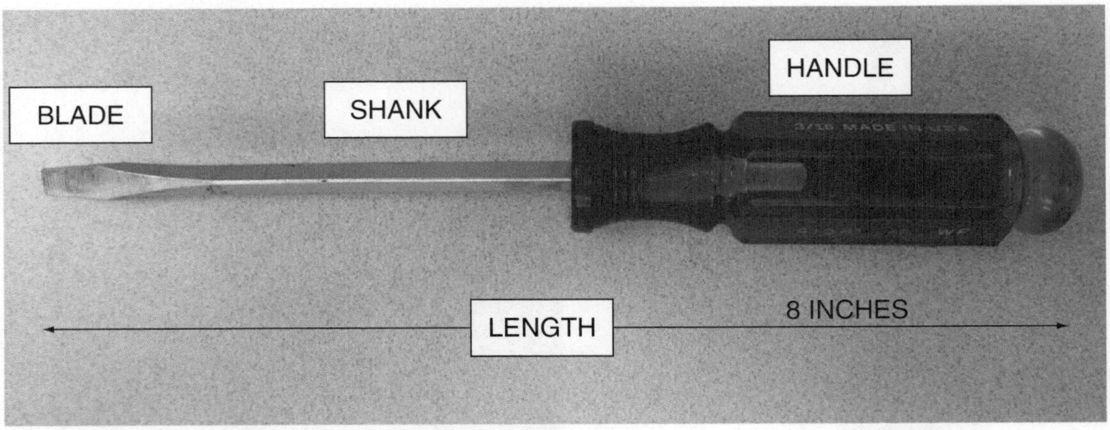

Figure 3-42 A typical 8-inch flat-tip screwdriver.

Screwdrivers

Many automotive components are held together with screws. A screwdriver is used to turn or drive a screw. There are several types of screws and screw heads. Consequently, there are several types of screwdrivers to drive them.

Standard Screwdrivers

Screwdriver A tool with a flattened, cross-shaped or star-shaped (torx) tip that fits into a head of a screw.

A standard **screwdriver** is the most common type. It is used to drive screws with a straight slot in the top. The main parts of the standard screwdriver are shown in **Figure 3-42**. The technician grips and turns the *handle.* The steel part extending beyond the handle is the *shank.* The part that fits into the screw head is the *blade,* which is often called a *bit.* The blade, or bit, is flat and ground at a right angle to the shank. The handle may be made from wood or plastic. The shank is designed to withstand a great deal of twisting force. The length of these screwdrivers is determined by their overall length. The larger the screwdriver, the larger its blade.

Screwdriver bits are available as socket drivers (**Figure 3-43**). These can be driven with the same drivers and attachments as socket wrenches.

Recessed-Head Screwdrivers

Recessed-head screwdrivers are made to fit recessed-head screws like Phillips, clutch, and Torx.

There are a number of screws with recessed-type heads. **Recessed-head screwdrivers** are required to fit these screws. The most common example is the *Phillips screwdriver* (**Figure 3-44**). This screwdriver is made to fit Phillips-head screws. Phillips-head screws

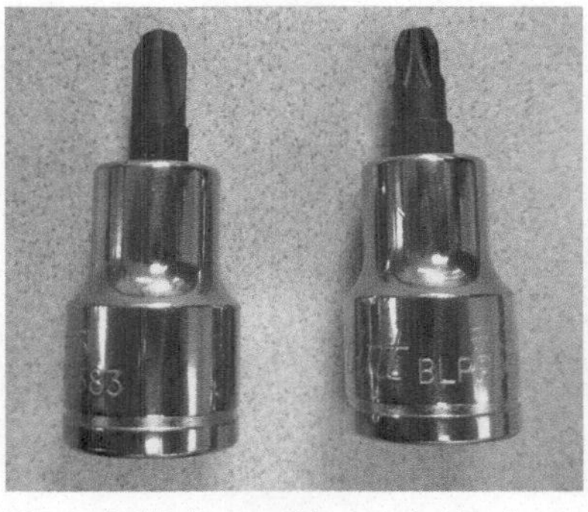

Figure 3-43 A screwdriver socket is almost always driven by a 1/4-inch ratchet in the automotive business. Larger sizes are available.

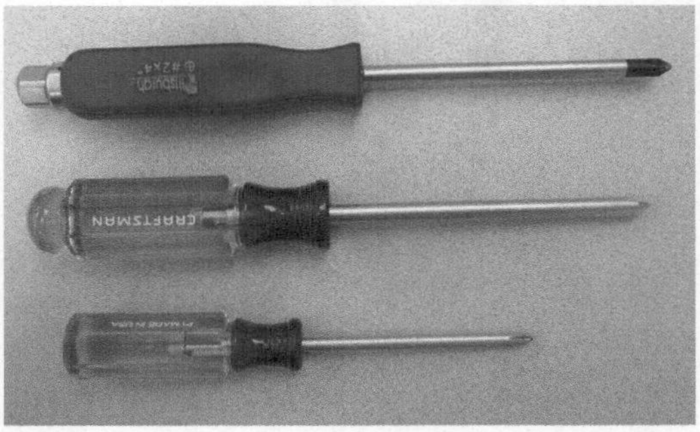

Figure 3-44 Typical Phillips-head screwdrivers. Note the different tip sizes along with the thinner shank and grip.

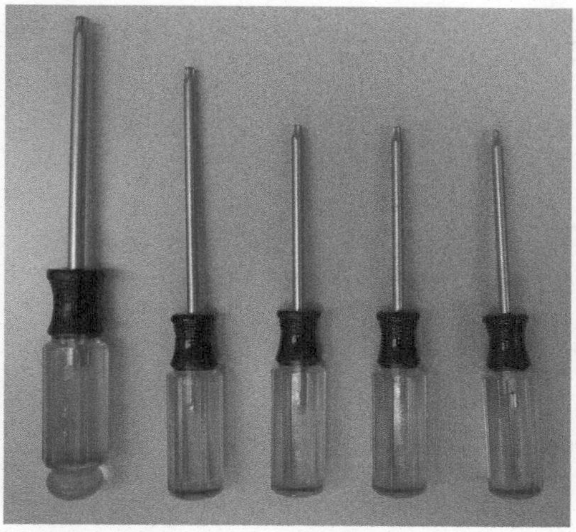

Figure 3-45 Torx-type fasteners are becoming more common on vehicle low-torque fasteners. They may be like the type shown or in a socket type driven by a ratchet.

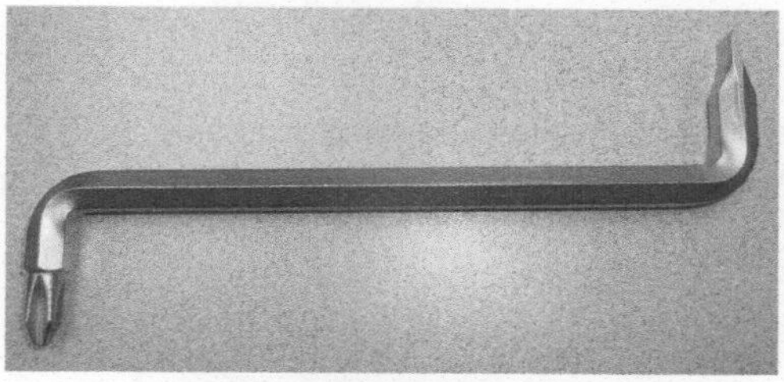

Figure 3-46 An offset screwdriver is handy in some dash or instrument work but may be awkward to use in tight places.

are used primarily to hold moldings and other trim usually on the body or interior of the car. The heads of these screws have two slots that cross at the center and do not extend to the edges of the heads. This design has the advantage that the screwdriver head will not slip out of the slots and scratch the finish. The blades are sized on a numbering system from 0 to 6, with 0 being the smallest and 6 the largest.

Another common recessed-tip screwdriver, called a *Torx-type screwdriver,* is shown in **Figure 3-45**. It has a six-prong tip and is often used to fasten many automotive parts. There are also *inverted Torx sockets* used to remove fasters shaped like the tips on the screwdrivers shown here.

There are many other types of recessed-head screws. Each requires a different type of screwdriver. Screwdrivers are also available that have interchangeable bits on the end of the shanks. Bits are available for each of these recessed-head screws.

Offset Screwdrivers

Offset screwdrivers (**Figure 3-46**) are designed for working in a tight space. The blades at opposite ends are at right angles to each other. The technician can change ends of the offset screwdriver after each swing and continue to turn the screw (**Figure 3-47**). These screwdrivers come in all the common blade or bit sizes.

Offset screwdrivers A tool with a flattened, cross-shaped or star-shaped (torx) tip at a right angle that fits into a head of a screw.

Hammers

Hammers are used in many different trades. Every hammer has two basic parts: a head and a handle. The handles are made from wood or plastic. The heads are made from varied materials for different jobs. The automotive technician will need to know how to use several types of hammers. A good technician knows when and how to use the correct type of hammer.

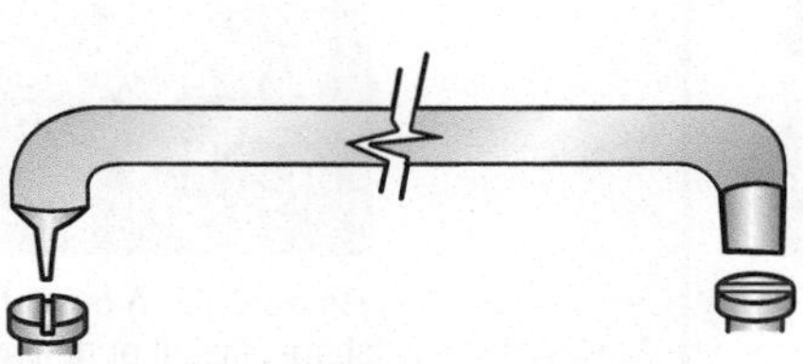

Figure 3-47 The tips of an offset may be reversed or right angled to each other.

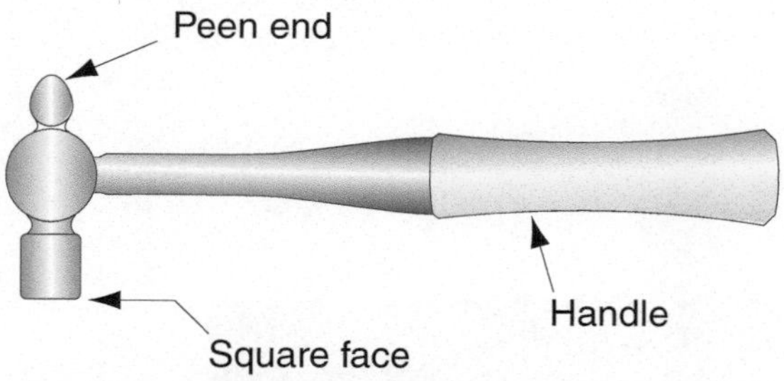

Figure 3-48 A technician's favorite tool, a ball-peen hammer. It is used to remove or install pin fasteners; shape metal; remove, install, or position some components; and drive metal cutters like the chisel.

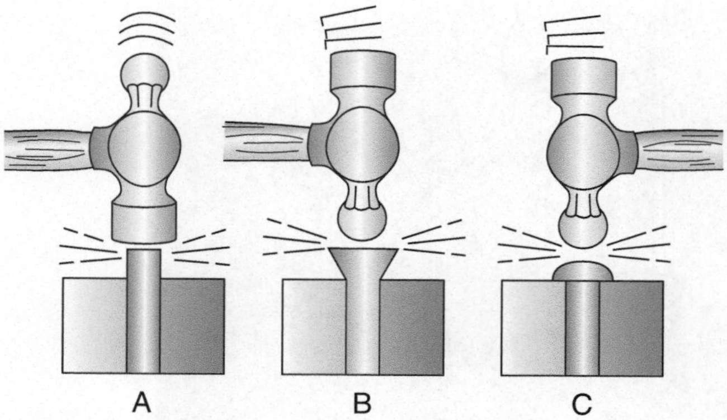

Figure 3-49 Using a ball-peen hammer to set a rivet.

Ball-Peen Hammers

Ball-peen hammer
A hammer that has two heads, one side is flat the other is rounded (peen).

The **ball-peen hammer** (**Figure 3-48**) is one of the most common hammers used by automotive technicians. These hammers have one face on the head that is square to the end of the handle. The opposite end is rounded and is called the *ball peen.* The square end is used for general hammering, such as driving a pin punch or chisel. The rounded ball peen is used to form or peen over the end of a rivet, as shown in **Figure 3-49**.

The hardened head of a ball-peen hammer should not be used to hammer on automotive parts. Because the head is harder than the part, the part will be damaged. Ball-peen hammers of assorted sizes are listed according to the weight of their head. Small ones weigh as little as 4 ounces; big ones weigh more than 32 ounces.

Plastic-Tip Hammers

Plastic-tip hammers
A hammer with a plastic head used to strike a surface without damaging it.

Plastic-tip hammers (**Figure 3-50**) have a head made from plastic. These heads are much softer than the ball-peen, steel-head hammers. They are used to align or adjust parts that might be damaged by a steel-head hammer. Plastic hammers will not damage metal parts because the plastic is softer than metal. Plastic can also absorb the force of the impacts instead of transmitting them to the metal parts. The technician should always remember to use a hammer that is softer than the material being hammered on.

Brass Hammers

Brass hammers
A hammer with a brass head use to strike an object without causing a spark or damaging it.

Hammers with brass heads (**Figure 3-51**) are softer than ball-peen hammers. **Brass hammers** are much heavier than plastic ones. They can be used when more force is required. Brass hammers are often used for driving pilot pins and other materials that will be harmed or distorted by steel hammers.

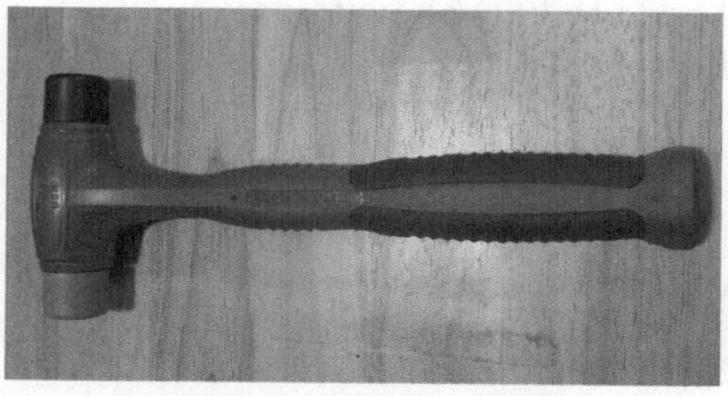

Figure 3-50 A plastic-head hammer can be used to position, remove, or install some lightweight components without damage to the components. It is never used to drive chisels or punches.

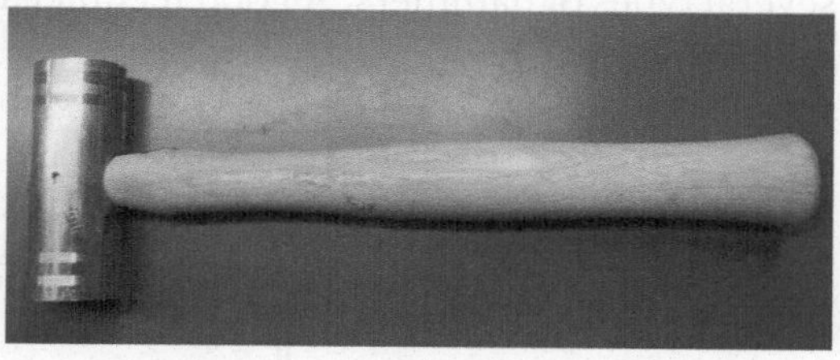

Figure 3-51 A brass-headed hammer is used to shape, install or remove, and position some lightweight components without component damage. It is never used to drive chisels or punches.

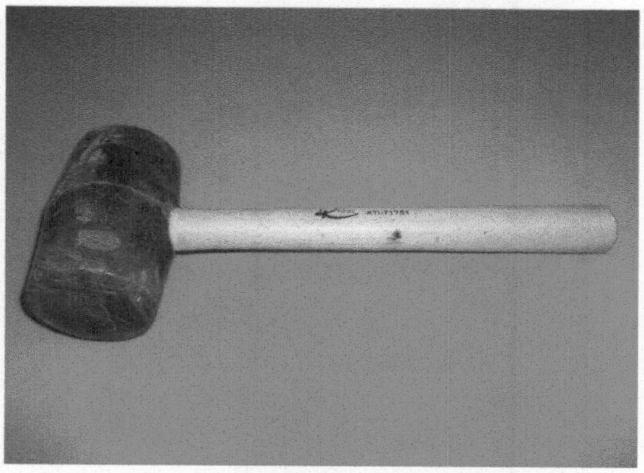

Figure 3-52 A rubber mallet is almost always used to shape metals. It is never used to drive chisels or punches.

Rubber Mallets

Rubber mallets (**Figure 3-52**) are softer than brass or plastic. These are used to seat hub caps, wheel covers, and other trim parts. They are soft enough so that they will not damage the finish of parts like wheel covers.

Rubber mallets A hammer with a rubber head softer than a plastic and brass hammer.

Punches and Files

Punches are used with ball-peen hammers to drive pins or to make the center of a part to be drilled. The most common types are the *pin punch, starter punch, center punch, cotter pin remover,* and *roll pin punches.*

Shop Manual
page 62

Pin and Starter Punches

Some automotive components are held together with pins or rivets. Two different kinds of punches may be needed to remove a pin or rivet (**Figure 3-53**). A **starter punch** is used to break the pin loose. Then a **pin punch** smaller than the hole is used to drive a pin out of a hole. A starter punch, also called a "drift punch," is made with a shank that is tapered all the way to its end. The tapered shank of a starter punch is stronger and will withstand greater shock from hammer blows than a pin punch. The starter punch is used to start the removal of pins and rivets (**Figure 3-54**).

Starter punch A short shaft punch that give you more control to break a pin loose.

Pin punch A punch with a very small tip used to dislodge rivets and pins.

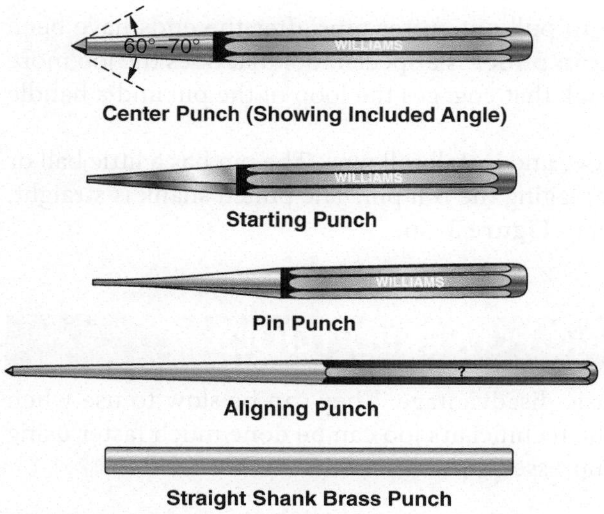

Figure 3-53 A selection of punches.

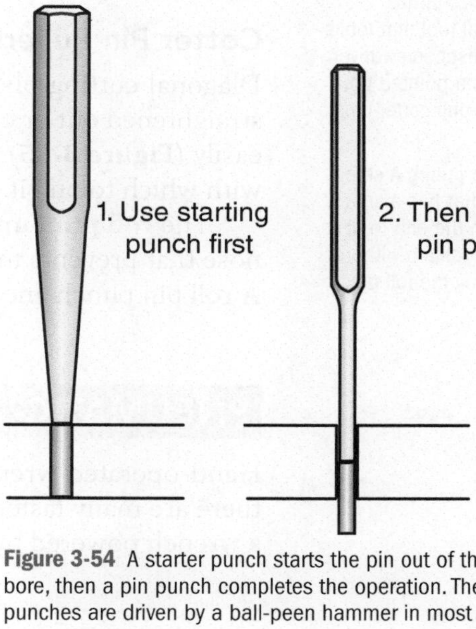

Figure 3-54 A starter punch starts the pin out of the bore, then a pin punch completes the operation. The punches are driven by a ball-peen hammer in most cases.

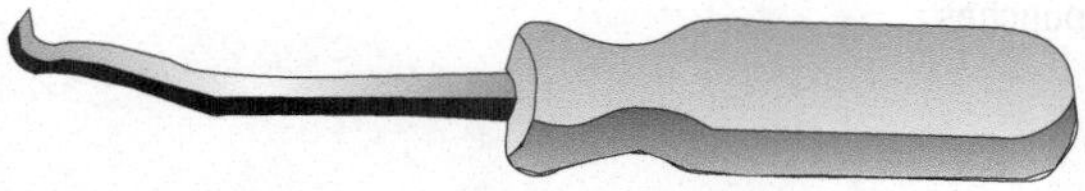

Figure 3-55 A cotter pin puller.

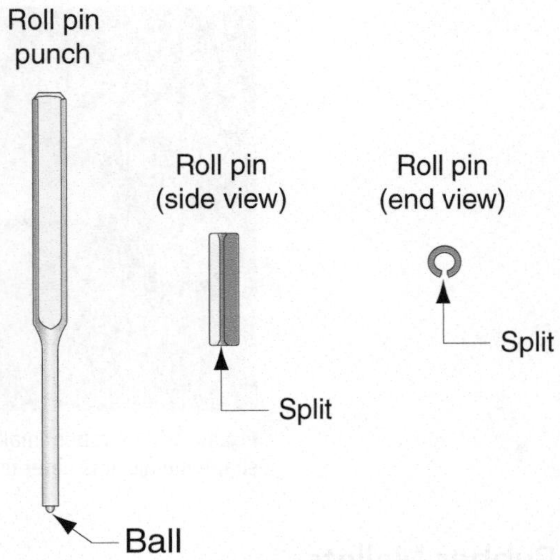

Figure 3-56 The ball on the roll pin punch fits into the hollow core of the pin. This allows the flat driving surface around the ball to drive the edge of the pin without deforming it.

Drifts are usually made of brass and are used to drive out hard or highly finished parts without damaging them. They are not always tapered. Long, tapered, steel drifts are often used to align two parts so that a bolt or screw can be installed.

Pin punches have straight shanks and are used to complete the removal of pins and rivets that have been started with a starter punch. The thin, straight shank of the pin punch is more likely to break under heavy blows from a hammer. The starter punch is used only to start a rivet or pin moving. The tapered design of its shank will prevent the starter punch from going all the way through. The pin punch is used to complete the job. Always use the largest starter and pin punches that will fit the hole.

A **center punch** is made of steel and used to mark the centers of holes to be drilled.

The **center punch** is a short, steel punch with a hardened conical point ground to a 90-degree angle. A center punch is used to mark the centers of holes to be drilled. If the center punch is used correctly, the depression it makes in the metal will guide the point of the drill bit and make sure the hole is drilled in the correct spot.

Cotter pin puller A special tool that looks like a offset screwdriver but with a pointed tip, used to pull cotter pins.

Roll pin punch A small punch that has a small point on the end to fit into the roll pin without damaging the roll pin.

Cotter Pin Pullers

Diagonal cutting pliers are often used to pull out cotter pins after the ends have been straightened out or cut off. The **cotter pin puller** is a special tool that does the job more easily (**Figure 3-55**). This tool has a hook that engages the loop of the pin and a handle with which to pull it.

The **roll pin punch** is used to remove and install roll pins. The pin has a little ball or nose that prevents mushrooming or damaging the roll pin. The punch shank is straight. A roll pin punch and roll pin are shown in **Figure 3-56**.

PNEUMATIC- AND ELECTRIC-POWERED TOOLS

Hand-operated wrenches have one basic disadvantage. They can be slow to use when there are many fasteners to remove. The technician's job can be done much faster using a wrench powered by electricity or compressed air.

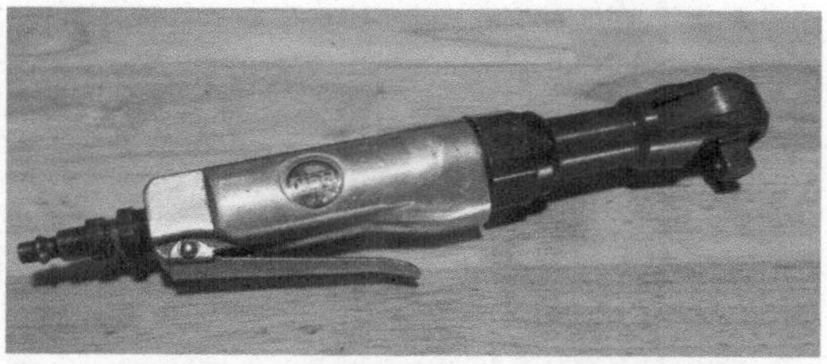

Figure 3-57 An air motor drives a reduction gear that in turn spins the socket driver. The rotation of the socket driver can be reversed for fastener removal or installation.

Air-Operated Ratchet Drivers

An **air-operated ratchet driver** (**Figure 3-57**) is connected to an air line. Pulling the trigger causes the air to rotate a socket attached to the drive on the wrench. A reversing lever allows the technician to loosen as well as tighten. Many air wrenches are designed with an impact feature.

Air-operated ratchet driver A air driven ratching wrench much like a hand ratchet.

Impact Wrenches

Shop Manual
page 50

An **impact wrench** (**Figure 3-58**) is an air-operated wrench with a powerful driver. An impact wrench not only drives the socket but also vibrates or impacts it in and out. The force of the impact helps to loosen a bolt or nut that is difficult to remove. They are often used to remove and replace wheel lug nuts.

Impact wrench An air operated impact gun that gives a rapid succession of sudden torque.

Most impact wrenches are not torque controlled. This means the technician must use a torque wrench to finish tightening after using the impact wrench.

Special heavy-duty impact sockets must be used with impact wrenches. Impact sockets are thicker and usually not chrome plated, as shown in **Figure 3-59**. Standard sockets cannot withstand the forces; they can break and cause serious injury.

Wheel Torque Sockets

Wheel torque sockets are commonly known as *torque sticks.* They are used with impact wrenches to deliver a specified torque to a fastener. They are similar to an extension in shape but are used to prevent overtorqueing a fastener while using an impact wrench for its speed. Tire stores are the primary users because of the high number of lug nuts that must be torqued each day.

Figure 3-58 An impact wrench is handy and quick for removing fasteners. It should never be used to torque a fastener. Always use an impact socket with this wrench.

Figure 3-59 Designed for use with an impact wrench, the impact socket is a softer material than the chrome sockets. It can be hand driven with a ratchet.

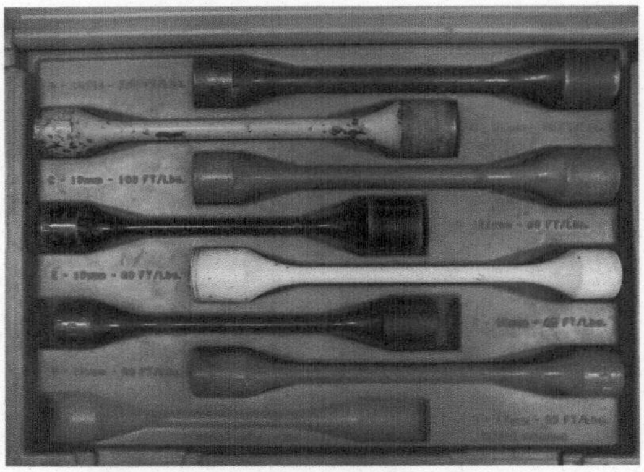

Figure 3-60 Color-coded torque sticks are designed to absorb torque above their designed limit. Some torque sticks have a fixed socket molded to one end sized to fit common wheel nut sizes.

Each stick is made for a certain torque and most toolmakers color them for that torque (**Figure 3-60**). The technician need not read the torque printed on the stick: just pick the right color for the desired torque and mount it to the wrench.

The torque stick will deliver the wrench's output up to the stick's limit. Once the limit is achieved, the wrench's output is absorbed by the stick and no torque is delivered to the fastener. However, a torque wrench is still the best method to torque fasteners.

AUTHOR'S NOTE A torque stick usually requires an impact wrench capable of attaining 250 ft.-lb. (339 N·m) of torque at 120 psi of air pressure. Less torque, or less air pressure, will not allow the torque stick to reach its designed torque limit.

Some tool manufacturers offer adjustable torque impact wrenches. This wrench can be adjusted to provide accurate torque to many types of fasteners. Others offer an impact wrench preset to a maximum torque. Impact wrenches of this type work best on larger fasteners such as wheel lug nuts.

METALWORKING TOOLS

Shop Manual
page 63

Many automotive repair jobs require the technician to do some metalworking. Jobs like drilling holes to install accessories or cutting sheet metal to repair body rust damage are typical examples. The common metalworking tools are described here.

Chisels

Chisels A long bladed hand tool with a beveled cutting edge.

Chisels are often used for cutting sheet metal, cutting off rivet heads, and splitting nuts that cannot be removed with a wrench. The most common chisel is called a *flat chisel* (**Figure 3-61**); these are often called "cold chisels." They are forged from round, square, rectangular, or hexagonal bars of tough carbon steel. A cold chisel is driven with a ball-peen hammer; the heavier the chisel, the heavier the hammer.

The *cape chisel* (**Figure 3-62**) is another common chisel. The cutting edge of a cape chisel is relatively narrower than the cutting edge of a flat chisel. The cape chisel is forged so that the cutting edge is slightly wider than the shank so that it will not bind when used to cut a narrow groove. Cape chisels are used for cutting narrow grooves such as keyways.

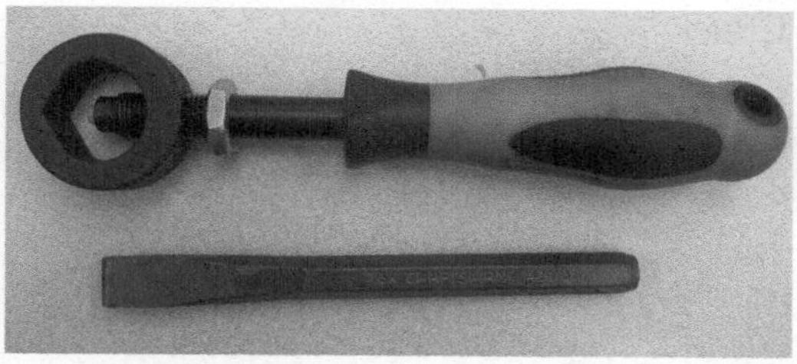

Figure 3-61 A cold chisel with a holder (top). The holder can be used to grip chisel or punches while helping to protect the hands from missed hammer blows.

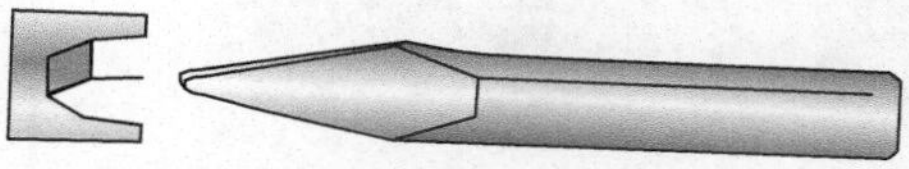

Figure 3-62 A cape chisel.

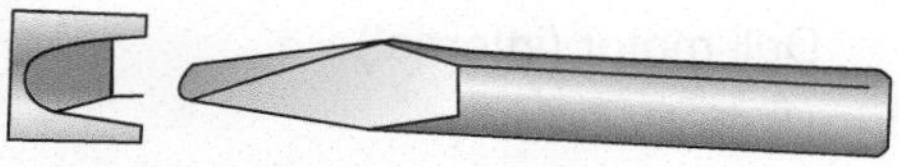

Figure 3-63 A round nose chisel.

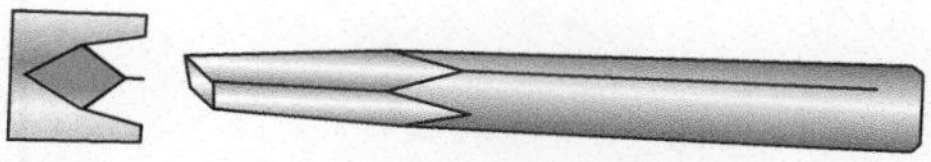

Figure 3-64 A diamond-point chisel.

The cutting edge of a *round nose chisel* (**Figure 3-63**) has only one bevel. It is used mostly for cutting semicircular grooves and inside rounded-off corners.

Diamond-point chisels (**Figure 3-64**) are forged with a tapered shank of square section that is ground on an angle across diagonal corners. This results in a diamond-shaped cutting edge. Diamond-point chisels are used for cutting V-shaped grooves and for squaring up the corners of slots.

Always wear eye protection when using a chisel. Make sure the part to be cut is held securely in a vise. Hammering sometimes curls over or mushrooms the upper end of a chisel (**Figure 3-65**). The technician must grind the head smooth because the chips may fly off a mushroomed head and cause injury.

Twist Drills

The technician will often use **twist drills** mounted or chucked in an electric drill motor to drill a hole. The four parts to a twist drill are shown in **Figure 3-66**. The end of the drill is called the *point*. The spiral portion is made up of the *body* and the *flute*. The part that

Twist drills A drill bit with a twisted body like an auger.

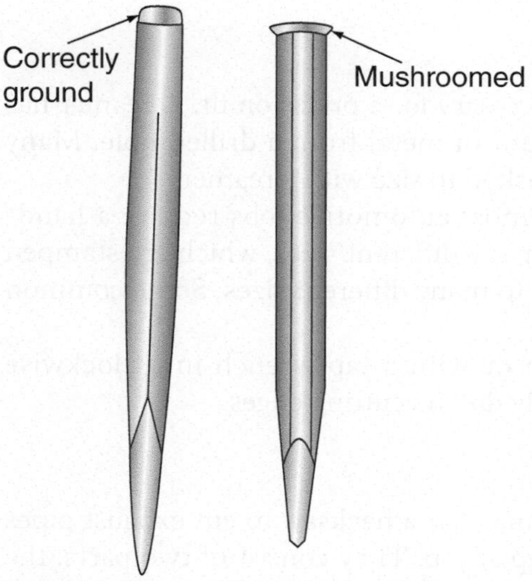

Figure 3-65 The mushroomed end of a chisel must be ground off.

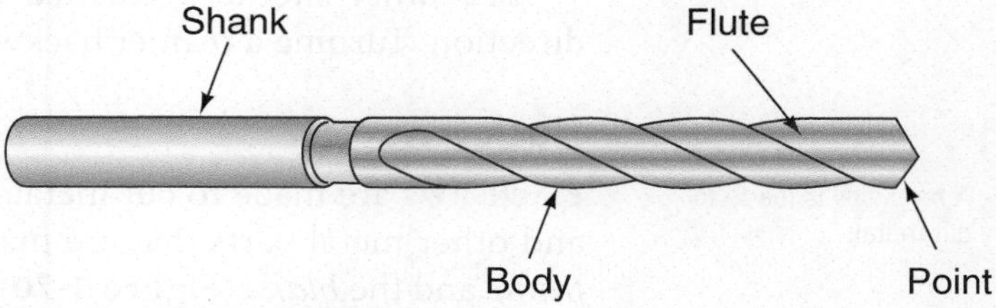

Figure 3-66 Parts of a twist drill.

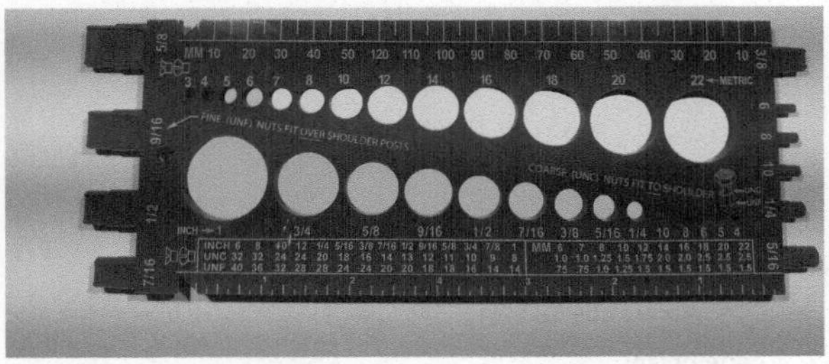

Figure 3-67 A drill gauge for measuring the size of a drill.

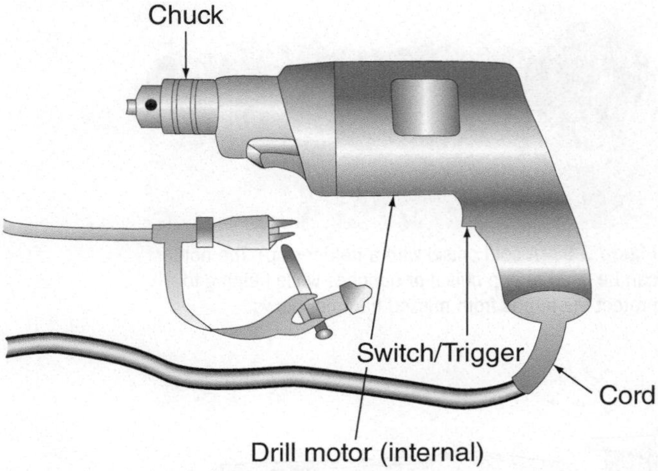

Figure 3-68 Parts of a portable electric drill motor.

fits in the electric drill motor is the *shank*. A straight shank drill is commonly used in portable drill motors.

Twist drills are made in four different size groups: (1) fractional sizes from 1/64 to 1/2 inch and larger in steps of 1/64 inch, (2) letter sizes from A to Z, (3) number sizes 1 to 80, and (4) millimeter sizes. Drill sets, or indexes, are sold in each size group. Drill size charts are available that list each drill and give decimal and metric equivalent sizes.

The size of a twist drill is stamped on the shank. But after a great deal of use, the stamp may be difficult to read. A drill may be measured with a drill gauge. The drill gauge shown in **Figure 3-67** is a metal plate with holes identified by size. The drill to be measured is placed in the holes until the hole size that best matches the drill is found.

A twist drill must be sharp to do a good job of cutting. The point of the drill is sharpened with a grinding wheel, usually in a special fixture.

Twist drills are most often used in portable electric drills (**Figure 3-68**). There are three common sizes of portable drills based on the size of the drill chucks. These are 1/4, 3/8, and 1/2 inch. The size of the chuck determines the largest drill size that will fit inside.

Make sure to exercise caution when using twist drill bits. If they are cracked or bent in any way, discard them. If they are dull, sharpen them using a special drill bit sharpening tool, not the regular bench- or floor-mounted grinder.

Reamers

A **reamer** has cutting edges used to remove a small amount of metal from a drilled hole.

A **reamer** is used when a very precise hole is necessary for a precision fit. A reamer has cutting edges designed to remove a small amount of metal from a drilled hole. Many automotive parts have bushings that must be finished to size with a reamer.

Machine-driven reamers are available, but most automotive jobs require a hand-driven reamer. Like drills, reamers are made in many different sizes, which are stamped on the shank. Reamers are also made that adjust to many different sizes. Some common types of reamers are shown in **Figure 3-69**.

A reamer should be turned with a wrench or with a tap wrench in a clockwise direction. Turning a reamer backward will quickly dull its cutting edges.

Hacksaws

A **hacksaw** is made to cut metal.

Hacksaws are made to cut metal. A technician may use a hacksaw to cut exhaust pipes and other metal parts that are made during a repair job. They consist of two parts: the *frame* and the *blade* (**Figure 3-70**). Almost all hacksaw frames are made so that they can use 8-, 10-, or 12-inch blades. The hacksaw blade installed in the frame is the part that

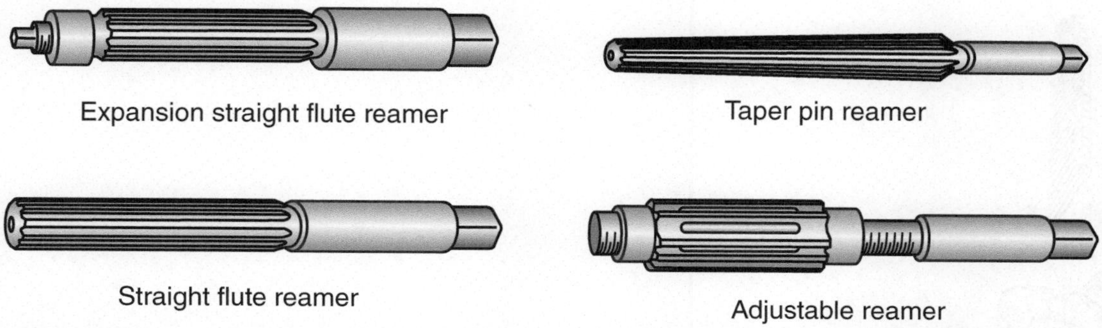

Figure 3-69 Reamers are used to fit a bore to exactly the right dimension for the component or fastener being installed. Reamers are most commonly used for piston pin bushings and certain suspension or steering bushings.

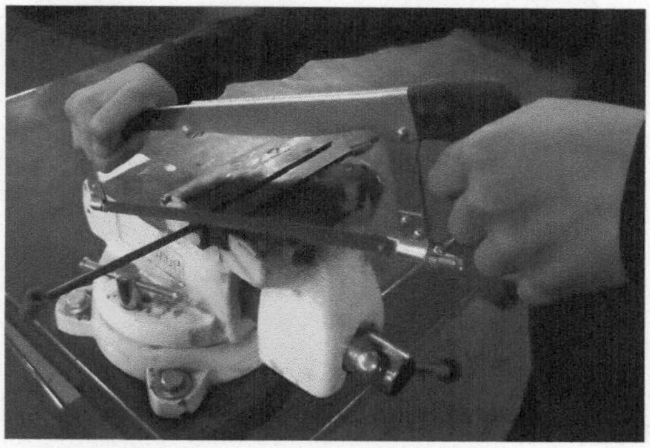

Figure 3-70 A hacksaw is used to cut metal only. It is commonly used on exhaust systems.

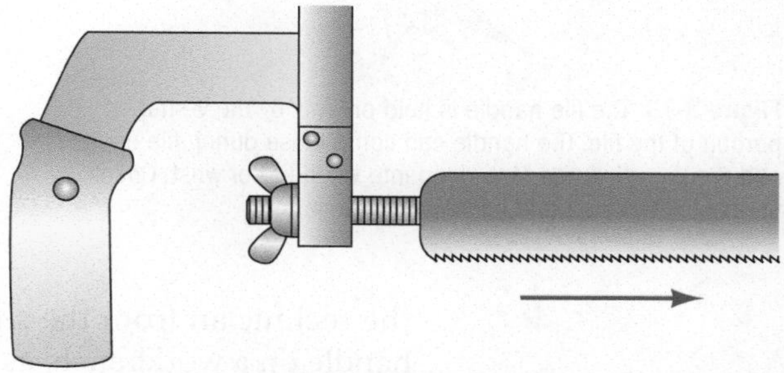

Figure 3-71 The teeth point forward or away from the handle.

does the cutting. Blades are made in different lengths and also with different numbers of teeth per inch (TPI). The common TPIs are 14, 18, 24, and 32. The number of teeth is marked on the hacksaw blade. For almost all automotive work, the 18- and 32-tooth blades are best. The 18-tooth blade is used for all sawing except thin metal, such as sheets or tubing, which should be sawed with 32-tooth blades.

When installing a hacksaw blade in a frame, always have the teeth pointing toward the front of the frame and away from the handle (**Figure 3-71**). Hold the saw with both hands and push forward and down to cut. Release the pressure to back the saw up for the next stroke. Take about one stroke per second.

Files

A **file** is used to remove metal for polishing, smoothing, or shaping. The parts of a file are shown in **Figure 3-72**. The *face* is the larger part of the file with the cutting teeth. The *tang* is the part that goes into the handle. The tapering part between the tang and the face is the *heel*. Files are made in different lengths, measured from the *tip* to the heel. The tang is shaped to fit into the handle. A handle must always be attached when filing to protect

A **file** is a hardened, steel tool with rows of cutting edges used to remove metal for polishing, smoothing, or shaping.

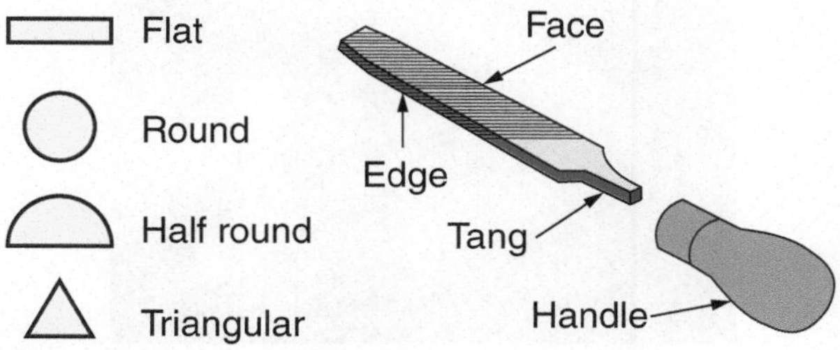

Figure 3-72 The parts of a file.

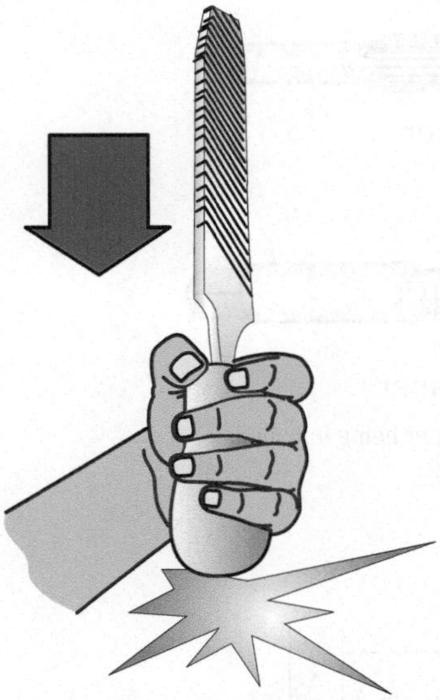

Figure 3-73 The file handle is held on only by the V-shaped portion of the file. The handle can come loose during file usage and the V portion can stab deep into the hand or wrist. Check the security of the handle often during use.

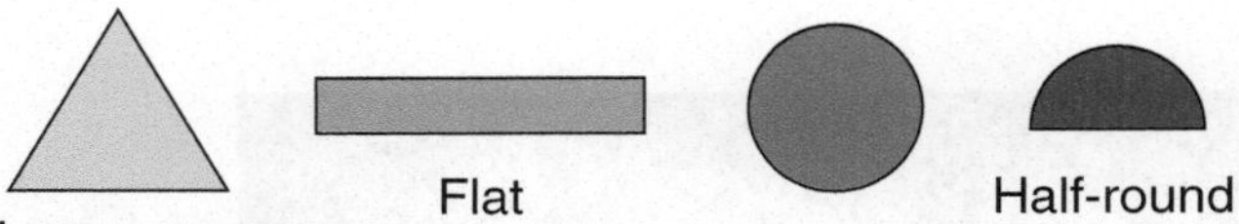

Figure 3-74 Different types of file shapes.

the technician from the sharp tang. The handle is set tightly on the file by striking the handle on a workbench, as shown in **Figure 3-73**.

Files with cutting edges that run in only one direction are called *single-cut files.* Files made with cutting edges that cross at an angle are referred to as *double-cut files.* The cutting edges may be spaced close together or wide apart. The wider they are spaced, the faster the file will remove metal. A file whose cutting edges are closer together will remove less metal and can be used to smooth or polish a metal surface. Files are also available in different shapes: flat, half-round, round, triangular, and square (**Figure 3-74**).

When using a file, grip the handle with one hand. Push down on the file face with the other hand. Because a file is designed to cut in only one direction, raise the file on the return stroke. Dragging it backward dulls the cutting edges. Mount small parts in a vise for filing. The correct way to hold a file is shown in **Figure 3-75**. When the teeth on the file become clogged with metal filings, remove them by tapping the file handle or brushing the teeth with a file card.

Thread Restorers and Chasers

Thread-restoring files A file used to repair thread on a damaged bolt.

Thread-restoring files are like other files in their appearance and use (**Figure 3-76**). The cutting edges run over and between the damaged threads, which are smoothed and straightened by the file's actions. Thread files will only *repair* damaged threads.

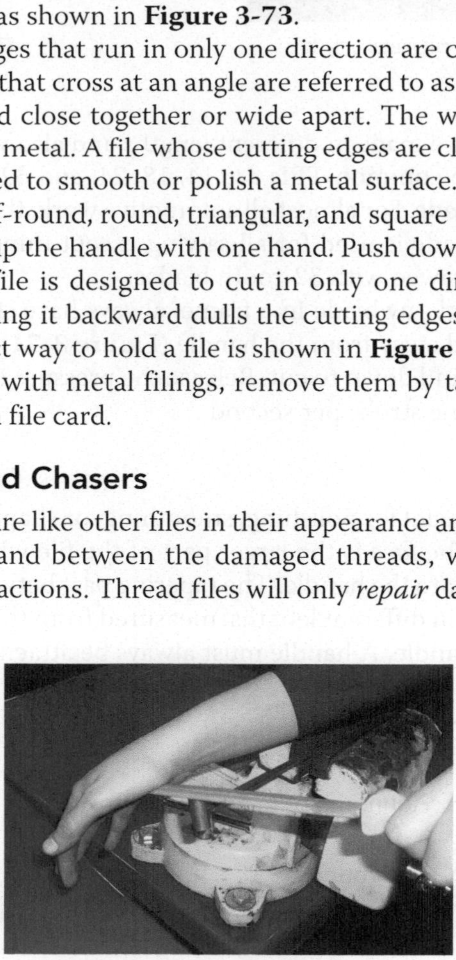

Figure 3-75 Correct way to hold a file.

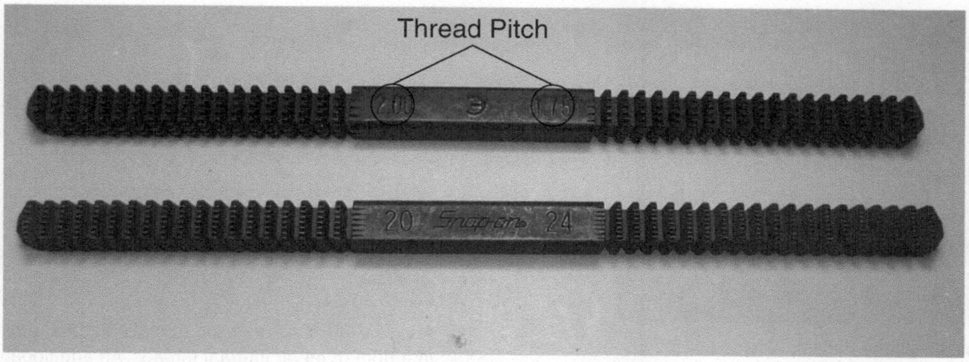

Figure 3-76 Thread files are used to straighten or clean threads. They cannot make new threads.

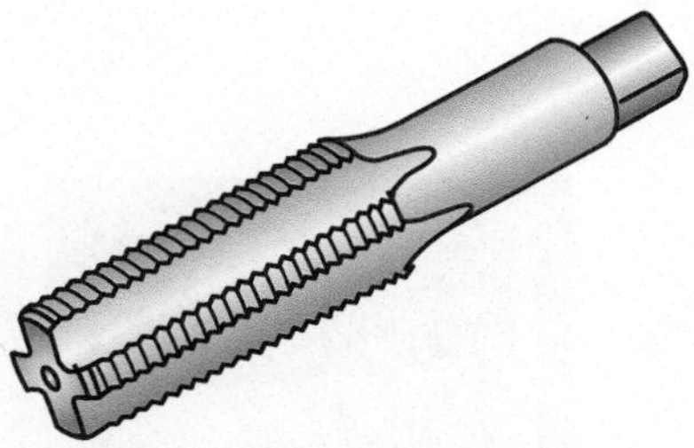

Figure 3-77 Taps are used to make internal threads in a bore or repair internal threads. An additional use is to clean internal threads. The tap's thread pitch and diameter must be matched exactly to the fastener being used.

Figure 3-78 A typical tap driver or tap wrench.

Thread chasers fit over the damaged thread like a threaded fastener. The chaser is tightened onto the damaged threads and then turned back and forth over them until they can be used. Like the file, chasers can only *repair* threads, not replace them. This type of thread repair tool does a better job and will correct more damage than a thread file. Chasers are also easier to use.

Shop Manual
page 64

Threading Tools

Many repair jobs involve repairing or replacing the threads on automotive parts. A **tap** is a metalworking tool (**Figure 3-77**) designed to make or repair inside threads. A tap of the correct size is installed in a holding tool called a tap wrench (**Figure 3-78**). The tap is then turned in the hole to make a new thread or to repair damaged ones. Taps are available for all the common thread sizes.

Tap A tool designed to repair or make inside threads.

A **tap** is used to cut internal threads.

Tap and Die Sets

Tap and die sets are designed to make threads instead of repairing them. However, careful use of taps and dies can be used to correct some minor thread damage without reducing the fastener's efficiency. A tap and die set is a combination of many different-size thread repair tools (**Figure 3-79**). The set generally includes special wrenches, an external die, internal taper taps, internal flat taps, and a pitch gauge. Sometimes special dies and taps for pipes are also included.

A tap is used for making or repairing internal threads. Internal threads are located inside a hole or nut. The taper tap is used for holes that extend completely through the material. Flat or bottom taps are designed with a flat tip to allow the tap to thread to the bottom of a hole (**Figure 3-80**). A **die** is a metalworking tool used to repair or make outside threads. A die is installed in a tool called a *die stock*. The die is then turned down

A **die** is used to cut external threads.

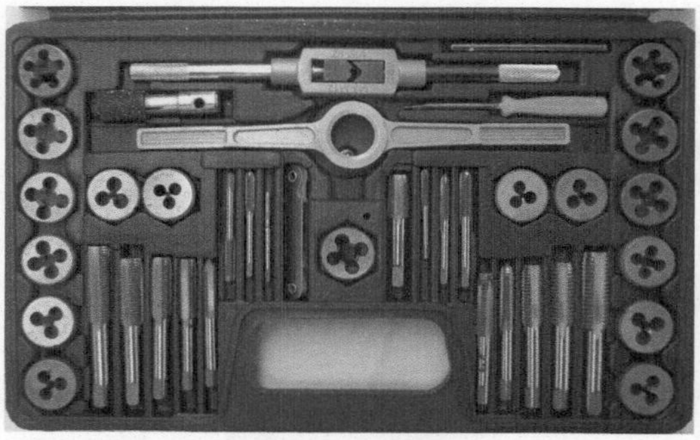
Figure 3-79 A typical tap and die set.

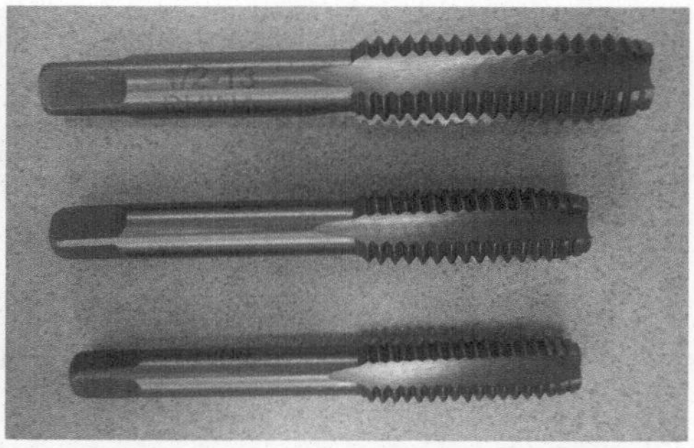
Figure 3-80 Taps are used to make internal threads in a bore or repair internal threads. An additional use is to clean internal threads. The tap's thread pitch and diameter must be matched exactly to the fastener being used.

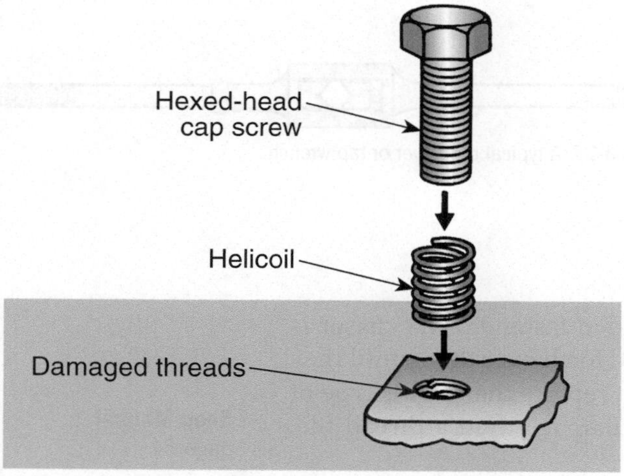

Figure 3-81 A helicoil is used to repair damaged threads.

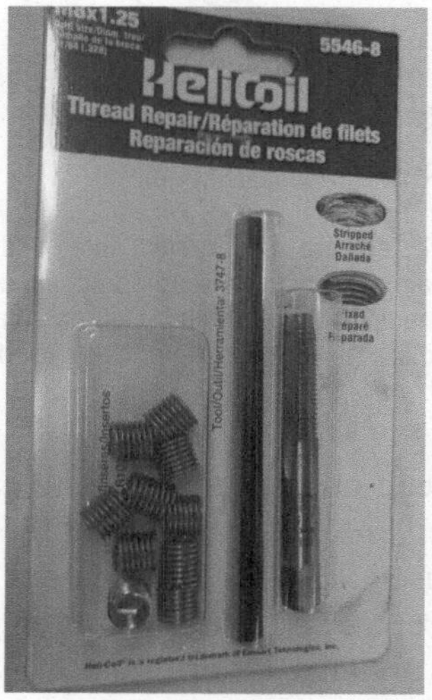

Figure 3-82 Helicoil sets can be purchased in packets as shown or a helicoil can be purchased separately. However, if purchased as individual coils, the correct drill bit, tap, and installation must be available.

over the part to make new threads or repair damaged ones. Like taps, dies come in a variety of thread sizes.

Helicoils

Helicoil A coiled wire type of threaded insert used to create inside threads.

Basically, a **helicoil** is a threaded device that can be threaded into a hole. Once positioned, it will provide internal threads for a bolt (**Figure 3-81**). The helicoil is used to make a major thread repair but retains the same diameter size and thread type. It must be used in conjunction with a drill and tap that match the size of the helicoil. A small, single-size helicoil set can be purchased for one-time jobs (**Figure 3-82**). It consists of a drill bit, the tap, and 5 or 10 helicoils. Larger sets can be purchased with several different-size helicoils and matching drill bits and taps.

A **screw extractor** is used to remove broken screws, bolts, or studs from automotive parts.

When a bolt or screw has broken off in a threaded hole, the technician must remove it. One tool for this job is a **screw extractor** (**Figure 3-83**). A hole of the proper diameter is drilled into the broken screw. A screw extractor is inserted into that hole. A wrench is

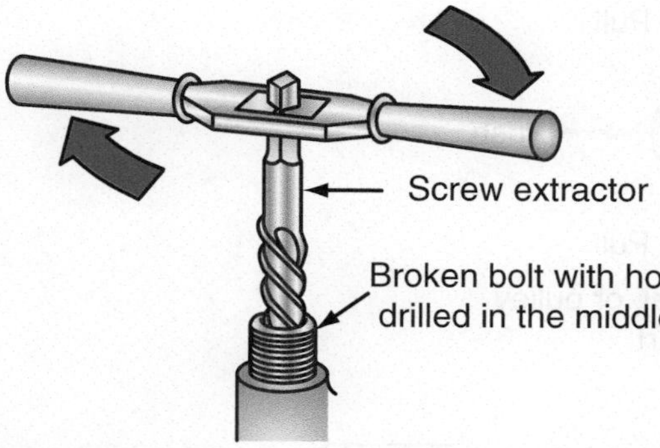

Figure 3-83 The extractor reverse (left-hand) flutes grip the fastener for removal. A hole must be drilled into the fastener for the extractor's insertion.

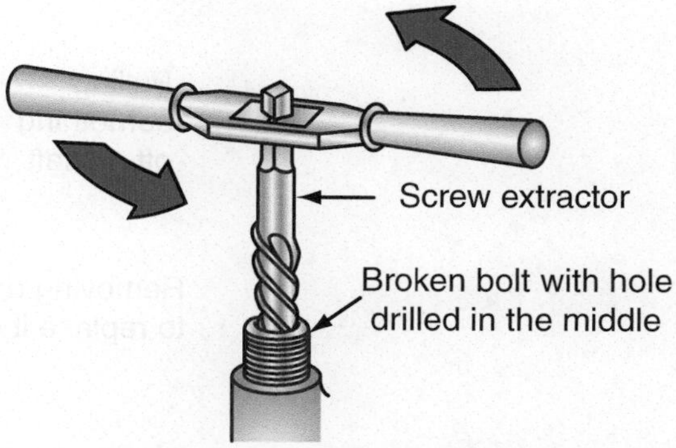

Figure 3-84 Using a screw extractor to remove a broken screw or stud.

used to turn the screw extractor counterclockwise. The reverse threads dig into the broken screw and allow the technician to unscrew it as shown in **Figure 3-84**.

Gear and Bearing Pullers

All technicians should be able to set up and use a **gear puller** or bearing puller. Most shops have a large selection of these pullers (**Figure 3-85**). These tools are designed for hundreds of common service jobs on the automobile. Although the number and variety of puller tools are huge, they are all designed to do three simple jobs (**Figure 3-86**).

A **gear puller** is a tool used to remove gears, bearings, shafts, and other parts off shafts or out of holes.

1. *Pulling something off a shaft:* A common service job is to pull a gear, bearing, or pulley off a shaft. These parts are made to fit tightly together. Rust and corrosion on the shaft can cause them to be very difficult to remove.
2. *Pulling something out of a hole:* Parts such as bearing cups and seals may be installed inside holes with a press fit. These parts would be very difficult to remove without a puller.
3. *Pulling a shaft out of something:* Shafts are installed in many transmission components. These parts are very difficult to remove except with the correct puller.

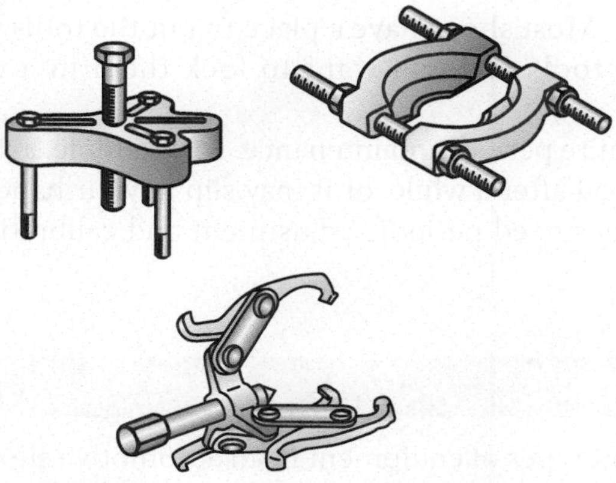

Figure 3-85 Pullers come in many shapes and sizes. Shown are the most common types used in automotive repair.

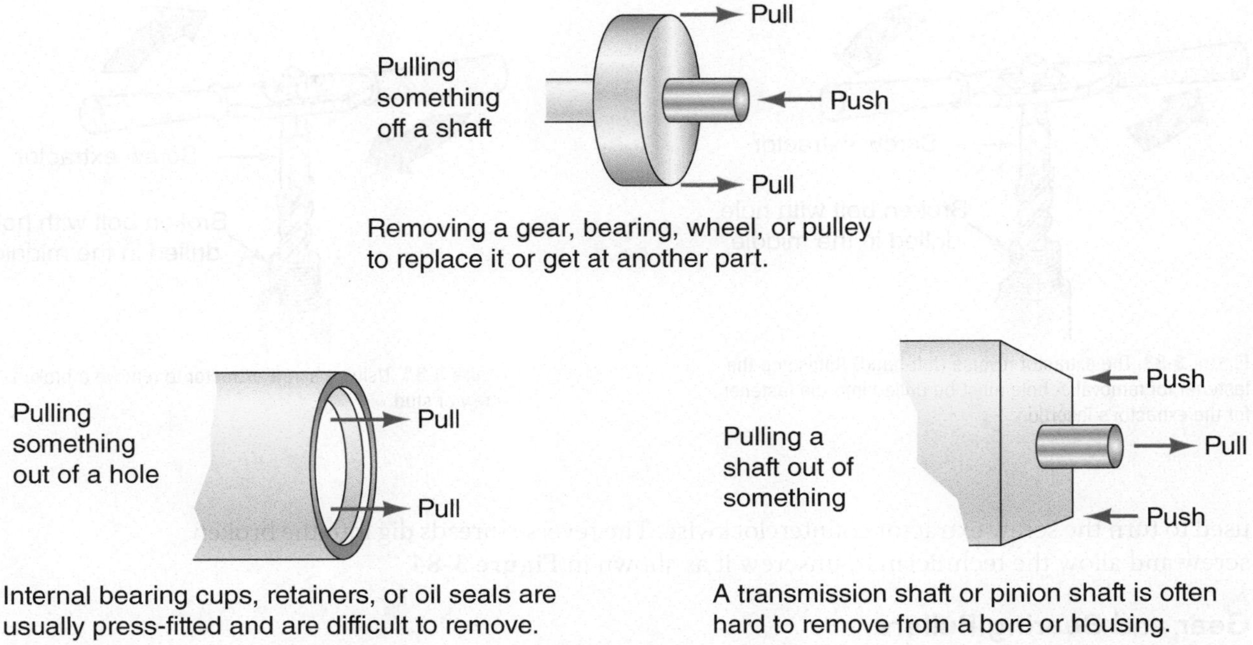

Figure 3-86 A puller can be used to remove a gear from a shaft (top), remove a bearing from inside a bore (bottom left), or remove a shaft from a bore (bottom right).

TOOL STORAGE, CLEANING, AND MAINTENANCE

Shop Manual
page 45

Each tool has its own set of procedures for keeping it in good shape. But the basic premise is that if you keep your tools in good shape, they will last longer, and you will be safer using them.

Cleaning your tools after using and before storing them in your toolbox will keep your toolbox clean, and it will prevent your hand from slipping the next time you use it. Keeping your hand tools clean also represents a good image to a watching customer. Some repairs are just going to be dirty, and customers understand that. But if you clean up your tools when you are done, it shows them that you care about your tools, and that sends an image that you care about keeping their car clean.

Safe and secure storage of tools is important. Make sure that you place all tools in a safe and secure place. Most shops have a place to put the tools where they are secure. If you own your own tools you will want to lock them in a toolbox at the end of the day.

Shop Manual
page 47

Some tools will require periodic maintenance. For example, an adjustable pliers needs to have the bolt tightened after a while, or it may slip in your hands when you absolutely need it. Torque wrenches need periodic adjustment and calibration to ensure that they are accurate.

SHOP EQUIPMENT

There are many different types of equipment in an automotive shop. They range between large 3/4-inch and 1-inch drive socket sets to automotive lifts capable of lifting up to 6 tons safely. Torches included in this section may be classified as equipment or as metalworking.

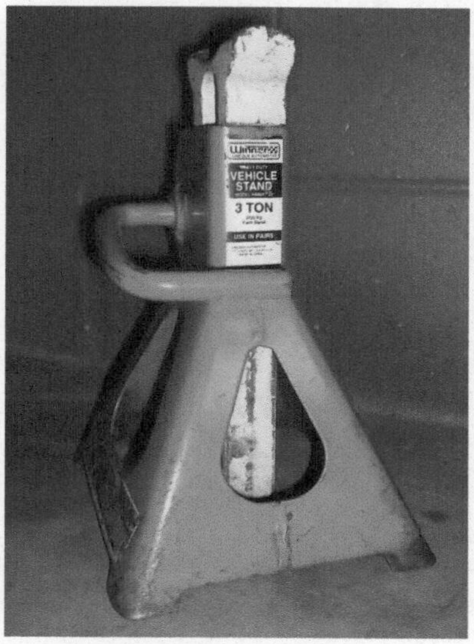

Figure 3-87 A 3-ton jack or safety stand is used to support a vehicle during undercarriage work.

Figure 3-88 A tall jack stand is commonly used to help steady a vehicle on a lift during removal of heavy components. It is also used to hold components in alignment for connections with other parts and for installation of fasteners.

Jack Stands, Wheel Blocks, and Jacks

Jack stands come in different heights with different load capabilities. The smallest can extend to about 24 inches in height and will support 2 to 4 tons of weight (**Figure 3-87**). Jack stands are used to support the vehicle or one end, so the technician can move under it on a **creeper.** Most under-vehicle work can be performed using this type of stand.

A **creeper** is a low, flat, wheeled platform on which the technician can lie and roll under a vehicle.

Other stands are much taller and designed to support a portion of a vehicle raised on a lift (**Figure 3-88**). For instance, a stand may be used to hold a rear-axle housing as the suspension system is being repaired. This type of stand is usually rated at a 2-ton capacity.

Wheel Blocks

Wheel blocks are placed forward and rearward of the wheels to prevent the vehicle from rolling. Chocks should always be used if only one end of the vehicle is being raised. Place the blocks forward and rearward of at least one wheel left on the floor (**Figure 3-89**).

Wheel blocks are also known as wheel chocks.

Jacks

The most common shop jack is a **floor jack** (**Figure 3-90**). This type of jack may be rated from 1 ton to 10 or more tons. However, the type found in most automotive shops is rated between 2 and 4 tons. Floor jacks are designed to roll as the vehicle is lifted. This keeps

Floor jack is designed to roll as a vehicle is lifted, thereby keeping the lift plate in position.

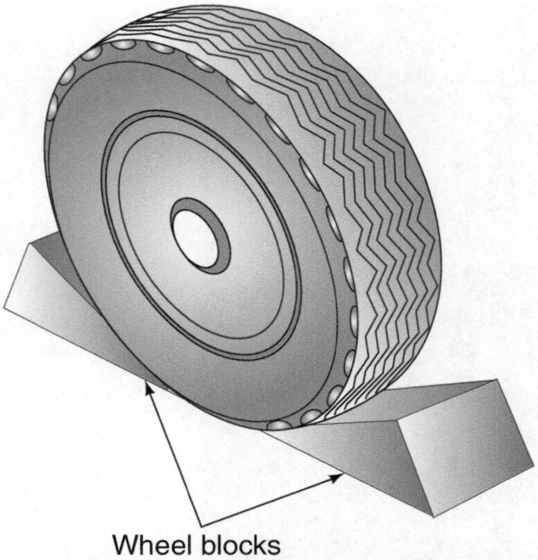

Figure 3-89 Blocks should be rearward and forward of the grounded wheels before jacking.

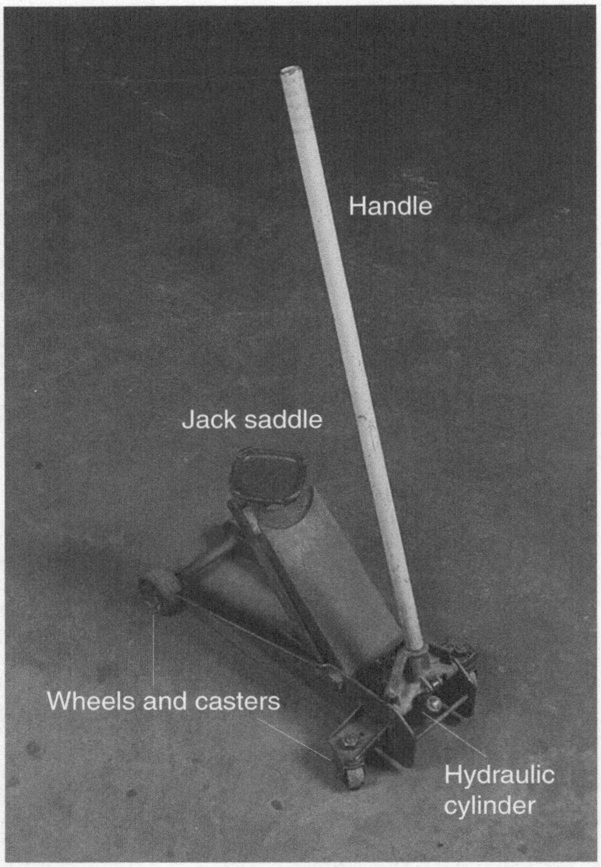

Figure 3-90 Floor jacks are designed to travel under the lift point as the vehicle is being raised.

the lift plate in position under the vehicle, but it *can* present a problem. On an incline, the jack and vehicle could roll down the slope. This is the reason for using wheel blocks.

Transmission jacks can be used under a lifted vehicle or with a vehicle on jack stands.

There are also **transmission jacks.** They can be tall enough to fit under a lifted vehicle or low enough to be used with the vehicle on jack stands (**Figure 3-91**). This type of jack has attachments to secure the transmission to the jack.

Lifts

Automotive lifts make the technician's work much easier. The vehicle can be raised so almost any under-vehicle repair can be easily accessed and accomplished. The entry-level technician will use a lift to change oil and inspect or replace tires.

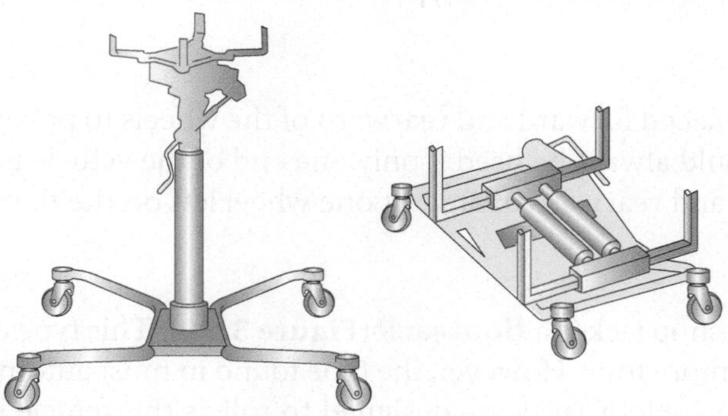

Figure 3-91 Transmission jacks are used to support the transmission during removal and installation.

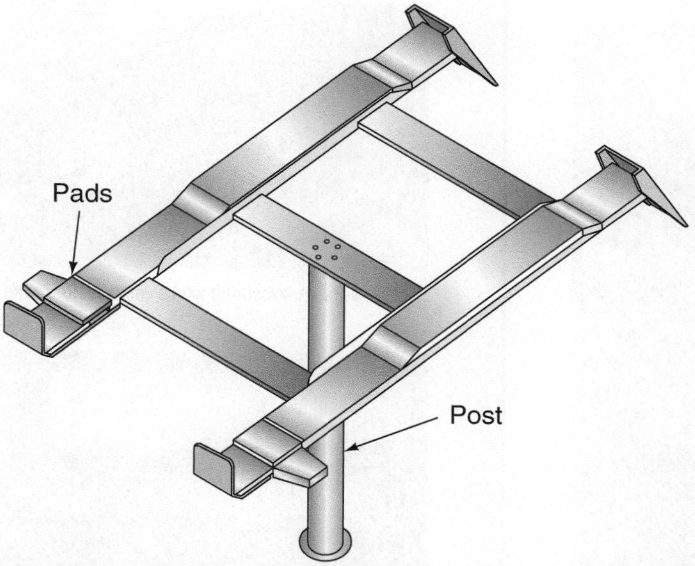

Figure 3-92 Single-post lifts have the post extending from and retracting into the floor.

Single Posts

Single-post lifts are older types of lifts. The hydraulic cylinder is buried in a floor cavity. The lift arms and the top of the cylinder present a blockage to an otherwise smooth floor (**Figure 3-92**). In addition, many of the under-vehicle components like the transmission and exhaust system are blocked by the arms when the vehicle is raised. The floorboard and exhaust can also be damaged if the lift arms and pads are not positioned correctly. There is usually only one safety lock positioned at the lift's full extension. Any position between *completely down* or *completely up* relies on the hydraulic and air control valves to hold the vehicle in place.

Single-post lifts a hydraulic cylinder lift that raises the vehicle from the center of the lift limiting access to the bottom of the vehicle.

Double Posts

The **double-post lift** is commonly known as the "above-ground lift" (**Figure 3-93**) but can also be found as in-ground lifts. It is designed to lift a vehicle and allow the technician to perform almost any repairs needed under the car. Many lifts of this type have the cross-over between the two posts at the top. This provides a clear work area under the vehicle, an advantage when compared to a single-post lift that takes up a lot of room where you need to work.

Double-post lift A hydraulic lift that raises the vehicle from the sides for the lift giving access to the bottom of the vehicle.

Most lifts of this type use two equalizing hydraulic cylinders to raise the vehicle. Cables or flat chains run over pulleys and attach to each cylinder, ensuring that the vehicle is lifted evenly. There are locks at different heights so the vehicle can be safely locked in place at several working heights.

Scissors and Drive-Ons

Scissor lifts are best used for brake or tire service. They fit under the vehicle and have limited height (**Figure 3-94**). As such, there is almost no room to work under the vehicle. Newer scissor lifts, however, are designed to allow full under-car repairs and are commonly fitted with slide plates at the rear and turntables at the front for wheel alignment. This newer lift can accommodate brake repair, but it may be awkward to use because of the runway width.

Scissor lifts A lift that vehicle can be driven on and it will lift the vehicle straight up.

Drive-on lifts allow the technician full access to under-vehicle components (**Figure 3-95**). Unless equipped with secondary jacks, the vehicle rests on its tires for the

Drive-on lifts A lift that a vehicle can be driven on and raised up with the wheels on the lift.

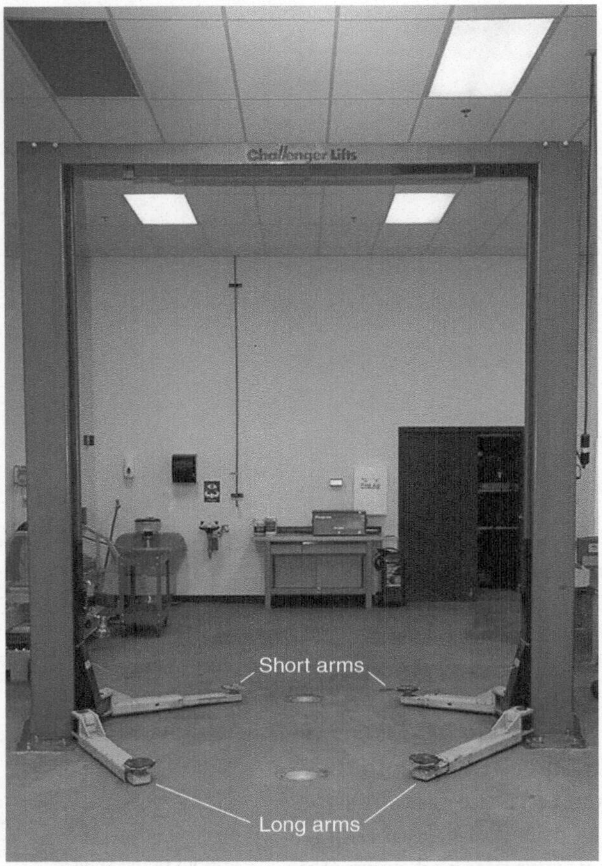

Figure 3-93 Double-post lifts with an overhead crossover leave a clean floor and work area.

Figure 3-94 A scissor lift works very well for brake and tire repairs.

Figure 3-95 Drive-on lifts are excellent work platforms for lower engine, transmission, and exhaust repair.

duration of the repairs. Drive-on lifts with secondary jacks are generally used for alignment. The vehicle can rest on its tires for alignment or the axles can be raised for access to the wheel and brake assembly.

Oxyacetylene Welding and Cutting Outfits

An **oxyacetylene welding and cutting outfit** is a system that uses oxygen and acetylene gas as a heating source for welding, cutting, and heating metal.

The **oxyacetylene welding and cutting outfit** is a common tool used in many automotive repair shops. This equipment can be used to weld parts such as exhaust systems. It can be used to cut metal to make repairs to body panels or to exhaust systems. Technicians often use the oxyacetylene outfit to heat parts to make them easier to remove.

The parts of the oxyacetylene welding and cutting outfit are shown in **Figure 3-96**. The oxyacetylene welding and cutting outfit uses two different kinds of compressed gases—oxygen and acetylene—to create heat. Each gas is stored in a steel cylinder. A regulator on the top of each cylinder is used to regulate the pressure of the gas coming out of the cylinder to the correct amount. Two gauges are attached to each regulator. One shows the pressure in the cylinder. The other shows the working pressure or the pressure going to the torch. Hoses connect the cylinders to the torch. The torch mixes the oxygen and acetylene in the proper proportions to create a heating flame. The flame is created at the tip of the torch and is hot to melt steel for welding or cutting.

Shop Manual
page 66

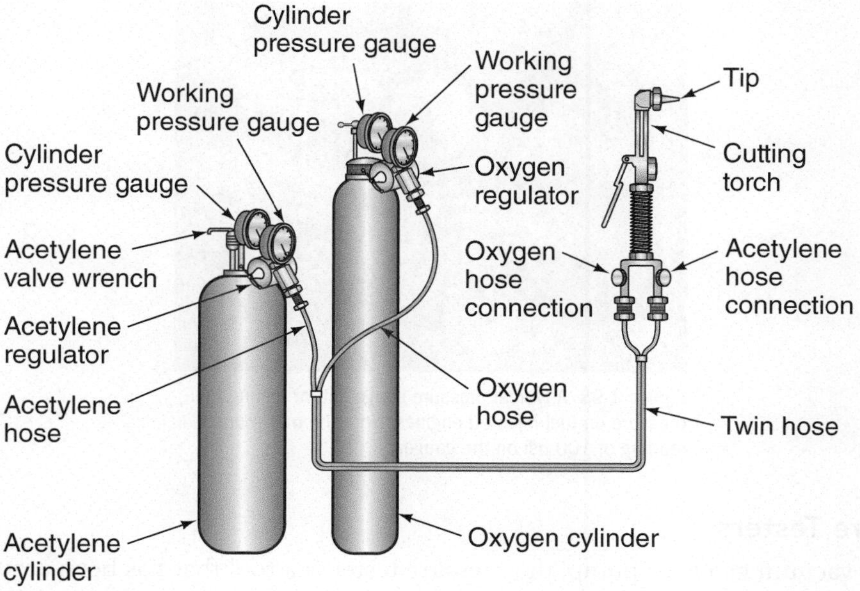

Figure 3-96 Parts of an oxyacetylene welding and cutting outfit.

TESTING TOOLS

Vacuum Testers

One of the best testers made for engine diagnosis is the vacuum gauge (**Figure 3-97**). Most automotive vacuum gauge testers have a small hand pump so the technician can supply a vacuum source to test vacuum motors and controls. The gauge itself can be used to diagnose the internal workings of an engine by reading the engine's intake vacuum under different operating conditions. Even on today's computer-controlled engine, the vacuum gauge is still one of the best tools to diagnose internal engine mechanical problems. The vacuum gauge or pump can be used to measure or apply vacuum to any vacuum-operated components on the vehicle, such as the blend door motors on heating and air-conditioning systems.

Figure 3-97 One of the oldest types of engine testers. This vacuum gauge can also be used to measure pressure, up to about 10 psi, for mechanical fuel pump testing.

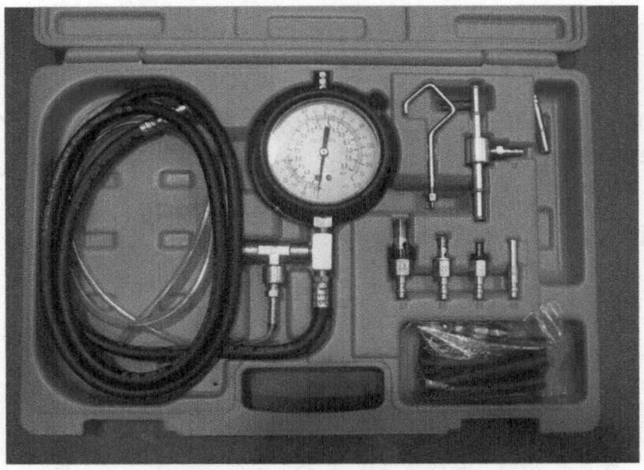

Figure 3-98 A typical pressure gauge set for testing fuel pressure on fuel-injected engines. Note the maximum reading of 100 psi on the gauge.

Pressure Testers

Like the vacuum gauge or pump, the pressure tester is a tool that has been around for a while (**Figure 3-98**). It is used to measure pressure in systems such as the transmission and fuel systems. About the only change to an automotive pressure tester is the ability to read higher pressure without damage.

Electrical Test Instruments

Electrical theory is covered in Chapter 7, "Automobile Theories of Operation." Electrical measurements have become a way of life with today's technician and are essential test devices. Some of the electrical terms used here are also explained in Chapter 5, where they are discussed within the context of electrical theory.

Test Lights

AUTHOR'S NOTE Test lights are now on the market that use light-emitting diodes (LEDs) and can be used to check electronic circuits. Before using one, ensure it can check electronic circuits without damaging them.

Even though the *test light* is old technology, it is still valid in many situations (**Figure 3-99**). One end, the *ground clip,* is attached to a metal portion of the vehicle. The other end, the *probe,* is used to probe the conductor. The test light will indicate voltage but not the amount. If voltage is present, the lamp will light. It is used on heavy-current circuits like the headlight and cooling fan motors.

Multimeters

High impedance is basically the same as high resistance. It is required to obtain an accurate measurement of electricity.

The multimeter is used to check almost any circuit on the car, but caution must be used when selecting one for purchase. It should be an automotive meter of **high impedance** and preferably digital (**Figure 3-100**). A high-impedance meter is used to protect electronic circuits during testing. A standard high-impedance multimeter will have a rating of 10 megaohms or higher. It will have two regular leads: black and red. Automotive multimeters have additional leads for temperature, engine speed, and other measurements exclusive to the automobile. The controls on most multimeters include selections for the three basic electrical measurements: voltage, current, and resistance. Some meters require the operator to select the range of measurements. The better and more expensive ones

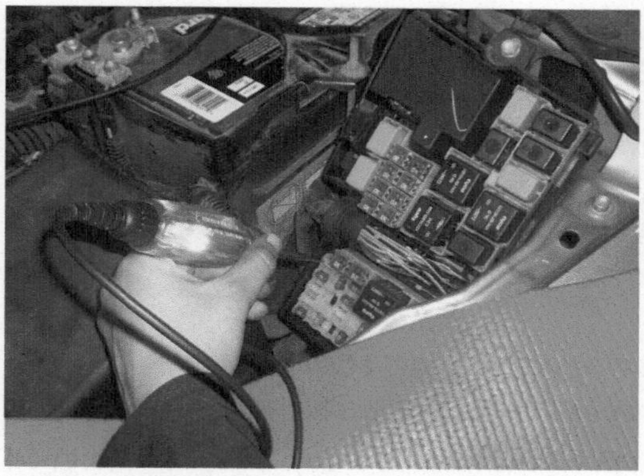

Figure 3-99 This test light is simple to use but should never be used on electronic devices. It will show that voltage is present, but not how much.

Figure 3-100 An automotive digital multimeter (DMM) is used to test electronic devices and some engine data. It can be used for simple voltage, resistance, and current measurements.

have automatic ranging. *Ranging* is the scale of measurement. For instance, if voltage is being read, the operator may have to select a 0–20-volt scale or a 20–100-volt scale. Automatic ranging meters do this automatically.

VAT 45

Automotive multimeters like a VAT 45 have selections for revolutions per minute, temperature, **frequency,** and others depending on the make and model of the meter. Additional leads are required to make the most of these measurements.

Frequency is the number of times an action or repeated action happens within a given time period. It is also known as "duty cycle."

There are meters that measure only resistance (ohmmeter), voltage (voltmeter), or current (ampere). Meters of this type are not common in the automotive field. There is one meter, however, that is common to almost every tire, parts, and repair facility. It is a *volts-amps-tester* (VAT) (**Figure 3-101**). The VAT 40 is dated technology because it is an *analog* type of meter, but many still exist and work as good as new. An analog meter is similar to the speedometer on a car where a moving needle points to a number for the measurement. Its replacement, the VAT 60, performs the same tests, but the display data is in a digital format. Both are used to test batteries and charging systems. This is one automotive meter that is capable of reading high current. On a starting system, the current may be as high as 400 to 500 amperes. Most multimeters can read no more than 10 amperes.

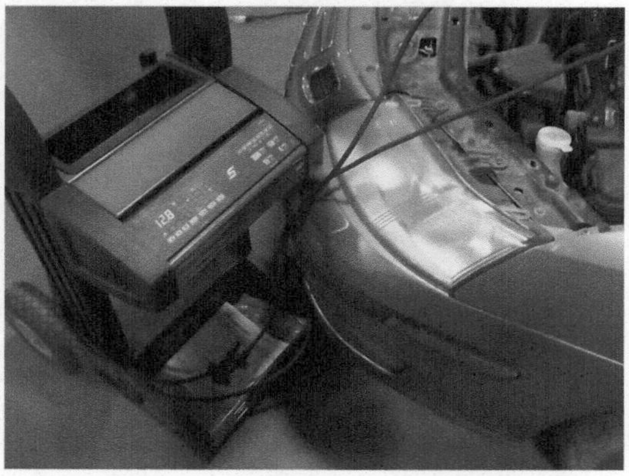

Figure 3-101 A volts-amp-tester (VAT). The top unit is a digital VAT 60. This unit has a battery charger mounted at the bottom of the frame and is sold as VAT 45.

AUTHOR'S NOTE None of the VAT testers will test a dead battery. There is a test with the VAT 40 that allows the battery to be charged for 3 minutes and then tested. This particular test does not mean the battery is good, but it can tell the technician if the battery will accept a charge. A battery should have at least 12 volts before testing to achieve a correct test result with a VAT 40 or VAT 60.

COMPUTERIZED TEST EQUIPMENT

Diagnostic trouble codes (DTCs) are numerical data that can be retrieved from the vehicle's computer with a scan tool. Each code designates a specific fault.

The **digital storage oscilloscope (DSO)** is a large, expensive piece of equipment. It is capable of storing and displaying enormous amounts of data in graph, wave, and numerical forms.

There are many different types of test equipment on the market today. They are designed to communicate with the vehicle's on-board computer or with a specific system. Most are known as *scan tools* or *hand-held lab scopes* (**Figure 3-102**). Computerized test equipment retrieves data and **diagnostic trouble codes (DTCs)** for the technician to use in diagnostic procedures. They will only collect data on electrical troubles. Mechanical problems will not show up as a problem but may cause a sensor to transmit erroneous information to the computer. This data may show as a sensor problem on the scan tool. Correct interpretation of the information is essential for correct diagnosis.

A larger, more expensive test instrument is the **digital storage oscilloscope (DSO)** (**Figure 3-103**). Shops with the customer base and funds to purchase a DSO can diagnose a fault down the last loose screw. The small, hand-held type will perform many of the functions of the larger scopes, but the large unit usually performs a more detailed readout of several data streams from the vehicle's computer at once. They are also quicker, which provides a better readout of what is happening within the vehicle's electronic systems. Most are equipped with pickup tubes to sample the exhaust and provide a total readout of four or five gases. By analyzing the amount of oxygen (O), carbon monoxide (CO), carbon dioxide (CO_2), hydrocarbons (HC), and nitrogen oxides (NO_x), gases within the exhaust provide the technician with information on the operation of the engine and its systems.

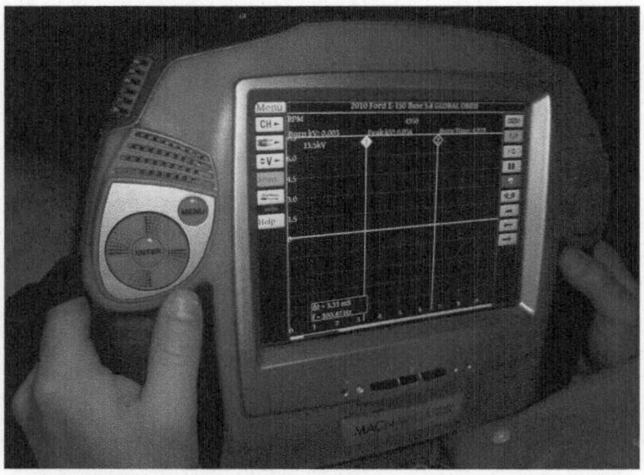

Figure 3-102 Shown is a common scan tool or graphing instrument used to collect electronic data from the vehicle's computers and assist in pinpointing the faults in the vehicle systems.

Figure 3-103 This engine analyzer is dedicated to more electronic graphing, but not the full digital data retrieval capabilities of other analyzers. Most shops have versions of both types.

SERVICE INFORMATION

Computerized Data Storage and Retrieval

Paper service manuals have been almost completely replaced by **computerized data** storage (**Figure 3-104**). Web-based internet sites are available for use by the technician. The data can be retrieved and printed quickly and easily. Most of the programs are user friendly and require only light training to get started. With time and some hands-on experience, you will be able to use these systems with ease. Fax, e-mail, internet, and direct satellite uplinks can also transmit computerized data. In addition, storage is not required except for the space required for the computer terminal. Dealerships use the manufacturer's Web site for their service information.

Computerized data is information stored on a computer or accessed through the internet.

There are aftermarket data systems such as Mitchell's On-Demand, AllData, and Identifix. These systems are the mainstay of independents, but some dealerships use these as a backup or when they are working on a vehicle that is not their represented brand.

All computerized repair information is now sold as a subscription to an online service. In the past, CD and DVD-ROMs were used to store the data. All that is needed is a subscription and a computer connected to the internet. All data suppliers offer this service, and many parts vendors such as advance Auto Parts, NAPA, and AutoZone have services that can be purchased on a short-term basis for the do-it-yourselfer (DIY) to perform some routine repairs at home.

The computer data system is less expensive after initial setup than purchasing the required paper manuals. The system is designed for a shop. With the correct software and computer hookup, the shop and even the technician can ask for, receive, and give technical assistance over the internet. Groups such as the Coordinating Committee for Automotive Repairs, the International Automotive Technician's Network, and the National Institute for Automotive Service Excellence provide online tips and conduct forums for automotive representatives to discuss repairs and the business in general. Many automotive businesses have internet sites along with libraries and schools that provide technical assistance and general information.

Service information identifies the vehicle, its major components, diagnostic checks and procedures, repair procedures, labor times, and much more. This data may be available in a paper book, on a computer hard drive, or through the internet. Even procedures that seem simple, such as changing a tire, may become a problem if the vehicle is equipped with devices to measure tire inflation. Either now or in the near future, everything on the vehicle will be monitored or controlled by electrical and electronic devices except possibly the operator. To repair almost any passenger vehicle requires some access to service

Figure 3-104 Paperback manuals are almost nonexistent in many automotive shops but are common for the do-it-yourself mechanic.

information. Some of the terms in this section have been covered earlier, so let's take a typical repair order and walk through the steps a technician follows to perform a "fixed right the first time" job.

An operator tells the service writer there is a shift problem with his six-speed automatic transmission. The writer gathers more details from the owner and starts a repair order. The assigned technician receives the repair order and realizes he has not worked on this type of vehicle but is somewhat experienced in transmission diagnosis. After determining the vehicle is safe to operate, it is taken on a short test drive to verify the customer's complaint. A problem definitely exists with the transmission shift pattern. But is it caused by the engine, transmission, or some other defect? This is where access to service information is important.

Using the car's vehicle identification number (VIN), the technician locates the type of engine, transmission, and other items that may pertain to this vehicle's problem. The next step is to check the repair history of this vehicle to see if similar repairs have been accomplished. No similar repairs were found except for a replacement of the vehicle speed sensor (VSS) several weeks earlier. Because the VSS has an impact on transmission operation, the technician makes a note and continues the research.

A search of recalls or technical service bulletins may point to the manufacturer having experienced this transmission problem with other similar vehicles. If any such reports are found, the technician's job may be half done at this point. Assuming no reports were found, the technician accesses the diagnostic procedures for this problem. He may retrieve DTCs from the vehicle computer and use them as a starting point for diagnostic steps. Usually on this type of problem, the "check engine" light will be illuminated, indicating a DTC. The technician can now pull up procedures under the heading "Testing with Codes" or a similar heading. This will lead to the diagnostic procedures that are most likely to find the fault. If no codes are present, go to the heading "Testing without Codes." This will lead the technician back to the "Codes" steps in many instances.

With this information, the actual repair can begin. The diagnostic steps will tell the technician what test tools to use and what the correct readings should be. The steps are arranged in a diagnostic tree, which is nothing more than a chart indicating "do this step, if the data is this; then go to next step or go forward to a specific step." This will eventually lead to the faulty part or system.

Another source of service information rests in the technician's head. If he understands how a typical modern transmission or transaxle works and how it interfaces with the other systems of the vehicle, then the problem is half solved. In this case, a problem with the engine may cause the computer to have the wrong data for proper shifting of the transmission. Or it could be a simple problem with thick, nasty transmission fluid that doesn't flow like it should. Either way, an understanding of what is supposed to happen can lead to a quick, accurate repair. Typical time for an average technician to access the service information and begin the diagnosis is perhaps 10 minutes.

Before assuming that all of the tools studied to this point are all that will ever be needed, dream on. There are probably hundreds of different types of tools and equipment needed for today's cars and trucks. Tools range from an exhaust wrench (a curved box-end wrench) to specialized screwdrivers and sockets, and from equipment providing 25 tons of force to remove bearings to noid lights that test electrical impulses.

There is an old saying in the automotive repair field: "You will owe the tool man money every day until you retire." The young technician will find this to be one of the truest adages he or she will ever encounter.

SUMMARY

- Wrench sizes are determined by the size of the bolt head or nut they fit. Sizes are given in either the U.S. (English) or metric system. Common wrench types are open-end, box-end, combination socket, and adjustable.
- Pliers are used as an extension of the technician's fingers. Common pliers are combination, channel lock, diagonal cutting, and vise grips.
- Screwdrivers are used to drive screws. There are many different types of screwdrivers to fit the many types of screw heads. Slotted screws are driven by the standard screwdriver. Recessed-head screws—such as Phillips, Torx, and clutch type—require a specific screwdriver.
- Hammers are used to drive tools or position parts. The ball-peen hammer is used for driving punches and chisels. Soft-face hammers—such as brass, plastic tip, and rubber—are made to protect the parts being hammered.
- Punches are used to remove parts held together by pins. Removing a pin requires both a starter punch and a pin punch. A center punch is used to make a mark for locating a hole to be drilled.
- Power wrenches are used for rapid removal or installation of bolts or nuts. Common power wrenches are the electric wrench, air-operated ratchet driver, and the impact wrench.
- Many repair jobs require the use of metalworking tools. Chisels are used for cutting metal. Twist drills are used to cut holes. Reamers are used to finish holes to an exact size. Metal is cut with a hacksaw. Files are used for removing, shaping, or polishing metal.
- Threads are made or repaired with several threading tools. Dies are used to make or repair outside threads. Taps are used to make or repair inside threads. If a screw is broken while still in a part, it can be removed with a screw extractor.
- It is important to clean and store tools when done using them. Some tools require periodic adjustments or calibration.
- Jacks that come with the vehicle for tire changing should not be used for any other repair work.
- Jacks are designed to lift only one end or one side of a vehicle.
- Lifts may be powered by air rather than hydraulics when air is used to operate under-floor controls for the hydraulic system.
- The most common electrical measuring instrument is the multimeter.
- Multimeters must be of high impedance to prevent damage to electronic circuits.
- VAT 40 and VAT 60 meters are used to test starting and charging systems and the battery.
- Large DSOs display more data than hand-held scanners and scopes. They can also be equipped to measure certain exhaust gases.
- Service manuals may be on paper or computerized.
- Computerized data banks can contain data for many years, makes, and models of vehicles.
- Web sites on the internet can provide assistance on vehicle repair or information on buying a vehicle.
- Computerized test equipment can interface with the PCM to display sensor and actuator data.

REVIEW QUESTIONS

Short Answer Essays

1. Describe how to determine the size of a wrench to use on a nut or bolt.
2. Explain how to safely pull and push on a wrench in a tight space.
3. What determines the size of a ratchet handle?
4. What is the purpose of a socket extension?
5. List the three main types of torque wrenches.
6. Describe how to safely use an adjustable wrench to loosen a nut.
7. Explain how to safely use an adjustable wrench to tighten a nut.
8. Explain why special heavy-duty sockets are used with an impact wrench.
9. What is the purpose of a die?
10. What three basic electrical measurements can be made with a high-impedance multimeter?

Fill-in-the-Blanks

1. The two different measuring systems for wrench sizes are ______________ and ______________.
2. A wrench with a box-end and an open-end is called a(n) ______________ wrench.
3. Torque can be measured in U.S. (English) units called ______________ - ______________.
   Torque can be measured in metric units called ______________ - ______________.
4. When using an adjustable wrench, care must be taken to apply the load to the ______________ jaw.
5. When using an impact wrench, be sure to use ______________ sockets.
6. A(n) ______________ hammer is used to drive a chisel.
7. The ______________ will measure resistance only.
8. A file should never be used without a(n) ______________.
9. A drilled hole can be finished to a specific size with a(n) ______________.
10. A broken screw can be removed with a(n) ______________.

Multiple Choice

1. Two technicians are discussing a wrench with "size 10" stamped on the handle.
   Technician A says this is a metric-size wrench.
   Technician B says the 10 indicates the maximum amount of torque this wrench can handle without breaking.
   Who is correct?
   A. A only
   B. B only
   C. Both A and B
   D. Neither A nor B
2. The use of a wrench to remove a tight nut is being discussed.
   Technician A says an open-end wrench provides better contact between the fastener and the wrench.
   Technician B says the box-end wrench is a better choice.
   Who is correct?
   A. A only
   B. B only
   C. Both A and B
   D. Neither A nor B
3. Two technicians are discussing which of two socket wrenches to use on a tight nut.
   Technician A says using a 12-point socket always provides twice as much contact surface between the fastener and socket.
   Technician B says a 6-point socket is best on a 6-point hex-head bolt.
   Who is correct?
   A. A only
   B. B only
   C. Both A and B
   D. Neither A nor B
4. Two technicians are discussing a torque reading that is specified in U.S. (English) units.
   Technician A says a low torque, such as 12 ft.-lb. is best applied with a box-end wrench.
   Technician B says there are 12 in.-lb. in a foot-pound.
   Who is correct?
   A. A only
   B. B only
   C. Both A and B
   D. Neither A nor B
5. A torque specification calls for a reading of 144 in.-lb.
   Technician A says this is approximately the same as 12 ft.-lb.
   Technician B says torque this large should always be converted to foot-pounds and applied with a heavier torque wrench.
   Who is correct?
   A. A only
   B. B only
   C. Both A and B
   D. Neither A nor B
6. A torque specification calls for a reading of 72 in.-lb.
   Technician A says applying 72 in.-lb. is best done with a box-end wrench.
   Technician B says this torque would probably be used on a small fastener.
   Who is correct?
   A. A only
   B. B only
   C. Both A and B
   D. Neither A nor B
7. Two technicians are discussing an Allen wrench with "8" stamped on the handle.
   Technician A says the wrench fits metric Allen head screws.
   Technician B says this Allen wrench will also properly drive a U.S. (English) Allen head screw.
   Who is correct?
   A. A only
   B. B only
   C. Both A and B
   D. Neither A nor B

8. Two technicians are discussing the use of an impact wrench.
   Technician A says heavy-duty impact sockets must always be used with an impact wrench.
   Technician B says standard sockets are designed to split open if used with an impact wrench to reduce possible damage or injury.
   Who is correct?
   A. A only
   B. B only
   C. Both A and B
   D. Neither A nor B
9. The removal of a pin from a part is being discussed.
   Technician A says a starter punch is used to begin the operation.
   Technician B says the final removal of the pin is done with a pin punch.
   Who is correct?
   A. A only
   B. B only
   C. Both A and B
   D. Neither A nor B
10. Lifts are being discussed.
    Technician A says dual-post lifts may extend from under the floor.
    Technician B says a drive-on lift is normally used for brake repair.
    Who is correct?
    A. A only
    B. B only
    C. Both A and B
    D. Neither A nor B

CHAPTER 4
FASTENERS AND MEASUREMENTS

Upon completion and review of this chapter, you should understand and be able to:

- Identify the United States Customary (USC) system of measurement.
- Identify the metric system of measurement.
- Convert measurements between the two measuring systems.
- Explain how to read rulers and tape measures.
- Understand the need for feeler gauges and their general use.
- Operate and read measurements made with micrometers.
- Operate and read dial indicators and calipers.
- Explain typical uses of vacuum and pressure gauges.
- Describe the USC threaded fastener measuring system.
- Describe the metric threaded measuring system.
- Identify the grade markings of threaded fasteners.
- Operate the common thread repair tools.
- Understand the purpose of common nonthreaded fasteners.

Terms To Know

Air gap
Atmospheric pressure
Balanced indicator
Bolt
Continuous indicator
Conversion factor
Decimal
Dial caliper
Dial indicator
Dowel pin
Endplay
Fasteners
Feeler gauge
Fractions
Gap
Grade markings
Index line
Jam hexagon nuts
Keys
Linear
Lock washers
Measurement
Meshed components
Meter
Micrometer
Nonthreaded fasteners
Nuts
Outside micrometer
Pitch gauge
Pressure
Rivets
Roll pin
Round feeler gauge
Ruler
Runout
Screws
Shank
Snapring
Splines
Standard gauge
Stepped feeler gauge
Stud
Tensile strength
Threaded fasteners
Threads per inch
Tolerances
Unified system
Vacuum
Vernier scale
Washers

INTRODUCTION

In the last section of Chapter 3, we discussed collecting repair data for the vehicle. Many times, the technician must collect information about the vehicle for comparison with manufacturer data. The method of collecting the data depends on what is being measured and what **measuring devices** are used. This chapter will discuss some of the most common automotive measuring devices. Your shop may send components out for machining based on a technician's diagnosis. If you do not check for excessive wear or damage, you may be wasting the customer's money by having repairs performed that weren't needed. Or worse, you may be repairing the wrong part because of a misdiagnosis. In Chapter 1 of the shop manual, we discussed the three Cs of a service repair order. Measuring something incorrectly could cause a misdiagnosis and not correct the original customer issue.

Automotive components are held together with **fasteners.** As a technician, you will spend much of your day removing and replacing fasteners. These small parts are very important. If fasteners are not used properly, the components they hold together can fail.

Many different types of fasteners are used on an automobile. Fasteners can be divided into two basic groups. Threaded fasteners use the clamping force from threads to hold parts together. **Nonthreaded fasteners** hold parts together without threads. Both types are discussed in the following sections.

Measuring devices An instrument use for measuring a physical quantity, amount or degree of something. A device that requires skill for proper use.

Fasteners A hardware device that mechanically joins two or more objects together.

Nonthreaded fasteners A fastener that does not have threads which includes pins, retaining rivets, push nuts.

MEASURING SYSTEMS

The **measurement** commonly used in the United States comes from the English system and is known as United States Customary (USC). Its scale is based on the inch (2.54 centimeters). USC measurements have been given terms for a certain number of inches. A *foot* equals 12 inches, a *yard* equals 36 inches or 3 feet, and so on up the scale.

Automotive USC measures **linear** distances in **decimals** (0.000) or **fractions** (1/2, 7/8) of an inch (**Figure 4-1**). Fractional measurements are usually used for overall vehicle length, tire or brake sizes, and other components where the **tolerances** are not required to be so precise. Internal engine measurements are expressed in thousandths (0.000) or ten-thousandths (0.0000) of an inch. The tolerances in tightly fitted and **meshed components** must be extremely precise to reduce wear and prolong the life of the vehicle. These types of tools and measurements are for experienced technicians who have experience with detailed engine overhaul.

The metric system of linear measurements is based on the **meter** (39.37 inches) or a portion of a meter. A kilometer (1,000 meters) is roughly five-eighths (5/8) of a mile (5,280 feet) (**Figure 4-2**). The meter can also be divided into smaller units. The most

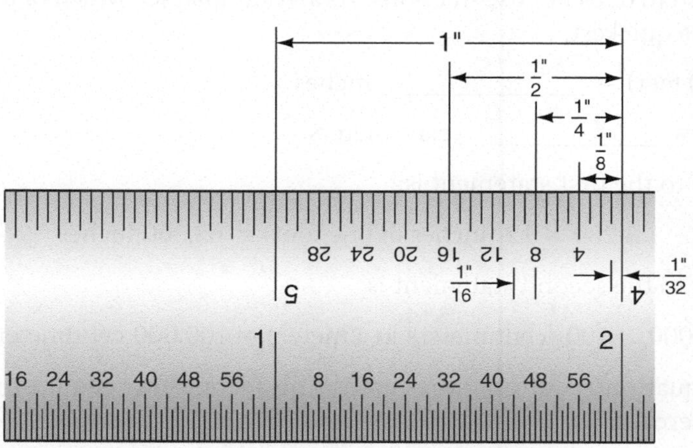

Figure 4-1 USC uses fractions of an inch to measure distances.

When recording **measurements,** always include the whole number before the decimal or fraction, for example, 0.002 or 1.05.

Linear measures are measurements of distance such as a mile or a meter.

Decimal numbers are expressions of whole numbers in hundreds, thousands, or ten-thousandths.

Fractions The resistance that one surface or object encounters when moving over another.

Tolerances set the limit that a component can be different from a perfect size.

Meshed components touch or engage each other all the time, such as two meshed gears.

A **meter** is the basic distance measurement of the metric system.

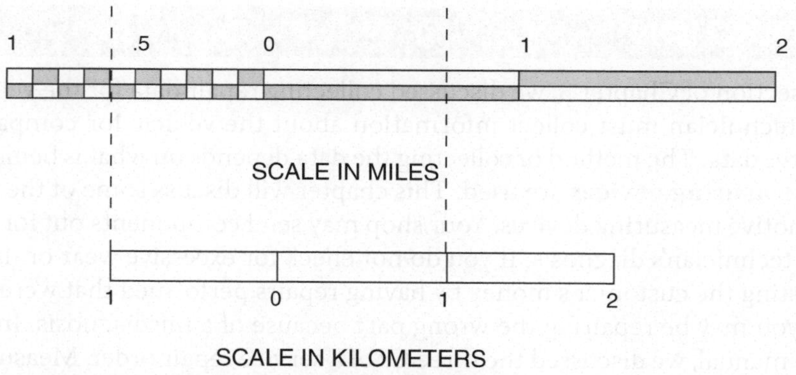

Figure 4-2 The metric system uses tenths as the base divider or multiplier.

Component	Torque*
Backing plate to axle flange	109 N·m (80 ft.-lb.)*
Brake hose bracket	45 N·m (35 ft.-lb.)*
Wheel cylinder to backing plate	23 N·m (17 ft.-lb.)*

*Note:Nottruetorquespecifications.

Figure 4-3 Most torque specifications give metric and USC data.

common units used on the automobile are *centimeter (cm),* or one-hundredth (1/100) of a meter, and *millimeter (mm),* or one one-thousandth (1/1,000) of a meter. Technicians and automotive designers use the millimeter to measure close tolerances in the same manner as when using the decimal units of inches.

Since the early 1980s, there has been an effort to change the USC system in United States to the metric system. Almost the entire world uses the metric system, and with global trade established it is becoming necessary to use a single system of measurement. Presently, most automotive measurements show the USC and its matching metric measurement together (**Figure 4-3**). For example, the speedometer is scaled both in miles per hour (mph) and kilometers per hour (kph). Another point that makes metric measurements a viable alternative to USC is the math required to change from one unit of measure to the other. With the USC system, the technician must divide or multiply by some hard-to-use numbers such as 12, 3, or 5,280. In addition, inches can be expressed as decimals or fractions. The metric system uses 10 for dividing or multiplying to change units of measure, and the system does not use fractions. As an example, see which of the following can be calculated the quickest.

One mile (5,280 feet) = ______________ inches

One kilometer = ______________ centimeters

The answer to the first statement is:

$$5{,}280 \times 12 \text{ (inches in one foot)} = 63{,}360 \text{ inches}$$

The answer to the second statement is:

$$1{,}000 \times 100 \text{ (centimeters in a meter)} = 100{,}000 \text{ centimeters}$$

The first equation requires some math. The second one requires the addition or subtraction of zeros.

In the automotive field, almost every vehicle and its parts are made to metric specifications. However, there are still many vehicles on the road based on the USC system, and those vehicles will be here for years to come. Eventually, the USC will be dropped, and the metric system will be used worldwide. Until then, it is necessary to convert some measurements from USC to metric or the reverse to properly perform daily tasks.

MEASUREMENT CONVERSION

At times, the data found may not be in the measurement format needed. This may happen because a technician who has been in the field for a while does not like the metric system or the tool being used only measures in one system. A prime example is brake disc thickness. Most new brake discs have the minimum thickness marked in millimeters (**Figure 4-4**). Most older shops use a brake disc caliper that measures in USC. To properly check the disc, the technician must first convert the metric data to a USC measurement. Sometimes a chart is readily available for quick conversion or the technician may have to perform some minor math. The following section will give a brief overview of converting measurements.

Figure 4-5 shows a chart to help make conversions. Notice that there is a **conversion factor** for going from the metric system to the USC and another for USC to metric. When making conversions, ensure that the correct conversion factor is being used. One of the most common conversions done in automotive repairs is converting inches to millimeters or centimeters or vice versa.

Conversion factors are used to change a measurement from one system to another measuring system.

A measurement that may require conversion is shaft **endplay** (**Figure 4-6**). This may be measured with a dial indicator, which may be marked in thousandths of an inch. However, the specification may be listed as metric. Using the data that follows, compute the measured endplay into a metric unit.

Endplay is the amount of horizontal movement of a shaft within its mounts.All automotive shafts must have some endplay.

Measured endplay is 0.0008 in.

Specified endplay is 0.2 mm to 0.3 mm.

Is the endplay within tolerances? If computed correctly, the answer is yes. Multiply the measured endplay (0.008 in.) by the conversion factor (25.40 mm). This equals 0.2034 mm, which falls within the 0.2 mm to 0.3 mm specified range. Try going in the opposite direction with the following information:

Shop Manual
pages 103–104

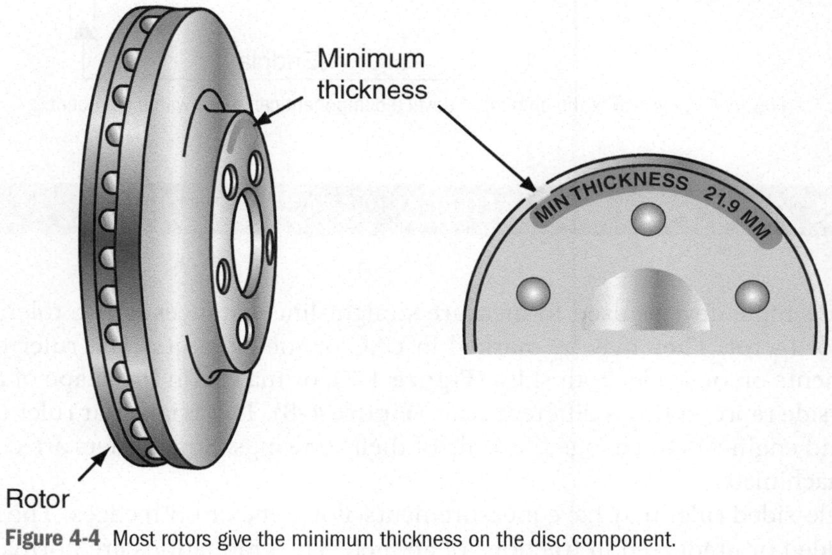

Figure 4-4 Most rotors give the minimum thickness on the disc component.

To Find	=	Multiply	x	Conversion Factors
millimeters	=	inches	x	25.40
centimeters	=	inches	x	2.540
centimeters	=	feet	x	32.81
meters	=	feet	x	0.3281
kilometers	=	feet	x	0.0003281
kilometers	=	miles	x	1.609
inches	=	millimeters	x	0.03937
inches	=	centimeters	x	0.3937
feet	=	centimeters	x	30.48
feet	=	meters	x	0.3048
feet	=	kilometers	x	3048.0
yards	=	meters	x	1.094
miles	=	kilometers	x	0.6214

Figure 4-5 Conversion charts make switching between measuring systems easy.

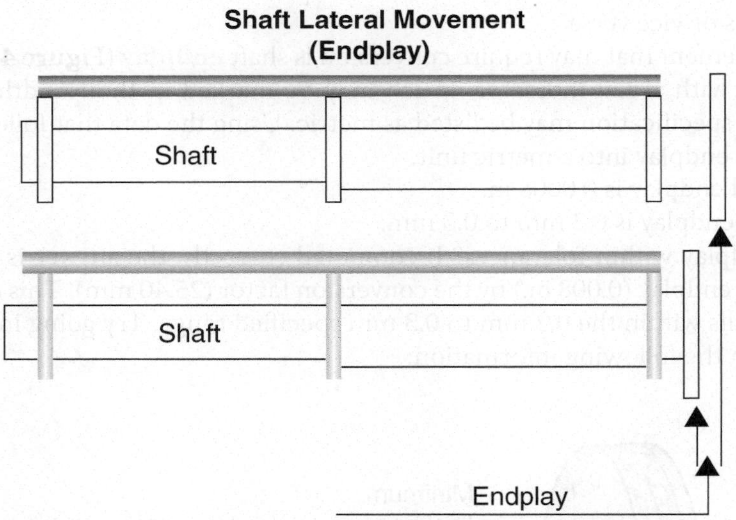

Figure 4-6 Endplay is the amount of travel a component can move within its mount.

RULERS AND FEELER GAUGES

Rulers

Rulers may have very precise graduations but still not be considered a precision instrument.

Rulers are simple devices used to measure straight-line distances where tolerances are not a major factor. They may be marked in USC or metric units. The ruler may have measurements on one side, both sides (**Figure 4-7**), or may be in the shape of a triangle with each side representing a different scale (**Figure 4-8**). This triangular ruler is used by drafters and engineers to change the scale of their drawings. Small rulers are commonly used by machinists.

A single-sided ruler may have measurements along one or both edges. The scale will be subdivided or graduated in fractions of an inch. The graduations are normally in 1/8 or 1/16 of an inch. Rulers with scales on both edges may have their second scale in 1/32

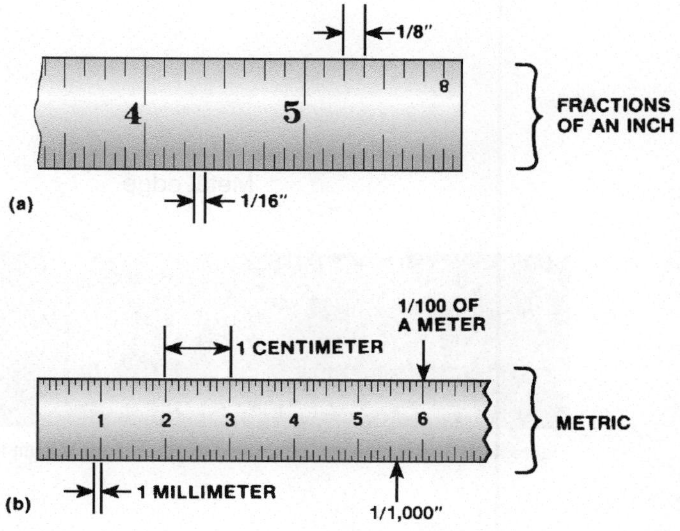

Figure 4-7 One scale has division of 1/16 of an inch. The other is 1/32 of an inch (a). The metric (b) is in centimeters and millimeter.

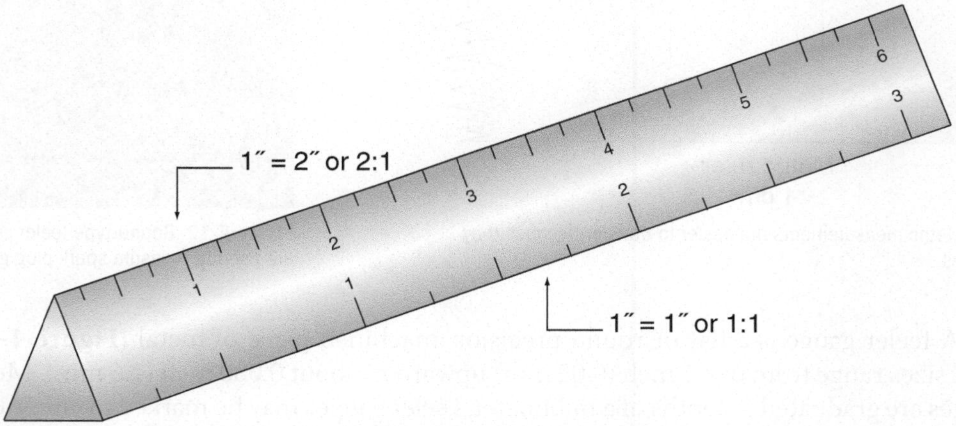

Figure 4-8 This type of ruler is used by drafters and engineers for blueprinting.

of an inch (**Figure 4-9**). USC rulers may have graduations along both sides; however, this is not a common practice. Most rulers have a thin metal strip embedded in the graduated edge to act as a straightedge for drawing or marking the work (**Figure 4-10**). However, the straightedge on a typical ruler is not considered to be an absolute precision tool or guide.

A metric ruler is made very similar to the USC version. The only real differences are the measuring system and overall length. Metric rulers are usually 20, 25, or 30 centimeters long. Metric rulers are graduated in millimeters, making it easier to interpret the measurement (**Figure 4-11**). There are no fractions on a metric ruler. Sometimes, a metric ruler will have the 0.5-millimeter marks labeled.

Feeler Gauges

At one time, **feeler gauges** were one of the most common measuring devices in an auto repair shop. Ignition points and spark plug gap were set using feeler gauges, as were other precision measurements.

Some **feeler gauges** are so thin, they must be used with extreme care to get an accurate reading and prevent damage.

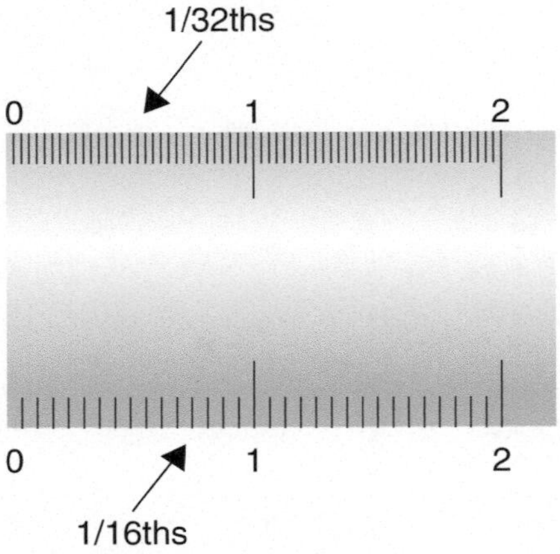

Figure 4-9 Rulers with 1/32-inch readings can be very hard to decipher.

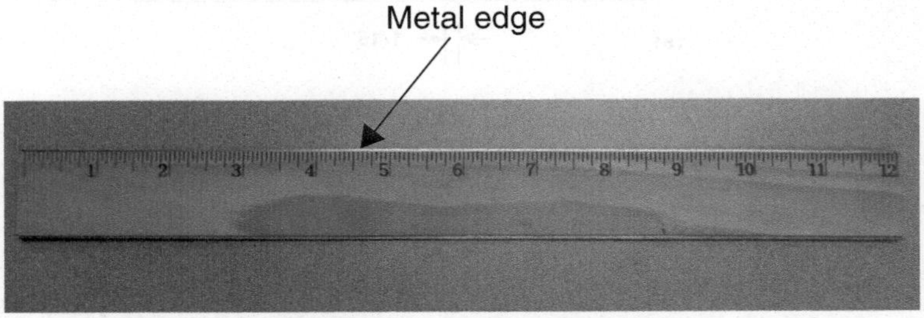

Figure 4-10 The metal edge is provided to help draw a straight line.

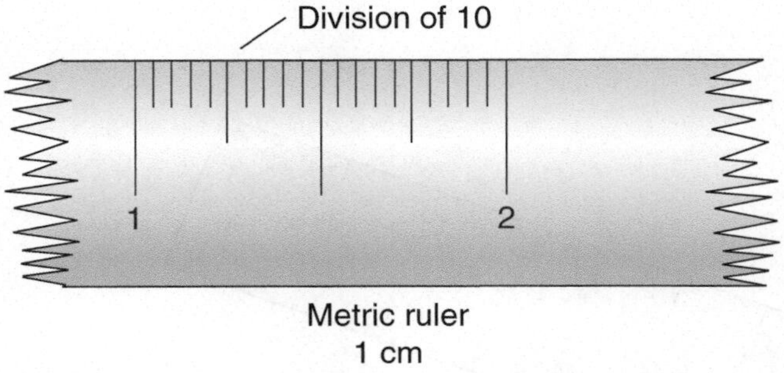

Figure 4-11 Metric measurements are easier to transfer because they use no fractions.

Figure 4-12 Round-type feeler gauges are used to measure spark plug gap.

Gap usually refers to the gap between two nonelectronic components.

Air gap usually refers to the space between two magnetic components on an automobile.

A feeler gauge is a flat or round precision-machined piece of metal (**Figure 4-12**). USC sizes range from 0.001 inch (0.02 mm) upward to about 0.080 inch (1.6 mm). Metric gauges are graduated in tenths of a millimeter. Feeler gauges may be marked in either USC or metric, but the widespread practice is to stamp each gauge with both measurements (**Figure 4-13**).

Usually the term "feeler gauge" refers to a set of 10 to 20 individual gauges (**Figure 4-14**). Feeler gauges are used to measure the distance between components. Flat gauges are used to check the **gap** between adjacent parts or measure the **air gap** between

Figure 4-13 Most feeler gauge sets are marked in USC and metric measurements.

Figure 4-14 Shown is a typical feeler gauge set that can measure most gaps.

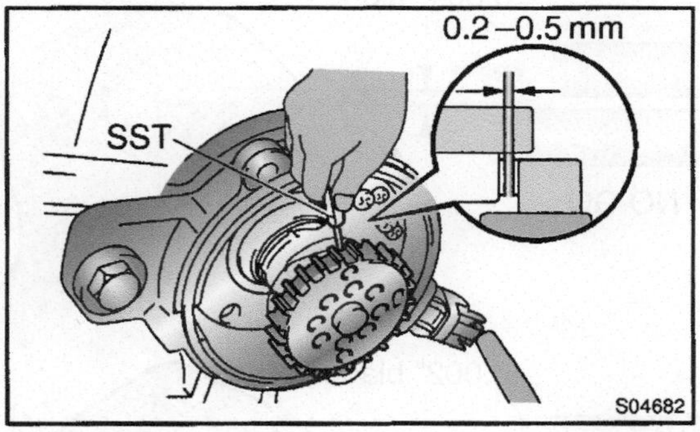

Figure 4-15 Air gaps should be checked with a nonmagnetic flat feeler gauge.

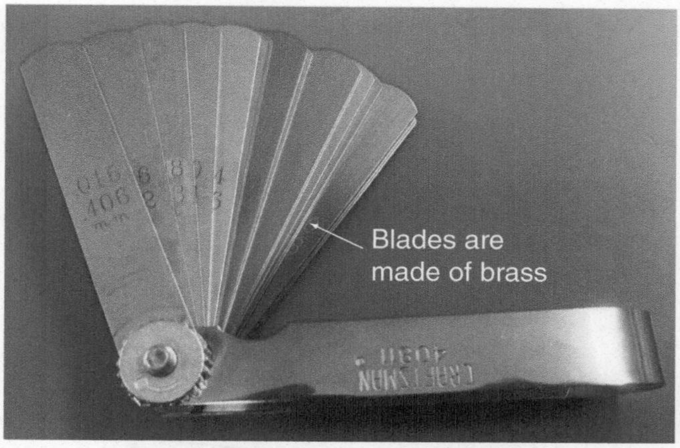

Figure 4-16 Typically, nonmagnetic feeler gauges are tan in color to indicate that the metal used is not steel or iron but usually brass.

electronic magnetic components (**Figure 4-15**). **Round feeler gauges** are used almost exclusively to measure spark plug gap. Usually, the gauges are made from steel, but special gauges may be made from a nonmetallic material, so they can be used to measure magnetic components air gap (**Figure 4-16**). Some gauges are built with an angle to make it easier to measure in tight places (**Figure 4-17**).

Round feeler gauges are made of precision manufactured wire attached to handle of some type.

The proper use of a feeler gauge is simple. However, the method of determining the point at which the measurement is correct relies on the technician's touch and feel. A feeler gauge is placed between two adjacent components. The object is to adjust the components until they are exactly *x* thousandths of an inch apart.

Instructions for the use of a feeler gauge state that it must slide back and forth between the two components until a *slight drag* is felt (Figures **4-18** and **4-19**). Technicians must develop the touch to find the slight drag required for a correct measurement.

In addition to regular feeler gauges, there are go/no-go gauges or **stepped feeler gauges.** This gauge has a tip that is 0.002 inch (0.05 mm) thinner than the rest of the gauge. If the tip slides into the space between the components and the remainder of the gauge will not slide in, the gap or space is correct. Two separate blades can be used to do the same check. For instance, a 0.002 inch (0.04 mm) can be used as the "go" gauge, and a 0.004 inch (0.010 mm) can be used as the "no-go" gauge (**Figure 4-20**).

Another name for a **stepped feeler gauge** is a go/no-go gauge.

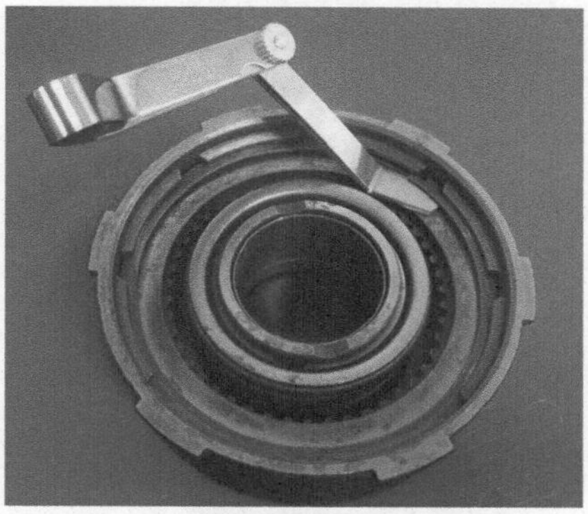

Figure 4-17 Angled feeler gauges assist in measuring gaps in tight places.

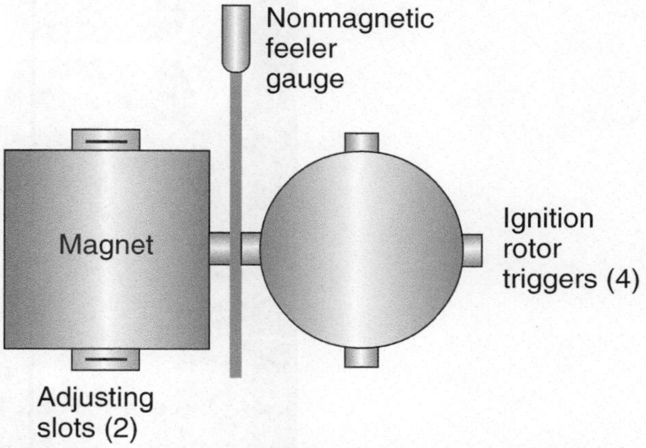

Figure 4-18 The gap must be adjusted until there is a slight drag on the feeler gauge.

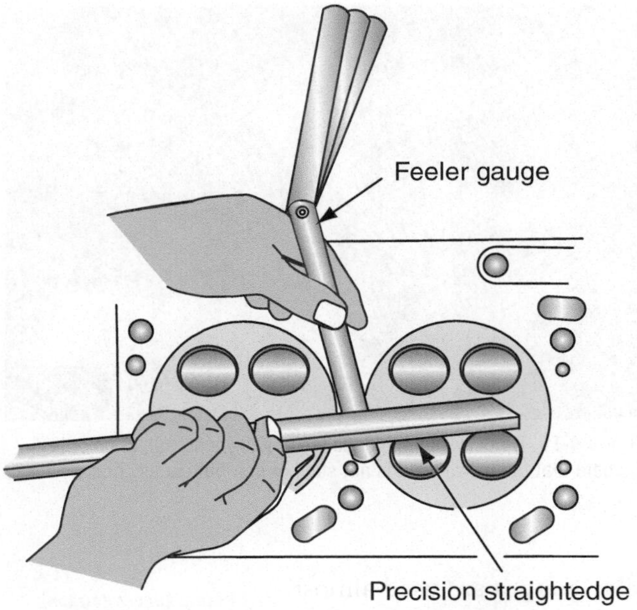

Figure 4-19 Use a straightedge and feeler gauges to determine warpage of the cylinder head.

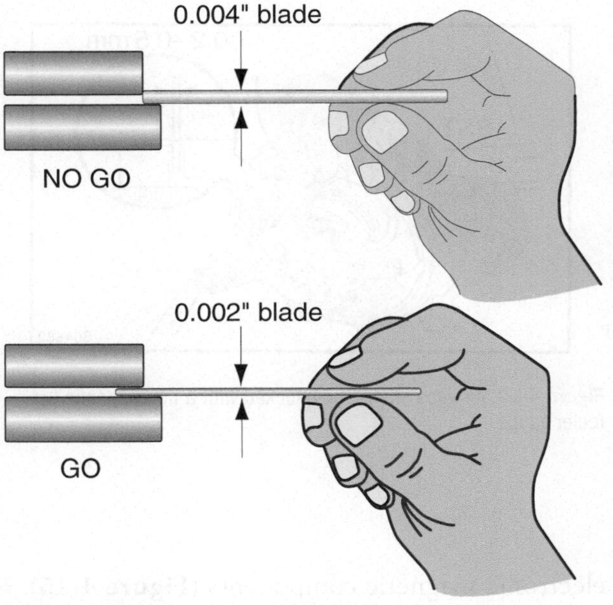

Figure 4-20 Two blades of different thickness with the correct dimensions can be used as go/no-go gauges.

MICROMETERS AND ASSOCIATED GAUGES

A **micrometer** may be referred to as a "mike."

Shop Manual pages 94–108

Micrometers are used to precisely measure the different shapes of a component. There are outside, inside, and depth micrometers, each built with a specific purpose. Each type has a measuring scale, an adjustable measuring face, a fixed measuring face, and a frame (**Figure 4-21**). Because of the limited space on the scales, micrometers are usually set up to measure either USC or metric. However, with the introduction of electronic and digital micrometers, the user can select either USC or metric measurements for the scale (**Figure 4-22**). It is suggested that manual micrometers be used for initial training so that an understanding of various measurements is achieved.

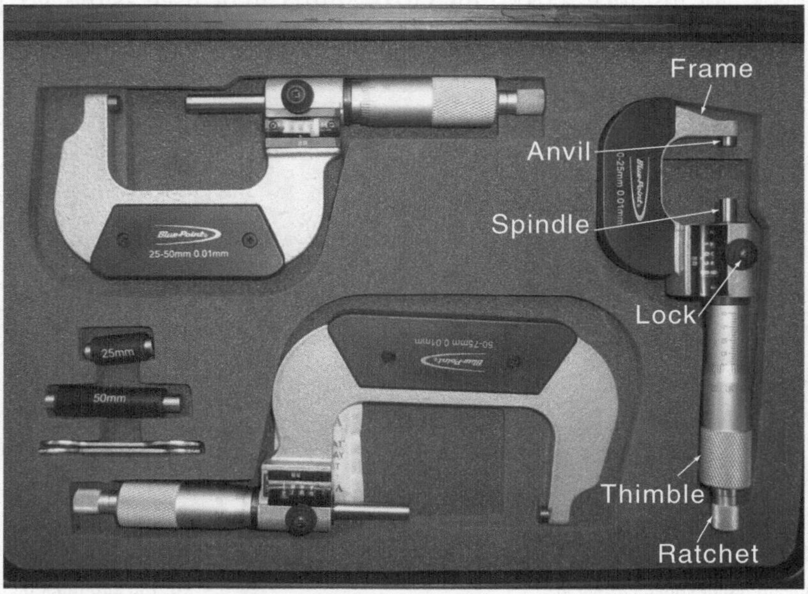

Figure 4-21 A typical outside micrometer set.

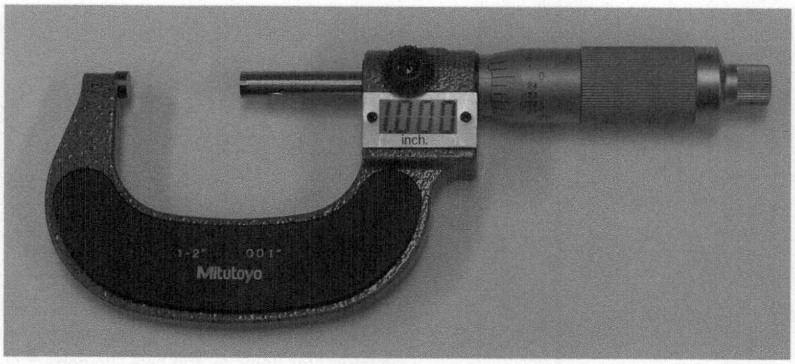

Figure 4-22 Electronic micrometers are quicker and easier to use and provide a push-button means for measurement conversion.

Outside Micrometers

An **outside micrometer** resembles a clamp that can measure linear distances. The *frame* supports the working components and is sized to measure within a limit, usually 1 inch or 25 mm. Metric scale will be discussed later. A 1- to 2-inch USC micrometer will measure a component between 1 and 2 inches in diameter or thickness (**Figure 4-23**). The *fixed face* or *anvil* provides a point from which the measurement is made (**Figure 4-24**). The other measuring face is on a *spindle* that can be extended to or retracted from the anvil by turning the *thimble* (**Figure 4-25**). The outer portion of the thimble is knurled for grip. At the outer end of most thimbles is a smaller, knurled protrusion that acts as a force release (**Figure 4-26**). The protrusion has a ratchet device that slips when the anvil and spindle are tight enough to make an accurate measurement. The ratchet also prevents excessive force from being applied to the spindle. The thimble rotates around a sleeve and extends from the frame in the opposite direction of the spindle. Most micrometers use some type of lock to hold the spindle and thimble in place (**Figure 4-27**). This allows a technician to remove the micrometer from the work to better see the measurement markings.

An **outside micrometer** is normally used in the automotive business to measure shaft external bearing journals or metal thicknesses.

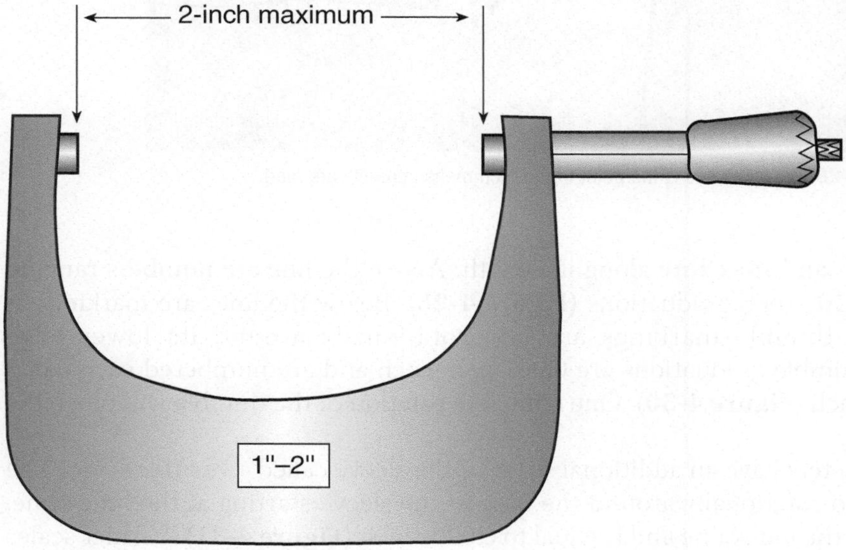

Figure 4-23 The micrometer frame shows the limits of its measurement.

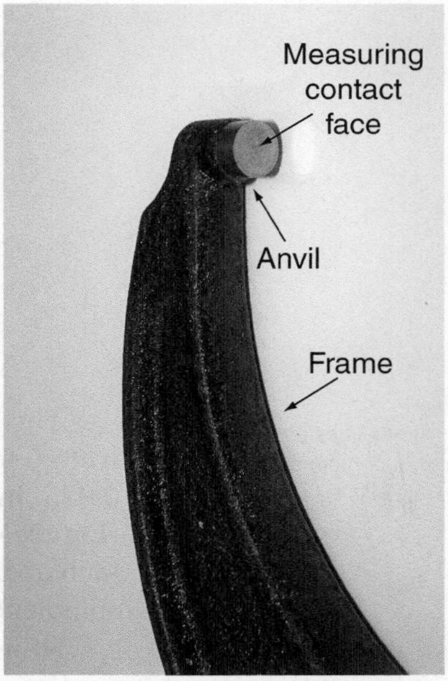

Figure 4-24 The anvil face is the nonmovable or fixed measuring face of a micrometer.

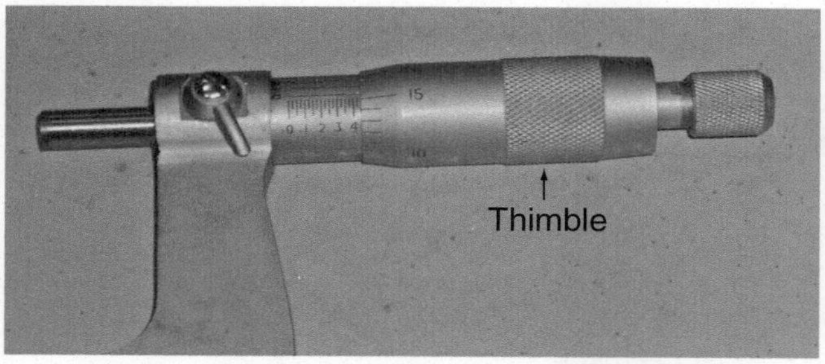

Figure 4-25 The thimble acts as a "wrench" to adjust the spindle.

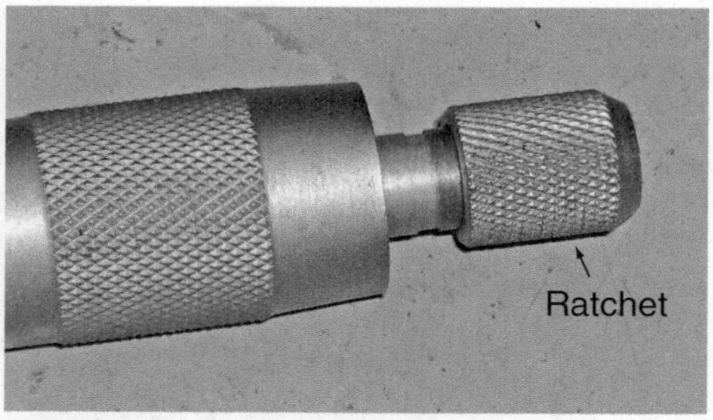

Figure 4-26 The thimble ratchet reduces the chance of micrometer damage and sets the correct contact force on the component being measured.

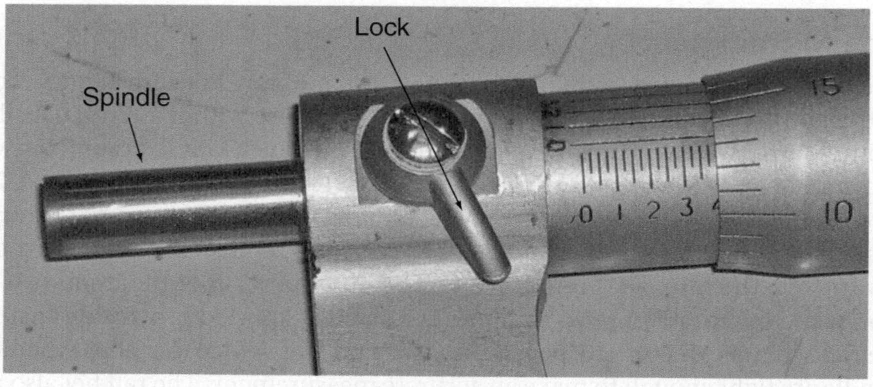

Figure 4-27 The lock fixes the spindle in place so the micrometer can be moved without changing the measurement reading.

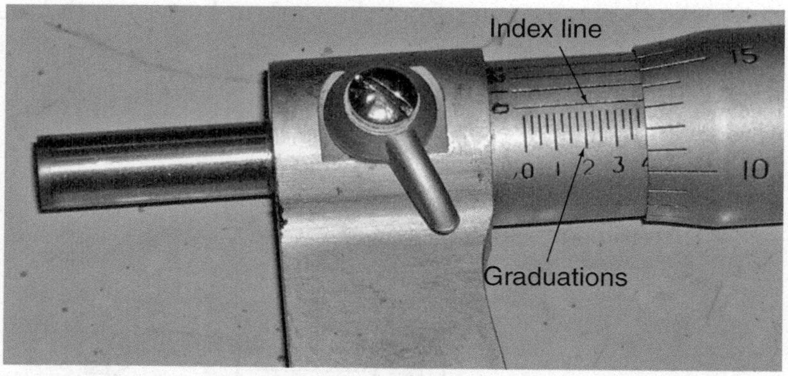

Figure 4-28 The index line is the point from which measurements are read.

Index lines are where two or more markings are aligned.

The sleeve has an **index line** along its length. Above the line are numbers ranging from 0 to 10 at 0.100-inch graduations (**Figure 4-28**). Below the lines are markings at 0.025 inch. The thimble markings are laid rotationally around its lower edge (**Figure 4-29**). Thimble graduations are 0.001 inch each and are numbered each 0.005 inch up to 0.025 inch (**Figure 4-30**). One complete rotation of the thimble will move the spindle 0.025 inch.

Vernier scale A fine measurement used to measure a ten-thousandths of an inch.

Some micrometers have an additional scale on the sleeve called a **vernier scale.** The vernier scale is laid rotationally around the base of the sleeve starting at the index line. Each line parallels the index line and is equal to 0.0001 inch (**Figure 4-31**). Vernier scales are used to read measurements in one ten-thousandths (1/10,000) of an inch.

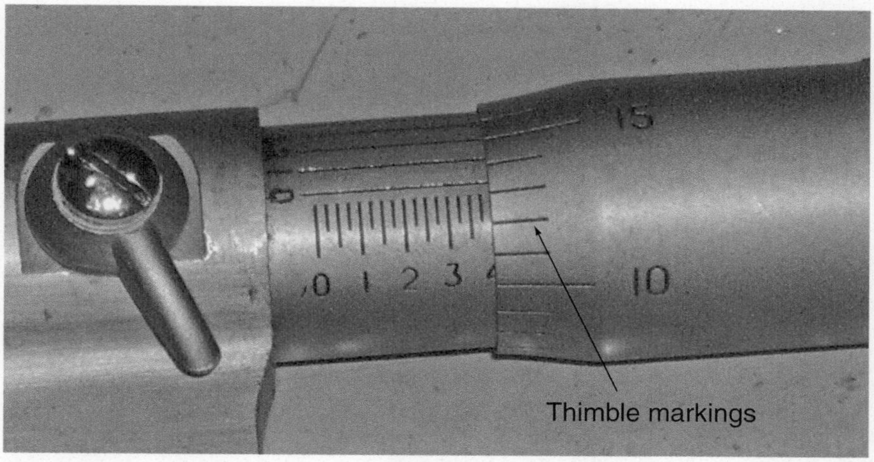

Figure 4-29 Each mark represents 0.001 inch and is numbered every 0.005 inch.

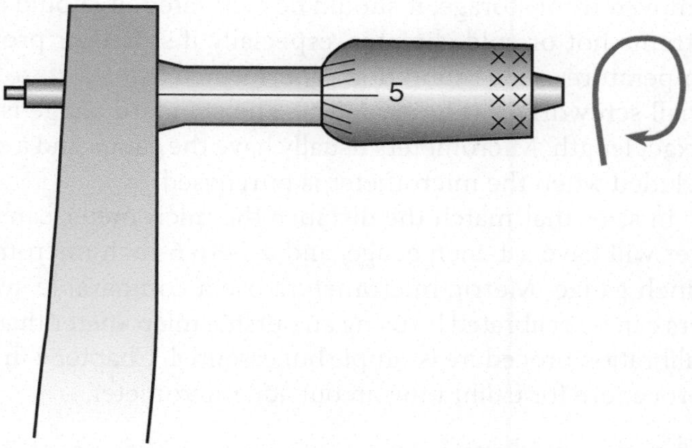

Figure 4-30 USC thimbles are graduated each 0.001 inch and numbered each 0.005 inch up to a total of 0.025 inch per complete revolution.

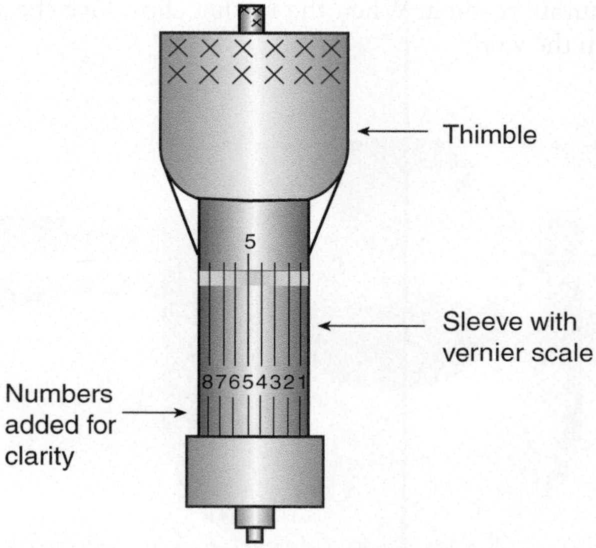

Figure 4-31 The vernier scale measures 0.0001 inch per line up to 0.0010 inch.

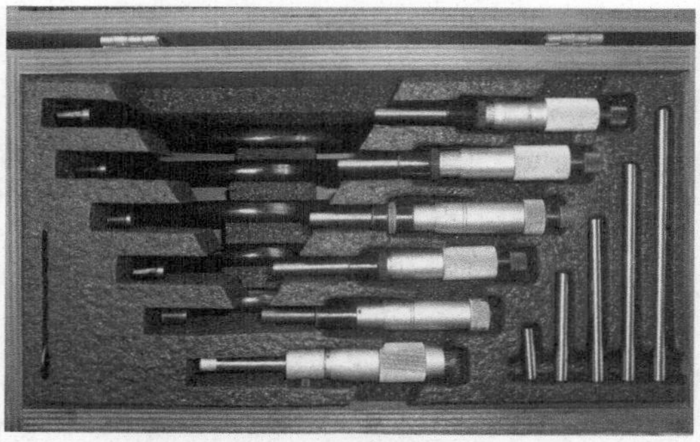

Figure 4-32 The five standards shown to the right of the case are used to calibrate outside micrometers from sizes 0–1 inch to 5–6 inches.

Micrometer Calibration

To have accurate measurements, the micrometer must be accurate itself. Any time the micrometer is removed from storage, it should be calibrated. It should also be calibrated when used in extreme hot or cold climates, especially if stored or protected at normal room or body temperatures. The calibration is performed using a **standard gauge** and a tool, usually a small screwdriver (**Figure 4-32**). The standard gauge is a shaft precisely machined to an exact length. Micrometers usually have the gauge and a small tool, usually a screwdriver, included when the micrometer is purchased.

Standard gauge A tool used to calibrate a dial caliper to insure proper measuring.

Gauges come in sizes that match the distance the micrometer can measure. A 1- to 2-inch micrometer will have a 1-inch gauge, and a 5- to 6-inch micrometer will be calibrated with a 5-inch gauge. Metric micrometers use a comparable system. Inside and depth micrometers can be calibrated by using an outside micrometer that has already been calibrated. The calibration procedure is simple but essential. Chapter 4 in the Shop Manual will explain the procedure for calibrating an outside micrometer.

Reading an USC Outside Micrometer

Reading any micrometer is easy after a little practice. Place the anvil against the work. Rotate the thimble to move the spindle to contact the opposite (180 degrees) side of the work (**Figure 4-33**). Use the ratchet knob to ensure sufficient contact is made with the work for an accurate reading. When the ratchet slips, lock the spindle and *remove the micrometer* from the work.

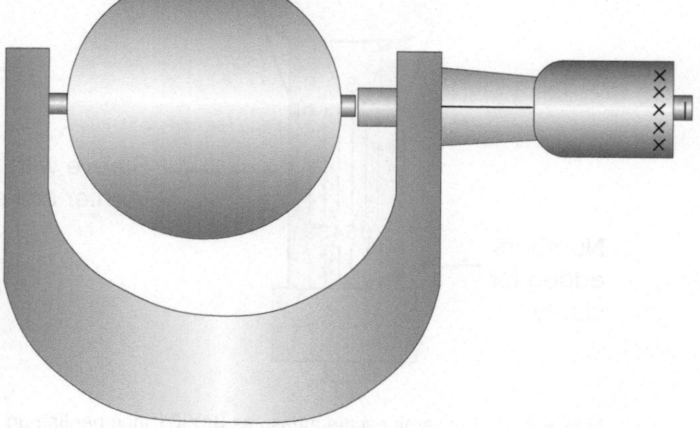

Figure 4-33 Ensure that the two measuring faces are opposite to each other.

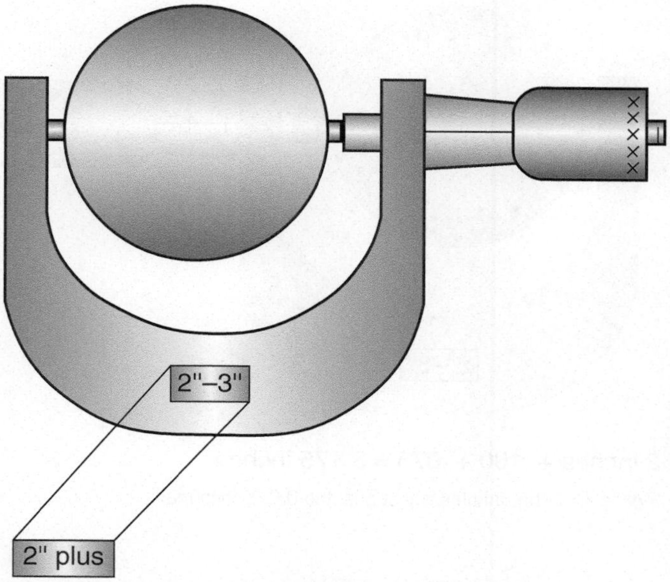

Figure 4-34 The frame contains the measurement limits of the micrometer.

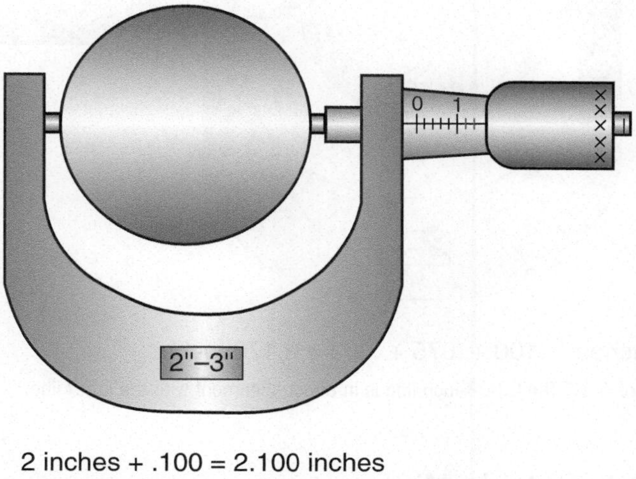

Figure 4-35 Use the highest visible number to start your decimal measurement.

For this example, the component is somewhere between 2 and 3 inches thick. This means a 2-to-3-inch micrometer is selected (**Figure 4-34**). With the micrometer removed from the work, the index line is checked first. Notice that 0 and 1 are exposed (**Figure 4-35**). Because 1 is the highest number exposed, the component is between 2.100 and 2.200 inches thick. Below the index line are three graduations past the 0.100 mark that are visible (**Figure 4-36**). Because each of these lines is equal to 0.025 inch, 0.075 inch is added to the 2.100 inches computed previously. The component is now at least 2.175 inches thick. Following the index line to where it disappears under the thimble, it is found that the third mark above zero on the thimble aligns with the sleeve's index line (**Figure 4-37**). This is the final number in the measurement. Because each line on the thimble equals 0.001 inch, 0.003 inch is added to the 2.175 inches from the last calculation. The final measurement for the thickness of this component is 2.178 inches.

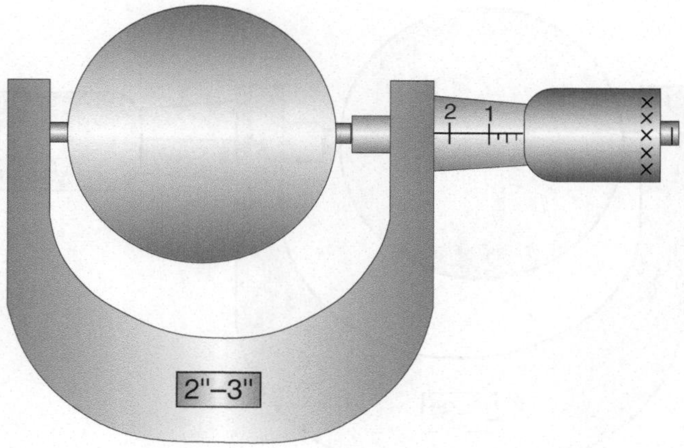

2 inches + .100 + .075 = 2.175 inches

Figure 4-36 The thimble is just past the 0.075-inch mark.

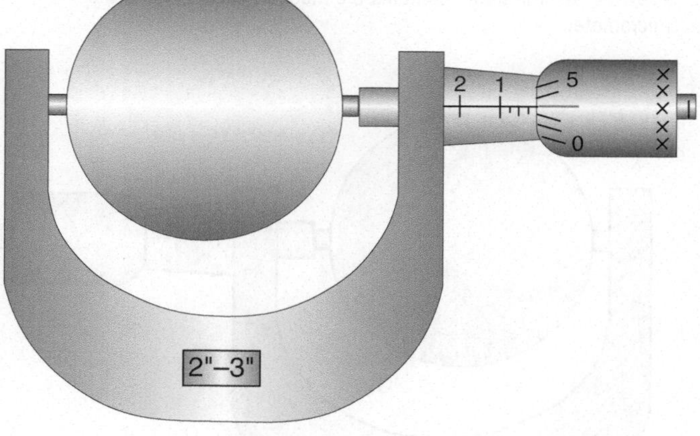

2 inches + .100 + .075 + .003 = 2.178 inches

Figure 4-37 The 0.003-inch line is in direct alignment with the index line.

Reading a Metric Outside Micrometer

A metric micrometer is laid out the same as a USC micrometer except for the size of the scales. Above the index line are 1-mm markings ranging from 0 to 25. They are numbered at each 5 mm (**Figure 4-38**). Below the index line are markings designated 0.5 mm each (**Figure 4-39**). The thimble is marked from 0 to 50 or half of a millimeter (**Figure 4-40**). Each graduation is equal to 0.01 mm and is numbered at each 0.05-mm mark (**Figure 4-41**).

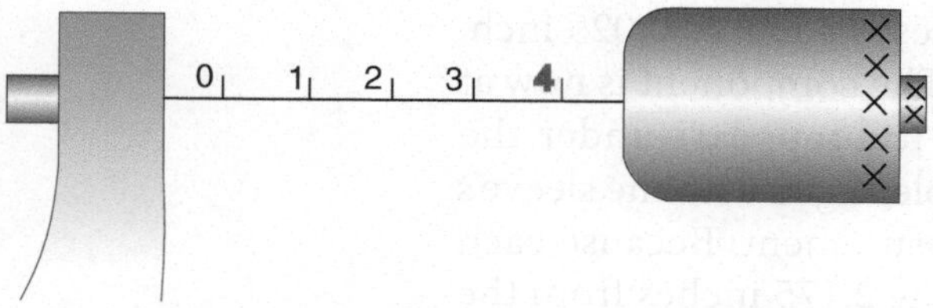

Figure 4-38 Each numbered line above the index line represents 5 millimeters.

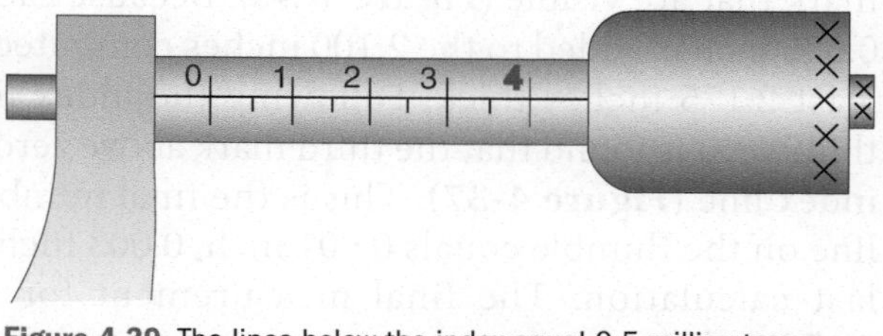

Figure 4-39 The lines below the index equal 0.5 millimeter.

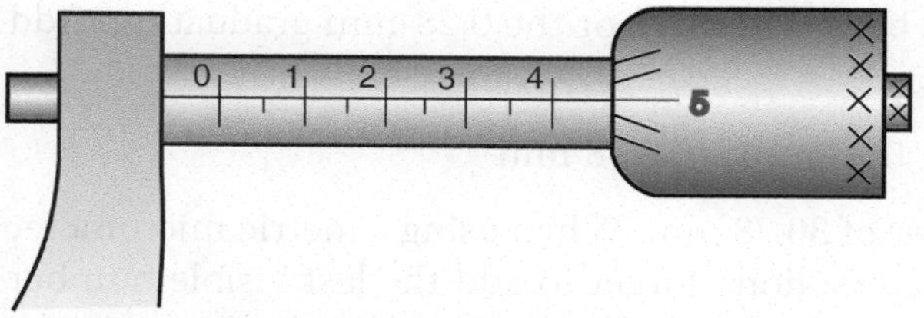

Figure 4-40 One rotation of the thimble equals half (0.5) of a millimeter.

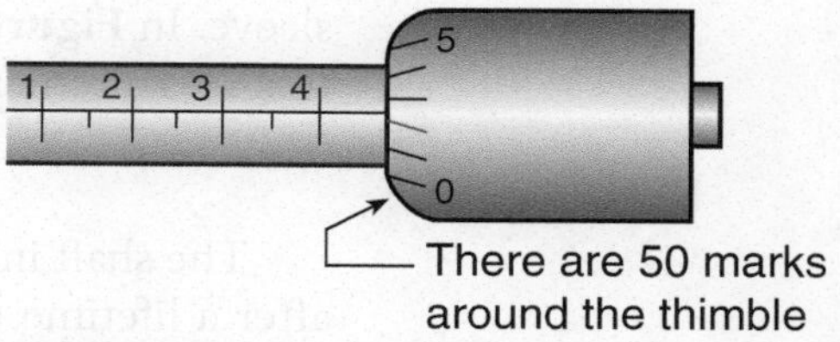

Figure 4-41 Each line on the thimble equals 0.01 millimeter for a total of 0.5 millimeter per thimble revolution.

Reading a metric micrometer is very similar to reading a USC micrometer. Like the USC micrometer, the frame has a numerical value based on the total distance or length the micrometer is capable of accurately measuring. The frame on a metric micrometer is sized for a total of 25 mm (i.e., 0 mm to 25 mm, 50 mm to 75 mm).

The first digit to the left of the decimal is the lowest number on the frame, just as with a USC micrometer. For this discussion, a 25-mm-to-50-mm micrometer will be used as the example. On a 25-mm-to-50-mm micrometer, this would initially be 25.00 mm. There are a few other differences between reading a USC and metric micrometer.

As stated, the first measurement (reading) is from the frame, in this case, 25 mm. Look on the sleeve's index line and read the last visible figure. **Figure 4-42** shows that the number 5 is visible. This is one change: Add the last visible number above the index line (5 mm) to the minimum size of the micrometer (25 mm). At this point, the shaft is at least 30 mm in diameter.

$$25 \text{ mm (frame)} + 5 \text{ mm (index)} = 30.00 \text{ mm}$$

The next step determines the first number to the right of the decimal. This number comes from the last visible marking on the index line under the sleeve. In Figure 4-42, the last visible mark under the index line is the 0.5-mm graduation. This is added to the previous reading, and we now know the shaft is at least 30.50 mm in diameter.

$$30 \text{ mm} + 0.5 \text{ mm} = 30.50 \text{ mm}$$

The last number in this measure is read from the thimble. Like the USC micrometer, find the graduation line on the thimble that most closely aligns with the index line on the

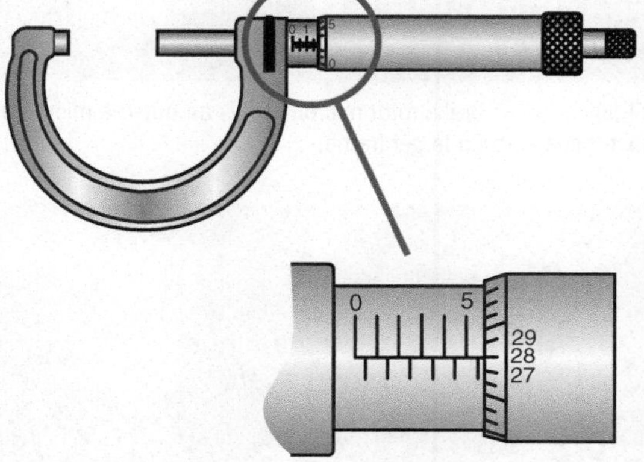

25 + 5 + .5 + 28 = 30.78 mm

Figure 4-42 Similar to the USC micrometer, the measurement on a metric micrometer is a matter of adding those numbers visible on the frame and index lines.

sleeve. In **Figure 4-42**, this appears to be the 28 mark or the 0.28-mm graduation. Add this measurement to the previous step:

$$30.50 \text{ mm} + 0.28 \text{ mm} = 30.78 \text{ mm}$$

The shaft in question has a diameter of 30.78 mm. When using a metric micrometer after a lifetime usage of USC micrometers, don't forget to add the last visible number above the index to the minimum measurement on the frame. This is a common mistake when an experienced technician begins using a metric micrometer as the instrument of choice for precision measurement.

Shop Manual
pages 93-97

Brake Micrometers

There are a couple of micrometers used almost exclusively by automotive technicians. They are used to measure brake rotor thickness and brake drum diameter. The brake rotor mike, as it is commonly called, works like any other outside micrometer except for its physical size, the size of measurement it can read, and the speed of the spindle movement (**Figure 4-43**). A standard 0.00-inch-to-1.00-inch (0.0 mm to 25 mm) or a 1.00-inch-to-2.00-inch (25 mm to 50 mm) micrometer will fit comfortably in the palm of an average hand. The brake rotor micrometer frame is larger and is generally easier to handle. In addition, the internal gears that move the spindle allow for a quick adjustment, which saves some time during a brake repair job. The markings and numbers on this micrometer are also larger and easier to read. This allows the technician to make quick and very accurate measurements of a brake rotor's thickness and the parallelism of its machined braking surfaces.

The brake drum micrometer or drum mike is an inside micrometer sized up in the same manner as the rotor micrometer (**Figure 4-44**). Like the rotor mike, the technician can make quick and accurate measurements of the inside diameter of a brake drum.

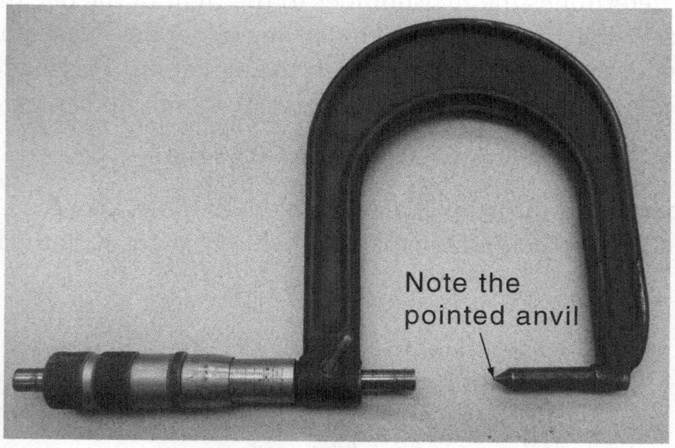

Figure 4-43 A brake rotor micrometer is an outside micrometer with faster gears and a larger frame.

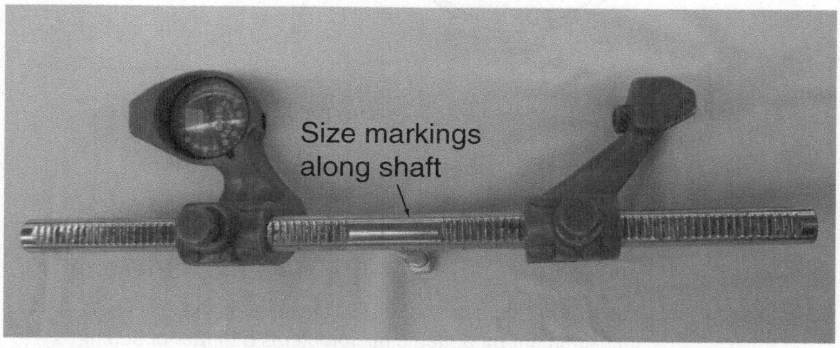

Figure 4-44 Dial readout drum micrometers are inside micrometers that, like the rotor micrometers, have faster gears and oversized frames.

DIAL INDICATORS AND CALIPERS

The **dial indicator** is used to check how far a component can move or is used to measure the distortion of a component, such as flywheel **runout.** A dial indicator must be mounted to a solid object for support during its use. The dial has a circular face with a scale around its edge (**Figure 4-45**). A moveable needle indicates the measurement much like a speedometer needle. Turning a knurled ring on the outside of the dial rotates the scale. There is a lock to hold the scale in place (**Figure 4-46**). The needle is geared to a moveable plunger extending from the bottom of the dial.

The scale can be in USC or metric measurements. The USC scale will use either 0.001 or 0.0001 inch. The scale size will be noted on the face. Metric scales are marked in 0.01 millimeter. The face may be **balanced,** meaning the readings extend from zero counterclockwise and clockwise until they meet at the midpoint of the face (**Figure 4-47**). A **continuous** type has the markings start at zero and extend clockwise around the scale back to zero (**Figure 4-48**). Care must be taken with both types when taking large

A **dial indicator** uses a plunger and gear to move a needle in proportion to plunger movement.

Runout is the amount of side-to-side movement in a rotating device. Runout may be referred to as "wobble."

Balanced scale indicators are usually used to measure runout.

Continuous scale indicators are normally used to measure endplay or lateral movement.

Figure 4-45 A dial indicator has a round dial with the graduations totaling 0.100 inch per needle revolution.

Figure 4-46 Once the scale is adjusted to zero, it is held in place by the lock during measuring.

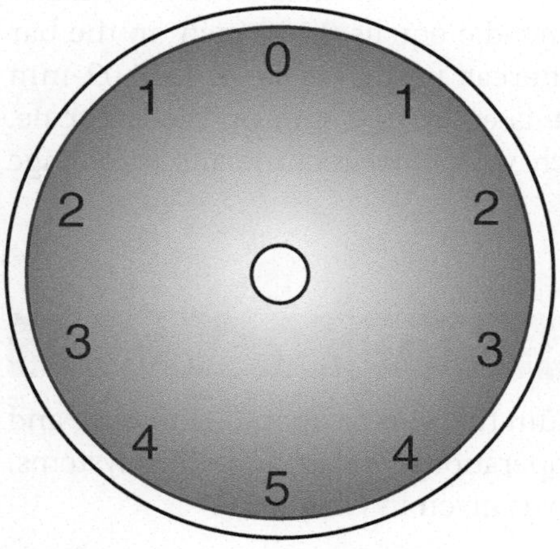

Figure 4-47 The balanced scale numbers meet at the midpoint of the face.

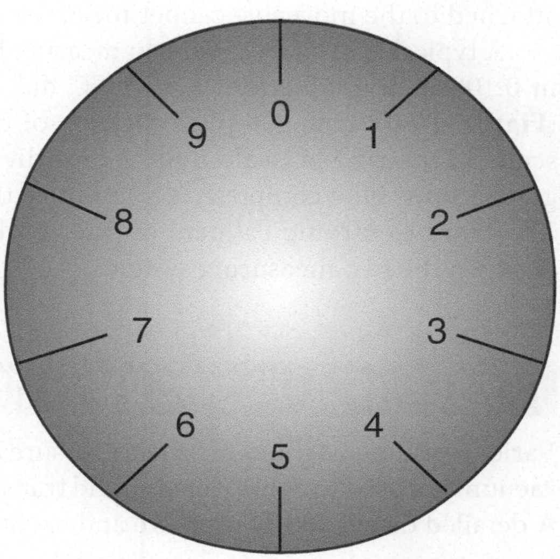

Figure 4-48 Numbers extend upward from zero on a continuous scale.

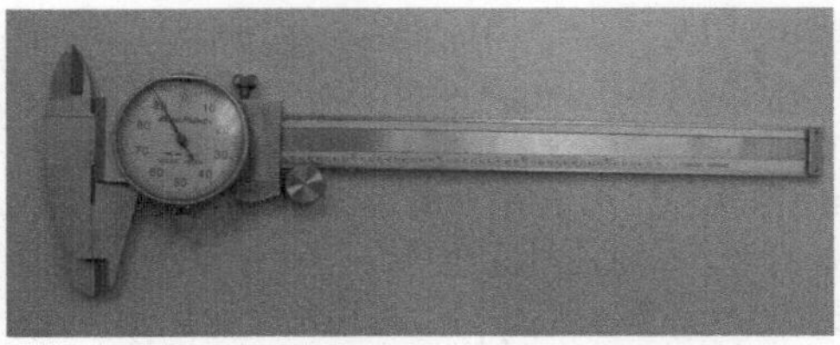

Figure 4-49 This nonelectric dial caliper is preferred by many technicians and machinists over the newer electronic version.

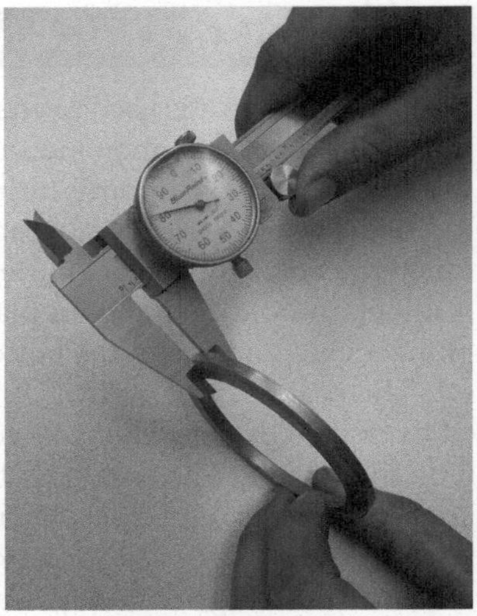

Figure 4-50 The dial face and scaling of a dial caliper are similar to the face of a dial indicator. The whole-inch measurements are read from the bar and added to the thousandths-of-inch readings from the dial.

measurements. The plunger may be designed to move the needle more than one revolution around the scale. A single revolution is equal to 0.100 inch (2.54 mm).

Dial indicators can also be used to check shaft endplay. Before setting up the dial, move the shaft as far as it can go in one direction. Mount the dial to indicate a zero reading at this point. The last step is to move the shaft as far as it will go in the opposite direction. The total movement will be shown on the dial scale.

Dial Calipers

Typically, a **dial caliper** in not used in an automotive repair shop but used extensively in automotive machining shops.

The **dial caliper** is a measuring device that can be used in place of almost all other automotive measuring devices (**Figure 4-49**). It can measure inside, outside, and depth dimensions. It has two pairs of jaws or calipers that fit inside or outside a component. The top half of each pair is fixed to the bar scale. The bottom half moves along the bar scale when the roll knob is turned. At the opposite end of the bar scale is the depth scale. The bar scale indicates the gross measurement of the caliper's movement. A dial indicator attached to the moveable caliper measures the small movement within its limits.

A typical USC dial caliper can measure between 0 and 6 inches. Its bar scale is divided in 0.100-inch graduations. The USC dial indicator measures in 0.001-inch segments (**Figure 4-50**). One complete rotation of the dial's needle equals 0.100 inch on the bar scale. A metric bar scale is in 2-mm divisions whereas its dial scale is in 0.02-mm graduations. One complete revolution of the needle is equal to 2 mm on the bar scale. There are electronic calipers on the market, which will allow technicians to change between the two measuring systems.

PRESSURE AND VACUUM MEASUREMENTS

Shop Manual
page 108

Various systems of the vehicle use pressure and vacuum theories to operate. Pressure and vacuum are used to control engine and transmission operations, brakes, and other systems. A detailed discussion of pressure and vacuum theory is given in Chapter 7.

Figure 4-51 The maximum amount of pressure a gauge must measure determines the graduation and numbering of the scale. This gauge is capable of measuring 100 psi.

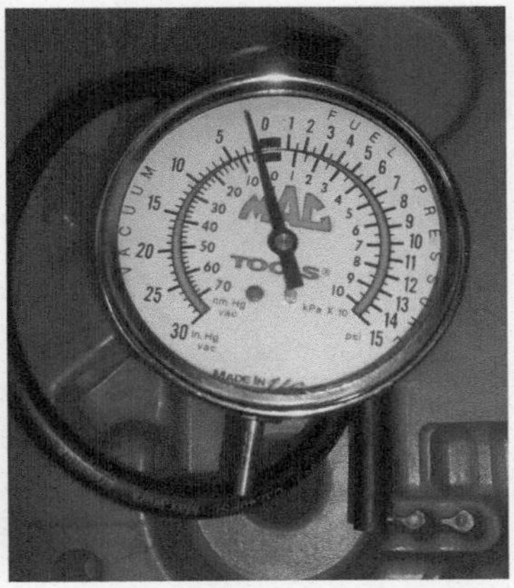

Figure 4-52 Almost all vacuum gauges used in automotive repair have a scale from 0 to 30 inches of mercury.

Pressure Gauges

Pressure is the amount of force being applied as measured in pounds per square inch (psi). If pressure is applied to a liquid in a closed, sealed system, the liquid can be used to transfer force. A hydraulic brake system is an example of pressure and liquid transferring forces.

Pressure is one of prime requirements in the fuel injection system.

Pressure gauges are fitted into the pressure line using adapters. The scales are marked (graduated) and numbered based on the size of the scale. A small-scale gauge may be marked each 1 psi and numbered each 5 psi, whereas a larger scale may be marked each 10 psi and numbered each 50 psi (**Figure 4-51**). When the gauge is connected, and the system energized, the scale will record the pressure being applied to the system.

Vacuum Gauges

Vacuum is properly defined as pressure below **atmospheric** and is generally measured in inches of mercury (in. Hg). This measure is based on how high (in inches) a vacuum source can raise a column of liquid mercury in a tube. Obviously, technicians cannot carry tubes of mercury around their shops. The vacuum gauge's mechanical system is geared to move a needle a specified distance based on the amount of vacuum in its inlet tube. Vacuum gauges are usually marked in the same manner as pressure gauges except the markings indicate inches of mercury (**Figure 4-52**).

Vacuum is essential to getting air into a gasoline engine cylinder.

Atmospheric pressure is considered to be 14.7 psi at sea level.

THREADED FASTENERS

Threaded fasteners are the most common fastener type. They use spirals called threads to wedge parts together. The common types of threaded fasteners are screws, bolts, studs, and nuts, and they are shown in **Figure 4-53**.

Threaded fasteners A hardware device that mechanically joins two or more objects together using a bolt and a nut.

Figure 4-53 Common types of threaded automotive fasteners.

A BIT OF HISTORY

Threads provide a mechanical advantage to hold parts together. The geometry behind threads was developed a long time ago, around 200 BCE. They were used by the ancient Romans to press grapes for wine.

AUTHOR'S NOTE Threads act like a lever that multiplies the applied force. This is the mechanical principle behind cranes, hand-operated presses, wine presses—and the high clamping forces applied by a wrench with minimum force.

Fastener Sizing and Torquing

For a fastener to work correctly, it must have the strength and size to handle the task. Fasteners are designed in different grade (strength) ratings for specific loads and applications. Replacement fasteners should meet the same requirements as the original item.

A light-duty fastener may be replaced with a heavy-class fastener in most cases. However, some industry and vehicle parts have a certain class of fastener designed to break or shear to prevent damage to the components. A typical example would be the shear pin in a winch. To prevent overloading a cable or chain that may result in very serious injury and great damage, the shear pin is designed to shear apart when its limits are reached. Replacing a shear pin that "broke too quickly" with a stronger pin has resulted in at least one death to the author's knowledge.

Screws

AUTHOR'S NOTE Technically, the term *screw* referred to in this section is correct for the fastener being described. However, if you ask any technician, and most automotive instructors, for a cylinder head's hexagon cap screw, you may not get a printable answer. In the automotive field, "screw" refers to a fastener driven by some type of screwdriver. A *bolt,* on the other hand, refers to almost any other threaded fastener holding parts together regardless of where or how they are installed. Generally, "bolt" will get you what you want, even when, technically, a bolt is used with a nut. (See the next section.)

The **screw** is one of the most common threaded fasteners. A screw fits into a threaded hole. Often there is a part without threads between the screw and the part with the threads, as shown in **Figure 4-54**. The threads on the screw engage the threads in the part. As the screw is turned, the two sets of threads work together to clamp the two parts together with a great deal of force.

Screw A sharp-pointed metal pin with threads running around it and a slotted or hex head.

Several types of screws are used to hold automotive parts together. The most common types of screws used in automotive parts are shown in **Figure 4-55**. The hex head cap screw is one of the most common types. It has a six-sided or hexagonal head and is driven with common automotive wrenches such as the box end, open end, combination, and socket. Hex head cap screws are used to hold large parts together.

Since the early 1990s, a specific type of screw has been used to hold high-pressure and high-force components together. The most significant are the screws that hold cast-aluminum cylinder heads to cast-iron engine blocks. This type of fastener is known as a torque-to-yield bolt. They bear no outward visible differences from other same-size screws. It is critical that they be used as engineered.

Standard screws hold adjacent components together but may not provide exact, all-around clamping force to the components. This was the cause for many cylinder head failures in the 1980s when this type of engine was being introduced. Torque-to-yield

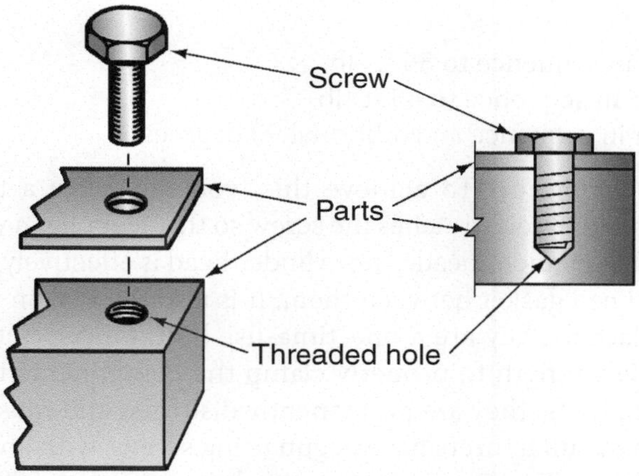

Figure 4-54 A screw fits through one part and into a threaded hole in a second part.

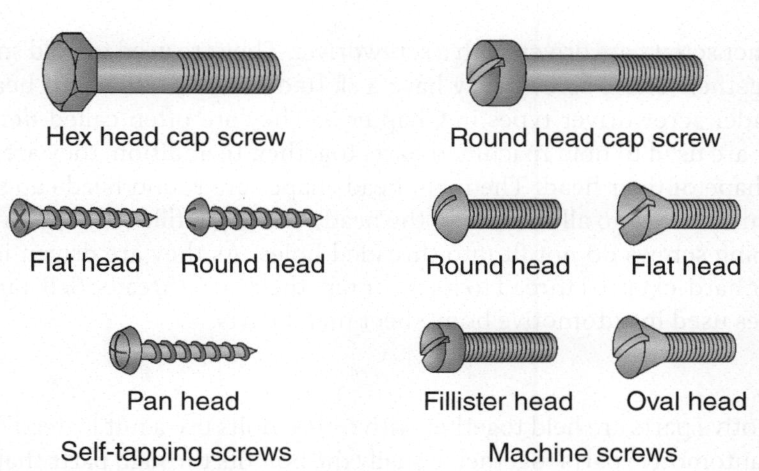

Figure 4-55 Common types of automotive screws.

Figure 4-56 A torque angle gauge allows you to accurately measure the degrees of turning.

screws helped to solve this engine problem. Torque-to-yield screws must be replaced when removed; they are not reusable.

The primary reason for replacing torque-to-yield screws is the method of tightening them to specifications. An example of a typical torque specification for cylinder head bolts would be:

1st step Tighten in sequence to 35 ft.-lb.
2nd step Tighten in sequence to 45 ft.-lb.
3rd step Tighten in sequence an additional 90 degrees.

The third step is the key. To achieve this, you must use a torque angle gauge (**Figure 4-56**). This torque step stretches the screw so that it applies a very specific amount of clamping force to the cylinder head. The cylinder head is effectively sealed to the engine block with a special head gasket between them. It is also the reason that torque-to-yield screws must be replaced. They are a one-time-use bolt. Once stretched, the screw no longer has the tensile strength to properly clamp the components together. When you remove these types of bolts, they are permanently distorted and must be discarded.

Recently, vehicle manufacturers have begun using screws with an Allen or Torx head, which require an Allen or Torx wrench, respectively, to turn them. These types of screws are used when space is limited or recessed into a hole making it difficult to use a standard wrench. Care must be taken when the screws "are broken loose" because if the wrench is of the wrong size or not fitted correctly into the screw head cavity, the corners will be rounded.

Many other screws are driven with a screwdriver. They are used to hold small automotive parts together. These screws may have a slotted head or a recessed head like those discussed under screwdriver types in Chapter 3. They are often called *machine screws* because they are used to hold machined parts together. In addition, they are often classified by the shape of their head. The main head shapes are round head cap screw, round head (with threads that go all the way to the head), flat head, fillister head, and oval head.

Self-tapping screws do not fit into threaded holes. As they are driven into the part, they use their hard-external thread to form, or tap, their own threads. Self-tapping screws are sometimes used in automotive body sheet metal parts.

Bolts A threaded fastener with external male threads. Bolts are very closely related to screws.

Bolts

Many automotive parts are held together with **bolts.** Bolts use a nut instead of a threaded hole to hold automotive parts together. Usually the bolt fits through parts that do not have threads. The bolt and nut are driven together to clamp the two parts together as shown

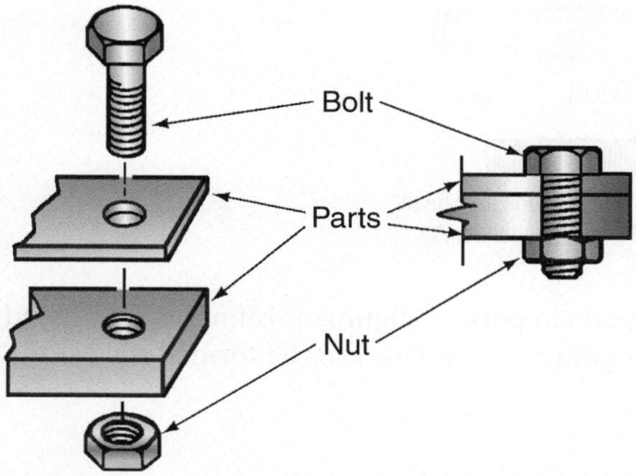

Figure 4-57 A bolt fits through both parts and into a nut.

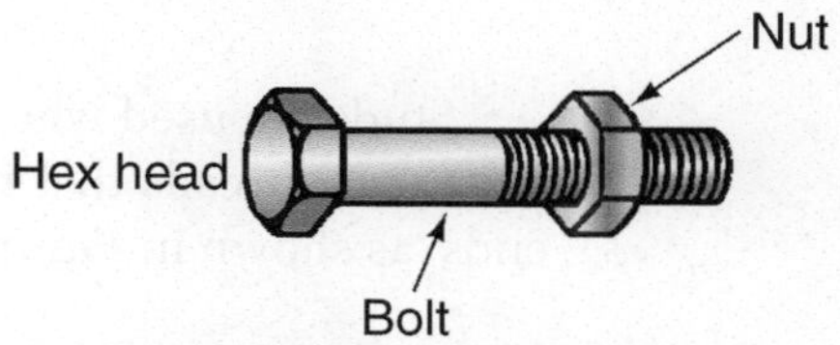

Figure 4-58 The hex head bolt is the most common fastener for automotive parts.

in **Figure 4-57**. Most bolts for automotive use have a hex head (**Figure 4-58**). The difference between this bolt and a hex head cap screw is that the bolt is used with a nut. The hex head cap screw is used in a threaded hole. There are bolts with heads that are not hex-shaped, but they are not often used in automotive parts.

Studs

A **stud** is a common type of automotive fastener that has no head. It has threads on both ends. One end of the stud fits into a threaded hole in the part as shown in **Figure 4-59**. A second part fits over the stud. Then the two parts are forced together with a nut on the other end of the stud.

Stud Threaded fasteners that are threaded on both ends and does not have a head.

AUTHOR'S NOTE Usually the threads at the opposite ends of the stud are different and may be of different lengths. This ensures the correct end is screwed into the threaded hole.

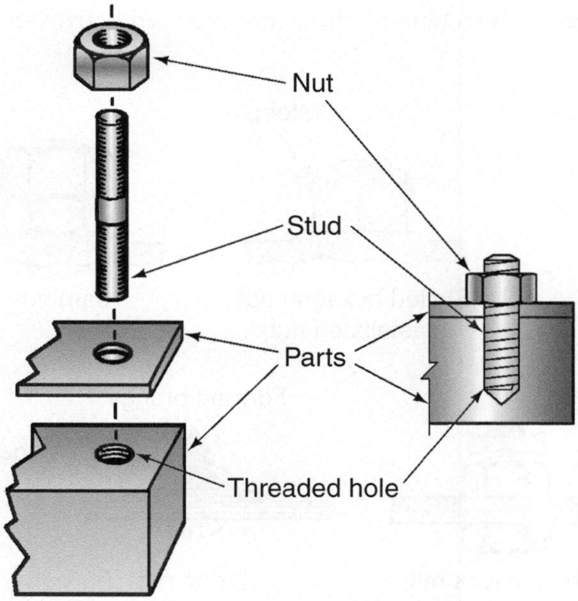

Figure 4-59 One end of a stud fits in a threaded hole and the other end has a nut.

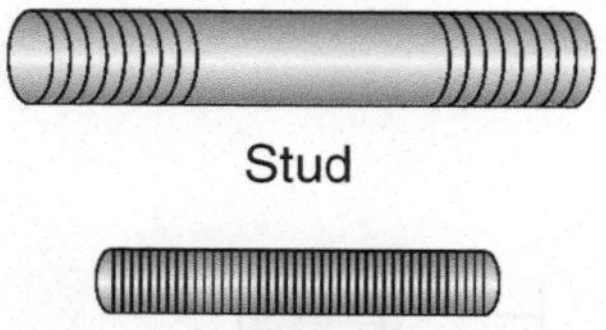

Figure 4-60 Stud threads may be at each end or along the entire length.

Studs are used when getting the two parts in perfect alignment is important. A stud may have threads that run all along its length or the threads may be formed only at the ends, as shown in **Figure 4-60**.

Nuts

Nuts CA hardware device that mechanically joins two or more objects together with internal threads.

Nuts are used with bolts and studs. Unlike other fasteners, they have their threads on the inside, called internal threads. Several types of nuts are used on automotive parts (**Figure 4-61**). The most common automotive nuts are hexagonal or *hex nuts.* Hex nuts have six sides and are made to fit box-end, open-end, or socket wrenches.

Several specialized nuts are available when parts may be subject to a great deal of vibration. Specialized nuts are also used where the part held by the nut is critical to safety. These parts might include the steering-wheel retaining nut or wheel spindle nut. The slotted or castellated nut is another important specialized nut. After the castellated nut is tightened onto the bolt or stud, a metal cotter pin is inserted through its slots and also through a hole that has been drilled in the stud or bolt. After the cotter pin is installed, it is bent around the nut. This method helps keep the cotter pin from vibrating or wigging loose.

Jam hexagon nuts A secondary thin nut used to jam or lock against a full size nut.

Jam hexagon nuts are thinner than regular hex nuts. These are used in pairs and provide double protection against loosening. The lock nut has inserts of fiber or plastic. The inserts jam into the threads as the nut is tightened. This prevents the nut from coming loose during vibration. A seating lock nut is a thin nut with a concave surface that flattens when it contacts the top of the regular hex nut. The concave action binds the lock nut to the threads and prevents it from coming loose. Spring nuts are made from thin spring metal. They have prongs that fit into a sheet metal screw thread and an arched base that pushes down on the part to be retained. These nuts are used in trim and sheet metal parts.

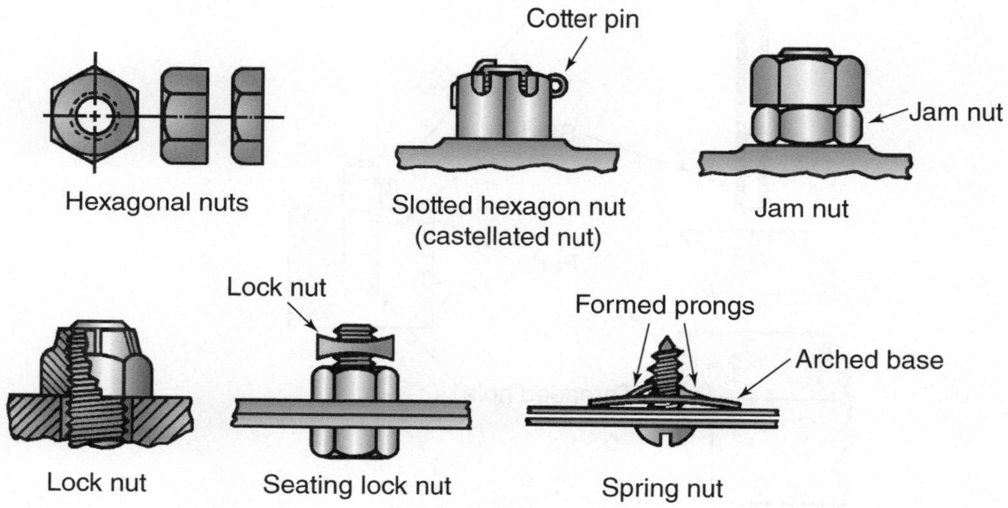

Figure 4-61 Common types of automotive nuts.

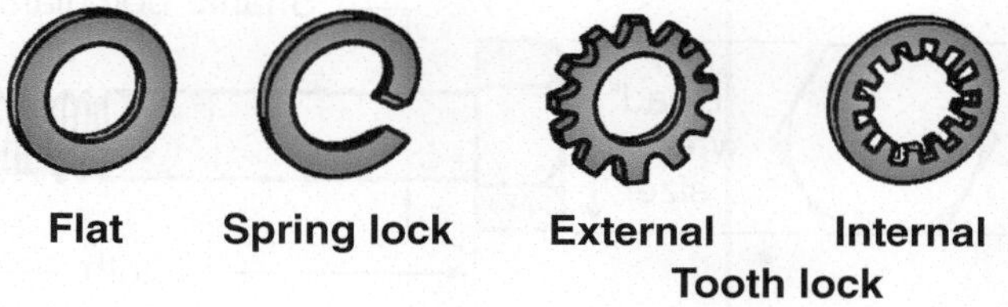

Figure 4-62 Common types of automotive washers.

Washers

Washers (**Figure 4-62**) are often used with other threaded fasteners for several reasons. First, they can help a nut distribute the clamping load on a part. They can also prevent a nut from coming loose. Finally, they can prevent a nut from digging into a machined surface.

Washers A thin plate with a whole that is used to distribute the load of a fastener.

A flat washer is often used between a nut and an automotive component or under the head of a screw or bolt. This washer helps spread out the clamping force over a wider area. It also prevents machined surfaces from being scratched as the bolt head or nut is tightened.

Lock washers are used to prevent nuts from vibrating or working loose. The spring lock washer has a sharp edge that will dig into a fastener or component surface. This prevents the fastener from coming loose. The tooth washer has either internal or external (or both) teeth that dig into a surface and prevent the nut from coming loose. The tooth washer is commonly used where an electrical terminal is fastened to a component, providing a better electrical connection and locking the fastener.

Lock washers A helical spring washer designed to lock by squishing flat between a nut and the mounting surface when tightening them.

U.S. (English) Thread Sizes

Threads on threaded fasteners are made to sizes specified in either the U.S. (English) system or the metric system. Older automobiles manufactured in the United States use only U.S. (English) system threads. Late-model U.S.-manufactured cars and import cars use metric system threads.

The U.S. (English) threads are manufactured to specifications called the **Unified System.** This system has two common types of threads: coarse and fine. Coarse means there are fewer threads in a given length of a bolt or screw; fine means there are many threads in a given length of a bolt or screw. They are designated as NC for National Coarse and NF for National Fine, as shown in **Figure 4-63**. The difference between fine and coarse threads is easy to see. Coarse threads are used in aluminum parts because they provide greater holding strength in soft materials. Fine threads are used in many harder materials, such as cast iron and steel.

Unified System A standard thread pitch mainly used in the United States.

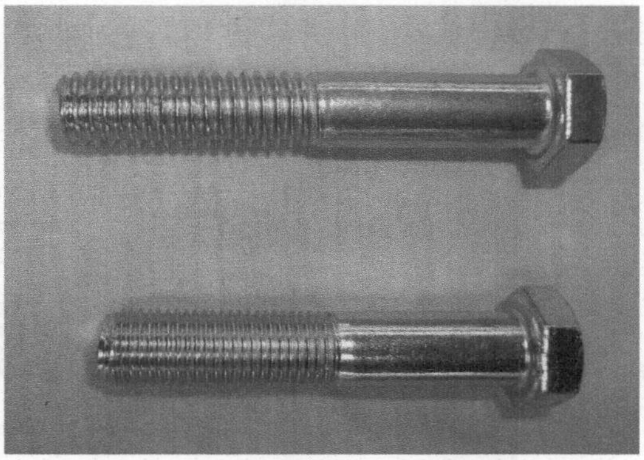

Figure 4-63 Fine and coarse thread bolts with the same diameter.

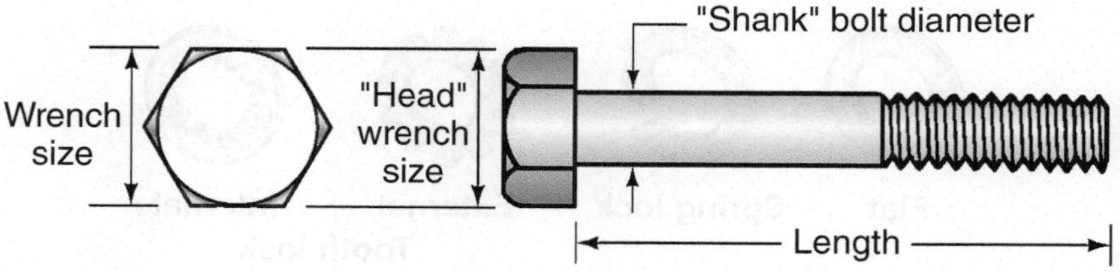

Figure 4-64 Typical measurements for a bolt or hex head cap screw.

Shank The thread-less part of a bolt between the threads and the head.

Threads per inch A measurement of the threads per inch also known as thread pitch.

U.S. (English) threaded bolts and hex head cap screws are described by a number of measurements. Three important measurements are head size, shank diameter, and overall length, as shown in **Figure 4-64**. The **shank** is the part of the bolt between the head and the threads. Its diameter is part of the bolt's identification specification. Shank diameter sizes are available in most of the fractional divisions of an inch. Common sizes include 1/4-, 5/16-, 3/8-, 7/16-, and 1/2-inch diameters.

The length of the bolt or cap screw is measured from the bottom of the head to the end of the bolt or screw. Bolts and hex head cap screws are commonly manufactured in 1/2-inch increments. Typical length sizes would be 1/2, 1, 1 1/2, 2 inches, and so on.

Head size determines what wrench fits the bolt or hex head cap screw. The shank diameter is not the wrench size. The head size is determined by the distance across the hexagonal flats on the head, as shown in Figure 4-64. A bolt with a 1/4-inch shank diameter usually has a head size of 7/16 inch. The common head sizes are shown in **Figure 4-65**.

The number of **threads per inch** is another important measurement in the Unified System. Each bolt and screw are identified by the number of threads in an inch, as shown in **Figure 4-66**. When a ruler is placed along the bolt, as shown in Figure 4-66, you can see that there are 20 threads in 1 inch. This bolt would be described as having 20 threads per inch (abbreviated to 20 TPI). When applied to a nut, these measurements refer to the size of the bolt that the nut fits.

These measurements are used to give a callout or designation for a threaded fastener. If you were to go to an automotive parts store, you would find the fasteners in drawers with labels like 3/8 × UNC × 1 1/2, as shown in **Figure 4-67**. The first measurement is the shank diameter of 3/8 inch. Each × means "by." The second measurement is the 16 threads per inch, which is in the Unified National Coarse category. Finally, the 1 1/2 means that the bolt shank is 1 1/2 inches long.

Common U.S. (English) Head Sizes

Head Size	Head Size
3/8" 7/16" 1/2" 9/16" 5/8"	1" 1 1/16" 1 1/8" 1 3/16" 1 1/4"
11/16" 3/4" 13/16" 7/8" 15/16"	1 5/16" 1 3/8" 1 7/16" 1 1/2"

Figure 4-65 Common U.S. (English) bolt head sizes.

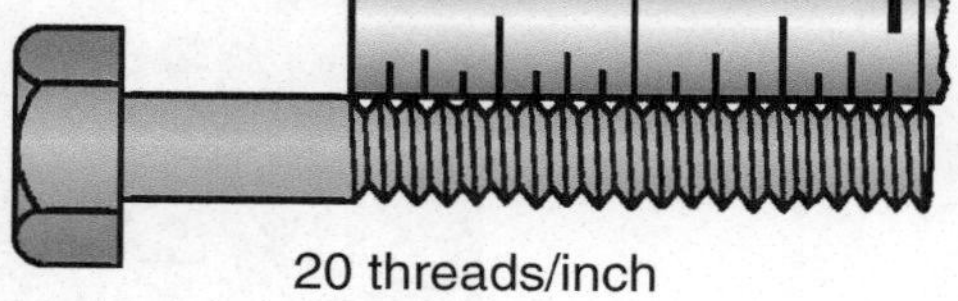

Figure 4-66 A bolt with 20 threads per inch.

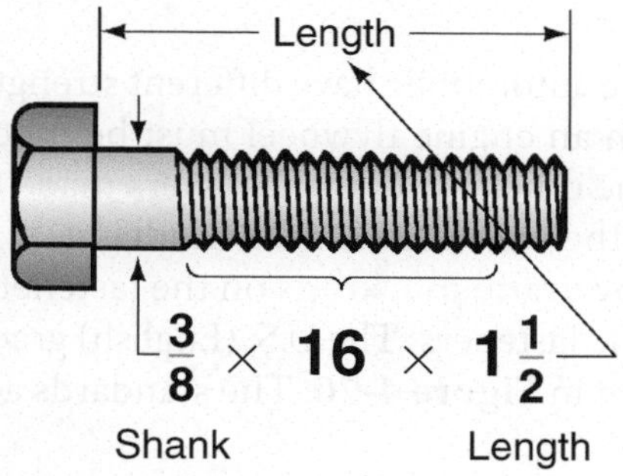

Figure 4-67 A complete bolt size designation.

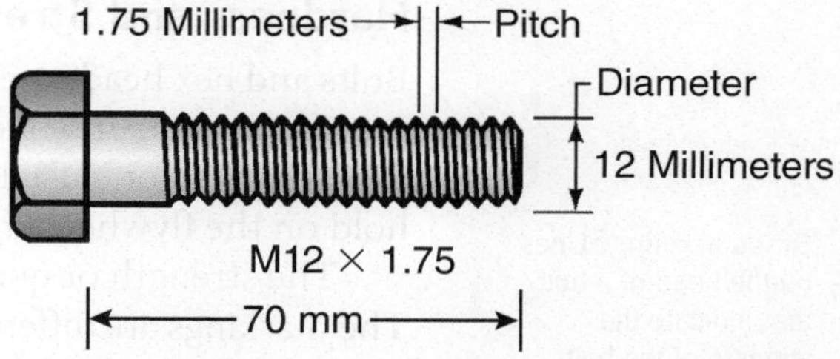

Figure 4-68 A metric fastener designation of a M12 × 1.75 mm × 70 mm bolt.

Metric Thread Sizes

Fasteners manufactured in metric sizes have the same basic measurements as those described for the U.S. (English) system, but they are given in metric units. A common metric system fastener might be M12 × 1.75, as shown in **Figure 4-68**. The M indicates that the fastener has metric threads. The first number is the outside shank diameter of the bolt, hex head cap screw, stud, or the inside diameter of a nut in millimeters. The second number after the sign × is the pitch. This is the distance between each of the threads measured in millimeters. After the second × is the length of the fastener's shank in millimeters.

As you can see, there are no abbreviations such as NC or NF to identify fine and coarse threads. In the metric system, the pitch number distinguishes between fine and coarse. For example, a fine thread metric bolt may be M10 × 1.0. A bolt of the same diameter with a coarser thread might be labeled M10 × 1.25. The larger pitch number indicates wider spacing between threads. The length of metric fasteners is measured in the same way as in the U.S. (English) system, but the measurements are given in metric system units.

Like the U.S. (English) system, the metric fastener head size is not part of the designation or callout. The head sizes are manufactured in the common metric units. The common head sizes are made to fit common metric wrench sizes. The common head sizes for metric fasteners are shown in **Figure 4-69**.

Common Metric Head Sizes	
Head Size	Head Size
9 mm	19 mm
10 mm	20 mm
11 mm	21 mm
12 mm	22 mm
13 mm	23 mm
14 mm	24 mm
15 mm	26 mm
16 mm	27 mm
17 mm	29 mm
18 mm	30 mm
	32 mm

Figure 4-69 Common metric head sizes.

Hardness and Strength

Grade markings Lines on the head of a bolt that indicate the strength of the bolt.

Bolts and hex head screws used for various parts of the automobile have different strength requirements. A hex head cap screw used to hold on an engine flywheel must be much stronger than one used to hold on a headlight. On the other hand, the fasteners made to hold on the flywheel are much more expensive than the ones used on the headlight.

The strength or quality of fasteners is identified by **grade markings** on the fasteners. The markings are different for U.S. (English) and metric fasteners. The U.S. (English) grade markings for bolts and hex head cap screws are shown in **Figure 4-70**. The standards are

GENERAL FASTENER MARKINGS AND PROPERTIES

ID Marking	Proof Load (fastener size)	Minimum Yield Strength (fastener size)	Minimum Tensile Strength (fastener size)
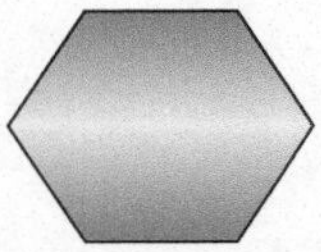 No Marking Grade 2	55,000 psi (1/4 to 3/4 in.) 33,000 psi (3/4 to 1-1/2 in.)	57,000 psi (3/4 to 1-1/2 in.) 36,000 psi (3/4 to 1-1/2 in.)	75,000 psi (3/4 to 1-1/2 in.) 60,000 psi (3/4 to 1-1/2 in.)
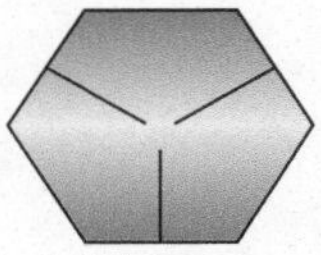 3 Lines Grade 5	85,000 psi (1/4 to 1 in.) 74,000 psi (1 to 1-1/2 in.)	92,000 psi (1/4 to 1 in.) 81,000 psi (1 to 1-1/2 in.)	120,000 psi (1/4 to 1 in.) 150,000 psi (1 to 1-1/2 in.)
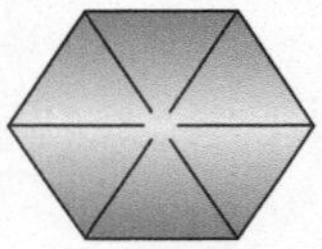 6 lines Grade 8	120,000 psi (1/4 to 1-1/2 in.)	130,000 psi (1/4 to 1-1/2 in.)	150,000 psi (1/4 to 1-1/2 in.)
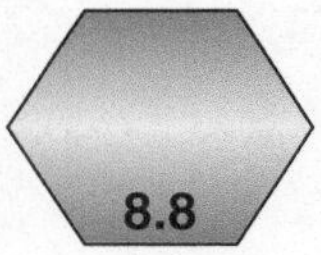 8.8 Class 8.8	85,000 psi (Up to 1-1/2 in.)	92,000 psi (Up to 1-1/2 in.)	120,000 psi (Up to 1-1/2 in.)
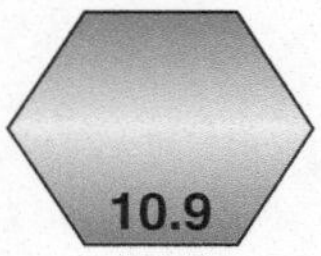 10.9 Class 10.9	120,000 psi (Up to 1-1/2 in.)	130,000 psi (Up to 1-1/2 in.)	150,000 psi (Up to 1-1/2 in.)

Proof load: Axial tensile load without material shows no evidence of permanent deformation.
Yield strength: Axial load before material shows a permanent deformation.
Tensile strength: Maximum load in tension during pulling or shearing action.

Figure 4-70 General grade and class markings for U.S. (English) and metric cap screws. The required minimum loads for each class or grade are also shown.

set by the Society of Automotive Engineers (SAE). The system uses marks on the head of the bolt or the hex head cap screw. Unfortunately, the system is confusing because the number of marks and the grade number do not agree. A grade 0, 1, and 2 bolt has no markings on the head and is not very strong. A bolt with three marks is called a grade 5 bolt and is much stronger than a grade 1 bolt. The strongest bolt for automotive use has six marks and is called a grade 8 bolt. The chart in Figure 4-70 also describes the material each grade is made from and the **tensile strength.** Tensile strength is the amount of pressure the fastener can take before it breaks.

Tensile strength The maximum load in tension that a bolt can stretch without breaking.

Metric fasteners use property class numbers to indicate their strength. These numbers are stamped on the head of the bolt or hex head cap screw. Typical metric bolt strength numbers are shown in **Figure 4-71**. The higher the property class number, the stronger the fastener. A 10.9 fastener is much stronger than one marked 4.6. A metric fastener without a number would be the same as a grade 0 U.S. (English) fastener. Metric studs have a marking system at the end of the stud that shows their property class number.

Caution

When replacing bolts, make sure that you are using the exact same bolt as you removed and as designed by the manufacturer. This means you need to check the pitch, diameter, length, thread, head size, and strength rating. Failing to do so will cause issues.

Nuts are also graded according to the same grading systems, as shown in Figure 4-71. Nuts must have the same grade as the bolts or studs with which they are used. English nuts have dots that represent grade markings. Three dots represent grade 5. Six dots represent grade 8. Metric nuts have numbers that represent strength. The number on the nut is its property class number. In both systems, the higher the number or the more dots, the stronger the nut.

Fastener Torque

Each threaded fastener on an automobile must be tightened just the correct amount. Untightened fasteners can loosen and cause parts to fail. Fasteners that are tightened too tightly can damage the part or the fastener and cause the parts to fail. Overtightening causes fasteners to stretch. These no longer have the strength they had before the overtightening.

Inch System		Metric System	
Grade	Identification	Class	Identification
Hex nut grade 5	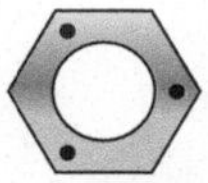 3 dots	Hex nut property class 9	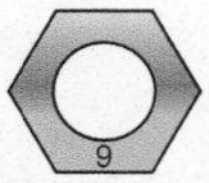 Arabic 9
Hex nut grade 8	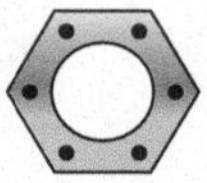 6 dots	Hex nut property class 10	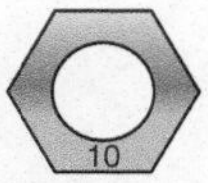 Arabic 10
Increasing dots represent increasing strength.		Can also have blue finish or paint dab on hex flat. Increasing numbers represent increasing strength.	

Figure 4-71 Grade marking system for U.S. (English) and metric nuts.

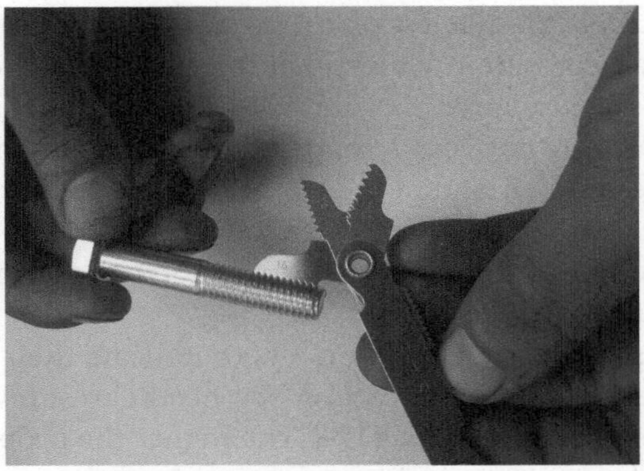

Figure 4-72 A pitch gauge is used to match threads for identification.

To avoid these problems, fasteners must be tightened with a torque wrench. There are torque specification charts for each important fastener in the service manual for the car you are working on. Always follow these specifications.

Caution

Never use a regular impact wrench to tighten a fastener. The impact wrench can quickly exceed the required torque and damage the fastener or part.

Pitch Gauges

There are two different thread systems in use, each with fine and coarse threads, which can cause a great deal of confusion. Metric threads cannot be used with English threads. Fine threads cannot be used with coarse threads. These distinct types of threads can make it difficult for the technician to tell one thread from another.

A thread **pitch gauge** (**Figure 4-72**) is used to identify fasteners. It has several blades with teeth. The thread size is written on the blade. By matching the teeth on the blade with threads on a fastener, the thread size can be determined (**Figure 4-73**). Pitch gauges are made for both metric and English threads.

Pitch gauge A tool used to measure the distance between threads on a bolt.

When threads are damaged by using the wrong sizes or starting a fastener incorrectly with a wrench, we call this *cross threading* or *stripping.*

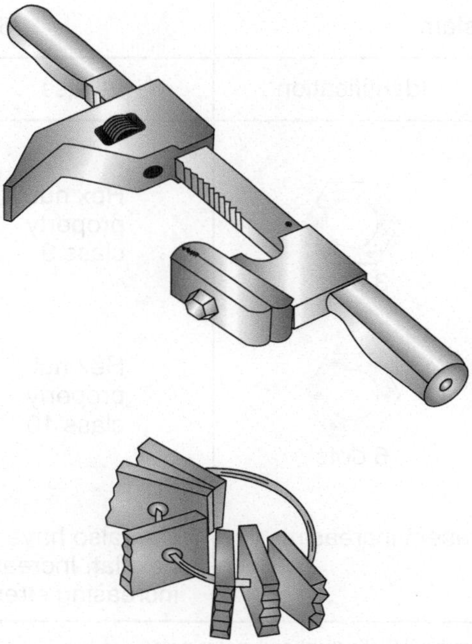

Figure 4-73 A thread restoring tool with insertable blades (lower).

AUTHOR'S NOTE Some unified threads almost match some metric threads. The key here is "almost." A case in point is that a 16 TPI thread almost matches the pitch of a metric 1.25. A technician in a hurry can mistakenly install the wrong bolt, such as a 1.25 into a 16 TPI hole. When the fastener is torqued, the threads on the bolt, nut, or hole will be damaged. Under low torque, this may not be obvious to the technician until the next time the fastener is loosened by vibration—or the next technician.

NONTHREADED FASTENERS

There are many types of nonthreaded fasteners on a typical vehicle. They range from simple press-in pins to shaft spines. All have the same purpose: to hold parts together in some manner.

Pins

A pin is often used to hold automotive parts together when they are not disassembled. A pin is a small, round length of metal. It is driven through holes in two parts to hold them together. There are several common types of pins, as shown in **Figure 4-74**. A **dowel pin** is straight for most of its length but tapered on the very end. This tapered part helps to position two parts that fit together. Tapered pins are tapered for their entire length. The taper locks into the part for a tight fit. A **roll pin** is not solid but rolled from a thin piece of metal. The split allows the pin to contract during installation. This action causes a spring effect, which holds the pin in place. It is used where strong holding power is not required.

Dowel pin A fastener that are use to join two or more items together. Short round rods that can be tapered, slotted, or grooved.

Roll pin A spring pin that secures the position of two or more parts together.

Snaprings

A **snapring** is often used to hold parts in place on a shaft. Because snaprings are used to retain parts in position, they are sometimes called "retaining rings." They are made from a high-quality steel that allows them to be expanded or contracted by a tool for installation. Once in place, the snapring "snaps" back to its original size.

There are two basic types of snaprings as shown in **Figure 4-75**. The internal snapring fits in a machined groove inside a hole. The external snapring fits in a machined

Snapring A retaining ring is a fastener that holds components or assemblies onto a shaft or housing/bore.

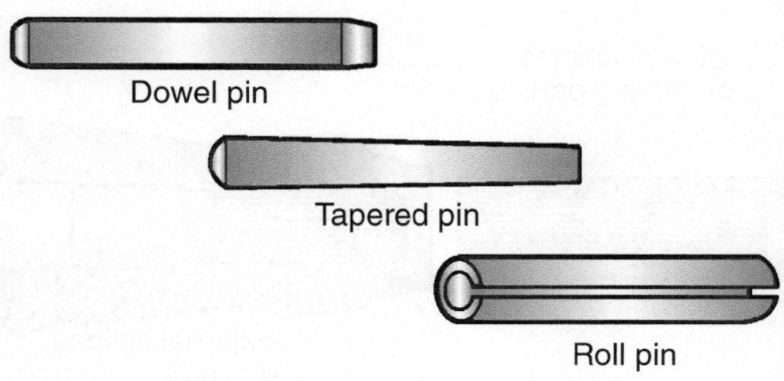

Figure 4-74 Three common types of pins.

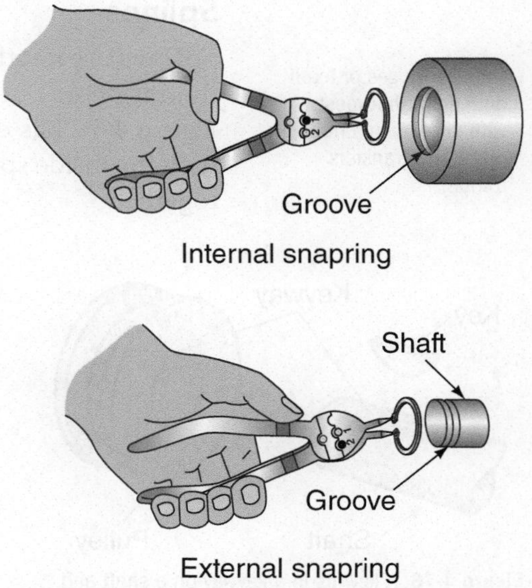

Figure 4-75 Internal and external types of snaprings.

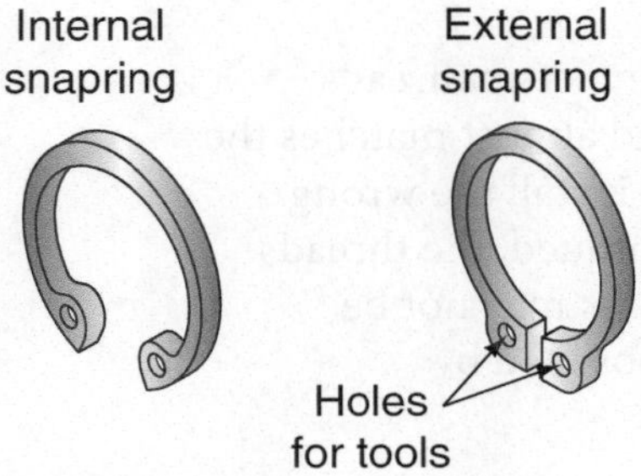

Figure 4-76 Internal and external snaprings with holes for a snapring tool.

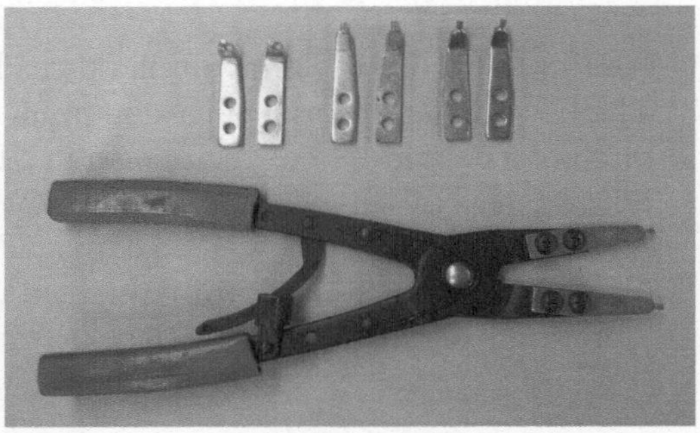

Figure 4-77 Snapring pliers with replaceable tips are handy and relatively inexpensive, but they may not be able to survive the force sometimes needed to remove snaprings.

groove on the outside of a shaft. The external snapring is expanded with retaining ring pliers for installation. The spring tension from the ring holds it in the groove. The internal snapring is compressed with retaining ring pliers to fit it in the groove. When the tool is released, its spring tension will hold it in position.

Many snaprings have small holes at the ends, as shown in **Figure 4-76**. The ends of snapring pliers are made to fit into these holes. Most snapring pliers come with an assortment of ends to fit assorted sizes of snap rings (**Figure 4-77**).

Keys

Key A small piece of metal that fits into a slot called a keyway.

Another common way to retain a part on a shaft is to use a **key**. A key is a small, hardened piece of metal. They are often rectangular, or half-moon-shaped. As shown in **Figure 4-78**, the key fits into a slot called a keyway. When in position, half the key sticks up past the shaft. There is a matching keyway in the part to be retained on the shaft. The part to be retained in Figure 4-78 is a pulley. The keyway in the pulley fits over the key, holding the two parts in position. A key of this type is commonly known as a *Woodruff key*.

Keys alone will not keep two components mated. Typically, a nut or bolt is installed to hold the two components in place. The installed key is usually the drive connection between a round shaft and its mated pulley or gear.

Splines

Splines Ridges or teeth on a shaft that mesh with grooves in a mating piece and transfers torque.

Splines are another way to hold parts together. These are long teeth that can be cut on the inside or outside of a part. A set of splines is shown in **Figure 4-79**. The shaft in Figure 4-79 has external splines. The matching part has splines on the inside. The part with the inside splines will slip over the splines on the shaft. The mating splines hold them together.

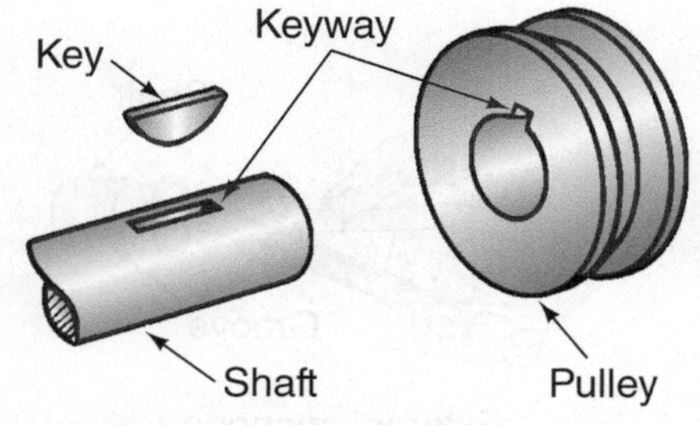

Figure 4-78 A key fits in a keyway on a shaft and pulley to hold the two parts together.

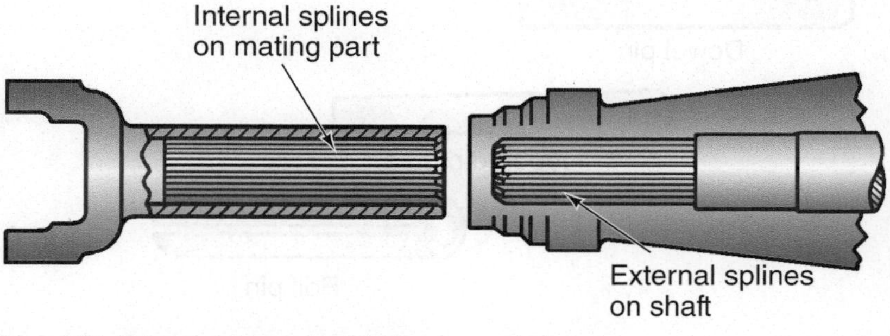

Figure 4-79 Splines can be used to hold a part on a shaft.

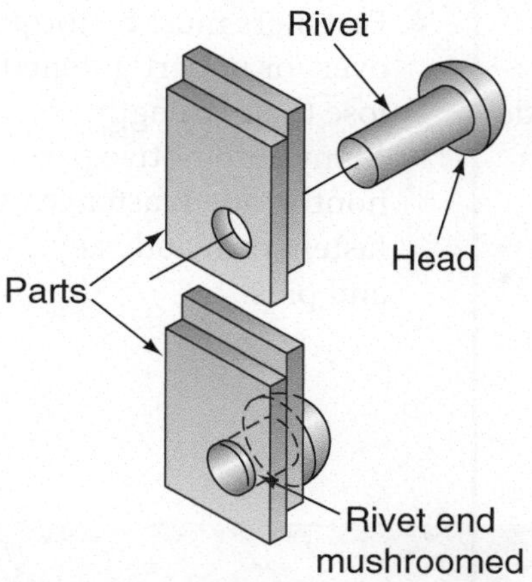

Figure 4-80 A rivet before and after installation.

In Figure 4-79, note that this fastener is not meant to hold two parts tightly together without movement between them. In fact, these two "minor" components are designed to slide back and forth over each other and thereby increase or decrease the distance between the two components.

Rivets

Rivets are used to hold parts together that are rarely disassembled. A typical use of rivets in vehicles is to hold frame members or the window regulators together. They may be made from soft material such as aluminum, but some are also made from steel. A rivet fits through a hole drilled in the two parts to be joined. The small end of the rivet is formed or mushroomed into a head with a rivet set or a ball-peen hammer. A rivet before and after installation is shown in **Figure 4-80**. A rivet is taken out by first removing the head with a drill or chisel. It may then be driven out with a punch.

Rivets A permanent fastener that consists of a smooth cylindrical shaft that when installed mushrooms and holds to object together.

SUMMARY

- The most common measurement system in the United States is the United States Customary (USC) or the English system.
- The base USC measurement is the inch.
- The base metric measurement is the meter.
- The metric system units are based on the number 10 and no fractions are used.
- USC and metric measurements can be converted to each other using a conversion chart and math.
- Rulers may be in USC or metric measurements.
- Feeler gauges are used to measure small gaps or distances between two parts.
- Micrometers are made for measuring the size of objects.
- Micrometers may be USC or metric.
- Dial indicators are used to measure very small measurements.
- Shaft endplay is usually measured with a dial indicator.
- Dial calipers can measure depth, outside diameters, and inside diameters.
- Pressure is measured in pounds per square inch (psi).
- Pressure and liquid can be used to transfer force.
- Vacuum is pressure below atmospheric pressure.
- Threaded fasteners use threads to hold automotive parts together. Common threaded fasteners include bolts, screws, studs, nuts, and washers.
- Threads are measured and classified according to the U.S. (English) and metric systems. U.S. (English) threads are made to standards of the Unified System.
- Unified System threads are classified according to the bolt shank diameter, length, and number of threads per inch.

- Metric system threads are classified according to shank diameter, length, and pitch.
- Both metric and U.S. (English) fasteners have grade markings to show fastener strength. U.S. (English) fasteners use marks on the bolt head to indicate grades. Metric bolts use property numbers to indicate grades.
- A pitch gauge can be used to determine threads per inch or thread pitch.
- Fasteners must be torqued to ensure they are not over- or undertightened. Overtightened fasteners lose their strength.
- Many automotive parts are held together with nonthreaded fasteners. Common non-threaded fasteners include keys, snaprings, rivets, splines, and pins.

REVIEW QUESTIONS

Short Answer Essays

1. Explain the measurement markings on a USC micrometer.
2. List and describe the different components on a micrometer.
3. Explain how stepped or go/no-go feeler gauges are machined.
4. Explain how to convert 3 inches to millimeters.
5. Discuss the components of a dial caliper.
6. Discuss the difference between pressure and vacuum.
7. List and discuss the differences between a USC and a metric dial caliper.
8. Describe the means for measuring vacuum.
9. Explain how a micrometer is placed to measure a shaft's outside diameter.
10. Explain the metric system of measurement.
11. Describe the difference between a bolt and a screw.
12. Explain the difference between a bolt and a stud.
13. Describe how the length of a bolt is measured.
14. Explain how to tell the strength of a U.S. (English) bolt.
15. Describe how to tell the strength of a metric bolt.
16. Explain how to tell the strength of a metric nut.
17. Explain the purpose of a pitch gauge.
18. Explain why a torque wrench should be used when tightening fasteners.

Fill-in-the-Blanks

1. The dial caliper is a measuring device that can be used in place of almost all other automotive ______________ ______________.
2. Some micrometers have an additional scale on the sleeve called a(n) ______________ ______________.
3. The sleeve on a micrometer has an ______________ ______________ along its length.
4. A 1- to 2-inch micrometer can measure a total of ______________ inch(es).
5. The vernier scale is used to read ______________ of an inch.
6. A stepped feeler gauge may be known as a(n) ______________ gauge.
7. Grade markings are found on the ______________ of a bolt.
8. The two types of snaprings are ______________ and ______________.
9. A measurement across the head of a bolt gives a(n) ______________ size.
10. A measurement from under the head of a bolt to the end of the threads gives a ______________ measurement.

Multiple Choice

1. *Technician A* says pressure and liquid are used to make hydraulic brakes systems work.

   Technician B says vacuum helps the engine operate.

   Who is correct?

   A. A only
   B. B only
   C. Both A and B
   D. Neither A nor B

2. Dial calipers are being discussed.

   Technician A says the bar scale may be graduated in 0.100 inch.

   Technician B says a bar scale may be graduated in 2 mm.

   Who is correct?

   A. A only
   B. B only
   C. Both A and B
   D. Neither A nor B

3. The grade markings on a U.S. (English) bolt head are being discussed.

   Technician A says the more marks, the larger the bolt.

   Technician B says the more marks, the weaker the bolt.

   Who is correct?

   A. A only
   B. B only
   C. Both A and B
   D. Neither A nor B

4. *Technician A* says 2.54 cm is equal to 1 inch.

   Technician B says 1 inch is equal to 0.3937 mm.

   Who is correct?

   A. A only
   B. B only
   C. Both A and B
   D. Neither A nor B

5. *Technician A* says a micrometer may be used to measure the wobble in a gear.

   Technician B says a micrometer can be used to measure the diameter of an engine cylinder bore.

   Who is correct?

   A. A only
   B. B only
   C. Both A and B
   D. Neither A nor B

6. The installation of a hex head cap screw is being discussed.

   Technician A says always start the screw by hand.

   Technician B says once the screw is started, it is best to use a speed wrench to tighten the *screw* to the torque point.

   Who is correct?

   A. A only
   B. B only
   C. Both A and B
   D. Neither A nor B

7. *Technician A* says pressure is generally measured in inches of mercury.

   Technician B says vacuum is generally considered to be any pressure less than atmospheric pressure.

   Who is correct?

   A. A only
   B. B only
   C. Both A and B
   D. Neither A nor B

8. *Technician A* says a drum micrometer is basically an inside micrometer.

   Technician B says a rotor micrometer is used to measure diameter of a brake rotor.

   Who is correct?

   A. A only
   B. B only
   C. Both A and B
   D. Neither A nor B

9. The installation of a hex head cap screw is being discussed.

   Technician A says it can be installed with an impact wrench.

   Technician B says it should be tightened with a torque wrench.

   Who is correct?

   A. A only
   B. B only
   C. Both A and B
   D. Neither A nor B

10. The use of washers is being discussed.

    Technician A says washers distribute the load on the part being tightened.

    Technician B says washers protect machined surfaces. Who is correct?

    A. A only
    B. B only
    C. Both A and B
    D. Neither A nor B

CHAPTER 5

ELECTRICAL AND ELECTRONIC SYSTEMS

Upon completion and review of this chapter, you should understand and be able to:

- Perform applications of electricity, voltage, current, and resistance.
- Describe how magnetism affects a circuit and components.
- Understand the types of electrical circuits.
- Explain how switches and relays work in a vehicle.
- Describe how automotive batteries work.
- Describe the components and basic operation of computer control in a vehicle.
- Explain the components and basic operation of the starting system.
- Identify the components and basic operation of the charging system.
- Identify the components and basic operation of the ignition system.

Terms To Know

Absorbed glass mat (AGM)
Actuator
Alternator
American Wire Gauge (AWG)
Ampere
Atom
Brushes
Butt connector
Circuit
Circuit breaker
Coil
Coil pack
Coil-near-plug
Coil-on-plug
Component locator
Conductor
Conventional theory of electricity
Corrosion
Current
Electromotive force (EMF)
Electron Theory of Electricity
Farad
Fuse
Fusible link
Ganged switch
Ignition timing
Input device
Insulator
Magnetic field
Maintenance-free battery
Multiplexing
Nickel-Cadmium
Nickel-Metal Hydride
Normally-closed (NC)
Normally-open (NO)
Ohm
One-wire return
Open
Output device
Parallel circuit
Pigtail
Potential differential
Relay
Secondary circuit
Sensor
Series circuit
Series-parallel circuit
Short finder
Short-to-ground
Short-to-power
Single-pole, single throw switch
Single-pole, double-throw
Specific gravity Stator
Switch
Terminal connectors
Twisted pair
Valence ring
Volt
Voltage
Weather-tight connectors
Winding
Wiring diagram

INTRODUCTION

The electrical system of an automobile can be both complex and simplistic at the same time. It is the one thing that has always managed to change throughout the year, and it will continue to be the dominant item in a car that changes. Once you understand the basics of electricity and the basics of the main components, you will be able to keep up with the changes in technology in the future. When working around electricity, it is always important to practice safety.

ELECTRICAL THEORY, MEASURING, AND CIRCUITS

Georg Simon Ohm (1787–1854) wrote the law regarding electrical theory and measurement. It states that it takes one **volt** to push one **ampere** of current through one **ohm** of resistance. A change in one of the three affects the other two values proportionally. In order to understand Ohm's Law, we must first look at **electricity.**

Ohm's law is stated like this. The voltage **across** a circuit is equal to the current **through** a circuit times the resistance of a circuit. These verbs are highlighted because they will help you understand how to measure electrical circuits when you get to the Shop Manual.

The basic element of any material is the **atom.** It is made up of three parts: neutrons, protons, and electrons (**Figure 5-1**). The center or nucleus of the atom is made of the neutrally charged neutrons and the positively charged protons in equal numbers. An equal number of negatively charged electrons travel in orbits around the nucleus much as the planets travel around the Sun. An atom that has equal numbers of electrons, neutrons, and protons is in balance. A balanced atom really does not produce any outside actions. If the atom can be unbalanced in some way, some type of energy can be generated and used. Splitting the atom creates enormous amounts of energy, either uncontrollably, as in the case of nuclear weapons, or in controllable situations like nuclear power plants. However, if only a portion of the atom is moved, and it basically remains intact, useable and controllable energy can be acquired.

The **volt** is the standard measurement of electricity.

The **ampere** is the standard measurement of moving electrons.

An **ohm** is the standard measurement of electrical resistance.

Electricity: The presence and flow of electric charge

Atom: The smallest unit of a chemical element

Conductors

Causing an electron to leave one atom and move to another creates a chain reaction in which an extremely large number of electrons are moving (**Figure 5-2**). At this point, electricity or electrical current is present and can be used if controlled. However, there

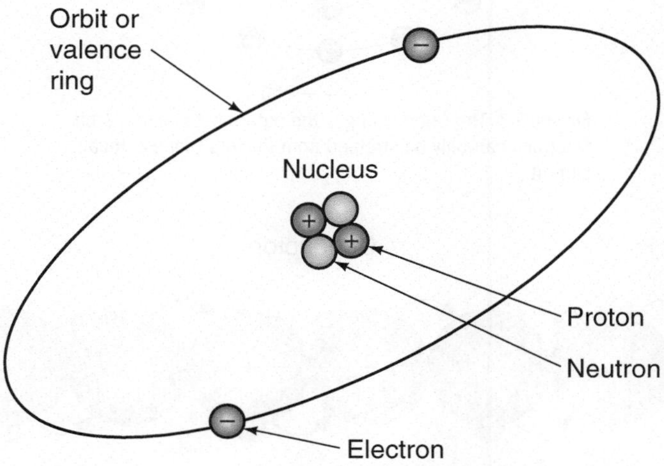

Figure 5-1 A balanced atom has equal numbers of protons, neutrons, and electrons.

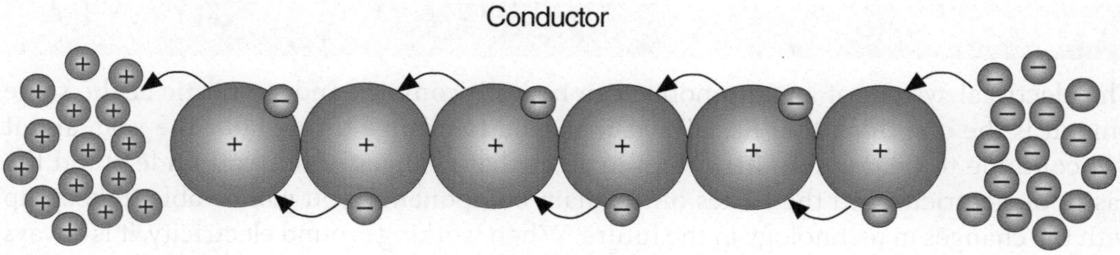

Figure 5-2 Electricity is the mass movement of electrons from negative to positive.

Valence ring: The outer path of electrons around the nucleus of the atom

A **conductor** is any material that will carry electrical current.

Voltage is a measurement of **electromotive force (EMF),** which is the difference between the number of free electrons at one end of the circuit compared to the other end or **potential difference.**

are only a few natural elements that can be used. In order for an element to be used for electricity, it must not have more than four electrons in the atom's outer orbit or the **valence ring.** The best element or **conductor** is copper (**Figure 5-3**). It has one electron in the valence ring. That electron can be forced out of orbit in a relatively easy manner. Elements with two, three, or four electrons in the valence ring can be used but require more voltage. For ease of understanding in this book, we will assume that the conductor is copper.

Voltage

An electron is moved by **voltage.** Voltage acts like the pressure in a water hose and is a measurement of **electromotive force (EMF).** EMF is produced or is available when there is a difference in the numbers of free electrons at one end of a circuit compared to the other end (**Figure 5-4**). This is known as **potential difference** expressed in voltage. If

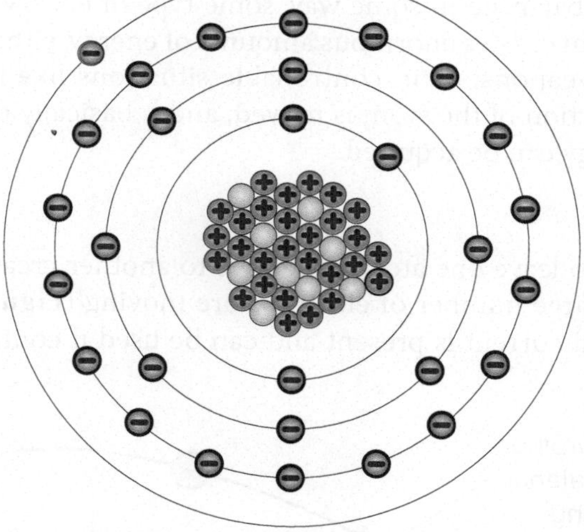

Figure 5-3 The valence ring is the outermost electron orbit. Electrons can only be stripped from this orbit for electrical current.

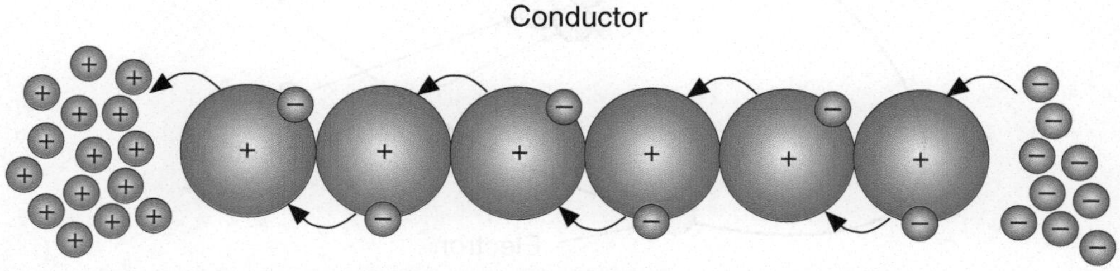

Figure 5-4 The difference between the number of electrons at the positive and negative sides causes a potential differential condition and creates voltage.

there is no potential difference between negative and positive, then there can be no voltage. An example is a driver trapped in a car under a live, 13,500-volt electrical line. Many people might assume that the driver is protected by the rubber tires that act as insulators. The actual situation, however, is that the driver and the vehicle have the same voltage as the electrical line, hence there is no potential difference . When the driver attempts to leave the vehicle and places a foot on the ground, he or she has created a conductor between the line and the ground. Since the ground will have very little voltage, the potential difference is enormous, and 13,500 volts of electricity will travel through the new conductor and usually result in the death of the driver.

AUTHOR'S NOTE Insulators are used to stop current flow. However, if enough voltage is present, almost all types of insulation can be overcome as the current can "jump a path" across it.

As voltage pushes an electron out of orbit, two things happen at once. The atom becomes unbalanced and tries to regain balance by attracting another electron. The moving electron attempts to attach itself to another atom. A chain reaction has started. All of the electrons will move from the negative (ground) toward the positive because unlike polarities, negative and positive attract each other. The movement of the electron is known as the **Electron Theory of Electricity.** When about 628 billion electrons pass a single point in one second, one ampere of current is available for use.

The **electron theory of electricity** is the actual physics of electricity in that electrons flow from negative to positive.

The Electron Theory of Electricity is what happens at the atomic level: electrons move from negative to positive. However, because of the way a vehicle is wired, the **Conventional Theory of Electricity** is applied. In this theory it is *assumed* that the electrons flow from positive to negative. One reason for this is that automobiles are wired as negative grounded: the battery's negative post, through its cable, is connected directly to the vehicle's frame and other metal components. In this way, the entire vehicle becomes a source of negative-polarity electrons. When performing a test for voltage, the technician is not actually looking for the moving electrons but for the positively charged atom or **ion** left behind. This type of wiring system is known as a **one-wire return.**

The **conventional theory of electricity** assumes that electrons flow from positive to negative and allows the automotive industry to use **one-wire return** circuity.

Current

Current is the amount of moving electrons expressed in an understandable term as amperes. A typical automobile has systems that use current ranging from milliamperes up to 30 amperes. Most electronic systems, like the engine computer and radio, use very low amperage while engine cooling-fan motors may use up to 30 amperes. The electrical current will continue to flow toward the positive until it meets resistance. It is wise to remember that voltage doesn't kill; however, current does.

Current is the number of moving electrons flowing between the ends of a circuit.

Resistance

Resistance is anything that uses voltage and is measured in ohms (**Figure 5-5**). Elements that are nonconductive are resistant and classified as **insulators.** The only way to use electricity is to create a resistance within a system.

Resistance slows or blocks current flow. It may be an electric working load or a solid **insulator.**

Resistance may be caused by **corrosion** on wires and connectors, small wire size, or it may be engineered into the system. Unwanted resistances like loose connections or corrosion disrupt the current flow in the system and devices may or may not work. Each electrical system is engineered to have specific resistance using specific amounts of current and voltage. Lamps, motors, and other electrical devices are engineered to use resistance and are known as loads. Loads are designed to use the electrical current for tasks. The lamp has a filament with a specific resistance (**Figure 5-6**). As current flows

Corrosion can be found most commonly on battery cables, usually resulting from a lack of maintenance.

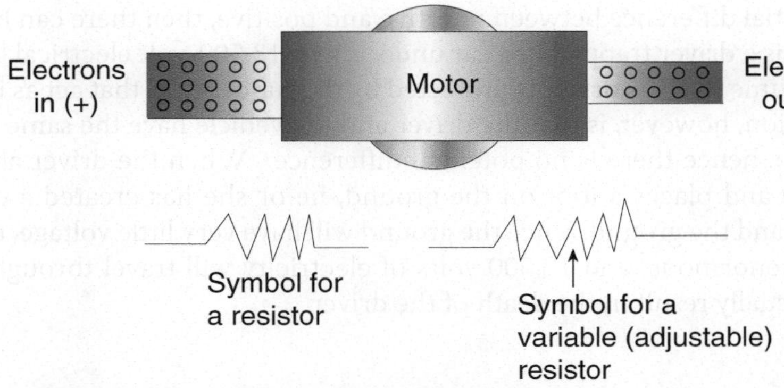

Figure 5-5 Resistance uses or stops current. Note the symbols for a resistor.

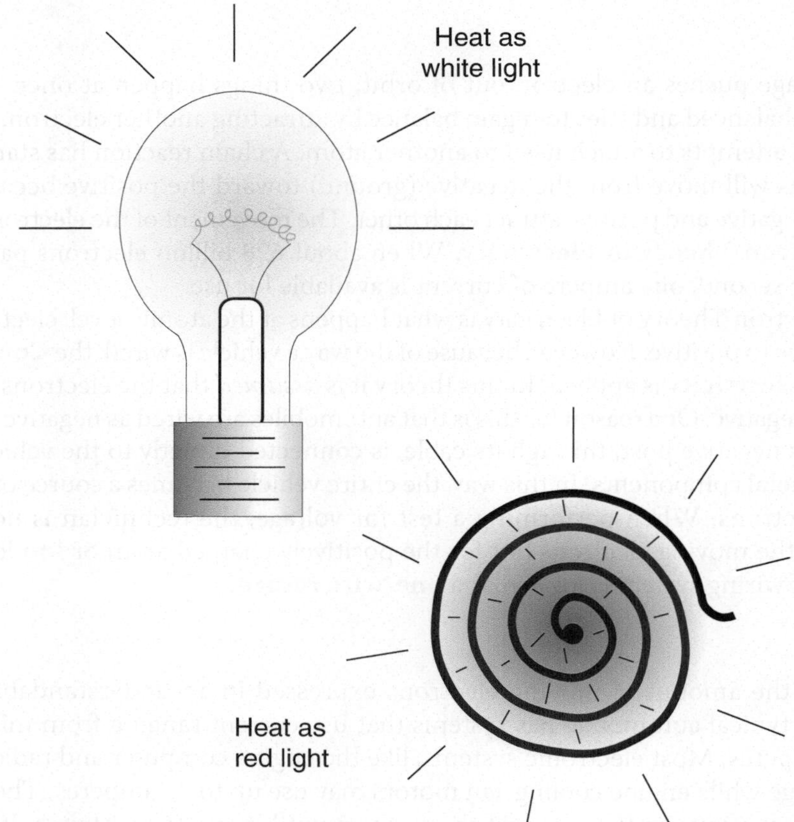

Figure 5-6 Depending on the type of resistance and circuit design, current can be used for light or heat.

Magnetic field may be natural, such as magnetic rock or something electrical. Materials that are magnetized by electricity are electromagnets.

A **winding** is one wrap of wire, rope, hose, or any item that can be wrapped around itself. Coiling a hose could be considered as several windings of the hose.

through the filament or resistance wire, it produces heat. The heat is seen as white light. An electric stove uses specific resistance coils for heating. The coils are heated to high temperature by current flow, which is usually seen as a red light. Electric motors must overcome the resistance of a shaft and attached devices to create mechanical work.

Magnetism

Any time current is flowing, a **magnetic field** is created around the conductor (**Figure 5-7**). If the conductor is wound tightly in a spool fashion around a shaft and each **winding** is insulated from the others, the magnetic fields overlap and increase the total magnetic field

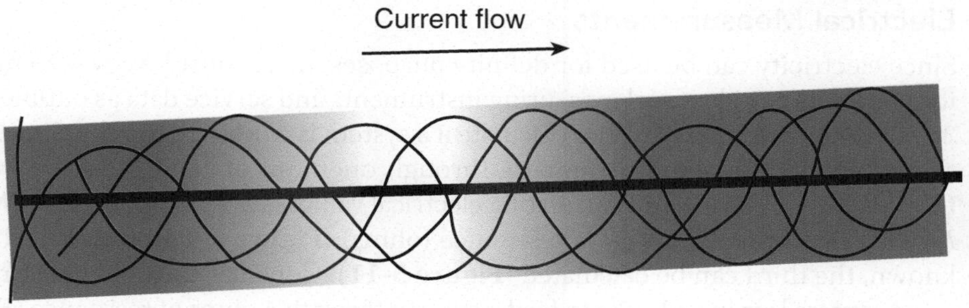

Figure 5-7 Current creates a magnetic field as it moves through a conductor.

around the shaft. This is known as a **coil** (**Figure 5-8**). Placing the shaft and its coil inside stationary magnetic fields makes the shaft turn. Stationary magnets produce the second magnetic fields. The positive and negative polarities of the shaft's coil fields attempt to move toward their opposing polarity of the stationary fields (**Figure 5-9**). The result is a spinning shaft, which can be attached to a fan or other device. The polarity of the shaft's magnetic field can be continuously changed to continue the rotation. Magnetic fields can also be engineered to produce electrical current. Devices that use or produce electricity will be covered in other chapters in this book.

Coil: One or more turn of a current carrying wire designed to produce a magnetic field

A BIT OF HISTORY

Michael Faraday (1791–1867) was an English chemist and physicist best known for his experiments with electromagnetism. Using the principle that electricity could be made by moving a magnet inside a coil of wire, he developed the first electric motor, generator, and transformer. He was responsible for several electrical terms in use today including *electrode, cathode, anode,* and *ion*. The unit of capacitance, the **farad,** is named in his honor.

Farad A unit of electrical capacitance or the ability to hold an electric charge.

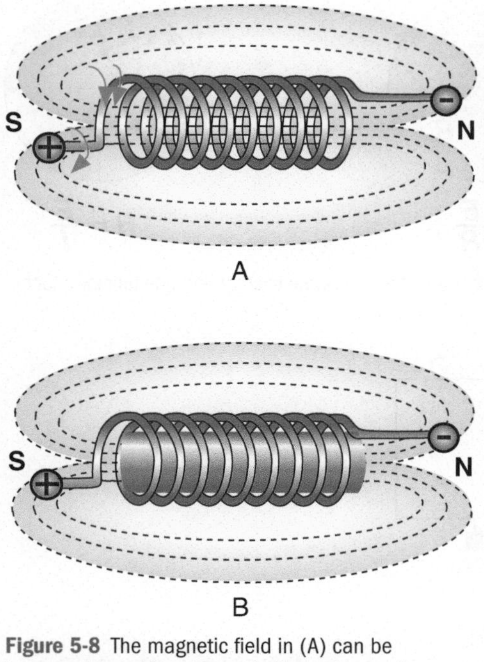

Figure 5-8 The magnetic field in (A) can be increased by adding more winding or increasing the current. Inserting an iron core (B) into the coil's, center concentrates the field identity.

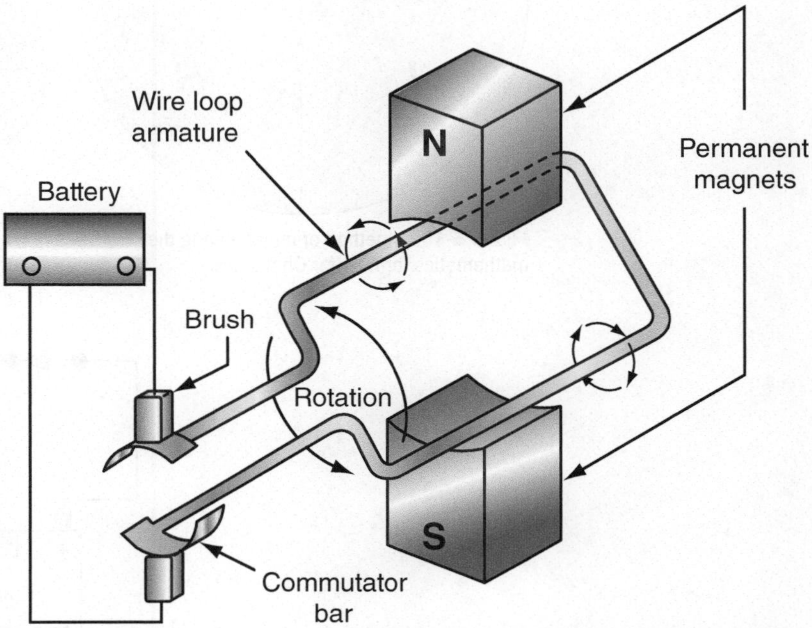

Figure 5-9 Electric motors use permanent magnets (pole shoes) and a switching polarity rotating magnet to cause shaft rotation. The shaft is attached to a mechanical device, such as a gear, to do mechanical work.

Electrical Measurements

Since electricity can be used for definite purposes, there must be a way to measure it for its control. Using electrical measuring instruments and service data as outlined in Chapter 3 and using Ohm's Law, we can calculate if a system is working properly. Since Ohm stated that one volt will push one ampere through one ohm of resistance, the mathematical formula for measuring and calculating electrical values is $E = IR$, where E is EMF (voltage), I is current (ampere), and R is resistance (ohms) (**Figure 5-10**). If two of the values are known, the third can be calculated (**Figure 5-11**). Consider the following problem.

A system is using 12 volts to feed a current through 4 ohms of resistance (**Figure 5-12**). The missing measurement, current, can be determined using Ohm's mathematical formula. Since $E = IR$, then $I = E/R$ or current is equal to voltage divided by resistance. In this case, $I = 12/4$ or I equals 3 amperes. If the system or device being checked uses 3 amperes to operate, then the system is functional. A device that uses more or less current would not work properly in this system. By the same token, if we measured this system with an instrument and found that only one ampere was being used, then the system is inoperative.

Consider a second situation.

A technician used a service manual to find that a system is supposed to have 12 volts supplied and 6 amperes of current (**Figure 5-13**). According to the formula, the working device should have 2 ohms of resistance ($R = E/I$). After taking measurements, it was found that only 1 ampere of current was present. Applying the formula again ($R = E/I$), the system appeared to have 12 ohms of resistance. Somewhere in the system there was too much resistance and the device did not work.

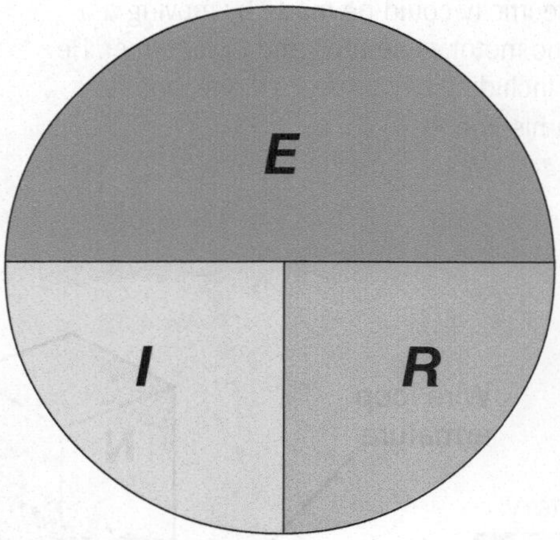

Figure 5-10 A method of remembering the mathematics formula for Ohm's Law.

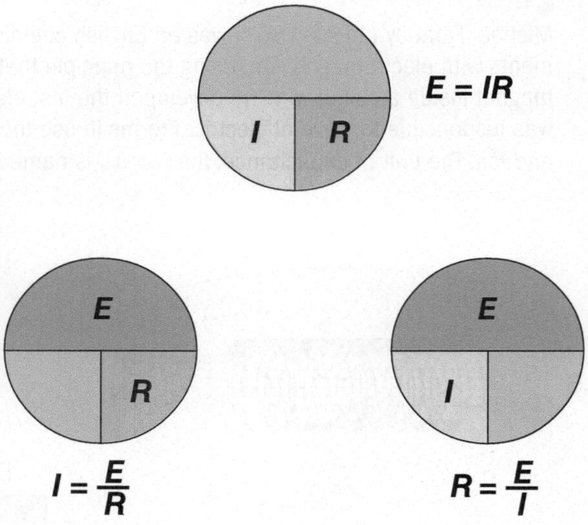

Figure 5-11 The equation can be used to find one unknown factor.

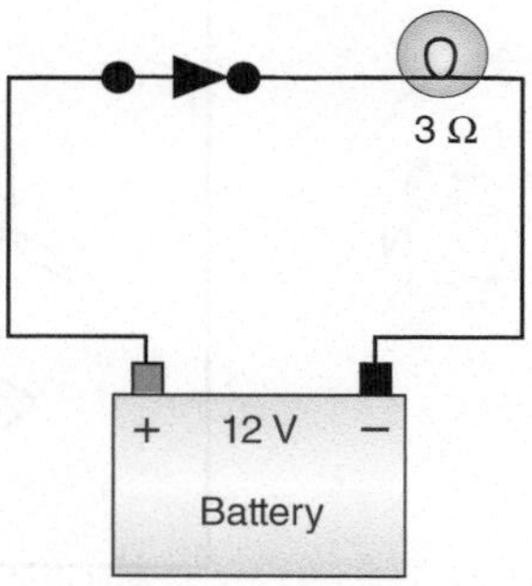

Figure 5-12 Simplified light circuit with 3 ohms of resistance in the lamp.

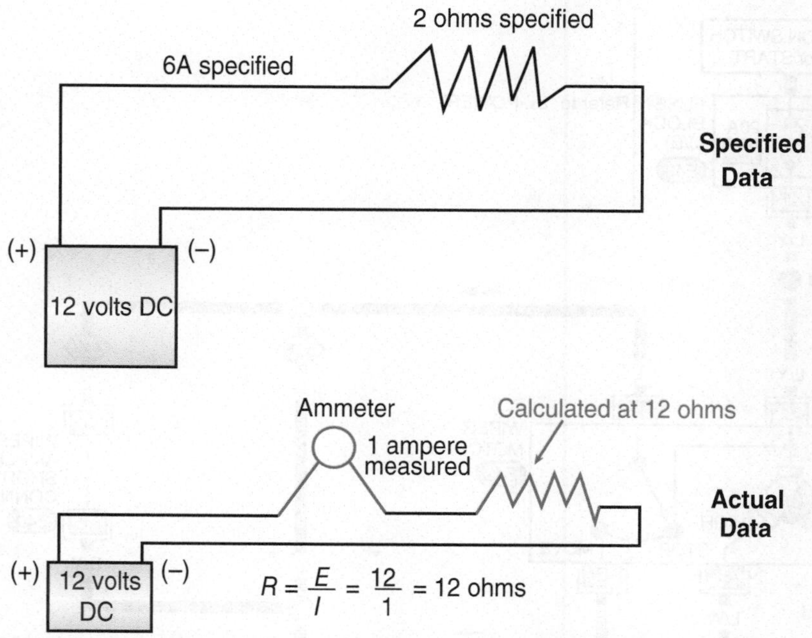

Figure 5-13 Measuring the actual voltage, current, and resistance of a circuit will assist the technician in diagnosis.

AUTHOR'S NOTE The resistance of many electrical devices increases when current is applied. This factor is taken into account by the electrical engineers. However, a failing device may increase the resistance enough to cause operating problems. Furthermore, a resistance test of this device, when shut down, may not reveal this increased resistance.

Circuits

Up to this point, we have discussed electrical systems. The systems comprise one or more electrical circuits. A **circuit** is the path that electricity follows from the negative side to the positive side. A circuit can work with a power source, conductor, and load. However, if those were the only things used, the operator would have to open the hood and connect the wires to the battery just to start the car. Such a procedure would have to be done for every device or system to be turned on or off. With this in mind, a viable working circuit must have a load, a conductor, a power source, protection and controls like the ignition or headlight switches (**Figure 5-14**). There are three distinct types of circuits.

A **circuit** provides a path for current to flow from the power source through the conductors and back to the power source.

Series Circuits

A **series circuit** provides one path for the current to travel. In order to work, the circuit must be intact from one end to the other (**Figure 5-15**). A defect anywhere in the circuit would affect the entire circuit. Older Christmas tree lights used series circuits. If one lamp went bad, then all of the other lamps would not work.

A **series circuit** provides one path for current flow.

There is only one true series circuit on the typical vehicle and that is the older starter control circuit. The current must flow from the battery, through the ignition switch, and finally through the park or neutral (clutch switch on manual transmission) to the starter relay. If the transmission is in a drive gear (clutch pedal up), then the park or neutral (clutch) switch is opened and the starter relay will not switch. Recent vehicle models have added a starter relay to the circuits. They are no longer a true series circuit.

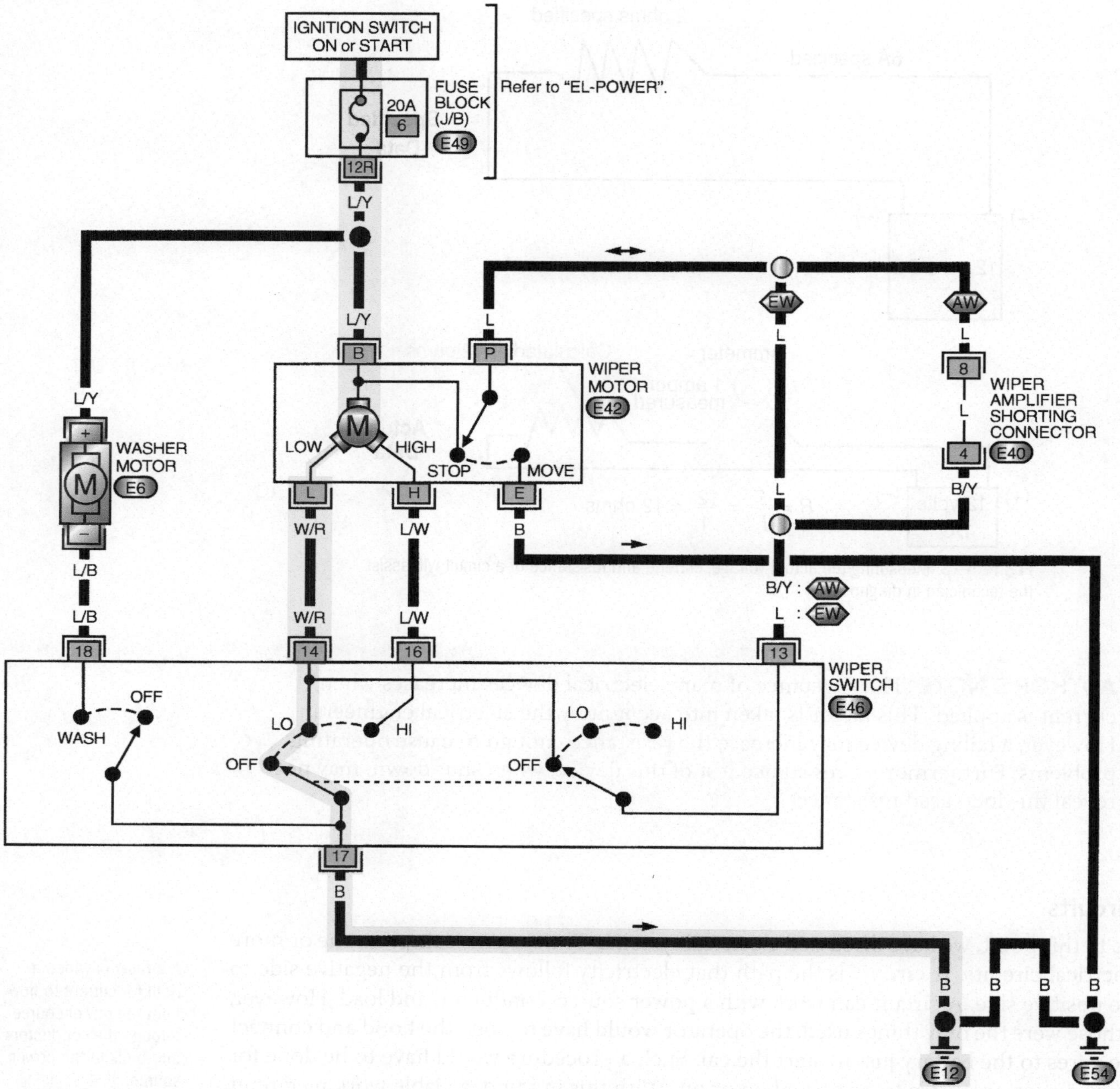

Figure 5-14 Current in this windshield wiper circuit flows from the battery through the ignition switch, the protection device, the wiper motor, and finally to the wiper switch. When the switch is moved to ON (low), current flows through the switch to ground. The switch here is on the negative side and is considered to be a grounding switch.

Parallel Circuits

A **parallel circuit** has multiple paths for current flow.

A **parallel circuit** provides more than one path to ground and is common on vehicles (**Figure 5-16**). One portion of the circuit can work while the others may not. Typical examples of this circuit on a vehicle are the exterior lights. In the case of the headlights, from the last control device in the system, the conductors branch into two legs or parallel circuits, one to each headlight. If one light fails, the other will still work. The brake lights, parking lights, and turn signals work in the same manner. Most of a vehicle's safety equipment, like the lights, are parallel wired to some extent.

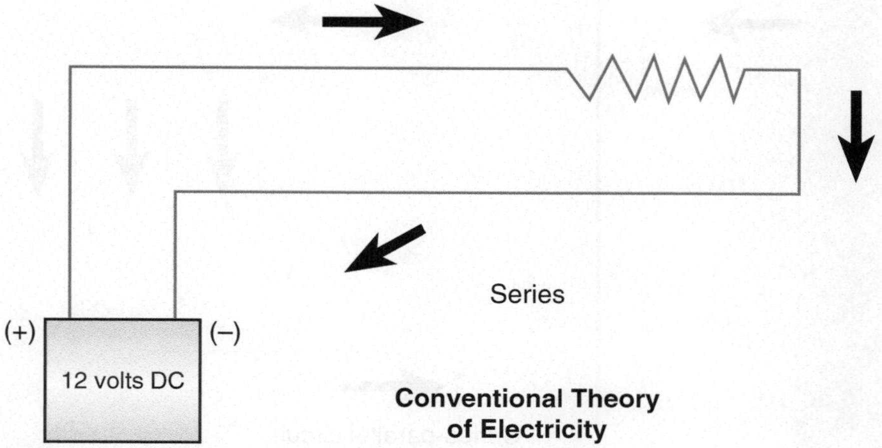

Figure 5-15 A series circuit has one path for current to follow.

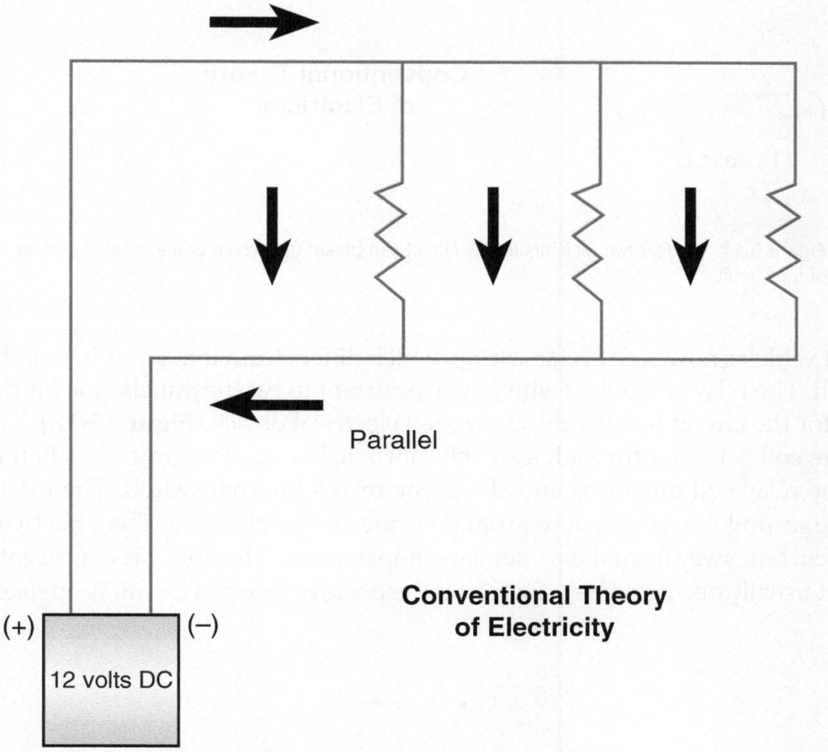

Figure 5-16 A parallel circuit has multiple paths to ground.

Series-Parallel Circuits

The most common type of electrical circuit on a vehicle is the series-parallel. Part of the **series-parallel circuit** is in series and other parts are in parallel (**Figure 5-17**). In effect, the series portion controls the parallel portion. Again, a typical example is the headlight system. As discussed in the last section, separate circuits feed the headlights from the last control device. That control device is usually an electric-mechanical switch known as a **relay.** A relay uses low current to switch a high-current circuit on or off. Many newer relays control multiple circuits. Before the relay is the headlight switch. The current flows from the power supply through the headlight switch, and finally to the relay. The switch controls the relay, which in turn routes power to the lights. When the switch is moved to the ON position, the relay closes and current moves from the power source through the relay to the headlights.

A **series-parallel circuit** typically has the series portion controlling the parallel portion.

An electronic switch that uses low current to control a high current is a **relay.**

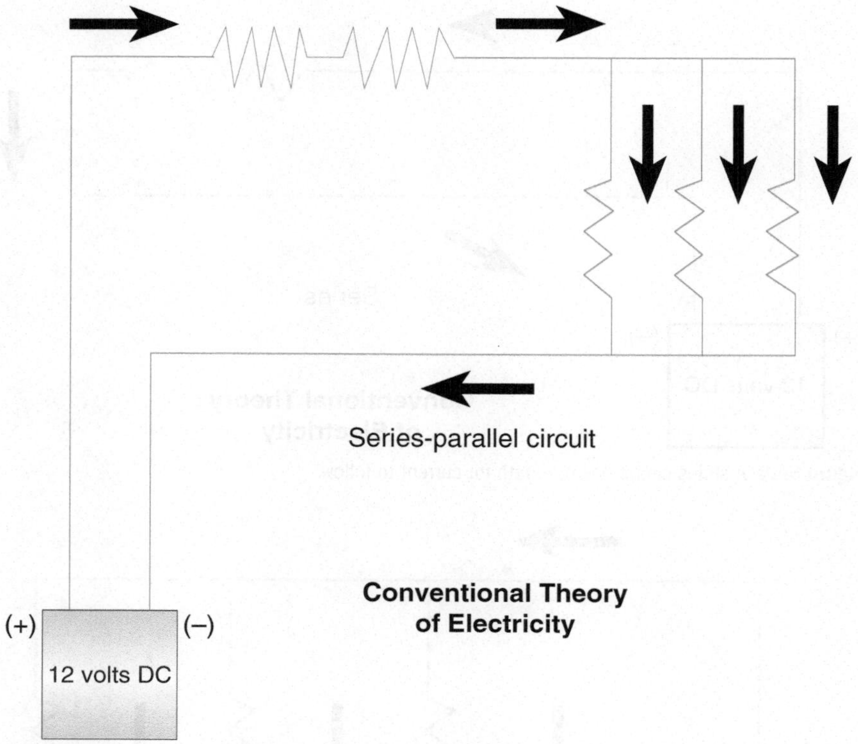

Figure 5-17 The first two resistors in this circuit can be set up to control the parallel portion of the circuit.

Most vehicles now use reverse wiring, which differs from that described in the previous paragraph. The relay is supplied with positive current to two terminals: one for the relay coil and one for the circuit feeding the controlled electrical device (**Figure 5-18**). The negative side of the coil is routed through a switch (horn, lights, etc.) to ground. When that switch closes, the relay coil functions and closes the relay's internal switch. The internal switch routes the second power circuit to ground via the electrical device. This effectively removes any high current away from the passenger compartment. The object is to prevent the power surge that usually occurs when a switch on the positive side of a circuit is opened or closed.

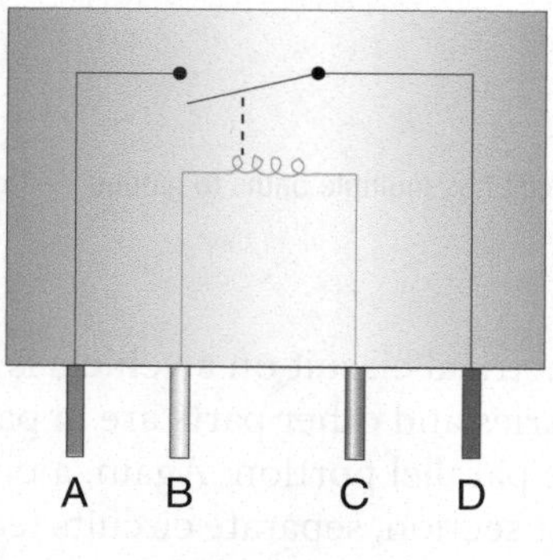

Figure 5-18 Current through the coil (b to C) will switch on the high-current circuit (A to D).

Circuit Protection

Circuits should be protected by some device that will fail if the circuit is overloaded. This protects other circuits and possibly the whole vehicle. There are three basic protection devices: **fuses, circuit breakers,** and **fusible links.** If a circuit protector fails, it is for a reason and the circuit must be checked to find the problem. Simply replacing the protector may damage the circuit or at the very least burn through it again.

Fuses, circuit breakers, and **fusible links** are devices most commonly found in automotive circuits.

AUTHOR'S NOTE Like all man-made things, circuit protection devices will fail from old age. Granted there is no mechanical "fair wear and tear," but the devices are heated and cooled by the current flowing through it. This weakens the wire, and eventually it will fail. If a protection device fails on an older vehicle, check the system for problems but consider that the protection device may have failed because of age.

A fuse has a small internal wire through which all of the circuit's current flows (**Figure 5-19**). When a circuit current exceeds the capability of the fuse, the wire burns through, thereby stopping the current. The fuse can be removed from its holder and ohmed checked for failure. Fuses are rated in different amperes ranging from 0.5 to 60. The larger ones, 30 to 60 amperes, are used in circuits such as charging units, fuel pumps, and cooling fans. The fuse may be a cartridge type or blade type. The cartridge has been replaced with the blade for convenience and reliability. A fuse is not serviceable and must be replaced when it fails.

Circuit breakers open because of high current and are rated like fuses (**Figure 5-20**). Some circuit breakers will reset after they cool off. Others must be replaced or manually reset. Breakers are better in some circuits because their design allows them to be slow acting compared to a fuse. This prevents an electrical surge from tripping the breaker. The problem with a circuit breaker is that most must be checked with an ohmmeter or multimeter. There are usually no visible signs of failure.

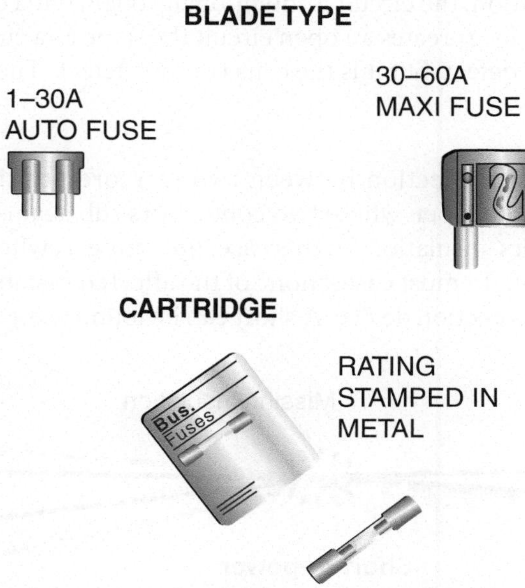

Figure 5-19 Fuses are made to be effective, but they are also inexpensive and easy to replace.

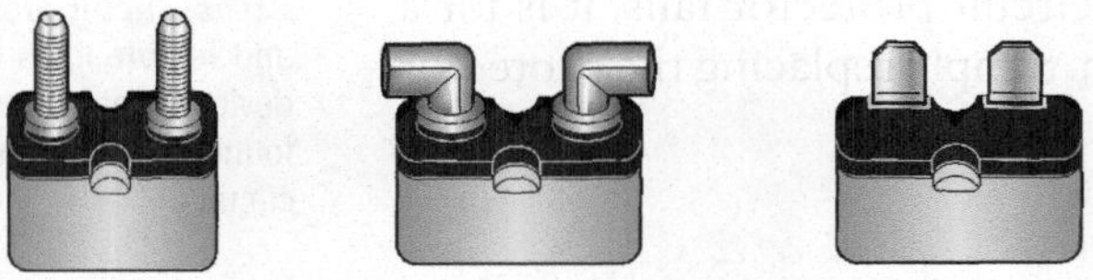

Figure 5-20 A circuit breaker can be used to protect a high-current circuit or used as the flasher for the turn signal.

Figure 5-21 This type of fusible link is being phased out but can be found on trucks, older vehicles, and similar industrial engines.

A fusible link is used on some heavy-current circuits. The link is a wire within a molded form (**Figure 5-21**). The wire is sized below the wire of the circuit and usually rated for low amperes. Excessive current will burn through the link. A major problem with this type of protection device is the diagnosis and replacement of it. Many times, the link fails without any visible signs. They are usually checked with a test light, voltmeter, or multimeter and may be placed in locations difficult to access. Fusible links must be replaced when they fail. Fusible links have been replaced on newer vehicles with high-rated blade fuses known as *maxi-fuses*.

Circuit Defects

A **short-to-power** is an unwanted connection between positive conductors of two different circuits.

A **short-to-ground** is an unwanted connection between a positive conductor and ground.

Open: An electrical circuit that is not complete.

There are only three possible defects in an electrical circuit. They are **open, short-to-power,** and **short-to-ground.** Any type of electrical device, including ones that use magnetism, can have one or more of the defects.

An **open** is a break in the circuit.

Open

An open circuit is an incomplete circuit where current will not flow. When a light switch is moved to the OFF position, the circuit is intentionally open, but a defect is an unwanted open. A burned-through fuse creates an open circuit the same as a cut conductor. A blown lamp is not the electrical defect, but it is the cause of that defect. The blown lamp is open.

Short-to-Power

A short is an unwanted connection between two or more positive (hot) conductors (**Figure 5-22**). This could happen where two conductors rub against each other, thereby weakening the conductors' insulation. In this case, operating a switch could cause two or more circuits to switch on. In most cases, none of the affected systems will work correctly and will probably trip a protection device. A short can be found using a test light, voltmeter,

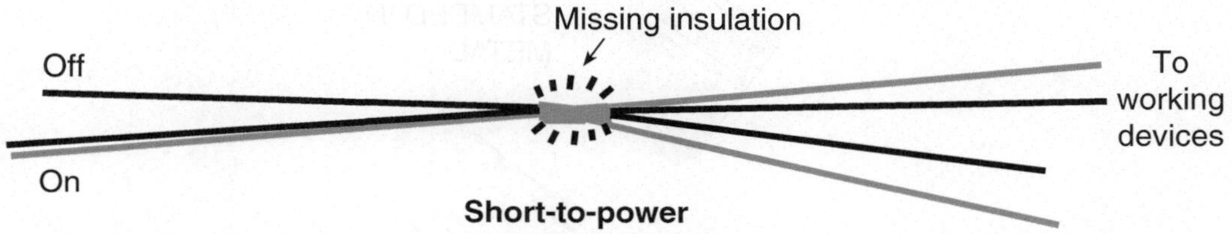

Figure 5-22 In this short, current is being directed into a circuit that is actually turned off. The result will be a blown fuse or poorly performing electrical devices.

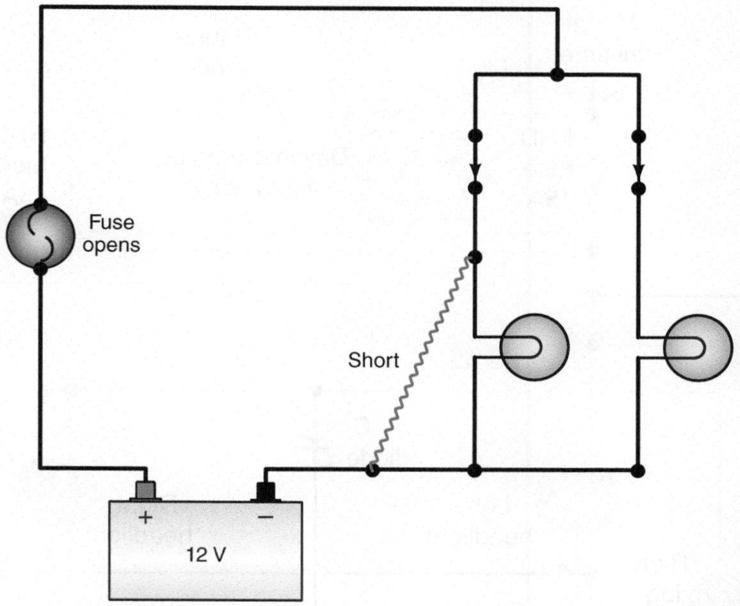

Figure 5-23 A grounded circuit.

ohmmeter, multimeter, or a short finder. A **short finder** is the common name for a tool that can be moved over a conductor for testing. A moving needle indicates current and that a magnetic force is present, and a stationary needle indicates no current or no magnetism. The same tool can be used to locate a grounded circuit.

A **short finder** is an electrical tool used to locate shorts in a circuit.

Short-to-Ground

A short-to-ground is an unwanted connection between a feed conductor (positive) and the ground of the vehicle (**Figure 5-23**). Since the ground is the negative side of the power source, and there will probably be no resistance at this time in the circuit, extremely high current will flow. The protection should trip to protect the circuit. If it does not, heat can damage the circuit very quickly or a fire could erupt, possibly causing damage and injury.

AUTHOR'S NOTE Prior to 1995, before automotive terms were standardized, there were three defects: open, short, and ground. Many technicians, however, still refer to any short as either a "short" or "ground" or any ground as a short or ground. They knew what they were doing, even though many they did not know the difference between the two terms and, quite frankly, really did not care. It was a "short" or a "ground."

ELECTRICAL SERVICE INFORMATION

To diagnose an electrical circuit or system, information must be obtained on how the circuit is placed in the vehicle. It may also be necessary to trace a circuit without tearing the vehicle apart to chase wires. There are two items to help the technician in this area. They are the vehicle's wiring diagram or schematics and electrical component locator.

Shop Manual
pages 148–149

Wiring Diagrams

The **wiring diagram** is a paper or computerized document that shows every circuit or a single system circuit on a vehicle (**Figure 5-24**). It also shows controls, switches, loads, and other devices in each circuit. A single circuit may extend over several pages or screens,

A **wiring diagram** is a drawing of multiple electrical circuits.

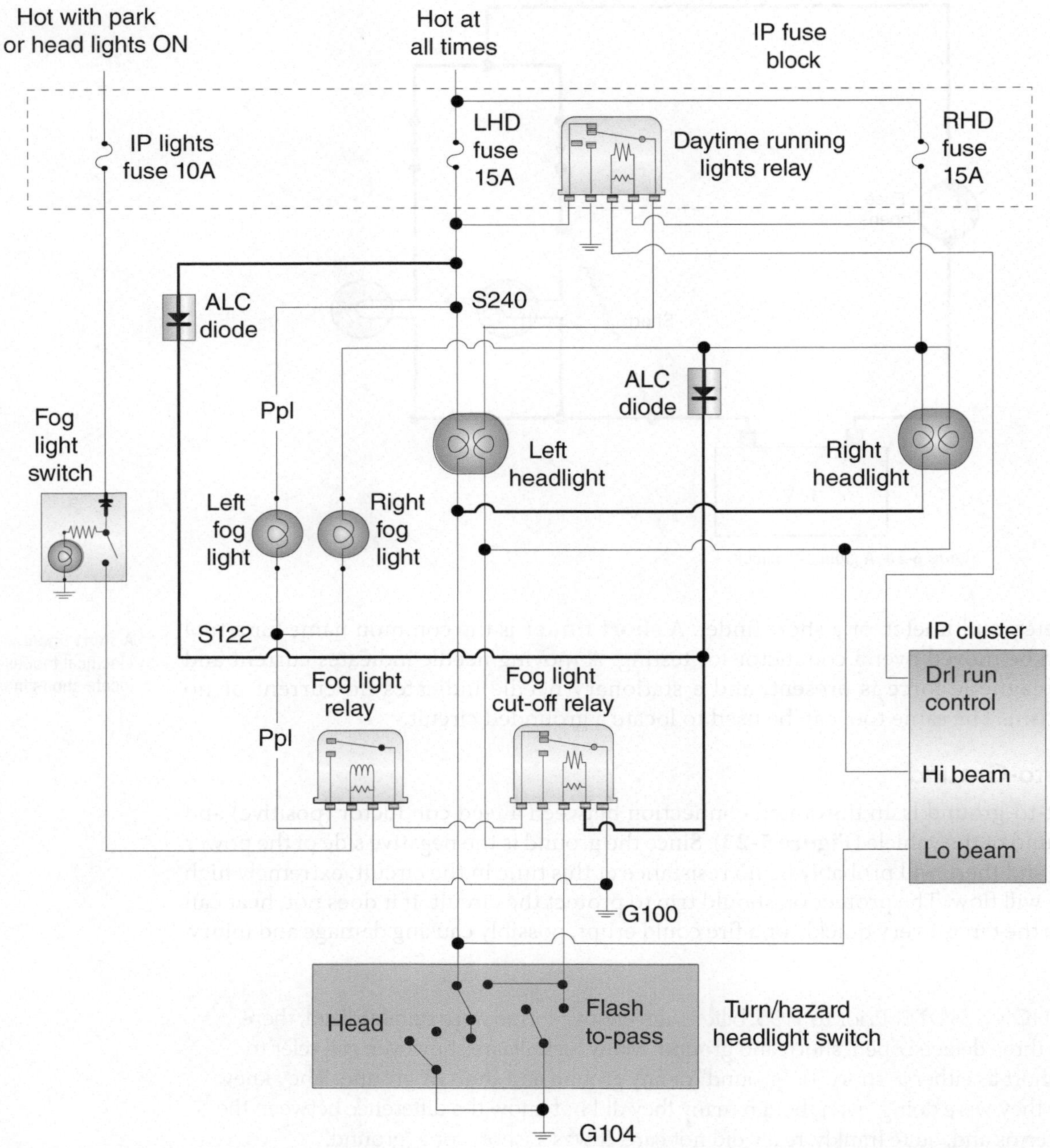

Figure 5-24 Circuit for a Chrysler antitheft system. Note the connection to the powertrain control module (PCM). A system fault may be related to a fault in one of the circuits in or connected to the PCM.

so tracing the circuit can be a time-consuming task. Wiring symbols are usually shown on the first page or screen of the wiring diagram section of the manual. Wire colors and sizes, along with circuit designators, are also shown on the diagram. Also known as schematics

Newer vehicle wiring diagrams usually have separate drawings known as system diagrams (**Figure 5-25**). They show the wiring for a single system on one page and usually include all the information needed to trace the power flow. However, if a system diagram indicates that the system is controlled by a relay or computer, it may be necessary to find the diagram for that relay or computer to determine what electrical signal(s) are needed to cause the system in question to turn on. For instance, two possible sensor signals needed to switch on the cooling fan relay are the engine temperature and the

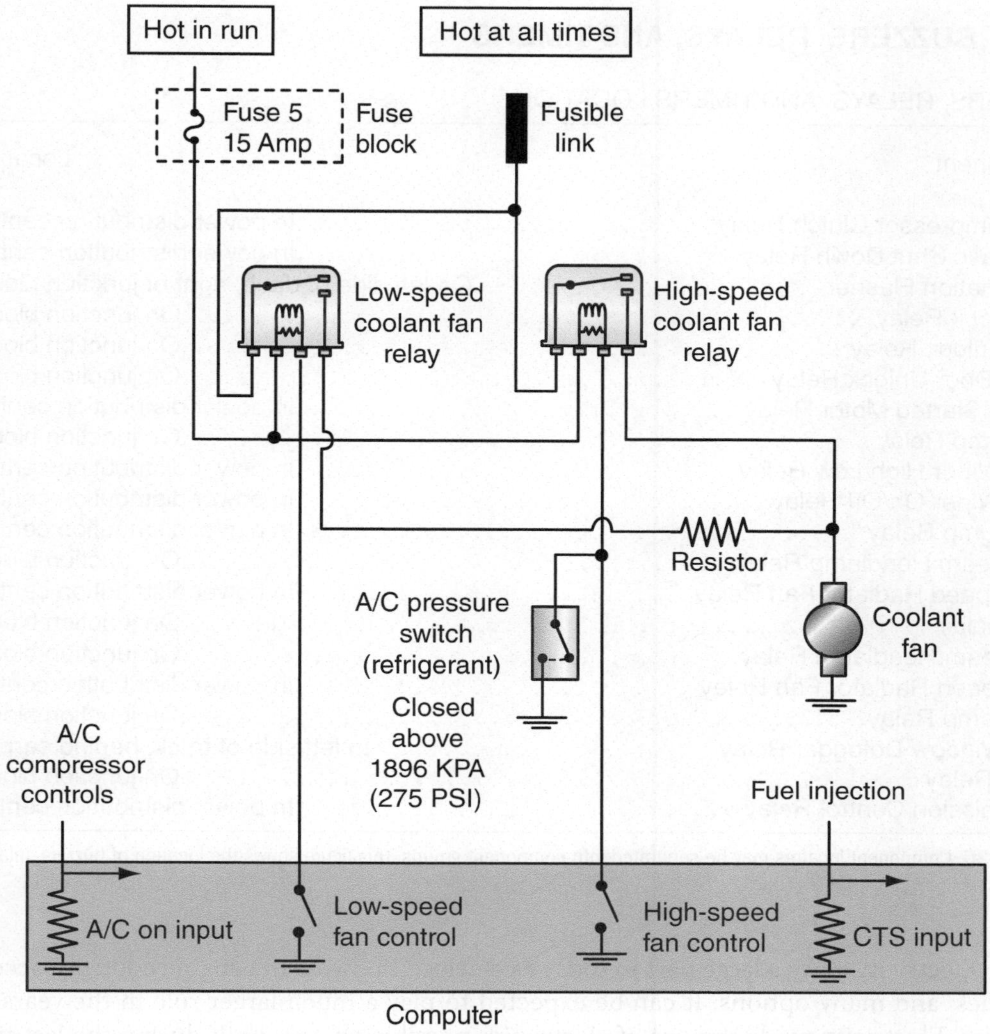

Figure 5-25 This system diagram shows the electrical components that control the coolant fan operation. They are the computer with low- and high-speed fan control and an independent A/C pressure switch.

air-conditioner compressor control circuit. If one of those signals is incorrect or missing, the cooling fan may not be energized, leading to engine overheating. Those signal wires may not be shown on the cooling-fan system circuit. Even using a wiring diagram locating a specific component or connection may be difficult.

Component Locators

A **component locator** shows the location of each electrical component on the vehicle. It can be a list of the items with their location spelled out, as shown in **Figure 5-26**. Some manufacturers use pictures or drawings of sections of the vehicle with the component highlighted in some manner.

A **component locator** shows the location of each electrical device in a vehicle including switches, splices, and connections.

A drawback to locators and wiring diagrams is their publication date. The manuals should be in the dealerships when their corresponding vehicle model is available for public sale. This means that last-minute changes may not be included and either the diagram or locator may be in error. This is not a real issue unless the manufacturers update their product during the model year instead of waiting for the next model. Even then, most changes are small and may only be a change in wire color. Experience is the best method of learning how to read wiring diagrams and component locators.

BUZZERS, RELAYS, AND TIMERS

BUZZERS, RELAYS, AND TIMERS LOCATION

Component	Location
A/C Compressor Clutch Relay	In power distribution center.
Automatic Shut Down Relay	In power distribution center.
Combination Flasher	On left side of dash, right of junction block.
Door Lock Relay	On junction block.
Door Unlock Relay	On junction block.
Driver Door Unlock Relay	On junction block.
Engine Started Motor Relay	In power distribution center.
Fog Lamp Relay	On junction block.
Front Wiper High/Low Relay	In power distribution center.
Front Wiper On/Off Relay	In power distribution center.
Fuel Pump Relay	In power distribution center.
High-Beam Headlamp Relay	On junction block.
High-Speed Radiator Fan Relay	In power distribution center.
Horn Relay	On junction block.
Low-Beam Headlamp Relay	On junction block.
Low-Speed Radiator Fan Relay	In power distribution center.
Park Lamp Relay	On junction block.
Rear Window Defogger Relay	In left side of truck, behind carpet.
Spare Relay	On junction block.
Transmission Control Relay	In power distribution center.

Figure 5-26 Component locators may be separated into component groups. This figure shows the location of buzzers, relays, and timers.

Electricity plays a large part in today's vehicles. It powers the engine controls, accessories, and many options. It can be expected to play a much larger role in the years to come. Electricity made the use of electronics possible on cars built during the last two decades of the twentieth century. The twenty-first century will see electronics expand greatly in the automotive industry.

SWITCHES, RELAYS, AND SOLENOIDS

Switches

Switch A device for making and breaking the connection in an electric circuit.

A **switch** is the most common means of providing control of electrical current flow to a component. Switches can be mechanical or electrical. Mechanical switches are easy to give example of. This would include any switch in the vehicle where you move or turn something and now current is flowing through it. Current then flows to the component.

The switch in **Figure 5-27** shows how a simple mechanical switch works. There is current flowing from the battery positive wire into the switch and when it is in the open position (as drawn), current stops flowing, but voltage is still available at the tip of the switch. Once the switch closes, the circuit pathway is completed and current will flow through the switch and through the load, then to ground. Let's take and apply this to a simple headlight circuit. **Figure 5-28** shows a wiring diagram for a simple headlight circuit with high and low beams. This is a series parallel circuit. The headlight switch only has one job to do. It is either on or off. This is called a **single-pole, single throw** switch. The

A **single-pole, single-throw** switch only has on/off functions.

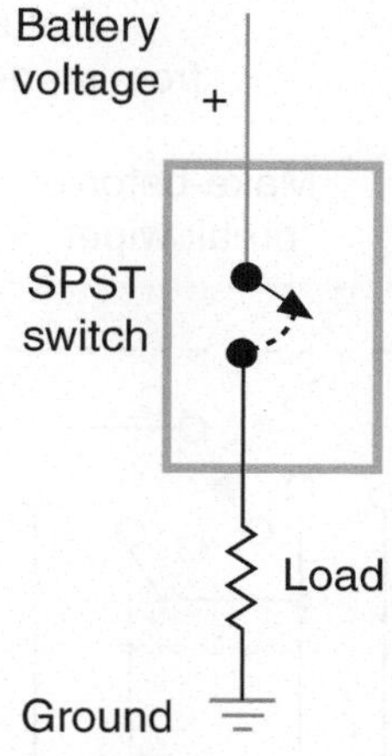

Figure 5-27 A simplified illustration of an SPST switch.

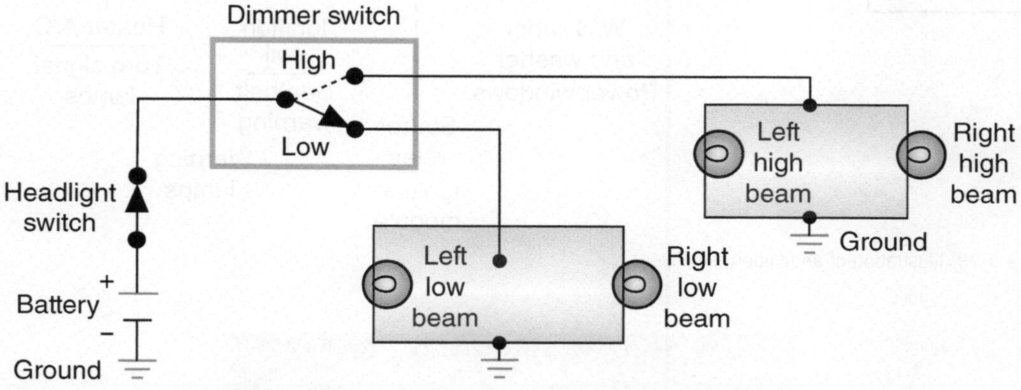

Figure 5-28 A simplified schematic of a headlight system using an SPDT dimmer switch.

dimmer switch controls the brightness of the lights (high or low beam) and can have two different positions, but both are "on." This is called a **single-pole, double throw** switch. Switches can also be classified as **normally open (NO)** or **normally closed (NC).** An example of an NO switch would be the seat belt switch. The seat belt reminder light is always on until you buckle your seat belt. An example of a **ganged switch** is an ignition switch (**Figure 5-29**).

When you look around an automobile you will find many types of switches that control many things on the car (**Figure 5-30**). However, **conventional** (mechanical) **switches** are not as prevalent on today's modern vehicles. Most switches on today's modern vehicles do not carry much current. They are simply switches that are there to tell the computer to carry out a function. These are still mechanical switches, but they are not directly connected to the component they control; rather they are an **input device** or **sensor.** A computer carries out the function of turning something on and off via an electronic switch. This is very common on many vehicles today.

A sensor can be a switch. There are many different types of sensors on the vehicle. A sensor changes the current in a circuit that a computer is watching. The computer then makes a decision and controls and **output device** or **actuator** (**Figure 5-31**).

A **single-throw, double throw** switch has two 'on' positions.

A **normally open (NO)** switch is open until it is activated.

A **normally closed (NC)** is closed until it is activated.

A **ganged switch** controls multiple circuits.

A **conventional switch** has only an on/off for a single circuit.

Input devices or **sensors** sends data to a computer on circuit conditions at a given time.

The computer uses input data to control **output devices** or **actuators.**

Relays

All circuits have some type of controlling device. Some circuits only use switches to control the flow of current, some circuits use a computer to control current flow, and

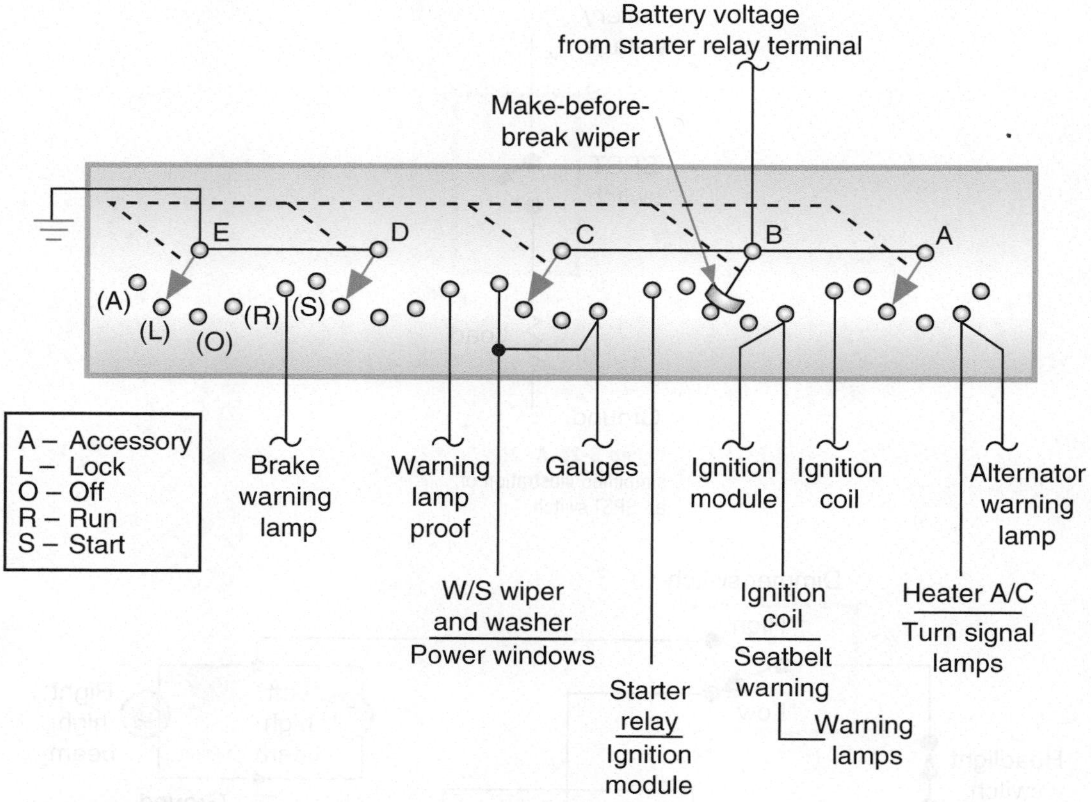

Figure 5-29 Illustration of an ignition switch.

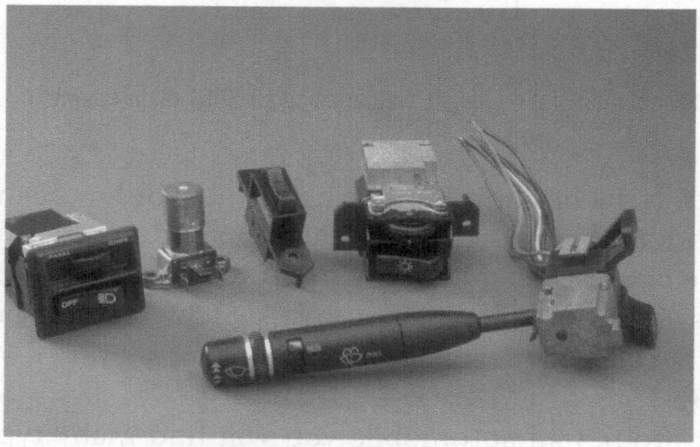

Figure 5-30 Common types of switches used in the automotive electrical system.

some circuits use a relay. Many modern vehicles use a combination of each of these devices to control the current flow.

A relay is an electromagnetic switch that consists of a coil that has very high resistance, thus drawing very low current. This low current is used to produce a magnetic field that will close a physical contact switch and allow high current to flow through a different circuit. A relay needs a parallel circuit to operate. It is usually found in circuits that need higher current. It is more desirable to have high current go through a relay than through a switch where your hands may come into contact with it, or the switch will be worn out because they are built for lower current and lower heat ranges. A relay also allows a design where the thicker wires (for high current) can only be used in short distances, thus reducing the chance of fire and current drop from high resistance due to increased length.

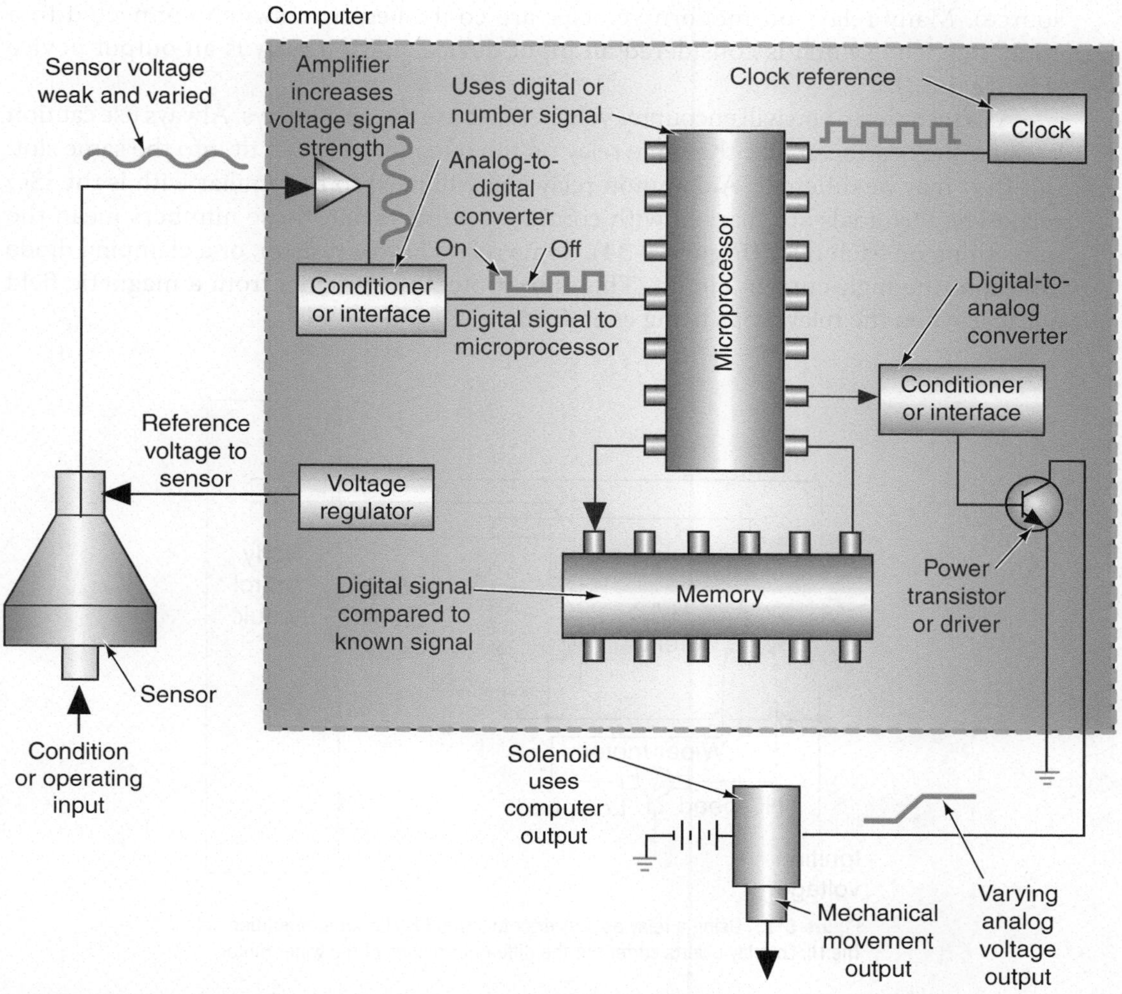

Figure 5-31 A microprocessor uses sensors or switches to control and determine what type of output to perform.

When current is applied to the coil by the low current side, it energizes. A basic relay is shown in **Figure 5-32**. The drawing shows a horn that is controlled by a switch. The relay is usually located under the hood, close to the horn or in the fuse box. Figure 5-32 is a three-terminal relay. A four-terminal relay would have two separate wires for the power source (high and low current) and then two separate output wires (ground and

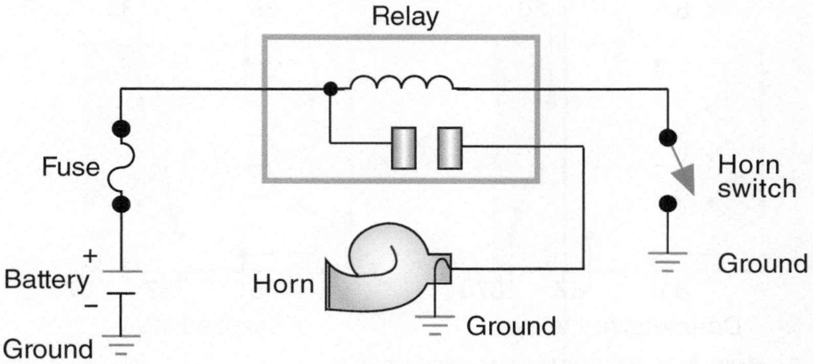

Figure 5-32 A relay can be used in the horn circuit to reduce the required size of the conductors installed in the steering column.

source). Many relays on modern vehicles are controlled by a switch connected to a computer. The switch is considered an input device and the relay is an output device (**Figure 5-33**).

As a technician you will encounter vehicles that use the same relays. Always use caution because they can seem like the same relay on the outside, and even fit into the same slot. But they may be different. A common relay you will need to be familiar with is the ISO relay. The terminals are marked with common numbers and those numbers mean the same thing on each relay (**Figure 5-34**). Some relays have a resistor, or a clamping diode built into the high-current circuit. This is to protect the circuits from a magnetic field collapse when the relay stops being energized.

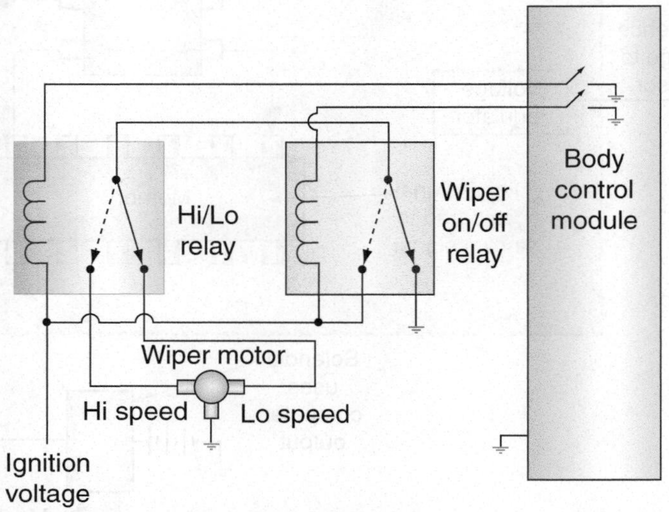

Figure 5-33 Using a relay as a diverter to control Hi/Lo wiper operation. The Hi/Lo relay diverts current to the different brushes of the wiper motor.

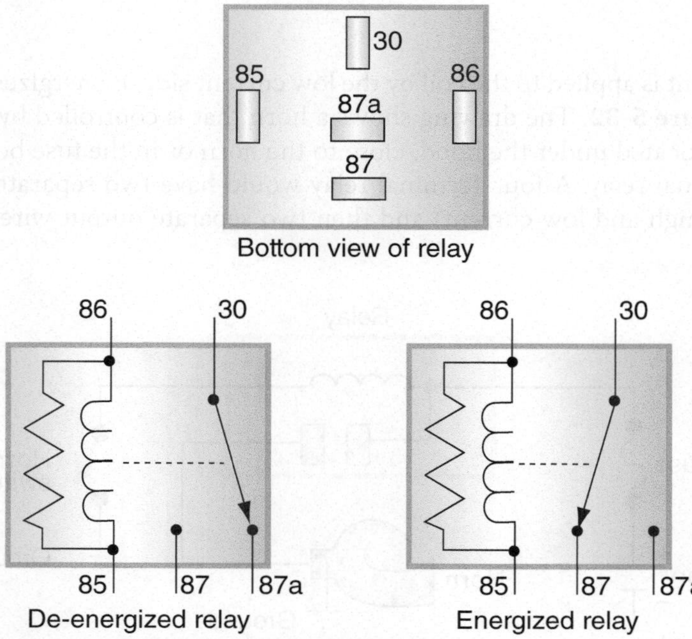

Figure 5-34 ISO relay terminal identification.

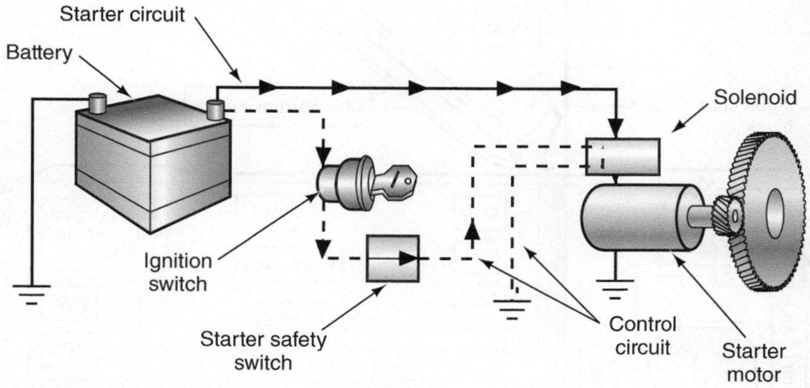

Figure 5-35 Major components of the starting system. The solid line represents the starting (cranking) circuit and the dashed line indicates the starter control circuit.

Solenoids

The best way to remember what a solenoid is or does is to think of it as a relay with a mechanical action. Starters have solenoids in their circuits, but they also have relays. The relay uses low current to control high current. A starter solenoid uses low or high current to control the flow of electricity and to physically move and engage the starter ring gear with the flywheel of the engine (**Figure 5-35**). Another example is a fuel injector. A fuel injector has a magnetic coil inside of it that is energized by the computer driver circuit. When it is energized, it magnetically lifts the injector nozzle to let pressurized fuel into the engine.

COMPUTER CONTROL SYSTEM

Many times, throughout this text we have mentioned how modern vehicles have many circuits controlled by a computer. It's true that this will make things more complex and difficult to comprehend. But you can be assured that after learning how a basic computer functions, everytime you have a computer issue, just remember the basics of operation and testing and you will be fine. Computer systems are somewhat of an advanced topic, so we will try to cover only the basics of operation, and then in the Shop Manual we will cover the basics of inspection and testing.

A modern computer is not really alone (**Figure 5-36**). It is usually combined with the function of other computers in the vehicle. Another thing to remember as you are learning is that almost all of the components on a modern vehicle have some sort of computer control or function. All of those circuits still have the same function as they did 50 years ago. The application hasn't changed, but the control of it has. It is still important to understand how a system works, and then just realize that there is a computer integrated into the same system.

Let's look at the basics of a computer. The main parts of a computer are the microprocessor, outputs, and inputs (**Figure 5-37**). The inputs are usually considered sensors. They allow data to come into the computer. An example of a sensor would be the oxygen sensor. The computer receives this data and makes decisions based on the data it receives from that sensor and other related engine control sensors. It then decides what to do to the outputs. In this case, one of the outputs is a fuel injector (which is an output).

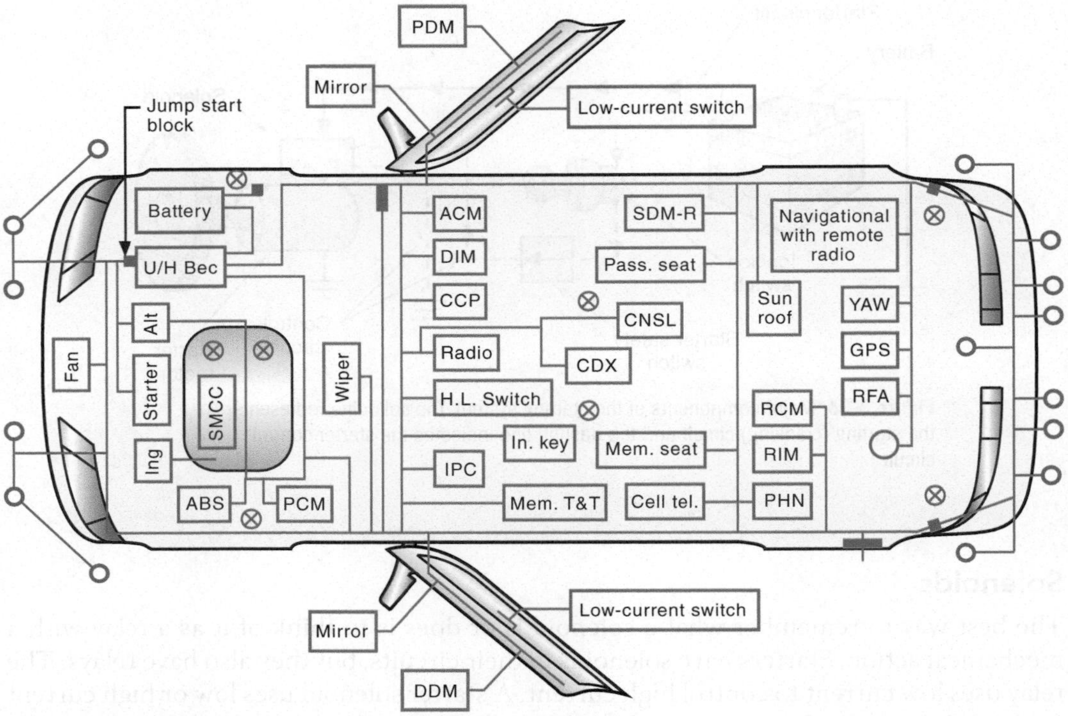

Figure 5-36 The electrical system of today's vehicle can be complicated.

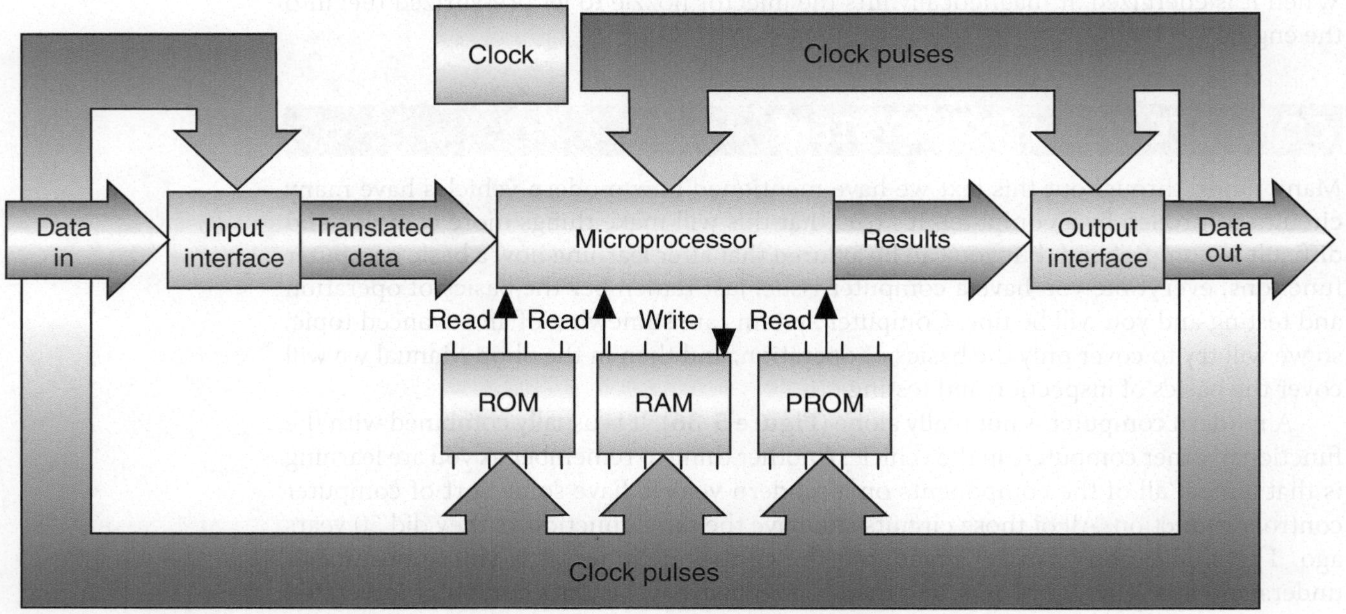

Figure 5-37 Interaction of the main components of the computer. All of the components monitor clock pulses.

Multiplexing is a method of connecting multiple electronic devices to send/transmit data.

On a modern vehicle with multiple computers, it is more efficient to operate each computer on a network, instead of operating each computer individually. This is called **multiplexing.** When a computer system is multiplexed, it uses a network to control the commands, inputs, and outputs of the system. It also allows the technician to access all of the computers from one diagnostic link connector (**Figure 5-38**).

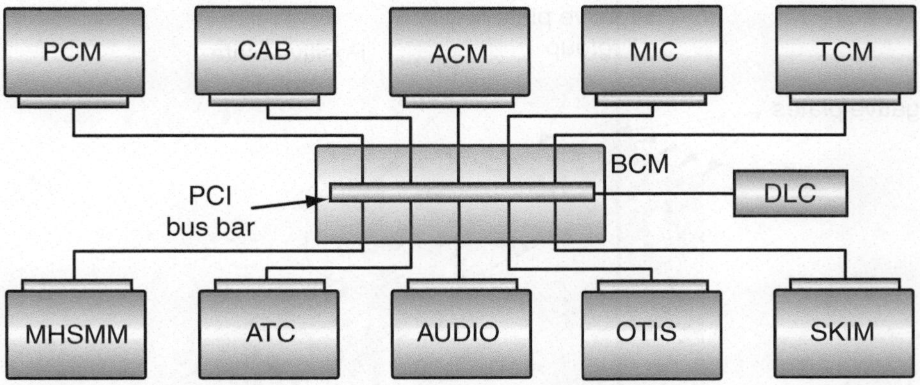

Figure 5-38 Automotive computers are networked together through multiplexing.

BATTERIES

An automotive battery (**Figure 5-39**) is an electrochemical device capable of storing and producing electrical energy. The source of power for everything in the car is the battery. Just like any other battery, when you use energy the voltage level and energy left in the battery will go down. The alternator's job is to replenish the energy removed from the battery. This is probably the single most important electrical concept!

The automotive battery has several very important functions, including:

1. It operates everything electrical and provides power for all electrical components.
2. It is a large enough storage device for times when the alternator's recharging current is less than the current demand on the battery from the vehicle.
3. It acts as a stabilizer of the voltage for the entire automotive electrical system.
4. It stores energy for extended periods of time.

Types of Batteries

Conventional batteries are composed of plates made of lead. An electrolyte acid is placed inside the battery to act as a chemical that will work with the plates to produce the conversion from chemical to electrical energy (**Figure 5-40**). Conventional batteries have positive and negative plates grouped together as a cell (**Figure 5-41**). Each cell is designed to produce 2.1 volts.

A conventional battery consists of six 2-volt cells in series (**Figure 5-42**). The cells are contained in a plastic case. This type of battery changes chemical forms when it is

Figure 5-39 Typical automotive 12-volt battery.

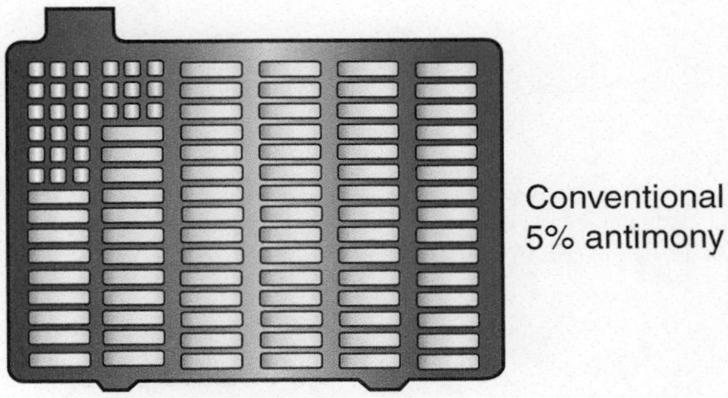

Figure 5-40 Conventional battery grid.

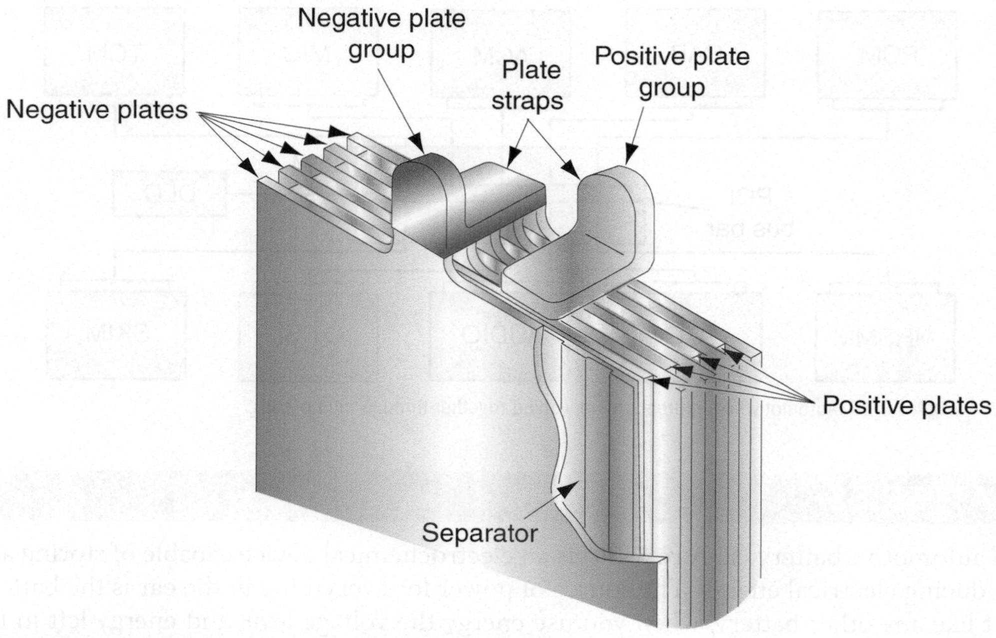

Figure 5-41 A battery cell consists of alternate positive and negative plates.

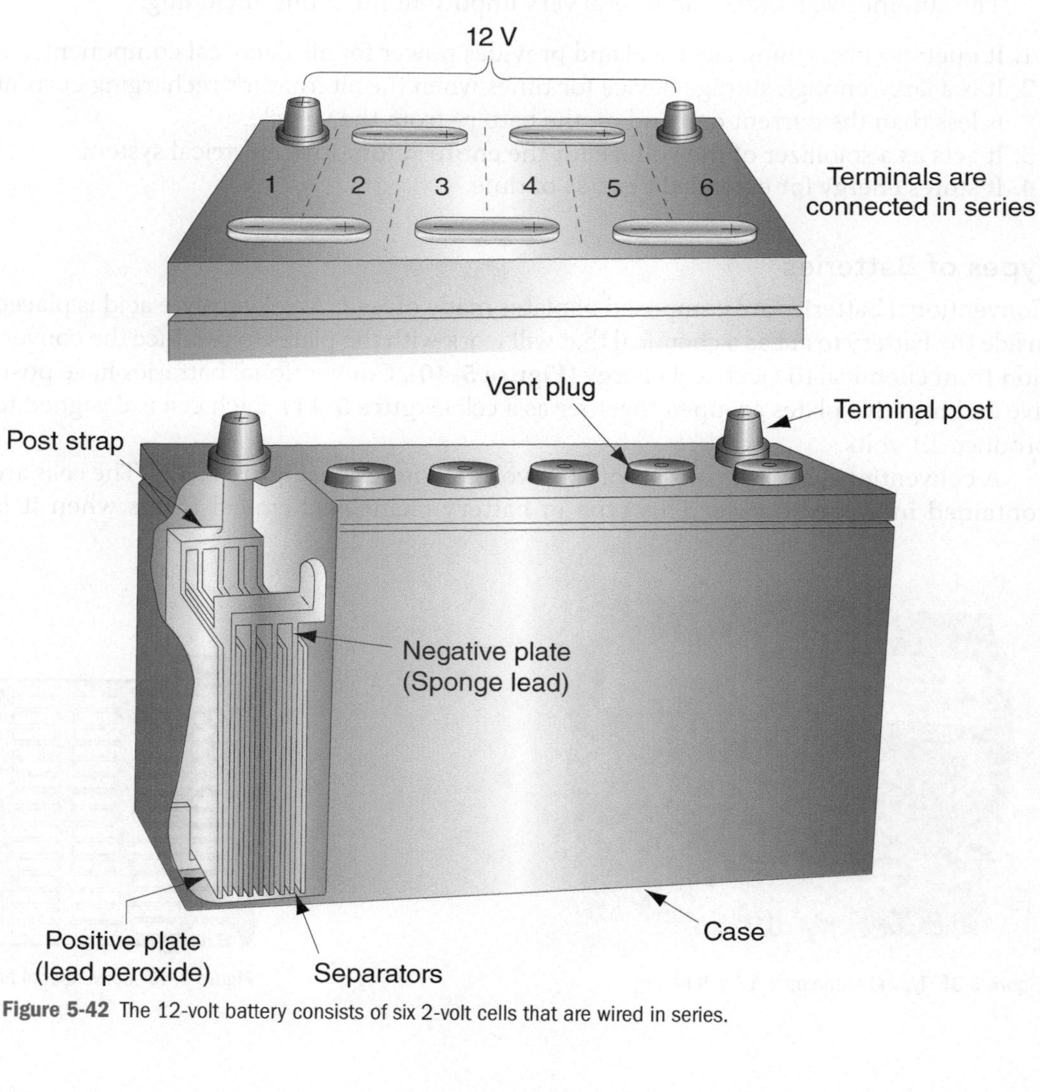

Figure 5-42 The 12-volt battery consists of six 2-volt cells that are wired in series.

discharging a recharging. A hydrogen gas is emitted when it is recharging. It is for this reason that you will find vents in different forms on the top of the battery (**Figure 5-43**). The solution of the electrolyte is a mixture of 64 percent water and 36 percent sulfuric acid. A chemical reaction takes place between the lead peroxide of the positive plates and the sponge lead of the negative plates. The electrolyte is the carrier that moves electrical current between the plates and the separators.

A battery's charge amount can be read by its **specific gravity**. The specific gravity of a battery is the weight of the fluid in comparison to the weight of water (for the same volume). Water has a specific gravity of 1.0 and anything that is less dense than water weighs less for the same volume. An example of this would be most lubricating oils. Most of the oils have a specific gravity of around 0.7. An automotive battery electrolyte solution has a specific gravity of about 1.265 when it is fully charged. The specific gravity of a battery can give you an indication of how charged a battery is, but it will not tell you if the battery can conduct electricity properly. The readings below are for a battery when the fluid is corrected to 70°F.

Specific gravity is the difference between the weight of water compared to a specific fluid.

Fully Charged:	1.265
75 percent Charged:	1.225
50 percent Charged:	1.190
25 percent Charged:	1.155
Discharged:	1.120 or less.

This conventional type of battery is very common and has only changed its technology slightly over the past 30 years, hence the term conventional. A **maintenance-free battery** (**Figure 5-44**) uses a different electrolyte solution containing lead calcium. A conventional battery vents hydrogen gas when it is recharging. Older designs had a lot of venting and after a period of time the fluid level would be low due to evaporation. A maintenance-free battery still vents hydrogen, but at a slower rate. Thus, much of the time a maintenance-free battery will be sealed and will not have removable caps to check the fluid level (**Figure 5-45**).

In theory, a **maintenance-free battery** requires no maintenance during its life.

Recombination and **absorbed glass mat (AGM)** batteries are the newest types of automotive batteries. Recombination batteries use a gel instead of an acid solution. This gel doesn't spill, and it doesn't splash around when the vehicle is driving over bumps. This type of battery is very common with racing vehicles due to its tough construction. The separators

Absorbed glass mat (AGM) batteries are becoming more common in small engine like motorcycles and garden tractors.

Figure 5-43 The vents of a conventional battery allow the release of gases.

Figure 5-44 Maintenance-free battery.

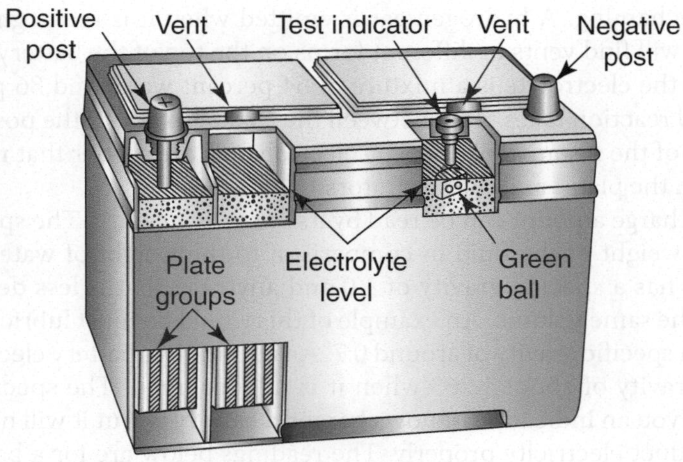

Figure 5-45 Construction of a maintenance-free battery.

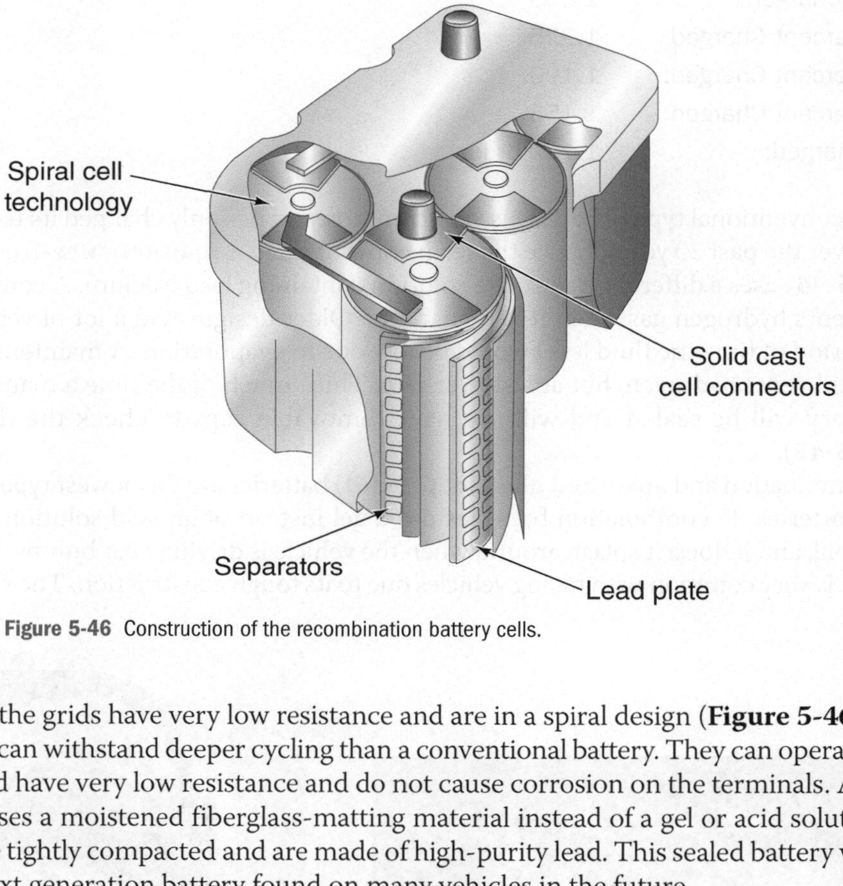

Figure 5-46 Construction of the recombination battery cells.

between the grids have very low resistance and are in a spiral design (**Figure 5-46**). These batteries can withstand deeper cycling than a conventional battery. They can operate upside down and have very low resistance and do not cause corrosion on the terminals. An AGM battery uses a moistened fiberglass-matting material instead of a gel or acid solution. The plates are tightly compacted and are made of high-purity lead. This sealed battery will likely be the next generation battery found on many vehicles in the future.

Nickel–cadmium (NiCad) and nickel-metal hydrid (NiMH) are used because of their high total output and battery life.

Hybrid electric vehicles (HEVs) typically use a **nickel–cadmium** (NiCad) or **nickel-metal hydrid** (NiMH) battery. These batteries are capable of high voltages and constant charging and recharging. Hybrid battery packs consist of many cells that combine to make a larger battery pack (**Figures 5-47** and **5-48**). Similar to a conventional battery where each 2.1-volt cell combines in series to 12.6 volts, except in a hybrid battery pack, the voltages are high, and a fluid is not used. The voltages on hybrid battery packs can range from 270 to 320 volts. High-voltage battery packs have more safety features than the lower voltage 12-volt batteries. The color orange is used on all high-voltage connectors and wires (**Figure 5-49**). This is to warn the technician not to touch these wires unless the

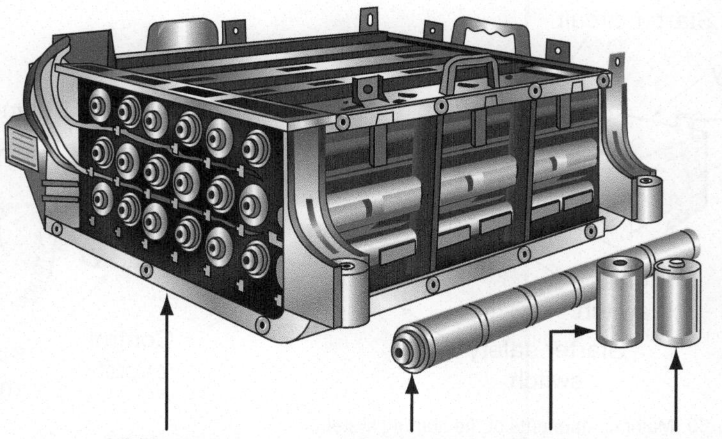

Figure 5-47 HV battery constructed of cylindrical cells.

Figure 5-48 Battery pack made of several prismatic cells.

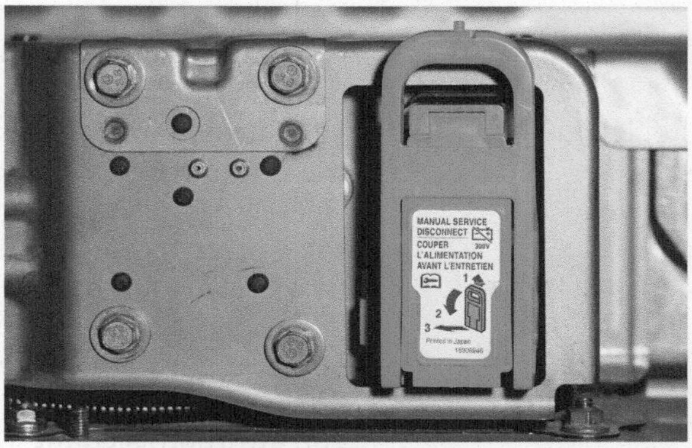

Figure 5-49 Service disconnect plug.

system is disconnected. Lower voltage hybrids (36-volt systems) use blue wires and connectors. These types of hybrids were only made for a few years and are not very common.

It is important to note that HEVs use the high-voltage battery for the traction and regenerative motors only. The remainder of the vehicle uses a standard 12-volt battery. The high voltage is unsafe and can cause an arc when a switch closes.

STARTING SYSTEM

The starting system consists of the following components (**Figure 5-50**):

Battery
Cable and wires
Ignition switch
Solenoid and relays
Starter motor
Flywheel or flexplate
Neutral safety or transmission range switch or sensor

The starting system works by converting electrical energy into mechanical energy. The internal combustion engine must be rotated before it will run under its own power.

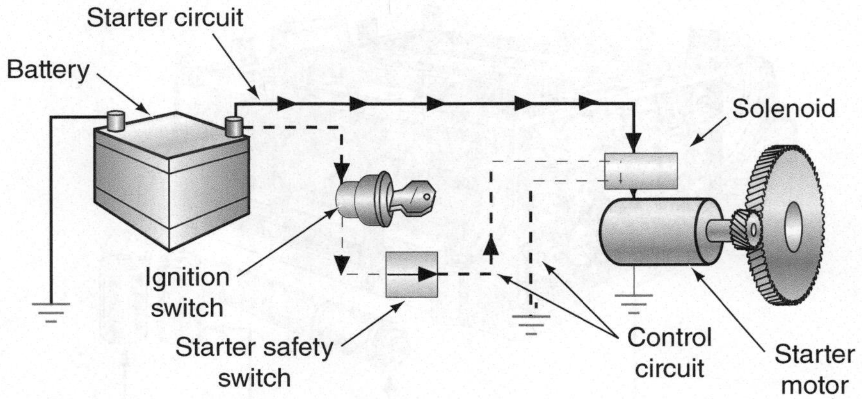

Figure 5-50 Major components of the starting system.

It will need to build up speed and momentum before the ignition and fuel system can successfully take over and run the engine. The starting system is a combination of mechanical and electrical parts working together to start the engine. The starting system is designed to change the electrical energy that is being stored in the battery into mechanical energy. For this conversion to be accomplished, a starter or cranking motor is used. The parts of a starter are shown in **Figure 5-51**.

The battery delivers electrical power to the starter solenoid when the driver turns the ignition key to the crank position (**Figure 5-52**). A starter rotates because of an electromagnet being energized, like a relay. The solenoid delivers battery power to the starter motor. The starter motor drive gear is engaged with the ring gear on the flywheel or flex-plate; the flywheel (or flexplate) is bolted to the engine's crankshaft. When the starter motor spins, it rotates the engine at about 200 to 250 rpm. This is fast enough to allow the four-stroke cycle to produce sustainable combustion and begin running under its own power. Many newer vehicles have computer modules and antitheft systems and auto start features installed on them. These computers are used to help control the engine's starting procedures. **Figure 5-53** shows a conventional starter circuit and **Figure 5-54** shows a computer-controlled starter circuit.

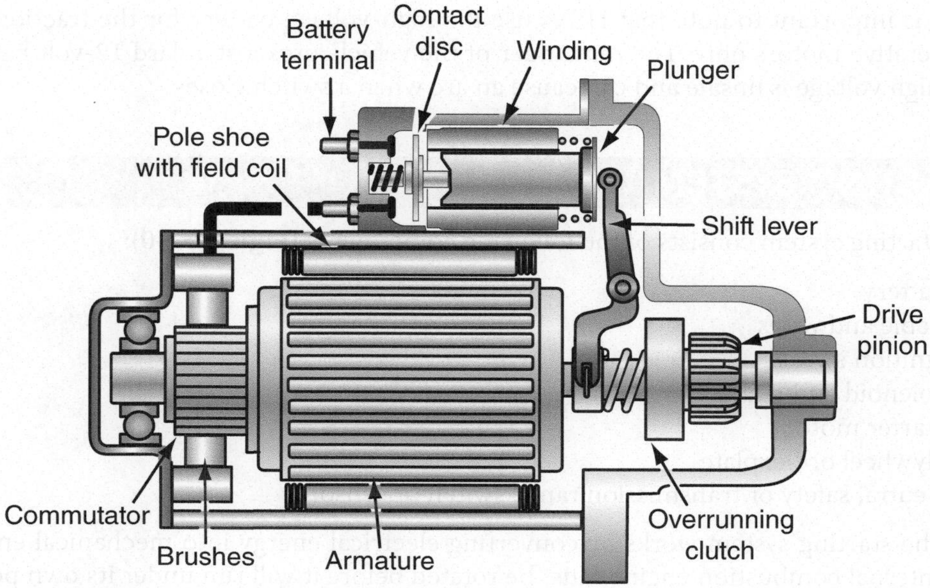

Figure 5-51 Starter and solenoid components.

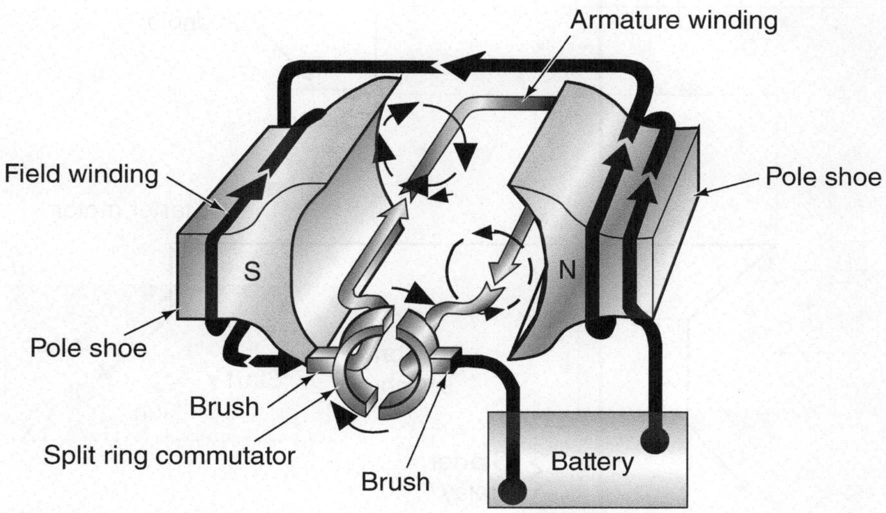

Figure 5-52 Simple electromagnetic motor.

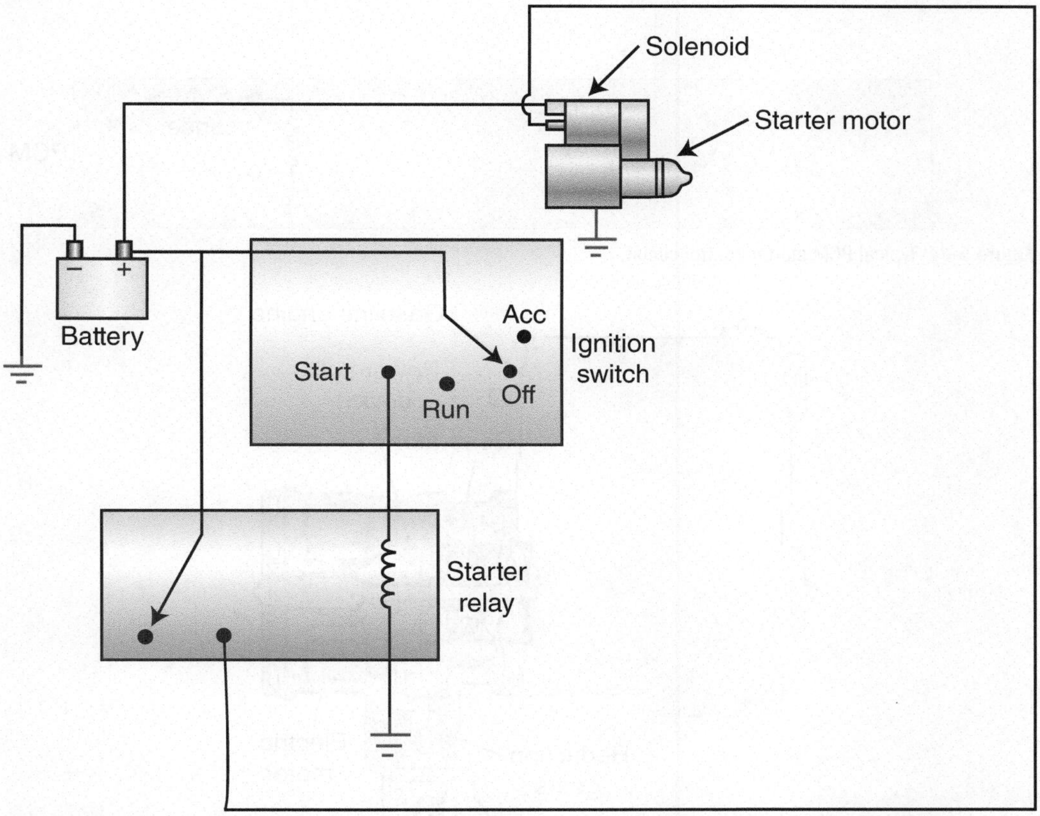

Figure 5-53 Starter control circuit using an insulated side relay to control current to the starter solenoid.

The flywheel is a heavy disk mounted at the end of the crankshaft to make the engine power pulses and vibrations more stable. It has a smooth surface in which the clutch assembly of a manual transmission is attached to. This smooth surface is like the surface of a brake rotor. A flexplate is a lighter version of a flywheel and is bolted to the torque converter of an automatic transmission. The ring gear around the circumference of a flywheel or flexplate allows the starter motor drive gear to turn the engine. Some flexplates and flywheels are used as part of a sensor to determine the speed of the engine.

The starting system of an HEV is different from the normal starting system (**Figure 5-55**). The high-voltage battery pack is typically used to start the gasoline engine.

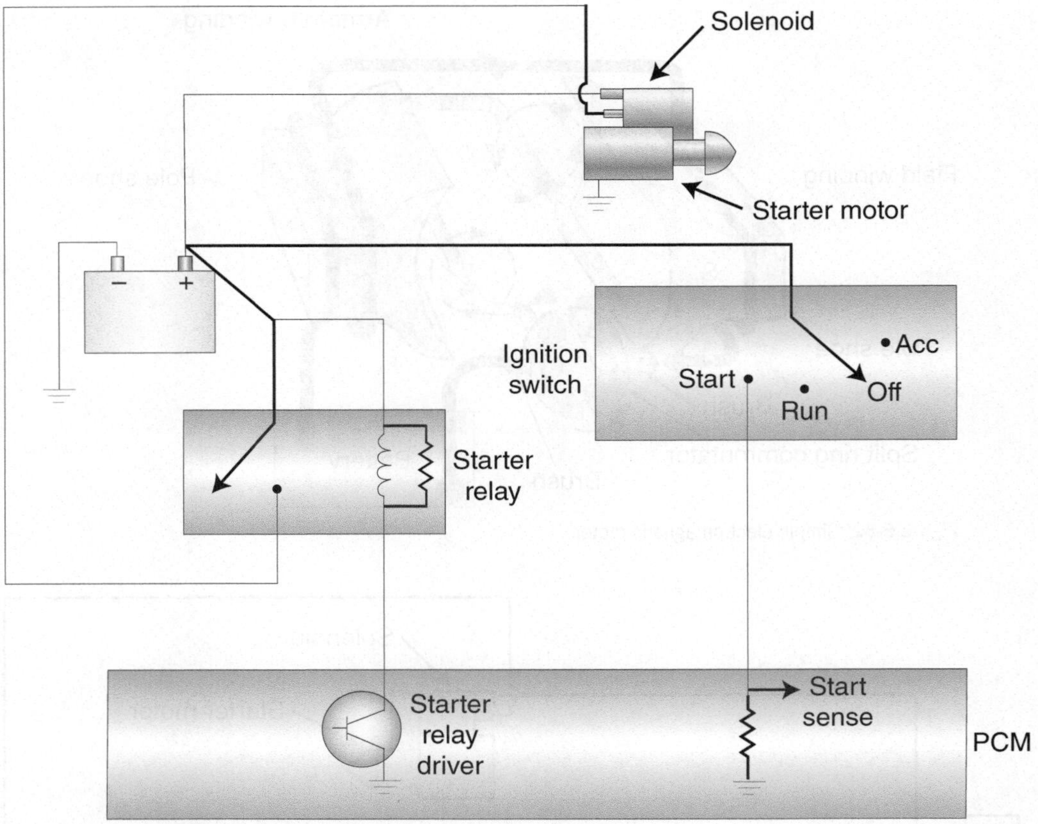

Figure 5-54 Typical PCM starter control circuit.

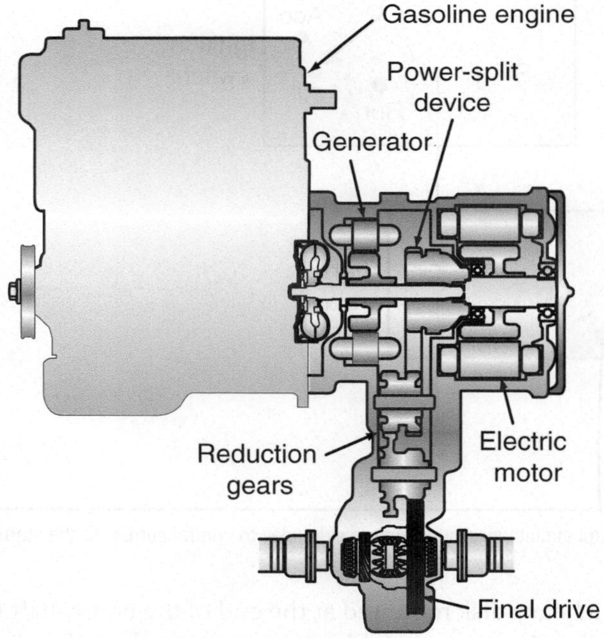

Figure 5-55 Hybrid power system with gasoline engine and electric propulsion motor.

The same set of electric motors that is used for propulsion is used in conjunction with a gear set to start the gasoline engine. An HEV's main computer will start the engine when the conditions are met. These conditions may be low voltage on the HEV battery pack, A/C demand, electrical demand, or acceleration demand. The starter on some modern gas and diesel engines are controlled by the computer, like an HEV. Special care must be taken when servicing these vehicles.

Charging System

The charging system consists of the following components (**Figure 5-56**):

Battery
AC generator
Drive belt
Cables and wiring harness
Ignition switch
Charge indicator light (or gauge)
Relays and fuses

An alternator converts mechanical energy into electrical energy. As the battery is used to power the vehicle, it will drain. The more power that is requested of the battery, the more the alternator will output and recharge the battery. And although modern vehicles use a computer to control the charging rate, the current charge rate on older systems was self-regulating.

A BIT OF HISTORY

In the early days of the automobile, the vehicle did not have a starter motor. The operator had to use a starting crank to turn the engine by hand. Charles F. Kettering invented the first electric "self-starter," which was developed and built by Delco Electric. The self-starter first appeared on the 1912 Cadillac and was a combination starter and generator.

An **alternator** (also called an AC generator) spins by the connection of the drive belt. Like a starter and a relay, it uses an electromagnet. A starter draws current into it and converts it to mechanical motion. An alternator draws a low amount of current and the mechanical movement moves the magnet through a magnetic field. This multiplies the low current coming into the alternator into a higher current. The voltage on an alternator is limited to about 14.4 volts. Anything higher than that would cause the battery to overcharge. The voltage output of the alternator must be slightly higher than the battery's

The proper name for an **alternator** is AC generator.

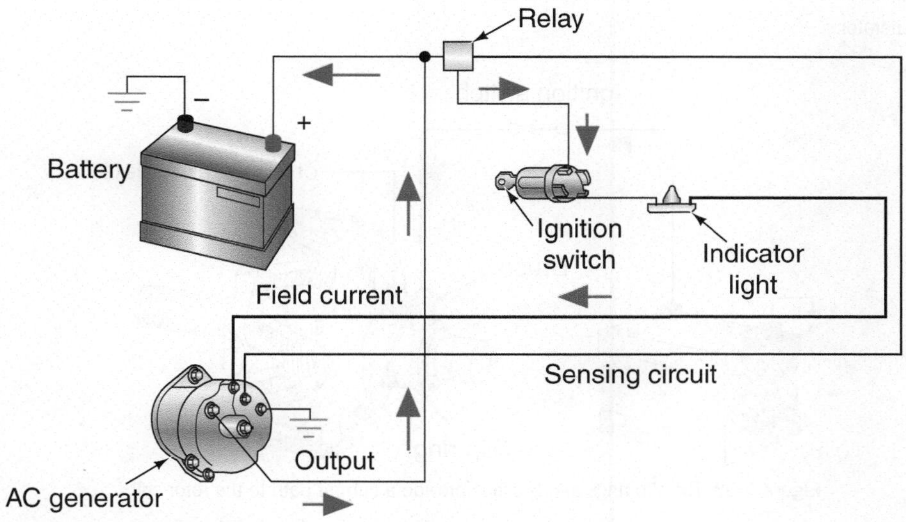

Figure 5-56 Current flow when the system is operating.

The **brushes** and **rotor** of an alternator work together to create the magnetic fields necessary to convert mechanical energy to electricity.

A **stator** is the place where the rotating magnetic fields are converted into AC current.

voltage. The difference in voltage is what allows the current to flow. Without this difference the battery will not be recharged. Alternators have many parts (**Figure 5-57**).

The flow of electricity starts at the battery and goes through the ignition switch, then into the brush (**Figure 5-58**). The **brushes** are sitting on the slip rings (which is a moving part). The current goes through the slip ring and into the **rotor,** thus creating a rotating magnetic field (**Figure 5-59**). The current leaves the rotor through the other slip ring and brush, and then heads to ground or the voltage regulator.

The rotating magnetic field then cuts through the windings of the **stator** (**Figure 5-60**). An easy way to remember these terms is that the rotor rotates, and the stator is stationary. The stator develops electricity because of the rotating magnetic field. There are three stator windings. The stator windings produce an alternating current. Since there are three stator windings, the voltage that is produced is three-phase voltage (**Figure 5-61**). The alternator develops an alternating-current voltage. But the battery can only be charged

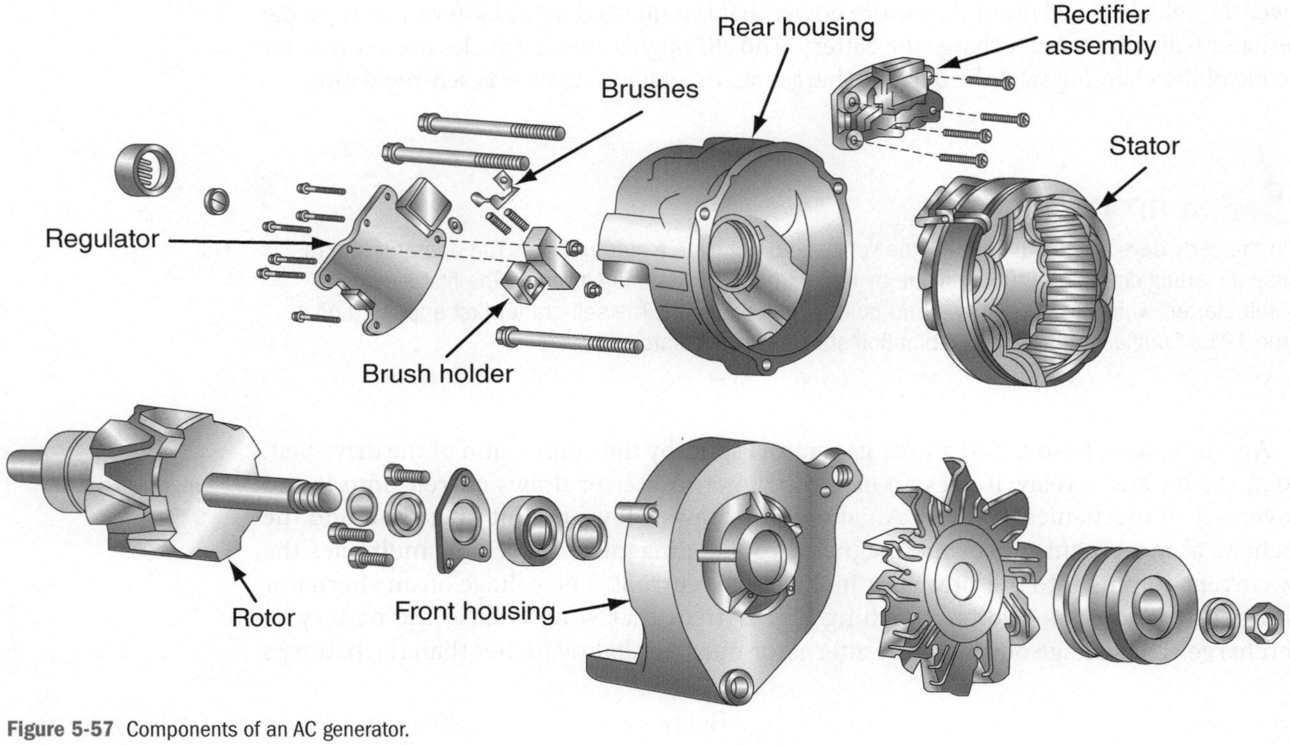

Figure 5-57 Components of an AC generator.

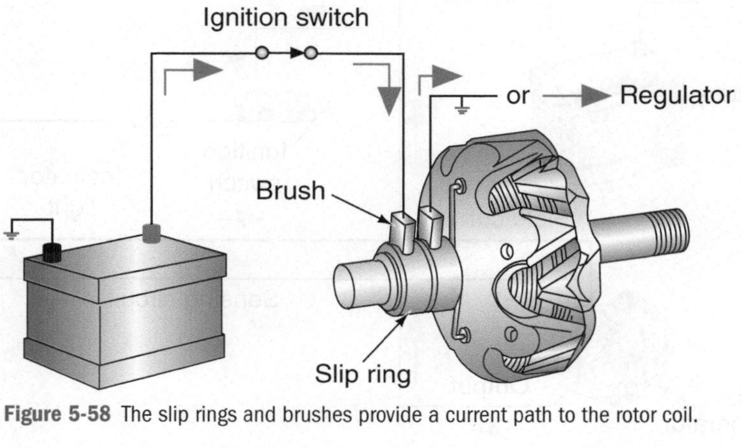

Figure 5-58 The slip rings and brushes provide a current path to the rotor coil.

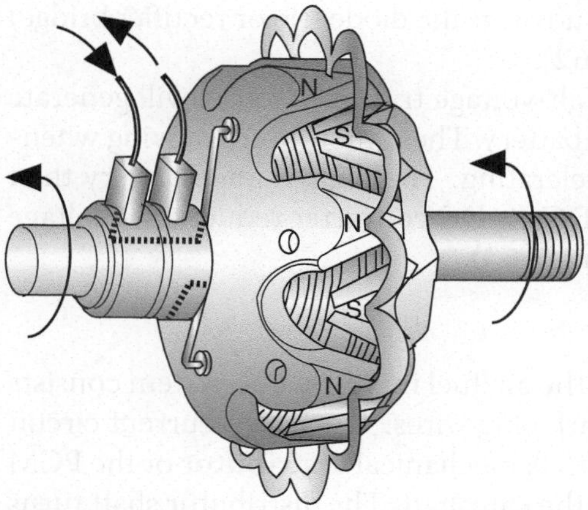

Figure 5-59 The north and south poles of a rotor's field alternate.

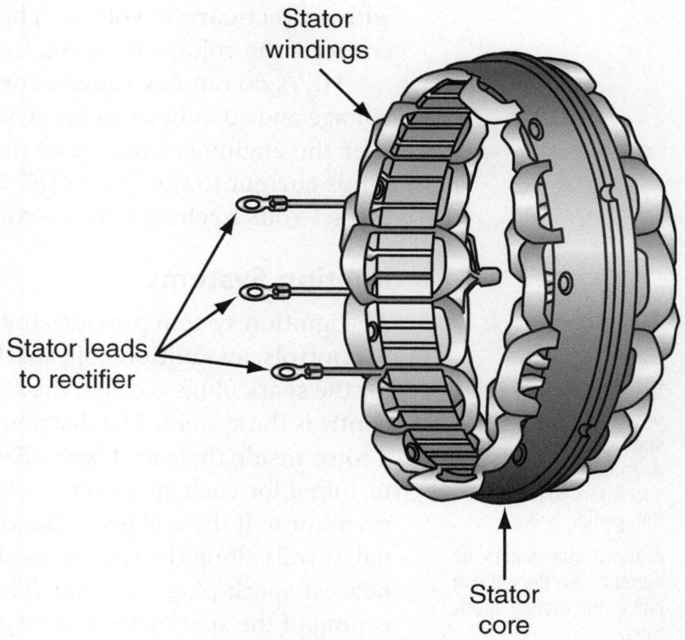

Figure 5-60 Components of a typical stator.

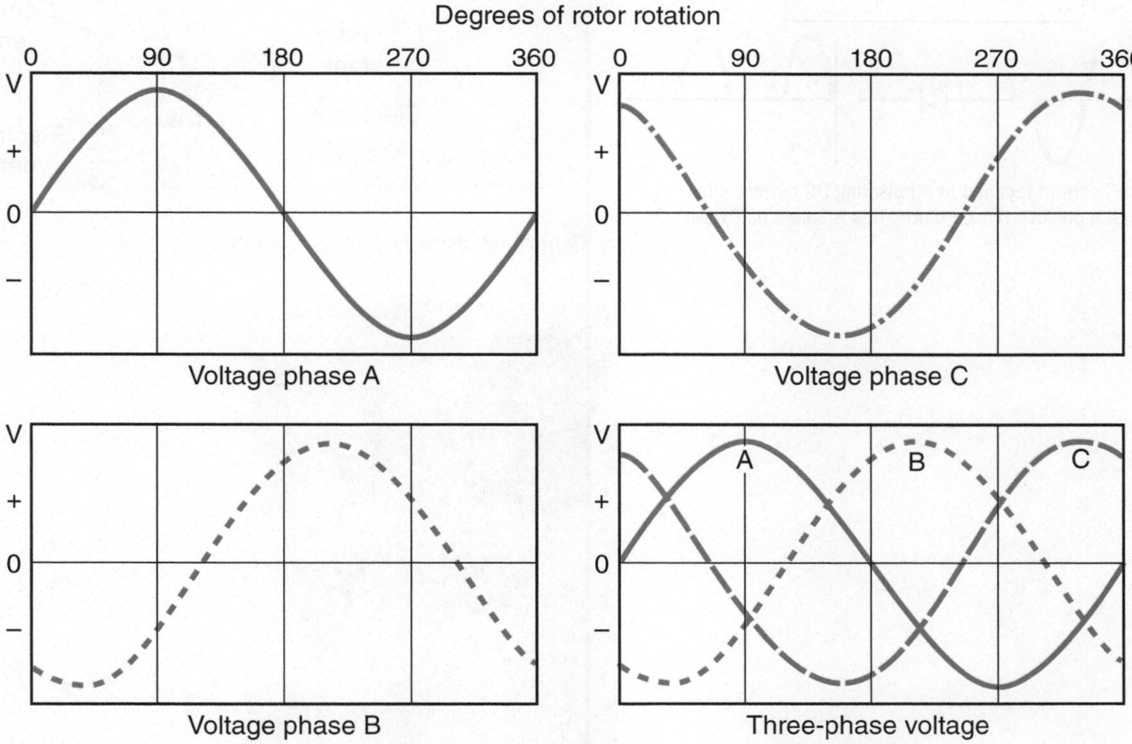

Figure 5-61 The voltage produced in each stator winding is added together to create a three-phase voltage.

with a direct-current voltage. The rectifier (also known as the diode trio or rectifier bridge) converts the voltage from AC to DC (**Figure 5-62**).

HEVs do not have alternators. Rather the high-voltage traction motors will generate voltage and current to recharge the high-voltage battery. These motors are moving whenever the engine is running or the vehicle is decelerating. The high-voltage battery then sends current to the DC-to DC-converter. The DC-to-DC converter reduces the voltage to 14.4 volts to charge the 12-volt battery (**Figure 5-63**).

Ignition Systems

The vehicle's **secondary circuit** consists of the spark plugs, plug wires, distributor, and the secondary windings of the ignition coil.

A **distributor** directs the current from the ignition coil to the correct spark plug.

The ignition system provides the spark to ignite the air/fuel mixture. The system consists of controls, an ignition coil, spark plugs, and spark plug wires. The heavy-current circuit for the spark plugs is called the **secondary circuit.** A mechanical **distributor** or the PCM controls the system. The distributor is geared to the camshaft. The distributor shaft turns a rotor inside the cap (**Figure 5-64**). The cap has an internal ignition coil terminal and a terminal for each spark plug. The rotor points at each spark plug terminal once in each revolution. If the coil is discharging, then the current moves through the cap's coil terminal, travels along the rotor's conductor, and bridges the gap between the rotor tip and the nearest spark plug terminal. The distributor has an advanced mechanism that adjusts timing of the spark based on engine speed and load.

Many distributor systems have gone the way of carburetors. They have been replaced with electronic systems that are faster and more accurate in controlling ignition spark. Some heavy equipment vehicles still use distributors.

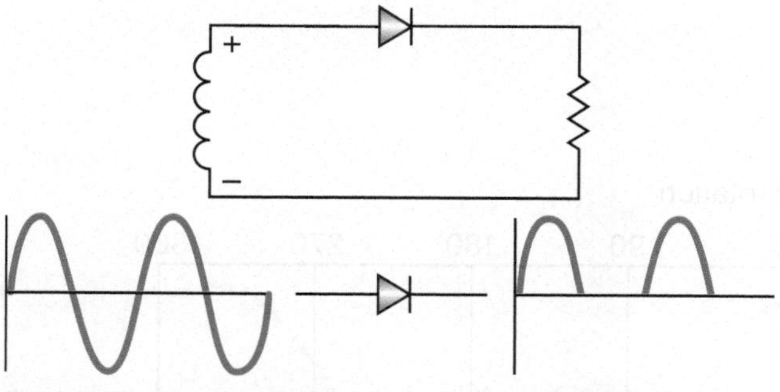

Figure 5-62 AC current rectified to a pulsating DC current after passing through a positive-biased diode. This is called half-wave rectification.

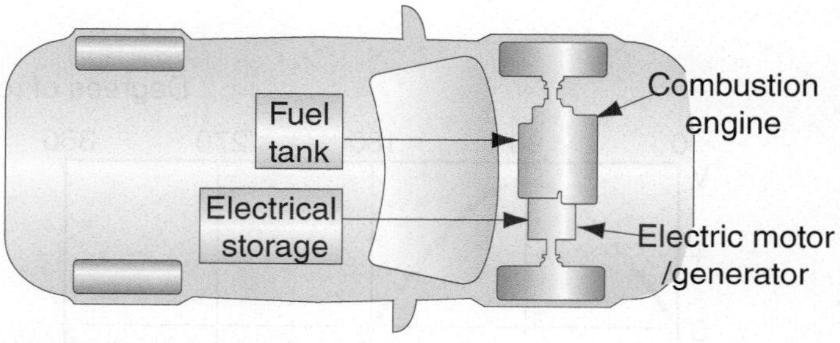

Figure 5-63 Parallel hybrid configuration.

Figure 5-64 An electronic distributor used on older vehicles.

Modern vehicles use electronics to control the ignition spark. The distributor has been replaced completely with an electronic package that determines crankshaft and camshaft position, engine load, engine temperature, and engine speed to determine **ignition timing** within each individual cylinder (**Figure 5-65**). This same data is also used to fire the fuel injectors and vary the time of injector opening.

The **ignition timing** must be adjusted because the load of the engine and the speed of the pistons change as the vehicle is driven.

The PCM performs all the calculations necessary based on engine, transmission, and other sensor input. This provides maximum fuel mileage, cleaner emissions, and more available power for any given engine application. Some of the vehicle sensors and actuators are shown in (**Figure 5-66**). Some of the terms and explanations listed there may be unfamiliar, but they are only offered at this point as an indication of the electronics used on an engine. Also, this figure does not list all the electronic devices, and some manufacturers use different terms for the same type device. Some of the devices are not used on some vehicles.

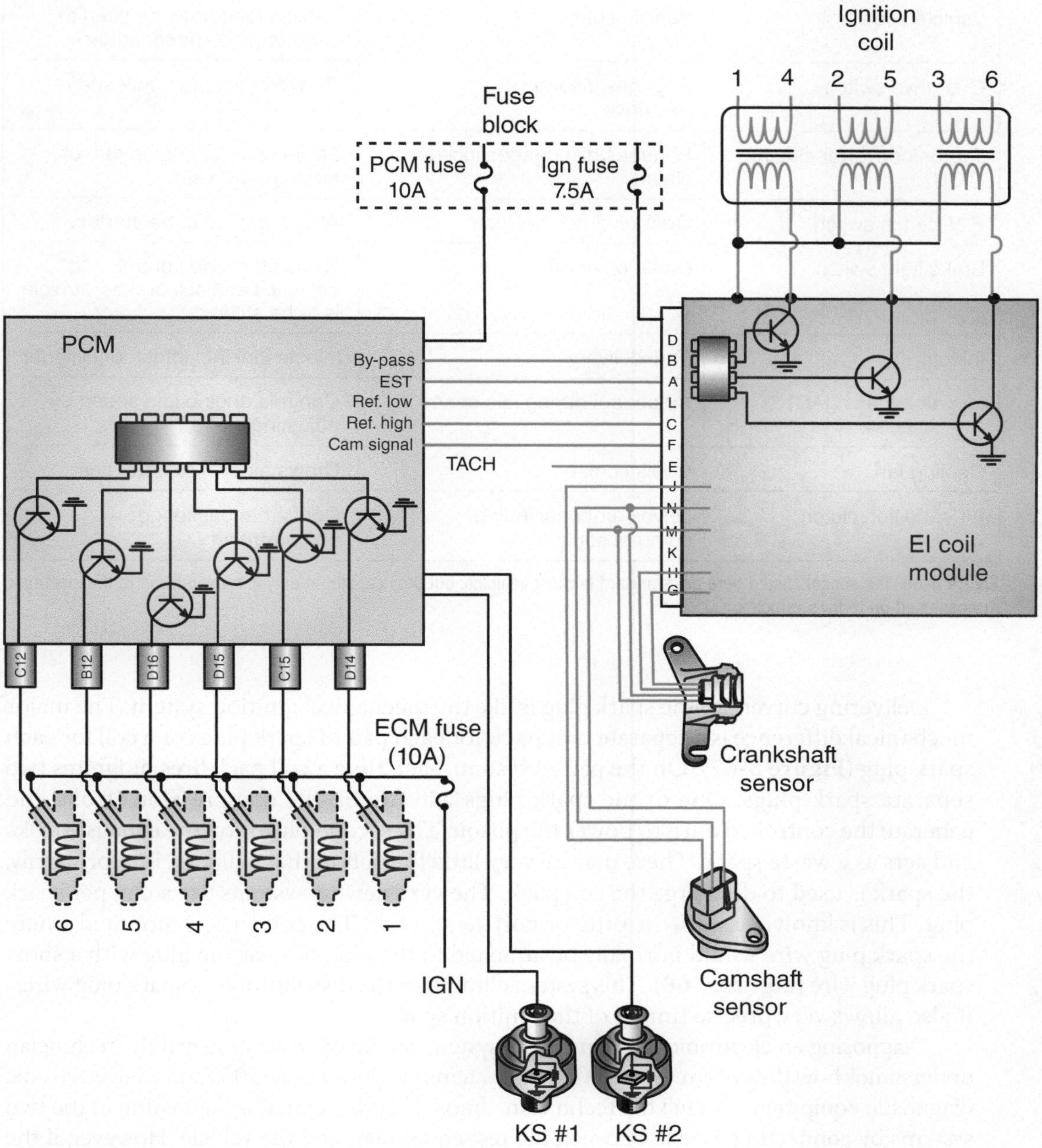

Figure 5-65 Note the sensors and control circuits in this ignition system.

Sensor/Actuator	Information	Determines
Throttle position (TPS)	Throttle opening	Driver's speed intention
Oxygen sensor (O_2)	Oxygen in exhaust gases	Air/fuel mixture ratio
Manifold absolute pressure (MAP)	Pressure in intake manifold	Engine load
Mass air flow (MAF)	Airflow in grams per minute	Engine load (may be used with or in place of MAP)
Engine coolant temperature (ECT)	Engine temperature	Engine temperature
Air charge temperature (ACT)	Intake air temperature	Intake air temperature
Vehicle speed	Vehicle speed	Vehicle load (may be used to send data to speedometer)
Overdrive switch	Engages/disengages overdrive	Provide better fuel mileage
Torque converter clutch (TCC)	Engages/disengages lockup clutch	Locks/unlocks engine output directly to driveline
P/N; clutch switch	Gear or clutch position	Allows engine to be started
Brake light switch	Brake on or off	Turns off cruise control, activates antilock brakes, vehicle is to be slowed
Injector	Fuel delivery	Injects fuel into intake or cylinder
Idle air bypass (IAC)	Air control during idle speed	Controls engine idle speed by adjusting air flow
Cooling fan	Cools coolant	Draws air over radiator fins
Air condition clutch	Drives air-conditioning compressor	Cool air to passenger compartment

Figure 5-66 The sensors listed here are standard in most vehicles, but they constitute only a small portion of the electronic devices used on today's vehicles.

A **coil pack** consists of a primary circuit control package and an ignition coil mounted over the package. The coil is in the secondary circuit.

Coil-on-plug means the coil is attached directly to the spark plug terminal.

Coil-near-plug has a coil for each spark plug and is mounted near the spark plug with a short connecting spark plug wire.

Delivering current to the spark plug is like the mechanical ignition system. The major mechanical difference is a separate **coil pack** for each pair of spark plugs or a coil for each spark plug (**Figure 5-67**). On the paired system, each time a coil pack fires, it ignites two separate spark plugs. One of the spark plugs will fire on the compression stroke and generate the controlled burn to power the piston. The second plug is on the exhaust stroke and acts as a waste spark. There may be very little fuel left in that cylinder, but, primarily, the spark is used to discharge the coil pack. The very newest systems use a coil per spark plug. This is known as **coil-on-plug** or **coil-near-plug.** The coil pack is mounted where the spark plug wire would normally be attached to the plug or near the plug with a short spark plug wire (**Figure 5-68**). This system eliminates the distributor and spark plug wires. It also allows very precise timing of the ignition spark.

Diagnosing an electronic ignition or fuel system can be done accurately if the technician understands how the system works. With the systems produced since 2000 and later, electronic diagnostic equipment can tell the technician almost everything that is happening in the two systems by connecting a cable between the test equipment and the vehicle. However, if the technician does not understand the operation of the systems, all that data is worthless.

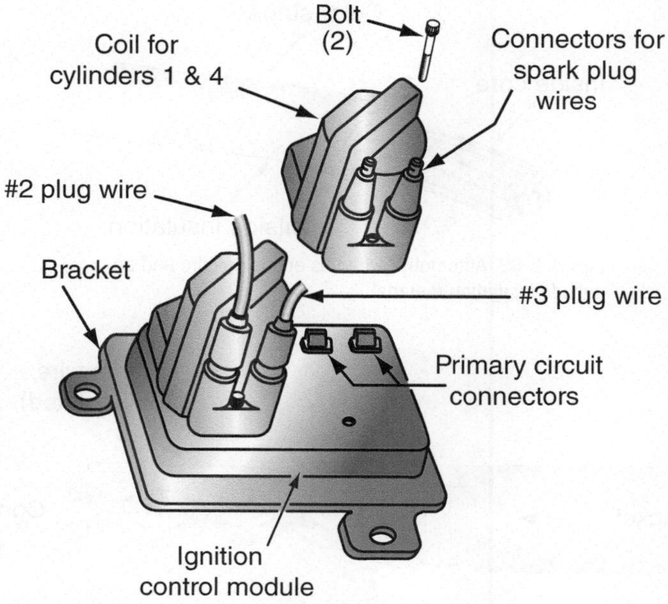

Figure 5-67 Two coil packs supply secondary voltage to this four-cylinder engine. Note the companion plugs, #1 and #4.

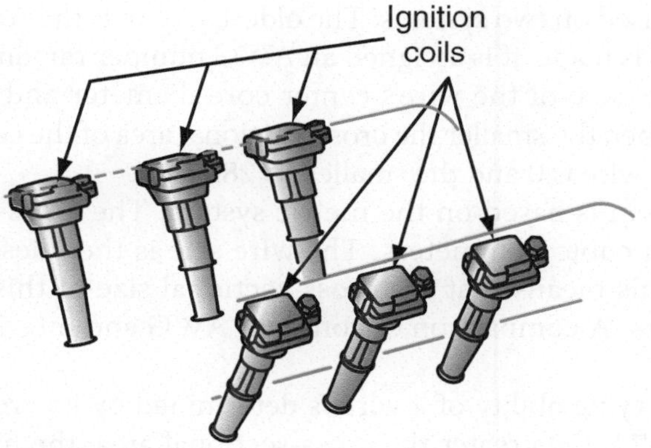

Figure 5-68 Typical coil-on-plug layout.

ELECTRICAL WIRE AND FASTENERS

A technician will be required to make repairs to the car's electrical system. This often means replacing damaged wires. Numerous types of wire and electrical fasteners are available to make these repairs. Wires, terminals, and connectors are the conductors of the electrical current within a vehicle.

When replacing a wire, always use a replacement that is at least as heavy as the one to be replaced. Automotive wire has two basic parts: an inside core of metal that conducts current flow and an outside of insulation material (**Figure 5-69**). The wire depicted in Figure 5-69 is a stranded cable. There are several wires laid side by side or woven around each other to form the cable. This increases the surface area of the cable, thereby reducing its resistance to electrical current. Solid wire cables are available, but generally only stranded wire is used in automotive applications. The sizing of the wire is the same regardless of whether it is stranded or solid. Computer communication circuits use a **twisted pair** wire. This helps reduce the chance of interrupting the signal from nearby wires (**Figure 5-70**).

Twisted pair wires are used as communication wires for computer circuits.

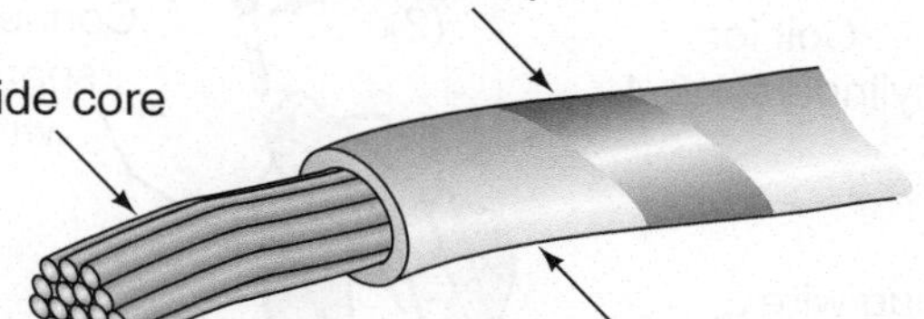

Figure 5-69 Automotive wire has an inside core and an outside insulation material.

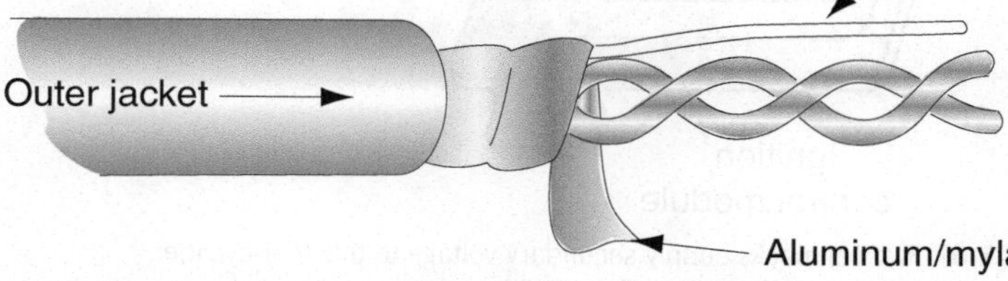

Figure 5-70 Twisted or shielded wire used in computer circuits.

American Wire Gauge: A standardized wire gauge system to determine the diameter of a round wire.

Wire sizes are based on two systems. The oldest system is the **American Wire Gauge (AWG).** When wire is made, it is assigned an AWG number ranging from 0 to 28. These numbers specify the size of the wire's center core diameter and cross-sectional area. The higher the number, the smaller the cross-sectional area of the center core. The largest cross-sectional core wire is 0 and the smallest is 28.

The newer system is based on the metric system. The cross-sectional area of the wire is measured in cubic millimeters. The wire size is then designated by a number such as 0.5 mm. This means that the cross-sectional size of this wire core measures 0.5 cubic millimeters. A comparison of common AWG and metric wire sizes is shown in **Figure 5-71**.

The current-carrying ability of a wire is determined by its cross-sectional area. As shown in **Figure 5-72**, the greater the cross-sectional area, the more current can flow through the wire. The smaller the cross-sectional area, the more resistance exists and

Wire Gauge Sizes	
Metric Size (cubic millimeters)	AWG Size
0.5	20
0.8	18
1.0	16
2.0	14
3.0	12
5.0	10
8.0	8
13.0	6
19.0	5

Figure 5-71 Comparison of AWG and metric wire core sizes.

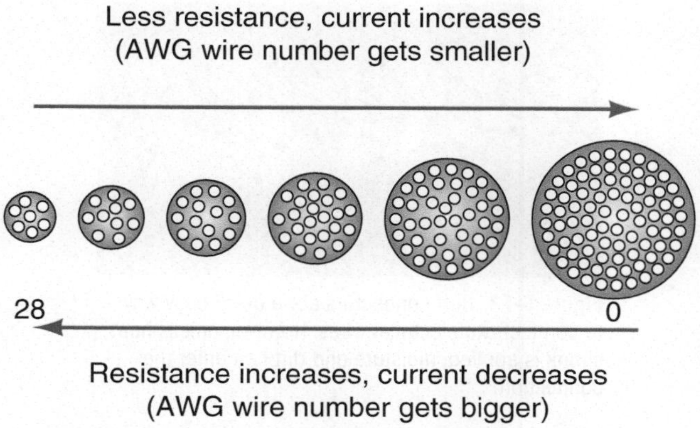

Figure 5-72 Current flow increases and resistance decreases as the wire core gets bigger.

the less current can flow. The wire size of a circuit is carefully determined by automotive electrical engineers to meet specific requirements. For example, large battery cables are often no. 1 or no. 2 gauge because a larger flow of current is needed. On the other hand, a no. 16 gauge (1.0 mm^2) wire is often used with a headlight because less current is needed. As you can see, when a wire is replaced, it must be the same size as the one it is replacing.

New wire is sold in spools that are clearly identified by AWG or metric size. Wire sizes in various circuits on the automobile are often identified on the wiring diagrams. Two procedures can be used to discover the size of an unidentified wire. First, a wire gauge is available that has a series of graduated holes. Numbers are printed on the wire gauge to correspond with the hole sizes. The center core of the wire to be identified is inserted into the various holes to determine the closest fit. If a wire gauge is not available, the diameter of the center core can be measured with an outside micrometer and then matched to a wire table to determine the size of the wire.

Wire insulations are colored so that they can be traced on circuits on the automobile. Colors on the wire are also printed on the wiring diagrams for the car. You should always try to use the same color as the original when replacing wire.

CONNECTORS

Soldering

Connecting two wires or a terminal to a wire is best accomplished using electrical-type **solder.** This type of solder adds little or no resistance to the conductor and is resin based. Regular solder is acid based, so the soldered area is cleaned as the solder is being applied. However, acid-based solder will damage the surface of the wire and may add resistance to the circuit. Only resin-based solder should be used for soldering wires together. Once soldered, the connection should be covered with heat-shrink material instead of tape whenever possible.

Solder: A low melting alloy used for joining copper wires

Butt Connectors

Butt connectors are used to connect two wires together (**Figure 5-73**). They are quick and easy to install but may add resistance to the total wire. Very-low-amperage circuits such as computer data lines should be soldered.

Butt connector: A round crimp connector with a crimp on each end

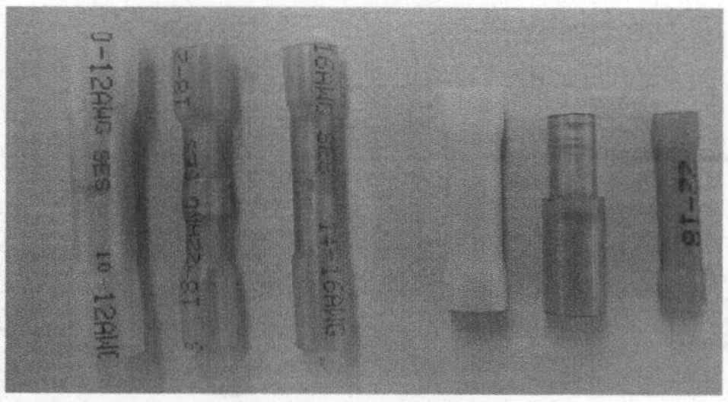

Figure 5-73 Butt connectors are a quick easy way to connect two electrical wires. However, unless heat shrink is applied, moisture and dirt can enter the connection.

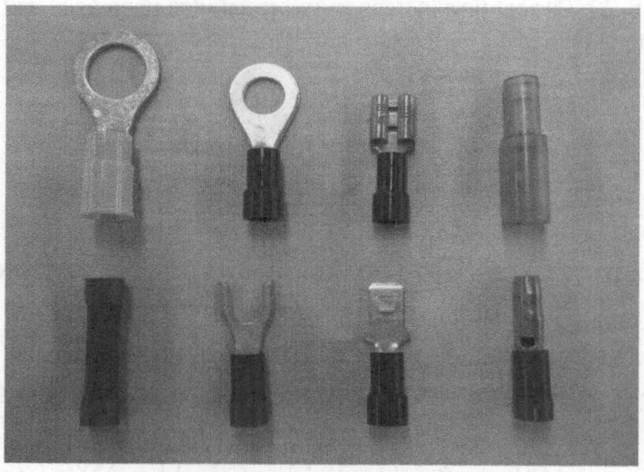

Figure 5-74 Terminals are installed at the end of the wire to form a way of connecting the wire to a permanent terminal like those on a switch or are mated with another terminal to connect two wires.

Crimp-on **terminal** and **butt connectors** are called "solder-less connectors."

Terminal Connectors

The **terminal connector** is a device fastened to the end of a wire that allows the wire to connect to or disconnect from a component (**Figure 5-74**). This allows components to be unplugged from the circuit without damaging the conductor or component. Most positive conductor terminals are covered in some type of insulator or installed in connectors to prevent shorting the circuits.

Molded and Shell Connectors

Molded and shell connectors come in many assorted sizes, shapes, and materials, depending upon their designed use. In areas of the automobile where there are several connections in one place, the designers have, in many cases, installed connectors that can be matched according to shape, size, and color (**Figure 5-75**). This reduces the possibility of crossing circuits. Connectors can be made for a single wire or any number of wires installed into one shell or mold.

Pig tail: A connector with a loose wire on one end

Weather-tight connector: A connector that can with stand exposure to extreme temperatures, moisture, harsh engine compartment fluids and chemicals

A molded connector imbeds wires inside a formed plastic or rubber coating (**Figure 5-76**). The wires cannot be removed from this connector. Replacement connectors with short wires can be purchased to replace the damaged connector. This part is called a **pigtail** and is made to exactly replace the original part, including the color of the wires.

Shell connectors are usually made of hard-formed plastic (**Figure 5-77**). With the right tool, damaged terminals can be removed, replaced, and reinserted into the shell. Shell connectors may have some type of seal to block moisture from entering the system. This is called a **weather-tight connector.** Most electronic circuits are connected this way.

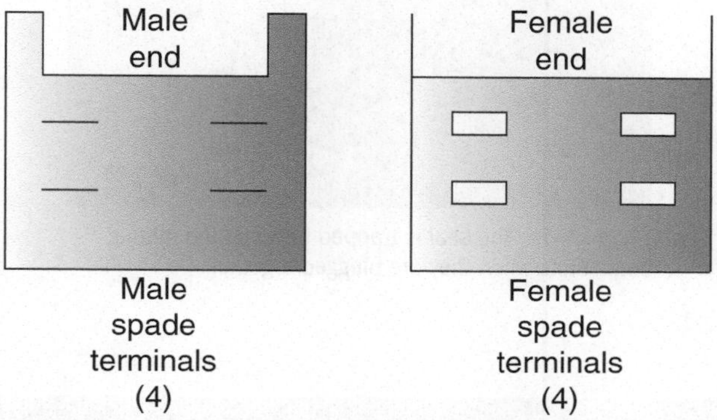

Figure 5-75 The male end will slide and lock into the female end.

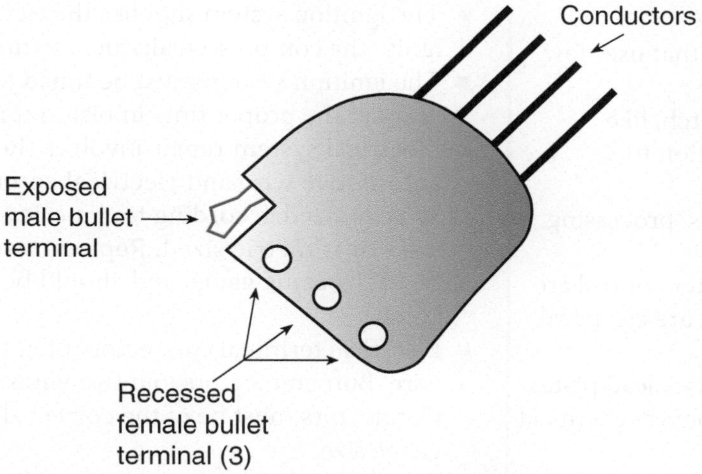

Figure 5-76 This is a typical molded connector. Note the bare terminal on the left. This is a four-wire trailer connector.

Figure 5-77 Multiple-circuit, hard shell connector.

Weather-Tight Connectors

Most of the electrical connections on late-model vehicles use the weather-tight design. This type of connector may be molded or a shell with rubber seals. As the two connectors are plugged together, the seal on one side is trapped and clamped between the two (**Figure 5-78**). The clamped seal prevents moisture and dirt from entering the connector.

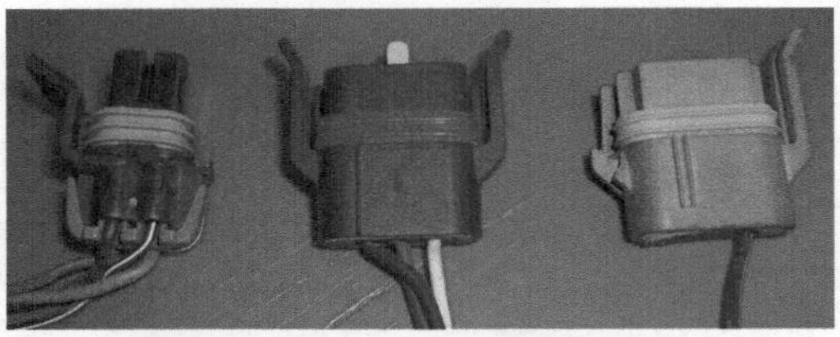

Figure 5-78 The seal is trapped between the mating connectors when they are plugged together.

SUMMARY

- There are several types of switches used in an automobile. Some examples of these are NO, NC, SPST, and SPDT.
- A relay is an electromagnetic switch that uses low current to control high current.
- A solenoid is an electromagnetic switch, like a relay, that causes a mechanical motion to occur.
- A computer system consists of inputs, processing, and outputs.
- Many of the circuits that use computer control are very similar in function to those that are electrical and not computer controlled.
- A conventional automotive battery uses lead plates and an acid electrolyte solution to convert chemical energy to electrical energy.
- The starting system converts electrical energy into mechanical energy.
- The charging system converts mechanical energy into electrical energy.
- The ignition system supplies the electrical spark to ignite the compressed air/fuel mixture.
- The ignition system must be timed to deliver the spark at the proper time in piston travel.
- Electrical system repair involves the use of automotive wire and electrical connectors. Wire is sized according to American Wire Gauge sizes or is metric sized. Replacement wires must be of the same gauge and should be of the same color.
- Electrical terminal connectors fit at the end of the wire. Butt connectors join two wires together. Connectors must be of the correct shape and wire gauge size.
- Weather-tight connectors are used to protect the interior of the connector from moisture.

REVIEW QUESTIONS

Short Answer Essays

1. List the process of how a switch works in a simple circuit.
2. Explain the difference between a normally open and a normally closed switch.
3. Describe the operation of a circuit with and without a relay.
4. Explain how a battery produces electrical energy.
5. Describe the difference between a conventional and maintenance-free battery.
6. List the components of a starting system.
7. Explain the flow of current in a starting system.
8. List the components of a charging system.
9. Explain the flow of current in a charging system.
10. Explain why a hybrid electric vehicle does not have an alternator.
11. Why must the correct wire gauge size wire be used when replacing wire?
12. Explain the purpose of soldering wires and terminals.

Fill-in-the-Blanks

1. A relay is an electromagnetic switch that consists of a coil that has very _______________ resistance and draws very _______________ current.

2. An automotive battery is a(n) ______________ device.

3. When a battery is recharging, it produces ______________ gas.

4. The electrolyte solution used in a conventional battery is ______________ percent water and ______________ percent acid.

5. A fully charged battery has a specific gravity of ______________.

6. The ______________ is the part of the starting circuit that acts like a relay but also has a mechanical movement to engage the starter.

7. The ______________ is a direct-current motor.

8. The ______________ generates electricity that is used to recharge the battery.

9. The ______________ converts AC to DC voltage inside the alternator.

10. An alternator converts ______________ energy into ______________ energy.

11. The two types of electrical connectors are ______________ and ______________.

12. The ______________ the inner core of a wire, the greater the resistance.

Multiple Choice

1. Which of the following describes the function of the starter solenoid?
   A. An electromagnetic device that transfers high current to the starter motor
   B. An electromagnetic device that moves and engages the starter
   C. Neither A or B
   D. Both A and B

2. The rotor in an alternator:
   A. Is a stationary component.
   B. Controls the speed of the alternator.
   C. Is touching the rectifier.
   D. Is a rotating component.

3. A hybrid electric vehicle uses a DC-to-DC converter to:
   A. Charge the 12-volt battery.
   B. Turn the headlights on.
   C. Power the alternator.
   D. Start the engine.

4. A flywheel is mounted to the:
   A. Crankshaft.
   B. Camshaft.
   C. Cylinder block.
   D. Transmission housing.

5. A fully charged conventional automotive battery has a specific gravity of:
   A. 1.460.
   B. 1.265.
   C. 1.225.
   D. 1.190.

6. The battery stores energy in which form?
   A. Chemical
   B. Electrical
   C. Mechanical
   D. Kinetic

7. Two technicians are discussing relay function. *Technician A* says that a relay can be controlled by a switch. *Technician B* says that a relay can be controlled by a computer. Who is correct?
   A. A only
   B. B only
   C. Both A and B
   D. Neither A nor B

8. A recombination battery:
   A. Uses a gel instead of acid.
   B. Contains flat cells and plates.
   C. Can operate only in the upright position.
   D. None of the above.

9. The source of electrical power for everything in the car is the:
   A. Alternator.
   B. Starter.
   C. Computer.
   D. Battery.

10. A computer circuit consists of:

    A. Inputs.

    B. Outputs.

    C. Processing device.

    D. All of the above.

11. Automotive wire size is being discussed.
    Technician A says the AWG number includes the insulation around the wire.
    Technician B says the larger the center core, the smaller the AWG number.
    Who is correct?

    A. A only

    B. B only

    C. Both A and B

    D. Neither A nor B

12. Connectors are being discussed.
    Technician A says terminals are used to connect two wires.
    Technician B says a butt connector can do the same job as soldering two wires. Who is correct?

    A. A only

    B. B only

    C. Both A and B

    D. Neither A nor B

CHAPTER 6

ACCESSORY LIGHTING AND ELECTRICAL SYSTEMS

Upon completion and review of this chapter, you should understand and be able to:

- Describe the different and overlapping features of convenience and comfort systems.
- Understand the warning systems found on most vehicles.
- Explain electrical circuits used with instrumentation and driver display.
- Describe interior and exterior lighting system.
- Explain the general operation of typical accessory systems.
- Describe the components and operation of the restraint system.

Terms To Know

Adjustable pedal system (APS)
Backlit
Bayonet
Bimetallic
Center high-mounted stop light (CHMSL)
Communication bus
Daytime running lights
Dimmer switch
Emergency flasher
Flasher
Global Positioning Satellite (GPS)
Grounding switch
Heads-up-display (HUD)
Instrument panel
Integrated circuit
Light-emitting diodes (LEDs)
Master switch
Microprocessor
Multifunction lever
Snapshot
Transponder
Trigger
Warning lights

INTRODUCTION

The increasing use of electricity and electronics in the automobile has provided an almost unlimited range of accessories and informational devices. The car or truck can be equipped with systems that memorize seat positions for a certain driver, adjust individual temperature controls, provide antitheft protection, and remote start the engine and heater on a chilly morning. More and better protection devices for the passengers are possible with electronics. With these advances comes an increasing need for electronically proficient technicians who can diagnose and repair sophisticated electronic circuits quickly and accurately. To the customer, the convenience of electronics brings the possibility of hugely increased repair costs.

This chapter, and its corresponding chapter in the Shop Manual, is not designed to make the entry-level technician an electrical or electronic repair expert. Instead, the chapters will give the reader a working knowledge of the various auxiliary systems' operations, some basic diagnostic testing, and the ability to perform basic maintenance.

Because some of the systems share common components and functions, a failure of one system or a component may affect engine and transmission operation.

PASSENGER COMFORT AND CONVENIENCE

Passenger comfort devices are items such as heat, air conditioning, heated and cooled seats, and others that make a passenger's ride in the vehicle enjoyable. They also keep the driver alert. Warning and information systems provide the driver with engine and vehicle operation data and indications of problems before they get too serious. Systems such as lighting, rearview mirrors, rear-window defrosters, and automatic ride control give the passenger a comfortable ride while the driver has an almost unrestricted view of the actions around the vehicle.

Convenience features are those that make the comfort items easier or simpler to use. Remote electric controls and heated outside mirrors eliminate the need to lower the windows on cold, rainy days to wipe them off or move them by hand. The driver can change seat positions during a long drive to relieve or at least postpone fatigue. There are driver seat systems offered today that will massage the driver's lower back. In the past, if a driver got lost, the solution was to use a map or ask directions. Today, the punch of a button displays the vehicle's position on a small screen or digital display. On some vehicles, a driver can call a live person for help with directions or in emergencies. In the old days, if a car broke down on a lonely, dark road, the driver and passengers either slept in the vehicle or walked.

Some modern vehicles have lane departure warning systems, adaptive cruise control, blind spot warning technology, and enough computer control for the vehicle to parallel park itself. Many vehicles can change the traction control system, wipers, and headlights when the weather changes. A few vehicles can even brake for you when an object is suddenly in front of the vehicle that could cause an accident. In fact, many of the modern technologies on vehicles are safety added features and accident avoidance. Vehicle communications is the area where the most technology has been developed. Systems can include cell phone and Bluetooth connections, internet browsing, traffic management, and so on.

All the features on a modern vehicle place a tremendous load on the battery, electrical system, and alternator. They also require technicians with good training and the intuition to know how to operate and repair many different systems.

INSTRUMENTATION

Instrument Panel: The surface in front of the driver, on which are arranged an array of dials, lights and gauges.

A **light-emitting diode (LED)** is a diode that gives off light when current passes through it.

Shop Manual page 222

Gauges and lights on the dash and information center are the means used to convey information to the operator. They are usually in an area of the dash referred to as the **instrument panel**. The gauges may be analog or digital. Analog gauges use a needle moving over a fixed scale to display the data much like the hands on a clock (**Figure 6-1**). Other systems use numbers or digits that change based on sensor input (**Figure 6-2**). The numbers are displayed by **light-emitting diodes (LEDs)**. Most analog gauges on modern vehicles are digitally controlled. Many consumers prefer to see the analog gauges.

Some vehicles use an illuminated band or bands against a fixed and printed scale. The bands act like an analog gauge but are LEDs that are turned on and off by a computer module based on incoming data. The bands can be programmed as different colors to attract the driver's attention. For instance, the band representing vehicle speed can be programmed to turn red for any speeds over 70 miles per hour (mph). It should be noted that certain red and amber warning lights are present in every dash regardless of the type of instruments used. There are more colored warning lights in today's instrument panel than just 10 years ago. They will be discussed in the section describing warning systems. When the ignition is turned on, all the dash warning lights should illuminate to perform what is called a bulb check. After the engine starts the warning bulbs should turn off.

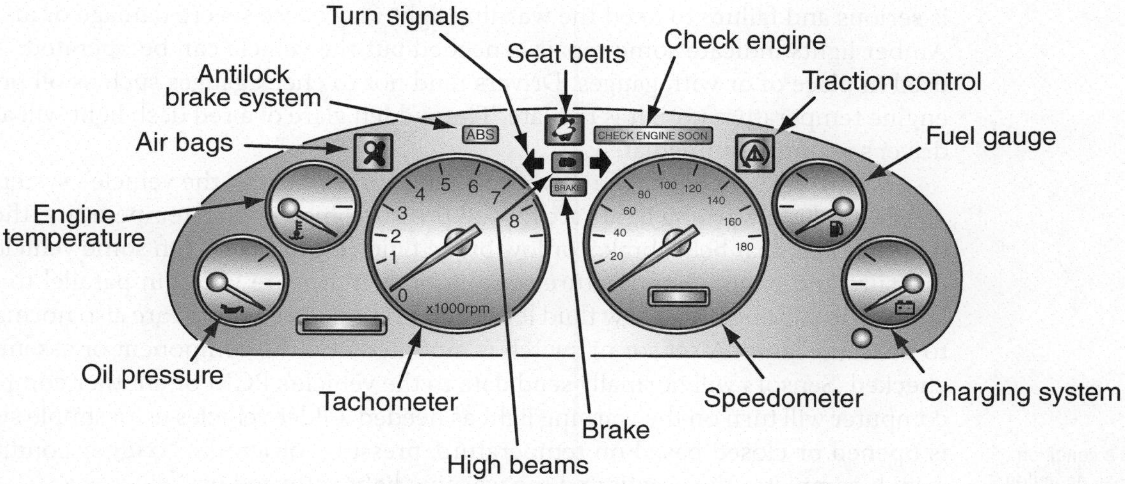

Figure 6-1 Typical gauge and warning indicator layout of an instrument panel.

Figure 6-2 All of the warning lights should illuminate during the bulb check.

Figure 6-3 Some of the same data shown on the instrument panel can be projected onto the windshield with HUD.

The **heads-up-display (HUD)** was adopted from military fighter aircraft. The information most commonly required is projected onto the windshield in the driver's line of sight (**Figure 6-3**). The information usually consists of vehicle and engine speed, oil and temperature conditions, and warning lights. In the future, the driver could possibly scroll through other data using steering wheel–mounted controls. The object of HUD is to keep the driver's eyes and attention on the environment around the vehicle while providing necessary vehicle data. Some instrumentation systems can display data on an overhead console.

The term **heads-up-display (HUD)** was adopted because the driver does not have to look down to read operational data.

WARNING SYSTEMS

Warning and information systems (see next section) would appear to be basically the same; however, they have distinct purposes. Warning systems are found on every vehicle and alert the driver to a fault or failure that could damage the vehicle or cause a loss of control. Information systems generally provide data that can assist the driver in some way.

Shop Manual page 222

Red and amber lights are used because of their visibility, relationship to danger, and eye-catching properties. Many times, gauges supplement the lights. Red means the danger is serious and failure to heed the warning light may cause severe damage or an accident. Amber lights indicate some repair is needed but the vehicle can be operated. Lights are used in place of or with gauges. Drivers tend not to check gauges such as oil pressure or engine temperature until it is too late. The sudden glare of a red dash light will attract the driver's attention immediately.

A warning light alerts the driver to a fault somewhere in the vehicle's systems.

Typical red warning lights are low oil pressure, high engine temperature, the alternator, door ajar, seat belts, brake, or low brake fluid (**Figure 6-4**). On some vehicles, the oil pressure and engine temperature sensors or switches are wired in parallel to the same lamp. The parking brake, low fluid level, and brake failure sensors are also normally wired to the same lamp. A sensor or switch is mounted into the component or assembly being checked. Sensors will normally send data to the vehicle's PCM or another computer. The computer will turn on the warning light as needed. Older vehicles use a simple switch that is opened or closed based on temperature, pressure, or another **trigger** condition. The switch completes the circuit and the warning light is turned on.

A **trigger** is a condition that will cause an action to be taken (e.g., low oil pressure or fire a spark plug).

The switch in some warning circuits is a one-wire **grounding switch** (**Figure 6-5**). In an oil pressure warning circuit, the switch may be held normally open (NO) or normally closed (NC) by oil pressure. When pressure drops below a specific setting and allows the switch to function, the circuit activates. The alternator light is switched on by low voltage in the sensing circuit. Mechanical switches may be used to control other warning lights.

A **grounding switch** is a switch that, when closed, completes the circuit to ground.

In the case of **Figure 6-6**, one warning light is used and will illuminate when any one of the three switches close. A grounding switch at the parking brake pedal usually switches on the red brake warning light (Figure 6-6). A fluid sensor in the brake's reservoir may also turn it on, as well as a switch in the brake's hydraulic system. Some vehicles have a system like this whereas others have one lamp for each circuit.

There are several amber-colored lights that alert the driver to a condition that is not life threatening or a problem that will not result in severe damage to the vehicle if corrective action is taken soon (**Figure 6-7**). One of the most prominent is the malfunction indicator light (MIL). It is switched on by the powertrain control module (PCM) based on data received from the engine and transmission sensors and is commonly known as the ***check engine light***. The light is a warning that some condition is present that interferes with the proper operation of the vehicle and should be corrected at the earliest possible time. This allows the driver to safely drive the vehicle until a repair shop can be reached. The lens of this light may reflect the words ***"Service Engine Soon"*** or ***"Check Engine."*** Many vehicles have another amber light that indicates a service is due or the light may indicate which service is required. Labeling on these lights may be ***"Change Oil"*** or

Figure 6-4 This lit warning light alerts the driver to a high engine temperature. Note the temperature gauge.

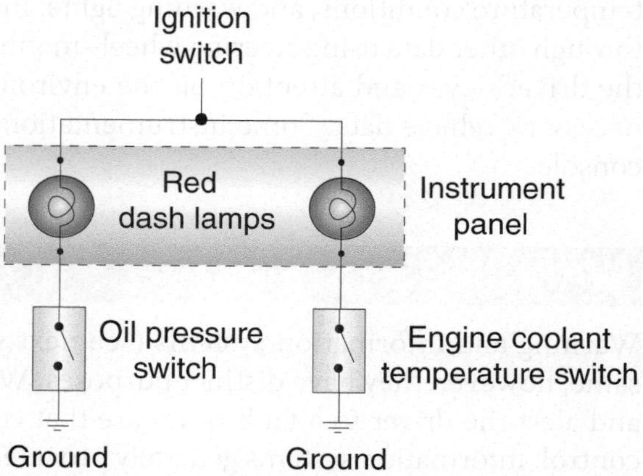

Figure 6-5 This red warning light may be switched on by either the oil pressure or the engine coolant temperature.

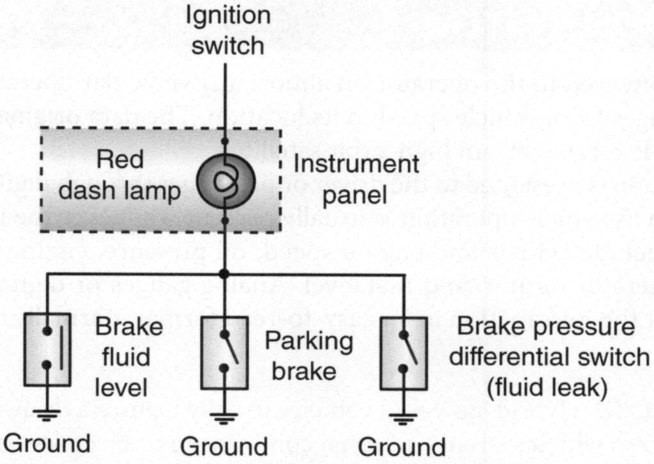

Figure 6-6 A typical red brake warning light may be switched on by a switch on the parking brake lever or pedal, low brake fluid in the reservoir, or a leak in one of the brake hydraulic circuits.

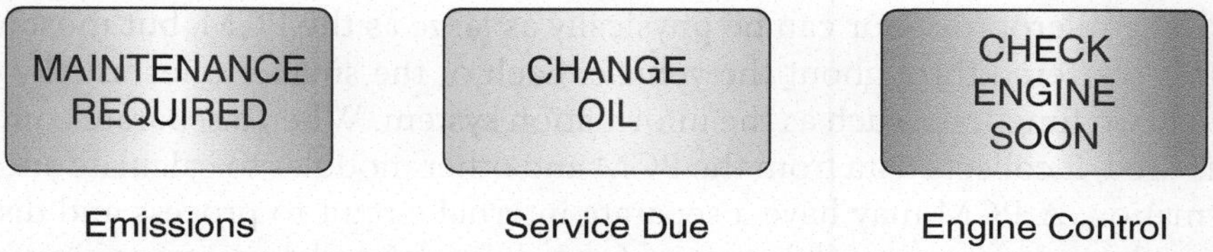

Figure 6-7 Three possible maintenance reminder lights. The one on the far right is required by the EPA and deals with engine and transmission controls that affect exhaust emissions.

"***Check Emissions***." Another prominent amber light is the warning light for the ABS. A lit ABS warning light indicates two things: the ABS has a fault or the ABS is deactivated. The regular (service) brakes will work fine and there are no safety concerns as long as the driver understands that the ABS will not function in an emergency stop. Most vehicle electronic systems that have some type of self-diagnosis capability can alert the driver to the fault. In many cases, the technician will need a scan tool and other electronic testers to isolate and diagnose the problem.

SERVICE TIP Although the amber ABS light or the red brake warning light may not indicate an immediate danger to the driver, if both lights are lit there is a severe problem involving the service brakes *and* the ABS. The fault(s) must be corrected before the vehicle is allowed on the road even if the brakes seem to be working.

Other amber lights include low fuel, traction control, overdrive off, electronic suspension, low tire pressure, supplemental restraint systems (SRSs) or airbags, shift now, and others either present today or that will exist soon. Most warning lights are backed up by an audible alert, usually a sharp beep or buzz that may not be heard if the radio or road noise is above "normal." There are also light and audible reminders of keys left in ignition, a seat belt not fastened, and a door or trunk ajar, and the newer systems with mapping directions can even alert you that a stop sign or obstruction is ahead, and it is time to apply the brakes.

INFORMATION SYSTEMS

Information is conveyed to the operator on almost any vehicular operation desired. The data available ranges from vehicle speed to its location. The data originates from sensors on the vehicle or is received from high-orbit satellites.

The information is presented to the driver or passenger through digital or text display. Information such as vehicle operation is usually present whenever the ignition switch is on. The basics include vehicle and engine speed, oil pressure, engine temperature, air conditioning generator output, and fuel level. Analog gauges or digital readouts using LEDs can present the information in an easy-to-read format, normally a number.

AUTHOR'S NOTE Hybrid-powered vehicles usually feature a dash-mounted LED screen that shows which system—internal combustion or electric—is powering the vehicle.

Microprocessors: An electronic component that is used by a computer to do its work.

A **communication bus** has all sensors, computers, and modules connected to shared circuits. In this manner, each computer and module can monitor data.

Newer vehicles may use a section of the PCM or a separate computer module to compute information. The 2008 model cars may have up to 80 **microprocessors** or computer modules. The microprocessor can be physically as large as the PCM, but most are small black boxes scattered throughout the vehicle. Each of the smaller ones usually controls only one or two functions, such as the information system. When a separate computer or module is used, it collects data from the PCM and other modules to calculate and present the information. A PCM may have a separate internal circuit to process and display the information; however, it is typical that separate modules share all data via a **communication bus.** A communication bus is like a computer network in a large office. The newest onboard communication protocol is *CAN* for *controller area network.* Each computer is wired into the network and can communicate and share data with all other computers in the system. Additional information about the vehicle and its surroundings may include:

- Outside and inside temperatures
- Continuous or accumulated fuel mileage
- Time of departure and expected time of arrival
- Miles left on the remaining fuel
- Mileage completed or left on this trip
- The location of the vehicle at any single time
- How to get from here to there
- Road conditions and suggested detours

Some of this information is displayed continuously or at the touch of a button (**Figure 6-8**). The operator can select the desired information and even select USC or metric measurements. Like operating data, this information can be displayed by analog or digital method, but digital is the most common. Many options offered on the early 2000 high-line models have become standard features on most vehicles by 2013.

Figure 6-8 An overhead console module usually displays "nice to know" information that is not essential to vehicle operation.

One option that is becoming more common came from military and space exploration. The **Global Positioning Satellite (GPS)** system is a network of satellites deployed in orbit around the earth. A vehicle-mounted receiver translates data from several satellites and calculates the exact location of the vehicle within a few feet. The resulting display provides the vehicle's position in an easy-to-read and easy-to-understand format. Using this information and an onboard mapping system, the operator can plan a route to an unfamiliar destination, navigate around an unexpected detour, or find out how to get back on route.

The satellite network known as the **Global Positioning Satellite (GPS)** system consists of hundreds of satellites that communicate with one another and transmitters and receivers on the Earth.

The future implications of GPS are not easy to define. The GPS is used by manufacturer-sponsored roadside assistance options. The OnStar® system, which is standard on many modern General Motors vehicles, allows the driver to request the vehicle's doors be unlocked (if proper ID is presented), provides breakdown assistance via cell phone, and alerts the OnStar network in the event of a vehicle emergency. In case of an airbag incident, an OnStar operator will attempt to contact the vehicle driver and offer assistance in notifying emergency personnel. This system, known as automatic crash response (ACR), is available because of signals sent from the vehicle via satellite and GPS locators to alert the system's human operators.

More sophisticated systems are in the research and developmental stages that will provide even more assistance or information with a lower consumer cost. At some point soon, repairs to this type of system will be routine provided the shop and the technicians have spent the time and money to continuously upgrade their equipment and training. This upgrade applies to all electronic systems now available or that will be available on future vehicles.

Many manufacturers of large trucks have a vehicle/dispatcher satellite transmitter/ receiver. If the truck breaks down, the driver calls the dispatcher and gives the code for that truck. The maintenance department communicates via satellite with the truck's PCM. The PCM transmits operating data and **snapshot** data back via the satellite. The technician selects the most probable repair parts based on that data, proceeds to the downed vehicle, and makes the necessary repairs. Vehicle down time is greatly reduced.

A **snapshot** is stored data that can be retrieved and studied later. It is usually data of a particular event at a particular time, such as a misfiring cylinder under heavy load. It may be referred to as a "movie."

Shop Manual
page 222

INSTRUMENT CLUSTER OPERATION

The instrument panel houses most of the gauges and warning lights. Most informational data can be displayed in an overhead console or near the midpoint of the dash (**Figure 6-9**). This keeps the small space in the panel clear for essential operating data. A printed circuit board, or **integrated circuit,** is located behind the instrument panel. This board supplies the necessary circuits between the gauges and warning lights to their individual sensors or computers. A printed circuit board is used to reduce or eliminate the number of wires needed to connect all the different components. A printed circuit board is basically a very early, crude model of an integral circuit. It has been replaced by much better versions that are smaller and capable of many more circuits.

An **integrated circuit** is a system of many transistors and other electronic devices on a special board. There may be thousands of those devices on one board.

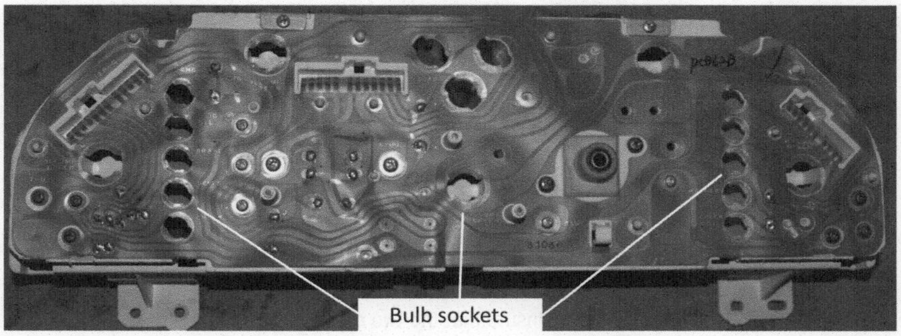

Figure 6-9 Printed circuit boards are used in instrument clusters. The thin plastic and copper circuit covers the back of the cluster and provides connections for all of the gauges and lights.

The panel's power and all other electrical connections to the remainder of the vehicle are made using one or two large, multiple-pin connectors. Large, molded, or shell terminals are plugged into the connector on the vehicle side. The panel side of the connector is an integral part of the board. The conductors are downsized and attached to the board's circuits using small soldered connections. Most integrated circuit boards are built off-site and installed in the vehicle much like installing an engine.

AUTHOR'S NOTE The original printed circuit boards installed in vehicles left a lot to be desired. They were fragile and susceptible to things such as cigarette smoke and tar, rough roads, and just plain abuse. They were almost impossible to repair at the technician's level and were usually replaced at a high cost to the consumer. But they did provide a base from which integrated circuits for the dash could be designed, developed, and built at a reasonable cost.

LIGHTING SYSTEMS

Shop Manual
pages 229–237

Most of the exterior lighting is required by federal and state regulations. The only mandated interior lighting is the instrument panel. Each of the mandated lighting systems, interior or exterior, must meet requirements for the amount of illumination generated and for positioning. Other lights may be provided for passenger comfort and convenience provided they do not distract from the mandated lights.

Lamp Specifications

The vehicle manufacturer specifies the type of lamp. There are many types used on a modern vehicle. Lamps for the marking, turn, brake, and headlight systems must meet illumination capacity specified by the federal and state regulations. Other lamps such as the courtesy lights are selected by type to provide the level of illumination needed and the mounting space available.

A **bayonet** connection is so named because of its push in, twist, and lock steps to make the connection.

The lamps are connected to the circuits using several types of connections. Most turn and brake lamps use **bayonet** housings to lock the lamp in place (**Figure 6-10**).

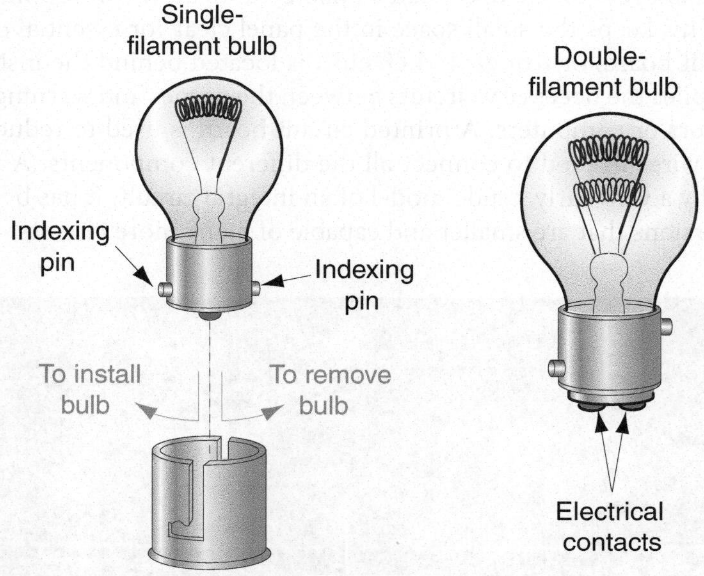

Figure 6-10 The offset pins ensure that the two contacts at the bottom of the lamp connect to the correct circuit.

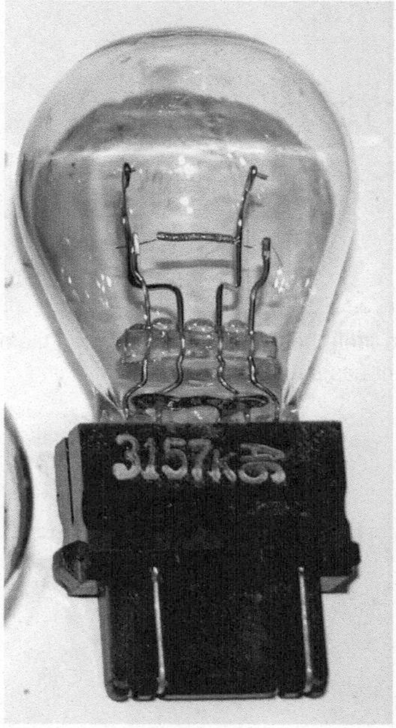

Figure 6-11 Slide-in lamp contacts are small wires folded up over the base. The female socket will have two brass strip contacts.

A single-filament lamp will have locking tabs across from each other. Dual-filament, lamp-locking tabs are offset from each other to prevvent improper installation. Headlight lamps use weather-resistant, hard-plastic shell connectors. Interior lamps may be the bayonet-type or the slide-in-type and are usually single-filament lamps (**Figure 6-11**). Instrument panel lamps may be bayonet, slide-in-type, or LED-type lights.

Exterior Lighting

The mandated exterior lights are:

- Headlights—one on each front corner of the vehicle, white, must provide high and low levels of illumination.
- Marking lights—one on each corner of the vehicle, amber in front, red in rear.
- Turn signal lights—one on each corner of the vehicle, amber in front, red or amber in rear.
- Brake or stop lights—one on each rear corner in the following driver's direct line of sight, red. A third brake light must be installed above and centered on the two lower brake lights.

Headlights

Headlight systems may use two or four lamps provided they meet the illumination and position requirements. The two-lamp system uses two filaments in each lamp (**Figure 6-12**). One filament is for the high beam, and the second is for the low beam. The high beam is the strongest and provides illumination over a higher area. This tends to blind oncoming drivers as the vehicles approach each other. The headlight system must be equipped with a switch for the driver to change to low beam without losing visibility of the road. The low-beam filament produces less illumination and is aimed to the right and low to provide coverage in the area nearest the vehicle and prevent blinding oncoming drivers (**Figure 6-13**).

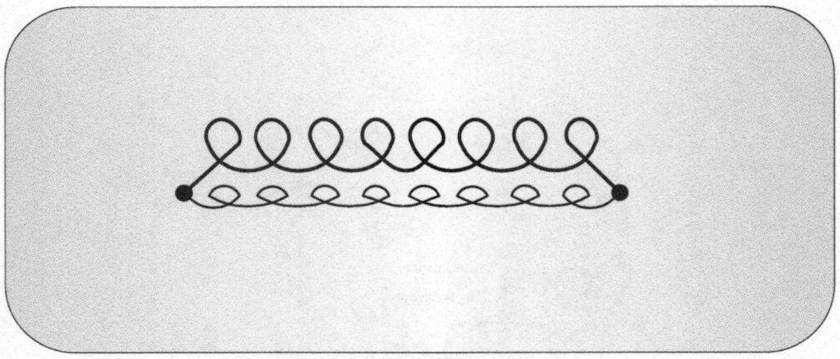

Figure 6-12 A two-lamp headlight system has two filaments in each lamp. Each filament has a different resistance.

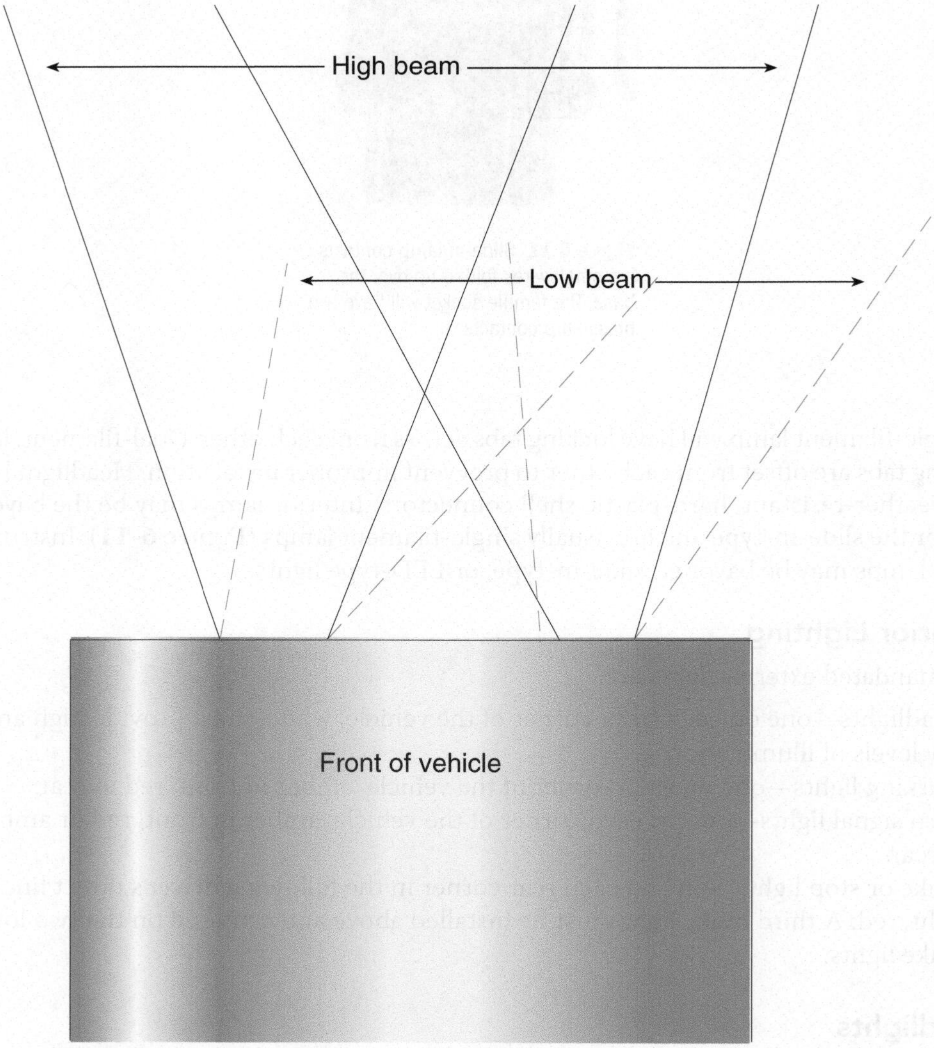

Figure 6-13 As indicated by the highlight, the low-beam illumination is aimed to the lower right and is not as bright as the high-beam illumination.

The aim of the headlight is set by federal regulations. The term **dimmer switch** was adopted because the headlights were much dimmer on low beam.

The **multiple function** lever is commonly referred to as the turn signal lever.

The four-lamp system has two headlamps at each front corner of the vehicle (**Figure 6-14**). Two inside or outside lamps are for high beams while the other pair provides low beam. In this case, the high/low beam switch usually turns the high beam pair on and off. The two low beam lamps are lit in both modes.

The high/low beam selector or **dimmer switch** is most commonly part of the **multifunction lever** mounted to the left of the steering column (**Figure 6-15**). This lever

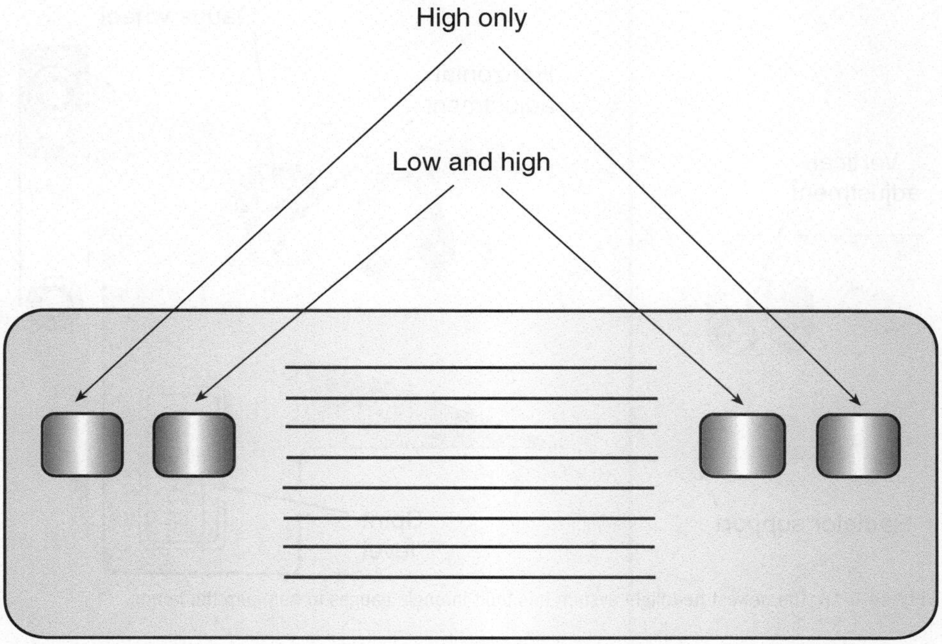

Figure 6-14 The smaller, four-lamp system allows the vehicle's grill and hood to be slightly lower and provides a wider field of illumination on high beam. This system is no longer used on the newer light vehicles.

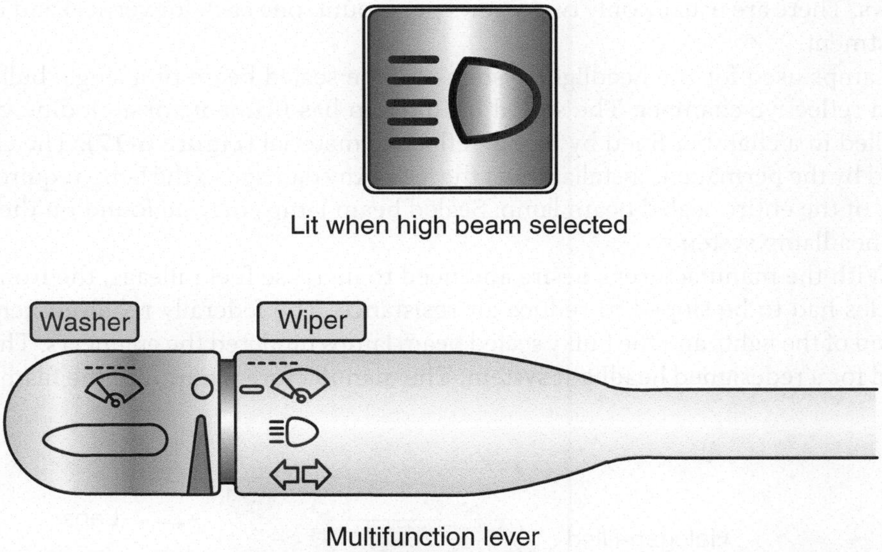

Figure 6-15 The high-beam headlight indicator is usually near the center of the instrument cluster and has a blue lens. Note the symbol for high beam on the multifunction lever (right side).

also controls the turn signal and may include wiper and cruise control switches. A blue light mounted in the top center of the instrument panel indicates that high beam has been selected. On older vehicles and some present-day heavier trucks, the high/low beam switch is mounted in the floorboard and is operated by the driver's foot. Automatic switching was tried on some of the more expensive vehicles in the past with mixed results. One of the problems then was the state of electronics. The sensor used to detect oncoming headlights was not always accurate, requiring the driver to make a panic changeover manually. Today's electronics are more reliable, and automatic high/low switching has reappeared.

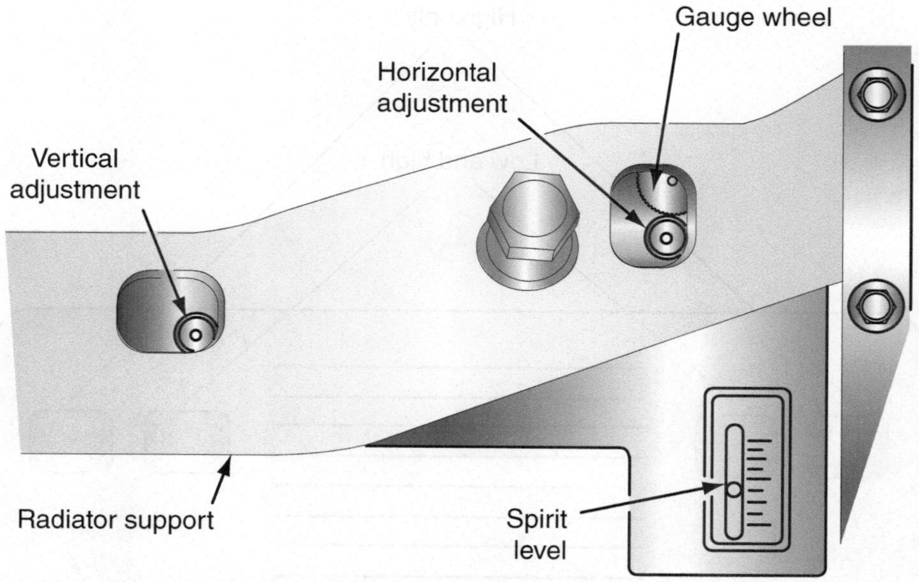

Figure 6-16 The newest headlight system has built-in angle gauges to help aim the lamps.

Headlamps must be aimed to prevent blinding other drivers and for the best illumination of the area ahead of the vehicle (**Figure 6-16**). In most cases, the lamps are installed into an adjustable mount. The mount is connected to the frame by springs and adjusting screws. There are usually only two screws per mount, one each for vertical and horizontal adjustment.

Lamps used for the headlight systems can be sealed beam or a single bulb installed into a reflective chamber. The sealed beam lamp has filaments or an iodine vapor bulb installed in a chamber lined by highly reflective material (**Figure 6-17**). The chamber is sealed by the permanent installation of the lens. Any damage to the lamp requires replacement of the entire sealed beam lamp. Sealed beam lamps may be found on the two- and four-headlamp systems.

With the manufacturer's desire and need to increase fuel mileage, the front of many vehicles had to be sloped to reduce air resistance. The federally required height above ground of the lights and the bulky sealed beam lamps hindered the engineers. The solution called for a redesigned headlight system. The manufacturers removed the filaments from

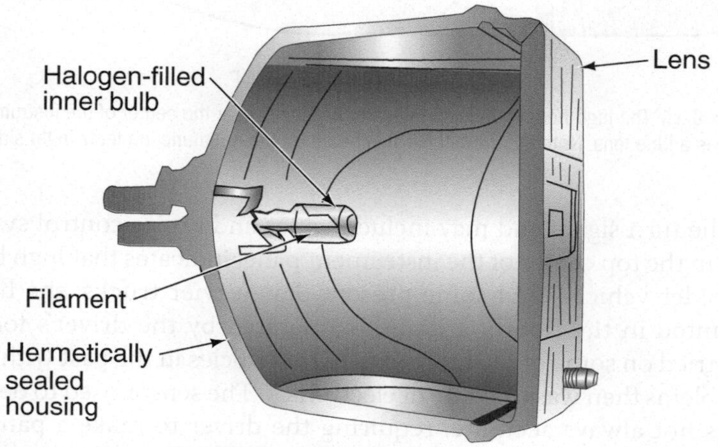

Figure 6-17 The interior of the sealed beam lamp is reflective and has the lamp or filaments mounted in the center of the cavity.

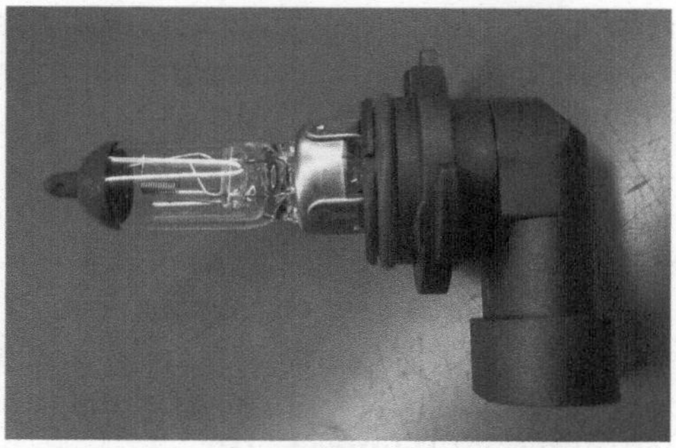

Figure 6-18 This small, high-intensity lamp actually provides more illumination than the older large bulb, but it is more a result of a better reflecting lens than the bulb itself.

the lens and reflector assembly. The assembly was shaped to fit the lower front contours of the vehicle while maintaining the height-above-ground requirements. A separate lamp was then installed into the rear of the assembly. The lamp is a high-intensity halogen bulb capable of producing comparable illumination as the sealed beam (**Figure 6-18**). Failure usually requires only the changing of the bulb but not the lens and reflector. Most vehicles now use this design and a two-lamp headlight system. Some newer vehicles are equipped with low-voltage headlights. The lights are on any time the engine is operating. This safety feature better highlights the vehicle to other drivers and pedestrians during daylight. They are referred to as **daytime running lights.**

Daytime running lights were first introduced in the United States by GM as a warning signal to other drivers.

Marking Lights

Marking or parking lights are used to identify the vehicle when visibility is low, but headlights are not needed. Marking lamps are usually single-filament in front and dual-filament in the rear. The lights are turned on and off with a separate set of contacts within the headlight switch. The front marking lamp is also used in most vehicles as the turn signal lamp. Most of the newer vehicles leave the marking lights on during headlight operations.

Fog and Driving Lights

Many vehicles are fitted with fog or driving lights or both. The two-light systems are designed differently and must meet separate requirements. Fog lights are mounted below the marker lights and have scattered-beam (diffused) lens. Generally, they are amber in color (may be clear) and are provided to increase near-distance visibility during fog or rain. The water droplets in fog or rain will reflect the headlight illumination back into the driver's eyes and basically blind him or her. Fog lights provide enough scattered illumination to show the outer side of the road and the center, in the hopes of preventing an accident. Fog lights may function only when the low-beam headlights are on and must be wired in series-parallel to the low-beam circuit. This prevents them from switching on singly or with high-beam lights. Fog lights must also be aimed to not blind the oncoming driver.

Fog lights may only legally work with the low beam headlights. Driving lights are designed to work the high beams.

Driving lights may function only when the high-beam lights are selected and are wired in series-parallel with that circuit. Again, this is done to prevent their operation unless the high-beam headlights are on. Driving lights can be very dangerous to an oncoming driver because of the lights' high illumination. Some driving lights are advertised to provide illumination up to 1 mile. Driving lights are mounted above the headlights and are most commonly found on pickup trucks. Although they have a viable off-road use, they are not worth much on the typical highway. In many states, it is illegal to use driving lights on public roads. In snow, rain, or fog conditions, they are totally useless and can present a danger to the driver and others on the road. Most driving lights are mounted by the driver using aftermarket components.

Turn Signals

A **flasher** may be electrical/mechanical or fully electronic to flash the turn signals.

A **bimetallic** strip has two pieces of metal laid side by side. One piece reacts to heat faster than the other one does.

The lamps in this system are used to communicate the driver's intentions to other persons on the road. Most turn signal systems use the front marking lamps and the rear brake lamps as the signaling devices. The turn signal switch is mounted behind the steering wheel on the column. It is controlled by a lever extending to the left from the column and has three positions: off, left, and right (**Figure 6-19**).

Moving the end of the lever downward supplies power from the fuse through a **flasher** to the left marking lamps at the front and rear of the vehicle (**Figure 6-20**). The flasher is a **bimetallic** circuit breaker. As current flows through it, the heat causes the breaker to

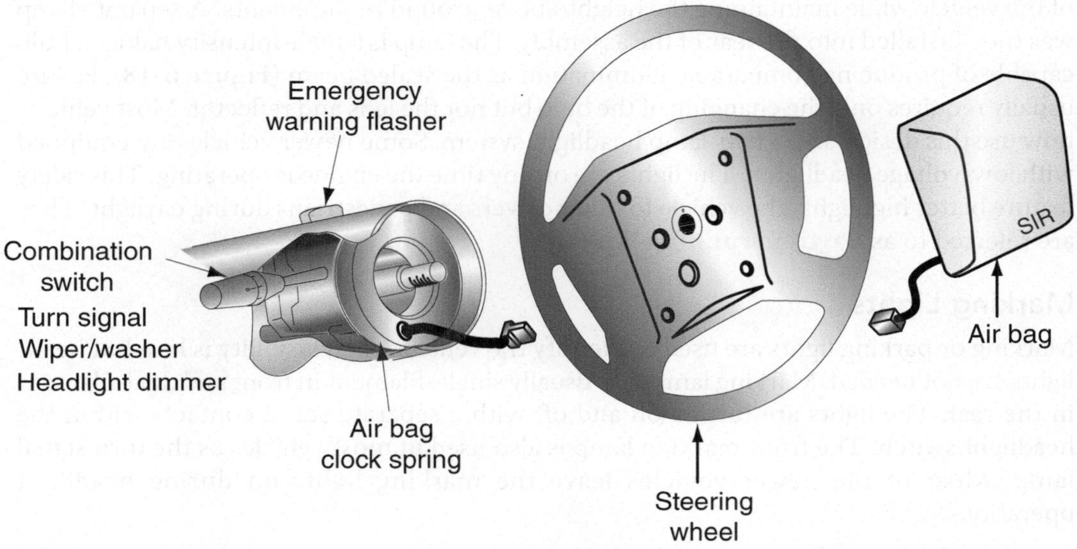

Figure 6-19 All standard modern vehicles in the United States have the turn signal lever on the left side of the steering column.

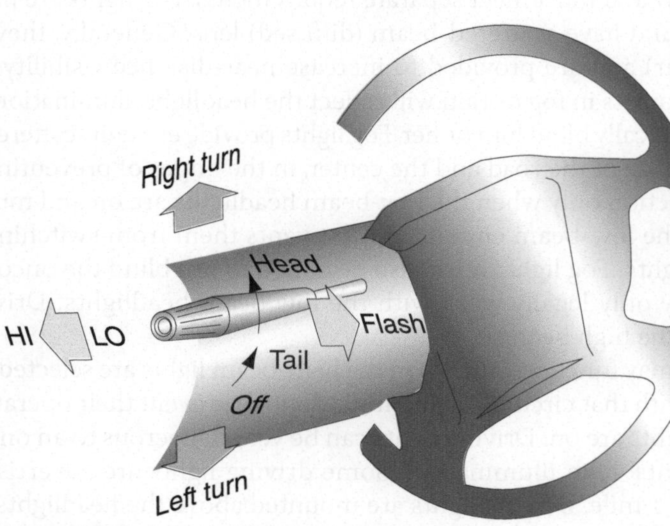

Figure 6-20 The turn signal lever can be moved up for right turns and down for left turns as shown here.

trip and open the circuit. As it cools, the breaker resets and closes the circuits to the lamps. This process cycles the flashers on and off between 0.5 second and 1.0 second on the average. The result is a flashing light at each end of the vehicle on the left side. The same process happens for a right selection except for the lamps that are flashing. Adding additional lamps such as those for a trailer can overload the flasher. Installing a heavy-duty flasher is usually required before adding additional lighting.

The rear turn signal lens may be amber or red. If the brake lamp is housed in the same reflective and lens assembly as the signal, the lens will be red. This is usually referred to as the single lamp system because the marker, turn, and brake use one lamp with dual filaments. The lens for a two-lamp system, which uses a separate lamp for turns, has amber for signaling and red for the marker or stop. Some vehicles are equipped with three lamps, one each for marking, turning, and braking. The single requirement is that the turn lens is amber or red and the marking and brake lights are red.

The **emergency flasher** or hazard lights work much the same as a turn signal. A separate switch is used to make all four turn signal lamps flash together. The four-way circuit bypasses the turn signal circuit and directly feeds each of the four lamps. Brake applications will override the four-way on the rear lamps to alert following drivers to the situation. Brake applications will not override regular turn signals. The marker lights, main brake lights, emergency flashers, and turn signal are all wired through the turn signal switch.

The **emergency flasher** must work with the ignition key in any position regardless of the operating conditions of the other lights.

Brake Lights

Brake lenses must always be red and have much higher illumination than the other lamps near them. A mechanical switch on the brake pedal rod controls the brake lights. Power is fed directly to the brake (stop) switch from the battery (**Figure 6-21**). When the brake switch is closed, current travels to the rear stop or brake lamps via the turn signal switch. If a turn signal is on, the flasher interrupts the current to that lamp. When the four-way circuit is operating, and the brakes are applied, the brake light current enters below the flasher and provides a normal steady current to the brake lamps.

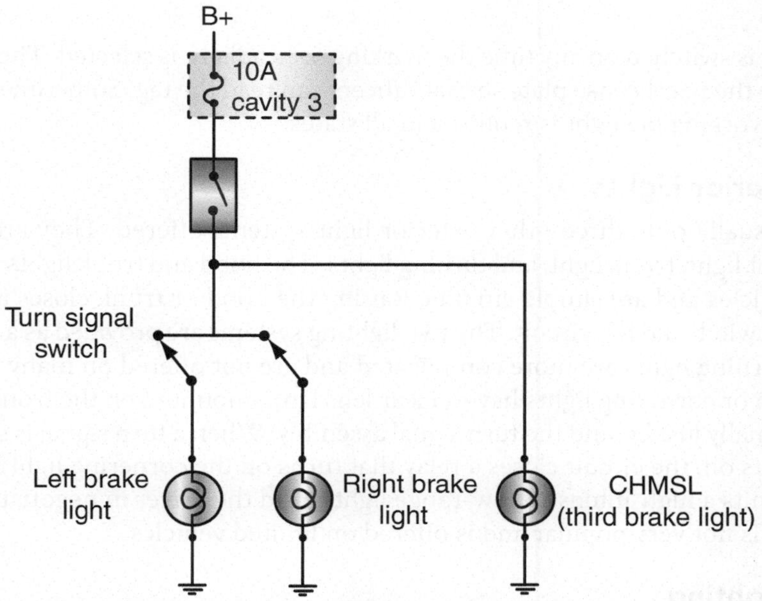

Figure 6-21 The third brake light, the center high-mounted stop light (CHMSL), is wired parallel to the other brake lights and will work even if there is a problem in the turn signal switch.

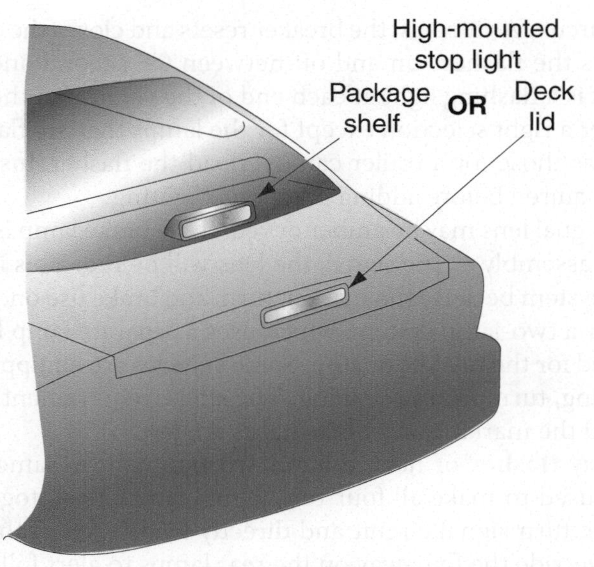

Figure 6-22 The CHMSL must be centered on the rear of the vehicle and not more than 3 inches below the bottom of the rear windshield.

The **center high-mounted stop light (CHMSL)** is a federally mandated third brake light, usually located on the lower rear window or on the trunk lid of a vehicle.

A third brake light was mandated by federal regulations to quickly gain the attention of a following driver. The third brake light is officially known as the **center high-mounted stop light (CHMSL).** It may be located on the lower edge of the rear window or mounted to the trunk lid (**Figure 6-22**). The third brake light is within the normal line of sight for a following driver. The lamp is wired into the brake switch output between the switch and the turn signal switch. In this manner, the light will function regardless of the turn signal switch or the other two brake lights. This light is not affected by the turn signal switch or the hazard light switch. There are some systems that have a flasher included in the CHMSL that will cause this light to flash. It is supposed to more readily alert the following driver, but in heavy, slow traffic it can become a nuisance to a following driver.

Tag Lights

The tag light is switched on any time the marking or headlight is selected. The light is used to illuminate the rear license plate so that officers can read the tag. Sometimes, two lamps are used. A working tag light is required in all states.

Other Exterior Lights

There are usually only three other exterior light systems offered. They are the engine compartment light, trunk light, and turning lights. The hood and trunk lights are provided on most vehicles and are simple circuits. Raising the hood or trunk closes a mercury or mechanical switch and the circuit. The two lighting systems are provided as a convenience item. The turning lights are more complicated and are not offered on many vehicles.

The turn or cornering lights have a clear lens lamp mounted on the front sides of the vehicle, normally just behind the turn signal assembly. When a turn signal is selected with the headlights on, the circuit closes a relay that turns on the cornering light for that side. The lamp emits a high-intensity, low-range light to aid the driver in negotiating a corner. This system is not very popular and is offered on limited vehicles.

Backlit means the light is coming from behind or from the side of the object being viewed.

Interior Lighting

Interior lighting is provided for entry or exit from the vehicle and illuminates the passenger compartment. The instrument gauges are **backlit** by twist-lock lamps into the rear of the panel. The lamps are positioned to light the gauges indirectly. The front side of the panel is coated with a nonglare finish. Contacts in the headlight switch turn on the lamps any time the marking or headlights are selected. The intensity of the illumination can be controlled by a rheostat mounted within the headlight switch or on a separate switch on the dash. The lights can be turned off with the rheostat switch without affecting the exterior lights. The headlight switch or the rheostat does not affect the warning lights in the instrument panel.

Included in the group known as the instrument lights are lamps to highlight various controls located on the dash. Most dash panels house the controls for heating and air conditioning, radio, blower, and onboard computer input keys. Lamps placed behind the control panels or in the button illuminate the markings and the buttons.

Dome and Courtesy Lights

The dome and courtesy lights are switched on when a door is opened (**Figure 6-23**). Most have separate switches to control individual lights when the door is closed (**Figure 6-24**). The dome light is mounted near the center of the roof. Courtesy lights, if equipped, are usually mounted on the kick panels, under the instrument panels, or on the doors. Rear courtesy lights are usually mounted to the left and right side of the passengers. In addition to the courtesy lights listed, reading or map lights may be positioned to shine toward each seat. This allows a passenger to read or search for an item without hindering the driver's front or rear visibility. Overhead dome lights tend to block the driver's vision through the inside rearview mirror.

Glove Box and Key Lights

The glove box light works like the hood and trunk lights. Opening the glove box door closes a mechanical switch. A small lamp is lit to illuminate the interior of the glove box. This is a common problem when tracing a drain on the battery. Items placed in

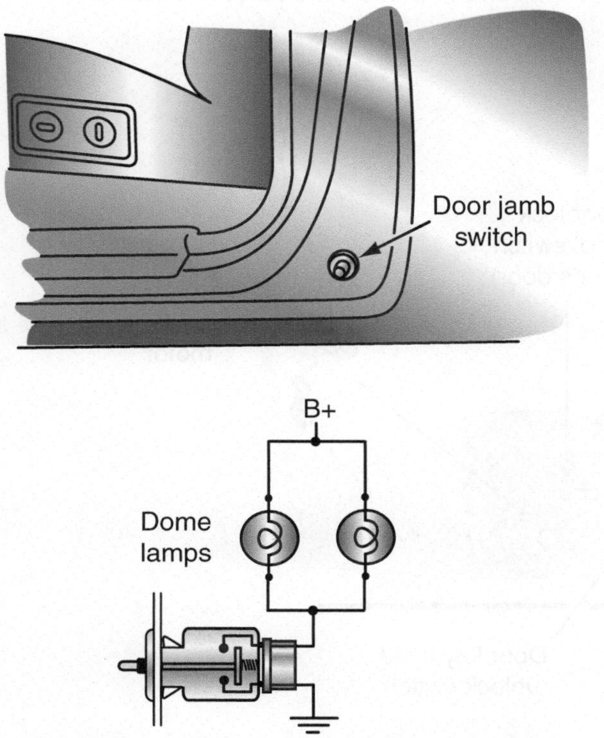

Figure 6-23 A typical location for a door switch. This switch will turn on the dome lights and the "door ajar" lamp if equipped.

Figure 6-24 Map or reading lights can provide a narrow field of illumination for use by the passenger. Similar lights may be installed over the rear seats.

the box may prevent the door from opening the switch, and the light slowly drains the battery overnight.

The slot for the key on the outside of the door and the ignition switch may have a small lamp, possibly an LED, for illumination. This helps the driver insert the key in the door quickly at night and helps prevent marring the door's finish. The ignition switch is in an awkward position for the typical interior light. Providing a small lamp near the switch helps the driver insert the key at night.

ACCESSORIES

Accessories include systems such as power door locks and windows, custom seat covers and carpets, trailer hitches, and many other much-liked but not-always-necessary items. In this section, we discuss power door locks and windows, antitheft devices, and power seats as examples of system operation. The next section will cover a major and expensive accessory: climate controls, commonly known as the heat and air conditioning system.

Power systems such as door locks, windows, and seats were designed to help the driver control security and provide comfort without distracting from driving conditions. The door locks have recently been incorporated into antitheft systems and remote control.

Power Door Locks

The electrical or mechanical mechanism for the locks is basically a reversible polarity solenoid or motor moving the manual lock mechanism. There is a solenoid for each door controlled by parallel-wired switches in the front doors (**Figure 6-25**). Operating either switch locks or unlocks all doors. A switch built into the key lock on the outside of the driver's door will also perform the same function. On some vehicles, one twist of the door key will unlock the driver's door and a second twist operates the other door locks.

Shop Manual
pages 239–240

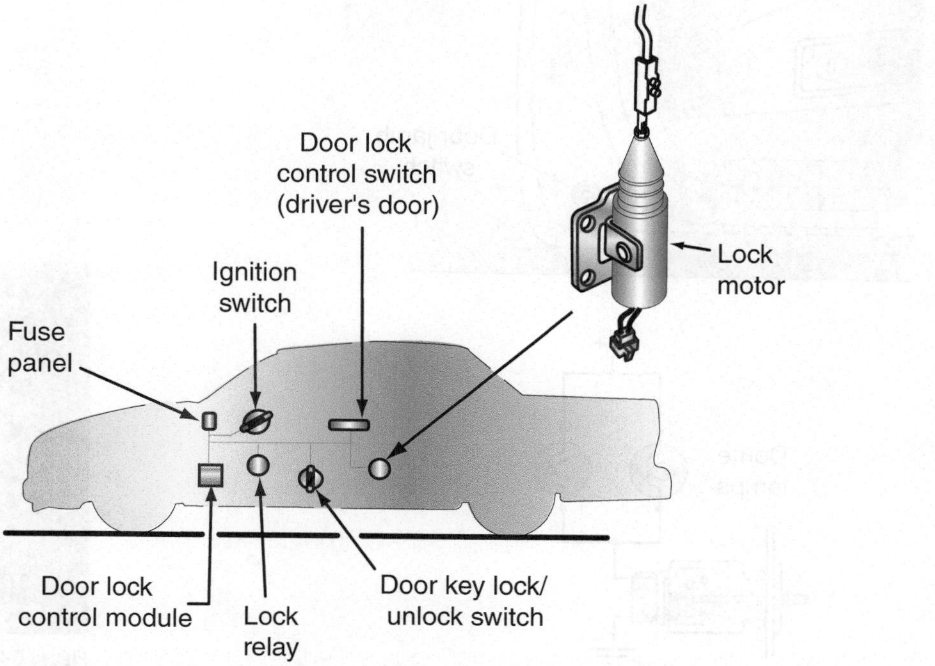

Figure 6-25 The power door locks may be controlled by the door-mounted manual switch, by the remote control, or by using the key in the exterior door lock.

Figure 6-26 This remote control or electronic key fob has a button to lock or unlock the door or trunk plus a "panic" button for the operator to use in case of emergency.

The solenoid is positioned to move the locking mechanism when activated. The solenoid can either push or pull the locking mechanism. The operation depends on the type of lock and the space available to install the solenoid or motor. Most vehicles have manually operated locks to back up the electric system.

With the availability of dependable electronics, the door locks can be activated with a small electronic remote-control device usually built into a key fob or the key itself. The Body Control Module (BCM) receives the remote signal and operates the locks. Antitheft security systems are connected to the (BCM) so that the alarm system is turned on or off whenever the locks are activated (**Figure 6-26**). Many vehicles have automatic systems that lock the doors when the vehicle speed reaches about 20 mph, or a drive gear is engaged or unlocks the doors when the park gear is selected. The locks must be unlocked with the key, switch, or the remote-control unit. A common fuse protects all the lock circuits.

Antitheft Systems

Antitheft systems range from a hidden toggle switch placed in the ignition circuit to ignition keys that must identify themselves to the PCM before the vehicle can be powered up. A toggle switch can be found and flipped on. This may prevent some thefts, but a professional thief will be gone with the vehicle in 2 or 3 minutes.

Most antitheft systems in use today rely on noise and flashing lights to alert anyone nearby that the vehicle is being stolen. However, once the thief is inside and the engine is running, the antitheft system will usually disarm, and the vehicle can be driven safety.

A later system uses an ignition key with an embedded resistor pellet. When the key is inserted into the ignition switch, a small current is sent through the key and pellet. If the resistor is present and correct, the current flows to the antitheft module to power up the vehicle.

In 1999, a new system was installed on some General Motors vehicles. Ignition keys containing small **transponders** are used with specific security codes. The system's control module was placed around the steering column under the column covers. A round antenna ran from the module and encircled the ignition key slot. The antitheft module and the PCM were programmed to recognize the key's security code and save it. This was done when the vehicle started for the first time at the assembly plant. It was also done at

A **transponder** will transmit its code by radio or radar when actuated by a specific signal from a radio or radar transmitter (interrogator).

the dealership for replacement of additional keys. The key's code was cross-matched against other selected components so that the key could not be used in another car. When a driver inserted the key into the ignition switch, the antitheft control module asked for identification from the key. The key transmitted its code and the antitheft control module compared it to the saved codes for the key and the selected components. If all codes were correct, the module signaled the PCM to switch on the electrical systems. Presently, multiple key codes can be stored in the module and PCM. If a key, control module, or PCM is replaced, the system must be reprogrammed or the vehicle will not start.

AUTHOR'S NOTE Aftermarket antitheft systems, including some dealer-installed units, offer some disarming problems during vehicle repairs. In some systems, if the battery is disconnected for extended periods of time, the security system will not disarm when repowered and the engine cannot be started. There may be a specific sequence of steps required to reset the system computer. In most cases, the instructions are contained in the operating manual for the system, and that manual cannot be located. Sometimes the owner may know of the sequence, or the data must be obtained from research on the internet or paper manuals.

Power Windows

Power windows use reversible electric motors to raise or lower the windows through gears, a punched plastic strip, or cables and pulleys. There is a window switch at each door with a **master switch** on the driver's side. The master switch may also have a lock switch that deactivates the other switches. The nondriver window motors are in parallel to the master switch.

A **master switch** controls multiple units from one location like the power window switch in the driver's arm rest.

The master switch consists of a switch for each window motor and the window lockout switch, if so equipped (**Figure 6-27**). The other switches placed on the other doors or panels will operate only that window. A common fuse or circuit breaker protects all the window circuits.

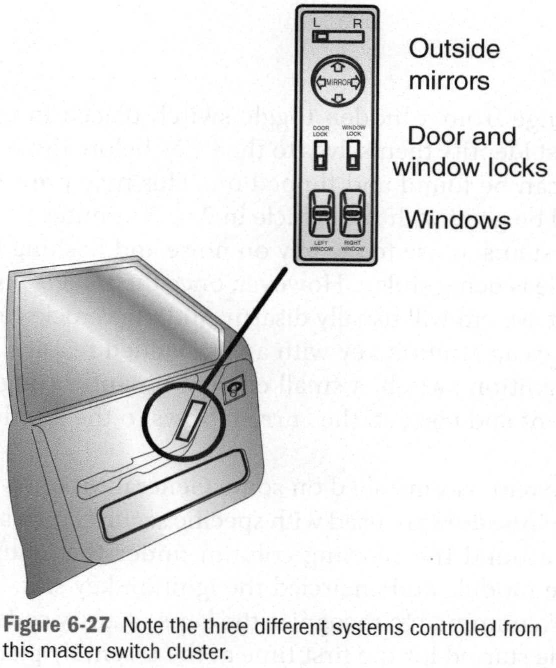

Figure 6-27 Note the three different systems controlled from this master switch cluster.

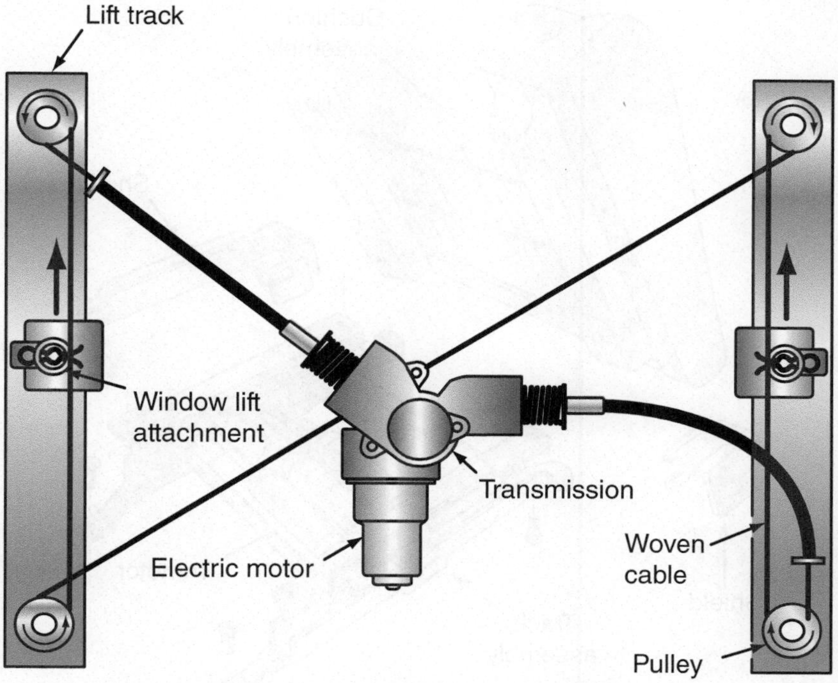

Figure 6-28 The part shown is a replacement for a cable-type window regulator.

The output gear moves the mechanical device that moves the window on the regulator. In a straight gear system, the output gear is meshed with a large, flat sector gear. The sector is attached to the arms or regulator that push up or pull down on the glass frame. Plastic strap (ribbon) or steel stranded cable and pulley mechanisms use the motor's output gears to pull the cable or ribbon that is attached to the glass frame (**Figure 6-28**). The ribbon is punched at even intervals and fits over the output gear. Two or three pulleys and a metal guide keep the ribbon or cable in contact with the gears and pulleys to prevent slippage.

Some vehicles are equipped with automatically lowering driver's-side windows. When the switch is operated under specified conditions, the window, once started, will lower automatically to its lowest position. This allows the driver to direct attention to the environment. A similar system allows for automatic window closing. This system is called an **"express"** window system.

Power Seats

Power seats are usually classified as four-way or six-way. Four-way seats can move electrically forward or backward, and they can tilt the backrest. Six-way types have the same movements plus tilting of the seat bench. The systems may be on each of the front seats or just on the driver's. Some newer luxury cars allow the passenger to adjust the positions of the rear seats.

The controls are usually mounted at the outside end of the seat bench, but may be mounted in the door or console. There will be two or three switches depending on the system. One will control the forward and rearward movement of the seat while another controls the tilt of the backrest. A third switch is used to control the motors to tilt the bench.

The actual movement of the seat components is done with reversible electric motors and gears, cables, or levers (**Figure 6-29**). Most systems use an output gear meshed with a toothed rail. The rail is attached to the seat component. As the gear turns, the rail and component are moved. The motors provide a great deal of torque for their size and energy usage and can cause damage to anything left under the seat.

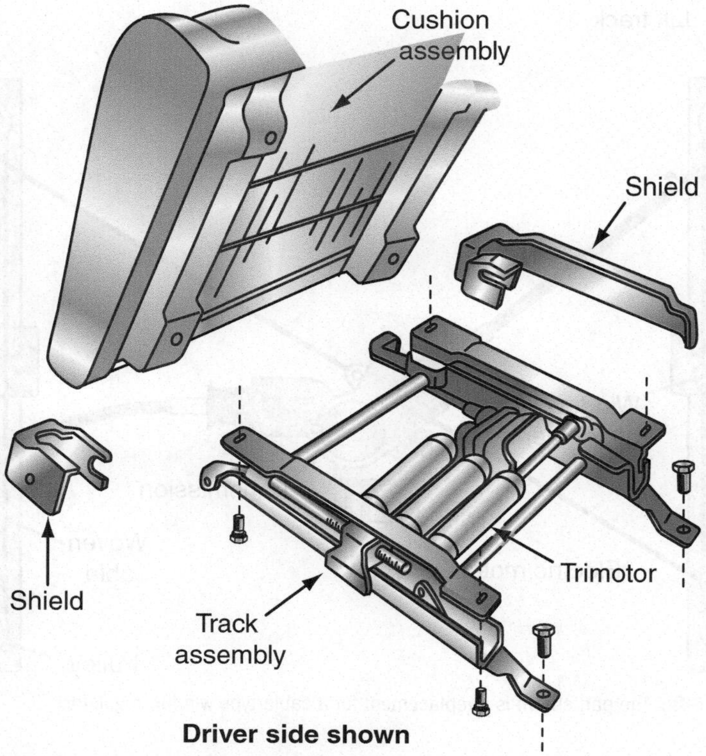

Figure 6-29 A typical power seat motor is one piece with all three motors mounted in one housing.

Many power seat systems are controlled by the BCM. Sensors on the undercarriage of the seat provide position data to the BCM. The BCM can be programmed to remember the data and relate it to a particular operator. Selecting the proper code activates the seat and returns it to the position recorded in the BCM's memory.

Power Adjustable Pedals and Steering Columns

For persons who are taller or shorter than the height of a "normal" person, reaching the various pedals on the floor can be difficult. Many times, the steering wheel position is not always comfortable for the driver. A taller person might have to move the seat back for pedal clearance but then would be required to reach for the steering wheel. A shorter person might slide the seat forward, but then he or she would be against or very near the steering wheel. The adjustable steering wheel, or the steering column, helps alleviate some of this, but it was not until the SRS was mandated for the driver that other, even more serious concerns were brought to light. The concerns associated with airbag deployment were emphasized during the late 1990s when some deaths and injuries were caused more by the airbag than the accident. The National Highway Traffic Safety Administration recommends the driver be at least 10 inches (25 centimeters) from the steering wheel. Safer airbags were developed, but something more sweeping was needed. Ford introduced **adjustable pedal systems (APS)** in some of its 1999 model vehicles.

APS can be mounted on a moveable platform or singularly mounted. The pedals can be moved up to 3 inches (8 centimeters) to accommodate the physique of most drivers (**Figure 6-30**). Screws or transmissions driven by electric motors move the platform or pedal along an axis close to the angle of the steering column. Singularly mounted units require a separate motor and transmission for each pedal. Some units are designed for

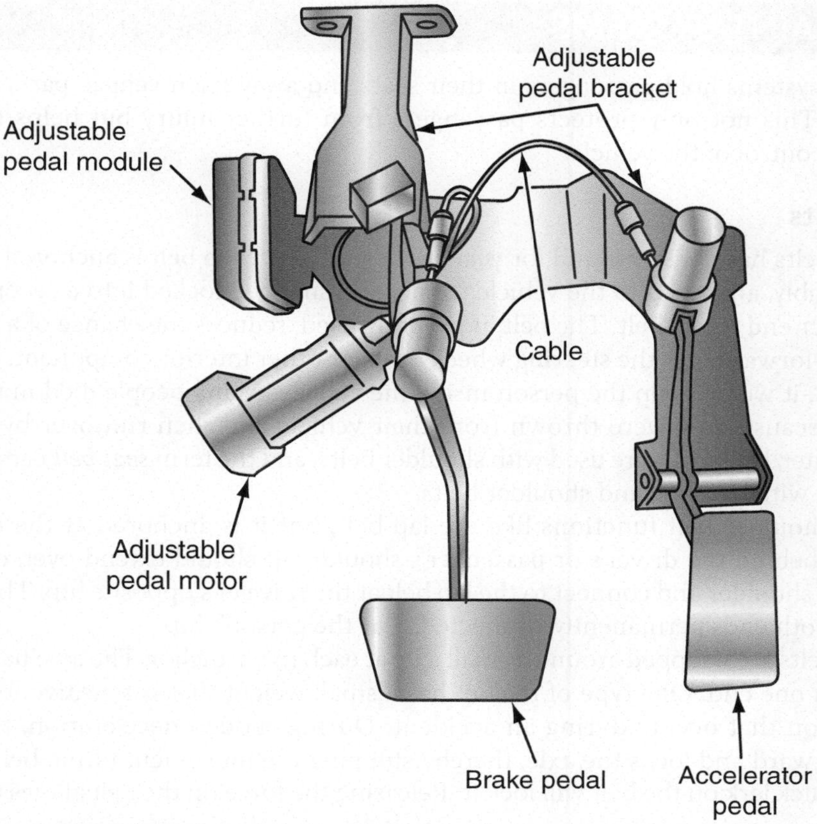

Figure 6-30 Adjustable pedals are offered in some high-line vehicles, with more to be offered in the next several years.

either linear or arcing motion. Linear motion moves the entire pedal mechanism in a straight line while the arcing motion moves the lower end closer to the driver with minimum movement of the pedal(s) pivot mount. The APS has been accepted by customers, generally with few problems to date. The original APS offered by Ford was deemed by drivers to have the pedals too close together. Ford has since corrected that problem through a recall and a new design for its later models.

A BIT OF HISTORY

APS is not new. It was introduced in 1910 to the public on a 1908 American Simplex. (Does anyone remember that one?) It was a mechanical system where the operator had to use a wrench to adjust the pedal. General Motors obtained a patent for a manually adjusted system in 1966, and the Dodge Viper offered a manual system in 1996. Viper is one of the few vehicles offering an APS for a manual-transmission-equipped vehicle.

The steering column can be lengthened, shortened, tilted, or all three—and, in most vehicles, the adjustable steering column works directly with the APS. Memory systems have been added so multiple drivers can use the same vehicle just by punching a code, and the pedals and steering column move to preset positions much like the memory seats. In fact, the controlling computer (BCM) will usually position the seats, pedals, steering column, and outside rearview mirrors.

RESTRAINT SYSTEMS

Restraint systems hold passengers in their seats and away from vehicle parts during an accident. This not only protects passengers from further injury but helps the driver maintain control of the vehicle.

Seat Belts

Shop Manual
page 240

Lap seat belts have been around for years. One end of the lap belt is anchored to a small axle assembly, anchored to the vehicle frame, and snapped locked into a second anchor at the other end of the belt. The belt, if properly used, reduces the chance of a passenger slamming forward into the steering wheel, dash, or other interior component. But, more important, it would keep the person inside the vehicle. Many people died in survivable crashes because they were thrown from their vehicle and then run over by the same vehicle. Later, lap belts were used with shoulder belts, and the term *seat belt* came to mean a restraint with both lap and shoulder belts.

The shoulder belt functions like the lap belt, but it is anchored at the top of the doorpost behind the driver's or passenger's shoulder. It should extend over, down, and across the shoulder and connect to the lap belt at the traveler's opposite hip. The seat belt now has both ends permanently connected near the person's hip.

The belts are wrapped around a small axle at each main anchor. The axle has a ratchet or lock on one end. One type of ratchet has a small weight that is sensitive to the rapid deceleration that occurs during an accident. During sudden deceleration, the weight swings forward and locks the axle, thereby stopping the movement of the belt. In some cases, a quick jerk on the belt will lock it. Releasing the force on the belt allows the axle to unlock.

There are two major operational problems with seat belts. The critical point to seat belt operation is that vehicle occupants must hand-connect their belts in almost every belt system. If belts are not connected, they do nothing to protect drivers and passengers. A second problem involves pregnant women. A seat belt must be positioned low on the abdomen (**Figure 6-31**) or injury to the unborn infant might occur. The belt will still work adequately even though it is set low. Seat belts will leave uncomfortable bruises and marks on passengers or drivers during an accident, but they will heal in a few days.

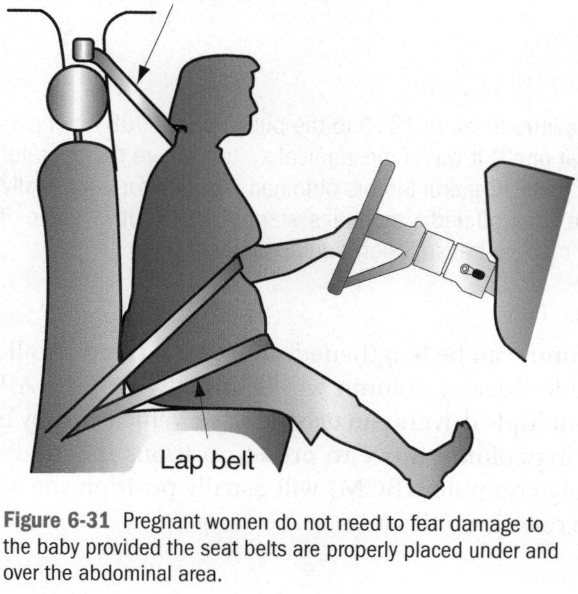

Figure 6-31 Pregnant women do not need to fear damage to the baby provided the seat belts are properly placed under and over the abdominal area.

Airbags

Supplement restraint systems (SRSs) are passive restraints requiring no action by the passengers or driver. They are generally referred to as airbag systems and are designed to keep anyone from hitting the front of the dash and the steering wheel. The original system included a single airbag mounted in the steering wheel. Later systems added a passenger airbag mounted in the right side of the dash. Unfortunately, airbags have caused the deaths of some motorists. A shorter person might sit closer to the steering wheel and, as a result, is too close to the deploying airbag. When a child is buckled into a child seat and placed in the front passenger position, the deploying passenger-side airbag could injure or kill the child. A public outcry directed at the National Transportation Safety Board prompted further research into a safer airbag system. Regulations and standards were adopted to allow vehicle manufacturers to reduce the force of airbag deployment. In addition, a switch was installed to turn off the passenger-side bag (**Figure 6-32**). The new regulations also allow the owner to request that the driver's airbag be deactivated under certain, specific conditions, such as when the regular driver is of smaller build and must sit closer to the steering wheel. On some vehicles, a weight sensor will determine if the airbag is to be disabled or not. This usually results in a dash indicator light stating the position of the airbag. The ideal solution is to reduce the force of deployment to meet individual needs. That option is available in some form now, but adjustable pedals help solve part of this problem.

The latest airbag system includes the two bags described in this section plus airbags mounted in the doors, head rests, and rear side curtains. The side bags are designed to reduce injuries during a crash to the side of the vehicle.

AUTHOR'S NOTE The federal and state governments now require that children be placed in the back seat in the proper restraint system, either a child seat or seat belts. In pickup trucks without a rear seat, the passenger side airbag can be deactivated either by a weight sensor or a manual key switch. However, to the extent possible using a child seat, seat belts, or moving the passenger seat away from the dash is better than turning off the airbag.

Airbag Operation

The airbag system relies on reliable electronics to function properly. It draws power directly from the battery and is turned on by switching the ignition to run. Impact or

Figure 6-32 The airbag deactivation switch was commonly used in pickup trucks to turn off the passenger airbag to protect children and infants.

arming switches are located behind the front bumper and inside the front doors if equipped with side bags.

A crash that is strong enough to trip (close) an arming switch signals the airbag control module to deploy the airbags. A frontal crash will fire both driver and passenger airbags.

The folded airbag is placed over a chemical that will produce a large volume of gas when ignited. When the control module ignites the chemical, it expands and rapidly inflates the airbag. The system requires a backup power source in case the battery is destroyed before the airbag can be deployed. This can cause serious injury or death if the system is not deactivated during repair work.

A capacitor is used as the reserve or backup power. The capacitor is charged each time the ignition is switched on and will input electrical current rated at about 36 volts. If the battery is destroyed and the capacitor loses input power, the capacitor will discharge. This provides enough power to operate the module and deploy the airbags. During normal operation, the capacitor slowly loses its charge after the ignition is switched off. During the drain down, the airbag can still be fired if the capacitor senses a sudden loss of power or receives a surge of electrical current.

Before beginning work inside the vehicle, the capacitor must be drained of electrical current. The acceptable method is to remove the negative cable from the battery and wait awhile. The amount of time is set by the airbag and vehicle manufacturer and is shown in the service manual.

To further protect the technician, it is suggested that the airbags be disconnected after the capacitor has discharged. There are two or more bright yellow plugs located near the bottom of the steering column under the dash or in the glove box. Once the yellow plugs are disconnected, the battery can be reconnected to supply power for testing of the failed components. When all the repairs are completed, and covers are installed, disconnect the negative cable again. The yellow plugs can now be connected without any danger of electric surge.

SUMMARY

- Passenger comfort devices are designed to make the driver and passenger comfortable during vehicle operations.
- Passenger conveniences are those items that make it easier for the operator or passenger to control vehicle components or provide some type of service to the passengers.
- Warning systems indicate to the driver that a dangerous or damaging condition in the vehicle may occur.
- Warning systems have bright red or amber lights and may have audible signals to alert the driver to some condition.
- Information systems provide data pertaining to vehicle operations.
- Instrumentation usually refers to the lights, gauges, and other devices that present information to the driver.
- Information may be conveyed in digital, audible, analog, text, or color displays.
- Most exterior lighting is required to meet federal and state regulations concerning illumination and positioning.
- Headlights may have either two or four lamps with high and low beams.
- Except for the turn signals, all the rear-mounted lights must be red.
- Brake lights override the rear four-way flashers but not the regular turn signals.
- A third brake light is mounted high to attract the attention of following drivers.
- The only interior light mandated by regulations is the lighting of the instrument panel.
- The instrument panel is finished with a nonglare coating.
- Most instrument panels use circuit boards.
- Each power lock circuit is parallel wired to the other lock circuits.
- One fuse protects all the locking circuits.

- Most power door lock systems are also part of the vehicle's antitheft system.
- Each of the power window circuits is parallel to each other and the master switch.
- The master switch can control all the windows plus lock out the other windows switches.
- Power seats can have four- or six-way positions.
- The ability to tilt the seat's bench is the difference between four- and six-way seat systems.
- Some seat systems can remember a designated position.
- Seat belts must be connected to protect the motorist.
- Supplemental restraint systems are known as airbags.
- A capacitor is used as backup power for the airbag system.

REVIEW QUESTIONS

Short Answer Essays

1. List and label the color of the lights or lenses required on mandated exterior lighting systems. Include options, if any.
2. Describe the major differences between four- and six-way power seats.
3. Explain how non-remote-controlled power door locks are operated from outside the vehicle.
4. Describe how a seat belt can reduce the chance of severer injury.
5. Explain how an express window switch operates.
6. Explain the process that makes the turn signal lamps blink.
7. Describe the purpose of the CHMSL.
8. Explain the purpose of the integrated circuit used in the instrument panel.
9. List and describe the purpose of the most common dash-mounted warning lights.
10. Explain how the high and low beams are selected and any light(s) used to alert the driver of the selection.

Fill-in-the-Blanks

1. The left, rear power window switch ______________ is with the ______________ ______________ master switch.
2. The power door locks are protected with a(n) ______________ ______________.
3. The ignition key slot may be illuminated with a(n) ______________ ______________.
4. The rear power door locks may be unlocked by turning the door key ______________ ______________ ______________.
5. The instrument panel is ______________ and finished with a ______________ coating.
6. The colored band type of gauge displays information similar to a(n) ______________ gauge.
7. HUD is used to project information onto the ______________ in the ______________ line of ______________.
8. Body computers can be programmed ______________ ______________ to for a designated ______________.
9. The dimmer switch can be mounted on the floor or within the ______________ ______________.
10. Supplement restraint systems (SRSs) are passive restraints requiring no action by the ______________ or ______________.

Multiple Choice

1. Interior lighting systems are being discussed.
   Technician A says the only mandated interior lighting is the instrument panel.
   Technician B says the courtesy and dome lights may be wired to the same door switch.
   Who is correct?
   A. A only
   B. B only
   C. Both A and B
   D. Neither A nor B

2. *Technician A* says the lamp or bulb can be replaced in the sealed beam headlight.
   Technician B says the need for better aerodynamics required the headlight lamps to be redesigned.
   Who is correct?
   A. A only C. Both A and B
   B. B only D. Neither A nor B
3. The rear lights on a vehicle are being discussed.
   Technician A says the turn signals may be amber or red.
   Technician B says three-lamp systems use dual-filament bulbs.
   Who is correct?
   A. A only C. Both A and B
   B. B only D. Neither A nor B
4. Rear dual-filament lamps are being discussed.
   Technician A says one filament is used for turn and marking lights.
   Technician B says one filament is used only for brakes.
   Who is correct?
   A. A only C. Both A and B
   B. B only D. Neither A nor B
5. Warning lamps on the dash are being discussed.
   Technician A says warning systems are found on every vehicle and alert the driver to a fault or failure that could damage the vehicle
   Technician B says when the ignition is turned on, all the dash warning lights should illuminate to perform what is called a bulb check.
   A. A only C. Both A and B
   B. B only D. Neither A nor B
6. Brake lights are being discussed.
   Technician A says a third brake light was mandated by federal regulations.
   Technician B says the CHMLS is usually located on the lower rear window or on the trunk lid of a vehicle
   Who is correct?
   A. A only C. Both A and B
   B. B only D. Neither A nor B
7. Headlamps are being discussed.
   Technician A says headlights are mandated lights.
   Technician B says headlights must provide high and low levels of illumination.
   Who is correct?
   A. A only C. Both A and B
   B. B only D. Neither A nor B
8. Power door locks are being discussed.
   Technician A says the electrical or mechanical mechanism for the locks is basically a reversible polarity solenoid.
   Technician B says only the driver's door lock will unlock or lock all the doors.
   Who is correct?
   A. A only C. Both A and B
   B. B only D. Neither A nor B
9. *Technician A* says information systems use data from the PCM to compute some information for display.
   Technician B says the information may be computed and displayed by a computer module linked the PCM.
   Who is correct?
   A. A only C. Both A and B
   B. B only D. Neither A nor B
10. Power seats are being discussed.
    Technician A says many power seats are controlled by BCM's
    Technician B says many power seats are controlled by PCM's.
    Who is correct?
    A. A only C. Both A and B
    B. B only D. Neither A nor B

CHAPTER 7
ENGINE SYSTEMS OPERATION

Upon completion and review of this chapter, you should understand and be able to:

- Describe the design of an internal combustion engine and its general operation.
- Understand the operation of a four-cycle internal combustion engine.
- Explain the purpose of bearings.
- Identify the different types, construction, and uses of automotive bearings.
- Name the primary differences between gasoline- and diesel-fueled engines.
- Explain the function and operation of the starting and charging system.
- Describe the function and operation of the lubrication system and engine oils.
- Identify the types of greases used in a typical light vehicle.
- Identify the different types, construction, and uses of gaskets.
- Recognize the types and uses of chemical automotive sealants.
- Identify the different types, construction, and uses of seals.
- Describe the function and operation of the cooling system and coolant.
- Explain the function and operation of the intake and exhaust system.
- Describe the function and operation of emission control system components.
- Discuss the function and operation of an engine computer.
- Describe the computer-controlled components on an engine.

Terms To Know

Actuator
Axial load
Analog
Antifreeze
Ball bearing
Bearing
Bearing journal
Block
Block configuration
Bore
Bottom dead center (BDC)
Boxer engine
Bushing
Bypass valve
Carbon monoxide (CO)
Catalyst action
Cavitation
Check valve
Chemical sealants
Clearance
Clutch position switch
Composites
Digital
Displacement
Dual overhead cam (DOHC)
Electrical
Electrolysis
Evaporative (EVAP) system
Exhaust gas recirculating (EGR)
Four-stroke
Hall-effect switch
Hydrocarbon (HC)
In-line engine
Lifter
Manifold absolute pressure (MAP)
Mass airflow (MAF)
Mechanical linkages
Muffler
Needle bearing
Offset in-line engine

Offset V-line engine
Oil Clearance
Overhead camshaft (OHC)
Overhead valves (OHV)
Oxides of nitrogen (NO_x)
Planetary gear set
Potentiometer
Pressure lubricated
Radial load
Resonator
Rocker arm
Rod bearing
Roller bearing
Room temperature vulcanizing (RTV)
Rotary
Secondary air injection
Sensor
Signal
Slant 6
Slant engine
Solenoid
Splash lubricated
Stroke
Switches
Synthetic motor oil and blends
Throttle position sensor (TPS)
Timing belt
Timing chain
Top dead center (TDC)
Transducer
Transmission range switch
Two-stroke
Valvetrain
Variable valve timing (VVT)
V-block
Vehicle speed sensor (VSS)
Venturi
Viscosity

INTRODUCTION

The powertrain has become the most complex system in the vehicle. The powertrain consists of the engine, transmission, and sometimes the differential. The engine itself has gone through significant changes since its original design, but its purpose is still the same. However, internal combustion engine operations have been refined internally and electronically to increase efficiency and power. There are many moving pieces in a typical vehicle, many of which work against or with others. Each moving part requires some type of lubricant and a means to retain that lubricant on or inside the assembly. Today's modern powertrains are more powerful, smaller, and more efficient than past engines. In this chapter, the use of bearings and assembly sealing will be discussed. In addition, common usage of each will be covered. The engine itself has more computerized sensors and actuators than the entire vehicle had in the mid-1980s. Modern diesel-fueled engines have just as many electronic devices as gasoline-fueled engines. Working on today's light vehicle engine is no longer a Saturday morning job in the backyard.

AUTHOR'S NOTE For the purpose of this text, the term *engine* refers to the assembly that produces power to operate the vehicle. An engine harnesses a chemical reaction that converts energy from one form to another (i.e., gasoline to heat). The term *motor* refers to an electrical device that uses power to perform a task.

ENGINE DESIGN AND COMPONENTS

Mechanical linkages are the piston, connecting rod, and crankshaft.

The basic engine design from the 1890s remains intact. It has been re-engineered, now made of better material, and designed to more exacting manufacturing standards, but it is still basically an air pump. When the starter is engaged, air is forced into the cylinders. Fuel is mixed into the airflow and lit with an electrical spark. The expanding gases (thermodynamic theory) supply the necessary power that pushes on the **mechanical linkages** in the engine (**Figure 7-1**).

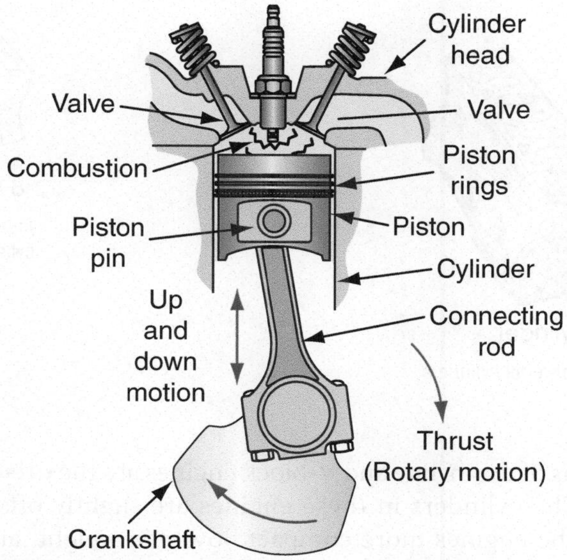

Figure 7-1 The ignition from the spark plug lights the air/fuel mixture and the piston (up-and-down motion) is forced downward onto the crankshaft, thereby turning the crankshaft (rotary motion).

An engine is classified using three criteria: **block configuration, displacement,** and **valvetrain.** Its operation is classified as **two-stroke, four-stroke,** or **rotary.** When an engine is referred to as gasoline or diesel, it is a reference to the fuel used and some small internal component differences.

Engine Block Design

The engine **block** starts as a large block of treated cast iron or aluminum. It is bored, drilled, machined, and polished to form the basic support for all the engine's operating systems, both internally and externally.

The block may be shaped in various configurations. An **in-line engine** has all the cylinders aligned in a series of vertical bores. The most common in-line blocks are four- and six-cylinder designs (**Figure 7-2**). More cylinders involve a longer and heavier block. Four-cylinder engines are available in many late-model vehicles. Six-cylinder, in-line, gasoline engines are used less frequently in modern cars because of the block's weight and size. They are most commonly used in light trucks. Many over-the-road trucks still use the in-line six and in some cases in-line eight or ten cylinders. **Slant engines** are in-line engines that are tilted to the side to allow for a lower overall height, a lower center of gravity, and a lower (more aerodynamic) hood. This design was used almost exclusively by Chrysler. It is an in-line six-cylinder that leans or slants to the right. It became known as the slant engine. This engine is still operating in many types of industrial equipment but not in cars.

Another block configuration is known as the **V-block.** The number of cylinders is divided into two rows set at an angle to each other (**Figure 7-3**). The angle between the rows of cylinders is 90 degrees or 60 degrees. The V-block allows for a lower hood line and for more cylinders to be added in a shorter, linear space. Then, the engine could be more powerful and contained in a smaller block. V-block engines also tend to create less vibration than a comparably sized in-line.

Block configuration is the shape of the block.

Displacement is the volume of air the engine can exchange in one complete engine cycle.

Valvetrain refers to the placement of the engine valves.

A **two-stroke** engine has a power stroke with each crankshaft revolution, whereas a **four-stroke** engine has one power stroke every two revolutions.

Rotary engines have no reciprocating pistons but use a rotor to move the air and fuel mixture and produce power.

The machined portions of a basic **block** are the cylinders, mounting deck for the cylinder head, lifter bores, and the saddles or bores for the bearings.

In-line engine A type in internal combustion engine having its cylinders arranged in a row.

A **slant engine** is the same as an in-line engine, but the cylinders are slanted and not upright.

V-block The V type engine has two rows (banks) of cylinders, usually forming an angle of 60 or 90 degrees between the two banks. see figure 7-3.

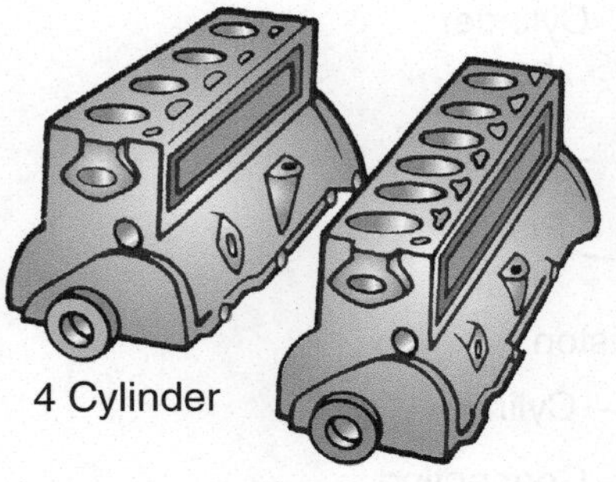

Figure 7-2 In-line engines have all cylinders in a straight line.

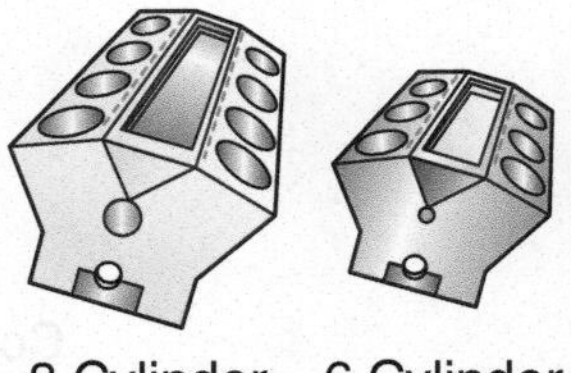

Figure 7-3 V-blocks have the cylinders in two equal rows.

An **offset engine's** cylinders are offset from the centerline of the block.

Boxer engine A horizontally opposed engine has the cylinders set horizontally at either side of the crankshaft. The cylinders are placed opposite each other, and they work lying down.

A simple **planetary gear set** has a sun gear that fits inside a planetary carrier. The carrier, in turn, fits inside a ring gear.

Rotary engine An internal combustion engine that works completely different way than a conventional piston engine. The crankshaft does not rotate.

Composites refer to engine components that are made of plastics and other materials resembling plastics.

Modified versions of the in-line and V-block engines are the **offset in-line** and **offset V-block engines.** The cylinders in these engines are slightly offset from each other. This design makes the engines more compact, lower in weight, and may increase the cooling efficiency.

Volkswagen made the horizontally opposed engine famous during its use in the original VW Beetle. The rows of cylinders are laid out 180 degrees apart (**Figure 7-4**). This provided a very low profile and air cooling. The engine was mounted in the rear of the vehicle. Subaru still uses a front-mounted, horizontally opposed engine in some of its models. Porsche used this configuration and called it the **Boxer engine**. Subaru uses this configuration in many of its current models.

Mazda uses another engine type for its RX7 and RX8 sports car models that does not use pistons or a cylinder to generate and transfer the combustion power to the crankshaft. Instead, two triangular rotors within matching oblong chambers intake, compress, and exhaust the air and fuel mixture (**Figure 7-5**). The rotors are attached to the crankshaft through a **planetary gear set,** which increases the force delivered by the rotors (**Figure 7-6**). This design is known as a **rotary engine**.

Fuel economy has caused some types of engines to be discontinued in many car models. The most common types of engines in today's vehicles are in-line four-cylinder, V-6, and V-8. V-8 engines have come and gone with fuel prices and consumer trends. The V-8 engine produces more horsepower, and new electronics and better engineering designs keep the fuel mileage and emissions within mandated limits, desirable for light truck owners. Engines are constructed of gray cast iron or aluminum. Some engine components are constructed using aluminum for the savings in weight and better heat transfer capabilities. A few research engines have been made of ceramic in an attempt to reduce weight and for durability. Ceramic material is expensive, but the research provides extensive data on how to use other materials. You will find that many engine components of modern vehicles are made of **composites.** They are lighter, can transfer heat, and are cost-effective to manufacture.

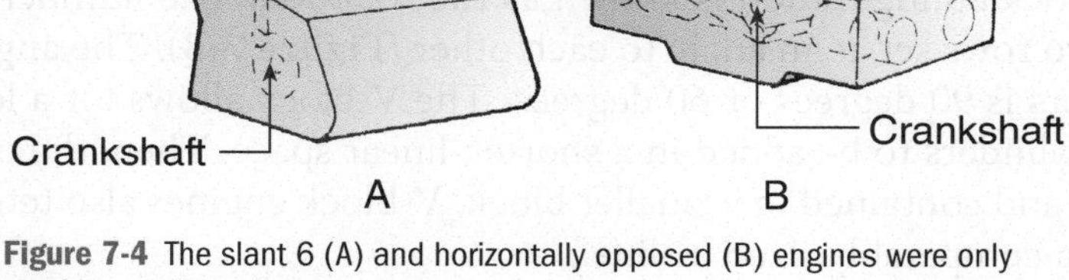

Figure 7-4 The slant 6 (A) and horizontally opposed (B) engines were only used in certain makes.

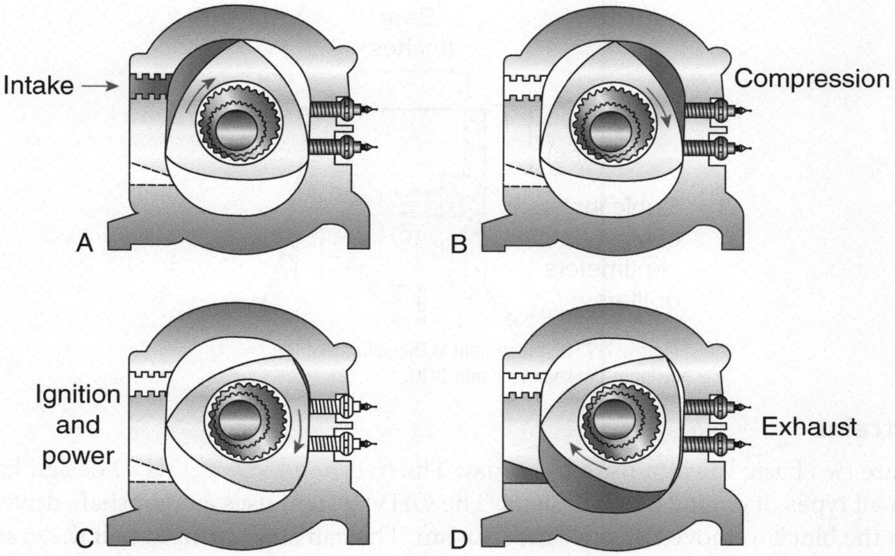

Figure 7-5 The rotary's oval chamber is shaped to create the same internal combustion conditions found in a piston-driven engine. The rotary engine is properly known as a Wankel engine.

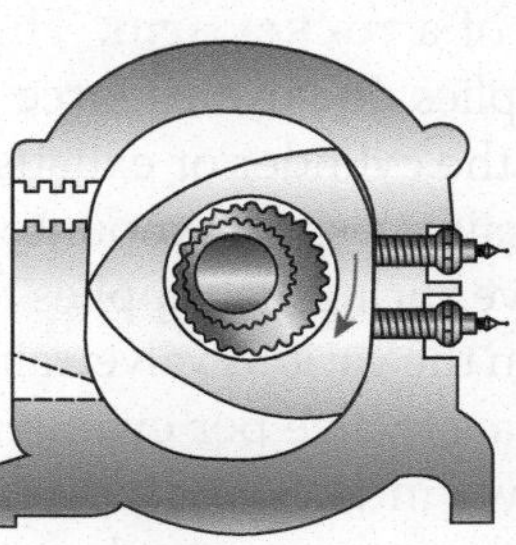

Figure 7-6 Planetary gear sets are also found in transmissions and transaxles. They are used to multiply and transfer torque.

Displacement

The amount of air an engine can move in one cycle is its displacement. **Displacement** is the total volume of the cylinders calculated by using the bore, stroke, and number of cylinders (**Figure 7-7**). The **bore** is the diameter of the cylinder. The **stroke** is measured from the top of the cylinder to the top of a piston that is resting at **bottom dead center (BDC).** The volume of a cylinder is equal to the square of the bore's radius multiplied by 3.1414 and then multiplied by the stroke. The volume is multiplied by the number of cylinders to find total engine displacement. Typical displacements of modern engines are 1.6, 2.0, 3.0, 3.8, 4.6, 4.7, and 5.3 liters (L). There are others that vary in displacement, but they generally fall within the ranges mentioned here. Engine size (displacement) is usually listed by both cubic inches and liters (imperial and metric systems). For example, a 6.0-L engine is also a 364-cubic-inch engine.

Engine **displacement** is the amount of air an engine can move through it. It may be measured in cubic inches displacement (cid), liters (L), or cubic centimeters (cc).

The **bore** of an engine is the diameter of the cylinder.

The **stroke** is the distance that the piston travels in the cylinder.

A Comparison—Old Versus New

During the 1950s and 1960s, high-performance, high-horsepower engines ranged from 350 cid (5.7 L) to 440 cid (7.2 L). Stock engines were capable of producing up to 400 horsepower and high vehicle speeds, but they were fuel-hungry and produced enormous amounts of harmful emissions. The newest engines can produce 300 horsepower with high fuel economy, high vehicle speeds, and very low levels of harmful emissions, all with a smaller displacement.

BDC The point at which the piston of an engine is nearest the lowest point or bottom of the pistons travel closest to the crankshaft. This is known as bottom dead center.

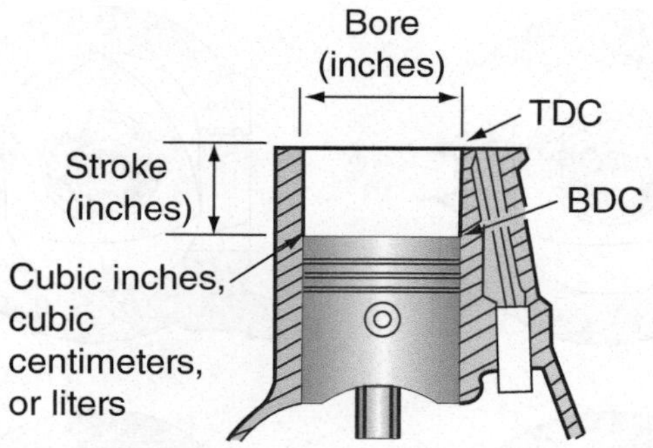

Figure 7-7 Displacement is the volume of the cylinder between TDC and BDC.

Valvetrains

There are two basic valvetrains in use today. The **overhead valve (OHV)** design has been used in all types of engine block designs. The OHV system uses a crankshaft-driven camshaft in the block to move the valve mechanism. The camshaft turns at half the crankshaft speed and must be timed to open and close valves in time with piston position (**Figure 7-8**). The camshaft pushes upward on a solid or hydraulic **lifter,** moving a pushrod. The push-rod applies upward force to one end of a **rocker arm.** The rocker arm has a pivot-mounted fulcrum, and the opposite end applies downward force to the valve (**Figure 7-9**). With the valve open, air and fuel can enter the cylinder or exhaust can exit. As the camshaft rotates, force is removed from the valvetrain. A spring mounted around the valve stem forces the valve upward and closed. The valve spring also applies force throughout the mechanism to maintain constant contact between the various valvetrain components (**Figure 7-10**). There is usually one intake and one exhaust valve per cylinder. Many newer, more powerful and fuel-efficient engines may have two intakes and two exhaust valves. This provides better flow characteristics for the air/fuel mixture and exhaust gases.

The **valvetrain** consists of valves, valve springs, a camshaft, valve-retaining devices, and may have valve **lifters**, pushrods, and **rocker arms**.

AUTHOR'S NOTE Many modern OHV engines are designed to work with variable displacement engine systems. These systems can shut down cylinders when they are not needed, thus increasing fuel economy and reducing emissions.

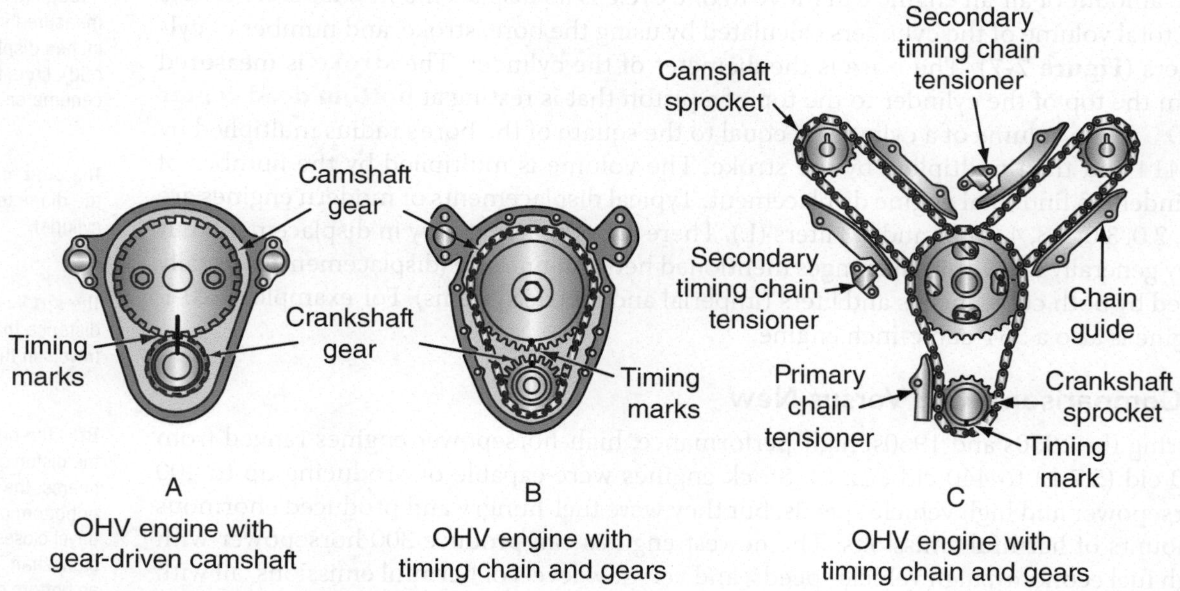

Figure 7-8 Three different types of timing mechanisms.

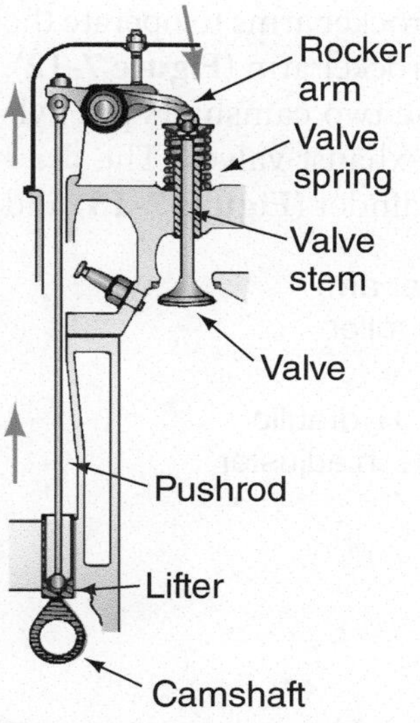

Figure 7-9 The rocker arm makes the connection between the pushrod and the valve stem.

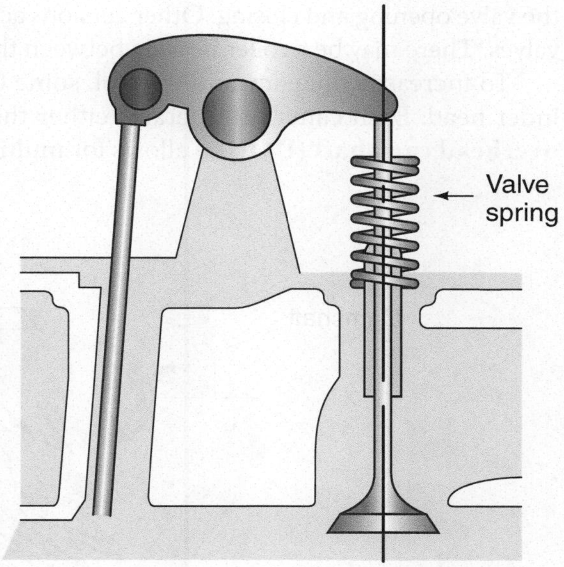

Figure 7-10 The valve spring applies continuous force on all components to prevent sloppy or noisy movement and returns the valve to the closed position.

Many of today's modern engines use a **camshaft** installed in the cylinder head with the valves. This is known as an **overhead camshaft (OHC)** engine. The camshaft may be mounted directly over the valve and exerts downward force directly onto the valve stem (**Figure 7-11**). This system eliminates the lifter, pushrods, and rocker arms, providing better durability, less maintenance, more efficiency, and less noise. Some versions of the OHC system have small, hydraulic, lifter-type mechanisms that absorb the vibration and shock of

Camshaft The mechanical component of an internal combustion engine that opens and closes the intake and exhausts valves.

An **overhead camshaft engine (OHC)** has one camshaft per cylinder head.

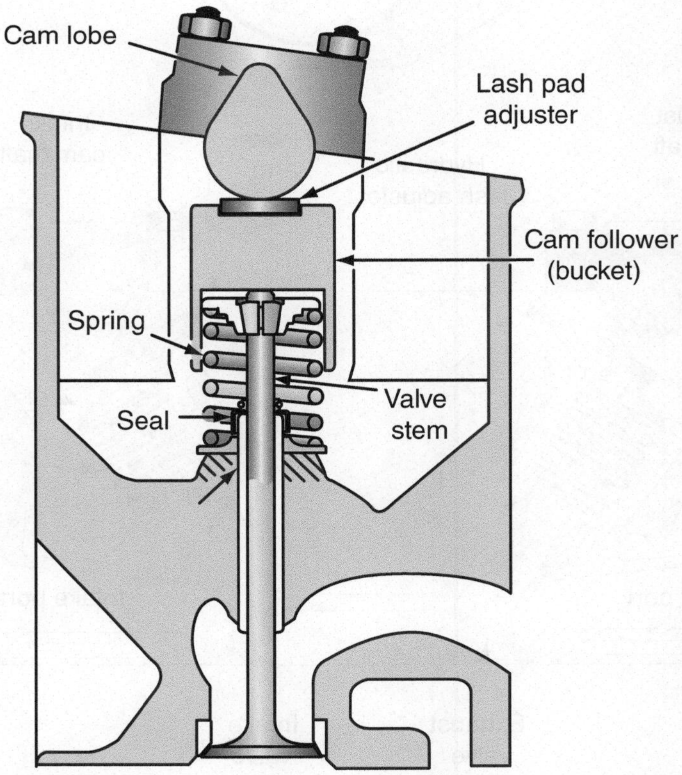

Figure 7-11 Overhead camshafts eliminated the lifter and pushrods.

the valve opening and closing. Other versions act directly on the rocker arms to operate the valves. There may be a roller bearing between the camshaft and rocker arm (**Figure 7-12**).

To increase efficiency and control, some OHC engines use two camshafts per cylinder head. Each camshaft operates either the intake or the exhaust valves. The **dual overhead camshaft (DOHC)** allows for multiple valves per cylinder (**Figure 7-13**) and

The term **dual overhead camshaft (DOHC)** does not represent a different valvetrain. It's a version of the OHC system.

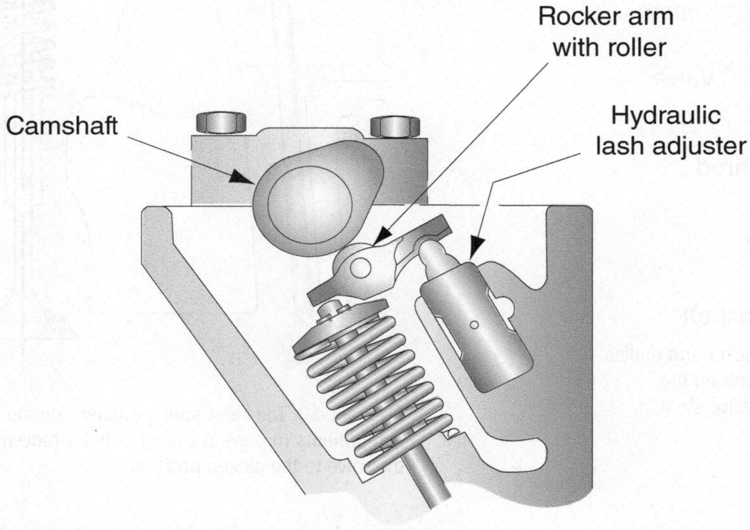

Figure 7-12 The roller bearings reduce camshaft wear and provide quieter operation.

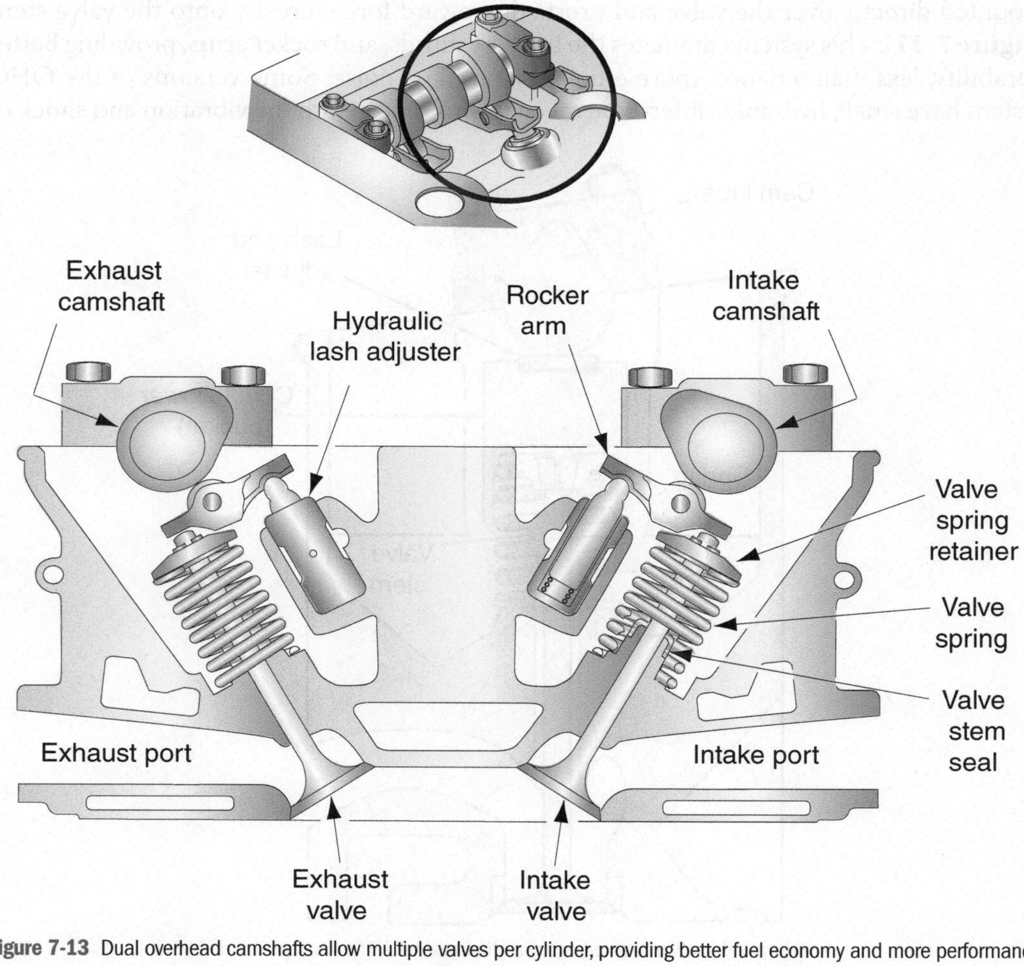

Figure 7-13 Dual overhead camshafts allow multiple valves per cylinder, providing better fuel economy and more performance.

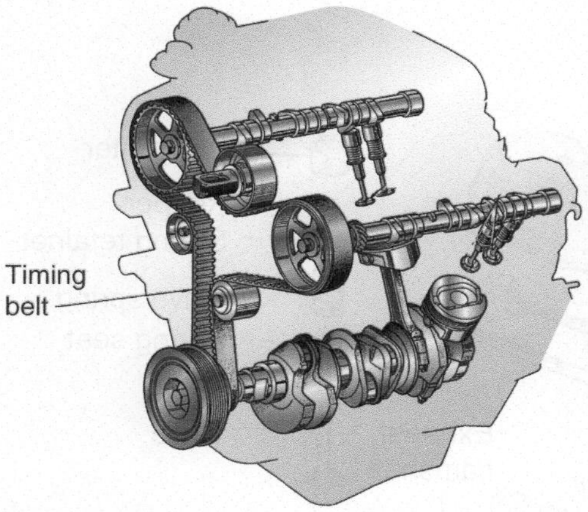

Figure 7-14 Timing belts are quicker and easier to maintain than chains.

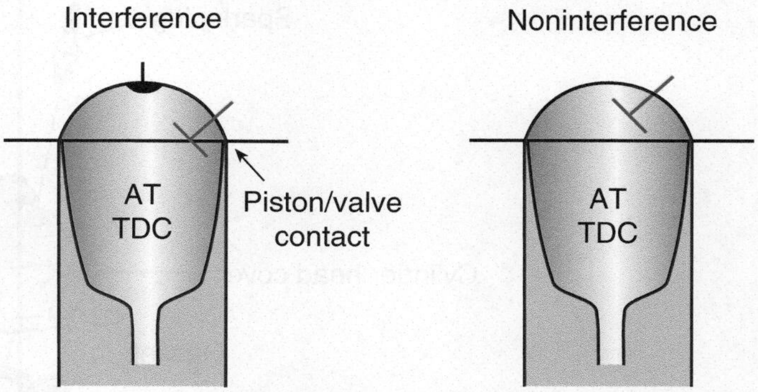

Figure 7-15 Noninterference engines keep the valve within the combustion chamber.

allows for independent control of the valves. Each camshaft is timed to each other and the crankshaft. Many modern DOHC engines use a technology called **variable valve timing (VVT).**

Variable valve timing (VVT) is a computer-controlled system that changes the timing of the valves by changing the timing and position of the camshaft.

The camshafts on an OHV engine are driven through meshed timing gears or a **timing chain** connecting the crankshaft and camshaft gears. When worn, both systems must be replaced as a set. The gears or chain are not readily visible, and noise or a breakdown is the first sign of a trouble. A chain and gear may drive the camshafts on an OHC system, but many use a flexible **timing belt.** The belt does not require lubrication, is quieter, and is replaced based on mileage or age (**Figure 7-14**). If the belt or chain is replaced at the recommended intervals, breakdowns because of failure become very uncommon. Timing chains tend to make a lot of noise before they break. Timing belts do not make noise before they break. Failure to replace the timing belt as scheduled can result in severe damage to the engine. If the timing belt breaks in an interference engine, a stuck open valve could hit a moving piston. A freewheeling engine will not do this.

The camshaft is connected to and driven by the crankshaft by a **timing chain** or **timing belt.**

An interference engine's pistons and valves operate very close to each other under normal operation. If not controlled (like when a timing belt breaks) the valve at full extension will hit the piston when it comes up to top dead center (TDC). The piston will break or bend the valve. Likely, the piston, cylinder wall, cylinder head, and valve will be damaged.

Noninterference or freewheeling engine valves open into the combustion chamber but do not extend to the point where it can touch the piston (**Figure 7-15**).

AUTHOR'S NOTE Owners who do not follow the replacement interval for timing belts may be forced to pay a lot of money for an engine repair. The problem usually results from the natural tendency of the driver to restart the engine when the timing belt is broken. The engine will not restart, and it only takes one revolution or less of the crankshaft to do serious internal engine damage.

A typical engine is an assembly of four major subassemblies or components. They are the cylinder block assembly, cylinder head, intake manifold, and exhaust manifold (**Figure 7-16**). The cylinder block assembly houses the pistons, connecting rods, crankshaft, and sometimes the camshaft and valve lifters (OHV only). The connecting rod is the component that connects the piston and crankshaft. These are all rotating parts. The crankshaft is then connected to the flywheel (or flexplate) and then to the transmission.

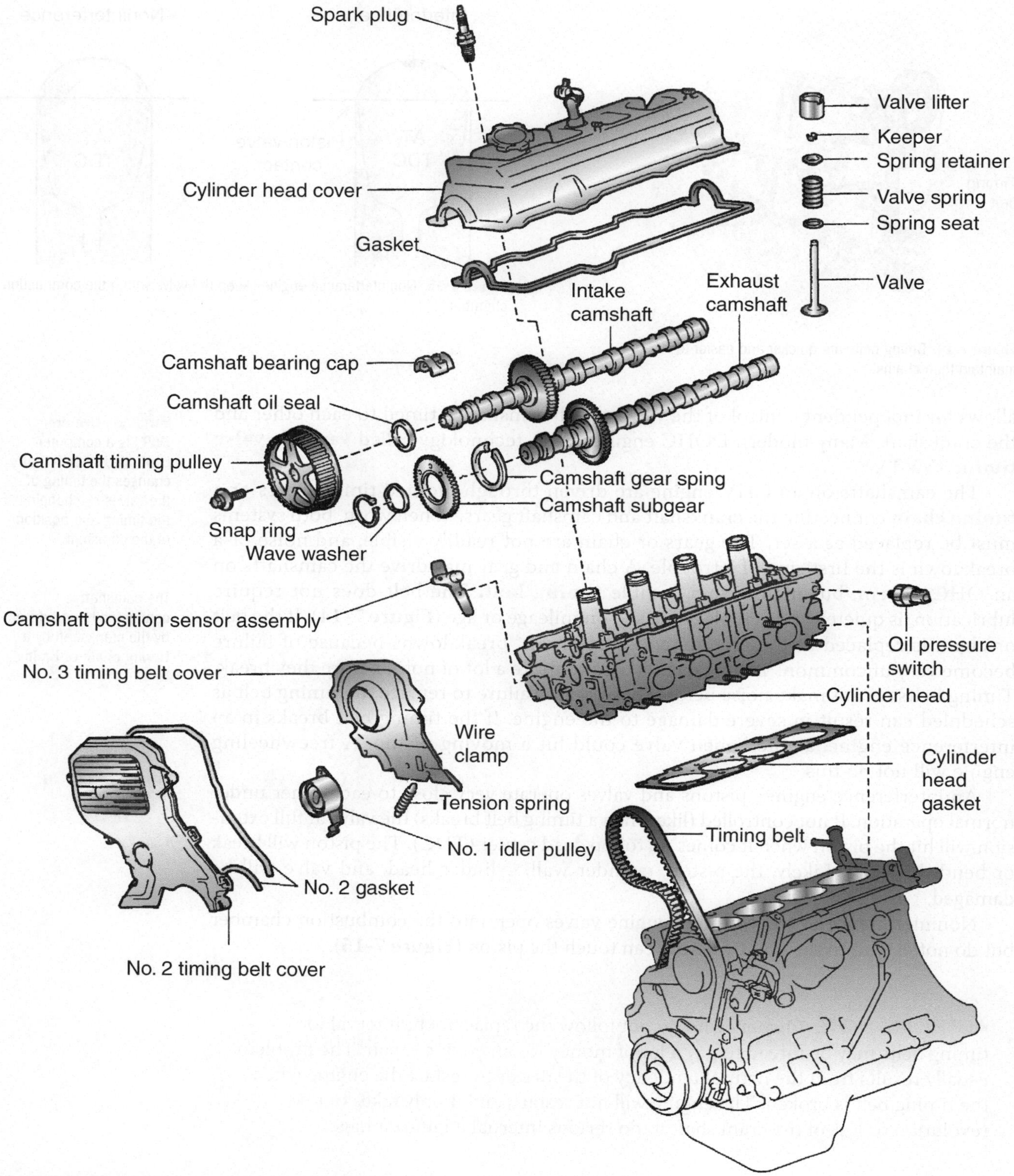

Figure 7-16 The engine is an assembly of hundreds of moving and stationary components.

An oil pump is mounted either into the block or suspended from the bottom of the block (**Figure 7-17**). External areas for mounting brackets of external components are machined on the block.

On top of the block is the cylinder head that houses the valvetrain. The valvetrain consists of the valve, valve spring, and any other components needed to move the valve mechanism. On an OHC engine, the camshaft is mounted in the head (**Figure 7-18**). In the cylinder head and block, passageways for oil and coolant flow exist within the component and cross over into the other components. There are external ports in the cylinder head(s) that match the ports on the intake and exhaust manifolds.

The two manifolds are mounted to the outside of the cylinder heads (**Figure 7-19**). An in-line engine has one of each. A V-block has two exhaust manifolds with one intake

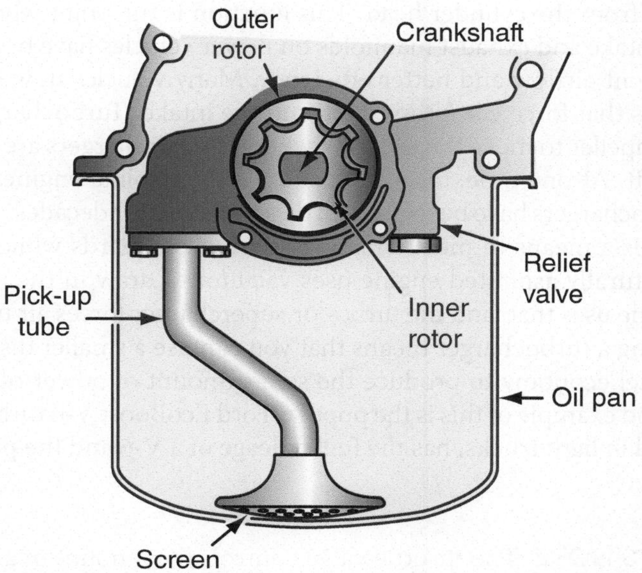

Figure 7-17 Some newer oil pumps are mounted internally and driven directly by the crankshaft.

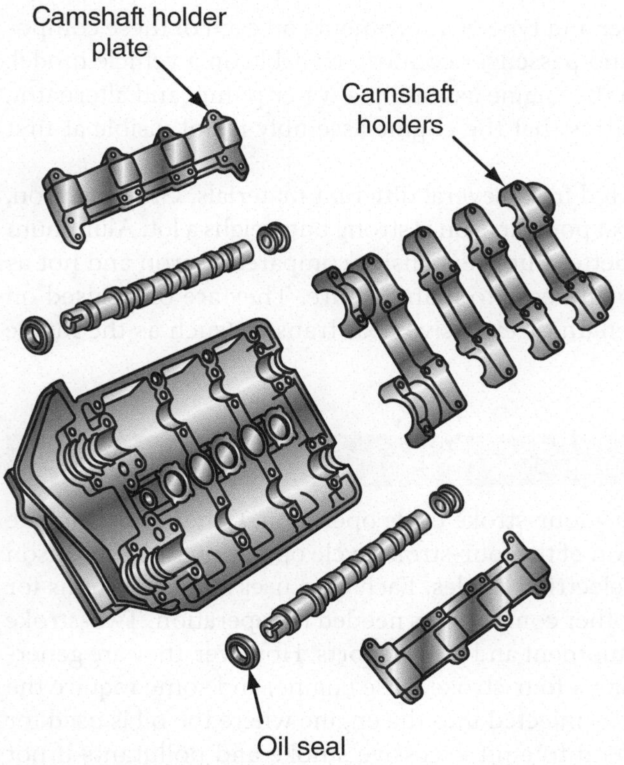

Figure 7-18 Overhead camshafts are mounted over or next to the valves.

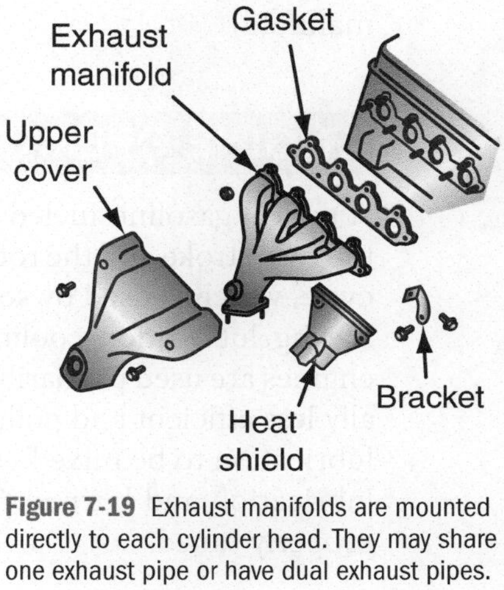

Figure 7-19 Exhaust manifolds are mounted directly to each cylinder head. They may share one exhaust pipe or have dual exhaust pipes.

manifold placed in the V between the two cylinder heads. The exhaust manifold has no moving parts but may have threaded holes for mounting emission-control devices. The intake manifold also has external mounting areas for attaching items such as an upper intake manifold, fuel delivery, and a throttle valve. The throttle valve controls the amount of air that will enter the intake manifold. Intake manifolds have runners that connect the air entrance to each of the cylinders. The runners are designed to deliver the same volume of air to each cylinder. Exhaust runners are provided for each cylinder, but they are joined at a point away from the cylinder head. This junction is the point where the exhaust pipe is connected. Intake and exhaust manifolds on newer vehicles have been designed to provide more consistent airflow and better efficiency. Many vehicles now have turbochargers or superchargers that force compressed air into the intake. Turbochargers use exhaust gases to drive an impeller to move the extra air, whereas superchargers are gear- or belt-driven by the crankshaft. At one time, racecars had the only gasoline engines using superchargers, whereas turbochargers have been used on diesel engines for decades. The recent use of these components is a means to meet emission and fuel standards while boosting the engine's output. A naturally aspirated engine uses vacuum to draw in the air. A turbo- or supercharged engine uses that, and the turbo- or supercharger forces air under pressure into the engine. Adding a turbocharger means that you can use a smaller displacement engine that gets better fuel economy to produce the same amount of power of a larger displacement engine. A good example of this is the popular Ford EcoBoost V-6 turbocharged engine. This engine, found in light trucks, has the fuel mileage of a V-6 and the power of a V-8.

AUTHOR'S NOTE The throttle valve controls the amount of air that enters the intake manifold. The intake valve(s) controls the amount of air that enters the combustion chamber.

It should be noted that the number and type of attachments on each of these components depend on the engine design and passenger comfort available on a vehicle model. It could be a very simple system with the engine assembly, a water pump, and alternator, or a system that is so loaded with extras that the engine assembly is not visible at first glance under the hood.

Engine components are constructed from several different materials. Gray cast iron, aluminum, and composites are the most popular. Iron is strong but weighs a lot. Aluminum is lightweight and can transfer heat better but is expensive compared to iron and not as strong. Plastic composites are cheap and easy to manufacture. They are often used on components that do not require strength or excessive heat transfer (such as the intake manifold).

ENGINE OPERATION

A typical gasoline-fueled engine uses a four-stroke cycle operation. Other types include the two-stroke and the rotary. A version of the four-stroke cycle operation is the Atkinson cycle, which is used by some hybrid electric vehicles. Each type uses similar systems for fueling, lubrication, cooling, and the other components needed for operation. Two-stroke engines are used primarily in light equipment and powersports. However, they are generally less efficient and pollute more than a four-stroke cycle engine, and some require the lubrication to be mixed with the fuel or injected into the engine where the oil is used for lubrication and burned. They also tend to emit excessive smoke and pollutants if not properly tuned.

Two-Stroke Operation

The two-stroke engine intakes and compresses the fuel/air mixture during the upward movement of the piston. The downward movement of the piston during the power stroke rotates the crankshaft and opens the exhaust valve. The lower portion of the piston's next upward movement not only starts clearing the cylinder of exhaust but also begins the intake of fresh air and fuel. The operation performed on this stroke is known as scavenging because of the method of clearing and filling the cylinder. Intake and compression are completed on one stroke while power and exhaust are accomplished on the next (**Figure 7-20**). Intake and exhaust valves are used to open or seal the cylinder. Power is produced on each revolution of the crankshaft.

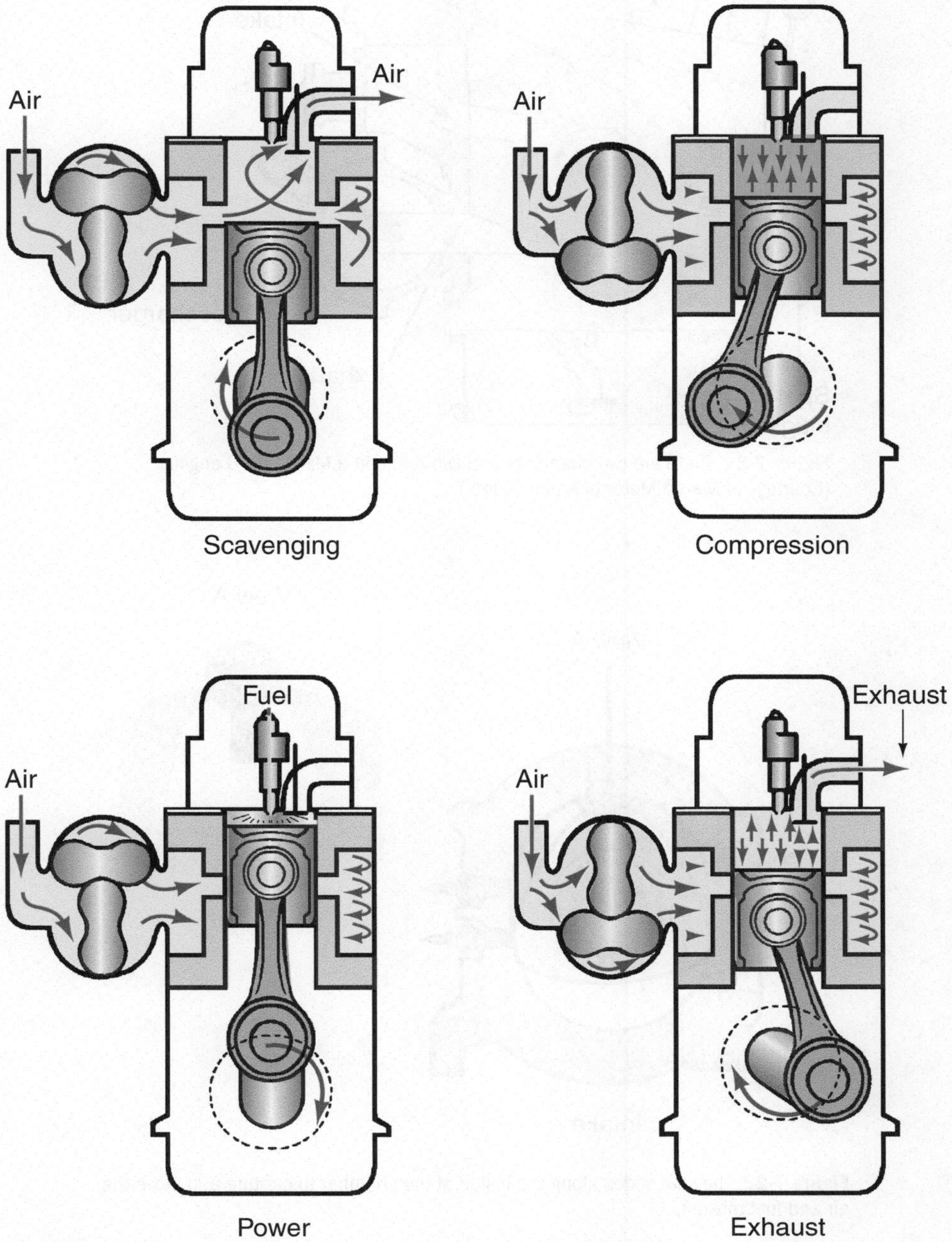

Figure 7-20 Some two-stroke engines use rotary valves to control the flow of the air and fuel.

Rotary Engines

In a rotary engine, the air/fuel mixture is compressed using a triangular rotor inside an oblong chamber (**Figure 7-21**). At each apex (corner) of the rotor, a seal keeps contact with the chamber wall and pushes the air/fuel ahead of it or is pushed by expanding gases (**Figure 7-22**).

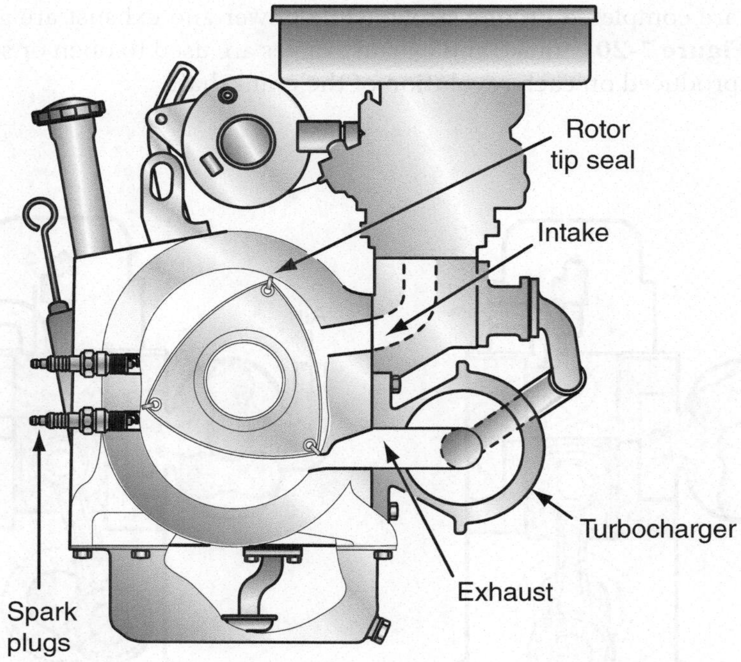

Figure 7-21 There are two chambers and two rotors in a Mazda RX-8 engine. (Courtesy of Mazda Motor of America, Inc.)

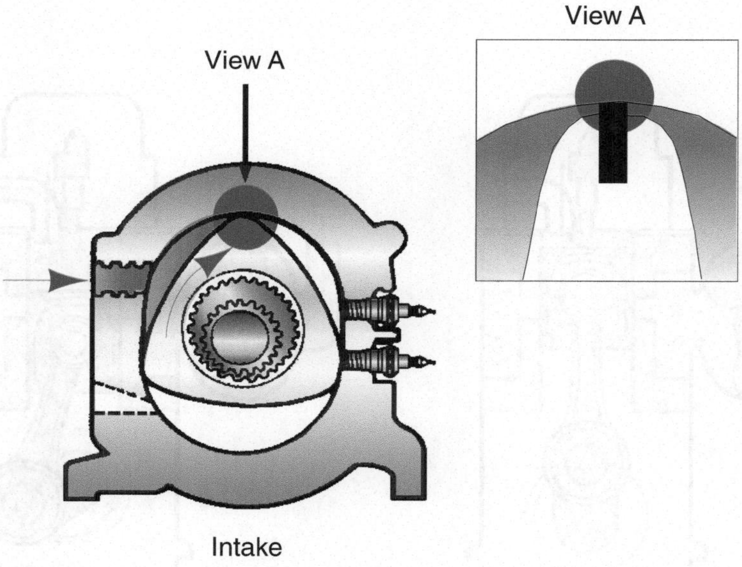

Figure 7-22 The seal slides along the inside of the chamber to capture and move the air and fuel mixture.

As the rotor turns, a seal forces the air/fuel mixture into a small area of the chamber. This reduced chamber volume compresses the mixture until ignition occurs. The expanding gases are directed against the seal ahead of the compression seal. This power is transmitted through the rotor to the crankshaft. As the rotor continues to turn, the gases enter a larger area where the spent gases are exhausted. As this area widens, a low-pressure area is formed under the intake valve. A new mixture is drawn in and the cycle is repeated. It should be noted that each side of the rotor and each seal act as the power, intake, compression, and exhaust component during each rotation of the rotor. Because the rotor always turns in one direction without the stop-and-start motion of a piston, the rotary engine can produce more horsepower with less wasted energy than a conventional reciprocating piston gasoline engine. However, the rotary engine does not produce the same torque as a piston engine of the same displacement and is considered less efficient. Rotary engines have the same problem with harmful emissions as piston engines.

AUTHOR'S NOTE In February 2006, Mazda introduced the Premacy Hydrogen RE engine in its hybrid vehicle. The Mazda RX-8 is not a hybrid vehicle, but it does offer a hydrogen-fueled engine known as the Hydrogen RE.

Four-Stroke Operation

The four-stroke engine is used in nearly every car and truck currently being produced. A power stroke is produced every two revolutions of the crankshaft. This engine requires four strokes (movements) of the piston to complete one complete cycle. One complete cycle occurs during two revolutions of the crankshaft (720 degrees).

To explain how a complete cycle—or four strokes—is accomplished, the crankshaft has to be rotated either by the starter during cranking or by the power of another piston. The starter is connected to the flywheel (or flexplate) by a gear, but only when the starter is engaged. The flywheel (or flexplate) is then connected directly to the crankshaft, and they rotate together. Without crankshaft rotation, all the pistons remain stationary. Each stroke—that is, one complete up or down movement of the piston—turns the crankshaft 180 degrees.

With a piston producing only one power stroke per complete cycle (four strokes), three of the four strokes use power produced by the other pistons power strokes. The compression stroke uses a lot of power to compress the air/fuel mixture. Remember this as the four strokes are discussed.

The first stroke is the intake stroke. During this stroke, the piston is pulled downward from **Top Dead Center (TDC)** by the movement of the crankshaft. During this movement, the camshaft opens the intake valve, which allows air and fuel to enter the combustion cylinder (**Figure 7-23**). The down movement of the piston causes low pressure (vacuum) above it within the cylinder. Atmospheric pressure (14.7 psi) forces the air past the intake valve into the space vacated by the piston. Fuel is mixed at the intake manifold's entrance or injected near the intake valve. Diesel and gasoline direct injection (GDI) engines inject the fuel directly into the cylinder at this point. When the piston is at or near the lowest point of its travel, the intake valve closes. At 180 degrees of crankshaft rotation—also known as BDC—the piston is forced upwards. This second stroke is called the **compression stroke** (**Figure 7-24**).

TDC The position of the cylinder when it is the farthest from the crankshaft or at the top of the pistons travel.

Compression stroke The stroke in an engine which the air/fuel mixture is compressed before ignition.

The force required to move the piston upward during the compression stroke takes away from the engine's total output. A few degrees of crankshaft rotation Before Top Dead Center (BTDC), the spark plug fires, starting the ignition of the air and fuel mixture

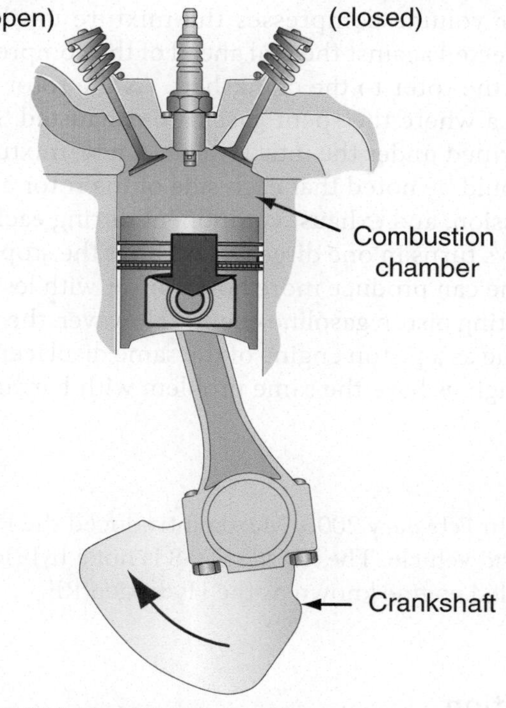

Figure 7-23 Air and fuel are drawn in during the intake stroke.

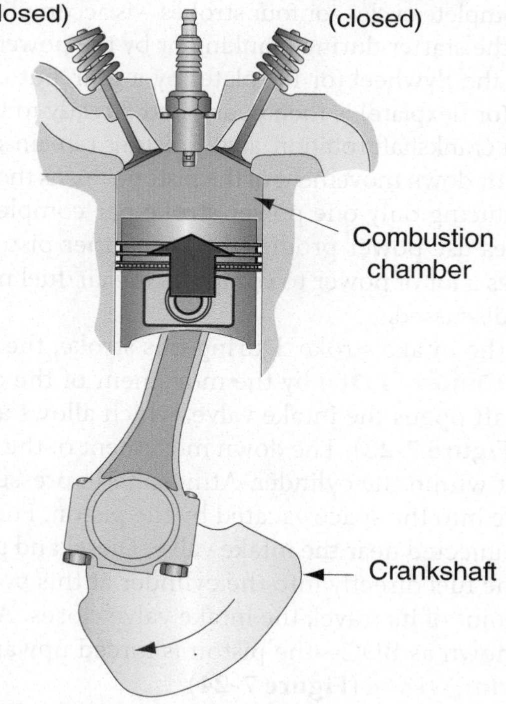

Figure 7-24 With the air and fuel trapped in the combustion chamber, the piston moves upward on the compression stroke.

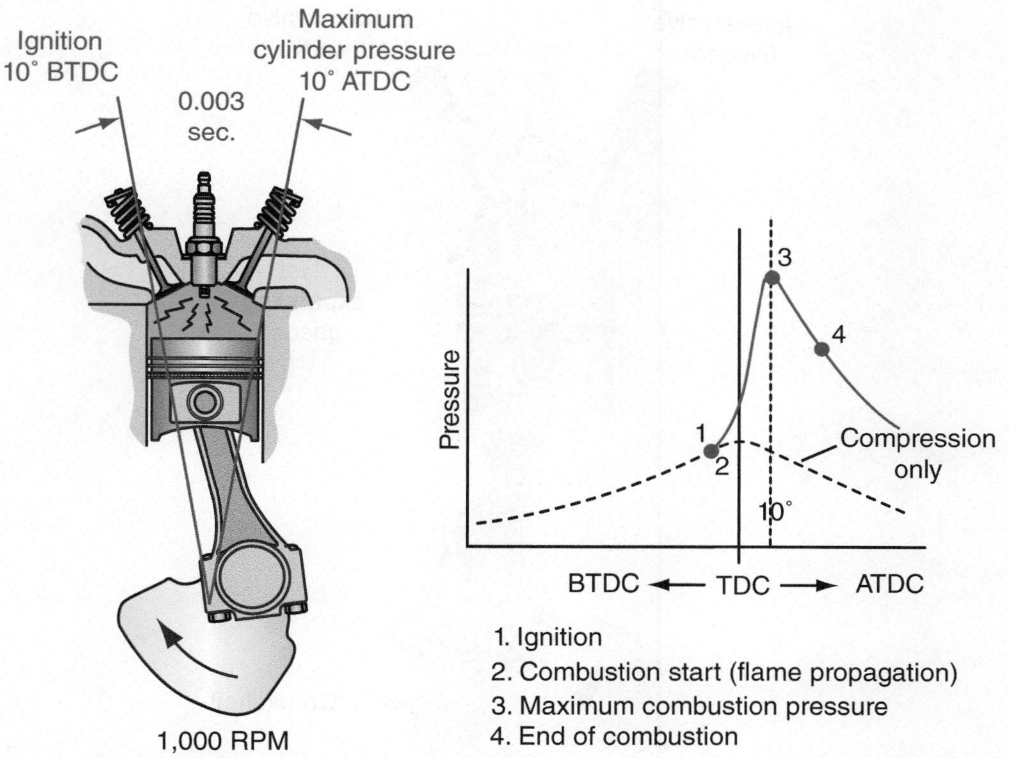

Figure 7-25 At idle, the spark plug fires about 10 degrees before the piston reaches TDC on the compression stroke.

(**Figure 7-25**). This allows the flame time to move across the combustion chamber in a controlled burn and have maximum force just as the piston arrives at TDC. A few degrees late or early in the firing of the spark plug can greatly reduce the power applied to the piston and can affect the emissions, drivability, and economy of the engine. So far two strokes, intake and compression, have taken place and one complete rotation of the crankshaft has been accomplished, but no power has yet been produced from the piston.

The third stroke is the power stroke. It is the only stroke where a single piston powers the crankshaft, other moving components of the engine, and the vehicle. The expanding gases force the piston down rotating the crankshaft (**Figure 7-26**). The crankshaft converts the linear (up/down) movement of the piston into the rotary motion needed for driveline action. The intake and exhaust valves are both closed completely during the compression and power stroke. As the piston nears, or just after, BDC the camshaft opens the exhaust valve. The crankshaft is now being driven by another piston forcing the highlighted piston upward again. The upward movement of the piston forces the spent exhaust gases from the cylinder into the exhaust system. Just before the piston reaches TDC, the camshaft begins to close the exhaust valve and begins to open the intake valve. This is called **valve overlap,** and it is designed to get a greater amount of fresh air and fuel into the cylinder more quickly when the piston begins the next intake stroke (**Figure 7-27**). At TDC center, the exhaust valve is completely closed while the intake is still in the process of becoming fully open. The cycle begins again.

Valve overlap The period during the valve timing where both the intake and exhaust valves are open.

The four strokes explained here add up to one complete cycle of one piston. This takes two revolutions (720 degrees) of the crankshaft to get one power-producing stroke. This single power-producing stroke is one reason that an internal combustion four-stroke engine is not considered to be extremely efficient either in power produced or fuel

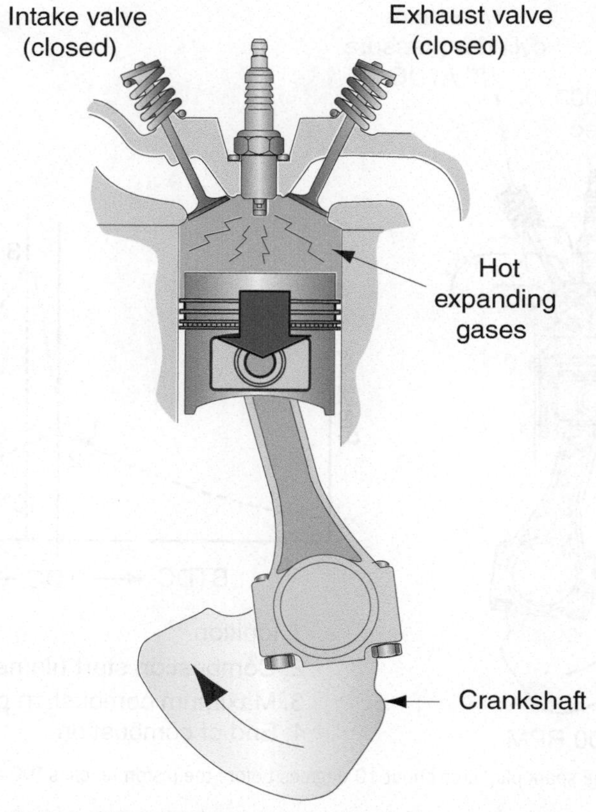

Figure 7-26 The hot expanding gases force the piston downward during the power stroke.

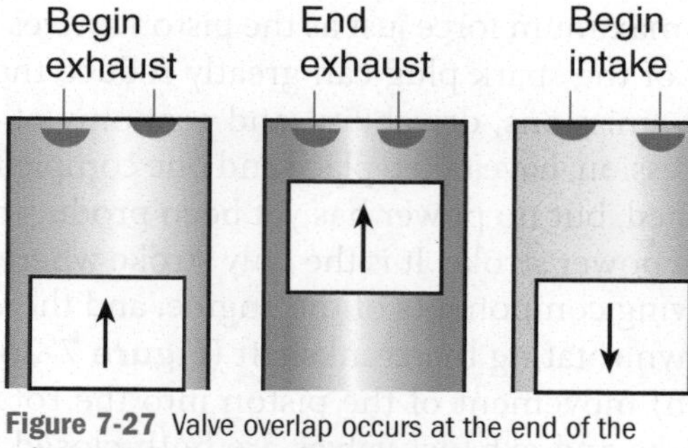

Figure 7-27 Valve overlap occurs at the end of the exhaust stroke and the beginning of the intake.

mileage. Most four-stroke engines are 20 percent to 30 percent efficient. **Figure 7-28** shows the entire four-stroke cycle.

The timing of the valves' opening and closing, the ignition timing, and the air/fuel mixture are critical to smooth engine operation. If any of the three is worn, broken, or out of adjustment, the engine will perform poorly and may be seriously damaged. Other operating systems of the engine can cause similar problems.

Newer technologies allow the engineers to design an engine that can change valve timing and valve opening, called VVT. The Powertrain Control Module (PCM) uses sensor data to change intake and exhaust valve timing and amount of opening (**Figure 7-29**). This helps reduce harmful emissions, increases fuel mileage, and increases power. Some engines use a system called displacement on demand (also known as active fuel management). This system controls the valves by closing the valves and stopping fuel flow to a few cylinders during low-load operating conditions. When you are cruising down the

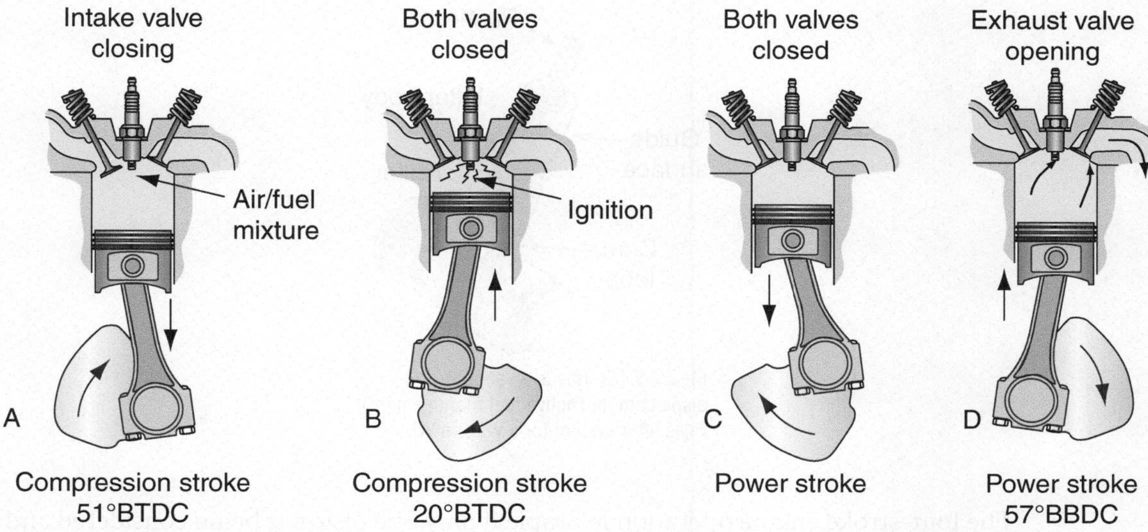

Figure 7-28 The four strokes of an automotive engine: (A) intake stroke, (B) compression stroke, (C) power stroke, (D) exhaust stroke.

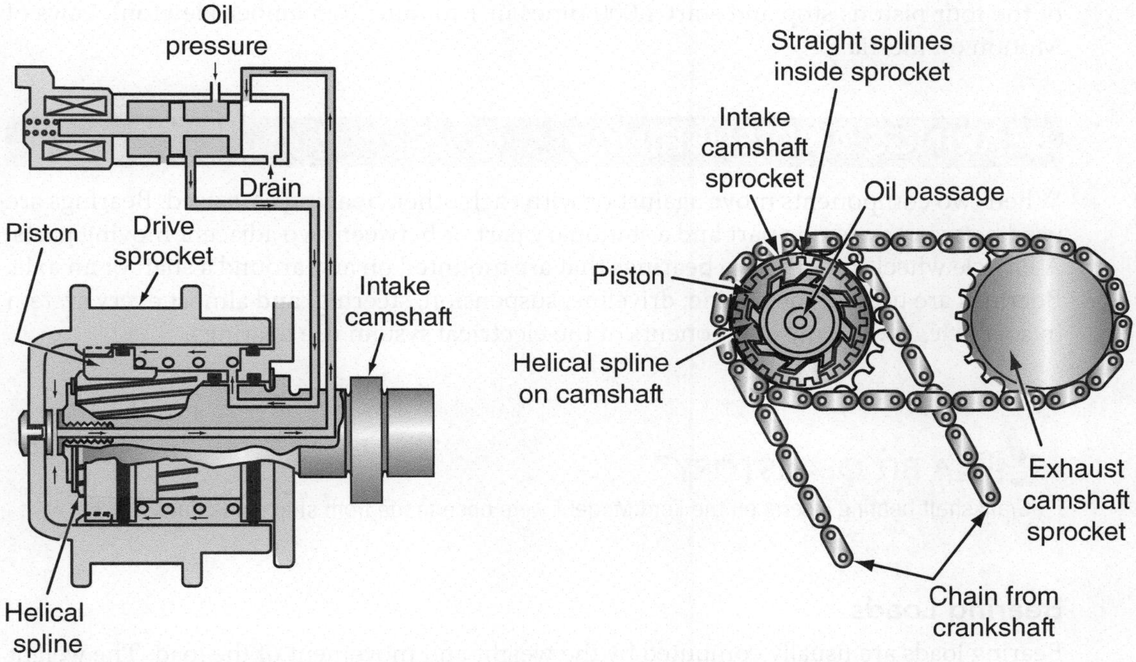

Figure 7-29 This is a cam phasing actuator; it is one of the main components of a variable valve.

highway you don't need the power of a V8, so four cylinders will shut off and the remaining four have enough power to keep it going. But when you accelerate, all eight cylinders are working (**Figure 7-30**).

Some of the engines offered since model year 2007 have virtually eliminated the mechanical controlling of the valvetrain by using electrical solenoids to control the opening and closing events of the valves. This provides an almost endless number of engine possibilities to reduce fuel usage, increase performance, and produce fewer harmful emissions. It also allows an engine to be more versatile in different driving situations and environments. Additionally, some vehicles can shut down selected cylinders when the engine is not under heavy load. A V-6 engine could operate just on two or four cylinders when cruising on flat roads with no load (Figure 7-30).

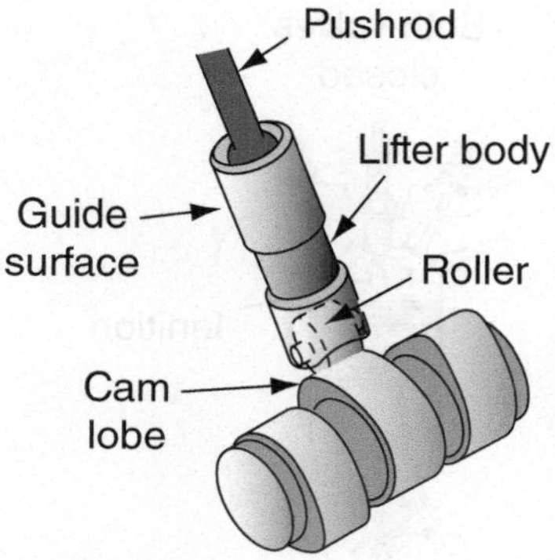

Figure 7-30 This is a variable displacement (active fuel management) valve lifter system for a V8 or a V6.

The four-stroke engine operation is simple if only one piston is being considered and the engine speed is low. However, when multiple pistons, multiple valve components, piston weight, piston speed, friction, timing, and other stress factors are considered, the operation is no longer simple. A four-cylinder engine running at 2,000 rpm will have each of the four pistons stop and start 4,000 times in 1 minute. Remember Newton's Laws of Motion on inertia.

BEARINGS AND BUSHINGS

Bearings reduce the friction, heat, and wear of moving components.

When two components move against or with each other, **bearings** are used. Bearings are used between a moving part and a stationary part or between two adjacent moving parts. A bicycle wheel spins on two bearings that are mounted on and around a shaft or an axle. Bearings are used in the engine, driveline, suspension, steering, and almost every system in a vehicle. Even some components of the electrical system use bearings.

A BIT OF HISTORY

Crankshaft bearing inserts on the Ford Model T were once made from strips of leather.

Bearing Loads

Radial load is produced by the component pressing at 90 degrees to the sides of the bearing.

Axial load is produced by the component pushing directly against the bearing from inside out or vice versa.

Bearing loads are usually computed by the weight and movement of the load. The weight is used to determine the size, placement, and type of bearing to be used. The movement is either **radial load** or thrust load. Thrust load is also known as **axial load** Radial load is the up-and-down movement of the load, whereas axial is the front-to-rear (in-and-out) movement of the load. It should be noted that the term *load* applies to not only the weight being supported but also the direction and amount of force against that weight. The force can be thermodynamics in the engine or centrifugal force acting against a turning vehicle.

Bearing Journals

Bearing journals are machined areas on a shaft (**Figure 7-31**). They have fine, smooth finishes to prevent damage to the bearing. Usually the area is hardened more than other parts of the shaft to withstand the loads. In most bearing failure cases, the journal will have to be machined and a smaller bearing used. There is a limit to the amount of machining before the shaft has to be replaced.

Figure 7-31 The shiny areas are the bearing journals on this crankshaft.

Bearing Inserts

Bearing inserts, also known as *plain bearings,* are most commonly used in the engine as **rod bearings** or **main bearings** (**Figure 7-32**). The piston rod's lower end wraps around a portion of the crankshaft (**Figure 7-33**). The rod and crankshaft move as a unit. The bearing allows them to work with reduced friction and increased life. The main bearing is fitted between the rotating crankshaft and the engine block.

Rod bearings A rod bearing is also called a plain bearing, connected to the connecting rod and the crankshaft which provides a smooth snug fit.

Main bearings hold the crankshaft in place within the engine.

A bearing insert is so named because it is fitted into a mounting component. A bearing insert is made from high-strength steel machined to a fine-surface finish, coated, and cut or stamped to the correct length and width. Bearing coatings may be of Babbitt, copper–lead alloy, or aluminum alloy.

Two insert pieces are needed to make one complete bearing that is curved to fit inside the rod and rod cap (**Figure 7-34**). One end is notched so the insert will lock into its mount. If the two halves of an insert are placed together like a circle and measured, it will be determined that the circle is not perfectly round (**Figure 7-35**). This is called *spread.* But when the two halves are fitted into a rod and rod cap, they are forced into a perfect circle. If they do not, the insert will damage the crankshaft and fail quickly. Rod bearings are machined to fit a specified diameter on the crankshaft. A small length at the end of

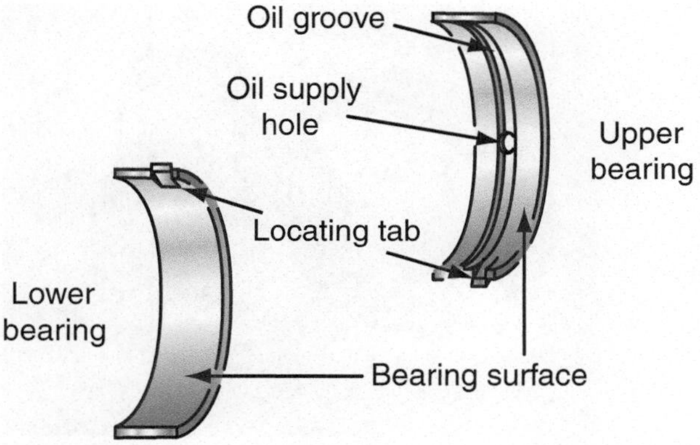

Figure 7-32 This type of bearing may be known as a bearing insert and is commonly used as a bearing for the crankshaft.

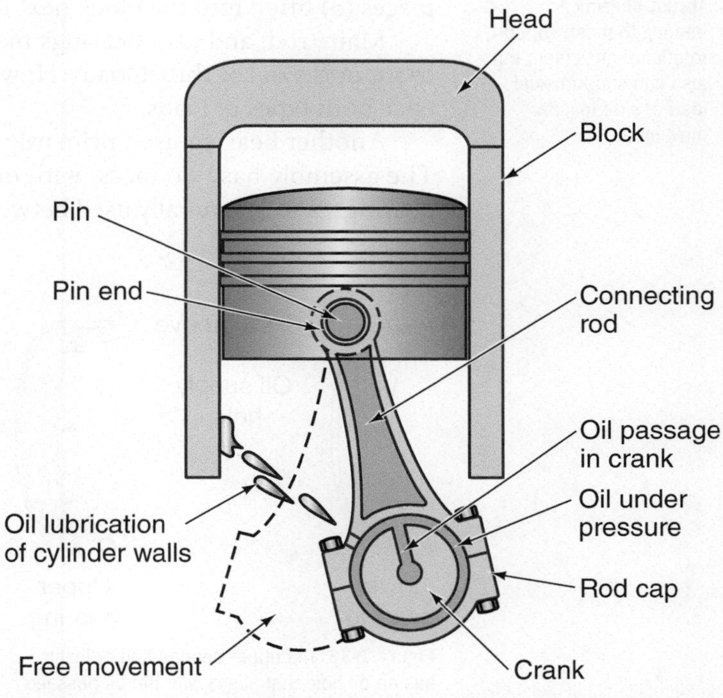

Figure 7-33 The lower end of the rod and rod cap holds a rod bearing in place.

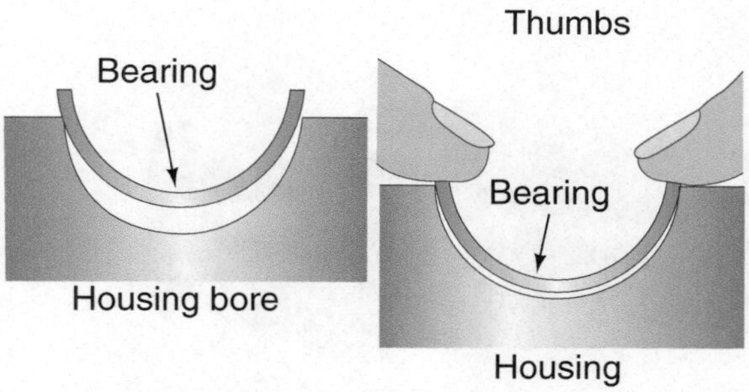

Figure 7-34 The bearing insert is pressed into the mount, usually a rod or main bearing cap, a rod, or the engine block.

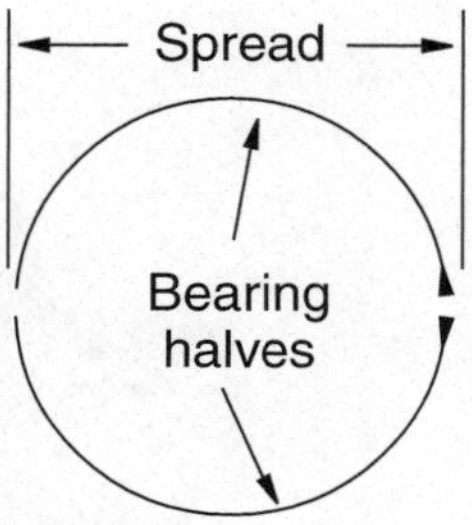

Figure 7-35 The two bearing halves are not in a perfect circle until installed. This is called bearing spread.

The area called **crush** is needed to allow for tolerances during bearing and rod manufacturing.

Main bearing The bearing that holds the crankshaft in place while still allowing rotation of the crankshaft.

Camshaft bearings for an OHV engine come with different outside diameters. The inside diameters are all the same.

Thrust bearing A bearing that can support rotational movement but also can support axial load or side to side movement.

each insert sticks halfway over the rod and cap. This is called **crush** and is used to fit the insert tightly within its mount. The bearings are lubricated by pressurized oil pumped through the crankshaft.

Main bearings are very similar to rod bearings in construction. They are usually a little wider and a little longer. One half fits into the main bearing cap and the other into the engine block. The top half of a main bearing has a lubrication hole that must be aligned with the oil galley hole in the block (**Figure 7-36**). In this manner, the bearing can be lubricated with pressurized oil. Main bearings also have a lock to keep them in place during operation.

Two other types of bearing inserts are the **camshaft bearing** and **thrust bearing.** Camshaft bearings are shaped like short, highly machined metal tubes (**Figure 7-37**). They are press-fitted into cavities of the engine block, and the camshaft is slid through from one end. Camshaft bearings have an oil hole that must align with the oil hole in the block. There are usually four or more bearings per camshaft. The camshaft operates the engine's valves.

Thrust bearings are fitted at the crankshaft and are used to limit the amount of end-to-end crankshaft travel. The bearing may be part of one of the main bearings or separate pieces (4) fitted into the block next to a main bearing (**Figure 7-38**).

Main, rod, and cam bearings mainly support radial loads. Thrust bearings are made to support axial or thrust loads. However, during normal operation, an engine will experience both types of loads.

Another bearing used primarily in transmissions is the *Torrington bearing assembly.* The assembly has two races, with one having a circular groove for a set of small needle bearings, and is generally used between two rotating parts (**Figure 7-39**). The Torrington

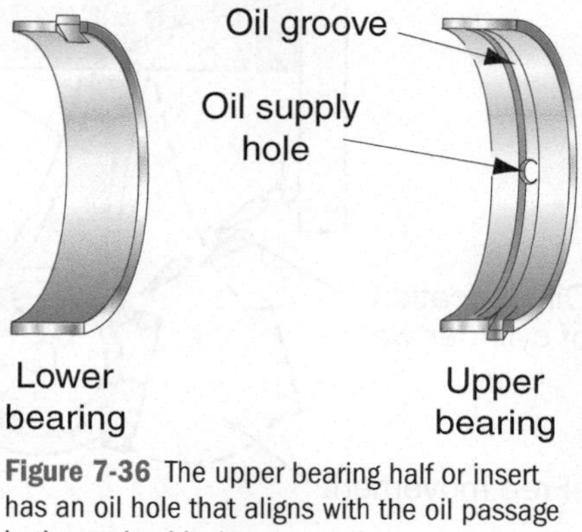

Figure 7-36 The upper bearing half or insert has an oil hole that aligns with the oil passage in the engine block.

Figure 7-37 Note the oil hole in each of the three camshaft bearings shown.

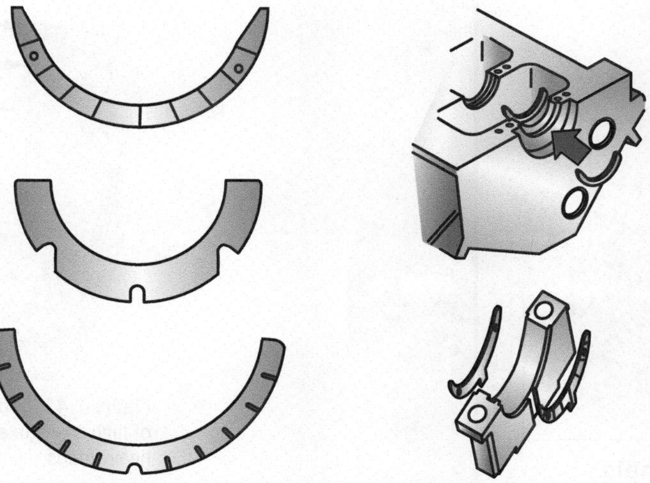

Figure 7-38 Thrust bearings are used to control or limit a shaft's forward and backward (in-and-out) movement within its mount. The figure illustrates the thrust bearings for a crankshaft.

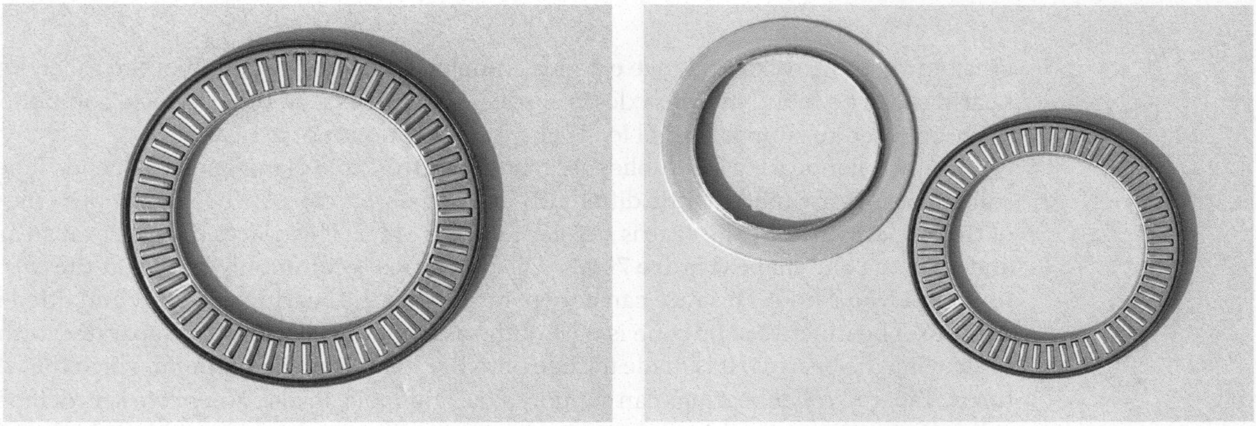

Figure 7-39 The Torrington bearing is often used as a typical bearing but may also be used to adjust shaft endplay.

bearing comes in different thicknesses to help adjust shaft endplay. Another type of bearing used in transmission is the thrust washer. It is used to adjust shaft endplay and works between two rotating parts or between a stationary component and a rotating part. The thrust washer is precision cut to a specific thickness out of strong steel.

The areas on the crankshaft where the rod and main bearings fit are called the bearing journals. A typical crankshaft for a four-cylinder engine will have five main and four rod-bearing journals (**Figure 7-40**). Most V-6 engines usually have four mains and three double-rod journals. A double-rod journal means that two piston rods are connected to the crankshaft side by side with a small space between them.

Roller Bearings

A **roller bearing** is a set of machined shafts of steel. The length and diameter of the shafts or rollers are determined by the size of the load that the bearing is expected to carry. The typical roller bearing in a vehicle is fairly small and has several rollers trapped in a cage or separator (**Figure 7-41**). The cage holds the rollers in a circle. An inner race is fitted through the circle's center and over a shaft. The roller bearing assembly can now rotate around the shaft with each individual roller turning within the cage. An outer race is fitted around the outside of the bearing and into the rotating component like a gear. The contact

All **roller** and **ball bearings** provide a rolling surface for the moving components so that friction and wear are reduced.

Figure 7-40 Crankshafts have a main bearing journal between each or between each pair of rod bearing journal(s) plus one at each end. There is a rod bearing journal for each piston or pair of pistons.

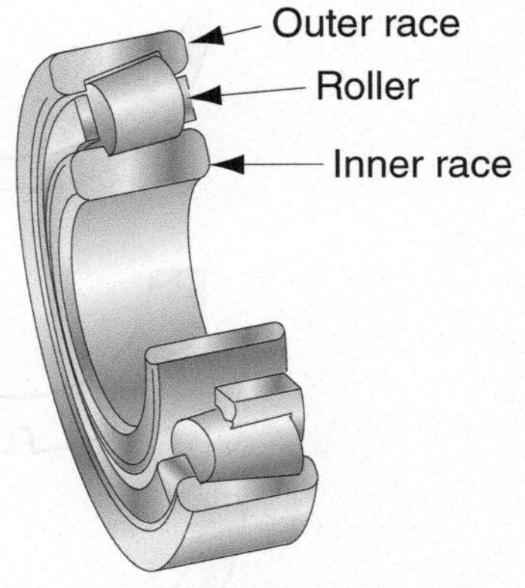

Figure 7-41 A roller bearing is made of high-strength steel and can support heavy loads.

area between the roller and race extends completely across the length of the roller. Roller bearings can be found in transaxles, electrical alternators, and other vehicle components. They are used to support radial loads and a small amount of thrust.

Many roller bearing assemblies are made with the same diameter at each end. Tapered roller bearings are made a little differently and usually support a wheel assembly. One end of the circle created by the cage is smaller than the other. This places the rollers at an angle and forms a cone shape (**Figure 7-42**). The outer race is commonly known as the *cup* and fits into a wheel hub. This race can be separated from the bearing assembly and fitted into the hub. The inner race or cone is part of the assembly and cannot be separated without destroying the bearing. It is made to slide onto the spindle or shaft around which the wheel turns. Tapered roller bearings can support axial and radial loads. When two sets of tapered roller bearings are used in the same component (like a wheel assembly), the bearings complement each other in handling the loads applied against the wheel assembly.

Needle bearings are not used for high-load conditions.

A third type of roller bearing is known as the **needle bearing.** Needle bearings are very small in diameter and may not be held within a cage. They fit into a bearing cup such as a driveline U-joint or into some older alternators (**Figure 7-43**). Improper disassembly of a component with needle bearings usually results in many needle bearings on the bench or floor, or they just disappear. If this happens, all of the bearings must be replaced. There can be no empty spaces between the needles. A needle bearing usually supports only radial loads.

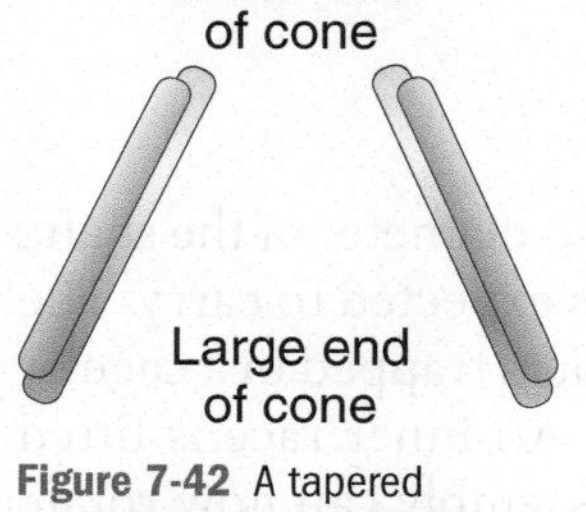

Figure 7-42 A tapered bearing forms a cone with the small end installed toward the inside of the component.

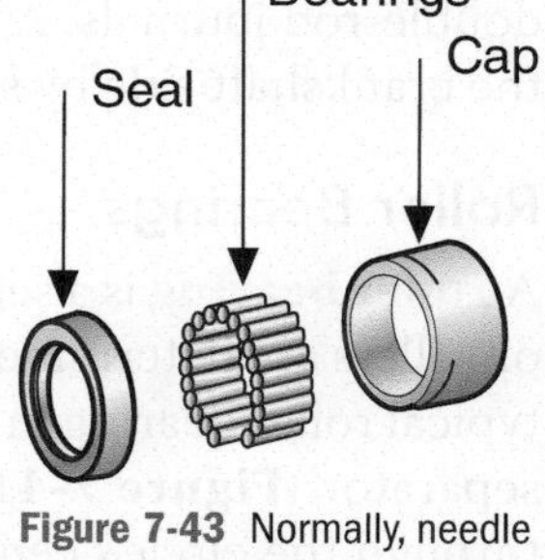

Figure 7-43 Normally, needle bearings fit inside a cup but may be placed in a cage like the roller bearing.

Ball Bearings

The construction of a **ball bearing** is like that of a roller bearing. The primary difference is the use of highly machined steel balls in place of rollers (**Figure 7-44**). The balls fit into a machined groove that keeps them aligned and distributes the load evenly to each ball. The two grooved sections form the cage that holds the bearing together and makes the inner and outer races. Contact between the ball and race is very small, but the bearing can support axial and radial loads well. Some bearings use two rows of balls to support loads better.

On the front wheels of many front-wheel-drive vehicles, tapered roller bearings have been replaced with **ball bearings.**

Shop Manual
pages 252-254

Lubrication

All components that come in contact with each other need some type of lubrication. The engine with its close-fitting parts, high speed, and high temperatures is no exception. Poor or no lubrication will be apparent within a few minutes of operation. Most damage from poor lubrication requires expensive repair or replacement of the complete engine.

The lubrication used today is a high-tech combination of lubricants and cleaning agents designed to reduce friction and improve fuel mileage. There are many different brands and combinations on the market today. Engine oil is rated by its **viscosity** (thickness) by the Society of Automotive Engineers (SAE). The American Petroleum Institute (API) creates a quality standard for each type of engine that oil manufacturers must meet. The oil is rated by numbers that refer to its viscosity (**Figure 7-45**). A common engine oil used in modern engines is rated as 5W30 by the SAE. 30 is the viscosity of the oil at a warmer temperature, and 5 is the viscosity of the oil at a lower temperature (like during winter, which is what the W stands for). An example of an oil that does not change its viscosity in the winter (colder months) would be SAE 30. Because of better manufacturing techniques, modern engines have much less clearance between moving parts and require thinner oil for lubrication. Some newer vehicles may require a 0W20 oil.

Viscosity is a rating of how well a fluid will flow.

The API rating of an oil refers to the quality and appropriate selection for the application of engine. Let's take an oil with an API rating of CG for an example. The letter C means that it is designed for a compression ignition (diesel) engine. The letter G means that it meets a newer standard than F. And F meets a newer standard than E, and so on. For a gasoline engine, an oil would have to be rated with an S. An example of this would be SL. Most likely, the newer an engine, the later the requirement of oil it will need. The newer oils will work backward on old engines, but older style oils will not work on newer

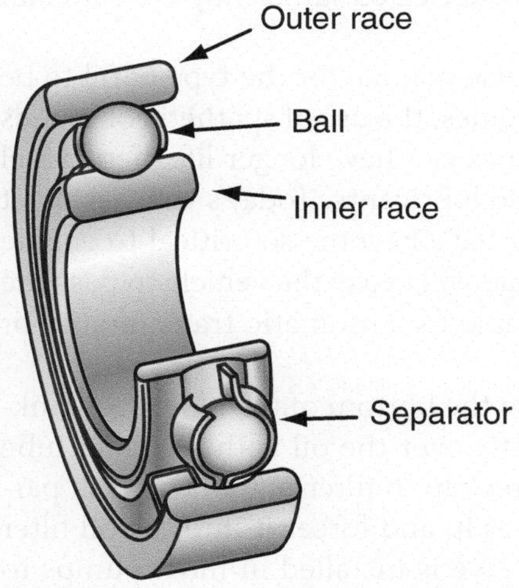

Figure 7-44 A ball bearing usually has the balls locked in a cage and comes in one piece. Ball bearings can support heavy radial (up and down) loads but may not be sufficient for heavy axial (sideway) loads.

Figure 7-45 Engine oil comes in several ratings, but only use the one specified by the manufacturer.

Figure 7-46 Synthetic oils are not made from natural products but are rated the same as natural petroleum oil.

engines requiring newer ratings. Some oils are combination oils and can work on both types of engines. Always check the service data or owner's manual for the correct type of oil to add to an engine. There are other types of oil ratings that exist, such as ILSAC (International Lubrication Standardization and Approval Committee) and ACEA (Association des Constructeurs Européens d'Automobiles—European Automobile Manufacturer's Association). As a technician, you should be aware of these.

Synthetic motor oils and blends are a chemical mix designed to be compatible with natural oils.

Consumers have many choices when it comes to engine oils. Synthetic oil is made from chemicals, does not use natural petroleum products, and is rated by thickness in the same manner as natural oils (**Figure 7-46**). **Synthetic motor oil and blends** are formulated so that they are compatible with the natural motor oils. In some cases, synthetic oils are better than natural oils; however, they are usually much more expensive than regular petroleum products. Synthetic and natural oil blends are a little cheaper than full synthetics but are more expensive than naturals.

One of the problems with lubricating an engine is the time during shutdown. The oil tends to run or drain back from engine components, resulting in a dry or nearly dry start-up where the moving components have little, if any, lubrication. This causes metal-to-metal contact and wear. Some synthetic brands, Castrol for one, offer a lubricant with magnetic tendencies. The oil has synthetic ester molecules that attach to the components and prevent drain back. In this manner, the moving components are lubricated sufficiently even after an overnight shutdown.

The main point to remember is to check the service manual for the type of oil to be used in the various systems of the vehicle. In some engines, the use of synthetic or blends may be a total waste of time and money. Other engines can have longer life, better fuel economy, and a quieter run with synthetics or blended lubricants. Today's vehicles must pass a federally mandated emissions test. Engine oils have become so critical to engine operation that sometimes a simple oil and filter change will cause the vehicle to pass the test. There are synthetic and blend lubricants available for automatic transmission or transaxles as well. Always consult the service manual.

A **check valve** is normally used to prevent fluid from reversing within a system.

A **bypass valve** is used to redirect fluid if a passageway becomes blocked or pressure is too high.

The engine oil is stored in an oil sump attached to the bottom of the block. A crankshaft or camshaft-driven oil pump is mounted directly over the oil with a pickup tube extending into the oil. The tube has a coarse screen pickup to filter out larger trash particles. The pump draws oil from the sump, pressurizes it, and forces it into the oil filter before it enters the engine's oil galleries. A **check valve** is installed in most pumps to prevent oil drain back during engine shutdown.

An oil filter removes most of the harmful particles floating in the oil. The particles are metal and carbon that result during normal engine operation. The filter adapter has a **bypass valve** that opens when oil pressure within the filter becomes too high (**Figure 7-47**).

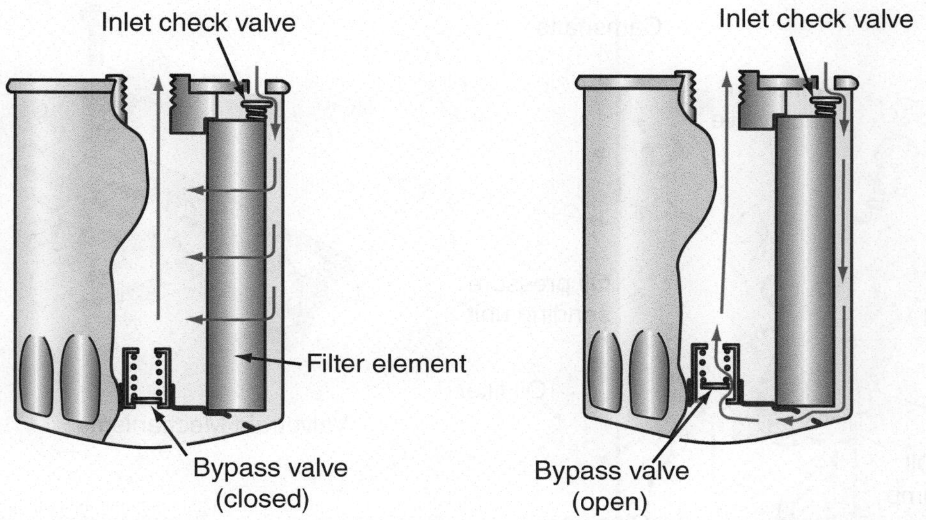

Figure 7-47 The bypass valve allows cold, unfiltered oil to flow to engine components during start-up and cold drive-away.

This allows oil to flow past the filter and into the engine. The valve is primarily designed to allow oil flow when the engine and oil temperatures are cold. However, if the filter is clogged or the oil is extremely dirty, the valve will open. In both cases, unfiltered oil is pumped into the engine. Damage could result.

Pressure lubricated is oil forced between two components by pressure.

Splash lubricated is oil randomly sprayed onto components by the actions of those components.

Clearance is the space between adjacent components.

The clearance between a bearing and its journal is called **oil clearance** and is usually about 0.0015 inch to 0.002 inch.

The oil moves under pressure into the block's oil galleries. Some engine components are **pressure lubricated** whereas others are **splash lubricated.** Pressure-lubricated components include the crankshaft bearings, connecting rod bearings, camshaft bearings, and some rocker arms (**Figure 7-48**). All the bearings operate under extreme conditions, and the pressurized oil is required to maintain a **clearance** between the bearing and the supported component. Splash lubrication is accomplished when the pump supplies a large pool of oil for the component to move through. The oil is splashed over the components before returning to the sump through large passageways or galleries. Some components using splash lubrication are camshaft drive gears and chains, valvetrain mechanisms, and cylinder walls (**Figure 7-49**).

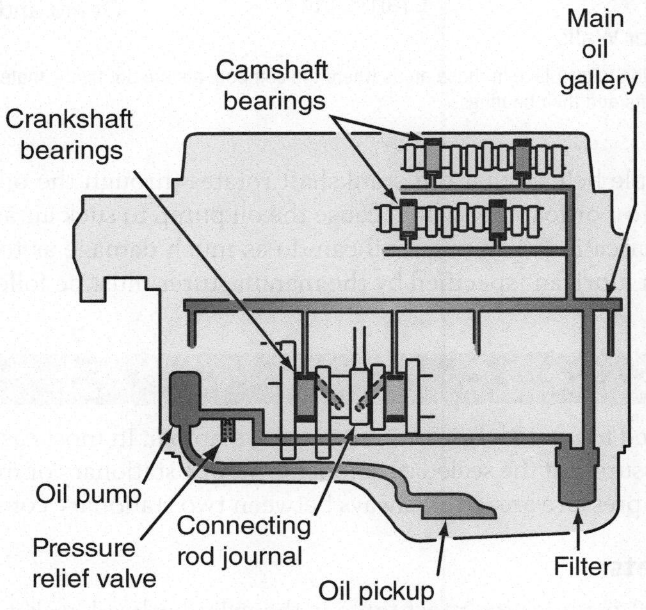

Figure 7-48 Pressurized oil is required for such mated components as the crankshaft and its bearings.

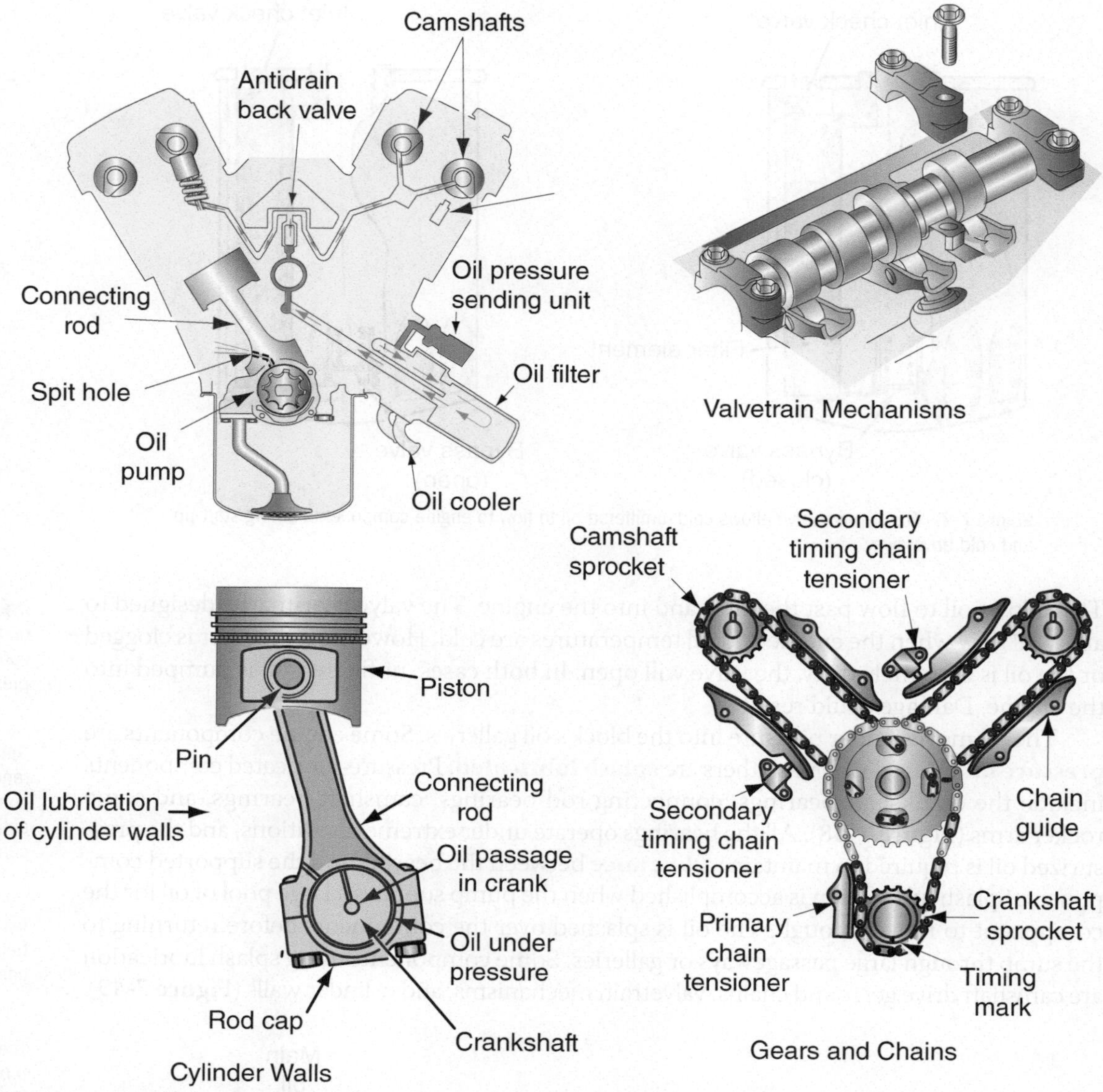

Figure 7-49 Splash lubrication takes place in those areas where the components are not tightly mated. The oil shown here is sprayed from around the shafts and their bearings.

Cavitation is the introduction of air into a fluid, in this case, air forced into the oil by the rotating crankshaft and connecting rod.

Some people believe that the crankshaft rotates through the oil in sump. If this did occur, **cavitation** or foaming would cause the oil pump to suck air and the engine would be without lubrication. Too much oil can do as much damage as too little oil. The type and amount of lubricant specified by the manufacturer must be followed.

GASKETS

Shop Manual
pages 256, 279–283

Gaskets are used to retain lubricants or to seal a chamber. In most cases, the lubricants are under low pressure, but the sealed components may be stationary or moving. A gasket seals high- and low-pressure areas and always between two stationary components or parts.

Head Gaskets

The gasket that is under the most force is the cylinder head gasket (**Figure 7-50**). This gasket seals the extremely high pressure of the combustion chamber and seals water and oil passages between the engine block and the head. It must accomplish this task under

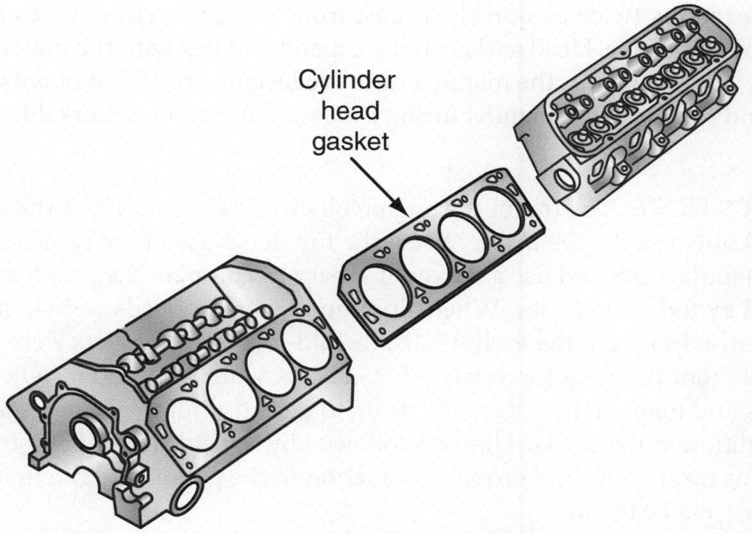

Figure 7-50 Cylinder head gaskets are one of the most critical and most stressed gaskets used on a vehicle. They must contain high temperatures and pressure while allowing adjacent components to expand and contract during temperature changes.

high and low temperatures, expanding and contracting components, different materials, and exposure to chemicals including acids.

A typical head gasket is composed of layers of different materials bonded together (**Figure 7-51**). Older head gaskets were stamped or embossed steel sheets. They worked well in many applications with cast iron blocks and cylinder head, but they are not recommended for vehicles produced after the early 1980s. The use of aluminum cylinder heads mounted on cast iron blocks produced some engineering problems on sealing this area.

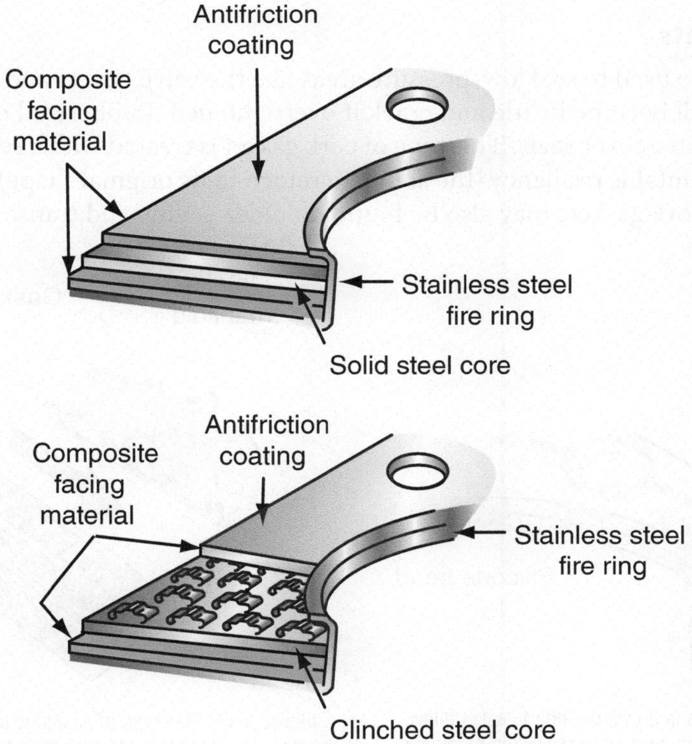

Figure 7-51 The antifriction coating allows the sealed components to slide over the gasket without damaging it.

Aluminum expands twice as quickly as cast iron. Also, current engines and heads are lighter and more flexible. Head gaskets must expand and flex with the materials above and below them. To correct this, the manufacturers redesigned the components, selected new materials, and created new manufacturing processes to produce workable gaskets.

> **AUTHOR'S NOTE** The top of the engine block and the bottom of the cylinder head are known as the *fire decks.* When the fire decks are of the same materials, design, manufacture, and use a gasket, it is easy to compare the problems presented by today's vehicles. When aluminum cylinder heads were first mated with cast iron blocks in the early 1980s, the old-style head gaskets were modified for the different fire deck materials. They did not work well, and it took the engineers and manufacturers several years to design a durable head gasket for fire decks of different materials. This was followed by several years of automotive technicians training for the proper installation of the gasket and the then-new torque-to-yield fasteners.

Composite refers to a mixture of materials to form a more durable gasket.

The modern head gasket has a steel core made of solid or clinched steel plate. A dense **composite** facing is placed on each side of the core and covered with a low- or antifriction material. Holes are punched through the gasket to line up with the engine cylinders and coolant and oil passages. The rim around each cylinder hole has a stainless-steel fire ring to help protect the gasket material from burning air and fuel in the combustion chamber. The antifriction coating is usually Teflon or a silicone-based material. Some types of head gaskets have a silicone bead placed around the oil and coolant holes to prevent liquid leakage (**Figure 7-52**).

Head gaskets are like intake and exhaust gaskets (**Figure 7-53**). They differ in construction and material and must perform well after being exposed to high temperatures and chemicals. Sometimes they are placed between two different metals and must be able to expand properly with both.

Other Gaskets

Cork gaskets are used to seal low-pressure areas like the valve covers and oil pans. When heated, cork will become brittle and crack if overtightened. Rubberized cork gaskets are often used as valve cover seals. This type of cork gasket is treated with a reduce-shrinking process and maintains resiliency (the ability to return to its original shape) when clamped and released. Cork gaskets may also be found on older engine and transaxle oil pans.

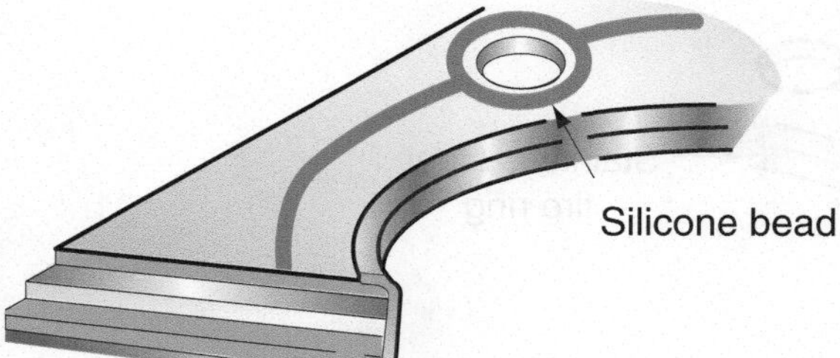

Figure 7-52 The smaller holes in a cylinder head gasket allow for movement of liquids, coolant, and oil between the block and cylinder head.

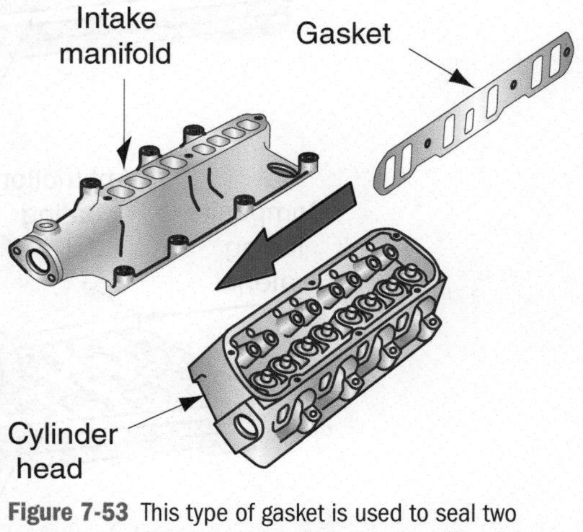

Figure 7-53 This type of gasket is used to seal two components that create high temperatures. (Courtesy of Fel-Pro Incorporated.)

Synthetic rubber gaskets are made from man-made materials. The material is usually neoprene, and the gasket may be reusable. For this reason, they are often used for valve cover gaskets because many valve covers require removal for routine maintenance (**Figure 7-54**). They may also be found as oil pan gaskets and in other areas.

Paper gaskets are made of treated paper and seal low-pressure, low-temperature areas between two stationary components. A typical application is between the different housings and covers on a manual driveline and some water pumps. Paper gaskets may be relatively thick for some applications, such as the water pump, or they may be as thin as a regular piece of paper.

Chemical gaskets are used on many new cars. They are installed by robotic equipment capable of laying an almost perfect 1/8-inch bead around the entire sealing area, thereby making a formed-in-place gasket. Because of this, they are difficult to use during repairs because a poorly laid bead will leak. Aftermarket part vendors may use cork or neoprene gaskets as replacement parts.

Gasket sealers are sometimes used to hold the gasket in place and help it seal. Care should be taken when using gasket sealers. Instructions should be followed, as many gasket manufacturers do not want any sealer on their gasket or only placed in certain amounts at specific places. The sealer may be applied with a brush or pressed from a tube. Gasket sealers are anaerobic chemicals, meaning they will cure only in the absence of air. The two components have to be fitted and torqued before the sealer will cure.

A gasket sealer or thread sealant may be required on some fasteners that extend into the cooling passages. The sealer is spread lightly over the fastener's threads to seal the area between the threads.

Shop Manual
page 280

Chemical Sealants

Chemical sealants include the sealing materials mentioned earlier and a popular sealer known as **room temperature vulcanizing (RTV)** (**Figure 7-55**). RTV is an aerobic sealant and may be known as silicone rubber. Aerobic means RTV cures when exposed to air. It can be used to seal two stationary components such as an oil pan and an engine block. It cannot be used as a head gasket or an exhaust manifold (high heat and pressure). It cannot be used on fuel systems either because fuel deteriorates RTV. RTV will set in about 15 minutes in warm temperatures and high humidity. It cures in about 24 hours. It will not set or bond to a gasket. RTV comes in different colors based on the application and temperature ranges.

Chemical sealants and sealers should be checked for their use with electronic engine controls.

RTV A silicone that begins curing as soon as it is exposed to moisture in the air.

> **AUTHOR'S NOTE** Do not place RTV on a gasket of any material unless directed to do so by the gasket installation instructions. The RTV forms a rubber-like bead and will not stick to cork and most other gaskets (**Figure 7-56**). A leak will form between the RTV and gasket.

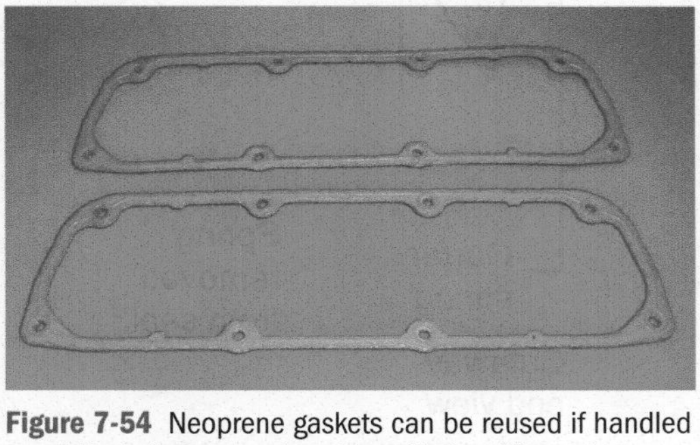

Figure 7-54 Neoprene gaskets can be reused if handled carefully during removal and installation. Shown is a typical valve cover gasket set.

Figure 7-55 Use RTV or another chemical sealant only where directed by the gasket manufacturer. Sealants applied improperly can prevent the gasket from sealing. (Courtesy of Fel-Pro Incorporated.)

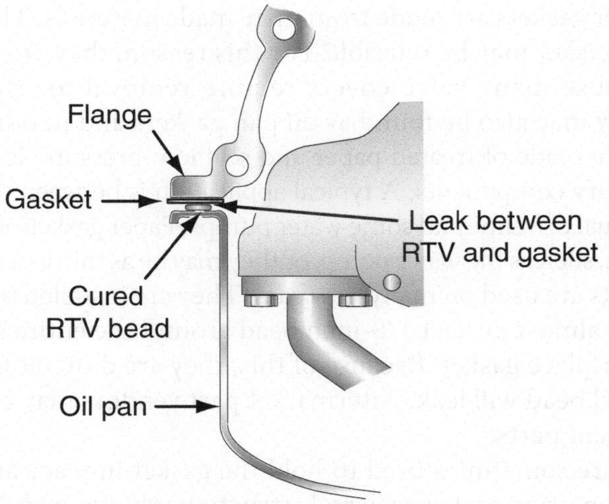

Figure 7-56 RTV will not stick to most gasket materials. Cured RTV is similar to rubber, so a leak can form between the gasket and the cured RTV bead.

As previously discussed, aerobic material cures in the presence of air. A second type of RTV is anaerobic, which cures in the absence of air. Anaerobic gasket material is best suited for use between two rigid components with less than a 0.015-inch gap. Such gaskets are commonly used to mate water pumps to engine block flanges.

Seals

Seals are normally used between moving components and their housings to retain liquids. Seals can be used in high-pressure and high-temperature areas and are usually made of butyl rubber or neoprene within a metal ring. The ring provides the connection to the stationary component.

Lip seal a circular seal encompassing a rotation shaft and preventing fluid leakage. A **garter spring** may be located inside the seal to help the lip seal better to the rotation shaft.

A **lip seal** is the type most commonly used on vehicles (**Figure 7-57**). The seal has a lip that is installed facing the lubricant. In this way, the pressure of the oil against the inner side of the lip forces the lip edge tightly around the moving part. The lip may have a **garter spring** behind it so it is better held to the shaft (**Figure 7-58**). Some seals have a smaller lip called a dust lip that faces away from the lubricant. It does as its name implies, keeping

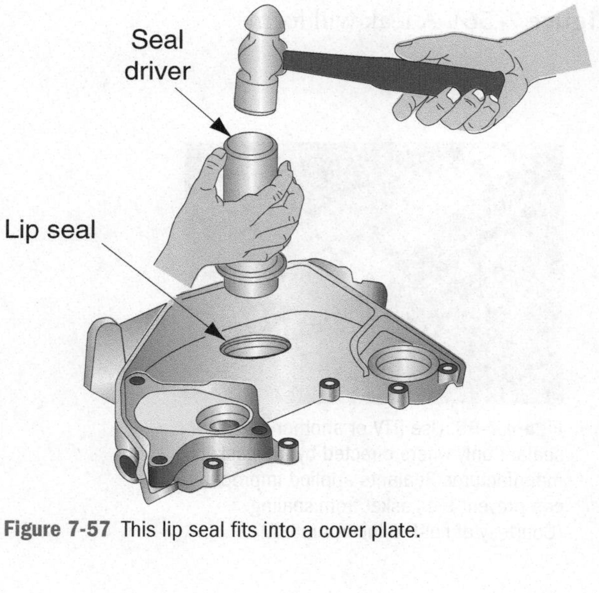

Figure 7-57 This lip seal fits into a cover plate.

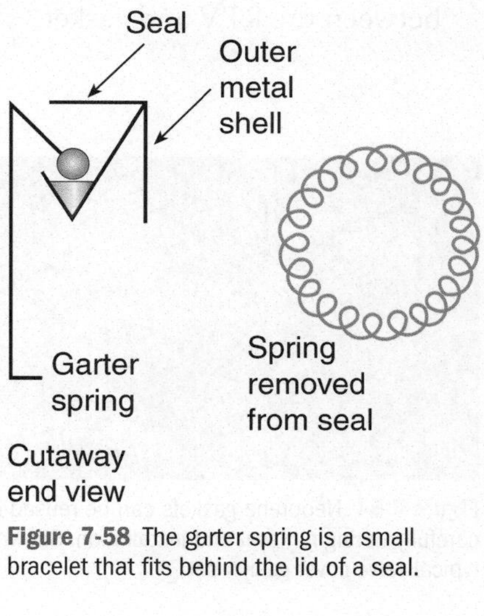

Figure 7-58 The garter spring is a small bracelet that fits behind the lid of a seal.

dirt from entering the cavity. Seals are found at the front and rear of the engine crankshaft, front and rear of a transaxle, around drive axles, and at the wheel or hub.

O-Rings

An O-ring is so named because of its shape and the manner in which it is used. The O-ring is shaped like a rubber bracelet without a latch (**Figure 7-59**). It fits into a groove, either in the moving component or in the stationary component. The ring extends above the groove and is pressed against the second component (**Figure 7-60**). O-rings are commonly used on high-pressure hydraulic fittings such as power steering, brakes, and air conditioning systems. A specially cut type, called a *square-cut O-ring*, is used on disc brake systems (**Figure 7-61**).

A third type of seal is used primarily to protect rather than seal a component. These seals are usually referred to as *boots*. A common one now in use is the boot for the CV joint. It does seal to some extent, but its main purpose is to keep debris and moisture from entering the joint. A common assertion is that if the CV boot is split, the CV joint will need to be replaced after about 8 hours of operation. Although this assertion is general in nature, it does point out how fast a component can fail when its seal or gasket fails.

Boot or dust boots are also used on the drum brake's wheel cylinders. In this case, they only protect the internal pistons and bore from dirt and brake dust. Other seals within the wheel cylinder perform the task of keeping the brake fluid inside the bore.

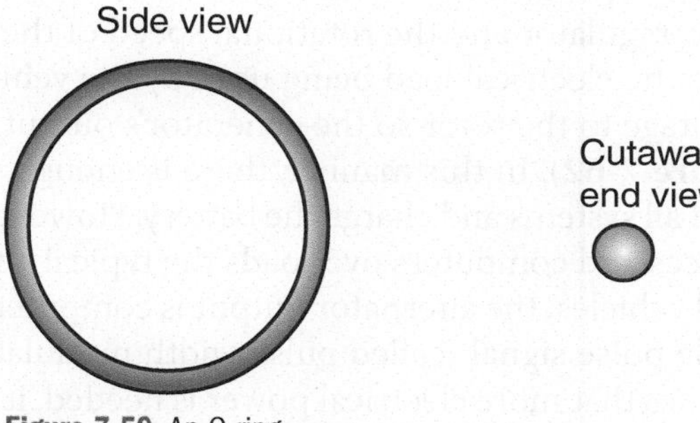

Figure 7-59 An O-ring.

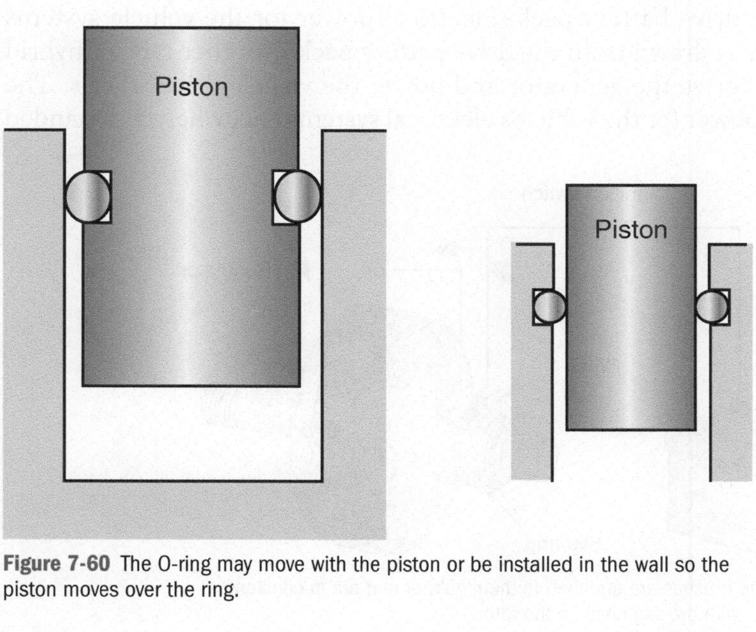

Figure 7-60 The O-ring may move with the piston or be installed in the wall so the piston moves over the ring.

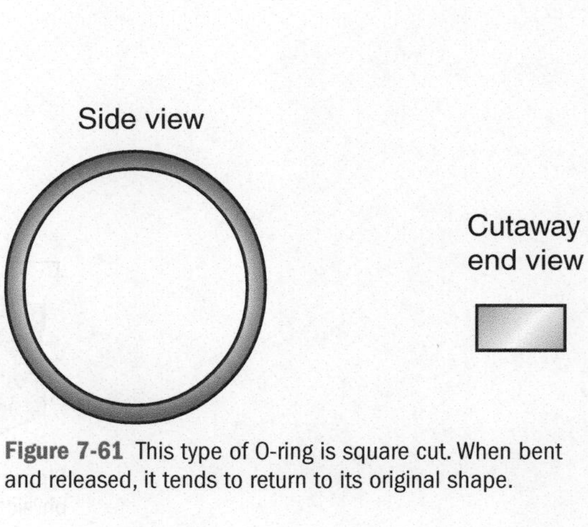

Figure 7-61 This type of O-ring is square cut. When bent and released, it tends to return to its original shape.

Starting and Charging Systems

Different circuits control the starting and charging systems. The starter controls are the ignition switch, a transmission range or clutch switch, and the relay. Many antitheft circuits are in series with or controlling power to the starter relay. The **transmission range switch** is in series with the ignition switch, starter relay, and the PCM. If the automatic transmission is not in park or neutral, the switch is opened and current cannot flow to the starter relay or PCM. A **clutch position switch** performs the same function on a manual transmission vehicle. If the clutch pedal is not fully depressed, the starter and PCM will not function. Some vehicles also have an antitheft system that will disable the starter or the ignition system if not disarmed by the correct security code.

A **transmission range** switch signals the computer which transmission gear is selected.

Clutch position switch A switch located under the dash that prevents you from starting the vehicle while the transmission is in gear.

The transmission range switch is like a manual position sensor and sends a signal to the PCM. It is sometimes referred to as the PRNDL (pronounced prindil) switch. Many of these types of switches are gate switches and only allow current to flow when the transmission is in park or neutral.

The charging system consists of the alternator, battery, wiring, and sometimes the PCM. The alternator (referred to as a generator by some manufacturers), as noted in Chapter 5, uses magnets (stator) around an electrically charged magnet (windings) on a rotor. The rotor is driven by a belt and the crankshaft. The amount of electricity allowed into the rotor windings determines the magnetic force of the rotor. The more current flows, the stronger the magnetic field. This rotating magnetic field moves through the magnetic field(s) of the stator, and each time a magnetic field is "cut" by another magnetic field, an electrical current is produced. The strength of the magnetic rotor field is controlled by the alternator's regulator and the rotational speed of the rotor.

The regulator senses the electrical load being used by the vehicle's electrical systems and adjusts the input voltage to the rotor so the generator's output from the stator is sufficient for the load (**Figure 7-62**). In this manner, there is enough voltage and current to allow for full operation of all systems and charge the battery. However, the addition of more and more electrical devices and computers overloads the typical alternator's capacity.

On some late-model vehicles, the alternator output is computer controlled. The computer will send a variable pulse signal (called pulse width modulation) to the alternator. When the PCM determines that more electrical power is needed, it sends a more frequent input signal to the alternator, thus causing it to output more current.

Hybrid electric vehicles may be the answer to this increasing electrical load. In general terms, a hybrid vehicle uses an internal combustion engine to drive a high-output generator that charges the large drive battery pack. Electrical power for the vehicle systems including the drive motors is drawn from the drive battery pack. Another type of hybrid vehicle uses the engine to drive the generator and power the vehicle drive wheels. The battery pack provides the power for the vehicle's electrical systems and, when commanded

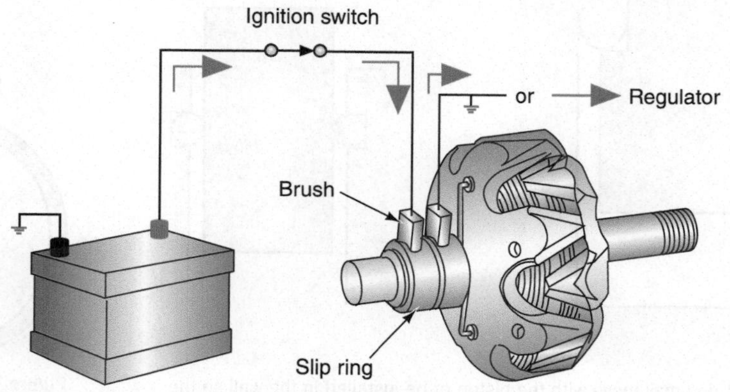

Figure 7-62 The brushes are attached to the regulator and are in constant physical contact with the slip rings on the rotor.

by the on-board computer, can assist in powering the vehicle. Both types have the ability to provide a constant electric current for all electric devices on the vehicle. The typical vehicle electrical system would still operate on 12 VDC. One or two AC/DC or DC/DC converters would reduce the voltage from drive battery voltage (typically 120 VDC) to the needed 12 volts. Another advantage of this system would be the precise management of current or amperes metered to sensors and actuators. This allows more exact measurement of conditions with a more exact response by the actuators.

Shop Manual
pages 287-290

FUEL SYSTEMS

Fuel systems deliver fuel from the tank to the mixing device, usually a carburetor or fuel injector. The fuel volume must be delivered at a specified volume and pressure. Older car engines use a mechanical pump that produces between 3 and 6 psi with a volume of 1/2 pint in about 30 seconds. This was good enough for carbureted engines. Fuel-injected vehicles use an electric pump that produces up to 60 psi and a similar volume. Diesel engines use an injector pump that produces from 1,800 to 15,000 psi and sometimes use a mechanical lift pump to provide a constant volume to the injector pump.

Mechanical Fuel Delivery Systems

Fuel is delivered to the mixing unit by a fuel pump. A mechanical pump driven by the camshaft is used on most carburetor systems. The pump operates from a camshaft-driven pushrod. Its fuel delivery pulses on and off as the camshaft turns. The carburetor is the mixing unit, which has a reservoir or bowl to hold sufficient fuel. The bowl is required due to pump pulses and the pump's inconstant pressure and volume. The mechanical pump delivers fuel to the carburetor on each pump stroke.

A **venturi** is the narrow portion of a larger bore.

The carburetor has various internal passageways to deliver fuel into the airflow moving through its **venturi** (**Figure 7-63**). The venturi is usually a large, vertical air passage

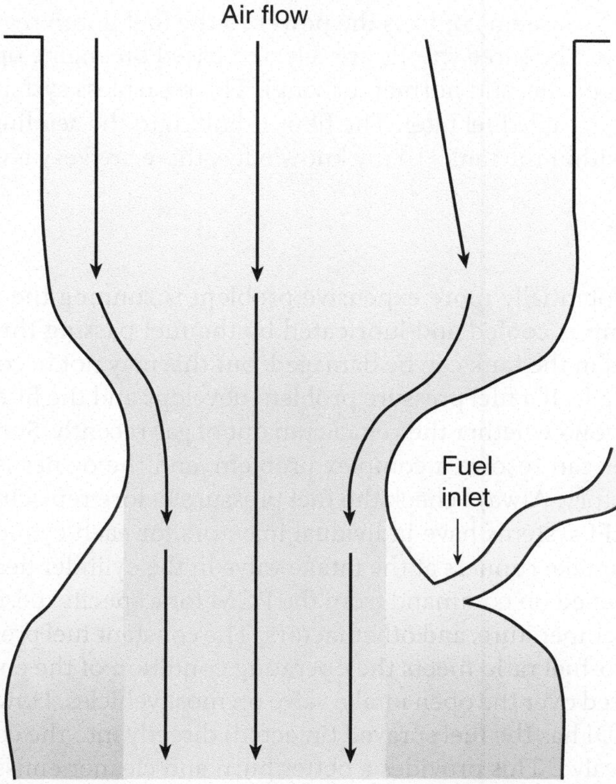

Figure 7-63 Low pressure is created by air moving rapidly past the recessed area.

near the center of the carburetor body. It opens directly into the top of the intake manifold. As the air flows past a narrow portion of the venturi, a low-pressure area is established at the point where the venturi widens again. In the bowl, atmospheric pressure pushes the fuel into the low-pressure area to mix with the air. The fuel passages are sized to deliver a certain amount of fuel depending on the volume of air moving through the carburetor.

Mechanical fuel systems are almost nonexistent in modern vehicles. However, there are still vehicles, powersports, and equipment in use today that rely on this type of fuel system.

Electronic Fuel Injection

The electronic fuel injection (EFI) system requires an electric fuel pump. This type of pump delivers fuel at a constant pressure and volume. The fuel may be delivered to injectors in a throttle body unit like the carburetor, individual injectors at each cylinder, or high-pressure injectors inside each combustion chamber. Throttle body injection has some of the same problems as the carburetor system. However, the pressure to the injectors can be regulated with an engine load-sensing fuel regulator to ensure a nearly constant air-to-fuel ratio at all engine speeds and loads.

Most vehicles have only one electric pump mounted inside the tank. There are two critical concerns with electronic fuel delivery maintenance. This type of system depends on a constant pressure and the volume of fuel. A clogged or restricted filter can cause numerous problems with engine operations. Sometimes, the filter is forgotten because some vehicle manufacturers do not require a typical service for 100,000 miles. The technician should keep this in mind when the vehicle comes in for an oil change. If the fuel filter is factory-installed and the vehicle mileage is more than 30,000 miles, it may be best to question the owner on the history of the fuel filter.

AUTHOR'S NOTE Chrysler, among other manufacturers, uses a returnless fuel injector system. The system controls the power to the fuel pump regulating the speed of the pump. The three speeds are selected based on engine operating conditions: start-up, idle, and normal (driving). The returnless system does not usually have a serviceable fuel filter. The filter is built into the sending unit and pump assembly within the tank. To my knowledge, there are very few problems with this filter.

A second and potentially more expensive problem is running the fuel tank dry. An in-tank electric pump is cooled and lubricated by the fuel passing through it. A pump running without fuel in the tank can be damaged, but this may not become obvious until a later date when it fails. If a fuel pressure problem develops and the filter is good, ask the owner if he or she knows whether the vehicle ran out of gas recently. Sometimes, a simple question and answer can resolve a complex problem, and the owner is usually the best source of vehicle history. Always check the fuel pressure before replacing the fuel pump.

Diesel-fueled engines use the heat of the compressed air to ignite the fuel.

Most modern EFI systems have individual injectors for each cylinder. The injectors are mounted in the intake runners at the intake valve in the cylinder head (**Figure 7-64**). The injectors are opened on command from the PCM for a specific period depending on engine speed, load, temperature, and other factors. The constant fuel pressure and volume ensure that the air-to-fuel ratio meets the operating condition of the engine.

The fuel is injected over the open intake valve on most vehicles. However, a new gasoline system called GDI has the fuel sprayed (injected) directly into the combustion chamber near the intake valve. This provides a better burn and cleaner emissions because the fuel is injected into the air being compressed in the cylinder. Some direct injection systems

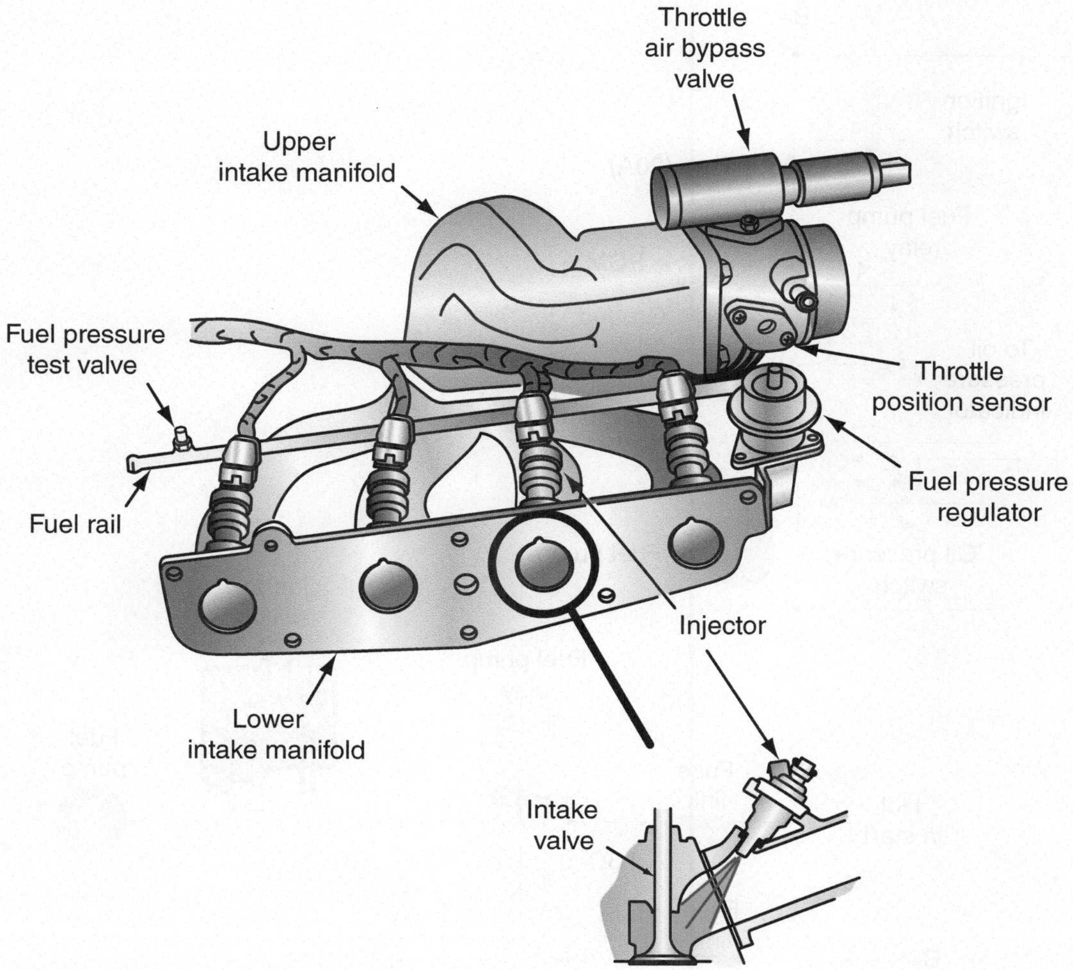

Figure 7-64 In MPI fuel systems, there is one injector per cylinder.

use an engine-mounted injector pump to pressurize the fuel. Direct injection with an injector pump is the fuel system used on all diesel-fueled engines.

Controls for the electrical fuel pump are the ignition switch, pump relay, and the PCM (**Figure 7-65**). Some vehicles use a separate switch in series with the relay and pump. This switch may be referred to as a rollover or inertia switch. Its purpose is to cut power to the pump if the vehicle is involved in an accident. Other systems use an oil pressure switch or engine speed sensor to signal the PCM that the engine is not running, or oil pressure has been lost. General Motors uses an oil pressure sensor to stop the pump if oil pressure is lost. A bypass circuit in the fuel pump relay will supply power to the pump during engine starting. Like the starter circuit, an antitheft circuit may be incorporated into the fuel pump circuit.

Shop Manual
page 286

The water pump may be mounted behind the timing belt cover, on the timing chain cover, or to the side of the block on a bracket.

Air for an EFI system is drawn into the engine in a manner like the carburetor. However, dirty air and fuel filters affect EFI systems more than a carburetor system. The best air/fuel mixture with current engine technology is 14.7 parts air to 1 part fuel, or a 14.7:1 ratio. A cold engine will need a rich mixture of about 13:1 for starting. A warm engine can operate on a leaner mixture of about 15:1 ratio or higher. The volume or amount of intake air depends on engine displacement, the fuel system, and the mechanical condition of the engine. A dirty air filter can enrich the fuel mixture to the point of flooding the engine by restricting the air intake.

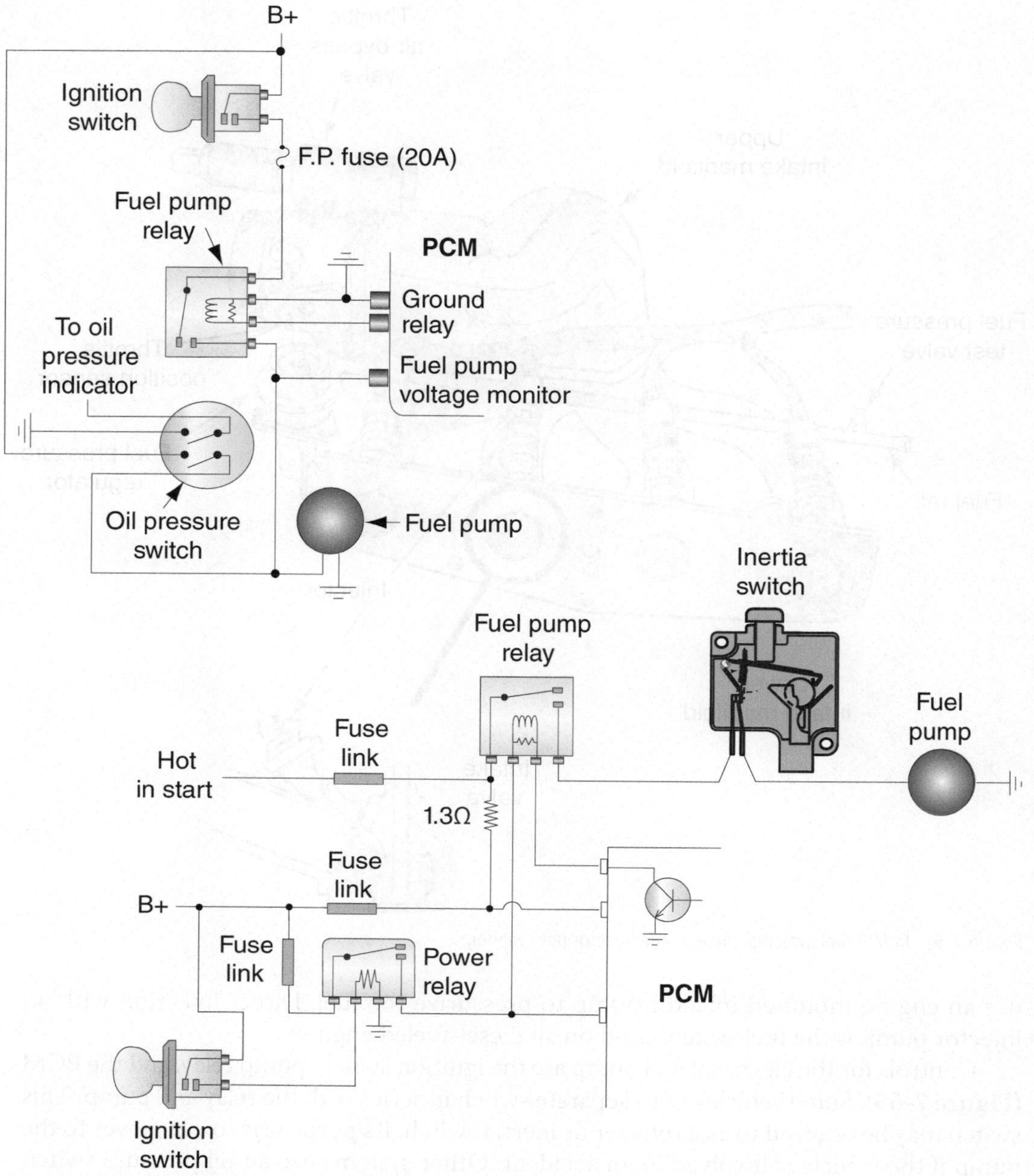

Figure 7-65 A typical starter control circuit.

COOLING SYSTEMS

Horsepower is produced from the expanding gases in the combustion chamber. High temperatures and excessive heat are created and must be dissipated. Some of the heat is absorbed by the engine's oil and carried to the oil sump. However, most of the engine heat is absorbed by the cooling system. Heat transfers from the combustion chamber to the metal of the engine and then into the coolant that circulates around the metal.

A crankshaft-driven water pump assists in moving hot coolant out of the engine and cool coolant back into the engine block. The coolant circulates through passageways in the block and cylinder head. Heat from combustion and friction is transferred to the coolant. The coolant leaves through a thermostat housing near the top of the engine. The thermostat is a temperature-sensitive valve mounted at or near the highest and hottest part of the engine (**Figure 7-66**). Some vehicles have the thermostat mounted at the top

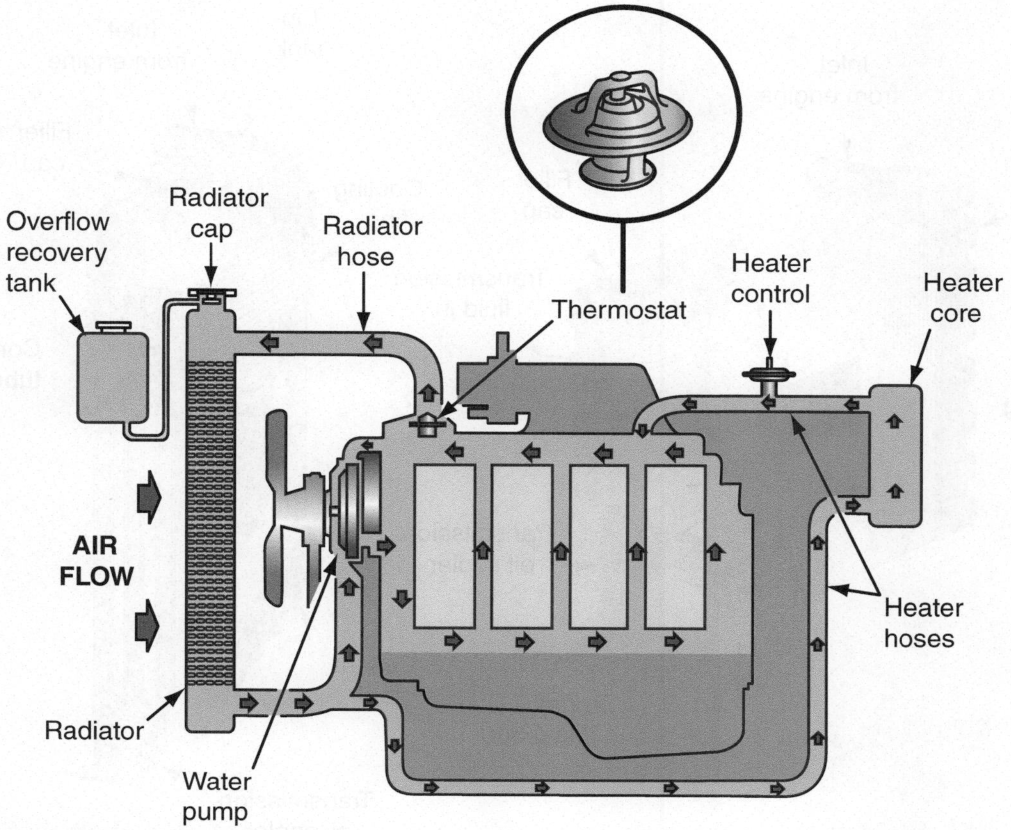

Figure 7-66 This thermostat is mounted slightly below the top of the engine. Some thermostats are mounted in the lower (return) hose near the bottom of the engine.

entrance into the radiator or at the exit from the radiator. During cold and warm-up conditions, the thermostat remains closed, forcing the coolant to circulate back into the engine block. By not cooling the coolant, the engine gets to operating temperature quicker. An engine is most efficient at operating temperature, so the quicker it gets there, the better. Most thermostats are fully open at about 190°F (88°C). As the temperature rises, coolant can flow through the upper hose into the top of the radiator.

The radiator is either a cross flow or a vertical flow. The two types are very similar except for the direction of coolant flow (**Figure 7-67**). The coolant moves through small tubes down or across to the radiator's exit. The tubes are finned to expose as much surface area as possible to the passing airflow. A fan draws air from the front of the vehicle over the radiator fins and exhausts the air over the engine block. As the coolant exits the radiator through the lower hose, the water pump picks it up and pushes it into the engine block.

Some engines use a cooling fan mounted on the water pump. These fans are either controlled with a thermostat inside them or they have an electronically controlled clutch. Most cooling fans are driven by electric motors controlled by a cooling fan switch mounted in a coolant passage or by the PCM (**Figure 7-68**). When overheating is sensed, the switch or PCM closes the fan relay. This control of the fan's operation is based directly on engine temperature and reduces the load placed on the engine and saves fuel by using the fans only when necessary. Some vehicles use two fans, which may be referred to as cooling fans 1 and 2 or as a cooling fan and an air conditioner fan. The second fan will turn on when the air conditioning is switched on to help move air through the condenser (a radiator for the air conditioning system). If only one fan is present, it will operate, whereas a two-fan system will have one of the two working. It should be noted that under extreme, overheating conditions, the PCM could switch on both fans while switching off the air-conditioning system.

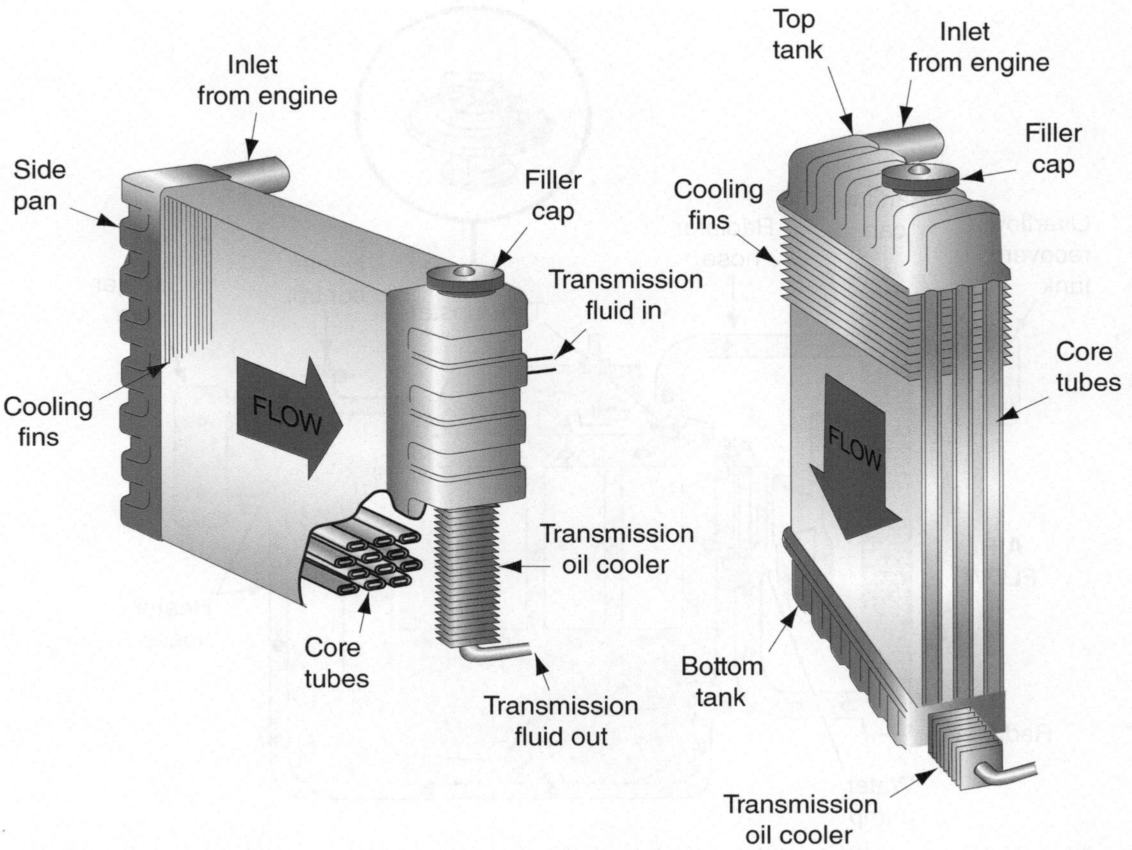

Figure 7-67 A typical downflow radiator on the right and a cross-flow one on the left.

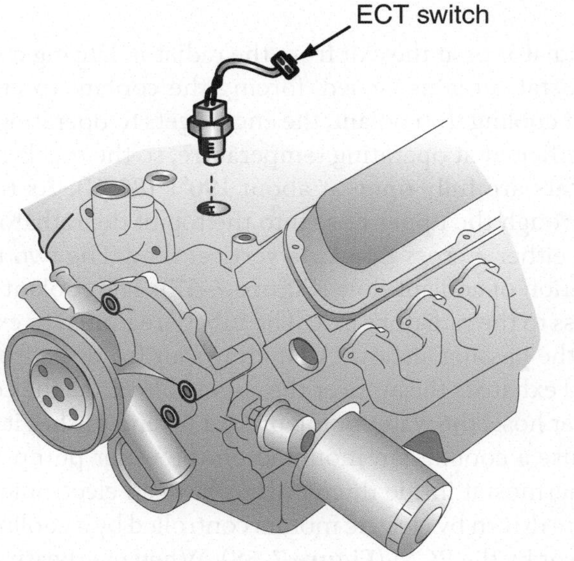

Figure 7-68 This engine coolant temperature sensor is mounted at about mid-block. Most are mounted near the thermostat housing.

Antifreeze is a chemical mixture that reduces freezing of a liquid, usually water.

Coolant

The coolant used in a cooling system is made up of a mixture of **antifreeze** and water. Antifreeze is a chemical compound added to water because water itself does not provide the proper characteristics needed to protect an engine. Water transfers heat quite well but freezes at 32°F and boils at 212°F. Engines must be able to operate above and below those temperatures. Coolant must protect the engine from freezing and overheating, but it also lubricates the water pump and protects against corrosion, rust, and **electrolysis.**

Electrolysis is the chemical process of decomposition between a liquid and metal producing a small electrical current.

The mixture of coolant and antifreeze will determine its temperature qualities. Adding coolant to water will make the boiling point go up and the freezing point go down. It is good practice for many technicians to put in a 60/40 mixture (60 percent antifreeze). This is close to the best mixture of 67/33 which will give the highest boiling point and lowest freezing point. But mixing too much antifreeze with water will have adverse temperature effects, and it will be difficult for the water pump to move around the engine. Heat will also have a tough time moving out of the engine because of the lack of water.

There are two main types of coolants used in modern vehicles. The traditional type of coolant is green in color and is an ethylene glycol which uses Inorganic Acid Technology (IAT)–based fluid. This type of coolant was used for many years in all vehicles. Very few new cars and trucks use it today, but there are a lot of vehicles you will see with this type of coolant in the system. It is easy to identify because of its color. Traditional IAT coolant lasts only 2–3 years before it starts to chemically break down and deteriorate.

Modern vehicles use long-life coolants that use a few different technologies and mixtures. They are easy to identify because they are any color except green. Some different colors that manufacturers have used are blue, purple, orange, red, yellow, pink, and others. These long-life coolants have different chemicals and technologies that allow them to be used for 5–6 years. Most of them have a service interval of around 100,000 miles. When coolant breaks down and degrades, it cannot protect the system as well. At this point the coolant must be flushed and replaced. The different types of coolant will be covered in more detail in Chapter 11, "Heating, Ventilation and Air Conditioning".

EXHAUST SYSTEMS

The original purpose of the exhaust was to clear the combustion chamber and quiet the noise of the engine. That same purpose remains, but some components of the exhaust system are now used to clean the exhaust gases of harmful emissions.

The exhaust manifold is attached to the cylinder head with a separate opening or port for each cylinder (**Figure 7-69**). The manifold runners run together at some point shortly after leaving the cylinder head. Some engines use a tuned exhaust manifold that has smooth runner (pipes). This design leads to better efficiency. A few newer Chrysler engines collect the exhaust gases at the cylinder head and then attach a short manifold or pipe to it.

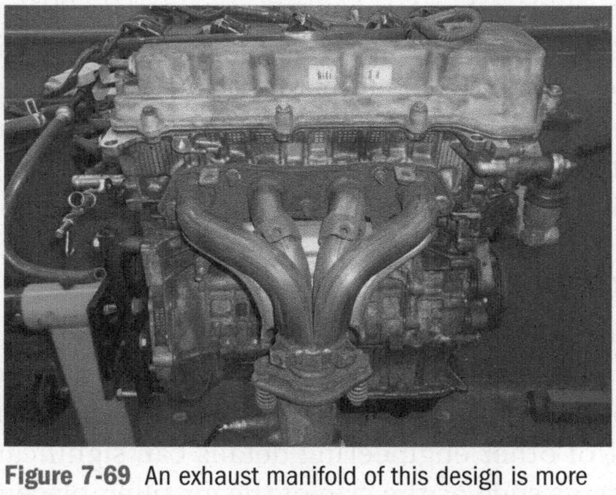

Figure 7-69 An exhaust manifold of this design is more efficient at gas flow. The design is commonly referred to as a lake pipe.

Mufflers and **resonators** are exhaust system components designed to reduce exhaust noises.

The exhaust pipe extends from the manifold to a muffler (pre-1974) or to a catalytic converter (post-1973). **Mufflers** and **resonators** are still used to quiet the exhaust noises but are positioned after the converter. Between the converter, muffler, and resonators is a tail pipe made of stainless steel. A section of the tail pipe extends from the last muffler to the rear of the vehicle.

AUTHOR'S NOTE A few years back, several children of the same family died while riding in the camper shell on their family's pickup truck. They were returning from a family outing less than 100 miles from home at night. The children were placed in the camper so they could sleep. During the trip home, the children succumbed to carbon monoxide poisoning. The cause was a broken tail pipe under the truck that allowed exhaust gases to enter and be trapped in the camper shell. During the safety inspection, inspect the exhaust system thoroughly.

For the best power and mileage, the exhaust must be tuned for the engine. Failure to exhaust gases quickly causes poor performance, particularly at higher engine speeds. Restricted or too-small exhaust pipes will not clear the combustion chamber for fresh air and fuel. A fully opened system with no noise or pollution devices is not necessarily the best means to achieve the most performance *unless* the system is designed or tuned to the vehicle and engine. In most areas, removal of noise-reducing devices is illegal. It is illegal everywhere in the United States to remove or disable the catalytic converter or other factory-installed emission devices.

AUTHOR'S NOTE A restricted exhaust creates symptoms that closely mirror the symptoms caused by loss of fuel delivery.

EMISSION CONTROL SYSTEMS

Hydrocarbons (HC), **carbon monoxide (CO)**, and **oxides of nitrogen (NO_x)** are harmful products of burning fossil fuels.

The burning of fossil fuels produces harmful emissions. Hidden within an engine's exhaust gases are three major chemicals or particles that harm the environment and living things. **Hydrocarbons (HC)** are particles of carbon left after ignition. HC is measured in parts per million of air. **Carbon monoxide (CO)** is a colorless, odorless, tasteless gas that can kill oxygen-breathing animals and people. **Oxides of nitrogen (NO_x)** are emitted in the exhaust when combustion chamber temperatures exceed 2,500°F (1,371°C).

At one time, six basic emission control systems were used on cars and light trucks. Electronics and engine designs have eliminated some of them even though allowable emission limits have been lowered. The first four are still in use today whereas the last two have been replaced with better technology. In addition to the operating systems, a great deal of emission control is accomplished during the vehicle's engineering and design stage.

Design and Construction

The design of the engine and its internal components and operation help determine the amount of harmful emissions produced. Even though no design alone can fully eliminate pollutants, the shape of the combustion chamber, valve timing, ignition timing, fuel and air delivery, and a host of other engineering details can significantly reduce emissions. These details involve items such as the size of the air filter, placement of the spark plug, and even the type of oil filter.

Catalytic Converters

Older vehicles use a two-way catalytic converter, which uses two metals, platinum and palladium, to reduce carbon monoxide and hydrocarbons through **catalyst action.** During the catalyst stage, the exhaust gases are heated to about 1,600°F (878°C) and the unburned fuel is burned off. The hydrocarbon (HC) molecules break apart, forming hydrogen (H) and carbon (C) atoms. This burning or oxidation of the exhaust also causes chemical changes in the carbon monoxide molecule by breaking it into oxygen (O) and carbon dioxide atoms (CO_2). Some of the oxygen combines with the released hydrogen to form H_2O, or water. The best exhaust gas mixture at the tail pipe is H_2O and CO. Modern vehicles use a three-way converter. It does the same as the two-way but includes rhodium to help reduce nitrogen oxides (NO_x) (**Figure 7-70**).

Catalyst action is a chemical reaction between two or more elements. One element (catalyst) causes a chemical change in the other element(s) without changing itself.

Shop Manual
page 286

Exhaust Gas Recirculating Systems

The sole function of the **exhaust gas recirculating (EGR)** system is to lower nitrogen oxide emissions. NO_x is formed in the combustion chamber when temperatures exceed about 2,500°F (1,371°C). The EGR valve allows some inert exhaust gases to enter the intake manifold and mix with the air and fuel (**Figure 7-71**). Because there is very little,

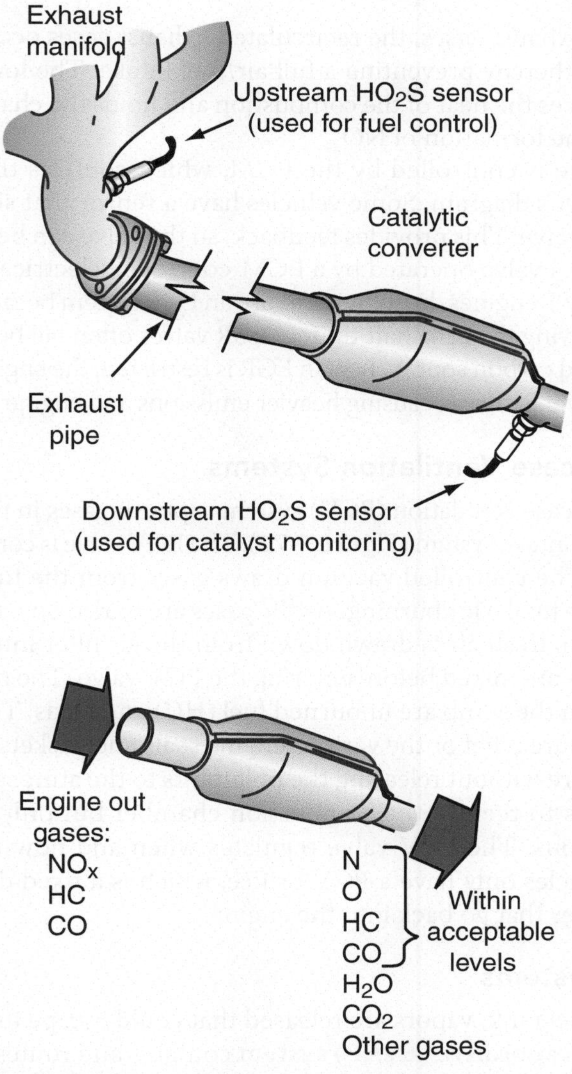

Figure 7-70 A three-way converter helps reduce HC, CO, and NO_x emissions.

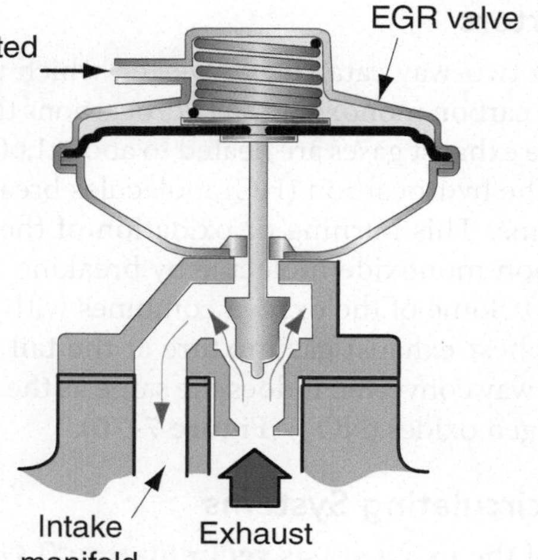

Figure 7-71 The exhaust gas recirculation valve helps lower combustion chamber temperature and reduces the formation of NO_x gases.

if any, fuel in the exhaust gases, the recirculated exhaust gases occupy space in the combustion chamber, thereby preventing a full air/fuel intake. The lower amount of air and fuel available reduces the heat of the combustion and holds the chamber temperature low enough to block the formation of NO_x.

The EGR valve is controlled by the PCM, which controls the amount of vacuum allowed to the valve's diagram. Some vehicles have a sensor that signals the PCM on the movement of the valve. This provides feedback, so the valve can be used more effectively. Other vehicles use a valve operated by a PCM-controlled electrical motor.

Some of the VVT engines do not need to use an EGR system because they can control the NO_x output by varying the camshaft timing. EGR valves often fail because they are a victim of exhaust gases and carbon soot. When an EGR is restricted, the engine's combustion chamber temperatures will increase, causing heavier emissions and engine drivability problems.

Positive Crankcase Ventilation Systems

The positive crankcase ventilation (PCV) system captures gases in the oil sump and routes them back into the intake system (**Figure 7-72**). The PCV valve is connected to a port on the intake manifold. The controlled vacuum draws gases from the top of the oil sump and feeds them into the intake for burning. As the gases are drawn out, a low pressure develops in the sump. Clean, fresh air is drawn down from the air filter into the sump, where the air and sump gases are mixed before entering the PCV valve. The most common harmful pollutants found in the sump are unburned fuel (HC) and acids. The oil sump must have some type of pressure relief or the various engine seals and gaskets will leak oil. The PCV relieves the pressure without releasing the pollutants to the atmosphere. The PCV allows unburnt HC gases to reenter the combustion chamber but only under certain engine operating conditions. The PCV valve regulates when and how much will reenter the engine. Some vehicles only have a PCV orifice, which is a fixed diameter that regulates the amount of gases that go back into the engine.

Evaporative Systems

An **evaporative (EVAP)** system captures and recycles vapors from the fuel system.

With fuel in the fuel tank, vapors are released that could escape into the atmosphere. To prevent this, the **evaporative (EVAP) system** contains and routes the fuel vapors to the intake manifold. The first component is the fuel tank filler cap. The cap is designed to allow air into the tank but prevents any vapor from escaping (**Figure 7-73**).

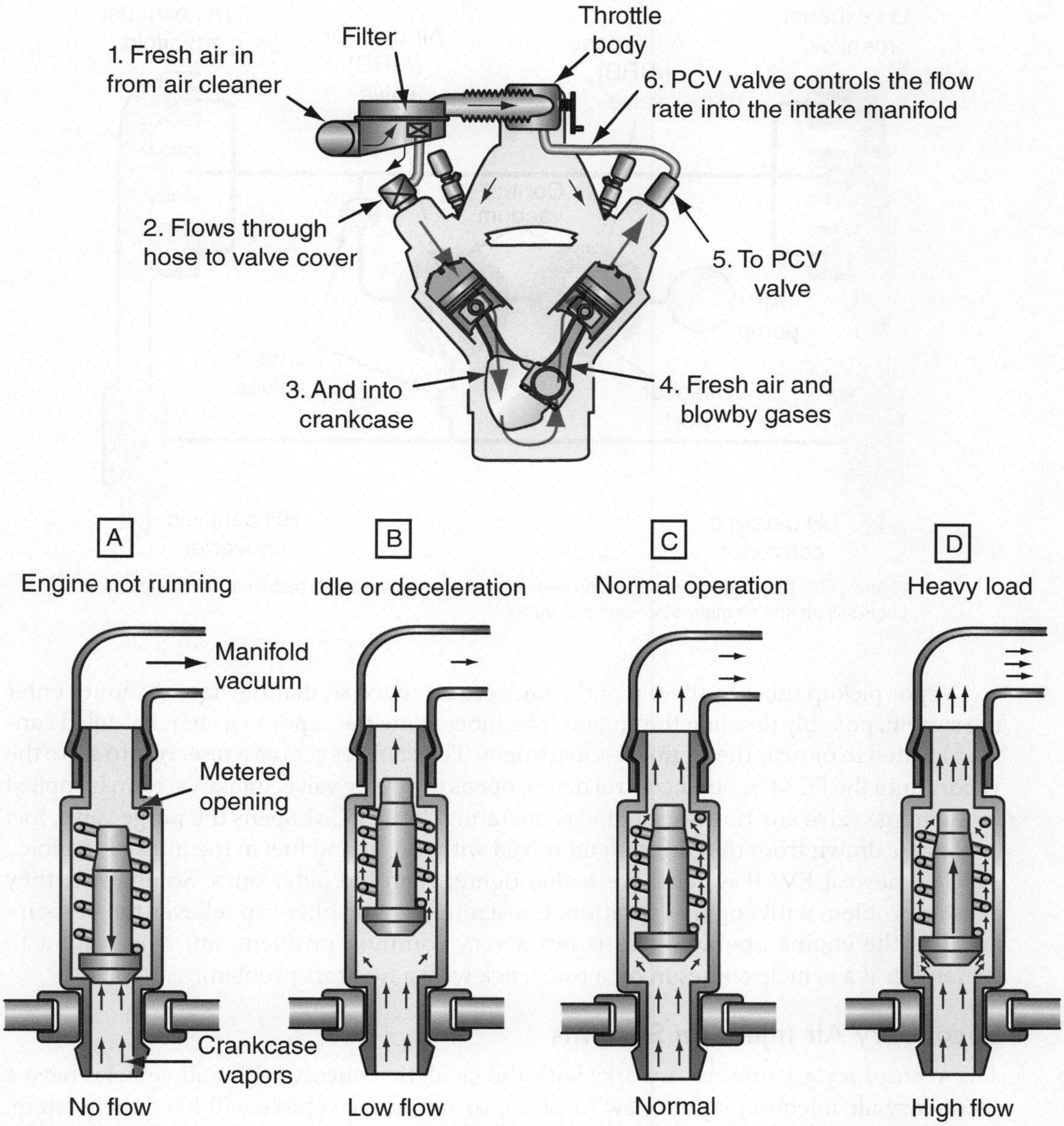

Figure 7-72 The PCV must be operational to reduce pressure in the oil sump and route harmful emissions back through the intake.

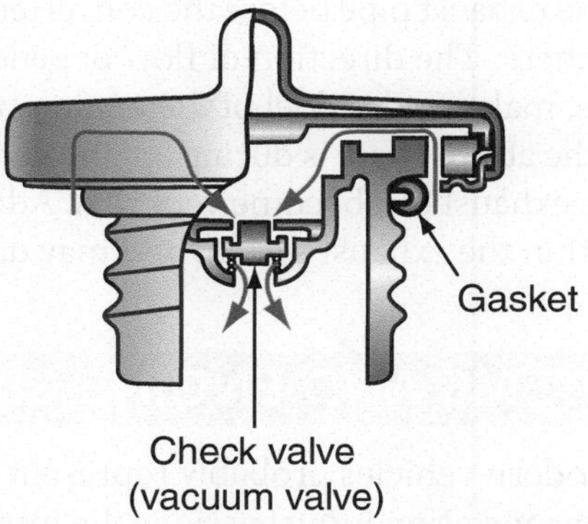

Figure 7-73 This is the only type of fuel cap allowed with the newest EVAP systems.

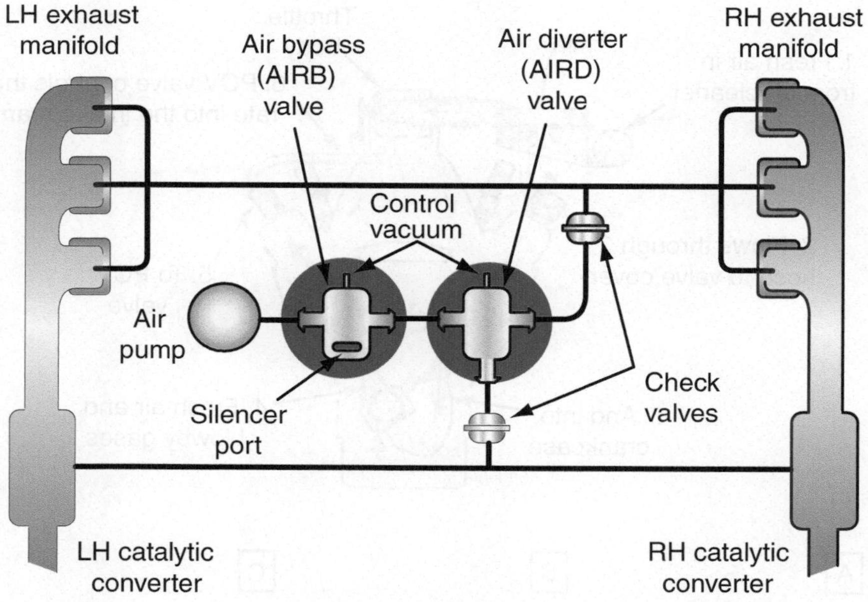

Figure 7-74 The secondary air injection system has been replaced with better engine controls but is still around on many older cars and trucks.

Vapor pickup tubes in the top of the tank are arranged so that the liquids cannot enter the system, possibly flooding the engine. The tubes route the vapor to a charcoal-filled canister located in or near the engine compartment. The canister acts as a reservoir to store the vapors until the PCM or other control device opens the purge valve. Intake vacuum is applied to the purge valve any time the engine is operating. If the PCM opens the purge valve, fuel vapors are drawn from the canister and mixed with the air and fuel in the intake manifold.

The newest EVAP systems are sealed tighter than the older ones. Sometimes, they cause a problem with engine operation. Loosening the fuel filler cap relieves the pressure and lets the engine operate. This is not a very common problem, but it is a point to remember if a vehicle comes in on a tow truck with a no-start problem.

Secondary Air Injection Systems

Secondary air injection forces fresh air into the catalytic converter to reduce exhaust emissions.

The **secondary air injection** works with the catalytic converter. Not all vehicles have a secondary air injection system law in place, so not every vehicle will have this system. Because the converter needs heat to work correctly, secondary air injection supplies the necessary oxygen to promote burning of any fuel in the exhaust. On most vehicles, this system is no longer used.A crankshaft-driven, high-volume, low-pressure air pump supplies the air (**Figure 7-74**). The air is directed by a valve mechanism. The air may be pumped into the exhaust manifold or the exhaust pipe before the converter, into the front of the converter, or diverted to the atmosphere. The direction of flow depends on the type of secondary air injection system; the type, make, and model of the engine; and engine load. The only time air should be diverted to the atmosphere is during engine deceleration. During deceleration, the air/fuel mixture in the exhaust can become very rich. Adding oxygen to the rich exhaust could cause a backfire within the exhaust system and may damage exhaust components.

COMPUTER SYSTEMS AND ENGINE CONTROL

Computer systems on modern vehicles probably represent the biggest and most comprehensive change to the transportation industry since the introduction of a credible internal combustion engine. Computer systems have provided more efficient vehicles to the public and can be helpful to technicians and repair shops. The only real drawback to vehicle computer systems is the fact that many technicians-to-be and in-field technicians either don't understand electronic systems or are afraid of them. If a technician can diagnose a headlight system, then she or he can diagnose a computer system. Different testers, procedures, and different values are present, but it is still electricity and it will follow certain rules.

Shop Manual
pages 287–290

The basic computer system requires you to understand that there are three main parts: inputs, processing, and outputs. Inputs come from sensors mounted on the engine and vehicle. They sense things, movement, temperatures, positions, and so on and send signals to the computer. The computer processes the information and makes decisions based on previous data and instructions in the programming. The computer then sends signals and current to the outputs. The outputs do something. Typically, the action is then watched by a sensor and the processes starts all over. A good example of this is how fuel is injected into the engine. After combustion, the remaining gases are measured by an oxygen sensor in the exhaust system. The oxygen sensor sends data in the form of a signal to the PCM. The PCM takes that data and compares it to the programmed data. It then adjusts the amount of fuel that goes into the engine by sending a signal out to the fuel injector. The Shop Manual highlights some tests that new technicians can do without causing damage or injuries.

AUTHOR'S NOTE Understanding the basics of computer systems and how they work on a vehicle are only the beginning of a career working with vehicles. Computer systems and computer control continues to get more complex and grow in terms of how many there are and what they control. My first automotive teacher told me years ago that vehicles would change significantly, and a student needed to know a lot about electrical and computers to remain successful. The most successful technicians today don't just understand computers and electricity; they embrace it as it constantly changes everything. Turns out, he was right.

Electrical is electricity flowing through a circuit and may be controlled by **electronic** signals from a computer.

Sensors

Before we start this discussion, let's define electrical and electronic circuits. **Electrical** refers to current flowing through circuits and components with a purpose of doing something, such as turning on a light bulb or motor. Electronic refers to the use of a computer or other regulation that controls the circuit to do similar things. In an electrical circuit, the control is a mechanical, like a switch. Both use the movement of electrons, voltage, and resistance to make a usable circuit. Circuits on vehicles can be electrical or electronic. More about electricity is covered in Chapter 5 and 6.

Sensors are devices that measure mechanical or electrical actions and sends the data to a computer.

Transducers convert energy from one type to another.

Sensors are electrical (and sometimes electronic) devices that measure some mechanical, chemical, or electrical action. There are many varieties, with many different names, on modern vehicles so we will discuss them as types and relate some as examples. Vehicle sensors are basically **transducers** that convert one form of energy to another. **Potentiometers** are one type of sensor.

A **potentiometer** converts mechanical movement into an electrical signal.

Potentiometers on a vehicle usually change mechanical action (movement) into an electrical voltage or a signal. The most prominent is the **throttle position sensor (TPS)** (**Figure 7-75**). As the driver moves the accelerator pedal and in turn moves the throttle (air) valve at the intake, the TPS mounted to the side of the throttle body measures this mechanical movement, converts it to electricity, and transmits a signal to the PCM. This is done by a wiper arm moving over a resistor wire inside the TPS. At idle speed, the signal voltage to the PCM is between 0.5 and 1.1 volts on most vehicles. This signal "tells" the PCM the operator has the throttle valve closed and gives the appropriate commands for

The **throttle position sensor (TPS)** measures the mechanical movement of the throttle and converts the movement to an electrical signal.

Figure 7-75 The TPS converts mechanical action into an electrical signal.

a smooth-running engine. There are usually three wires to a TPS: One is the reference voltage (REFV or VREF) of either 5 or 7 volts, one is the ground for the sensor, and the last is the signal return to the PCM. As the operator opens the throttle (depresses the pedal), the wiper arm moves, resistance between the VREF and the signal return conductors change, and the voltage signal increases (**Figure 7-76**). At wide-open throttle the

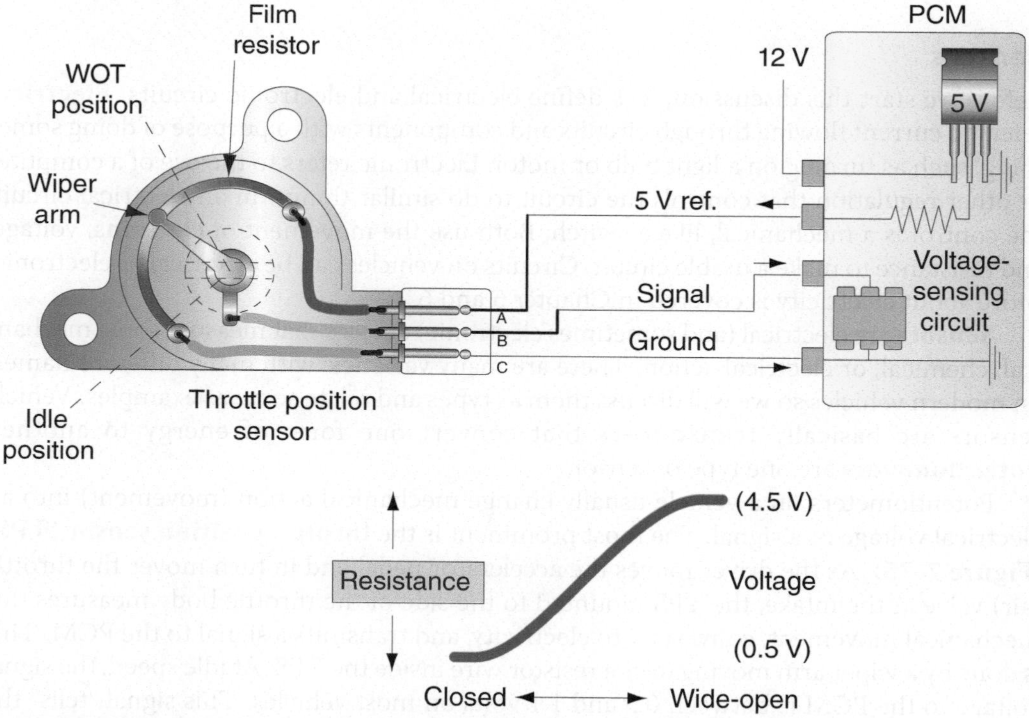

Figure 7-76 The TPS measures of infinite voltage between maximum and minimum on a resistor wire.

voltage is about 4.5 volts. The PCM sends the proper commands to change the air/fuel mixture, spark timing, and may even cause the transmission to shift gears. The most modern version of this sensor is the throttle position motor or sensor. It incorporates an electric motor as the throttle instead of using a cable. In this case, the sensor is simply a check and balance for the motor controlled by the same computer watching the sensor's return signal.

Another type of sensor is the alternating-current (AC) generator usually called a permanent magnet (PM) generator, which is nothing more than a very small version of the vehicle's generator. The signal voltage will be very low, but its operating principles are the same as the vehicle's generator. Magnetic teeth are attached to a mechanical device like a brake rotor or axle shaft. The sensor has an electrical magnet energized by current from the PCM. As stated in Chapter 5, as each tooth passes the sensor magnet, an electrical voltage or current is produced and transmitted back to the PCM. One sensor of this type that has been used for many years is the **vehicle speed sensor (VSS),** which does exactly what its name implies—it measures vehicle speed. The second common AC sensor is the wheel speed sensor used in antilock brake systems (ABS). They measure the rotational speed of the wheels, so the ABS controller can make braking decisions and issue commands in a panic stop.

Vehicle speed sensor (VSS) converts the rotary motion of an axle or wheel to an electrical signal for the computer to determine vehicle speed.

A **Hall-effect switch** is a type of pulse generator signal that either measures the mechanical speed of a device or the position of certain components. Hall-effect switches have been used for years as crankshaft and camshaft position sensors. These two sensors signal the PCM on how fast the shafts are rotating and when to command the fuel injectors on or off and when to fire the spark plugs. Many modern engines use an AC or PM sensors at the wheels with Hall-effect switch sensors. The main reason for this is the on or off type of signal generated by the Hall-effect switch as opposed to the constantly changing signal generated by AC generators. The Hall-effect switch is also grounded when it is mounted. Computers must respond to an on or off signal. If a sensor produces a wave signal, it must be converted to a digital on or off signal before being processed. A Hall-effect sensor already outputs a digital signal that does not need conversion. A following section deals with types of signals.

A Hall-effect switch measures the speed of position of a component and converts that data into an electrical signal.

AUTHOR'S NOTE It is important not only to understand how the sensors on today's vehicles work but to truly understand electricity and electronics because the type of technology that the sensors on future vehicles will use will likely be different than the ones today. The only thing constant is change.

The **manifold absolute pressure (MAP)** sensor measures the vacuum or absolute pressure in the engine's intake and converts that to an electrical signal (**Figure 7-77**). This signal informs the PCM on how hard the engine is working (engine load). As the throttle is open initially, air rushes into the intake and the pressure increases to near atmospheric (vacuum drops). The computer interprets this signal as a high engine load and holds the fuel injectors open longer and retards the spark timing. As the vehicle speeds up and the engine load decreases, the intake pressure drops (vacuum rises) and the computer changes the injector on time and may even cause the transmission to upshift. A **mass airflow (MAF)** sensor measures the volume of the air leaving the air filter en route to the intake (**Figure 7-78**). This is another indication of engine load. The MAF sensor may also be designed to measure the temperature and density of the incoming air. This makes the MAF more sensitive to engine load and the type of air entering the engine (i.e., the colder the air, the higher its density, hence more oxygen is available for thermodynamics). The PCM can now give more precise commands to the fuel and ignition systems.

A **manifold absolute pressure (MAP)** sensor converts air pressure into an electrical signal used by the computer.

A **mass airflow (MAF)** sensor converts air flow and density to an electrical signal.

Figure 7-77 The MAP converts pressure measurements in the intake manifold and generates an electrical signal. This signal is a measurement of engine load.

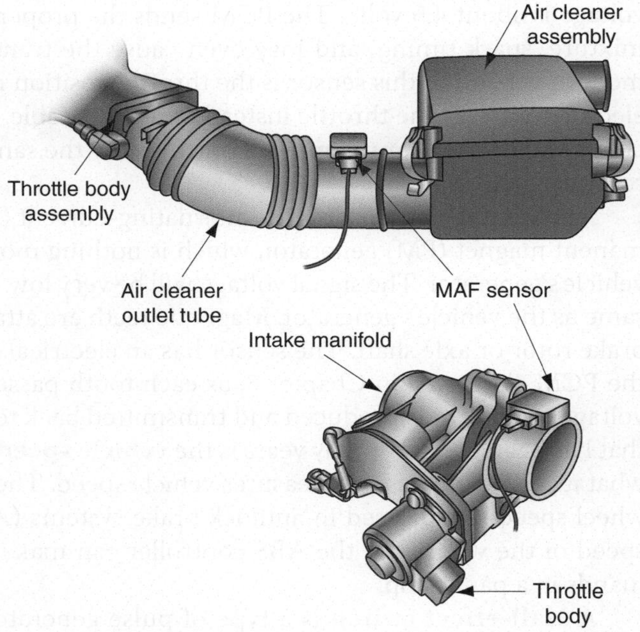

Figure 7-78 The MAF is mounted in the air intake duct between air filter output and the throttle valve.

Switches are mechanical or electrical sensors to signal when a device is either on or off.

Switches are the most common sensors on a vehicle. Switches may be operated by the driver or electronically controlled. A switch gives an on or off signal to the PCM and other computer modules. Modern vehicles have computers everywhere and they are used to control almost everything. Switches used as computer inputs can be as simple as a window switch and as complicated as an electronic command switch in the automatic climate control. In some instances, this signal does nothing but tell the computer what the operator wants. When the air conditioning button is pressed on, a signal is sent to a pin on the PCM labeled "AC Request." The air conditioning will not switch on until the PCM looks at certain other signals, like the engine coolant temperature, refrigerant pressure, vent position, and throttle position. If all conditions meet the required parameters, the PCM sends power to the compressor, and it is energized. If one of the required signals is missing or out of limits, then there is no air conditioning, no matter how hot the vehicle's interior gets. For example, the PCM will not switch the A/C on if the engine temperature is too high, there isn't enough refrigerant, or one of the temperature sensors is malfunctioning.

There are many more sensors on today's vehicles. Only a few were discussed here, but all sensors fall into one of the categories mentioned. As you continue with your automotive education and training, you will learn all the sensors. Each manufacturer does certain sensors a little differently, so it is important to remember the basics of how a sensor works.

Actuators

Actuators are mechanical work components controlled by computer command.

Solenoids are computer-controlled mechanical devices commonly used to regulate fluid flows.

Actuators are electrical devices that do the actual work commanded by the PCM or another computer. They range from fuel injectors to cooling fans to lighting. Most are **solenoids** or electric motors. A solenoid is like a relay. It is an electrical device that changes electrical energy into mechanical movement. Probably the most common is the fuel injector (**Figure 7-79**). On command from the PCM, the injector coil is energized, creating a magnetic field. This field acts on a metal, moveable core that is part of the injector pintle. The core's movement opens the injector, allowing pressurized fuel to enter the airflow. When the injector is switched off, a spring forces the core to move backward, shutting off the fuel. The second most common actuators on most modern vehicles are the shift solenoids in the transmission. They control fluid flow to and from transmission

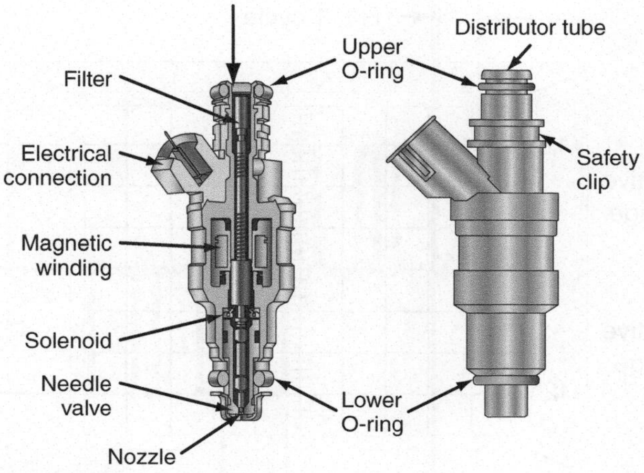

Figure 7-79 A fuel injector.

shift valves. Other solenoids may be used to operate power door locks or the blend doors on the climate-control systems.

Electric motors are used everywhere on a vehicle. They can be used to operate the throttle valve, power windows and locks, seats, blower and cooling fans, air conditioning blend doors, and the starter motor. They range from the heavier motors such as those for power seats and fans to small, palm-size ones used as blend door actuators, and operate based on commands from their controlling computers.

Signals and Commands

Signals, in this discussion, are those electrical currents generated by a sensor and transmitted to a computer. The signal may be **analog** or **digital.** Analog signals would be like those generated by an AC (PM) generator where the voltage goes from zero to a positive voltage back to zero then to a negative voltage (**Figure 7-80**). This is a signal that a computer has a problem interpreting. Usually an analog signal is sent to a signal converter where it is changed into a digital signal. Digital signals are usually interpreted as on or off (**Figure 7-81**). Based on the computer's programming, the "on" signal may be considered as "high" or "yes." The "off" signal can be considered as "low" or "no." The signal is used for the computer to make decisions and issue commands.

A **signal** is data sent to/from a computer in an electrical form.

An **analog** signal changes from positive to negative voltage at a given rate.

Digital signals switch between on and off voltages.

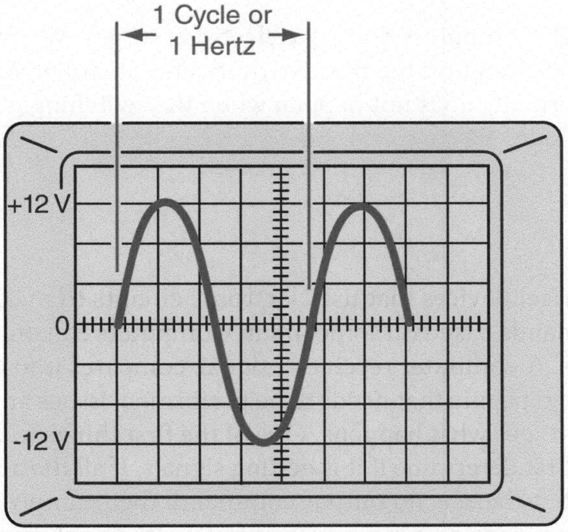

Figure 7-80 Note that an analog signal moves from positive to negative voltage. This is known as a sine waveform.

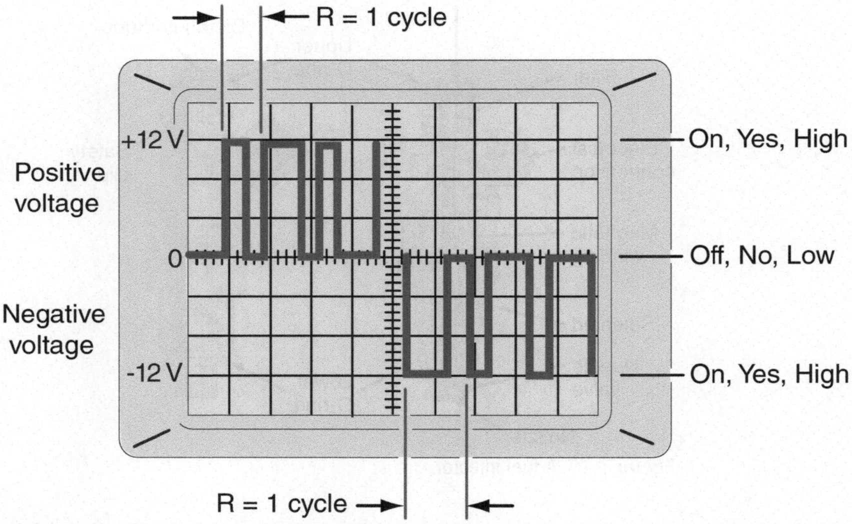

Figure 7-81 Digital signals can be positive or negative voltage. The ones shown here are listed as a rectangular waveform. The differences between the widths indicate the amount of time the signal was in the on (or high) position.

AUTHOR'S NOTE Computer decisions are not like those produced by the human brain. A computer decides by comparing input signals to installed programming and issues commands according to the program instructions. If something is not correct, the component will simply not work. Sometimes a diagnostic trouble code (DTC) will be present or even a check engine light will illuminate, and sometimes nothing will happen.

Commands are simply electrical signals sent by a computer to an actuator to do something. A simple command may only ground a relay coil to close a switch that turns on the cooling fan, or it may switch the fuel injector on. Usually the commands constitute an on or off signal to the actuator and are usually found on the negative side of the circuit.

AUTHOR'S NOTE Computers don't like electrical surges. When opening and closing an electrical circuit on the positive (hot) side, an arc or surge is usually generated. This normally does not happen when the switching is done on the negative (ground) side.

Computers

Computers are electrical devices that use electronic circuits to make programmed decisions and issue commands based on some input. Computers can only be as smart as their human programmers. A computer receives a signal, compares it to internal instructions, selects something very specific that needs to be performed, issues an electrical command, and then sits back and sees what happens. One of the first things to check when diagnosing a computer is to first determine if it is getting signals. If all the appropriate signals are present and correct but there is no output command, then, simply put, the computer is bad. Automotive technicians do not repair computers; they only replace them.

Automotive computer systems are here to stay. They will continue to become more powerful and will be associated with more mechanical components and systems. Already the mechanical linkages for the throttle and the transmission shift valves have been eliminated on some **drive-by-wire** vehicles. Some steering systems use electrical motors to assist the driver's effort in steering. Some of these computer systems come from military development and research into electric-powered vehicles and others come from racing, college competitions, and smaller companies. But most of the development of vehicle computer systems still occurs at the manufacturer level.

Drive-by-wire is a term used in the automotive industry which means that a component is controlled by a computer-regulated system and is NOT operated by cables or mechanical linkage.

SUMMARY

- The major systems of a vehicle are: electrical, engine, driveline, steering, suspension, climate control, frame, and body.
- The two-stroke engine produces a power stroke for each revolution of the crankshaft.
- Air and fuel are drawn in during the compression stroke of the two-stroke engine.
- The four-stroke engine completes intake, compression, power, and exhaust strokes in one cycle.
- Valves are used to open and close ports in the cylinder head.
- Valvetrains are either OHV or OHC.
- OHC engines may have two camshafts per cylinder head.
- The lubrication system cools, lubricates, and cleans the internal components of the engine.
- Synthetic lubricants and blends are made from chemicals rather than natural petroleum products.
- The starting system provides power to rotate the crankshaft, which draws air and fuel into the cylinders.
- The charging system charges the battery and supplies the electrical current to operate the vehicle.
- EFI systems require an electrical fuel pump.
- Air is drawn into the intake manifold by engine vacuum and atmospheric pressure.
- The cooling system removes the heat from the engine block and cylinder head.
- To get fresh air and fuel into the combustion chamber, an exhaust system must quickly route the exhaust gases from the chamber.
- The catalytic converter reduces CO, HC, and NO_x emissions.
- HC emissions are reduced using the PCV and EVAP systems.
- Computer systems monitor and control many mechanical devices on modern vehicles.
- Sensors provide input signals to the computer.
- Automotive sealants include gaskets, chemicals, and seals.
- Automotive bearings may be inserts, rollers, or the ball type.
- Bearings must support axial and radial loads, but they may be designed for one specific type of load.
- Generally, automotive grease is classified as either L or G.
- L grease is for suspension and A is for wheel bearings.
- A special grease is used on CV joints and brake systems.
- Gaskets are used between two stationary parts.
- Gaskets can be made to withstand high temperature and pressure.
- Gaskets can be very thin and made from paper.
- Gaskets can be made from a bead of chemical compound.
- RTV is a chemical sealant used only on stationary components.
- Seals are usually used between a moving component and a stationary component.
- Seals are installed with the lip facing the lubricant.
- O-rings are special-purpose seals.
- A rubber boot may seal the component from dirt and water.

REVIEW QUESTIONS

Short Answer Essays

1. List the four strokes of an internal combustion engine, and explain the actions that happen on each stroke.
2. Explain how the lubrication system works.
3. List and describe the actions of the cooling system and its components.
4. Explain the purpose of the exhaust system.
5. Describe the purpose of the sensors.
6. List the areas where RTV may be used.
7. Explain how head gasket material is protected from flame and chemicals.
8. List and describe the uses of the two major classification of automotive grease.
9. Describe the construction of a tapered roller bearing.
10. List the components of a roller bearing.

Fill-in-the-Blanks

1. The EGR system is designed to help reduce ______________ emissions.
2. On the compression stroke, the valves are ______________.
3. The exhaust system quiets engine noise and helps ______________ ______________.
4. The cooling fan is used to draw air over the ______________, ______________.
5. L-type grease may fail if used on ______________ ______________.
6. High pressure within the oil pan may cause the crankshaft ______________ to leak.
7. The head gasket has holes punched through it to allow passage of ______________ and ______________.
8. Gaskets may use a(n) ______________ to hold it in place.
9. A seal is installed with the dust lip ______________ from the lubricant.
10. An O-ring is installed in a(n) ______________ in the moving component.

Multiple Choice

1. A traditional coolant can be identified by the color:

   A. Red.
   B. Green.
   C. Purple.
   D. Orange.

2. *Technician A* says it is illegal to remove the muffler from the exhaust in all localities. *Technician B* says the catalytic converter was installed on most 1974 cars. Who is correct?

   A. A only
   B. B only
   C. Both A and B
   D. Neither A nor B

3. The lubrication system is being discussed. *Technician A* says the camshaft may drive the oil pump. *Technician B* says the pressure relief valve is used to bypass oil around the oil filter. Who is correct?

   A. A only
   B. B only
   C. Both A and B
   D. Neither A nor B

4. *Technician A* says the battery can operate the vehicle's electrical systems for a short time. *Technician B* says the starter motor draws current from the AC generator. Who is correct?

   A. A only
   B. B only
   C. Both A and B
   D. Neither A nor B

5. Which of the following choices of oil means that it is used only with a gasoline engine?

   A. CF
   B. API
   C. SL
   D. SAE

6. The head gasket is being discussed. *Technician A* says the steel ring is placed around the cooling passage. *Technician B* says silicone may be used to seal the oil passages. Who is correct?

   A. A only
   B. B only
   C. Both A and B
   D. Neither A nor B

7. *Technician A* says a gasket can be used to replace a seal.
   Technician B says a chemical can be used to replace a gasket.
   Who is correct?
   A. A only
   B. B only
   C. Both A and B
   D. Neither A nor B

8. *Technician A* says lubricant pressure can be used to better seal the lip to the shaft.
   Technician B says a garter spring is used to better seal the dust lip.
   Who is correct?
   A. A only
   B. B only
   C. Both A and B
   D. Neither A nor B

9. RTV is being discussed.
   Technician A says RTV is an aerobic sealant.
   Technician B says the absence of air cures RTV.
   Who is correct?
   A. A only
   B. B only
   C. Both A and B
   D. Neither A nor B

10. Gasket sealers are being discussed.
    Technician A says a sealer can be used to replace the gaskets.
    Technician B says sealers are cured by the absence of air.
    Who is correct?
    A. A only
    B. B only
    C. Both A and B
    D. Neither A nor B

11. *Technician A* says anaerobic chemicals cure when the parts are assembled and torqued.
    Technician B says aerobic chemicals cure completely in about 24 hours.
    Who is correct?
    A. A only
    B. B only
    C. Both A and B
    D. Neither A nor B

12. The use of O-rings is being discussed.
    Technician A says O-rings should be used in low-pressure systems.
    Technician B says a square cut O-ring is used on drum brakes.
    Who is correct?
    A. A only
    B. B only
    C. Both A and B
    D. Neither A nor B

13. Bearings are being discussed.
    Technician A says needle bearings are a type of roller bearing.
    Technician B says flat roller bearings are used to support a wheel.
    Who is correct?
    A. A only
    B. B only
    C. Both A and B
    D. Neither A nor B

CHAPTER 8

TRANSMISSIONS AND DRIVELINE SYSTEMS

Upon completion and review of this chapter, you should understand and be able to:

- Describe the purpose of the driveline.
- Explain gear ratios and how they change the torque ratio.
- Understand the operation of a manual transmission/transaxle.
- Understand the purpose of the clutch.
- Explain the basic operation of an electronically controlled transmission.
- Explain the basic operation of a Continuously Variable Transmission (CVT).
- Understand purpose of the torque converter.
- Describe the general operation of an automatic transmission.
- Identify the components of drive shafts and axles.
- Describe the purpose and operation of the differential and final-drive assemblies.

Terms To Know

Bands
Bell housing
Blocker rings
Cluster gear
Clutch pack
Compound gears
Constant mesh
Differential
Differential carrier
Driveline
Drive gear
Drive shaft
Driven gear
Extension housing
Final drive
Flywheel
Friction disc
Gear
Impeller
Input gear
Input shaft
Lockup clutch
One-way clutch
Pilot bearing
Pilot bushing
Pinion gear
Planetary carrier
Planetary gearset
Pressure plate
Release bearing
Reverse idler gear
Ring gear
Roller
Rotary flow
Servo
Shifter rails
Side gear
Simpson planetary gearset
Slave cylinder
Slip joint
Spool valve
Sprag
Stator
Sun gear
Synchronizer
Torque converter
Transmission housing
Turbine
Turbine shaft
U-joint
Vortex flow
Yoke

INTRODUCTION

The engine produces the power that drives the wheels. However, a direct connection between the engine and drive wheels is not a satisfactory arrangement. There has to be a method of delivering high torque to get the vehicle moving and then lowering the engine speed once underway. The transmission and other components of the driveline accomplish that task. For this chapter, the term *transmission* refers to a rear-wheel-drive (RWD) transmission and a front-wheel-drive (FWD) transaxle. The term *differential* refers to the RWD differential and the FWD final drive. Specific operational and design differences will be noted for each, but traditionally a transaxle is a transmission that contains the final drive in the same case.

PURPOSE AND TYPES OF DRIVELINE

Shop Manual
pages 313–314

The word **driveline** is used here in lieu of *drivetrain*. Both refer to the components that transfer engine output to the drive wheels.

The **driveline** delivers the power of the engine to the drive wheels. A transmission is used on most RWDs whereas a transaxle is used with FWD vehicles. Some vehicles, like the Porsche, have an engine located near the middle or rear of the vehicle and use a transaxle to drive the rear wheels. A Chevrolet Corvette has an engine in the front, and the transmission is mounted near the rear drive wheels.

The primary difference between the transmission and the transaxle is the method of delivering transmission or transaxle output power to the drive wheels. The transmission uses a driveshaft, which drives the input gear of a differential. On RWD vehicles, the differential turns the torque line 90 degrees and drives the axles (**Figure 8-1**). The axles are attached to the drive wheels.

A transaxle has an integral final-drive assembly. The final drive performs the same function as the differential except the transaxle output is delivered directly to the input gear of the final drive. There is no drive shaft in this arrangement; however, there are driving axles (**Figure 8-2**).

The transmission and transaxle use gears to change the engine's output torque. Ideally an engine is most efficient at its peak torque. The job of the transmission is to try to stay in that range (**Figure 8-3**). A manual transmission requires the driver to select, through

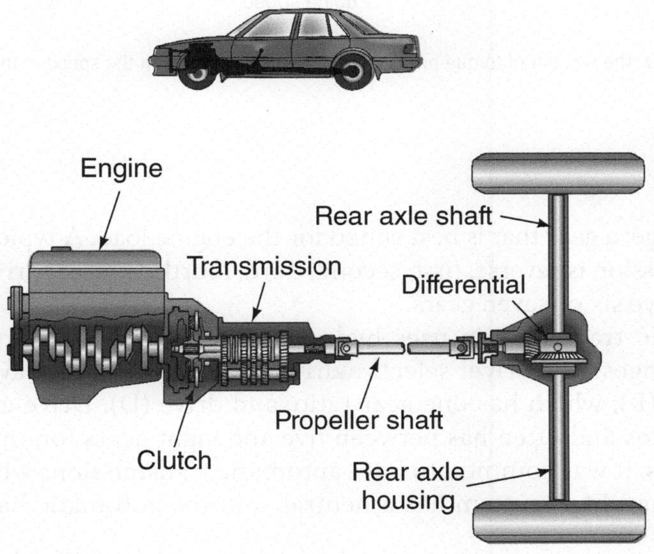

Figure 8-1 The drivetrain components for a typical rear-wheel-drive vehicle.

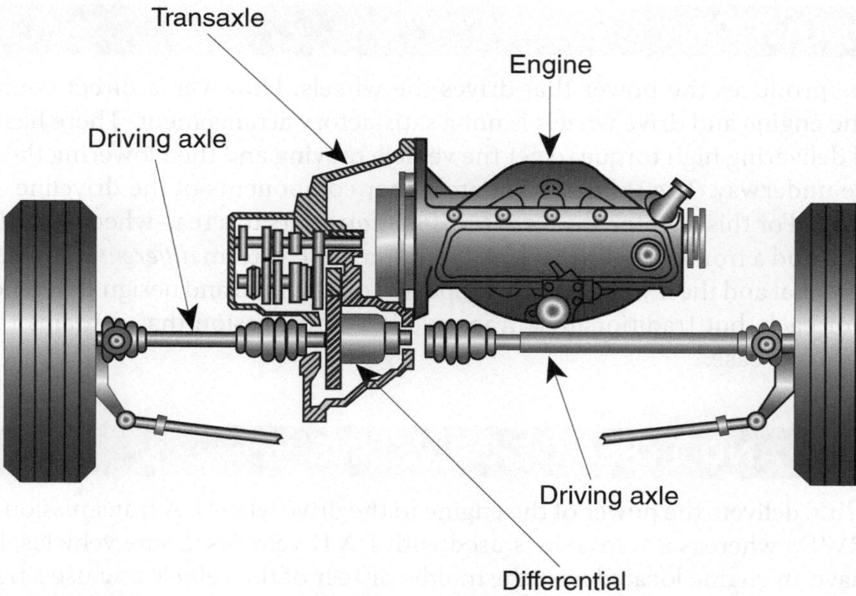

Figure 8-2 The main components of a typical transaxle assembly.

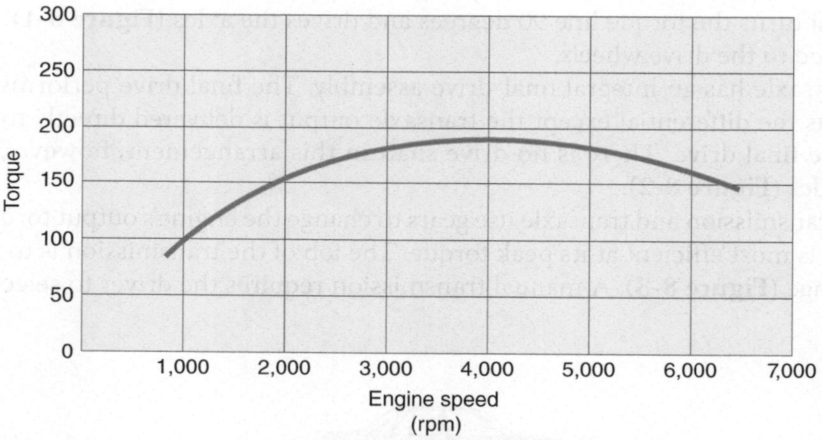

Figure 8-3 The amount of torque produced by an engine varies with the speed of the engine.

mechanical linkage, a gear that is best suited for the engine load. A typical gear setup for a manual transmission is reverse, first, second, third, fourth or direct drive, and fifth gear. Some vehicles have six or seven gears.

An automatic transmission uses hydraulics, valves, sensors, and actuators to achieve gear changes. The driver selects a drive range and drives away. A typical drive range is reverse (R), which has one gear ratio, and drive (D). Drive usually has three forward gear ratios and often has between five and eight gears (on modern vehicles). On older vehicles, it was common to have automatic transmissions with two, three, or four forward gears. Each system has a neutral, and the automatic has a position for park (**Figure 8-4**).

Figure 8-4 A six-speed automatic transmission.

GEARS AND GEAR RATIOS

Different gears and gear ratios are used to harness the torque of the engine as it is needed for different loads. The transmission and differential provide the means to do this.

Gears

Gears in a transmission are normally circular-toothed components. The teeth of one gear mesh or interact with another gear. The diameters of the gears determine if the input torque is increased or decreased. If torque is increased, the speed is decreased and vice versa. A simple gearset consists of two gears working together (**Figure 8-5**). The gear delivering the torque into the gearset is the **drive gear.** The gear delivering the torque out of the gearset is the **driven gear.** Gearsets of three or more gears are called **compound gearsets** and may have more than one gear ratio within the set. Gear teeth may be on the outside of the gear, external, or on the inside of the gear, internal. Internal gears are usually found in planetary gearsets. Planetary gearsets are most commonly found in automatic transmissions but may be used for fifth or sixth in a manual system.

Gear A toothed wheel that increase the force needed to push or pull something.

Drive gear Usually refers to the timing drive gear that receives forces from the power output then transfers power to the driven gear.

Driven gear The member or a pair of gear to which motion and power are transfered by the drive gear.

Compound gearsets A number of gear fixed together usually differ in size and number of teeth.

Ratios—Torque and Speed

Ratios are computed by dividing the driven gear by the drive gear. A gearset with a 1-inch drive and a 2-inch driven has a gear ratio of two to one or 2:1. If the drive was delivering 10 ft.-lb. of torque, the driven would be delivering 20 ft.-lb. of torque to the output. The example here shows a doubling to the available torque (**Figure 8-6**).

Using the same example, we can show the speed ratio for this gearset. Speed ratios are computed by dividing the drive by the driven, in this case, one is divided by two or 0.5 to 1. The driven gear will turn half a revolution to each complete revolution of the drive. If the drive turned at 10 rpm, the driven would only turn 5 rpm. In some vehicles, this would be low gear: high torque with a low speed.

Direct drive gear in most vehicles has the transmission output shaft turning at the same speed as the input. The gear ratios between low and high are determined by the gears

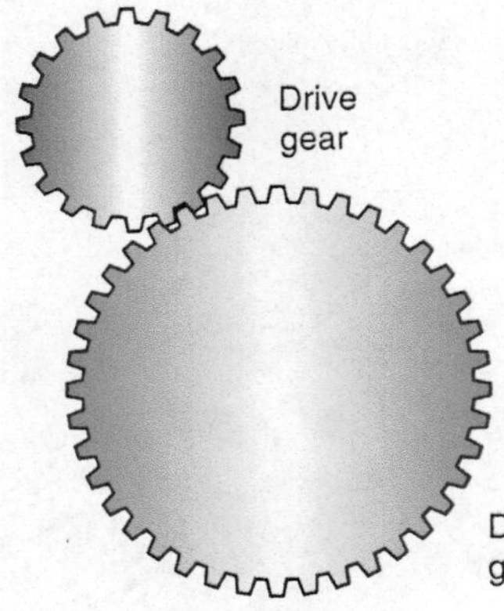

Figure 8-5 A small drive gear and a large driven gear increase the torque and decrease the speed.

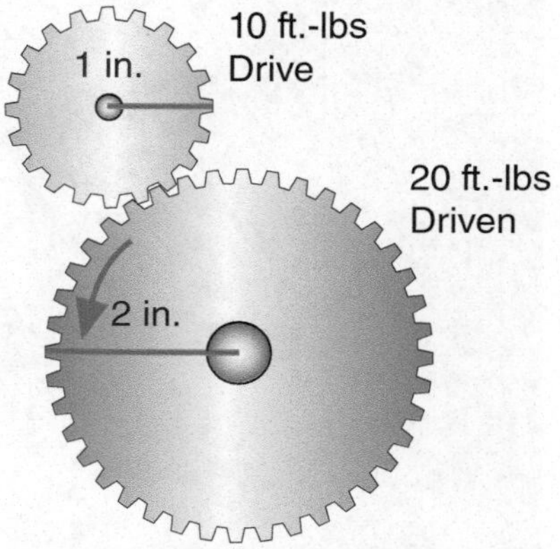

Figure 8-6 The 2:1 gear ratio in this gearset doubles the amount of torque available at the output.

being used. An overdrive gear would have a larger drive gear than the output gear (**Figure 8-7**). If the overdrive drive gear is 2 inches in diameter and the output is 1 inch, divide the drive by the driven for a gear ratio of 2:1. In this case, each time the drive gear completed one revolution, the output gear would rotate two times.

Note that this seems to be the same gear ratio mentioned in the first paragraph of this section. Both are 2:1. However, in that instance the ratio caused an increase in torque but a decrease in speed. The situation noted in this paragraph causes a decrease in torque and an increase in speed. When calculating gear ratios, always remember that a large gear driving a smaller gear will always result in a speed increase with a reduction in torque.

Note that the high torque reduces the speed by the same proportion. This is true of any gearset. If the torque is high, the speed will be low. As the vehicle nears cruising speeds, different gearsets are selected to provide a high speed, which in turn lowers the amount of torque delivered to the drive wheels and lowers the engine speed.

In **Figure 8-8**, the calculation shows what happens when the transmission and driveline create torque multiplication and speed reduction. This is necessary to get the vehicle up to speed and to operate efficiently in changing driving conditions.

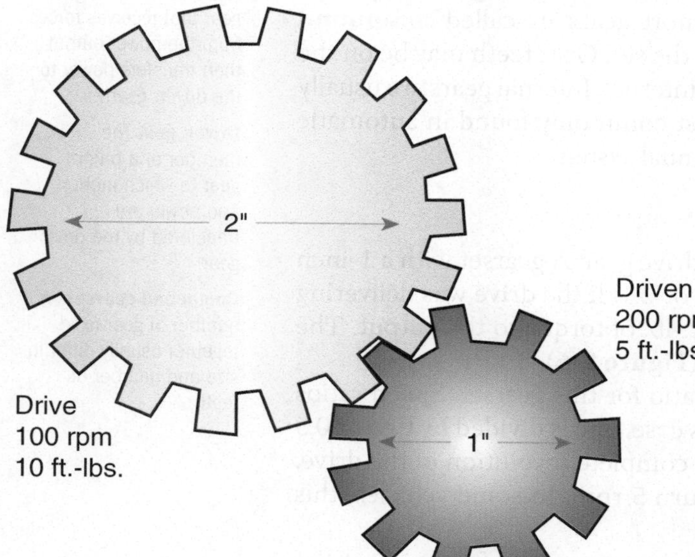

Figure 8-7 Using the small gear as the driven gear will result in higher output speed but less torque.

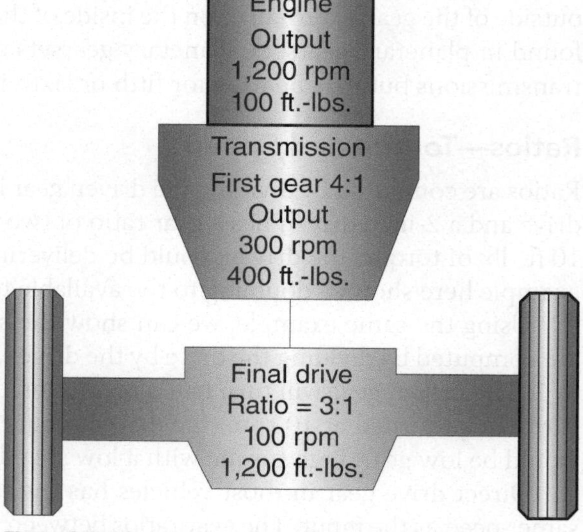

Figure 8-8 The torque multiplication and subsequent speed reduction of a typical vehicle's drivetrain.

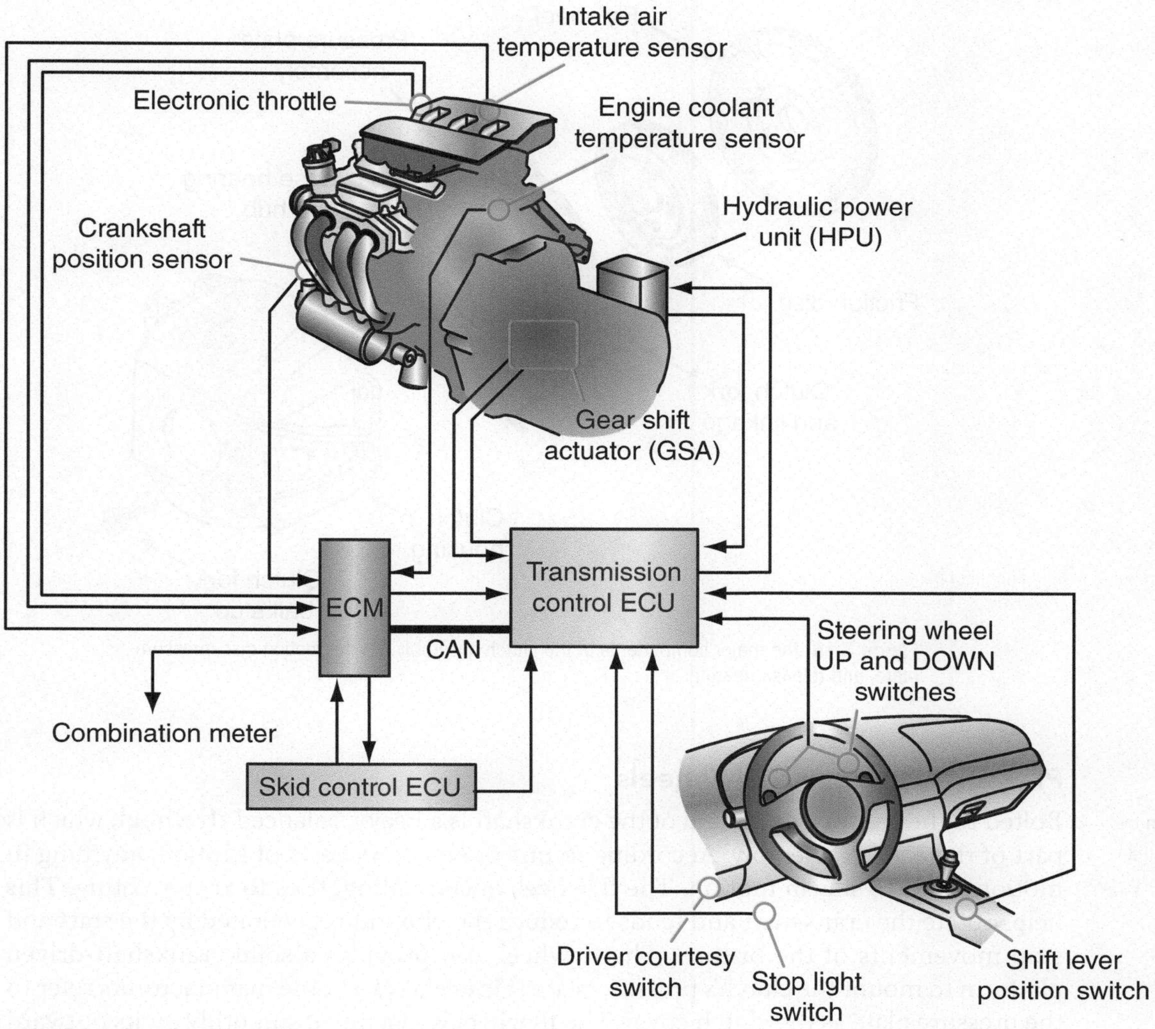

Figure 8-9 Typical computer control of a transmission.

Electronic Control of Transmissions

Shop Manual
page 319

Modern vehicles use a heavy amount of electronic control in transmissions. Transmission operation is dependent on engine-operating and vehicle-speed conditions. Before computer control, transmissions were operated based on hydraulic and vacuum controls that gave the transmission its control mechanically. Today's modern transmission contains several computer sensors and actuators inside it and relies on sensor inputs from the engine and driveline (**Figure 8-9**).

CLUTCHES

To apply the engine power smoothly to the transmission or transaxle, there has to be a device to couple and uncouple the two components. The most common clutch used in a typical vehicle is a manual clutch assembly for a manual transmission and hydraulic clutches within an automatic transmission. Hydraulic clutches will be discussed with automatic transmissions. The manual clutch assembly is made up of three subassemblies and the linkages (**Figure 8-10**). The subassemblies are the **pressure plate, friction disc,** and **release bearing.**

The **pressure plate** assembly is used to clamp the clutch disc to the flywheel.

The **friction disc,** once clamped, provides the connection between the engine and the transmission.

The **release bearing** provides a point of contact between a stationary component (linkage) and a moving component (pressure plate assembly).

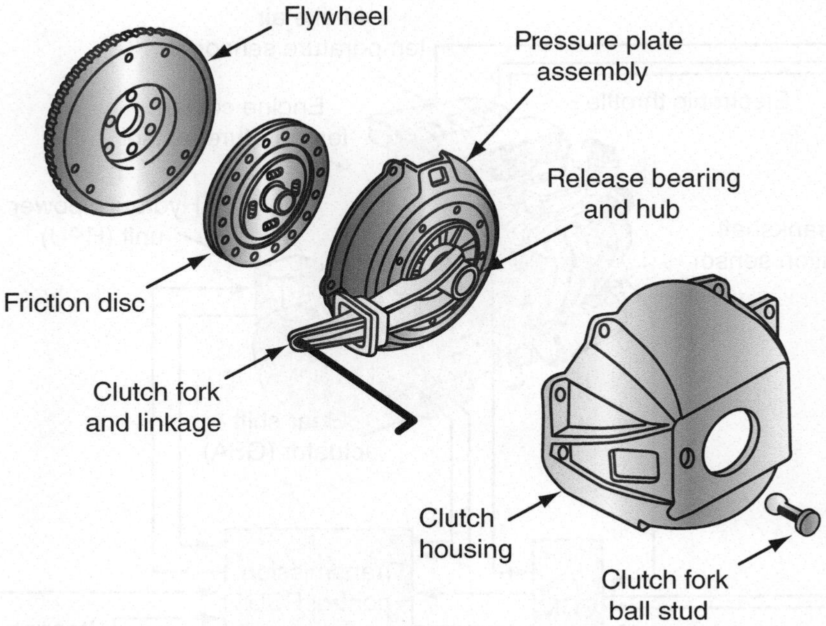

Figure 8-10 The major components of the clutch assembly are the friction disc, pressure plate, and release bearing.

Pressure Plates and Flywheels

The **flywheel** is driven by the starter motor during cranking to spin the crankshaft.

Bolted to the rear or output end of the crankshaft is a heavy, balanced **flywheel,** which is part of the engine assembly. According to one of Newton's Laws of Motion, anything in motion tends to stay in motion. The flywheel, once rotating, tries to keep rotating. This helps rotate the crankshaft and tends to reduce the vibrations generated by the start and stop movements of the pistons. The flywheel also provides a solid crankshaft-driven platform to mount the clutch's pressure plate (**Figure 8-11**). Some manufacturers refer to the pressure plate as the clutch cover. The flywheel is machined smoothly on its outward or rear face.

A steel frame supports the pressure plate, springs, and release levers or fingers (**Figure 8-12**). The frame is bolted to the flywheel. The pressure plate is finely machined on the flywheel side. High-tension springs apply force to hold the machined side firmly against the flywheel. This force clamps the friction disc between the flywheel and pressure plate. The clutch is now engaged, and engine power is transferred to the friction disc.

Release levers extend from pivot points along the outer edge of the frame up and toward the center (**Figure 8-13**). If force is applied against the inner tips of the levers, the levers' other ends pull the pressure plate away from the flywheel and friction disc. The friction disc spins free and no power is transferred. The clutch is disengaged.

Friction Discs

The friction disc is an assembly of springs and friction materials installed between the pressure plate and flywheel (**Figure 8-14**). There are two friction areas, one on each side of the plate. Between the two opposing friction areas are cushion springs that help absorb the shock of engagement. Mounted around the center of the plate are torsion springs. The torsion springs are designed to reduce the sudden rotation impact of engagement. In the center of the plate is a splined hole that fits the splines of the transmission's input shaft. When the clutch is engaged, and the friction disc is rotating with the flywheel, power is transferred to the transmission at this point.

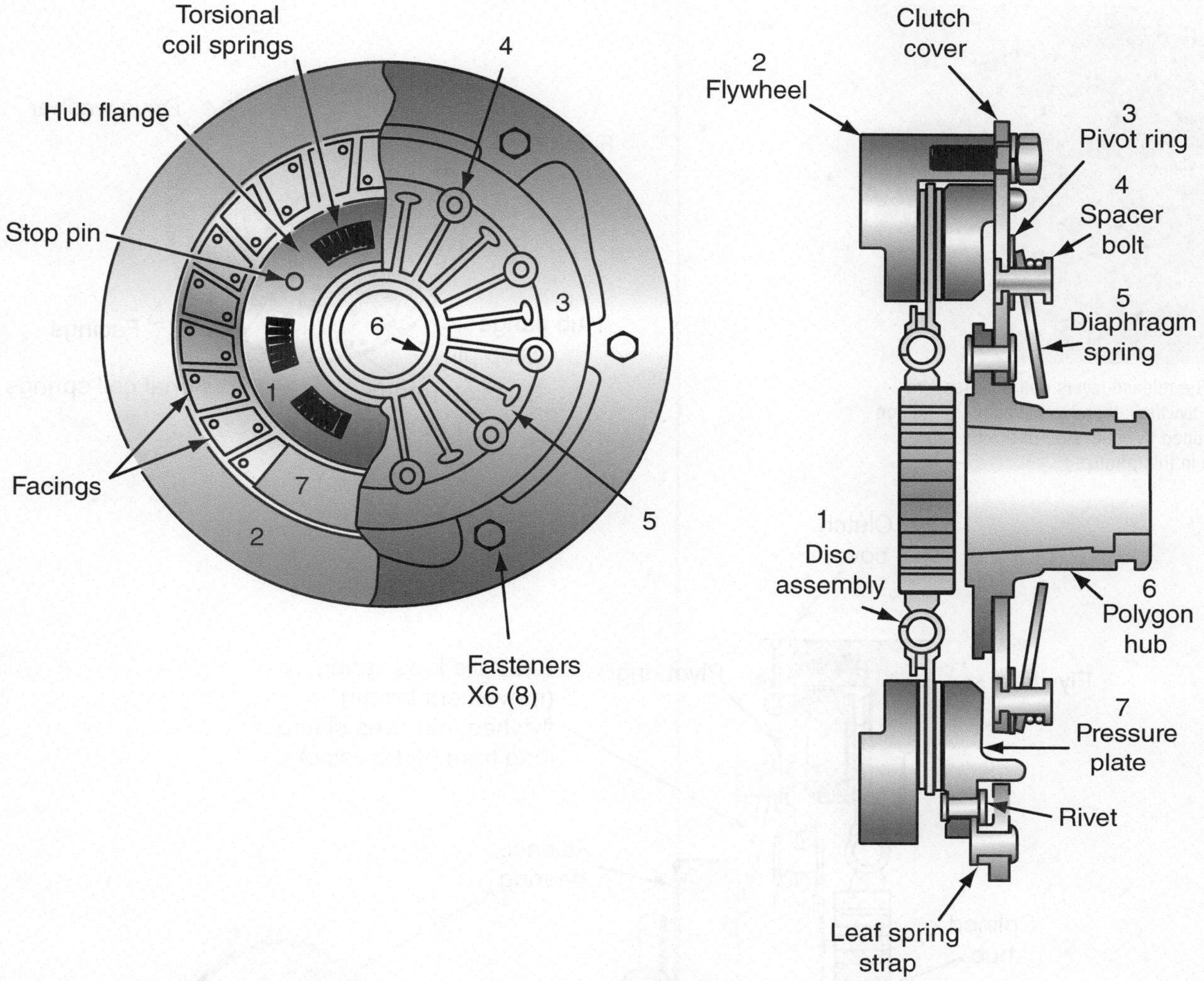

Figure 8-11 The pressure provides the clamping force to hold the friction disc firmly to the flywheel.

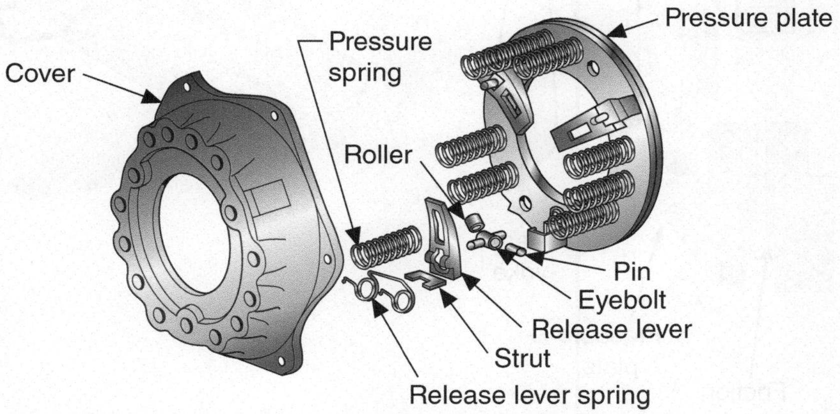

Figure 8-12 The pressure plate is moved rearward (disengaged) by the levers and clamped forward (engaged) against the flywheel by springs.

Release and Pilot Bearings/Bushings

The release bearing, like all bearings, provides a point of contact between stationary and moving components (**Figure 8-15**). The outer shell of the bearing snaps into a clutch fork and moves forward and backward on a guide extending from the transmission.

Figure 8-13 The release levers pull the pressure plate from the flywheel, thereby releasing the friction disc. The machined flywheel side of the pressure plate is visible in this photo.

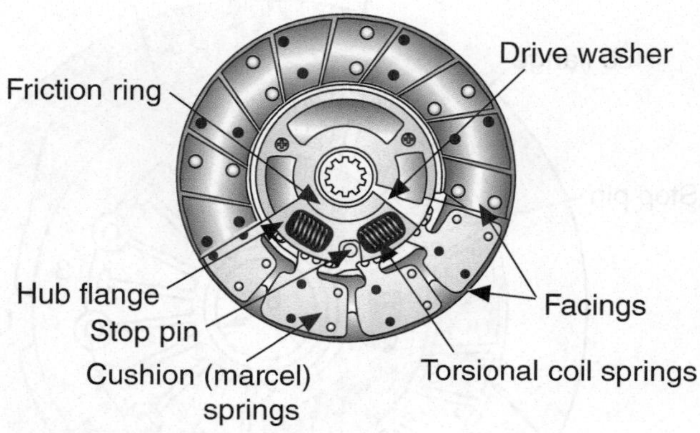

Figure 8-14 A clutch friction disc.

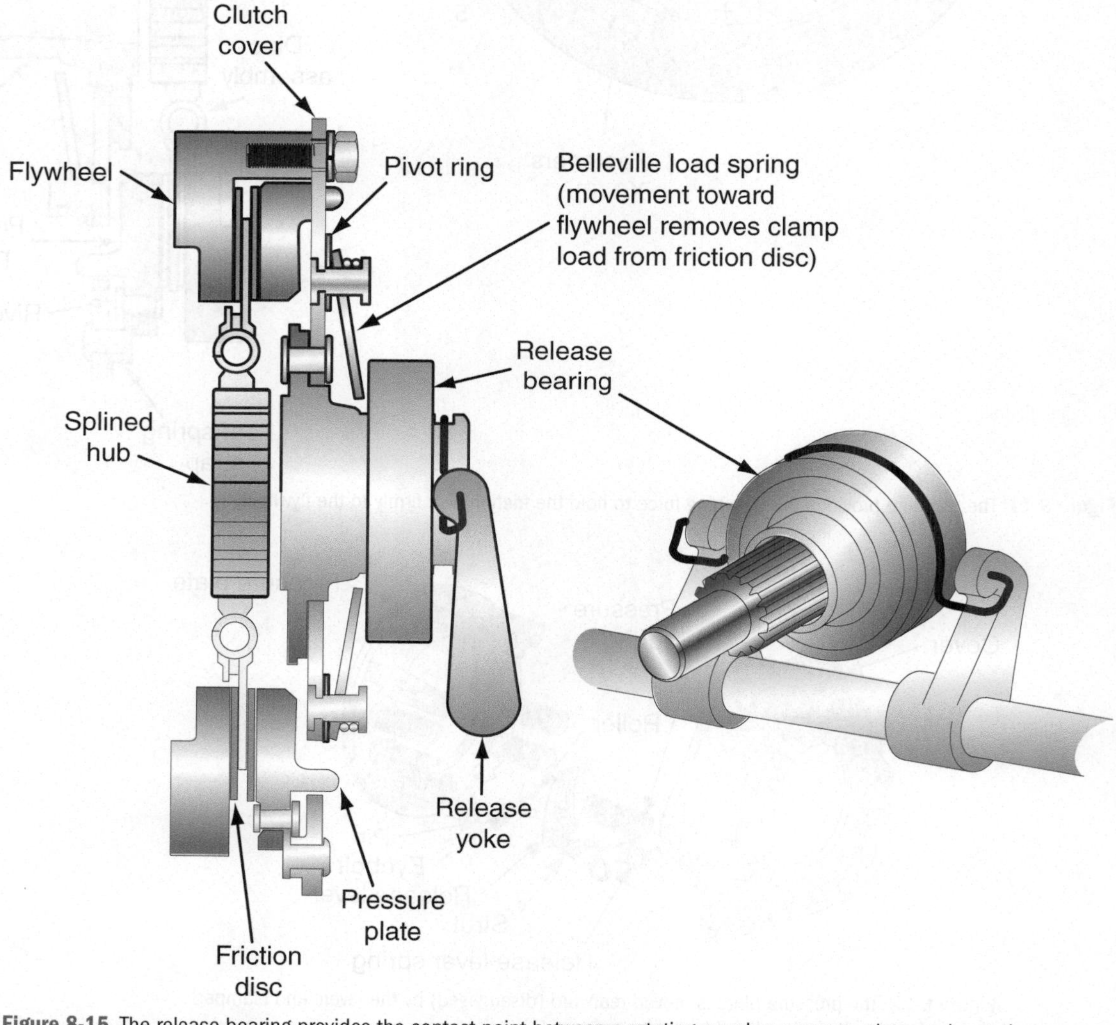

Figure 8-15 The release bearing provides the contact point between a rotating member, pressure plate, and a stationary member, the release fork.

The bearing shell does not rotate. The rotating portion is mounted within the shell and rotates when it contacts with the release levers. When the clutch is engaged, most release bearings totally break contact with the release levers and spin to a stop. Other systems use a continuously running bearing that maintains contact with the release levers at all times. This eliminates any free movement in the linkages. The bearing is sealed and permanently lubricated.

A **pilot bearing** or **pilot bushing** is installed in the rear of the crankshaft. The bearing or bushing is used to support the forward end of the transmission's input shaft. The pilot bearing is sealed and permanently lubricated. A pilot bushing is made of bronze and needs no lubrication.

A **pilot bearing** or **bushing** is not used with torque converters.

Mechanical Clutch Linkages

Older cars and many present-day trucks use a manual linkage. The linkage starts at the clutch pedal and extends to the clutch fork at the **bell housing.** The linkage is a system of rods and ball joints that transfer the linear force of the driver's foot through the floorboard and around obstacles to apply a linear force to the fork (**Figure 8-16**). (In contrast, the latest systems use a cable in place of rods and ball joints.) The fork extends through the side of the bell housing and moves the release bearing. There are usually several springs to ensure that the clutch linkage returns to the disengaged position when the pedal is released. When the driver releases the clutch pedal, the clutch is engaged. Mechanical linkage usually requires periodic adjustment to compensate for clutch wear.

The **bell housing** may be part of the molded transmission case or it may be bolted to the front of the transmission case.

Hydraulically Operated Clutches

New cars and many light trucks use a hydraulic system to apply the clutch (**Figure 8-17**). This requires less force from the driver, removes any free travel, and normally uses a continuously running release bearing. The hydraulically controlled clutch uses a small master cylinder to apply hydraulic pressure against a piston in the slave cylinder. The **slave cylinder** is mounted to the outside or inside of the bell housing. The piston pushes a small rod that moves the clutch fork. A release spring pulls the slave piston to the release position when the clutch pedal is released. Hydraulic controls do not normally require adjustment for clutch wear. The hydraulic system keeps the release bearing in contact with the pressure plate fingers as the clutch disc wears.

The **slave cylinder** converts hydraulic pressure from the clutch master cylinder to mechanical action to operate the clutch.

Some manufacturers use an internal slave cylinder mounted inside the bell housing. The release bearing guide is assembled as a two-bore hydraulic cylinder. The inner bore

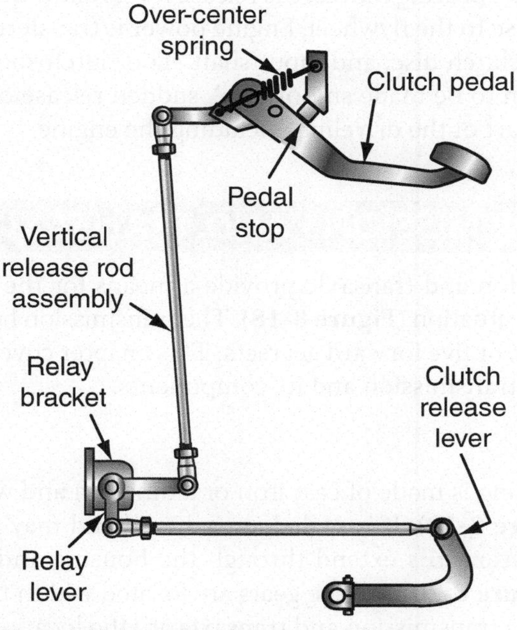

Figure 8-16 Mechanical linkages are still in use, particularly on heavier vehicles.

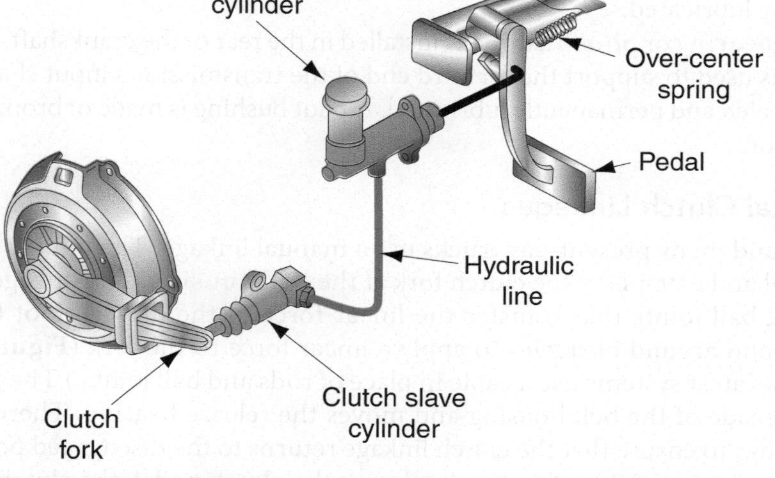

Figure 8-17 The hydraulic clutch control functions much like a hydraulic brake system.

fits around the transmission's input shaft's bearing retainer. Inside the outer bore is the slave cylinder's piston, which is shaped roughly like a doughnut. Fluid applied behind the piston directly applies the driver's force to the release bearing.

Clutch Operations

The clutch is released when the driver applies force to and pushes down on the clutch pedal. Through linkage or hydraulics, the release bearing is forced forward against the inner ends of the pressure plate release levers. The levers pull the pressure plate rearward, releasing clamping force on the friction disc. The connection between the flywheel and the transmission input shaft is broken and no engine power is transferred to the driveline.

When the clutch pedal is slowly released upward, the linkage allows the release bearing to move rearward. As the bearing moves, the release levers allow the pressure plate springs to clamp the friction disc to the flywheel. Engine power is transferred to the transmission through the flywheel, clutch disc, and input shaft. The clutch should be released slowly to allow the connection to be made smoothly. A sudden release of the clutch pedal may cause damage to any part of the driveline, including the engine.

MANUAL TRANSMISSIONS AND TRANSAXLES

The manual transmission and transaxle provide a means for the driver to select a gear ratio that best suits the situation (**Figure 8-18**). The transmission houses a reverse gearset, neutral, and three, four, or five forward gearsets. This chapter covers the basic five-speed, synchromesh, manual transmission and its components.

Housings

Shifter rails Rails that connect the shift lever to the shift fork.

Shop Manual page 314

The transmission housing is made of cast iron or aluminum and will usually consist of at least two pieces (**Figure 8-19**). It is sealed at each end and may have a gasket between individual pieces. **Shifter rails** extend through the housing and are connected to the driver's shift lever (**Figure 8-20**). All the gears are located within the housing. The major differences between the transmission and transaxle are the location and shape within the vehicle. Both use a similar gear layout and linkage and operate in about the same manner.

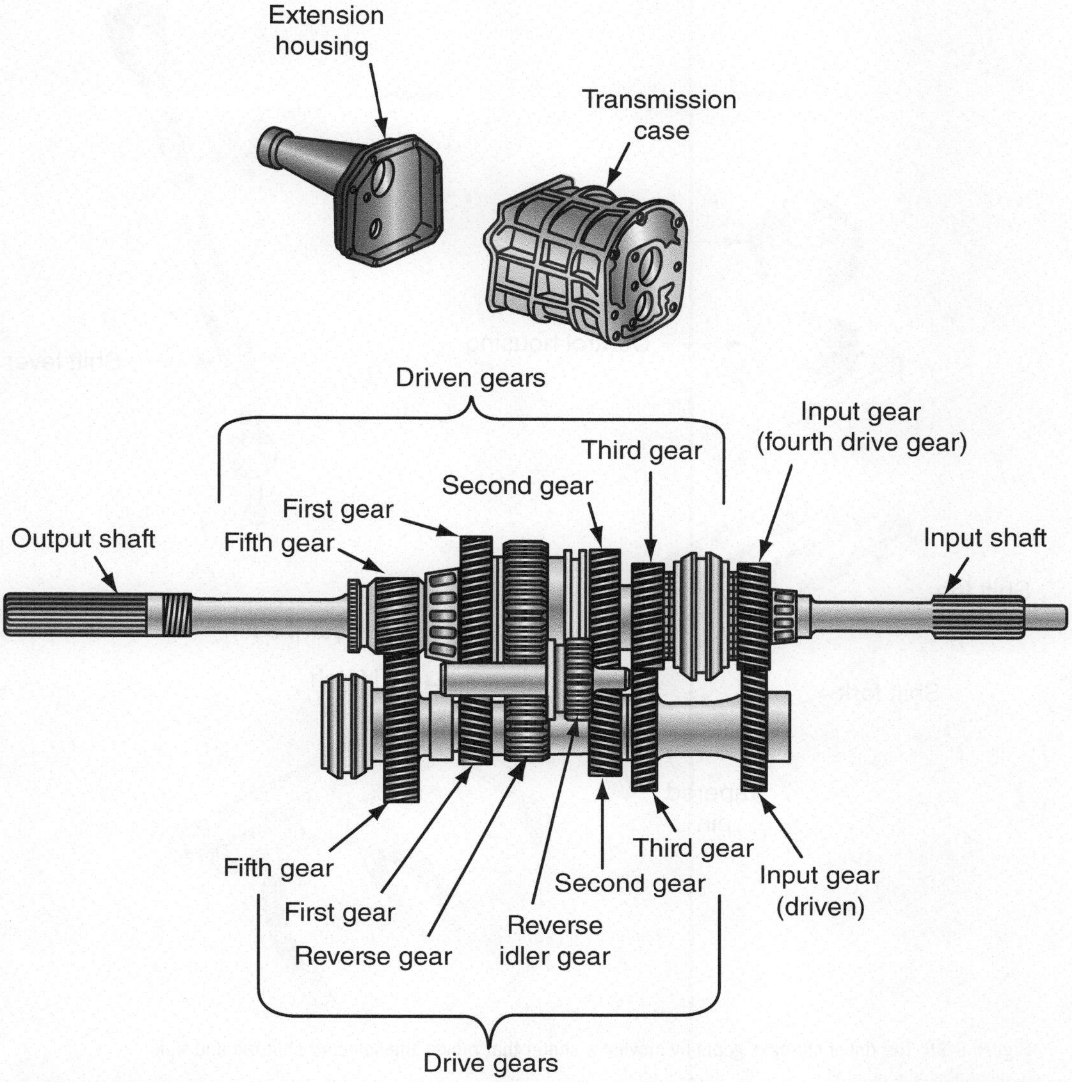

Figure 8-18 The different sizes of drive and driven gears allow the driver to select a gear ratio suitable for either torque or speech.

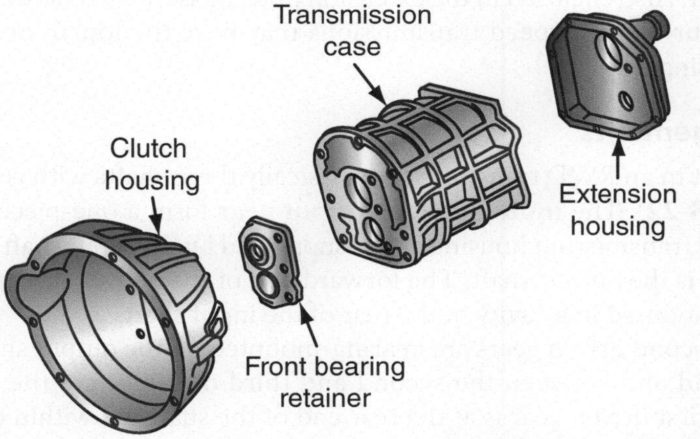

Figure 8-19 Both RWD and FWD transmissions usually have two main housings or cases.

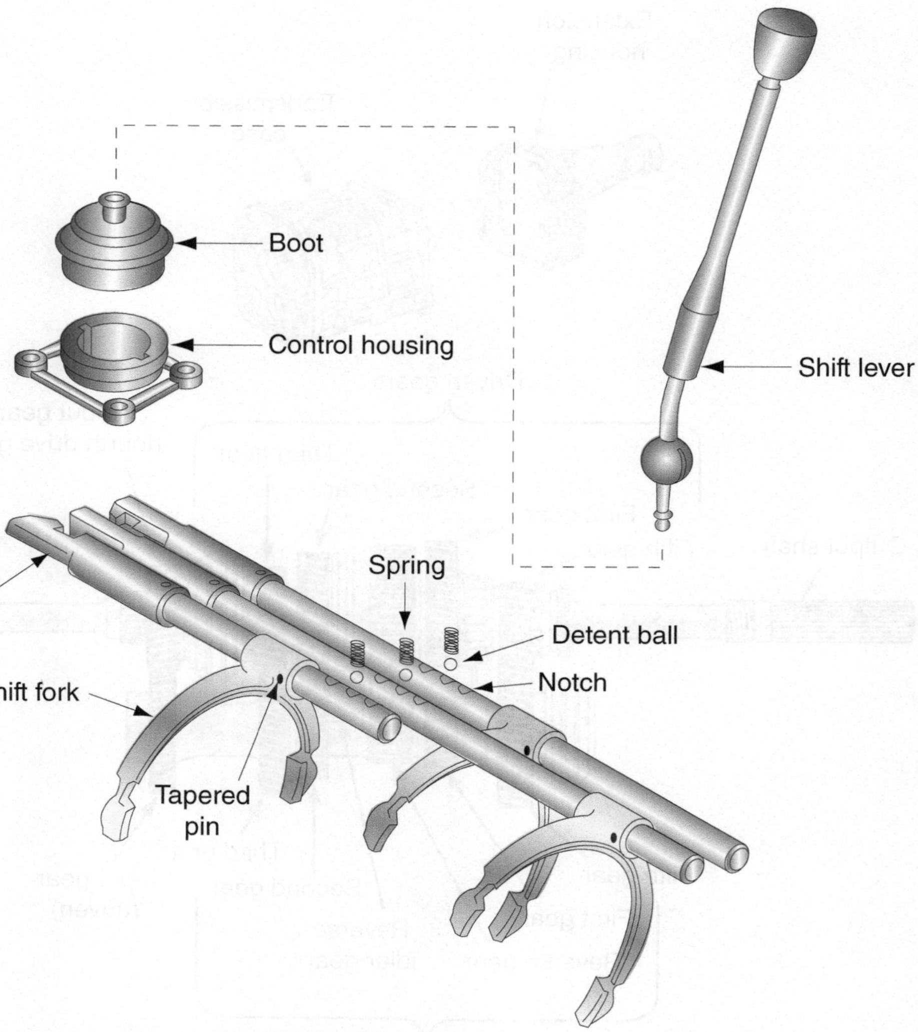

Figure 8-20 The driver changes gears by moving a shifter that moves the selected shift rail and fork.

Transmission housing Also known as the bellhousing is the portion of the transmission that covers the flywheel and clutch or torque converter.

Extension housing Also known as a tail housing it is fastened to the tail of a transmission or transfer case.

The nose of the **input shaft** extends through the friction disc and into the pilot bearing.

The **input gear** is a forged part of the input shaft and is the first drive gear in a manual transmission.

Synchronizers are used to adjust the speeds of the mating gears, so they can be engaged without damage.

The two pieces of the housing are the **transmission housing** and the **extension housing** or tail shaft housing. All the gears and some of the linkage are in the transmission housing (**Figure 8-21**). The extension housing extends to the rear and covers the tail of the output shaft. Also enclosed in the extension housing is the speedometer drive mechanism. Some four- or five-speed transmissions may have the fourth or fifth gear in the extension housing.

Gear Arrangements

The gear layout in an RWD transmission is basically three shafts with gears placed along them (**Figure 8-22**). The **input shaft** and **input gear** form a one-piece unit. It extends from inside the transmission housing and is supported by the input shaft bearing. Behind the input shaft is the output shaft. The forward end of the output shaft is inserted into a pilot bearing mounted in a cavity in the rear of the input shaft.

First and second driven gears are bearing-mounted on the output shaft. Between the first and second and between the second and third or direct are the **synchronizers.** Normally, the first driven gear is at the rear end of the shaft but within the transmission housing (**Figure 8-23**). The reverse driven gear is directly in front of first or just behind first with second normally just forward of reverse. Some systems have a portion of the first

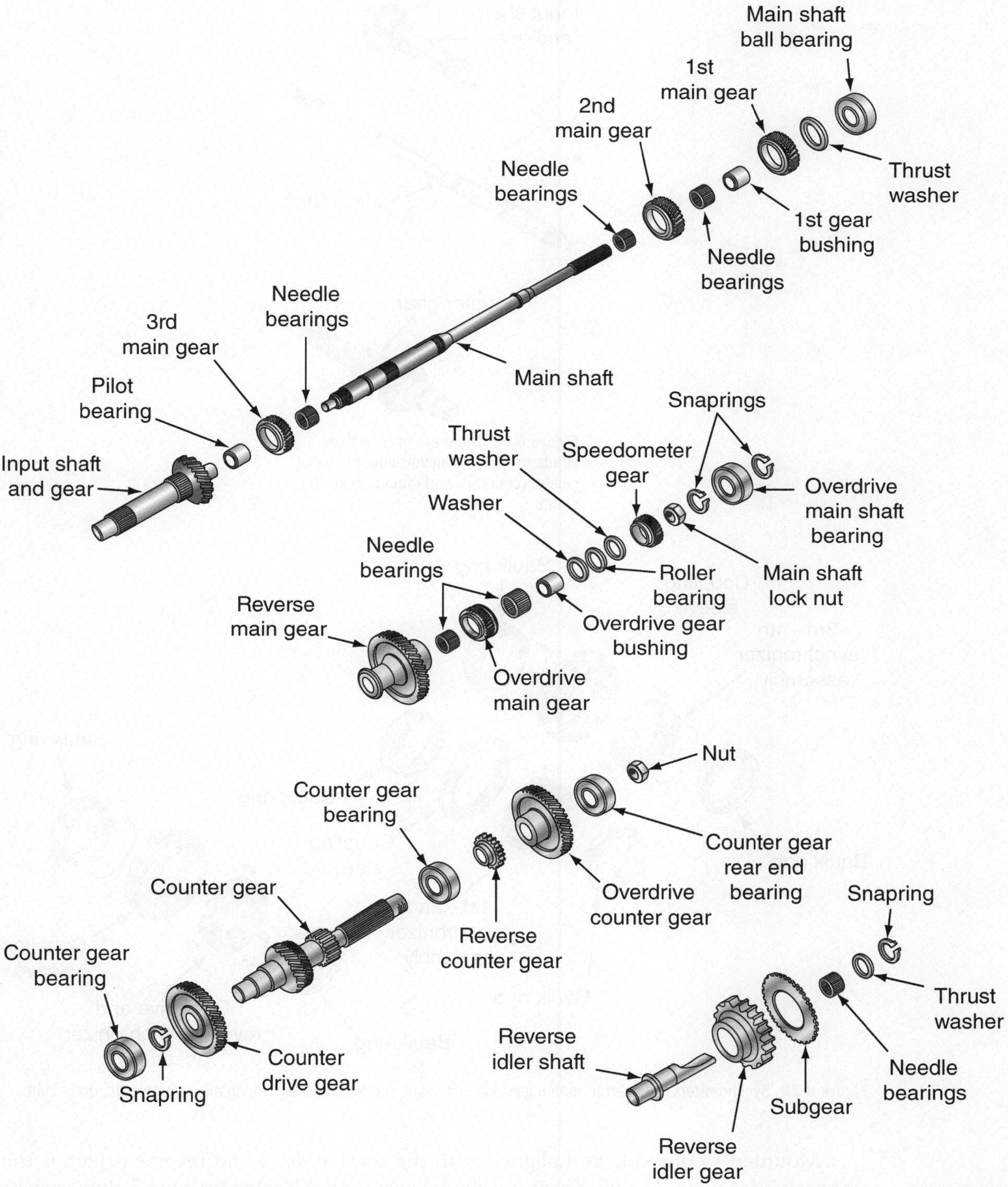

Figure 8-21 All of the gears are located inside the transmission case or extension housing.

or second synchronizer machined into a reverse driven gear. There is no actual third gear, but direct or third is obtained by locking the output shaft directly to the input shaft. Each forward gear is in **constant mesh** with a gear within the cluster gear. The first, second, and third gears turn freely on the shaft unless locked by the synchronizer. Fifth gear is usually outside the main transmission housing with fifth drive on the counter shaft and fifth driven on the output or main shaft.

Constant mesh means the mating gears' teeth are running together constantly.

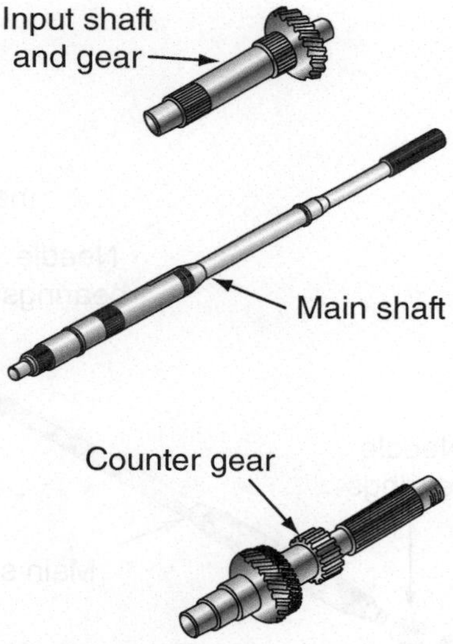

Figure 8-22 There are usually three shafts in a RWD transmission: the input, cluster (counter), and main or output shafts.

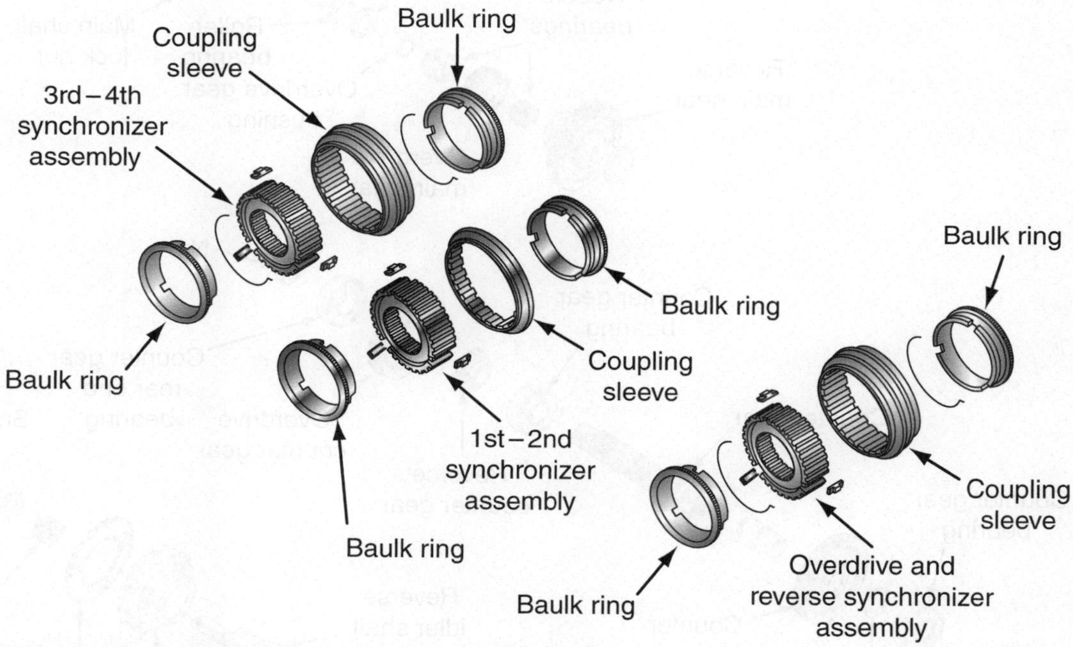

Figure 8-23 Synchronizers are used to match gear speeds inside the transmission, which allows a smooth, quiet shift.

The **reverse idler gear** changes the direction of rotation but does not change the gear ratio.

Mounted to one side and aligned with the reverse drive and reverse driven is the **reverse idler gear.** The idler gear is a sliding gear in that it must be moved along a shaft to mesh with the two reverse gears. It is not a synchronized gear. The vehicle should be stopped, and the clutch disengaged before selecting reverse. Once in place, the idler is driven by the reverse drive gear and drives the reverse driven gear (**Figure 8-24**). The idler gear rotates in the opposite direction of the drive gear. This causes the input rotation to change direction, and the output shaft rotation is reversed.

Between first and second gears and between third and fourth gears are synchronizer assemblies (**Figure 8-25**). They are splined to the output shaft and are used to make gear changing easier and quieter. They lock the driven gear to the output shaft.

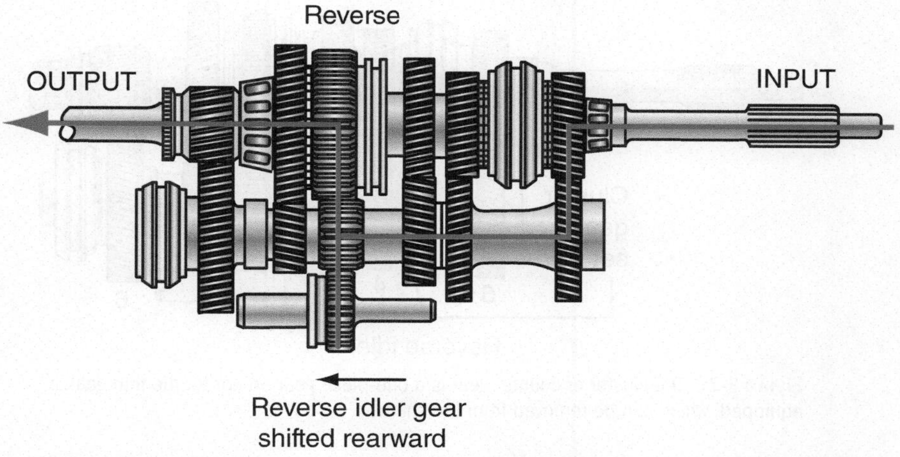

Figure 8-24 Reverse power flow is from the input gear through the cluster gears, the reverse drive, the reverse driven, and out the main shaft.

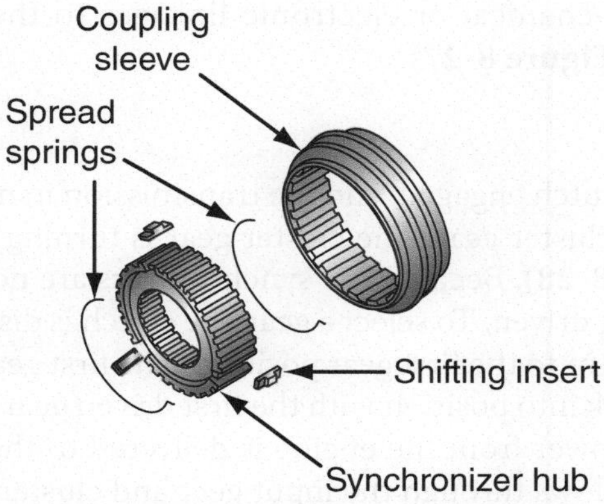

Figure 8-25 The synchronizer hub is splined to the output shaft (normally) whereas the sleeve can move over and lock onto the driven gear teeth.

The synchronizer assembly consists of a hub, sleeve, two **blocker rings,** three centering locks, and lock springs. The hub is the portion splined to the output shaft. The sleeve moves along external splines on the hub. The sleeve also has teeth to engage the driven gear. The blocker rings engage the gear first and is used to bring the gear speed to synchronizer speed. The center locks and springs lock the sleeve in neutral and drive the blocker rings. A shifting fork fits over the synchronizer sleeve.

Blocker rings make the first contact with a gear and initially allow the gear to slip a little. As the blocker ring is forced against the gear, the connection becomes tighter and the gear speeds up to blocker-ring speed, allowing a smooth, quiet gear change.

A **cluster gear** cannot be repaired, and if it is damaged there is probably damage on meshing gears.

Cluster Gears

The **cluster gear** is a one-piece unit. There are usually five molded gears in the cluster, which is mounted on bearings or a counter shaft (**Figure 8-26**). The forward gear (from the front of the transmission) is meshed with the transmission's input gear. The third and fifth gears in the cluster mesh drive the second and first driven gears on the output shaft. The fourth cluster gear is the reverse drive gear. Notice that the reverse drive and reverse driven are not meshed. Also notice that the entire cluster gear rotates any time the transmission's input shaft is rotating. This means that all gears meshed with the cluster gears are also turning. The transmission in Figure 8-26 shows the gear marked *"6"* is the fifth drive gear. It spins free on the cluster or counter shaft until it is locked to that shaft with the fifth gear synchronizer. The fifth driven gear is locked to the output shaft but can only deliver torque after the fifth synchronizer is engaged.

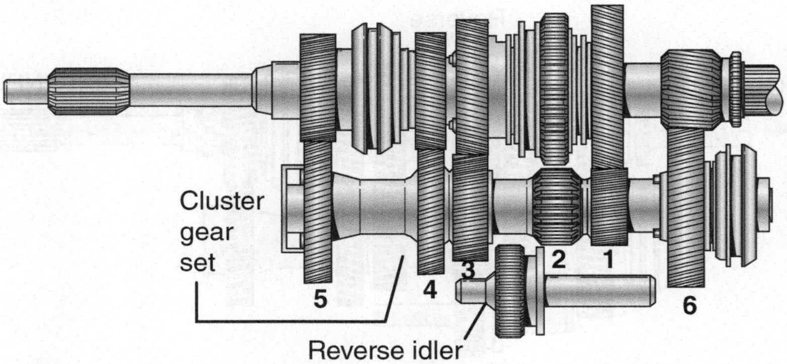

Figure 8-26 The cluster or counter gear is a one-piece gear except for the fifth gear, if equipped, which can be removed from the shaft.

Linkages

The shift lever is operated by the operator and may be mounted on the floor or the steering column. Most new vehicles have the lever on the floor. The movement of the lever is transmitted through mechanical or electronic linkages to the shifting fork in the transmission or transaxle (**Figure 8-27**).

Operation

With the engine running, clutch engaged, and the transmission in neutral, the input shaft is turning and driving the cluster gear. The cluster gear is turning the first, second, and third driven gears (**Figure 8-28**). Because the synchronizers are not locked to the gears, the output shaft is not being driven. To select a gear, the clutch is disengaged, and the shift lever is moved by the operator to the first gear position. The first gear synchronizer sleeve is pushed by the shifting fork into position with the first driven gear (**Figure 8-29**). As the clutch is slowly engaged, power from the engine is delivered to the transmission's input shaft and gear. The power flows through the input gear and cluster gear to the drive first gear. With the synchronizer in place, the power flow continues through the synchronizer assembly and rotates the driven gear and the output shaft. First gear has a gear ratio that provides high torque and low speed. The other forward gears are selected in the same manner. In direct drive or third gear, the 2–3 synchronizer locks the input and output shaft together giving a gear ratio of 1:1. In reverse, the two synchronizers are in neutral and the reverse shifting fork slides the reverse idler gear into position between the drive and driven reverse gears (**Figure 8-30**).

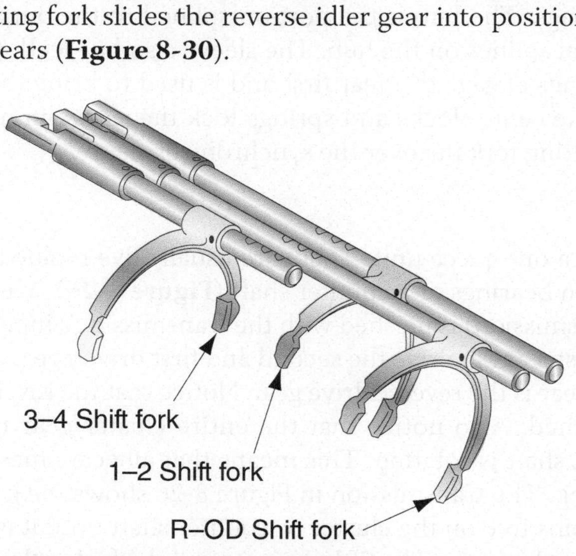

Figure 8-27 The shift forks fit into grooves on the synchronizer sleeves and are pinned to the shift rails.

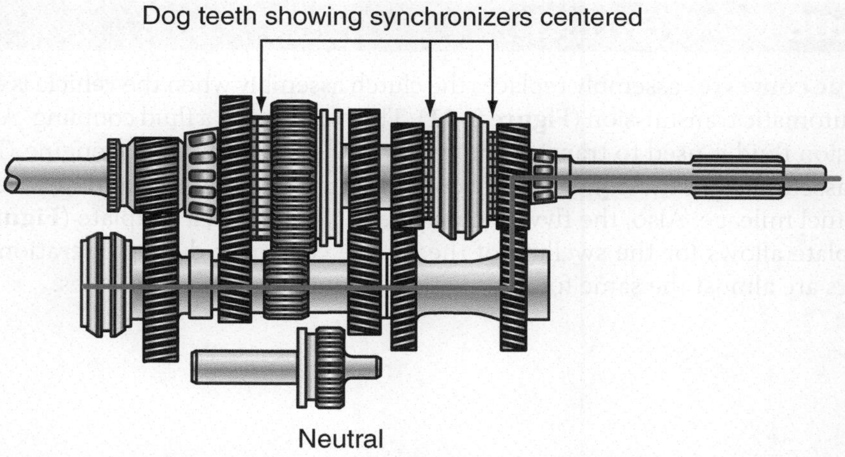

Figure 8-28 In neutral gear with the clutch engaged, all of the gears inside a transmission or transaxle are spinning except for the reverse idler gear.

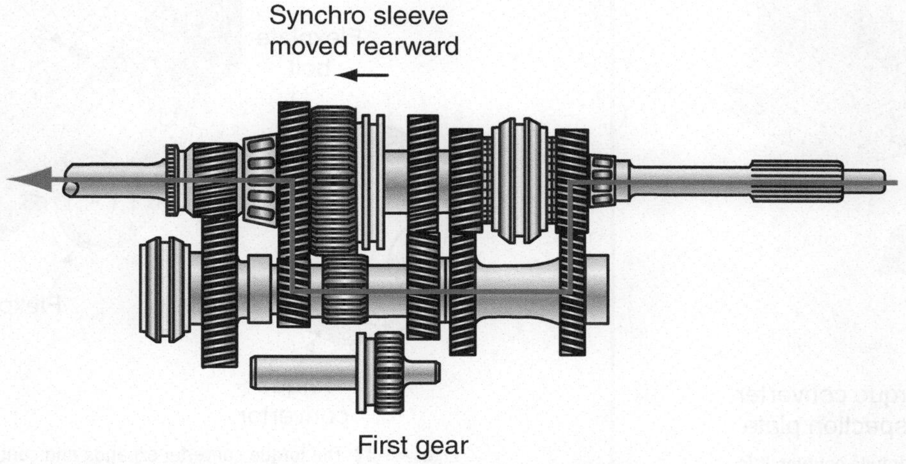

Figure 8-29 First gear power flow is from input through the cluster, first drive and driven, through the synchronizer and the main shaft.

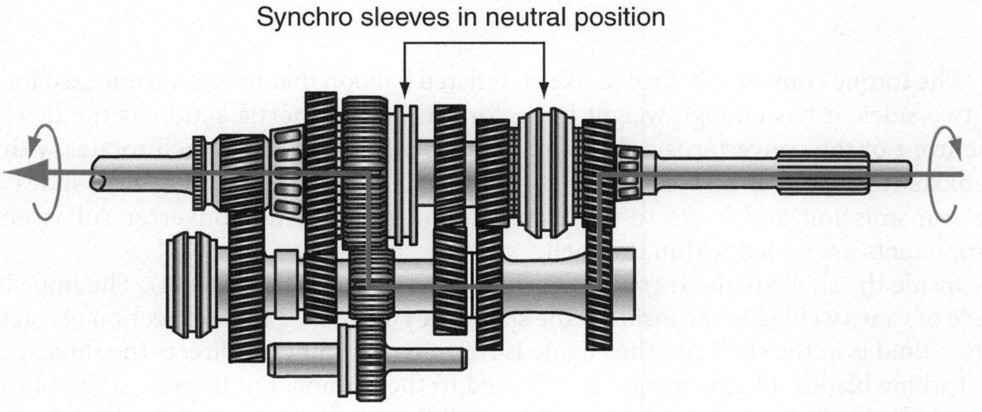

Figure 8-30 This reverse driven gear is a forged part of the 1–2 synchronizer. Note how the reverse idler gear has been moved forward to mesh with both the reverse drive and the reverse driven gears.

TORQUE CONVERTERS

Although there is no crankshaft pilot bearing or bushing used with **torque converters**, there is usually a cut out in the crankshaft, so the torque converter can expand.

The **torque converter** assembly replaces the clutch assembly when the vehicle is equipped with an automatic transmission (**Figure 8-31**). The converter is a fluid coupling. Automatic transmission fluid is used to transfer and multiply the torque from the engine. There are four subassemblies in the converter, three required for operations and another one to improve fuel mileage. Also, the flywheel may be replaced with a flexplate (**Figure 8-32**). The flexplate allows for the swelling of the torque converter during operation. Torque converters are almost the same for automatic transmissions and transaxles.

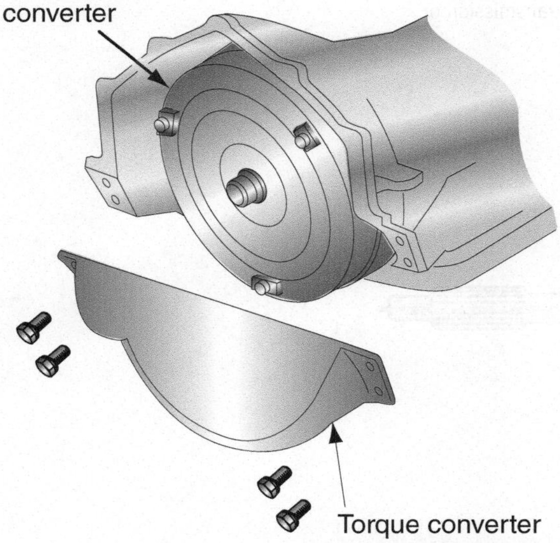

Figure 8-31 The torque converter is a hydraulic coupler. It is bolted to the flexplate and splined to the transmission pump.

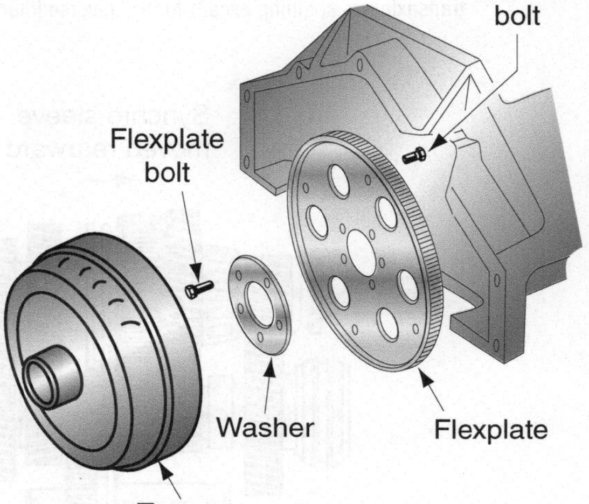

Figure 8-32 The torque converter expands and contracts during normal operation. The flexplate is flexible to absorb this ballooning action.

Impeller Vanes or blades that directs the transmission fluid against the turbine blades creates torque.

The **turbine**, **stator**, and lockup clutch are not visible without cutting the two halves of the converter apart.

To see **rotary flow**, put water into a bowl and use a spoon to make the water flow around the inside of the bowl.

The torque converter is shaped like an inflated balloon that has been squeezed inward on two sides. It has enough weight to perform the same inertia action as the flywheel. The front of the converter is bolted to the flex plate, so the outer shell rotates with the crankshaft (**Figure 8-33**). The rear of the shell has an extended lug that slides into the transmission and locks to the transmission pump. The converter subassembly components are sealed within the shell.

Inside the shell are the **impeller, turbine,** and **stator** (**Figure 8-34**). The impeller is made of vanes welded to the inside of the shell. They are angled in the direction of rotation. When fluid is in the shell and the engine is running, the impeller directs the fluid against the turbine blades. Engine torque is delivered to the turbine, but because a fluid instead of a mechanical connection is used, some or all the torque is lost and slipping occurs. If the vehicle is stopped with the engine idling, slipping is the desired condition. The fluid flow is classified as a **rotary flow** (**Figure 8-35**).

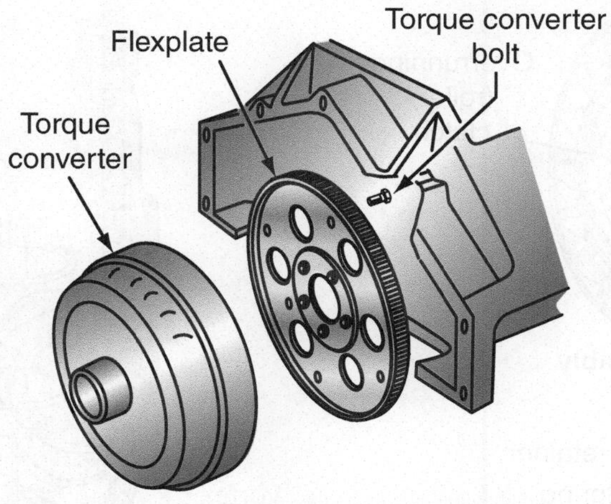

Figure 8-33 The flexplate is bolted to the crankshaft, and the torque converter is bolted to the flexplate.

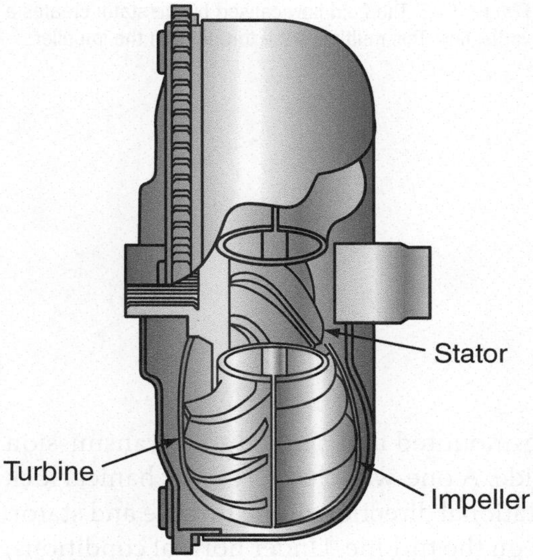

Figure 8-34 Internally, there are four major components of a modern torque converter: impeller fins welded to the converter housing, turbine mated to the transmission turbine (input) shaft, a one-way clutch known as the stator, and a torque converter lockup clutch for a mechanical connection between the converter housing and the turbine shaft.

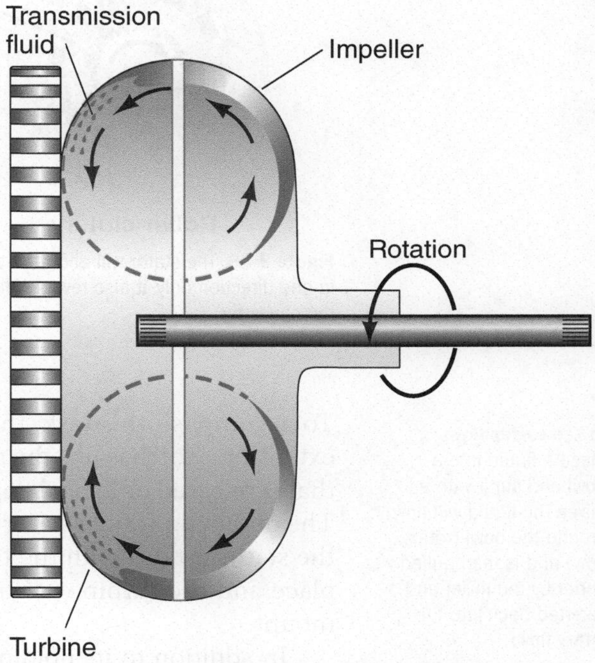

Figure 8-35 Rotary flow causes the turbine to spin.

The turbine is bearing-mounted and free of a mechanical connection with the converter's outer shell. The turbine is splined-mated to the input shaft of the transmission. In some makes, this shaft is referred to as the **turbine shaft.** Blades on the turbine face against the rotation and opposite to the impeller vanes. As fluid from the impeller strikes the face of the blades, the turbine is rotated in the same direction as the outer shell but with much less speed and force. At this point, the engine torque is delivered to the input shaft of the transmission. However, the turbine could turn backward if a holding device was not in place.

Turbine shaft Transmits the torque to the output shaft and redirects the fluid leaving the turbine to the hydraulic system.

The stator is the holding device. The stator is a **one-way clutch** that allows the turbine to rotate in one direction only (**Figure 8-36**). If the turbine attempts to rotate backward, the stator locks and the turbine stops moving or reverses its rotation.

One-way clutch A unit that rotates in one direction and locks in the opposite direction.

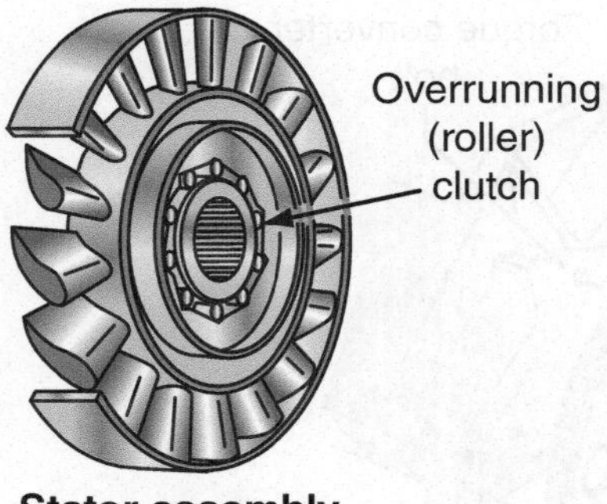

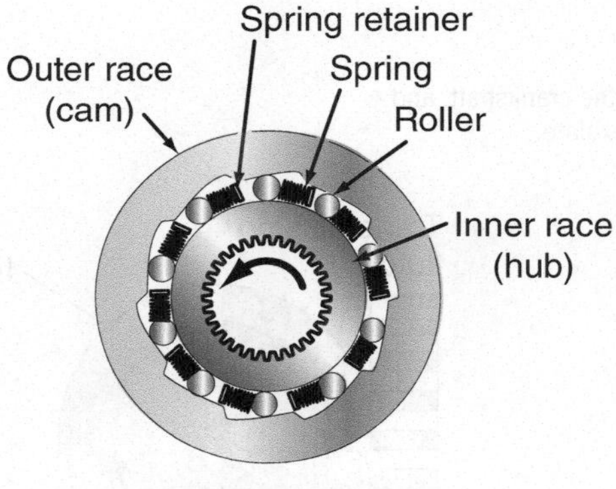

Figure 8-36 The stator will allow the turbine to spin in one direction only. It also reverses fluid flow within the converter.

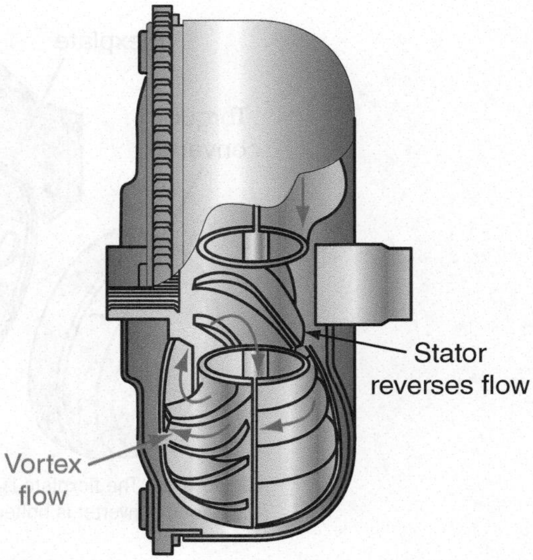

Figure 8-37 The fluid flow caused by the stator creates a vortex flow. This multiplies the torque from the impeller.

To see **vortex flow,** place a liquid into a bowl and stir it with a mixer. The liquid will flow around the bowl (rotary flow) and is then pulled under by the mixer and inserted back into the rotary flow.

Without the stator and vortex flow, there would not be enough torque to power the geartrain.

A **lockup clutch** locks the turbine to the outside of the torque converter and is not applied until the vehicle is in high gear and engine load has decreased. The transmission should be designed to work with a lockup torque converter.

To accomplish this task, the stator is bearing-mounted and splined to a transmission extension, which is like the release-bearing guide. A one-way clutch is a mechanical lock that is released or locked depending on the rotational direction of the turbine and stator. The outer edge of the stator locks into a cavity on the turbine. Under normal conditions, the stator rotates with the turbine. If the turbine moves backward, the stator locks into place and the turbine and stator are prevented from moving by the stationary stator mount.

In addition to its holding task, the stator has blades that redirect the fluid from the turbine. The stator blades are angled in the opposite direction of the turbine blades. As the fluid strikes the stator blades, it is forced back against the turbine blades. This creates higher pressures and increases the torque between the impeller and turbine. The fluid flow at this point is a **vortex flow** (**Figure 8-37**).

The impeller, turbine, and stator are all required for torque converter operations. The next component may or may not be included in a particular torque converter. It is the **lockup clutch** (**Figure 8-38**). The lockup clutch may be mechanical, hydraulic/mechanical, or electrical/hydraulic. The clutch, when operated, locks the turbine to the outer shell of the torque converter. When this occurs, the turbine and the input shaft move at the same speed as the crankshaft similar to direct drive in a three-speed manual. Usually, the lockup clutch is not applied until the vehicle is in direct (high) gear and the engine load has decreased. Obviously, if the lockup clutch was applied at a low speed, there would be no slipping within the torque converter and the engine would lug or stall. Every torque converter manufactured since the early 1990s has a lockup clutch of some type.

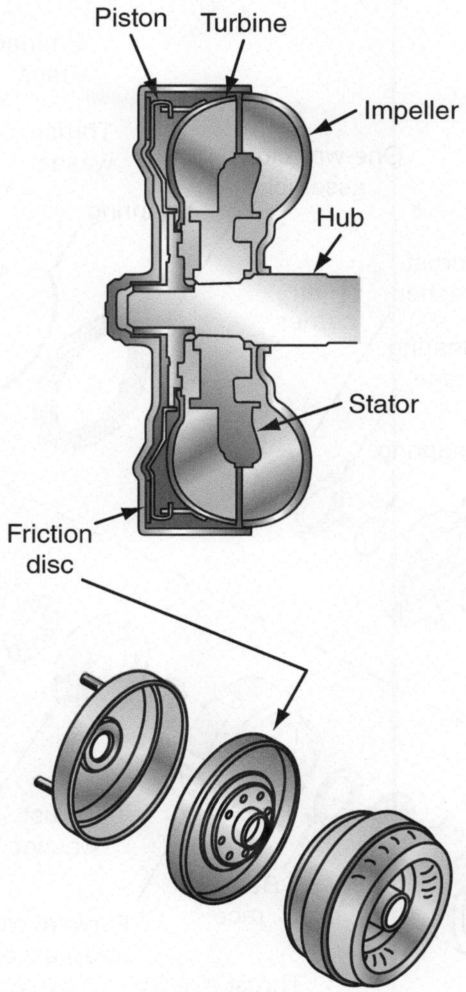

Figure 8-38 The TCC mechanically links the turbine to the converter housing. This provides direct drive from the crankshaft to the transmission turbine shaft.

AUTOMATIC TRANSMISSIONS AND TRANSAXLES

Automatic transmissions and transaxles shift up and down through the forward gear ranges using hydraulic clutches, a valve body, and various controls. The driver selects park, reverse, neutral, and a forward range using a shift lever like the one used for manual transmissions. A hydraulic pump driven by the outer shell of the torque converter supplies the necessary fluid volume and pressure to operate the transmission.

Shop Manual
page 314

There are many different transmission designs based on the Simpson geartrain. A basic Simpson system uses two hydraulic clutches, two bands, two planetary gearsets, an overrunning clutch, and a valve body (**Figure 8-39**). This chapter discusses the Simpson geartrain design because it is basic and the simplest to understand at this point.

A **clutch pack** is the term used for hydraulic clutches in automatic transmission. The clutch pack performs the same function as the clutch on a manual transmission but is internal to the transmission and applied hydraulically.

Hydraulic Clutches

A hydraulic clutch or **clutch pack** is composed of a piston assembly, a drum, and a number of steel and friction clutch plates (**Figure 8-40**). The center hub of the drum fits around a gear or shaft and provides the mounting area for the piston. The piston is a thin, aluminum plate with a seal around the outer and inner edges. The piston is held to the

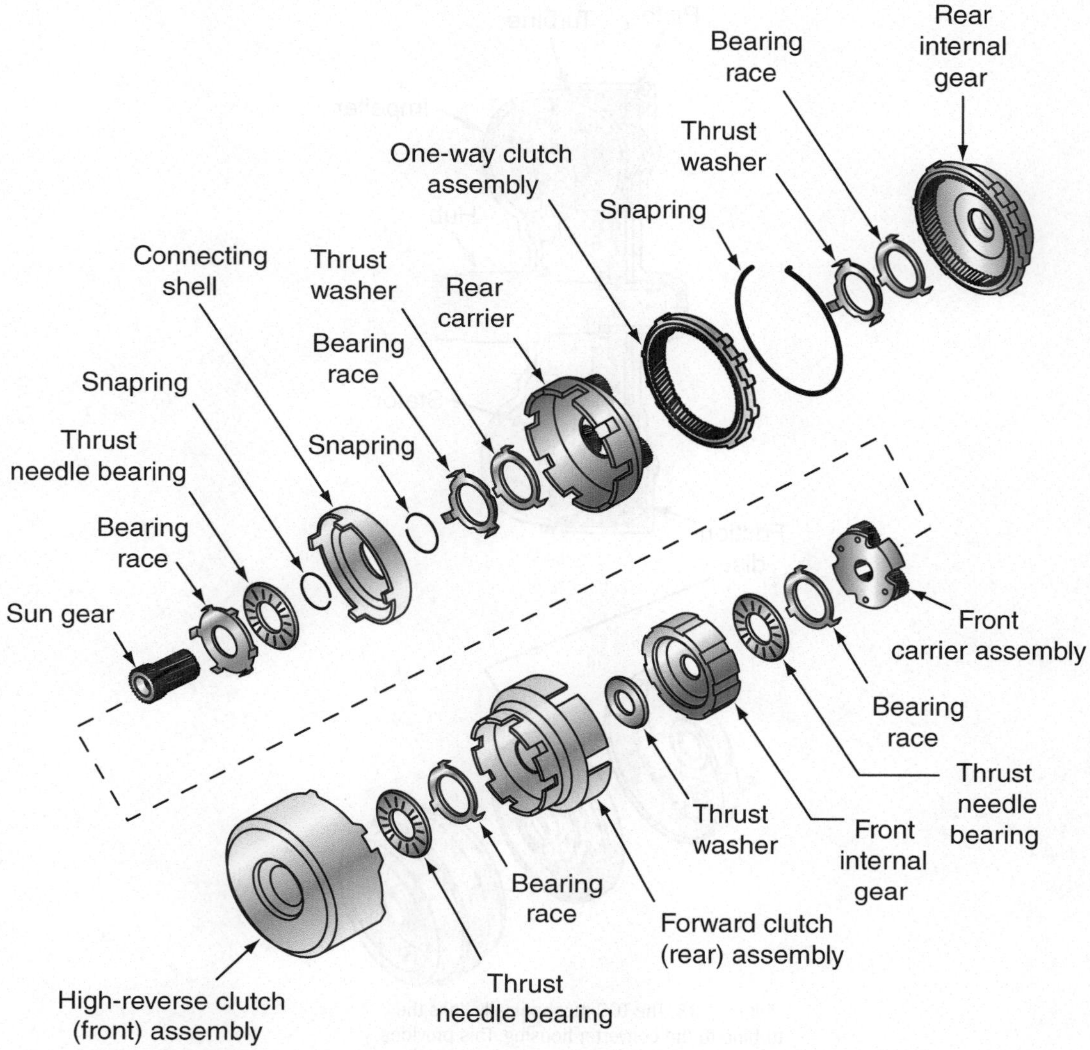

Figure 8-39 The Simpson geartrain in its simplest form uses two hydraulic clutches, two planetary gearsets, two holding bands, and a one-way mechanical clutch.

bottom of the drum's bore with return springs and a retainer. Above the piston and within the bore are alternating steel and friction clutch plates. The steel plates have lugs at the outer edge to lock into grooves on the drum. The inner edge of the clutch discs is splined to fit either a gear or shaft. The plates are held in place loosely with a retaining snapring.

As pressurized fluid is directed behind the piston, it moves forward and clamps the friction and steel plates together (**Figure 8-41**). This operation is like the manual pressure plate clamping the clutch's clutch disc to the flywheel. In hydraulic clutches, input power may be driving the clutch discs or the friction drum. As the plates are clamped together, they begin to turn together, and the power flow is directed through the clutch pack from input to output.

Steel **bands** that wrap around sections of the gear train and connect to the housing.

A **servo** is used to convert hydraulic pressure to mechanical force.

Bands

Bands are wrapped around the outside of the clutch drum and act as a holding device. The band is anchored at one end while the other end is attached to a hydraulic **servo** (**Figure 8-42**). When fluid is applied to the servo, the band is applied and prevents the clutch drum from moving. If the clutch is not applied, the power flow passes through the

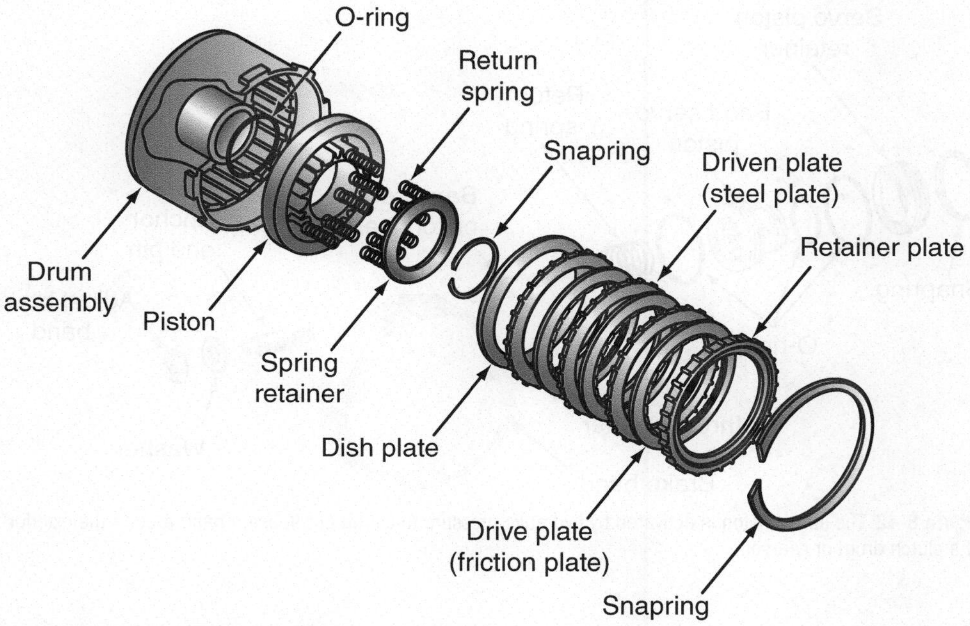

Figure 8-40 A typical hydraulic clutch assembly or clutch pack.

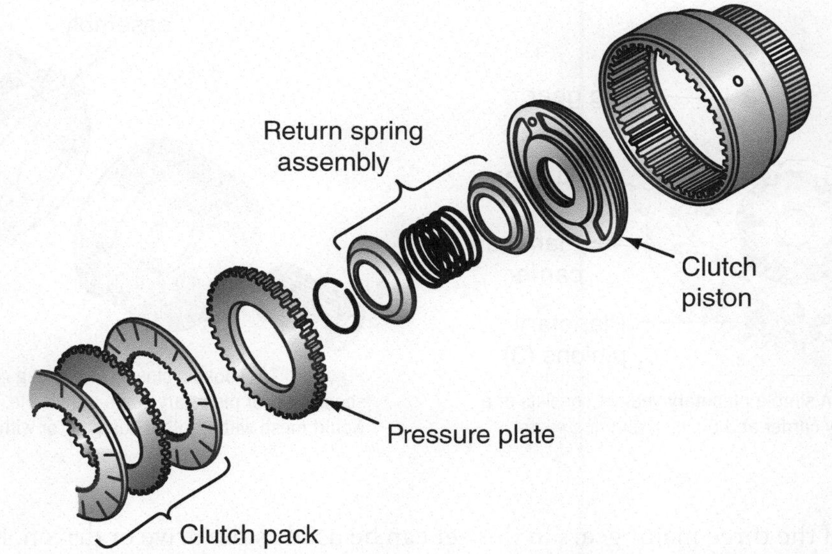

Figure 8-41 The hydraulically activated piston clamps the friction disc together similar to the pressure plate clamping the manual friction disc to the flywheel.

clutch pack without any action. If the clutch is applied when the band is also applied, the gear or shaft splined to the clutch's clutch disc is held and prevented from rotating.

Planetary Gearsets

A Simpson **planetary gearset** has three gears. The outer gear, called a **ring gear,** is an internal gear. The gear teeth are on the inside of the ring. Fitted and meshed inside the ring gear is the **planetary carrier.** The planetary carrier is a set of three or four pinion gears mounted to a platform or carrier. The carrier is designed to allow the pinion gears' teeth to extend past both edges of the carrier. This allows the gears to mesh with the ring gear around the carrier and the **sun gear** mounted through the center of the carrier (**Figure 8-43**). The external sun gear extends completely or partially through the planetary carrier and meshes with the pinion gears (**Figure 8-44**).

The two **planetary gearsets** in a Simpson geartrain are used together to obtain the correct gear ratio for a given situation.

Ring gear A large gear in the differential of an automobile that is driven by the propeller shaft pinion.

Planetary carrier Connects the center of the two gears and rotates to carry one gear around the other.

Sun gear The gear in the center of a Simpson planetary gearset.

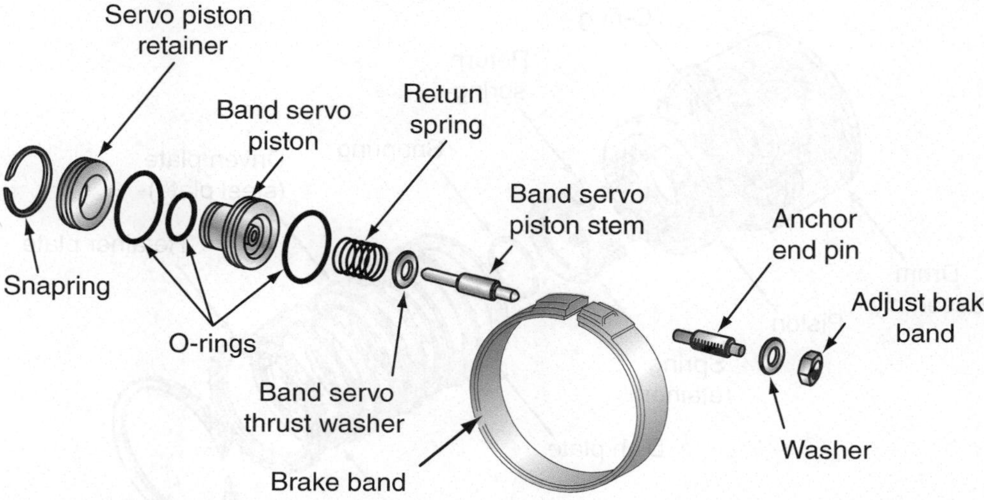

Figure 8-42 The servo piston is activated by hydraulic pressure to clamp or release a band around the exterior of a clutch drum or gearset.

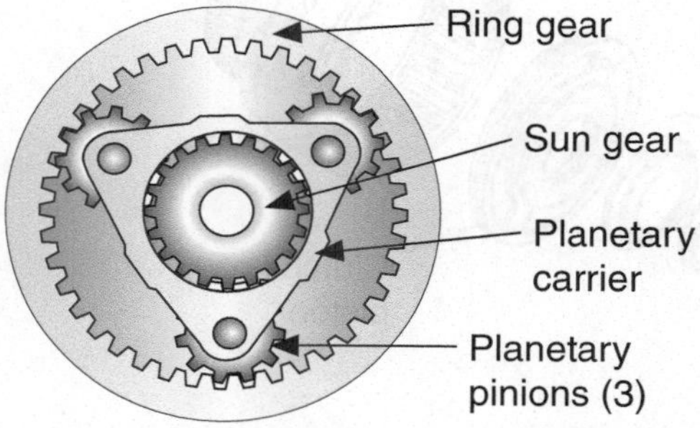

Figure 8-43 A simple planetary gearset consists of a sun, planetary carrier and pinions, and ring gears.

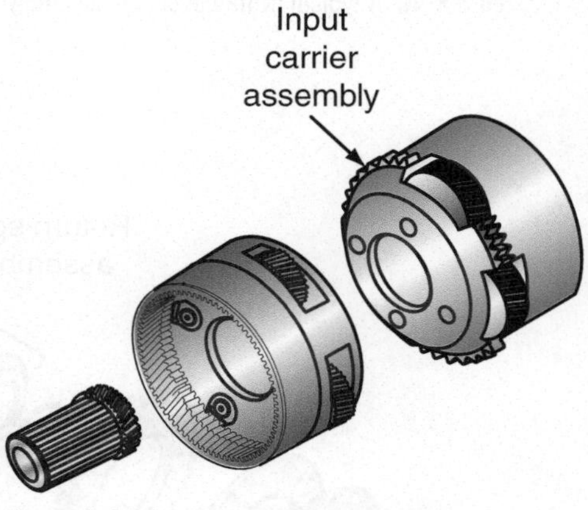

Figure 8-44 Looking closely at the ring gear, a second set of planetary gears are visible. This set would mesh with another sun gear or with a shaft.

Any of the three major gears in this set can be used as the drive or driven. However, for the set to work, one of the three must be held in some manner so the other two have a point to rotate around. For a quick example, use the following scenario. The ring gear is mechanically connected to the clutch pack drum. The planetary carrier is splined to an output shaft and an input gear drives the sun. The sun gear is also splined to the clutch discs of the clutch. The band is applied, which holds the drum and ring gear in place. The clutch is released, freeing the clutch discs. As power is applied to the sun, the sun rotates the pinion gears within the carrier. Because the ring gear is held in place, the pinion's gears "walk" around the inside of the ring gear, thereby turning the planetary carrier. The carrier now turns the output shaft. The power flow moves from the input, to the sun, through the planetary gears, to the carrier, and to the output shaft (**Figure 8-45**).

Because the sun gear is smaller than the carrier, there is an increase in torque and a reduction in speed. The pinion gears are not counted in computing the ratio because they are used as idlers. If the power flow is reversed using the carrier as the input and the sun as output, the torque would decrease, and speed would increase.

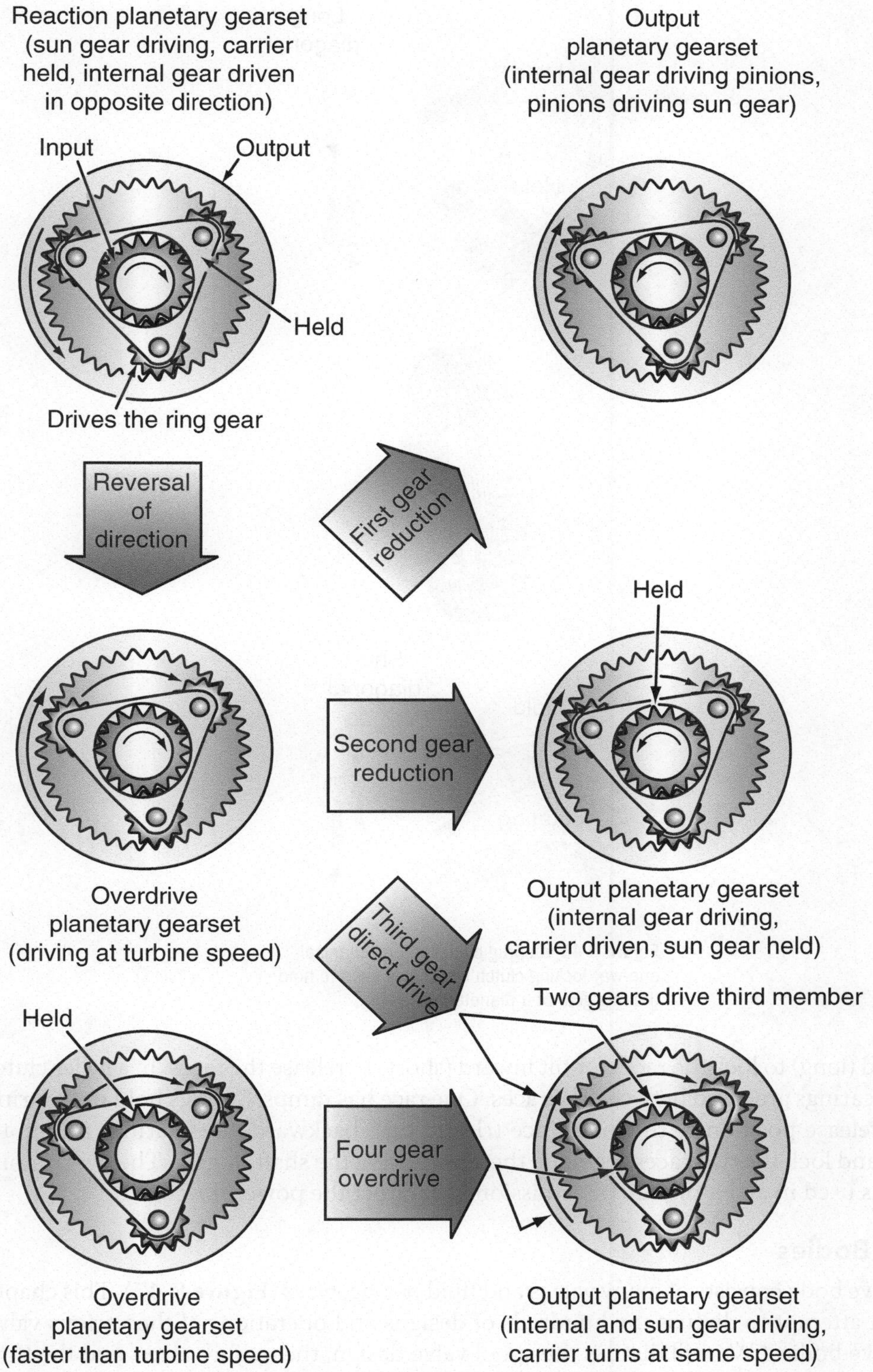

Figure 8-45 The use of planetary gearsets provides a variety of gear ratios in a small, lightweight package.

Overrunning Clutches

The overrunning clutch in most Simpson geartrains is a mechanical clutch. The inner ring or race is splined to a gear or shaft, and the outer race is locked to the transmission housing or some other anchor. There are two basic types of overrunning clutches: **sprag** and **roller** (**Figure 8-46**). On a sprag clutch, between the two races, there are diagonals that can tilt

A **roller** overrunning clutch is much simpler in operation than a **sprag** clutch.

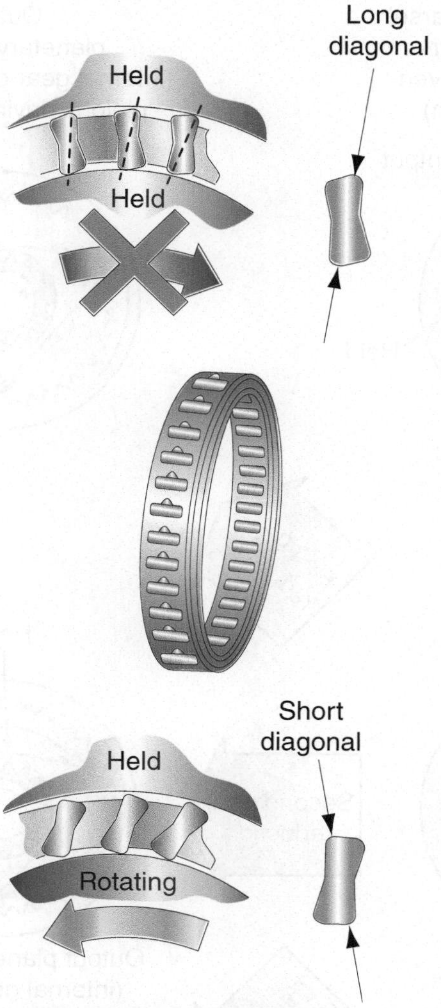

Figure 8-46 A sprag clutch is a mechanical one-way locking clutch commonly used to hold the ring gear of a planetary set.

outward (long) to lock the races, or tilt inward (short) to release the races. In a roller clutch, roller bearings are fitted between the races. One race has ramps. Springs hold the bearings in the release position. If the inner race tries to turn backward, the bearings roll up the ramps and lock the two races together, thereby locking the shaft or gear. The overrunning clutch is used in an automatic transmission to redirect the power flow.

Valve Bodies

The valve body consists of many valves and fluid passageways (**Figure 8-47**). This chapter will not attempt to discuss the hundreds of designs and operations of the various valves and valve bodies. We will discuss the spool valve design, the major valves, and the three major controls.

Spool Valves

On a **spool valve,** fluid flows through the grooves when they are aligned with a hydraulic port.

The **spool valve** is a highly machined steel rod that fits tightly and smoothly into a highly machined bore in the valve body. The rod is grooved at different points along its length. The width of the grooves may be different. The remaining high points on the rod are called lands (**Figure 8-48**). There are fluid passageways entering the valve body bore at different points. With the spool valve inserted into the bore and a land moved directly over a fluid passageway,

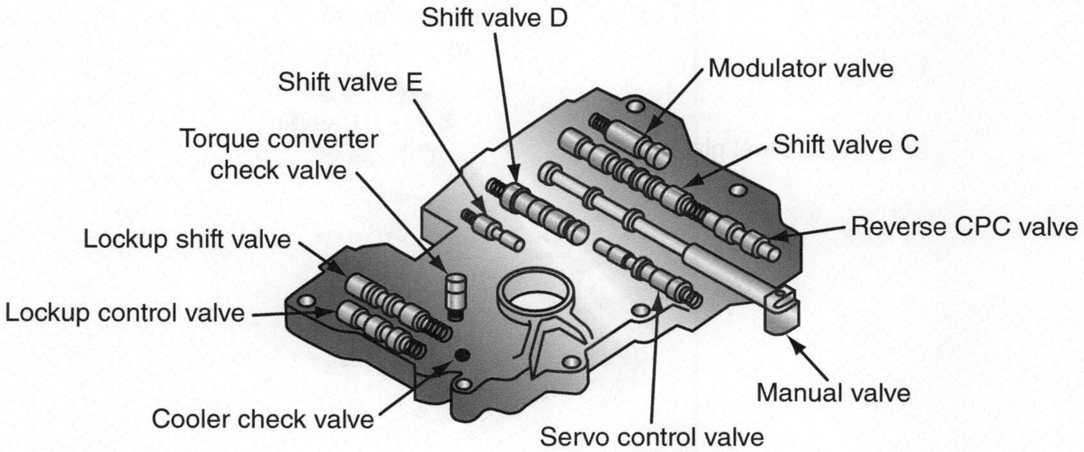

Figure 8-47 The valve body holds many different valves and their springs. They control fluid to and from hydraulic elements in the transmission and regulate fluid pressure.

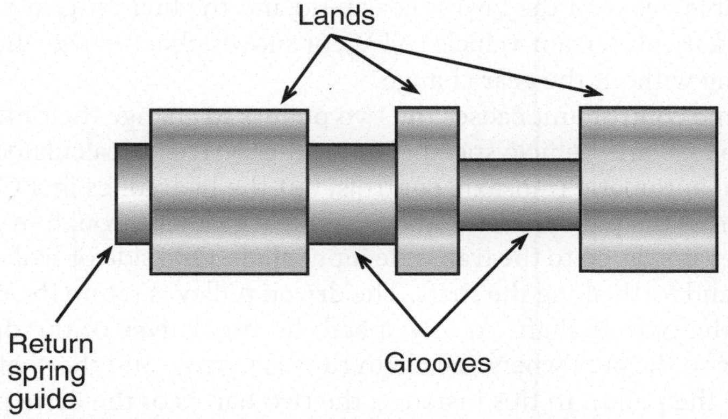

Figure 8-48 The spool valve is used to block or open fluid passageways in the valve body passages.

the fluid is blocked. Moving the spool until a groove lines up with the passageway, fluid flows through the groove and enters another passageway. In this manner, the fluid can be directed to or cut off from clutches, servos, other valves, or a control device. Springs or fluid move the spool valves. The major valves are the 1–2 and 2–3 shift, reverse, pressure regulator, pressure booster, and the manual valve. The operator controls the manual valve.

CONTINUOUSLY VARIABLE TRANSMISSIONS AND TRANSAXLES (CVT)

Many modern vehicles, including many hybrid electric vehicles (HEV), with an automatic transaxle or transmission are equipped with a CVT. They use hydraulics and electronic controls like the typical types but have very distinct differences in other areas of operation. Simply put, the CVTs don't use gearsets to change ratios. In fact, the CVT doesn't shift gears except for reverse and forward.

A typical CVT has manually selected reverse, park, neutral, and forward gears. The first three are almost like all automatic transaxles, but the real change is in the forward selection. In forward gear, the transaxles start in a lower ratio and high torque with the ratio increasing as the vehicle comes to speed. This changing of torque and speed output is basically done by two pulleys, one drive and one driven, split across into two halves each and connected by a steel belt (**Figure 8-49**) or a multiple-link chain. An infinite number

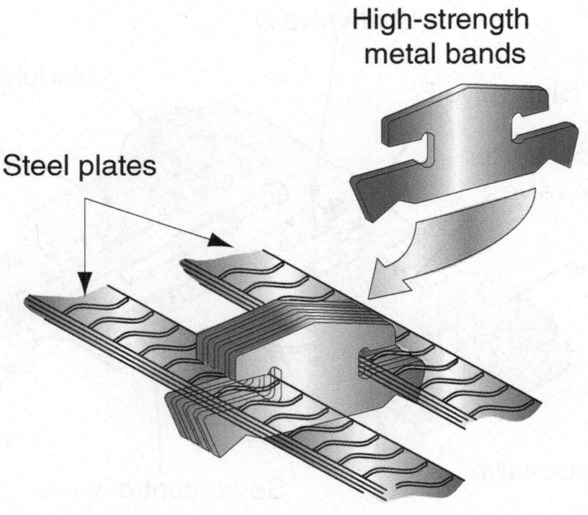

Figure 8-49 The construction of the steel belt used in most CVTs.

of ratios is possible between the lowest gear ratio and the highest gear ratio. If you have ever driven a go-kart, all-terrain-vehicle (ATV), or snowmobile, you get the same sensation of ratios changing without the gear change.

The electronic control unit causes the two pulleys to change their internal diameters based on engine load and vehicle speed (**Figure 8-50**). This is calculated very similar to all other types of automatic transaxle controls, but the hydraulics in a CVT unit is used to split or compress the pulleys instead of engaging a gearset through hydraulic clutches. The driven pulley is splined to the transaxle input shaft. One side or half of the pulley can be moved back and forth along the shaft. The driven pulley is set up the same way except it is splined to the output shaft. At low speed the two halves of the driven pulley are pressed together so the area where the chain runs is narrow and the chain rides near the outside edges of the pulley. In this instance, the two halves of the drive pulley are pulled apart so that the chain rides low. This system consists of a small pulley driving a large pulley with a torque increase at the output shaft.

Shop Manual
pages 324–328

As the vehicle speed increases and the engine load is reduced, the drive pulley halves are forced together, allowing the chain to ride higher and creating a bigger pulley. At the same time, and in direct proportion to the inward movement of the drive pulley halves, the sides of the driven pulley are forced apart creating a smaller pulley. This reduces the torque ratio and has a corresponding increase in the speed ratio. This system, when controlled by a well-programmed, onboard computer with accompanying sensors and actuators, allows for no "physical" feel of an up or down gear shift; provides innumerable ratios within the maximum fixed limits; provides almost instant response to engine load and vehicle speed; and provides a quiet and comfortable ride to the driver and passengers. This system also reduces the total number of parts that can wear out.

Drive shaft A rotating shaft which transmits torque in an engine.

Differential A device that splits the engines torque two ways allowing each output to spin at a different speed.

U-joint Also known as a universal joint is a device that allows one or more rotating shafts to be linked together which allows the transmission of torque.

Slip joint Serves to adjust the length of a shaft when demanded by axle movement.

Yoke A strap that holds a bearing in place and connects to the output shaft.

DRIVE SHAFTS AND DIFFERENTIALS

The **drive shaft** assembly provides the connection between the transmission's output shaft and the differential in the rear axle housing. The **differential** allows the drive wheels to rotate at different speeds when the vehicle changes direction.

Drive Shafts

A typical drive shaft assembly has two **U-joints**, a **slip joint,** and the drive shaft tube (**Figure 8-51**). The drive shaft is a hollow, balanced, steel pipe designed to withstand high rotational forces. Each end has a **yoke** to mount the U-joints. The rear U-joint connects

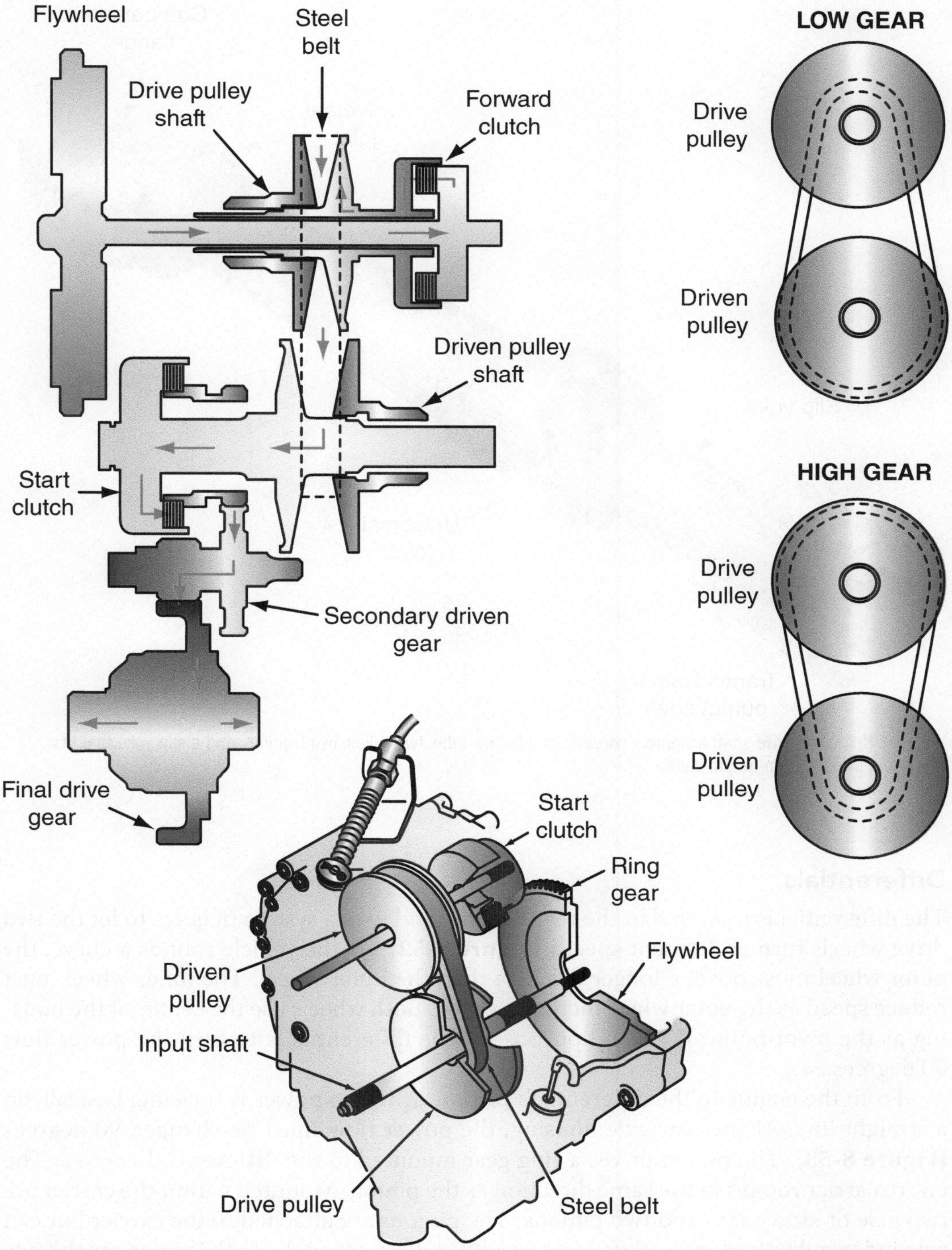

Figure 8-50 In low on this CV transaxle, the drive pulley is narrow, creating a larger pulley while the driven is spread apart, creating a small pulley. The pulleys change sizes based on engine load and vehicle speed.

the drive shaft to the companion flange on the differential. The front U-joint mates the drive shaft to the slip joint. The slip joint is internally splined to fit over the output shaft of the transmission. The slip joint can slide in and out on the output shaft to allow for linear distance changes between transmission and differential. When the rear wheels go over a high spot, the differential moves upward, in effect getting closer to the transmission. The opposite happens when the wheel drops into a hole. As the distance changes, so does the angle between the transmission and differential. The two U-joints rotate within their mounts to adjust for the different angles.

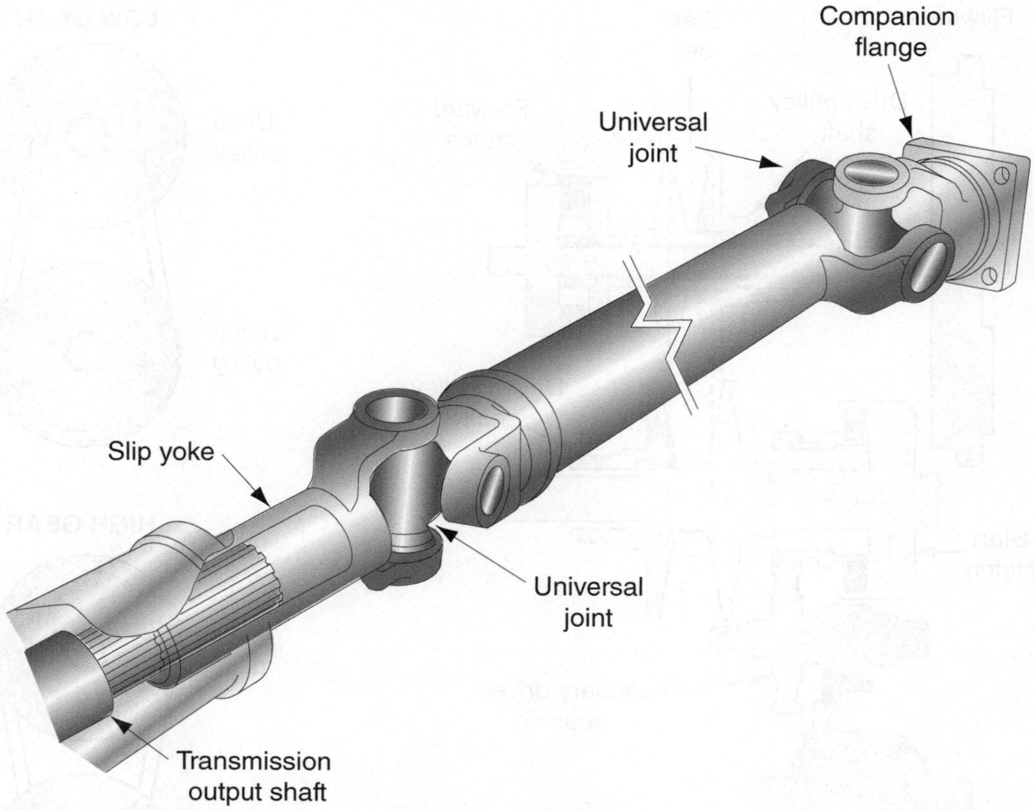

Figure 8-51 The drive shaft assembly consists of a hollow tube, two yokes, two U-joints, and a slip yoke that fits over the transmission output shaft.

Differentials

The differential is mounted in the rear housing and uses a system of gears to let the two drive wheels turn at different speeds (**Figure 8-52**). As the vehicle rounds a curve, the outer wheel must travel a longer distance than the inner wheel. The inner wheel must reduce speed as the outer wheel increases speed. Both wheels use the center of the housing as the pivot point. A second purpose of the differential is to turn the power flow 90 degrees.

The input **pinion gear** is a long gear with spiral teeth. The pinion gears in the carrier are small, round gears with 10 to 14 teeth.

Differential carrier A case that holds the differential gears.

Side gear Also commonly known as spider gears is a set of gears that allow the rear wheels of a vehicle to turn a different speeds.

From the engine to the differential's **pinion gear,** the power is traveling basically in a straight line. At the rear axle housing, the power flow must be changed 90 degrees (**Figure 8-53**). The pinion drives a ring gear mounted to the **differential carrier.** The entire carrier rotates in the same direction as the pinion. Mounted within the carrier are two axle or **side gears** and two pinions. The pinions are attached to the carrier but can spin independently of each other. Meshed with and at right angles to the pinion are the side gears. A solid axle is splined through the center of each side gear. As the vehicle moves in a straight line, the differential carrier rotates and carries the pinion with it. Because the pinions are meshed with the side gears, they rotate with the carrier and turn the axles. The pinions are not rotating or spinning at this time. During a vehicle turn, the inner side gear slows down, and the two pinions begin to spin. The outer side gear speeds up for two reasons: the inner wheel's action and the actions of the pinions. The two side gears change speed at the same rate. For example, if the inner one slows 5 rpm, then the outer one must speed up 5 rpm.

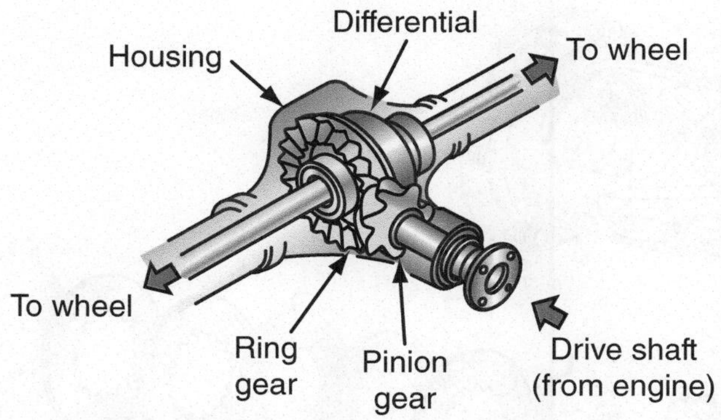

Figure 8-52 The final drive gears turn the power flow 110 degrees and the differential gears allow the drive wheels to rotate at different speeds.

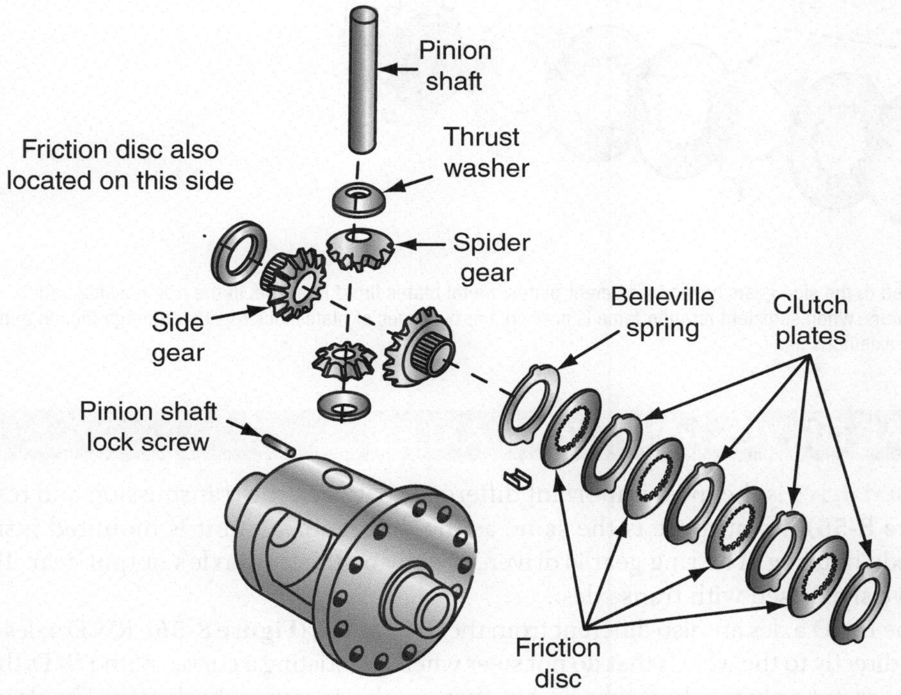

Figure 8-53 Locking differentials are used to manage power between a slipping wheel and a wheel with traction.

The differential carrier is mounted on bearings and must be adjusted for proper pinion gear and ring gear backlash. Some differentials may have a clutch that can lock or slow an axle (**Figure 8-54**). This is known by various names but is a *locking differential.* Under normal circumstances, the differential allows the wheels to speed and slow as needed. However, if a wheel is spinning, more power is sent to that wheel. The result is no vehicle movement. The locking differential uses a clutch to lock the spinning axle, so the power is sent to the wheel that has the traction. If wheel has enough traction, the vehicle will move. As the traction improves on the spinning wheel, the power is more evenly divided between the two wheels.

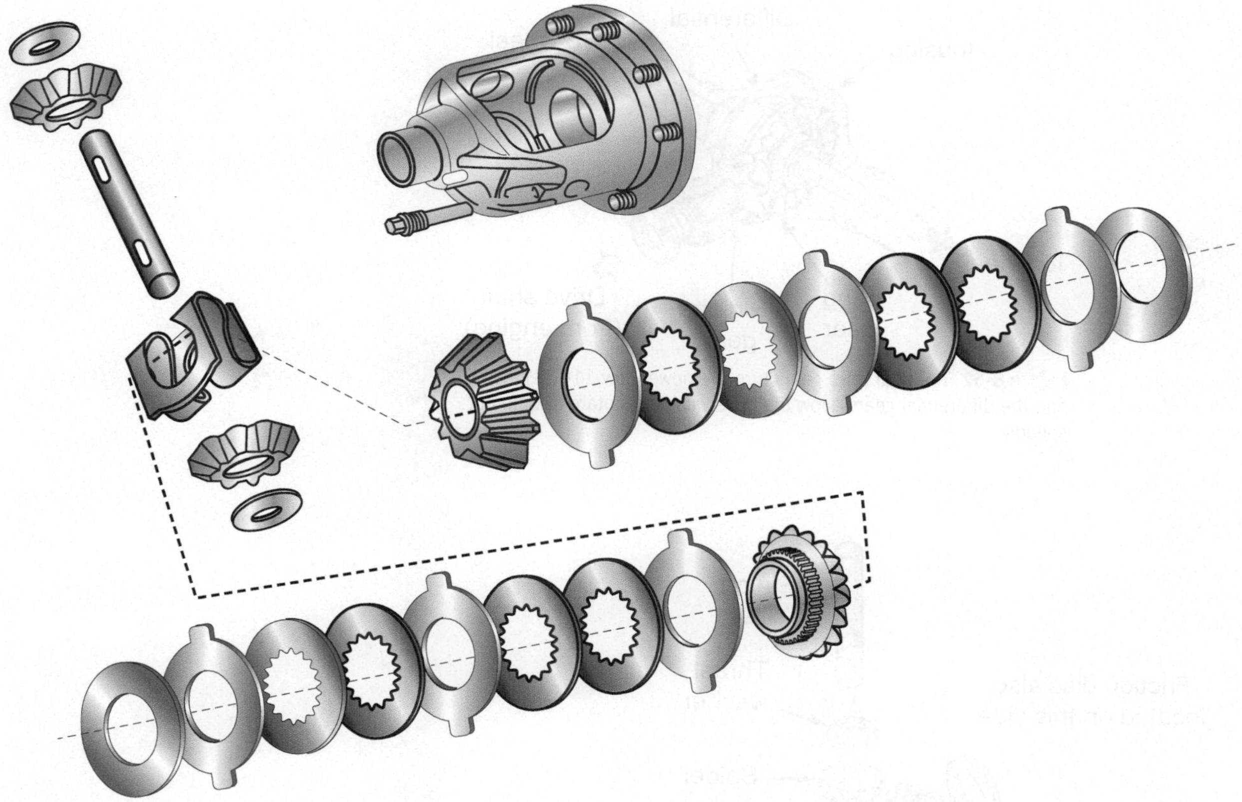

Figure 8-54 The clutches connected to the side gears have two different plates: metal plates fitted into slots in the axle housing and friction plates splined to the side gears. When sufficient rotation force is applied, the two types of plates lock together through friction and effectively lock the side gear to the axle housing.

FINAL DRIVES AND DRIVE AXLES

Shop Manual
page 315

Final drive The last set of gears in the drive train.

The **final drive** is the most important difference between the transmission and transaxle (**Figure 8-55**). Its purpose is the same as the differential, and it is mounted inside the transaxle housing. The ring gear is driven directly by the transaxle's output gear. There is no drive shaft used with transaxles.

The FWD axles are also different from the RWD axles (**Figure 8-56**). RWD axles deliver power directly to the wheels that do not steer when negotiating a curve. With FWD, the front wheels are not only the drive wheels, but they are the steering wheels also. The drive axles must transfer the power to the wheels on sharp turns while moving over road irregularities. To do this, each axle has two constant velocity (CV) joints. The outer or outboard CV joint allows the wheel to be steered and lets the wheel and suspension travel over the road (**Figure 8-57**). The inner or inboard CV joint also allows for the changing vertical angles and adjusts for the linear distance between the final drive and the wheel (**Figure 8-58**). This joint performs the same function as the U-joint and slip joint on the drive shaft.

HYBRID ELECTRIC VEHICLE TRANSMISSIONS AND TRANSAXLES

There are a few different types of transmissions used on HEVs. Early HEVs from Honda used an integrated motor assist (IMA) transmission. This was just a motor that was placed between the gas engine and a standard transmission (**Figure 8-59**). The motor was able to help power the vehicle and accelerate it up to speed. During light acceleration, it will work without the gas engine running, and during heavy acceleration, it will work in combination with the gas engine.

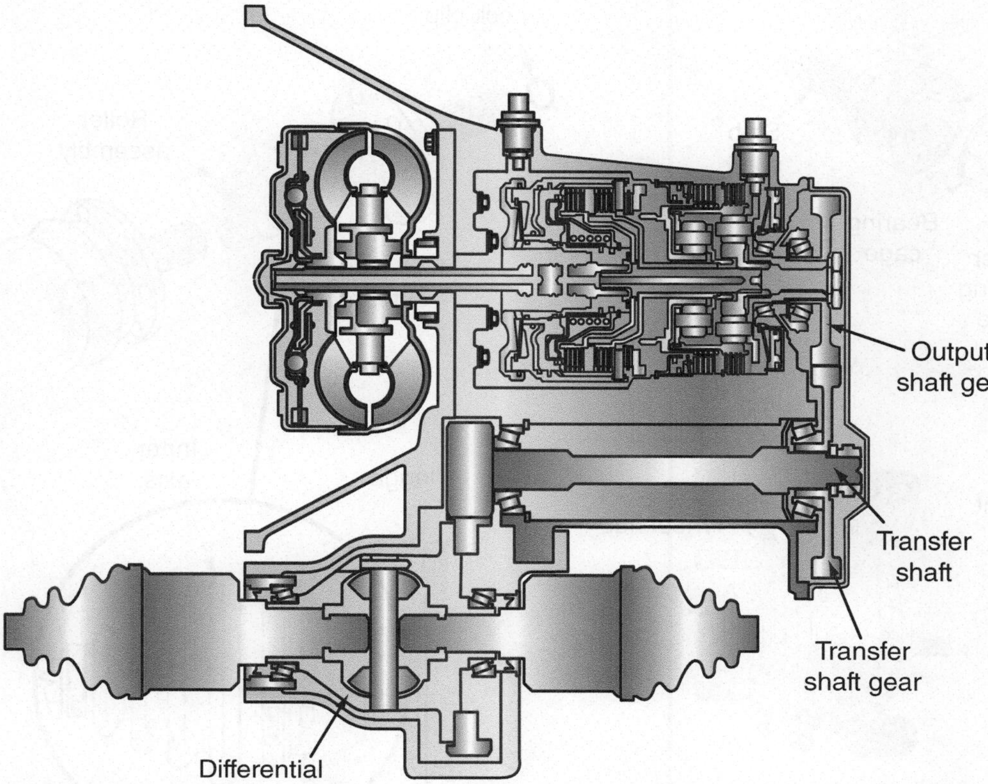

Figure 8-55 The final drive in a transaxle allows the drive wheels to turn at different speeds but may not change power flow direction on certain transaxle designs.

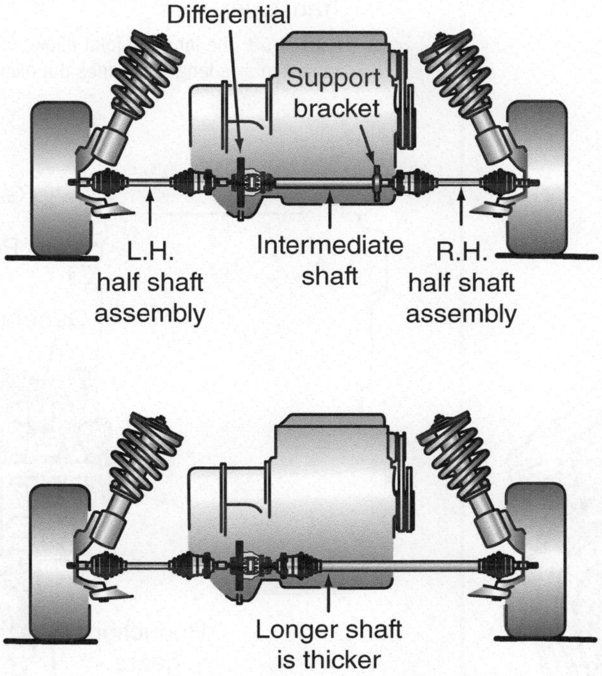

Figure 8-56 Generally, the right- and left-side FWD drive axles are of different lengths and may have additional mounts on the longer one.

The Toyota Prius uses two electric motors and a power split device in combination with a planetary gearset to control its transmission (**Figure 8-60**). The advantage of having two motors is that they can operate like a CVT transmission but only using electromagnetism, not belts.

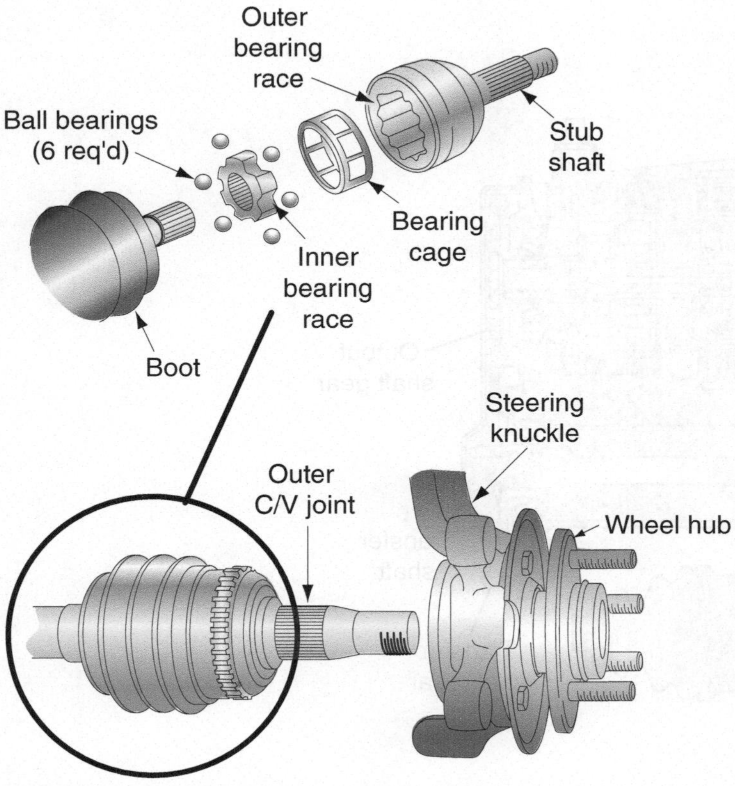

Figure 8-57 The outer CV joint allows for horizontal and vertical angle shifts, but it does not allow for changes in drive axle length.

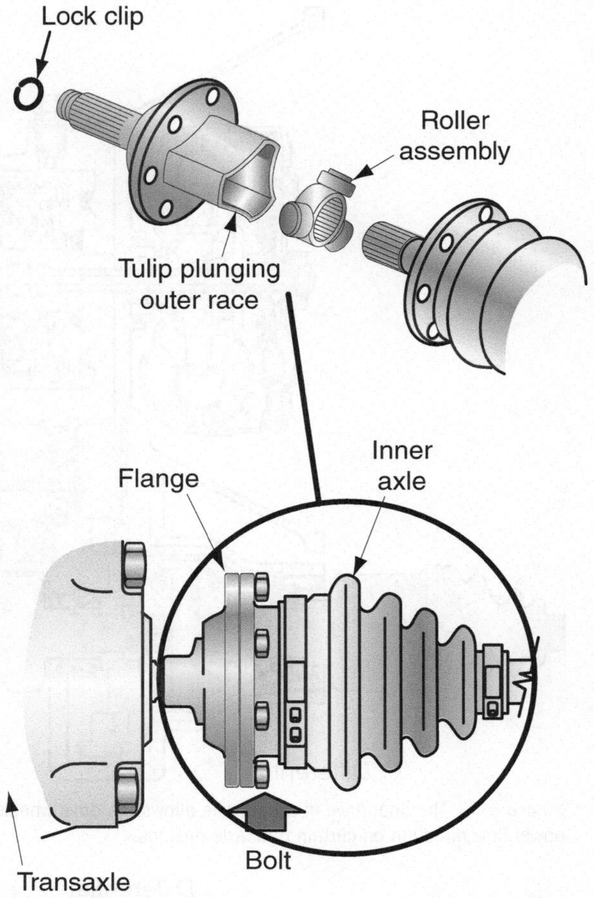

Figure 8-58 The inner CV joint allows for vertical angle shifts and changes in axle length but does not allow horizontal angle shifts.

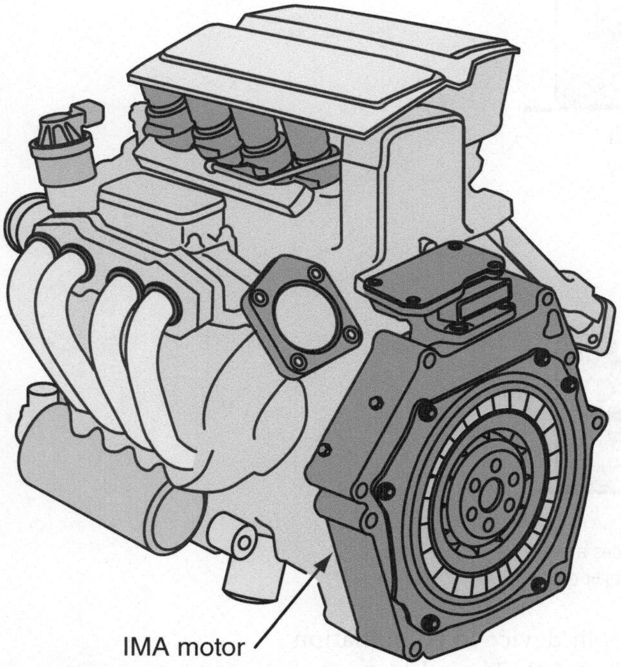

Figure 8-59 The IMA motor in Honda hybrids is mounted between the engine and transaxle.

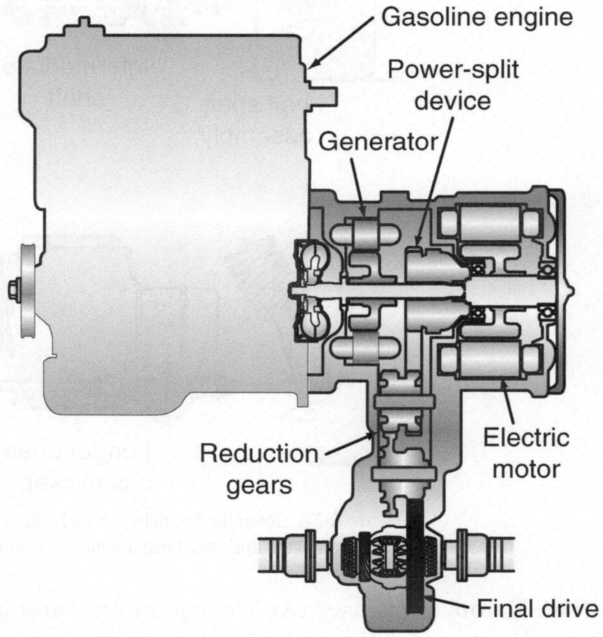

Figure 8-60 Hybrid power system with gasoline engine and electric propulsion motor.

TRANSMISSION FLUIDS

All transmissions need fluids to operate. Fluids are used to lubricate and/or cool the transmission. Some HEVs do not use the fluid to lubricate; the fluid is used to cool the transmission. In this case, the fluid is like engine coolant and has a similar system. Manual transmissions can use a heavy gear oil to lubricate the gears by splashing it up onto the gears. In this type of transmission, the fluid level is typically below the gears. Some manual transmissions (in smaller cars) use a lighter weight oil (like automatic transmission fluid) and a pump to lubricate the gears.

Automatic transmissions use an automatic transmission fluid under pressure. The fluid lubricates but also shifts the transmission and transfers movement when under pressure. There are many types of automatic transmission fluids used in vehicles today; each manufacturer has a specific requirement based on the type of clutches, metals, and shift characteristics of their transmission.

SUMMARY

- The driveline is used to transfer engine torque to the drive wheels.
- Gear ratios are used to match the engine load with the vehicle speed.
- A clutch provides the driver a method to separate the engine from the driveline.
- The pressure plate clamps the clutch disc to the flywheel for transferring engine torque to the transmission.
- Manual transmissions or transaxles provide the necessary gear ratios to move the vehicle.
- The driver makes the gear selection in a manual transmission or transaxle.
- Torque converters are used to connect the engine to automatic transmission or transaxles.
- The impeller drives the turbine.
- The stator acts as a one-way clutch.
- A driver with an automatic transmission or transaxle selects the gear range.
- The automatic transmission or transaxle selects the forward drive gear automatically.
- The Simpson geartrain uses two hydraulic clutches, two gearsets, two bands, and an overrunning or one-way clutch.
- The bands are hydraulic or mechanical-holding devices.
- A typical drive shaft has two U-joints.
- The differential turns the power flow 90 degrees and allows the drive wheels to turn at different speeds during turns.
- Final drives are used with transaxles in place of differentials.
- FWD axles use an outer and an inner CV joint.
- CV transaxles have an infinite number of ratios within their upper and lower limits.

REVIEW QUESTIONS

Short Answer Essays

1. Explain how a hydraulic clutch operates.
2. Describe the purpose and use of the torque converter stator.
3. Describe the two fluid flows in the torque converter during operation.
4. Explain how shifting is controlled in an automatic transmission or transaxle.
5. List and explain the purpose of each component of a drive shaft assembly.
6. Describe the purpose of the two CV joints commonly found in a drive axle.
7. Explain the purpose of transmission bands.
8. Explain how power is transferred from the flexplate to the transmission's input shaft.

9. Describe the actions of the 1–2 synchronizer during shifting from first to second gear.

10. Explain how ratios are changed in a CV transaxle.

Fill-in-the-Blanks

1. The _______________ _______________ has three gears: an outer gear called a ring gear, an internal gear called a planetary carrier, and a sun gear.

2. A hydraulic _______________ is composed of a piston assembly, a drum, and a number of steel plates.

3. The _______________ _______________ is used to block or open the passageways in the valve body passages.

4. The _______________ delivers the power of the engine to the drive wheels.

5. The pressure plate provides the _______________ _______________ to hold the friction disc firmly to the flywheel.

6. Bolted to the rear or output end of the crankshaft is a heavy balanced _______________.

7. The pinion drives a ring gear mounted to the _______________ _______________.

8. Using the small gear as the _______________ gear will result in _______________ output speed but less torque.

9. The torque is increased if the _______________ gear is smaller than the _______________ gear.

10. If the drive pulley on a CVT is spread apart and the driven pulley is completely closed, then the torque output is at its _______________ ratio.

Multiple Choice

1. Automatic transmissions are being discussed. *Technician A* says the band drives the clutch drum. *Technician B* says a basic Simpson geartrain uses three clutches. Who is correct?

   A. A only
   B. B only
   C. Both A and B
   D. Neither A nor B

2. A drive gear with 9 teeth and a driven gear with 27 teeth results in a gear ratio of:

   A. 33:1
   B. 3:1
   C. 3.3:1
   D. 1:.33

3. FWD axles are being discussed. *Technician A* says the outboard CV joint provides vertical and horizontal movement. *Technician B* says the inboard CV joint provides vertical and steering movement. Who is correct?

   A. A only
   B. B only
   C. Both A and B
   D. Neither A nor B

4. Manual transaxles are being discussed. *Technician A* says the reverse idler and reverse gears are connected with a synchronizer. *Technician B* says the reverse idler is moved with a shifting fork. Who is correct?

   A. A only
   B. B only
   C. Both A and B
   D. Neither A nor B

5. Differentials are being discussed. *Technician A* says the carrier lets the drive wheels travel at different speeds. *Technician B* says the pinions do not spin when the vehicle is traveling in a straight line. Who is correct?

   A. A only
   B. B only
   C. Both A and B
   D. Neither A nor B

6. Hydraulic clutches are being discussed. *Technician A* says the piston is applied by spring force. *Technician B* says the clutch discs are splined to a gear or shaft. Who is correct?

   A. A only
   B. B only
   C. Both A and B
   D. Neither A nor B

7. Final drives are being discussed. *Technician A* says the ring gear is driven by the transmission output shaft. *Technician B* says power is delivered to the final drive by a drive shaft. Who is correct?

   A. A only
   B. B only
   C. Both A and B
   D. Neither A nor B

8. Automatic transaxles are being discussed.
   Technician A says the transaxle pump is driven by the torque converter.
   Technician B says the shift valves control shifting.
   Who is correct?
   A. A only
   B. B only
   C. Both A and B
   D. Neither A nor B

9. Drive shafts are being discussed.
   Technician A says the slip joint corrects for distance changes between the transaxle and differential.
   Technician B says U-joints allow angular changes in the drive shaft's movement. Who is correct?
   A. A only
   B. B only
   C. Both A and B
   D. Neither A nor B

10. Planetary gearsets are being discussed.
    Technician A says that if the ring is driven and the carrier is the drive, the torque will decrease.
    Technician B says if the sun is driven and the carrier is held, the ring is the drive.
    Who is correct?
    A. A only
    B. B only
    C. Both A and B
    D. Neither A nor B

CHAPTER 9

TIRES, SUSPENSION, AND STEERING

Upon completion and review of this chapter, you should understand and be able to:

- Describe how the Laws of Motion affect the operation of the suspension and steering systems.
- Explain the relationship between vehicle frame, suspension, and steering.
- Explain the different suspension systems and components.
- Understand the purpose and operation of springs and torsion bars.
- Understand the purpose and operation of shock absorbers and struts.
- Explain the operation of electronic suspension systems.
- Discuss the different steering systems and their general operation.
- Explain the operation of electric steering systems.
- Explain the operation of four-wheel steering.
- Describe the different tire ratings and markings.
- Recognize purpose, components, and operation of the tire pressure monitor systems (TPMS).
- Understand the alignment angles and wheel alignment.
- Explain the difference between static and dynamic tire balancing.

Terms To Know

Active body control (ABC)
Air spring
Alignment machine
All-season treads
Aspect ratio
Automatic ride control (ARC)
Bias ply
Camber
Carcass
Caster
Center link
Center of gravity
Chapman strut
Coil spring
Contact patch
Curb height
Dependent suspension
Dropped center safety rims
Dynamic balance
Hertz
Idler arm
Independent suspension
Indirect TPMS
Jounce
Leaf spring
MacPherson strut
maximum
Metering
Monoleaf spring
Mud and snow tires
Parallelogram
Payload
Pitman arm
Pneumatic
Rack and pinion
Radial ply
Rebound
Recirculating ball and nut
Ride height
Run-flat tires (RFTs)
Sector shaft
Setback
Setback angle
Shock absorber
Sidewall
Spring air valves
Sprung weight
Static balance
Strut
Swag bar
Sway bar
Tie-rod

Tire bead
Tire diameter
Tire Performance Criteria (TPC)
Tire pressure monitor system (TPMS)
Toe
Toe-in
Toe-out
Toe-out-on-turns
Torsion bar
Tread
Tread width
Uniform Tire Quality Grading (UTQG)
Unsprung weight
Variable rate springs
Wheel sensors
Worm gear

INTRODUCTION

The tires, suspension, and steering systems of a vehicle all work together to follow the road surface for the purpose of maintaining traction and control of the vehicle over a variety of road conditions, while also provide a smooth ride for the passengers. Even with all the tires, suspension, and steering systems in good working condition, they also need to be aligned with the vehicle's centerline for the vehicle to operate safely and properly. The tires and wheels are the only point of contact between the vehicle and the road. The tires and wheels become an important link between the vehicle's suspension and steering systems. All the systems must be in alignment and working properly to provide the best quality and control of the vehicle.

LAWS OF MOTION

In Chapter 12, "Engine performance systems," we will discuss some theories of physics that apply to vehicle design and operation. One theory was Newton's Laws of Motion. The laws are based on an object, in the absence of outside forces, performing in each manner, such as moving at the same speed in a straight line. As in most systems on a vehicle, there are almost always some outside forces at work.

Obviously, the portion of the Laws of Motion discussed do not consider the outside forces that move, stop, or turn an object. For example, friction tries to stop or slow an object, whereas a force greater than the mass (weight) of the object is required to move or accelerate it. When designing a vehicle that must work within the Laws of Motion and other physics, related items like mass, force, torque, friction, and speed must be considered. The suspension and steering systems have components that are almost constantly moving, stopping, or changing direction during vehicle operation. The suspension and steering systems must be able to overcome the straight physics of motion so the vehicle can be controlled.

FRAME, SUSPENSION, AND STEERING RELATIONSHIP

Proper operation of the suspension and steering systems depends on a straight and true frame. The frame, full or unibody, must be strong enough to support the suspension and steering as they operate over the road and under different load conditions. A bent or warped frame may affect the steering to the point that the driver may have problems keeping the vehicle under control. A weak frame can flex as it is stressed when changing steering and suspension angles. A weak frame may also allow the suspension system to transmit the shock to the body, break mounting points, and create steering problems. Frames and unibody structures that are misaligned may not be clear until some other component fails or begins to show excessive wear.

Shop Manual
pages 352-353

AUTHOR'S NOTE A small 2004 "compact" truck on which we had installed four new tires was experiencing a vibration problem at cruising speed. The truck had not been abused, it was primarily transportation to and from work, and had no vibration problem until the installation of the new tires. Rebalancing and finally installing another set of new tires did not solve the problem. After consultation with the local dealership, a technical service bulletin (TSB) was found that indicated a twisting in the frame that could occur even during normal driving conditions. This problem was most prominent when hard, long-mileage tires were installed. Although we were unable to correct the frame condition, after installing four new "softer" tires (more like the factory originals) the vibration was no longer noticeable.

Contact patch The area of the tire that makes contact with the road surface.

Pneumatic The use of compressed air or gas to operate a tool or object.

Filling a balloon with air from the lungs makes that balloon **pneumatic**.

Note that, when a wheel is mounted on a vehicle, the place for the lug nuts appears to be deep inside the wheel opening. This would be a **dropped center safety rim**.

WHEELS AND TIRES

Wheels

The tire is the contact between the vehicle and the road. The area where the tire contacts the road is called the **contact patch**. The contact patch is where the tire supports the weight of the vehicle. Because the tire is **pneumatic** (air-filled), a wheel or rim is needed to support it. Most wheels are made of steel or an alloy of magnesium and aluminum. Steel wheels are made by welding the rim to the centerpiece. Magnesium and aluminum wheels are cast molded. Passenger car wheels are listed as **dropped center safety rims.** The rim flange is set above the center (**Figure 9-1**). Holes around the center fit over studs fitted

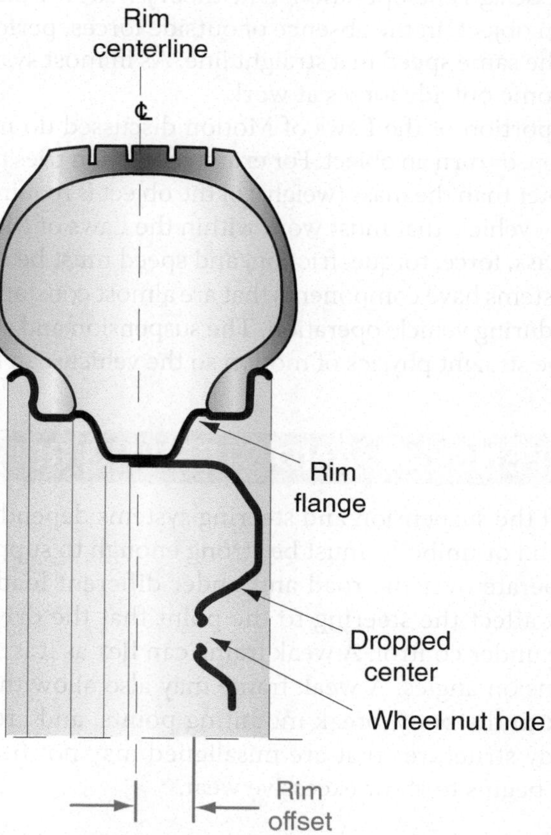

Figure 9-1 Shown is a dropped center rim. This places the mounting portion to the right of the vertical center line.

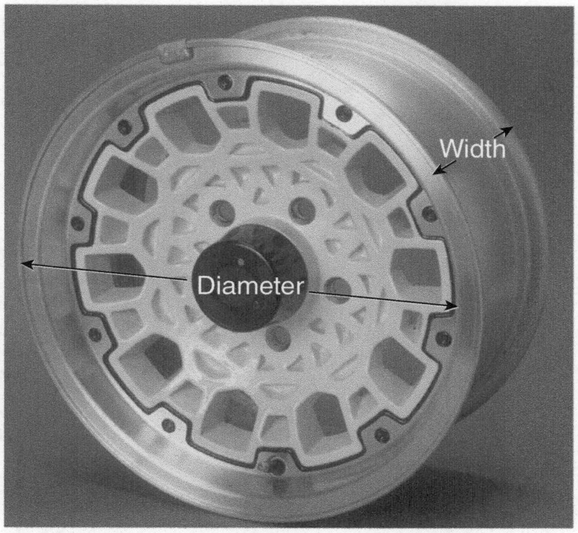

Figure 9-2 The diameter and width measurements of a wheel.

into the hub, and lug nuts are used to mount the wheel to the hub. Wheels are measured in diameter (across the wheel) and width (between edges) (**Figure 9-2**). The tire is held to the wheel by air pressure that forces the **tire beads** over the wheel's center humps and seals them against the rim flanges. The two tire beads are tightly wrapped steel wires formed into loops slightly smaller in diameter than those of the rim flanges. Tires are matched to the wheel by diameter and width.

Tire bead The edge of the tire that sits on the wheel or rim.

The present-day tire is a result of research and development conducted at the racetrack and at tire-manufacturing facilities. There are all types of tires, from the large racing slicks used for drag racing to the small ones for lawn mowers and go-karts. Passenger car tires range from 14 inches to 21 inches in diameter. There are several types of **tread** and **sidewall** designs to meet almost all road and driver situations. Tires can be selected for different types of weather, payload, and driving conditions. Some vehicle manufacturers have normal, sport, or touring tires that can be placed on the same model car for different driving styles. Tires come in two basic configurations: **bias ply** and **radial ply**.

Tread is the molded portion of the tire that touches the road. It is supported by the **sidewall,** which extends from the bead to the tread area.

Bias Ply

Bias ply tires are no longer used on most vehicles. They can still be found on larger trucks and trailers. The fleet service technician may deal with bias ply tires regularly. The word *bias* refers to the direction the cords are laid under the tread (**Figure 9-3**). Cords run from one bead to the other, forming the sidewall and supporting the tread. Bias ply tires have the cords installed at an angle to the radii of the tire and tread. The bias ply tire is stiff and can carry heavy loads. But as the wheel tilts during a turn, the bias sidewall picks up the inner portion of the tread from the road, thereby reducing traction and increasing wear. Bias tires are not fuel efficient and can be rough riding because of the lack of flexibility in the sidewall.

Today **bias ply** tires are usually associated with heavy load carriers like semi-trucks.

Radial ply tires are found on all modern cars and most light trucks. Some heavy load carriers are required to have radial ply tires on the steering axle.

Radial Ply

The **radial tire** is currently the most popular tire in use today. The cords are laid 90 degrees to the tread and parallel to the tire's radii (**Figure 9-4**). This provides a softer ride, better fuel mileage, and better traction. The radial sidewall stiffens the tread while maintaining a flat tire/road contact. As the wheel assembly tilts during a turn, the tire's sidewall flexes instead of pulling the tread from the road. The soft sidewall flexes more readily over various road conditions. The tire's flexing sidewall results in a smoother ride and less rolling resistance. The lower resistance increases the vehicle's fuel mileage.

Bias ply tires on heavy load carrier have straight treads.

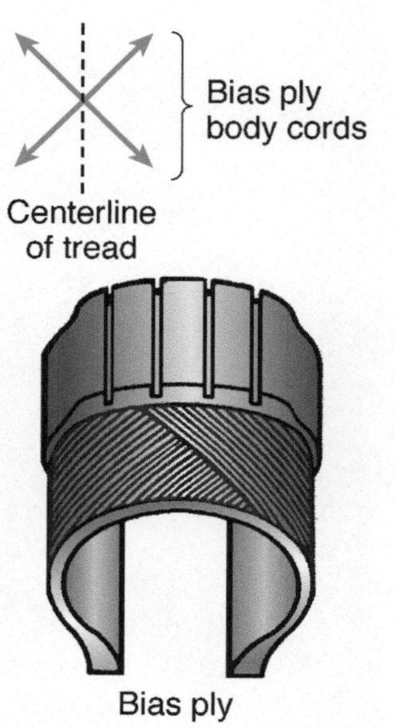

Figure 9-3 Bias cords make a stiff sidewall but tend to lift the tread during cornering.

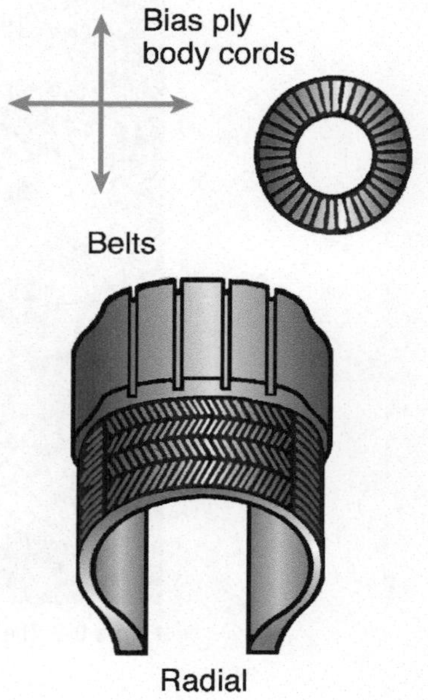

Figure 9-4 Radial cords cause a softer sidewall, allowing the tread to remain flat on the road.

The greatest drawback of the radial tire design is the lack of driver education on its inflation characteristics. With a bias ply tire, it is obvious when the tire is low in air pressure. However, with the soft sidewall of a radial tire, the tire tends to look a little underinflated even when pressurized to 35 psi. Many drivers do not recognize when the tire is low. As a result, many radial tires are worn out prematurely.

Run-Flat Tires (RFTs) have strengthened sidewalls to support a vehicle for an extended distance in the event of tire damage. Some have temporary self-sealing capabilities.

A tire that became available from several manufacturers in 1997 is commonly known as the ***run-flat* tire (RFT)**. It has extra strong sidewalls that will support the vehicle for up to 125 miles (**Figure 9-5**). These tires are usually mounted on vehicles equipped with electronic TPMS to alert the driver of low tire pressure, because a low tire condition is not as visible or noticeable when driving. Another type of RFT has self-sealing capabilities,

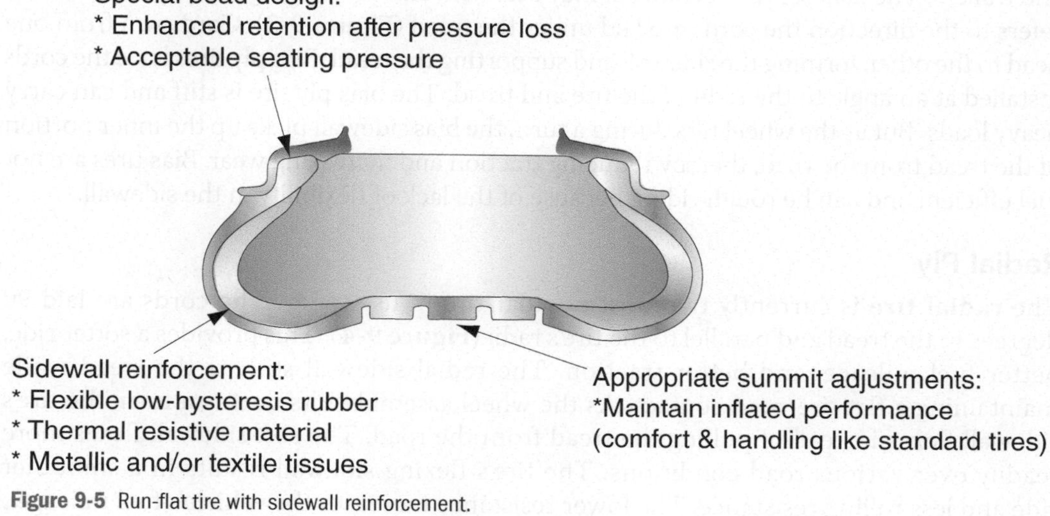

Figure 9-5 Run-flat tire with sidewall reinforcement.

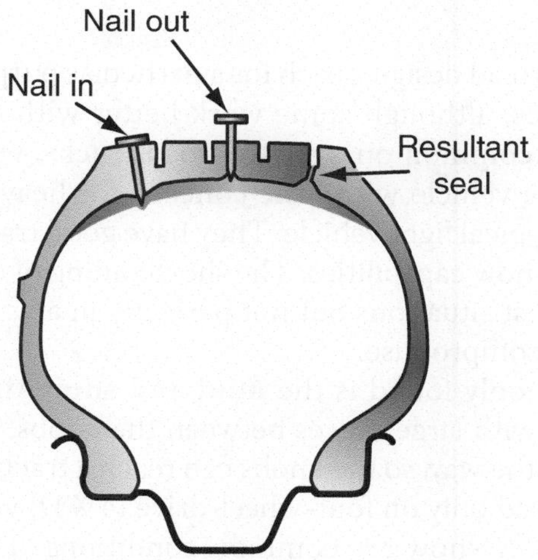

Figure 9-6 Puncture-sealing tire.

but it is not designed to run forever with a puncture. The tire was designed to seal the hole and hold air until a repair facility could be reached (**Figure 9-6**). With this tire, there is no loss of control that often happens during a sudden deflation of the tire.

Tire Construction and Labeling

The tire is made up of several layers of different materials, each designed for a specific task. The cord layout was discussed in the last section. Other major components of the tire are the belts and tread.

Belts

Belts of reinforcing materials are placed under the tread for better support. The belts are laid under and run in the same direction as the tread (**Figure 9-7**). The belts help hold the tread in position on the **carcass** and give better tread/road contact. The belts commonly used today are made from layers of steel wires. Belts are used in bias and radial tires.

The **carcass** is the completed tire minus the tread.

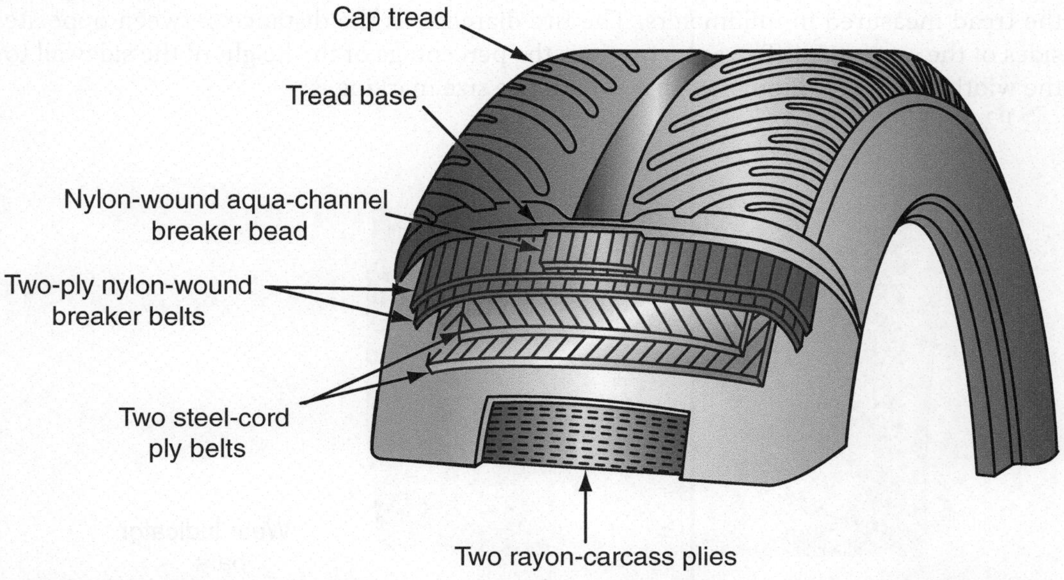

Figure 9-7 Note how the belts support the tread. This particular tire has a water (aqua) shedding channel supported by a special belt.

Treads

There are many different tread designs, each for a particular purpose. They can be molded to bias or radial carcasses, although some work better with one type than the other. Straight treads are most common on over-the-road trucks, which require traction for starting and stopping the vehicle with little concern for heavy cornering. **All-season treads** are made for the typical light vehicle. They have good traction on dry or wet roads and have some mud and snow capabilities. The shortcoming of the all-weather tread is its ability to work well in most situations but not perfectly in any. Like most equipment on the vehicle, it is a design compromise.

All-season treads are standard equipment on almost all standard cars and light trucks.

Another tread commonly found is the **mud and snow tire.** The tread consists of knobby lumps of rubber with large spaces between the knobs. This allows the mud and snow to be pushed out of the way, so the knobs can reach a traction surface. The mud and snow tread is generally used only on four-wheel-drive (4WD) vehicles and in those areas of the world where mud or snow are common conditions. The northern parts of the United States have tire treads designed for snow only. Although this tire works well in snow, it, like the mud and snow tread, is loud and wears quickly on dry pavement. A common solution in areas with regular snowfall is to use an all-season tread and install tire chains when necessary.

Mud and snow tires are used occasionally in the southern United States and very common during winter in the northern United States and Canada. None of the tires mentioned here work well in icy conditions.

A special tire is also available for ice. This tire has a mud/snow tread with small steel spikes protruding from the tread area. The spikes dig into the ice for traction. However, the spikes will not last long when driven on pavement without ice.

Each tire is equipped with tread wear indicators (**Figure 9-8**). Several solid strips of rubber run across the width of the tread area near the bottom of the grooves. When the tread wears to a depth where the indicators become readily visible, the tire should be replaced.

Tire Labeling

Each tire is identified by markings etched or molded into the sidewall. The markings include the manufacturer, tire brand, tread size, and test certifications. The most recognized marking is the tire size.

The tire is sized in three measurements: **tread width, tire diameter,** and **aspect ratio,** measured with the tire properly inflated. The size of the tire is one of the determining factors in its load capability. The tread width is the distance between the two edges of the tread measured in millimeters. The tire diameter is the distance between opposite sides of the same bead. The aspect ratio is the percentage of the height of the sidewall to the width of the tire (**Figure 9-9**). A typical tire size marking is:

P245/65R17

Tread width is the measurement of the actual width of the tread, not just the part that contacts the road.

Tire diameter is the size of the large opening in the tire, and the wheel must match that diameter.

The **aspect ratio** is the ratio of the sidewall to the tread width expressed as a percentage.

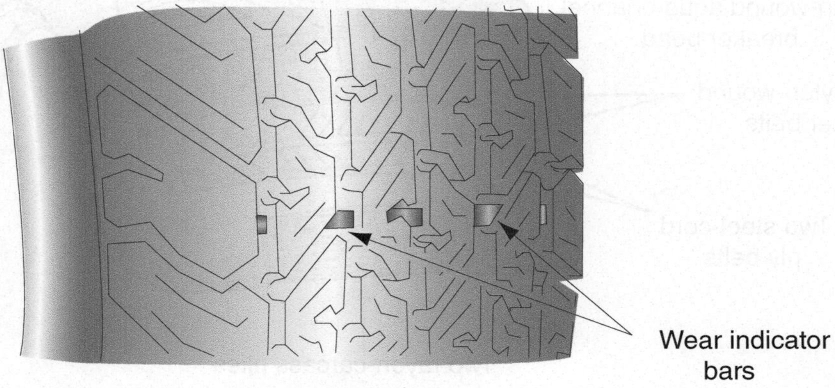

Figure 9-8 Wear indicators are a quick, visible method of determining tire wear.

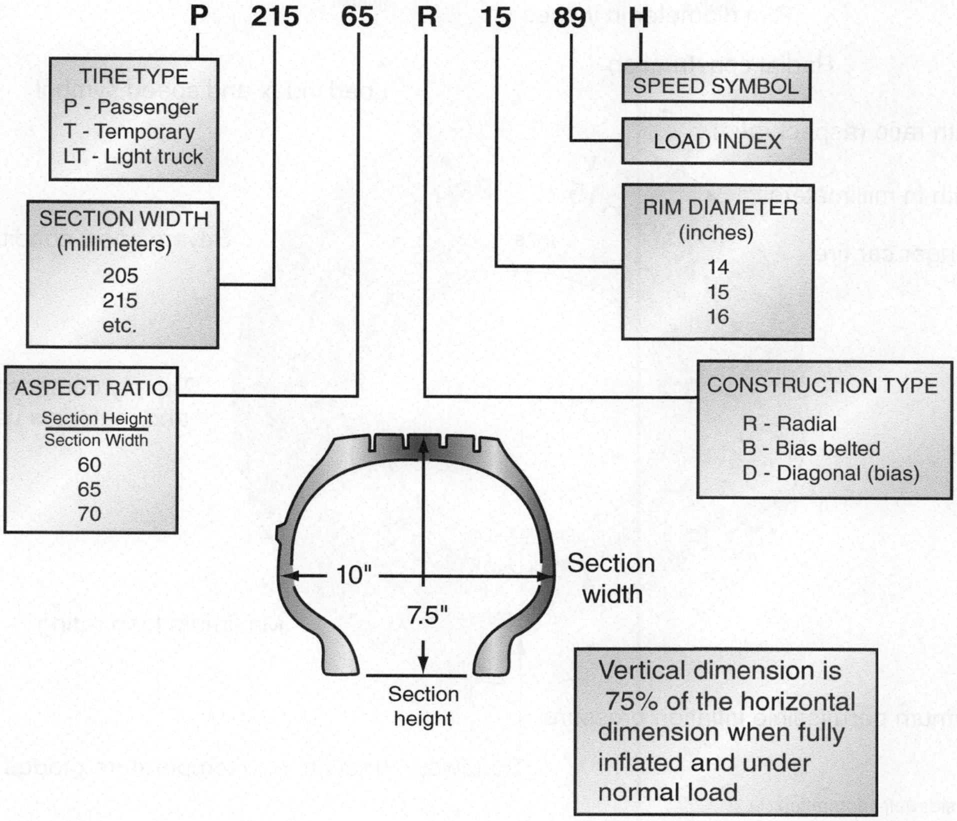

Figure 9-9 The aspect ratio is the height of the tire as a percentage of tire width. The measurements apply when the tire is at recommended pressure, cold, and loaded by the vehicle.

The first group, 245, means the tread is 245 millimeters wide. The letter "P" prefix designates a passenger car tire. Often SUVs and light trucks will use passenger style tires with the "P" designation. This does reduce the towing capacity of the vehicle. Instead of a "P"-type tire, some trucks use a "LT" tire. This designation means the tire is made of a heavier construction and is suited for "Light Trucks." The number 65 indicates the sidewall height is 65 percent of the tire's width, or about 159 millimeters. The letter R indicates a radial construction tire. The last group, 17, is for a 17-inch diameter. This size tire fits a 17-inch rim with a width of about 6 to 7 inches. The tire could be fitted to a wider or narrower or wider rim, but the beads would not seal correctly and might result in a blowout due to improper tire loading.

A **Tire Performance Criteria (TPC)** number is molded into the sidewall near the tire size (**Figure 9-10**). This number indicates the tire's traction, noise, handling, rolling ability, and endurance characteristics. Replacement tires should have the same TPC number.

Other markings include the **Uniform Tire Quality Grading (UTQG),** which outlines the tire's traction, tread wear, and temperature resistance. The grades are shown by letters and numbers stamped on the sidewall. Tread wear is based on a percentage of 100. Percentages range from 50 (poor) to 540 (long) relative wear. Resistance to skidding on a wet surface (traction) and temperature are listed by letters AA, A, B, or C, with AA being the best wet-surface traction and lowest temperature, and C being the worst wet-surface traction and highest temperature. High-performance tires can be rated for speed using the codes F (50 mph) to Z (149 mph). The codes would show up as part of the tire size markings such as 225/70R 16 Q for a tire rated at 100 mph (**Figure 9-11**).

The **Tire Performance Criteria (TPC)** is a number, molded in the sidewall, that lists important information about the tire.

Uniform Tire Quality Grading (UTQG) is a system developed to provide the customer a way to compare, select, and rate tires.

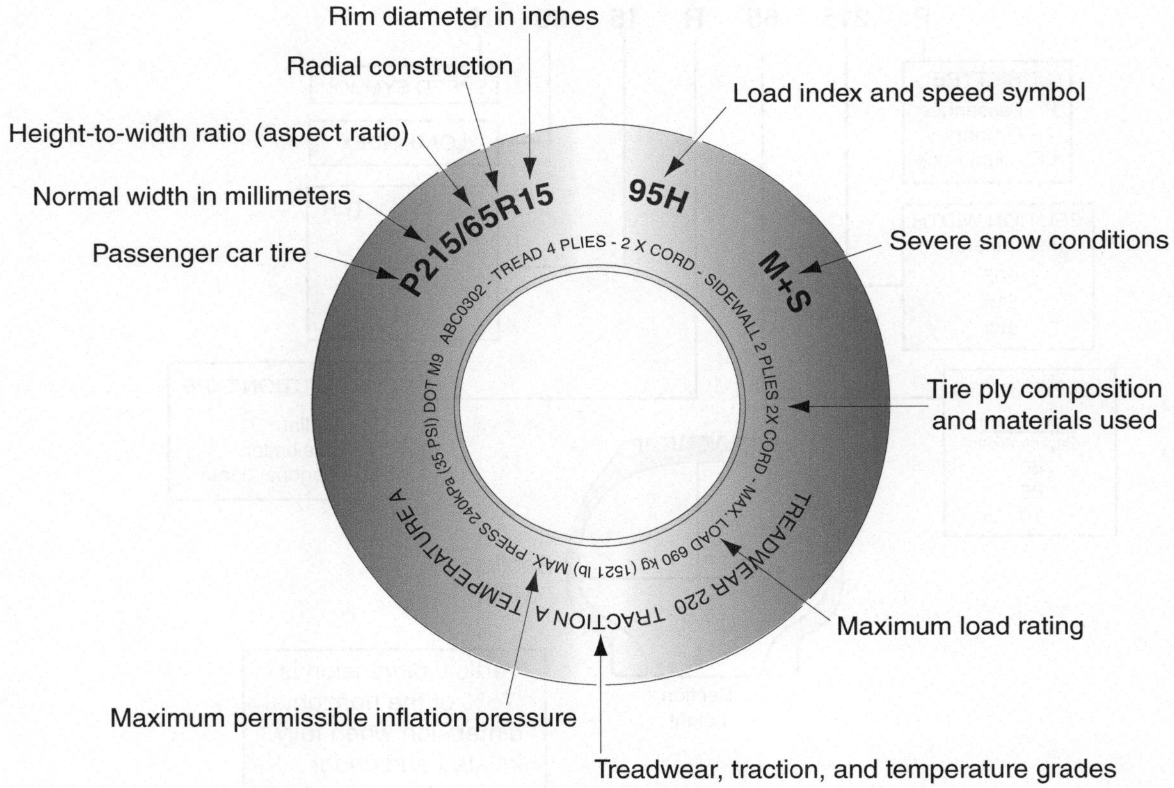

Figure 9-10 Tire sidewall information.

TIRE SPEED RATINGS

Letter Symbol	Maximum Speed
F	50 mph
G	56 mph
J	62 mph
K	68 mph
L	75 mph
M	81 mph
N	87 mph
P	93 mph
Q	100 mph
R	106 mph
S	112 mph
T	118 mph
U	124 mph
H	130 mph
V	149 mph
Z	+149 mph

Figure 9-11 Tire speed ratings chart.

The Federal Department of Transportation requires certain information to be placed on the tire. Examples of some of the required information are:

- Tire size
- Maximum permissible inflation pressure
- Maximum load rating
- Generic name of each cord material in sidewalls and tread
- Actual number of plies in sidewall and tread if different from sidewall
- Tubeless or tube type as applicable
- Radial as applicable
- DOT manufacturing code listing the tire manufacturer, factory location, and date of manufacture

AUTHOR'S NOTE In late 2008, tire problems concerning the shelf life of unsold "new" tires came to light. Accidents involving tires in which their composition degraded over time resulted in injuries and lawsuits. To help reduce this problem, locate the tire data stream located just around the tire's bead and find the last four digits. The digits represent the week and year of the tire manufacture date. The first two digits indicate the week of the year. For example, the digits 0307 indicate the tire was manufactured in the third week of 2007. Tire storage life is estimated at about 2 years, but some tire dealers and repair shops sell and install tires older than 2 years. A deteriorated tire can explode violently, causing loss of vehicle control and possible serious damage or injury. Although such explosions are rare and involve only certain types of tires, the installer should nevertheless check the "new" tire's manufacture date before installing it. In particular, check slow-selling tires, such as those designed for 13-inch rims, for their manufacture date. Because only a few vehicles use 13-inch tires now, they tend to be surplus and overstock rather than of **recent manufacture**.

Air Pressure and Nitrogen in Tires

Many new car and tire dealerships offer customers a choice of adding regular air or nitrogen to their tires. The air you breathe (and the "regular" air that an air compressor uses) contains 78 percent nitrogen, 21 percent oxygen, and about 1 percent inert gases. So, when a technician suggests filling a tire with nitrogen, they mean using an airline source from a special nitrogen generator and compressor or precharged tank.

Nitrogen molecules are larger than oxygen molecules (**Figure 9-12**). Therefore, a tire inflated with nitrogen has a lesser chance of leakage through the sidewall, valve stem, or bead. Oxygen carries moisture, which causes rust and corrosion. Oxygen's volume also increases and decreases significantly with temperature, whereas nitrogen does not. This is a major factor when the temperature swings. Tires filled with nitrogen are easily identified by the green valve stem cap.

When checking the air pressure in a tire, you must always refer to the tire placard (**Figure 9-13**). On most vehicles, it is found in the door jam. But other uncommon locations can include the glove box, trunk, or center counsel. The air pressure rating on the tire itself is the **maximum** inflation rating. This is not the correct pressure to achieve. The correct setting, as found on the tire placard, is specific to that vehicle given the weight, performance category, and designed ride quality. The pressure is always checked when the tire is cold (in other words, not right away after driving it). When you are driving, the tire creates friction and heat. This results in a higher pressure.

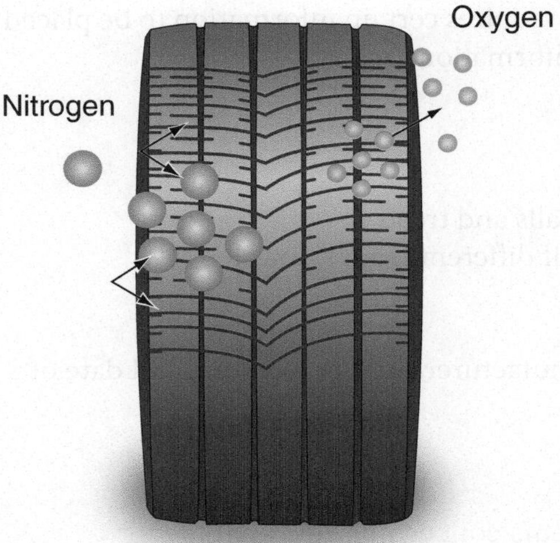

Figure 9-12 Oxygen and nitrogen leaking through tire walls and tread.

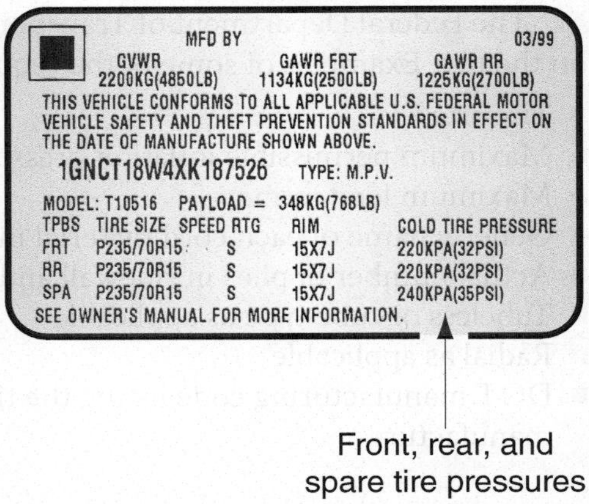

Figure 9-13 Tire placard.

Always follow the tire placard specification. Overinflating the tire causes the tire to be stiffer, the tread contact patch to be smaller, and the rolling diameter to be greater. Because the tire is stiffer, it also is more susceptible to damage from a pothole or bump in the road. Underinflating the tire causes the shoulders of the tire to be in contact with the road. This causes friction in an area where it is not designed to be. This will rapidly wear the shoulder or sidewall of the tire and cause it to blow out.

TPMS An electronic system designed to monitor the air pressure inside the pneumatic tires.

Using a spare tire that is oversized/undersized compared to the original tire will affect the **indirect tire pressure monitoring system (TPMS)** more than the **direct** system.

Tire Pressure Monitors System (TPMS)

A **tire pressure monitoring system (TPMS)** is an electronic air pressure monitoring and reporting system found on all new vehicles (**Figure 9-14**). It is a federal requirement that all vehicles sold after September 1, 2007, have some type of TPMS. There are two types of system in use, the **indirect** and the **direct** system. An **indirect TPMS** system uses the already existing wheel speed sensor data from the ABS system (**Figure 9-15**). An underinflated tire will rotate faster than a normally inflated tire because the rolling

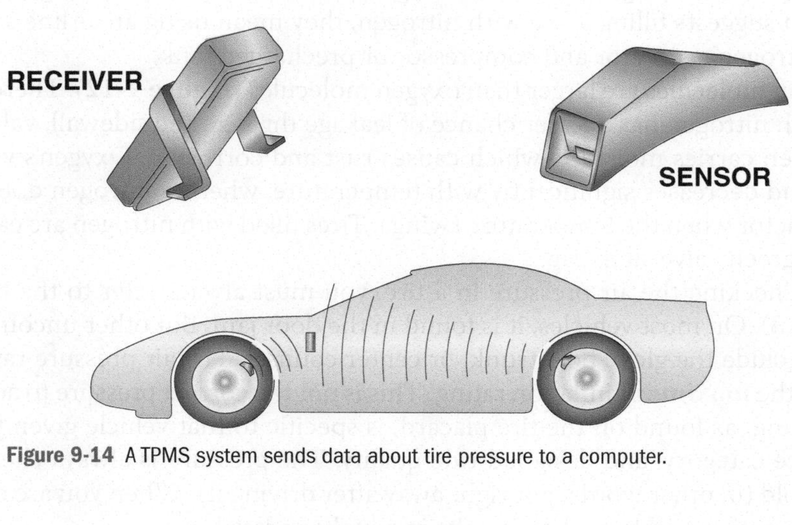

Figure 9-14 A TPMS system sends data about tire pressure to a computer.

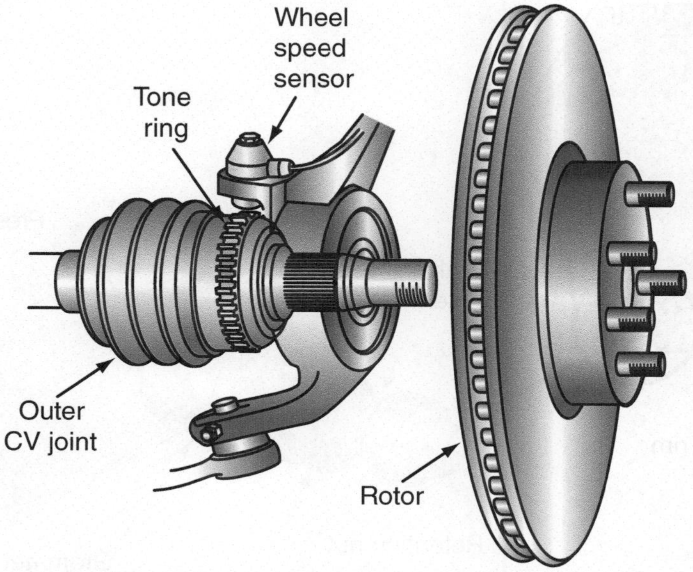

Figure 9-15 This is one type of wheel speed sensor. The other is sealed in the hub bearing.

diameter is less. The ABS computer notices this and sends an error signal to the instrument cluster (a malfunction light) for the driver to see. One main disadvantage to this system is the situation when all tires have low pressure. An indirect system only measures the difference in speed, not the actual pressure.

A direct TPMS system uses a sensor mounted in the tire/wheel assembly. Some TPMS sensors are mounted directly to the wheel with a band clamp (**Figure 9-16**). The more common type of TPMS sensor is the valve stem sensor that incorporates a valve stem with a pressure sensor attached to it (**Figure 9-17**). The valve stem style can be identified by either the metal construction of the stem and cap or the slightly different look if it is a rubber stem. The TPMS sensor monitors tire pressure and alerts the driver to a low-pressure condition. Some systems only illuminate a dash indicator warning lamp,

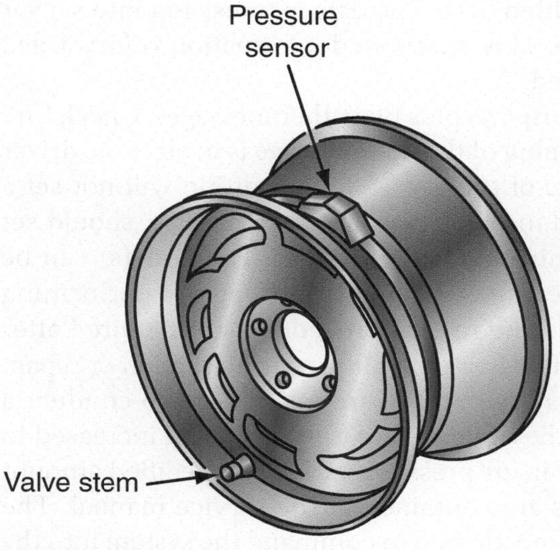

Location	Color code
Right front	Blue
Left front	Green
Right rear	Orange
Left rear	Yellow

WARNING: Pressure sensor inside tire. Avoid contacting sensor with tire changing equipment tools or tire bead.

Service note: Pressure sensor must be mounted directly across from valve stem.

Figure 9-16 Tire pressure sensor.

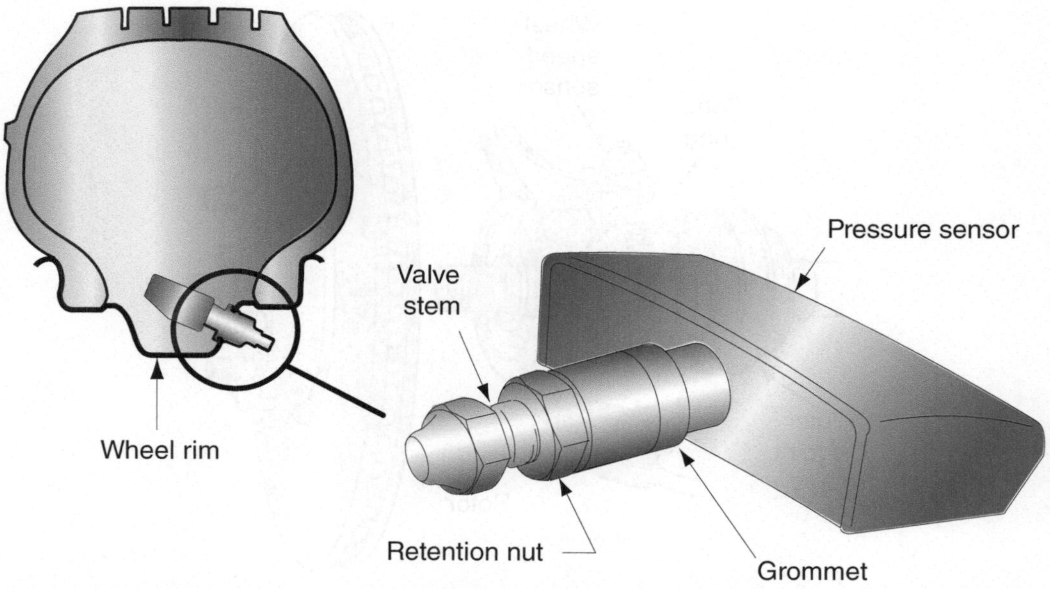

Figure 9-17 Some TPMS sensors can be identified by their metal construction, but some use a rubber valve stem.

while others can inform the driver of which tire is low and what the actual pressure is. The main components of a General Motors TPMS system consist of a driver information center (DIC), passenger door module (PDM), four pressure sensors, and the serial data circuit. Other manufacturers use similar system operations and similar components.

The pressure sensor mounted on each wheel transmits a radio signal to a module in the vehicle. The sensor has an internal roll switch opened and closed by centrifugal force or the lack of that force. When stationary, the switch is open and in a stationary mode. In this mode, the sensor monitors tire pressure every 30 seconds and transmits that data to the module every 60 minutes. When the vehicle is moving (tire is rolling), the switch is forced closed, putting the sensor into drive mode. In this mode, the pressure is sampled every 20 seconds and the data transmitted every 60 seconds. Each vehicle manufacturer performs the reporting data slightly different. It is important to refer to the service manual for specifics on each vehicle.

The PDM is the receiver for the transmitted data. The data is translated into sensor location, sensor mode, and the tire pressure. This interpreted information is forwarded to the DIC where it is processed and displayed.

The DIC provides tire pressure reading displays plus two other messages: Check Tire Pressure and Service Tire Monitor. The meaning of the first message is to alert the driver to a low-pressure condition in one or more of the tires. This condition will not set a diagnostic trouble code (DTC). The second message, Service Tire Monitor, should set a DTC and is indicative of a problem within the TPMS system. The code(s) can be retrieved with a scan tool. The procedures for servicing, replacing, or performing a relearn are covered in the service manual. The relearn procedures are required after tire rotation, after replacing a faulty sensor, and after a low air pressure situation or repair. In the case of the General Motors TPMS system, there are two means to conduct a relearn. One is tire pressure changes when the tire pressure is lowered and increased to force the sensor to remeasure. The changes in air pressure must be a specified amount in each direction. The specific instructions are contained in the service manual. The second method requires a special TPMS diagnostic tool to command the system into the relearn procedure.

AUTHOR'S NOTE I came across a technician who didn't perform relearn procedures after tire rotations. A customer's tire developed a leak (from a nail). The TPMS system told him that the left front tire was low, so he added air to it at the gas station. The TPMS warning indicator and lamp did not go out, so he continued to add air. After a while he drove the vehicle back to the repair shop. When they checked the vehicle, the left front tire had 80 psi in it and the left rear tire had 10 psi in it. Had the technician performed a relearn procedure, he would have saved himself from embarrassment and a very dissatisfied customer. He learned a good lesson the hard way.

SUSPENSION SYSTEM OPERATION

The suspension supports most of the vehicle's weight and absorbs most of the various shocks from the bumps and holes in the road. But a second and just as important purpose is the control that a properly functioning suspension system offers a driver. A suspension system allows the wheel assembly to move upward and backward or down and forward as it maneuvers over the road. This angle of movement absorbs some of the shock and reduces stress on the suspension and steering components.

Shop Manual pages 358–362

The mass of the vehicle is divided into two weights. The **sprung weight** is the mass supported by the springs and usually includes everything mounted above the axles (**Figure 9-18**). It is important to remember that only the springs support the sprung weight. **Unsprung weight** or *mass* includes components such as the axles, brakes, tires, shock absorbers or struts, and the springs themselves. Vehicle engineers try to reduce unsprung mass by reducing the weight of the wheels and suspension components. When the unsprung weight is reduced, the ride quality is increased, and the shock absorber does less work. For this reason, you will see a lot of unsprung components made of lightweight materials. Aluminum wheels are now being used on many vehicles instead of steel ones. Tires weigh less because of the different sizes being used today. Another way to reduce unsprung mass is to use an independent suspension system and aluminum control arms.

Sprung weight is all vehicle weight supported by the springs.

Unsprung weight is all vehicle weight not supported by the springs including the springs themselves.

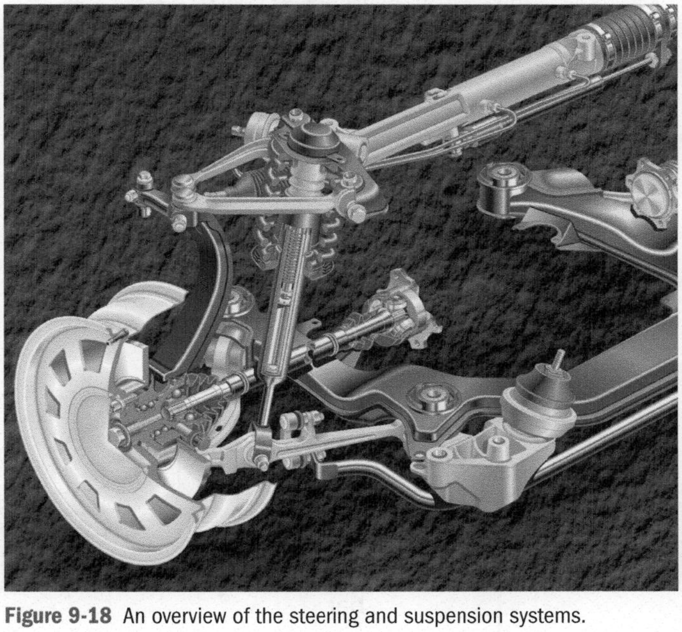

Figure 9-18 An overview of the steering and suspension systems.

Jounce generally occurs when the wheel hits a high spot like a rock.

Rebound is downward movement after a jounce.

A **shock absorber** or **strut** is designed to absorb shock and control spring movement.

Metering is the controlled flow of a liquid or vapor.

Sway bars and **stabilizer bars** help the driver control vehicle stability during turns.

The **center of gravity** is the balancing point of an object.

When the tire moves up (**jounce**) or down (**rebound**) over the road surface, a shock or jar is transmitted through the suspension system (**Figure 9-19**). For the most part, the shock is absorbed by the flexing of the tires and springs before it reaches the passenger compartment. However, a spring tends to rebound back and forth until its energy is expended. Because most roads do not have many long, flat surfaces, the vehicle would bounce up and down continuously unless the spring action was controlled.

A hydraulic cylinder known as a **shock absorber** or **strut** is used to control the motion of the springs. This cylinder allows the spring to give or extend enough to absorb most of the road shock. It controls further spring movement by **metering** fluid within the cylinder.

Ideally, the passengers would never feel the irregularities of the road, but the weight of a typical vehicle makes it practically impossible to eliminate all rough conditions. In addition to absorbing road shock, the suspension also helps control body lean during turns to help maintain vehicle control. **Sway bar** or **stabilizer bar** are used to reduce lean (**Figure 9-20**). As the vehicle is turned, its **center of gravity** and weight shifts toward the

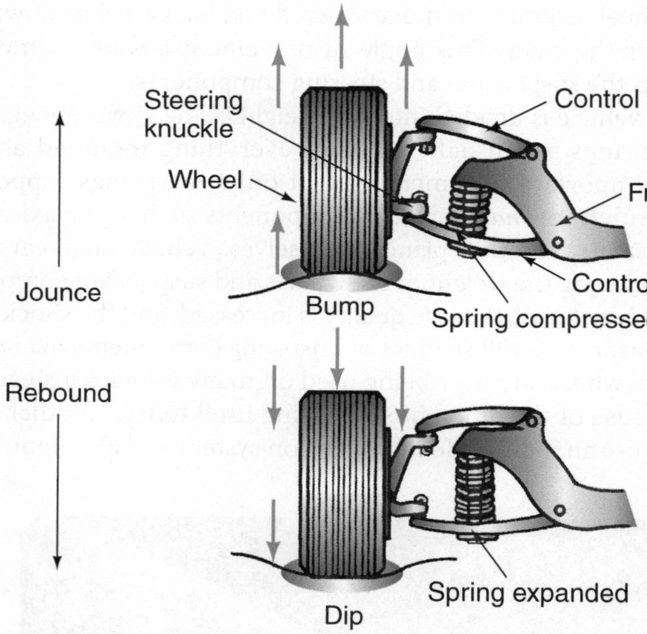

Figure 9-19 Most of the shock generated by the jounce and rebound of the wheel assembly is absorbed by the springs.

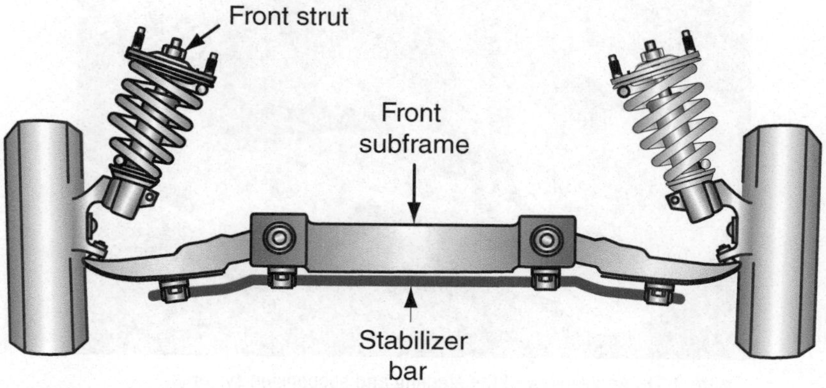

Figure 9-20 The sway or stabilizer bar is mounted at each end to the frame or body and helps control body roll.

outside of the turn, making the vehicle unstable. The springs and sway bar must be stiff or strong enough for the driver to retain control. Some vehicles have a form of the sway bar mounted at the rear of the vehicle.

The suspension system comprises all the factors involved so the driver experiences a smooth ride and controls the vehicle at the same time. Newer and more expensive suspension systems use electronics to control **air springs** for a smoother, more controllable vehicle.

Air springs are basically tough balloons that can be used to replace metal springs.

TYPES OF SUSPENSION SYSTEMS

Suspension systems can be generally classified as **dependent** or **independent.** Today most dependent systems are found on heavy trucks, primarily because of their rear-wheel drive and **payload** capacity. The most popular suspension for cars is a four-wheel, independent system whereas most light trucks use an independent system in the front and a dependent system on the rear axle.

Suspension systems are called either **dependent** or **independent** based on how much of the weight on one wheel is delivered in some manner to the other wheels.

Payload is the amount of weight a vehicle is designed to carry including the operator. The term "gross payload" includes the total weight of the vehicle, including all passengers and the payload.

A **coil spring** is a rod of steel shaped in a spiral.

A **torsion bar** acts like a coil spring in that it cushions road shock for a more comfortable ride.

Independent Suspensions

An independent suspension is designed to allow one wheel to follow road contours without affecting the operation of the other wheels (**Figure 9-21**). Independent systems allow for a smoother ride while providing good support for the vehicle. Older independent systems used a spring and control arm.

A **coil spring** is placed between one control arm and the frame to absorb road shock and hold the tire to the road surface (**Figure 9-22**). Some vehicles use a **torsion bar** in lieu of the spring (**Figure 9-23**). A shock absorber is attached to the frame and to the control arm. It helps control the movement of the spring.

Dependent Suspensions

Dependent suspension uses a solid or rigid connection between the two wheels on an axle. This is usually the rear-drive axle assembly. Many 4WD vehicles also have a rigid axle housing at the front. In a dependent system, any movement of one wheel is transmitted to the opposite wheel. In most cases, **leaf springs** are used in this suspension system to support the sprung weight of the vehicle. Some dependent systems use coil springs instead of leaf springs. Leaf springs are attached at one end to the frame and run over or under the axle (**Figure 9-24**). Shock absorbers are mounted between the axle and the frame.

The strength of a leaf spring can be changed by adding or removing leaves. A shock absorber cannot support weight because of its valve design.

Leaf Spring Strips of heavy metal slightly curved upward and clamped together one above the other.

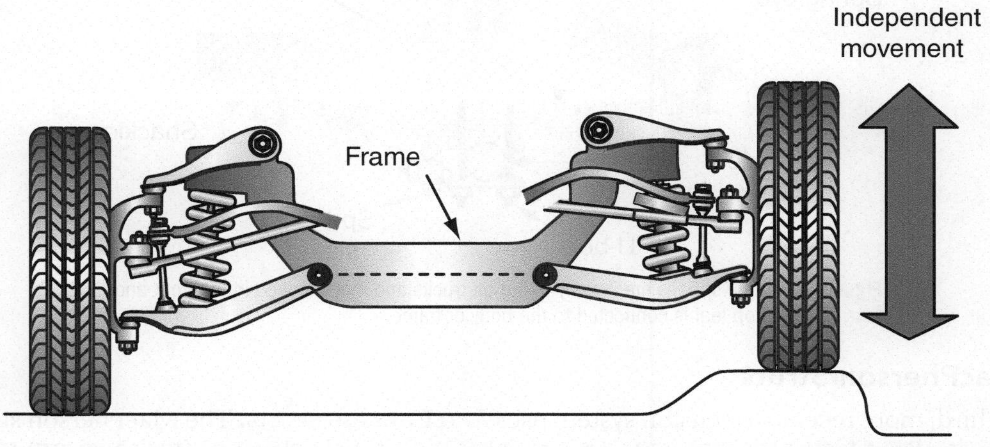

Figure 9-21 The control arms attach the steering knuckle to the vehicle and allow for the vertical movement of the wheel. One of the arms will support the spring and one end of the shock absorber.

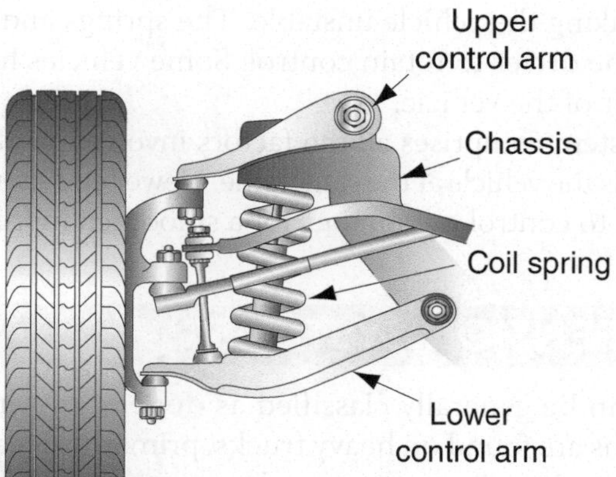

Figure 9-22 The coil springs support the weight of the vehicle and dampen road shock. In this illustration, the spring is retained between the lower control arm and chassis.

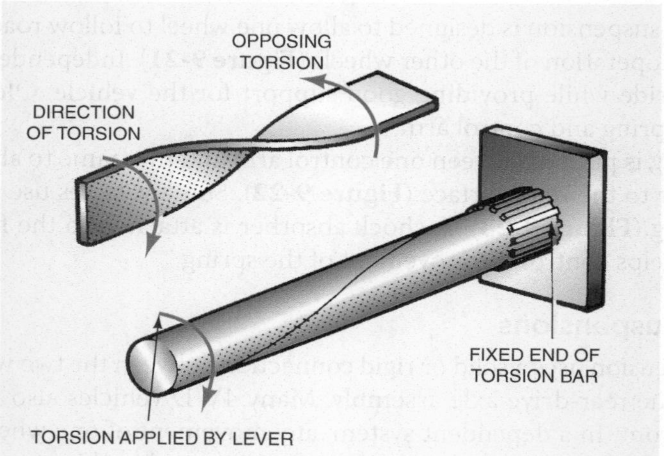

Figure 9-23 Twisting of the torsion bar acts like a coil spring.

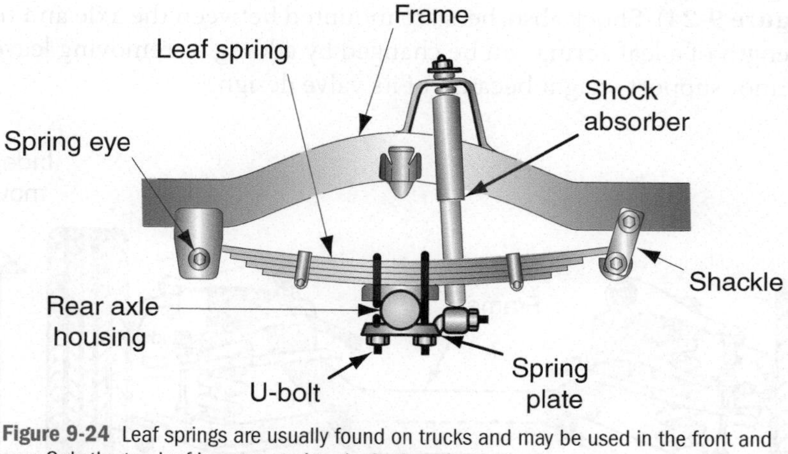

Figure 9-24 Leaf springs are usually found on trucks and may be used in the front and rear. Only the top leaf is connected to the body or frame.

MacPherson struts are lighter than a coil spring and control but are almost maintenance free and can be replaced much more easily.

MacPherson Struts

A third, more recent suspension system uses **MacPherson struts.** The MacPherson strut is basically a shock absorber with mounting brackets for a coil spring (**Figure 9-25**). The strut assembly also eliminates the upper control arm by using a rotating mount at the top of the strut. The MacPherson strut assembly provides better support for the vehicle,

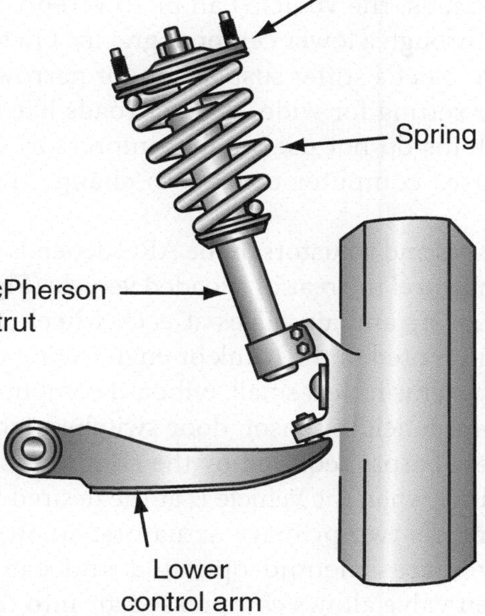

Figure 9-25 The strut performs the same function as a shock absorber but also provides mounting for the spring and eliminates the upper control arm.

controls the wheel assembly better, and reduces the weight of the vehicle. MacPherson struts can be found on all wheels but are most common on the front wheels.

An **automatic ride control (ARC)** system requires some events to occur in sequence, such as the ignition key turned off and the door opened, before the computer commands an action. Do not fault this system until the required events are performed in order.

Electronic Suspensions

In recent years, **automatic ride control (ARC)** has become popular on many cars and SUVs. There are several different names for a computer-controlled suspension system. Often it is joined with other computer control systems, such as roll stability and antilock brake systems. Most of these are independent suspension systems. Using sensors, an air compressor, air valves, and a computer, the strength of the individual springs can be adjusted. Air can be added to or removed from the air spring (**Figure 9-26**).

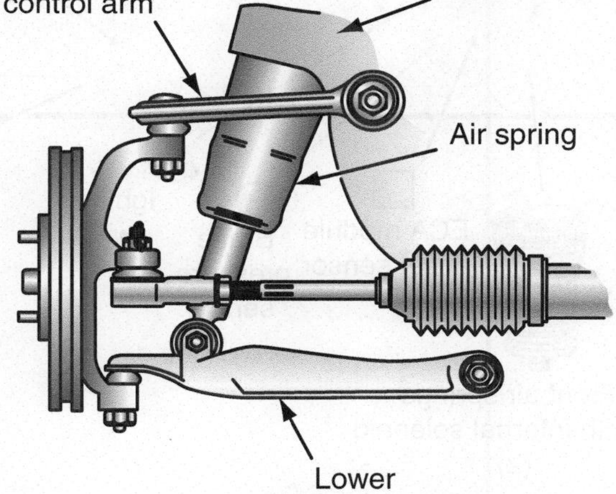

Figure 9-26 Air springs are heavy-duty balloon-type bags that can be inflated or deflated to raise or lower the vehicle.

The onboard suspension computer typically is programmed to raise the vehicle during low speeds and stops. The vehicle is raised at the halt, so the door can clear any road curbing. At cruising speeds, the vehicle can be lowered to reduce air resistance and provide better control through a lower center of gravity. On some automatic suspension systems, the driver can select a stiffer suspension for narrow, winding roads (for better vehicle feel) or a softer setting for wide, straight roads like the interstate road system. Some ride control systems do not use an air compressor; rather they use a magnetic shock absorber that uses computer control to change the valve setting and shock rebound rate.

The number of sensors and actuators in the ARC depends on the type of system. Some ride control systems only level the rear of a loaded vehicle (**Figure 9-27**). Fully automatic systems have height sensors and air valves at each wheel. There is also an electrically driven air compressor mounted in the trunk or engine compartment.

All the automatic systems have a small, onboard computer (**Figure 9-28**). It collects and analyzes data from the height sensor, door switches, ignition switch, brake switch, speed sensor, and other sensors required by the computer's programming. The height valve signals the computer when the vehicle is at the desired height.

Spring air valves are computer-controlled actuators to allow the air spring to inflate or deflate.

The computer controls two primary actuators: **spring air valves** and the compressor. The air valves are solenoid-operated and can only be open or closed (**Figure 9-29**). An open valve allows compressed air into the spring or allows air out. The compressor can only be switched on or off by the computer. If the compressor is operating and the air valves are open, the spring will inflate, thereby raising the vehicle. The air-controlled system can be switched on and off by a manual switch. Sometimes this switch is found in the trunk and must be turned off before lifting the vehicle in the air.

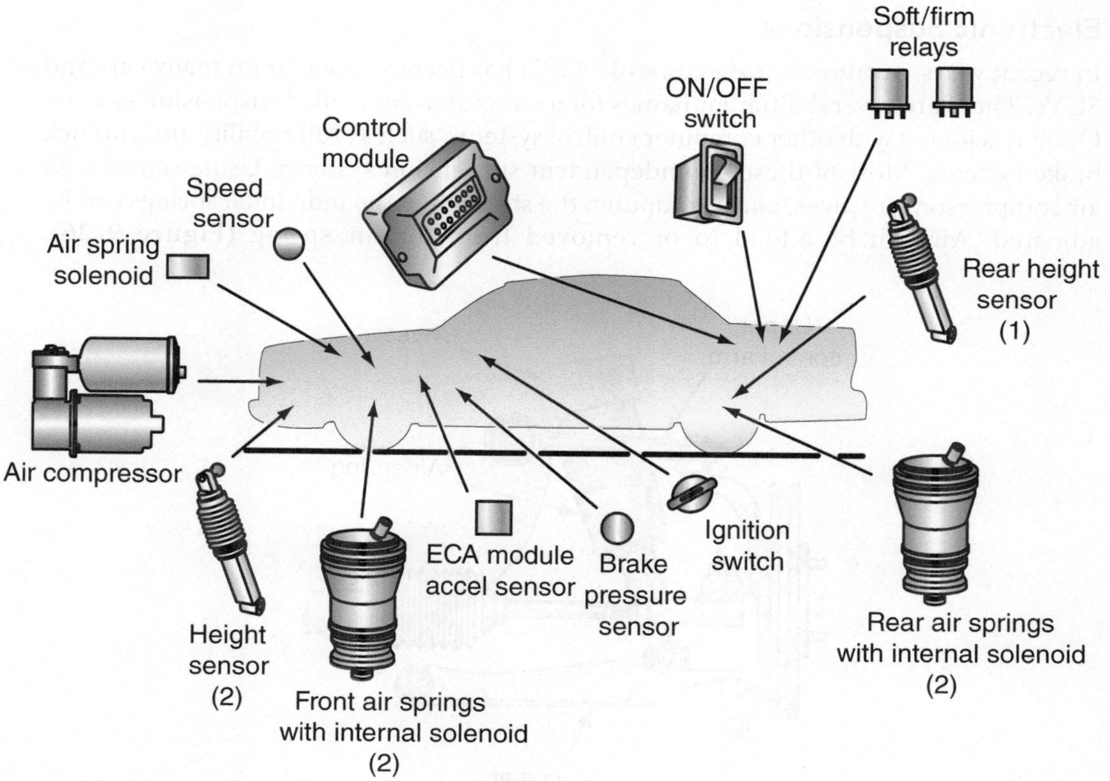

Figure 9-27 The illustration shows the layout of a fully automatic ride control (ARC).

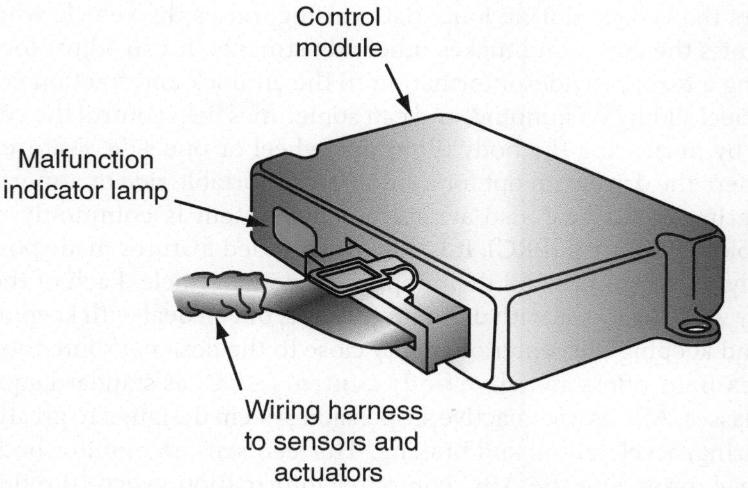

Figure 9-28 This computer is similar in appearance to other vehicle computers but only has the programming for ride control.

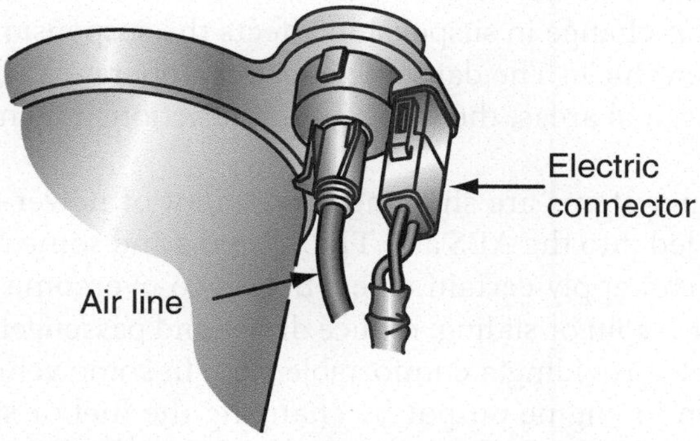

Figure 9-29 The air valve can be commanded by the ARC computer to allow air to enter the air spring or release air from the spring.

ELECTRONIC SUSPENSION SYSTEMS

There are several electronic suspension systems. Each deals with a particular type of suspension and its purpose. The earlier one came from an aftermarket idea for light trucks and so-called performance cars. This is the rear load leveler system. The idea came from adding air shock absorbers that can raise the rear of the truck when it was loaded. This assisted in steering control and ride. It was also used on some cars to gain more acceleration from one stop light to the next. In that case, it didn't work very well, plus the passengers took a beating on any road that wasn't smooth.

Automatic load leveling grew from this attempt to level the loaded vehicle. In place of the "air shocks," air springs were installed at the rear axle. When the vehicle was loaded, height sensors transmitted this information to a basic computer module. The module switched on the compressor and opened the air valves at the springs. When the rear of the vehicle reached its level height, the valves were closed, and the compressor shut down. When unloaded, the computer would open the air valves to deflate the air springs until the vehicle was level again. This system is available on many vehicles today but is not that common. It has largely been replaced by more versatile and better systems.

One of those newer systems is termed the ARC system, but *automatic ride control* is also the accepted term for many variations. The ARC provides leveling capability at each wheel, lowers the vehicle during long, flat cruising, raises the vehicle when the door is opened (to miss the curb), and makes other adjustments. It can adjust for the tilt of the vehicle during a turn, provides information to the antilock and traction control systems to control wheel sliding or spinning, and can sometimes help control the vehicle on high-speed turns by measuring the body tilt at one wheel or one side. Features such as ride selection, where the driver can opt for a smooth comfortable ride or a sport's ride with its better cornering ability, are also available. This system is commonly referred to as programmable ride control (PRC). It is ARC with added features made possible with the newest family of electronics and their applications in a vehicle. Each of these systems is used in many vehicle models with different features, but all deal with keeping the vehicle's body level and keeping the center of gravity close to the designer's intention.

Active body control (ABC) is a computer-controlled system that increases vehicle stability and reduces accidents.

Hertz is a measurement of how many times a function cycles in 1 second. In the 5-hertz instance, the suspension can change up to five times per second.

Mercedes-Benz offers an **active body control (ABC)** as standard equipment on its SL and CL classes. ABC is a semiactive suspension system designed to greatly reduce body roll in cornering, acceleration, and braking. Thirteen sensors monitor body movements and body level, providing the ABC computer information every 10 milliseconds. The computer commands four hydraulic servos mounted on four spring struts near each wheel. The strut is connected in parallel to a hydraulically controlled adjusting cylinder. The adjusting cylinder can be lengthened toward the strut, thereby changing the suspension length. This change in suspension affects the suspension and damping effect of road surface on the vehicle. The damping effect can operate at a frequency of up to 5 **hertz.** In other operational areas, the ABC system functions much like other automatic ride systems.

Active suspension systems are showing up on a lot of newer-model vehicles. This system has been wedded into the ABS and TRC systems and some steering systems. This "total" active system can apply certain wheel brakes to overcome cornering problems, prevent or reduce wheel spin or sliding, reduce driver and passenger fatigue, and increase driver control, all while providing a comfortable ride. In some vehicles, the systems can command a reduction in engine output by changing the fuel or spark control. *Today's Technician: Automotive Suspension and Steering,* Fifth Edition, and *Automotive Brake Systems,* Fifth Edition, texts provide in-depth studies of the active systems. Both texts are published by Delmar Cengage Learning.

SUSPENSION SYSTEM COMPONENTS

Shop Manual
pages 351–352

The components of the suspension undergo a great deal of stress during operation. They must keep the tires on the road and maintain vehicle stability during all road and driving conditions. The suspension components are designed for durability with a minimum amount of maintenance.

Coil Springs

Coil springs are commonly used on the front of passenger cars and light trucks. They have also been used on the rear of many vehicles, particularly on the smaller types. They come in various strengths according to the type of vehicle and its maximum payload. The spring is made of a high-tempered steel rod twisted into a coil. The diameter, overall length, and manufacturing treatment of the rod determine the strength or rate of the spring.

The wire in a **variable rate** spring may be tapered, truncated cone, double cone, or barrel-shaped.

Variable rate coil springs are made so the spring can support more weight per inch of compression (**Figure 9-30**). Some variable springs use the same-diameter wire and unequally spaced coils. Other types have a wire machined in different diameters. The smaller-diameter portion of the wire supports lighter loads. The vehicle tends to stay more level, even under moderate loads with variable-rate coil springs. This type of spring action provides a better ride at light payloads and better support as the vehicle is loaded.

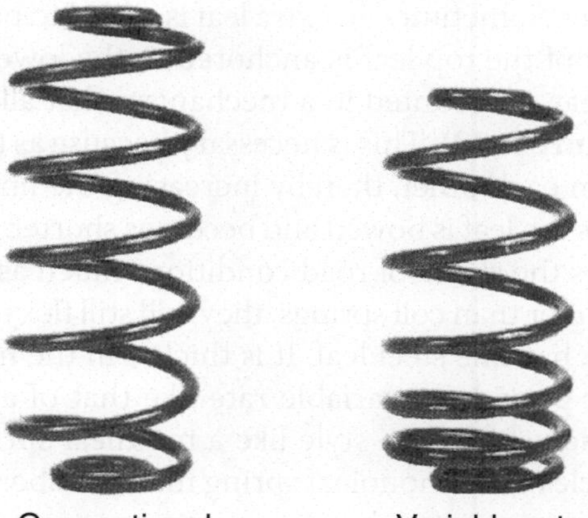

Figure 9-30 The variable-rate spring provides more vehicle control as the load becomes heavier.

Air springs have been used for many years on over-the-road trucks. The springs are large airbags that can be inflated or deflated to support different payloads (**Figure 9-31**). This ability to control the spring's capacity makes it possible to employ electronics to control a vehicle's ride. The bags are made of very tough, thick, flexible material like rubber. There is actually minor wear on the bags unless they are underinflated. The main points of wear occur at the top and bottom where the bags are mated to the vehicle.

Leaf Springs

Leaf springs are basically a set of long, curved, metal strips or leaves manufactured to bend at a certain rate. A leaf spring typically has three to five leaves. The leaves are placed on top of each other with the longest on top. Clamps are used to hold the stack of leaves in

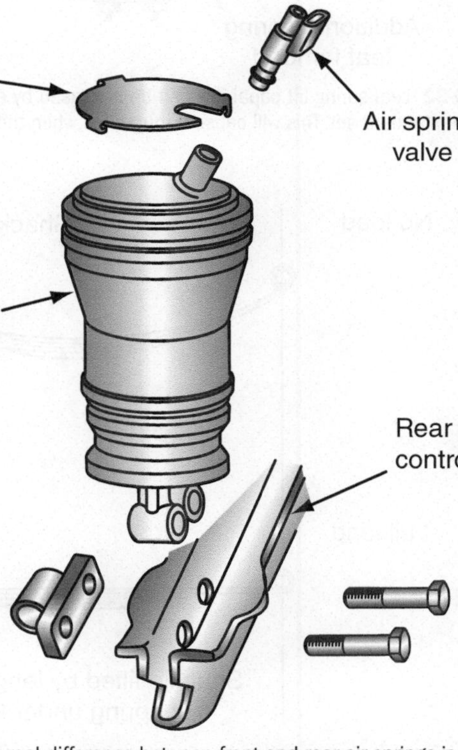

Figure 9-31 The only real difference between front and rear air springs is the method of mounting the spring to the control arm and body or frame.

alignment with each other. Sometimes an extra leaf is added to better support heavier loads (**Figure 9-32**). One end of the top leaf is anchored to the lower side of a frame rail. The other end, usually the rear, is mounted in a mechanism that allows the top leaf to extend or shorten in length (**Figure 9-33**). This is necessary because as the leaf flattens under load, its ends move away from each other, thereby increasing the linear length of the leaf. The opposite action occurs as the leaf is bowed and becomes shorter. The flattening and bowing of the leaf spring absorbs the shock of road conditions much as coil springs do. Although leaf springs are usually stiffer than coil springs, they will still flex until all energy is expended.

Monoleaf springs are usually not found on standard passenger cars and light trucks.

A **monoleaf spring** has one steel leaf. It is thicker in the middle and tapers down to each end. The tapering provides a variable rate like that of a variable coil spring. The spring may mount to the vehicle in a style like a multileaf spring or it may be mounted transversely. Some vehicles use a monoleaf spring made of fiberglass. Shock absorbers are used with leaf springs the same as coil spring systems.

Leaf and coil springs are designed and manufactured to support the expected weight of the vehicle and its payload. However, standard coil springs tend to be slightly less supportive of the vehicle during turns. Coil springs tend to offer a more comfortable ride. Standard leaf springs are harsher in their movements, resulting in a rougher ride for the

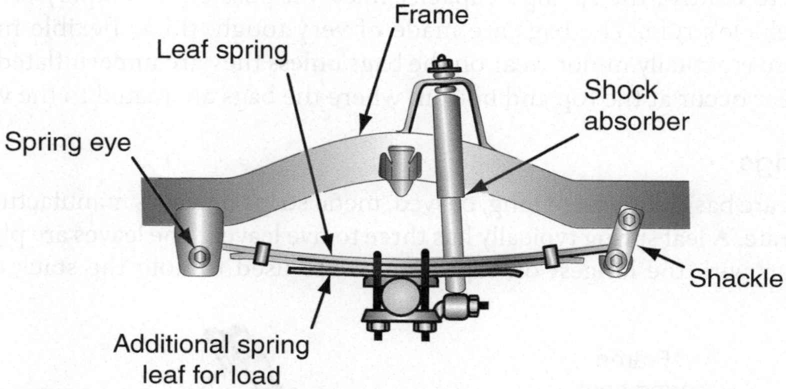

Figure 9-32 Leaf spring lift capability can be increased by adding an additional leaf under the standard set. This will cause a rough ride when the vehicle is not loaded.

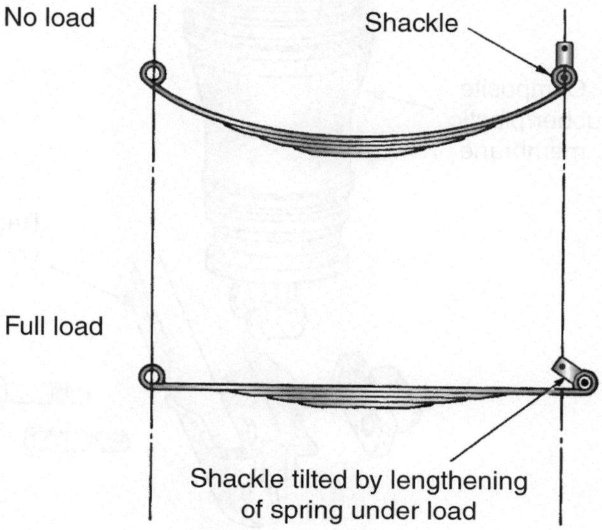

Figure 9-33 The rear shackle is pushed to the rear as the spring is flattened under load.

passengers and load but affording a little better support during a turn. Both types can be designed to provide a smoother or firmer ride. However, the typical spring is usually a compromise between the two requirements.

NASCAR and other racing vehicles use specially designed and manufactured coil springs on all four wheels. Each spring is designed for certain race conditions such as the type of track, condition of the track, and where the spring is mounted. Springs of this type are expensive and are impracticable for everyday use.

Torsion Bars

Torsion bars are long, round, steel bars that have been manufactured to twist and then return to their original condition. This type of suspension is very common on light trucks and SUVs. Torsion bars are anchored to the frame at one end and splined to the lower control arm at the other (**Figure 9-34**). As the control arm moves, the torsion bar is twisted and absorbs the shock of the movement. A typical torsion bar is usually stiffer

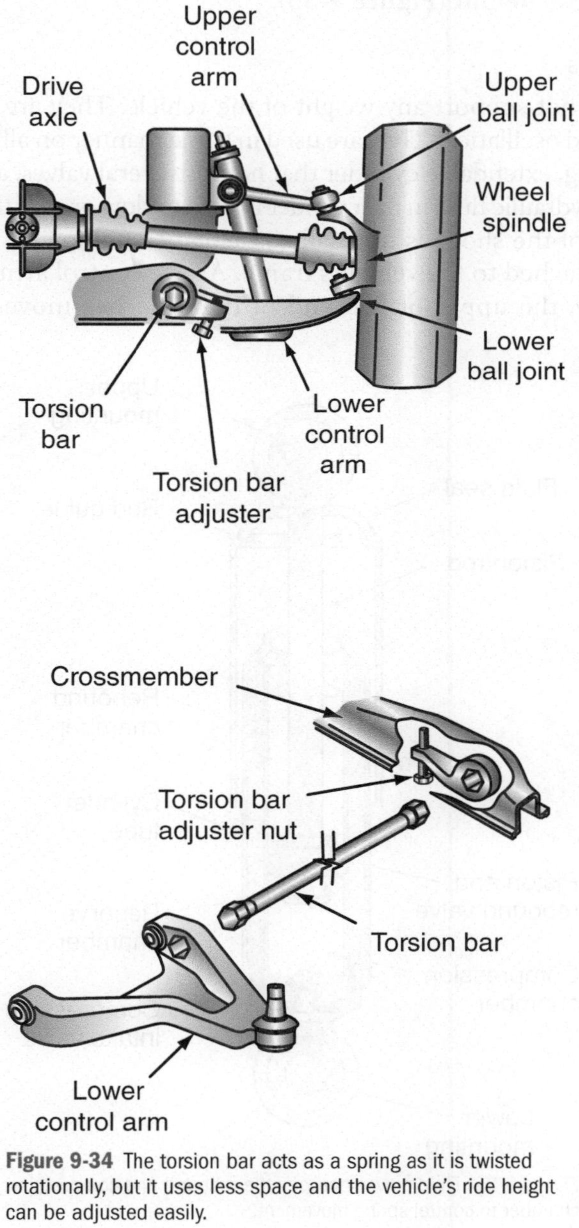

Figure 9-34 The torsion bar acts as a spring as it is twisted rotationally, but it uses less space and the vehicle's ride height can be adjusted easily.

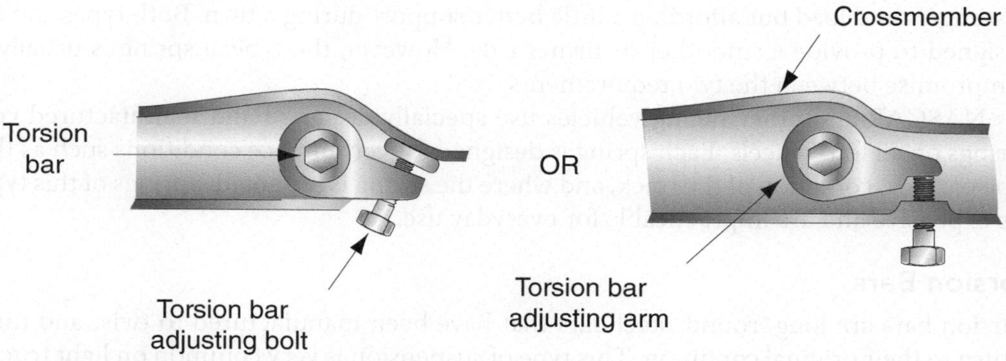

Figure 9-35 This is a torsion bar adjustment mechanism commonly found in light trucks and cars.

The **ride**, also known as **curb height**, is the distance an "at rest" vehicle sits above the ground. It is measured at specific points on the vehicle.

than a coil spring but less stiff than a leaf spring when used in similar conditions. Many vehicles with torsion bar suspensions have an adjustment capability to manually raise or lower the vehicle's **ride height** (**Figure 9-35**).

Shock Absorbers

Shock absorbers do not support any weight of the vehicle. They are designed to control spring movement and oscillation. They are used in some manner on all suspension systems. The absorber is a long, extendable cylinder that houses several valves, a piston, and a piston rod (**Figure 9-36**). Hydraulic fluid in the cylinder is used to slow or stop the piston movement.

The lower end of the shock is normally attached to the lower control arm or axle. The upper end is attached to the vehicle's frame. As the control arm or axle moves with the wheel assembly, the upper or rod end of the absorber moves in and out of the

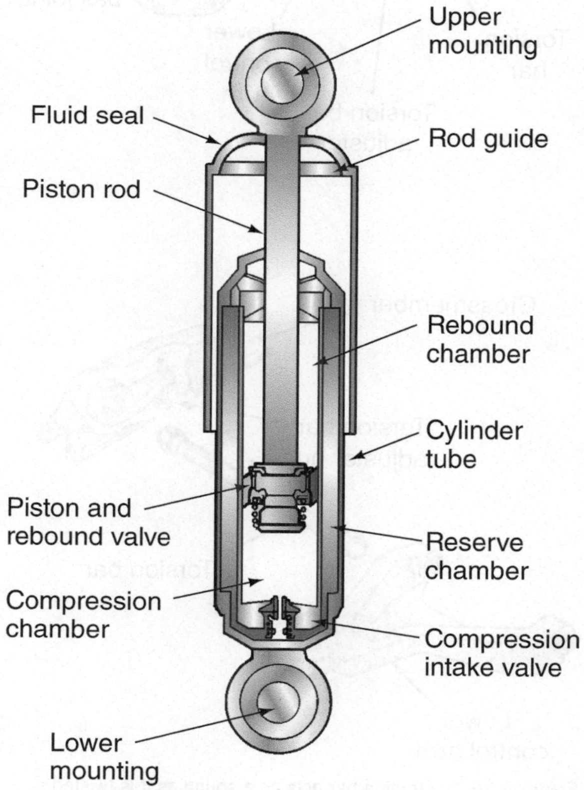

Figure 9-36 Fluid moves between the rebound and compression chamber to control spring movement.

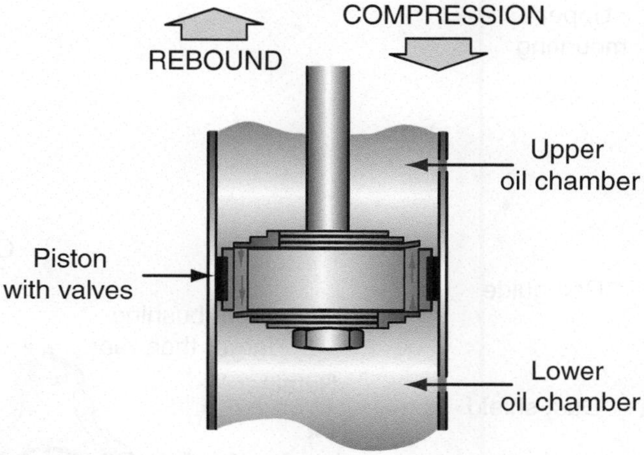

Figure 9-37 The metering valves in a shock absorber or strut control the flow of hydraulic fluid between the upper and lower chambers.

cylinder. Inside the housing at the lower end of the rod is a valve assembly mounted within a piston. There is hydraulic fluid on both sides of the piston. The fluid must move from one side of the piston to the opposite side as the piston/rod moves up and down. The piston valve assembly meters the fluid's rate of movement (**Figure 9-37**). The resistance of the fluid against the piston assembly absorbs the energy of the spring. This restricts the oscillations of the spring and controls the up-and-down motion of the vehicle.

A problem with earlier shock absorbers was the delayed initial action of the absorber. The cylinder could not be filled with fluid. The fluid did not restrict piston travel until it was pressurized and forced to move through the valves. Newer shocks have pressurized nitrogen gas pumped into the cylinder after the fluid has been installed (**Figure 9-38**). This eliminates the initial lag, and the restricting action of the fluid is instantaneous upon any spring movement.

Control Arms

Control arms are the connection between the vehicle's frame and the steering knuckle. Control arms make it possible to use independent suspension systems. They may be located at each wheel or just on the two front wheels. On the rear of some vehicles, *trailing* or *radius arms* connect the wheel assembly to the frame. This type of arm will not keep the tire's tread flat on the ground. Other suspension components must be used with trailing or radius arms to properly support and guide the wheels.

The control arms are mounted with rubber bushings at the frame and extend outward from the frame to the steering knuckle (**Figure 9-39**). The inner bushings allow the control arms to pivot up and down with the movement of the wheel. The SLA system has the shorter arm at the top. As the wheel moves upward over a hump, the shorter arm draws the top of the wheel inward. The same action happens when the wheel drops into a hole. This arc movement keeps the tire tread flat on the road. The two arms are designed to complement each other as the wheel moves over various road conditions.

If a coil spring is used, it is mounted between the frame and one of the control arms. This arm is the load-bearing arm and may be either upper or lower. At the ends of the control arms are ball joints that attach to the steering knuckle. The ball joints provide a pivot point for the steering knuckle to rotate during steering and while moving the wheel assembly over the road. Control arms used at the rear of the vehicle perform basically the same as front arms, although a steering operation is not usually present on the rear wheels.

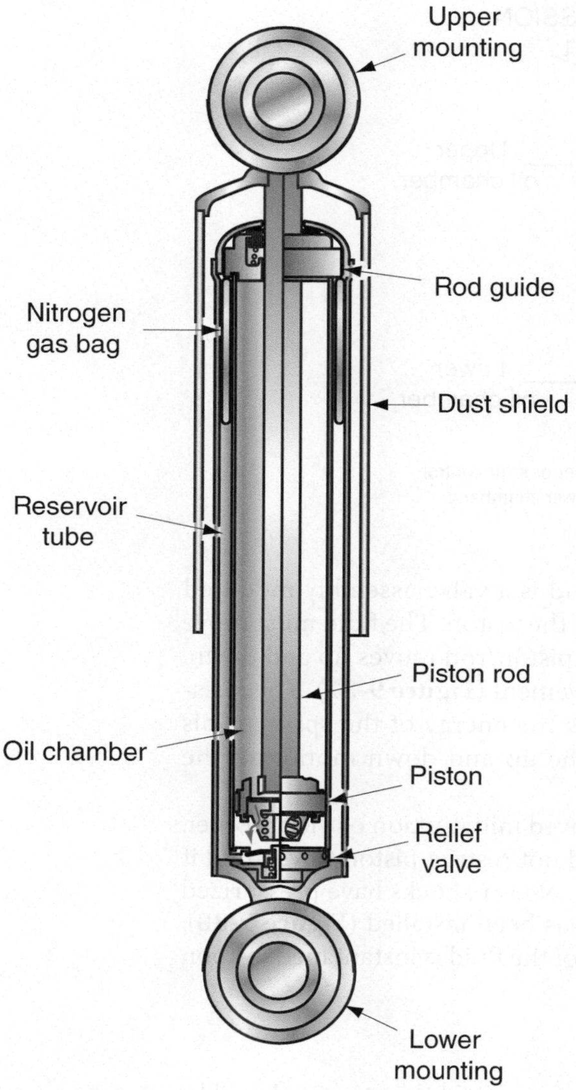

Figure 9-38 Older shocks with unpressurized oil had a lag time before they would damper shocks. Adding pressurized nitrogen gas to the top chamber pressurizes the hydraulic fluid, causing instant reaction to road shocks.

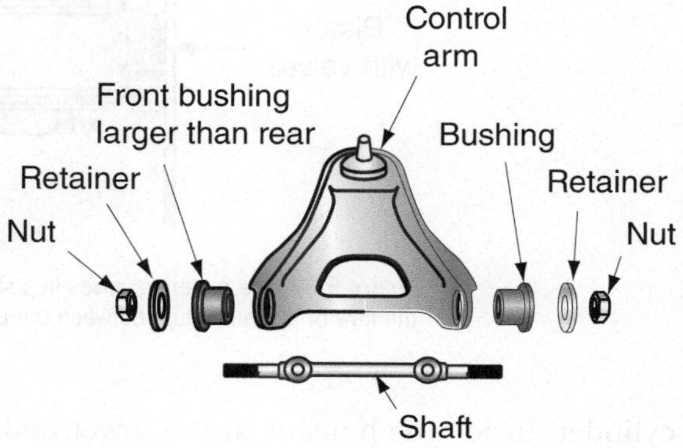

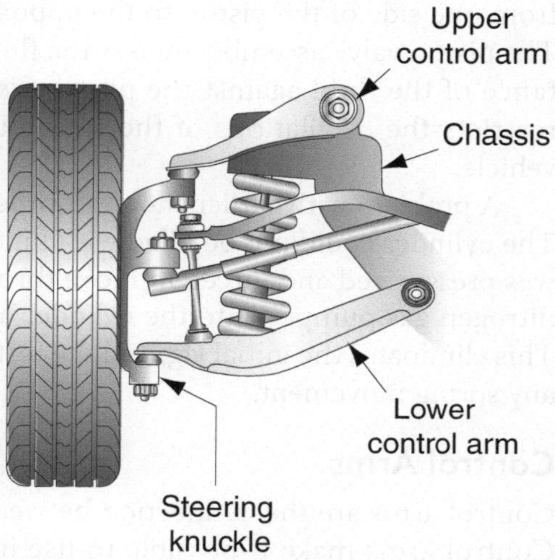

Figure 9-39 Control arm bushings lessen the shock transmitted from the control arm to the body or frame.

Struts

The MacPherson strut assembly houses a shock absorber internally and a coil spring mounted on the exterior (**Figure 9-40**). The coil spring mounted on the strut is not considered part of the strut assembly. The shock absorber action of the strut is very similar to that of a regular shock absorber. The primary advantage of the strut is its use as the upper control arm and weight reduction.

AUTHOR'S NOTE When a strut is used on the rear wheels of an independent suspension, the strut may be referred to as a Chapman strut. It is named for Colvin Chapman, a well-known engineer for Lotus. The first application of the Chapman strut was on the Lotus 12.

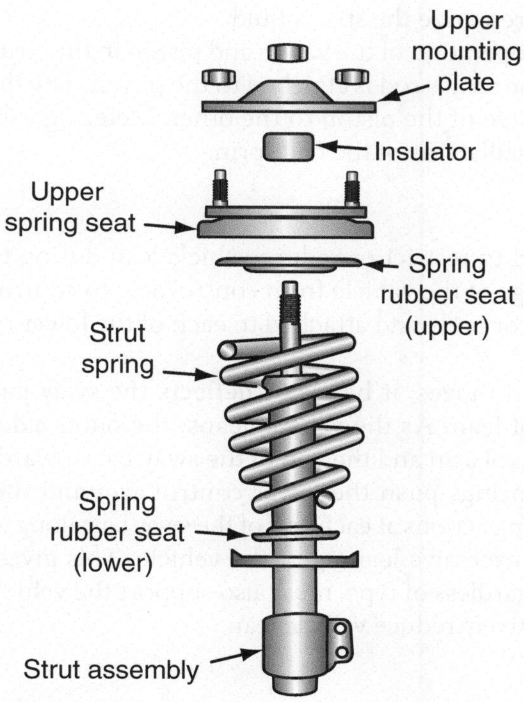

Figure 9-40 The strut acts like a shock absorber, provides mountings for the coil spring, and eliminates the upper control arm.

Because it is mounted to the frame or body with a rotatable bearing at the top, the entire strut turns with the steering knuckle (**Figure 9-41**). The strut and spring are kept in alignment with the knuckle in this manner. The road shocks and steering stresses are more easily absorbed when all components are kept in line. A smaller and lighter-weight

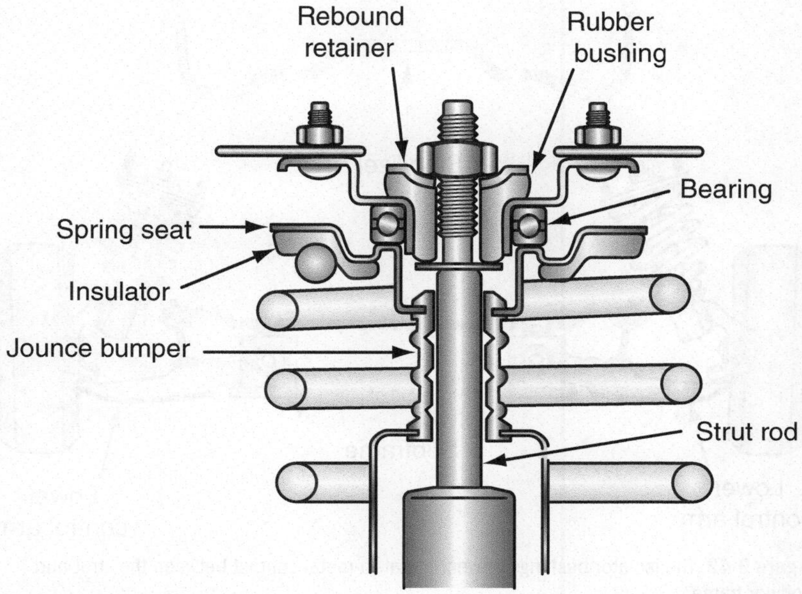

Figure 9-41 The strut's upper mounting mechanism allows the strut to rotate with rubber cushions to seat the coil spring and reduce shock.

lower control arm can also be used. In addition to the suspension advantage, better steering control can be maintained for the same reasons. Pressurized nitrogen gas is also added to the strut to pressurize the strut's fluid.

The assembly and operation of the valves and piston in the strut is very similar to that of a shock absorber. The strut's rod is attached to the piston. The fluid within the cylinder must move from one side of the piston to the other. Metering valves control the rate of flow and restrict the oscillations of the coil spring.

Sway Bars

Sway bars are designed to restrict or reduce vehicle lean during turns. The bars extend across the undercarriage of the vehicle from control arm to control arm. The sway bar is anchored to the frame or body and attached to each of the lower control arms with links (**Figure 9-42**).

As the control arm moves, it bends or deflects the sway bar. The bar resists and reduces the amount of lean. As the vehicle leans, the outer side of the vehicle drops, pushing the lower control arm and the end of the sway bar upward. The inner side of the vehicle lifts and the springs push the lower control arm and the end of the sway bar downward. The opposing actions at each end of the sway bar, along with the manufacturing process, help prevent excessive leaning of the vehicle. This gives better control to the driver. The springs, regardless of type, must also support the vehicle as it leans. The sway bar alone cannot effectively reduce vehicle lean.

Ball Joints

The ball joint is the connection between a control arm and the steering knuckle. It has a stud that is ball-shaped on one end and is threaded from near the ball to the end of the stud. The ball fits into a molded socket (**Figure 9-43**). The threaded end fits through a hole in the steering knuckle and is secured with a nut or through bolt. Some ball joints are fastened to the steering knuckle with the threaded end through the control arm. The ball can rotate within limits to allow vertical and rotational movement of the steering knuckle assembly. Each control arm has a ball joint. The ball joint is one of the few suspension items needing routine maintenance.

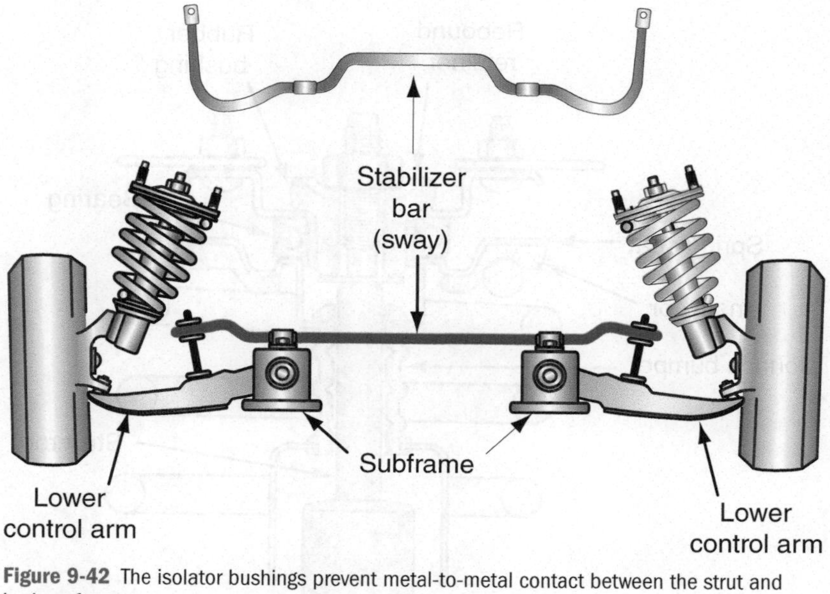

Figure 9-42 The isolator bushings prevent metal-to-metal contact between the strut and body or frame.

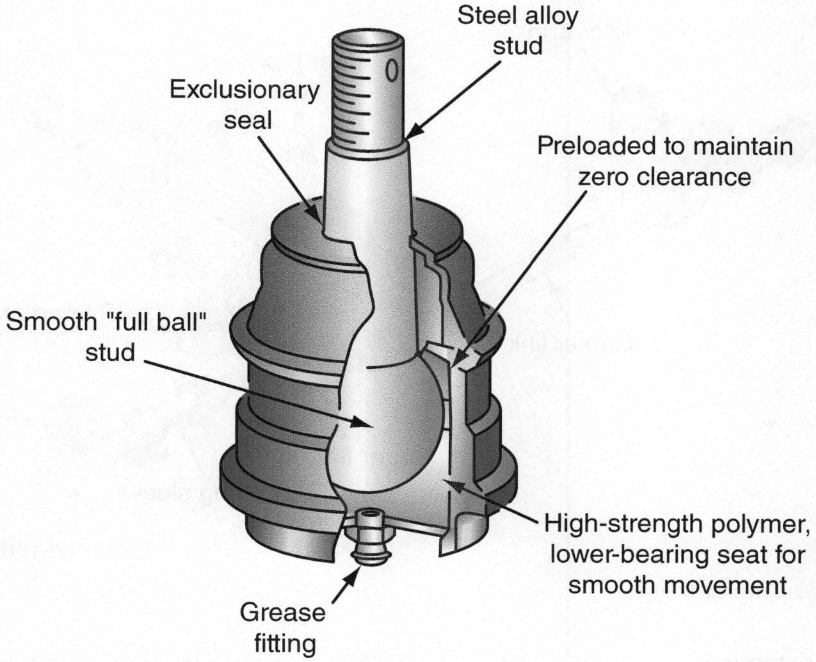

Figure 9-43 The ball joints allow the wheel assembly to vertically move up and down due to the road surface and allow the wheels to be steered horizontally.

TYPES AND COMPONENTS OF STEERING SYSTEMS

The steering system provides the driver with the means to steer the vehicle. There are two types of steering systems: **parallelogram** and **rack and pinion.** Power boost systems are available on both.

Shop Manual
pages 362–368

Parallelogram Used on car and trucks because all the pivot points are paralleled. the steering linkage consists of pitman arm, idler arm, tie-rod ends and a center link.

Rack and Pinion A compact steering system that uses a gearwheel (pinion) and a flat tooth bar (rack) that is connected to the steering wheel and tires.

Parallelogram System and Components

The parallelogram steering system has been in use for many years; its use on light vehicles and trucks is limited on modern vehicles. The linkage in this system is designed to remain parallel to the control arms during all steering movements (**Figure 9-44**). There are several ball-joint-type devices connecting different parts of the linkage. The driver is connected to the steering linkage and gear through the steering wheel, steering shaft, and steering gearbox. The steering wheel and most of the shaft is mounted within the passenger compartment, whereas the linkage and gearbox is located in the engine compartment or under the vehicle.

The steering wheel allows the driver to move the linkage to steer the vehicle. Steering wheels on manual steering systems are larger in diameter than the ones used on a power-steering system. This gives the driver some mechanical advantage during low-speed steering. The horn control button and airbag are located within the steering wheel. Some newer vehicles have switches for cruise control and other accessories on the steering wheel. The steering column supports and houses the steering wheel and shaft along with any switches and wiring mounted on or in the column (**Figure 9-45**). The steering shaft extends through the column to the steering gearbox. There are a few variations of this system that use more or less joints.

Recirculating ball and nut Also known as a worm and sector or steering gear is a steering mechanism commonly found in older vehicles.

The most common type of steering gearbox in the parallelogram system is the **recirculating ball and nut.** Large steel balls are placed in grooves on the nut and

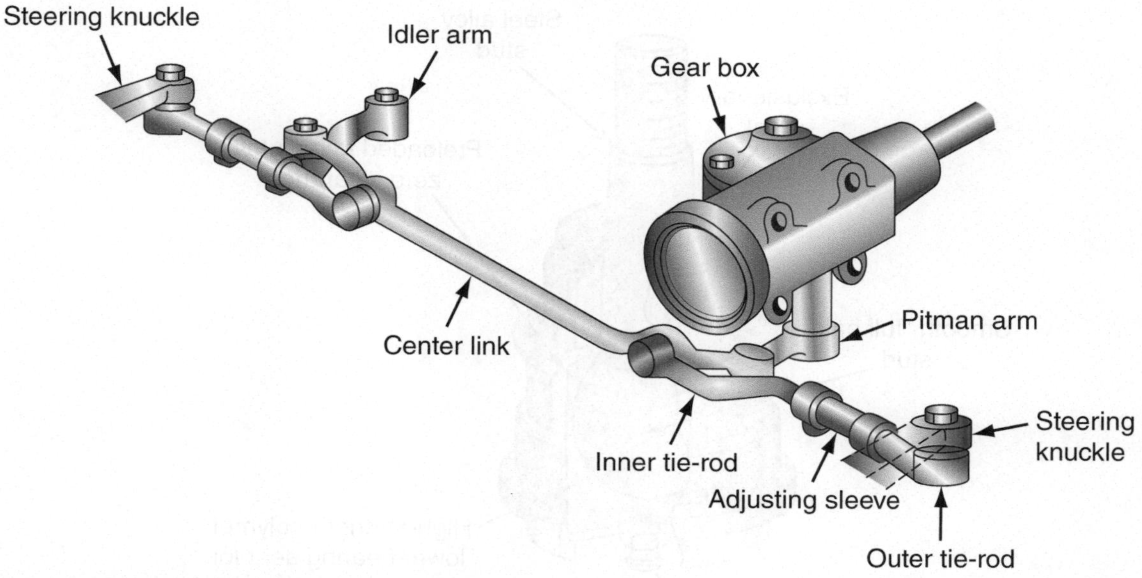

Figure 9-44 Parallelogram steering linkage.

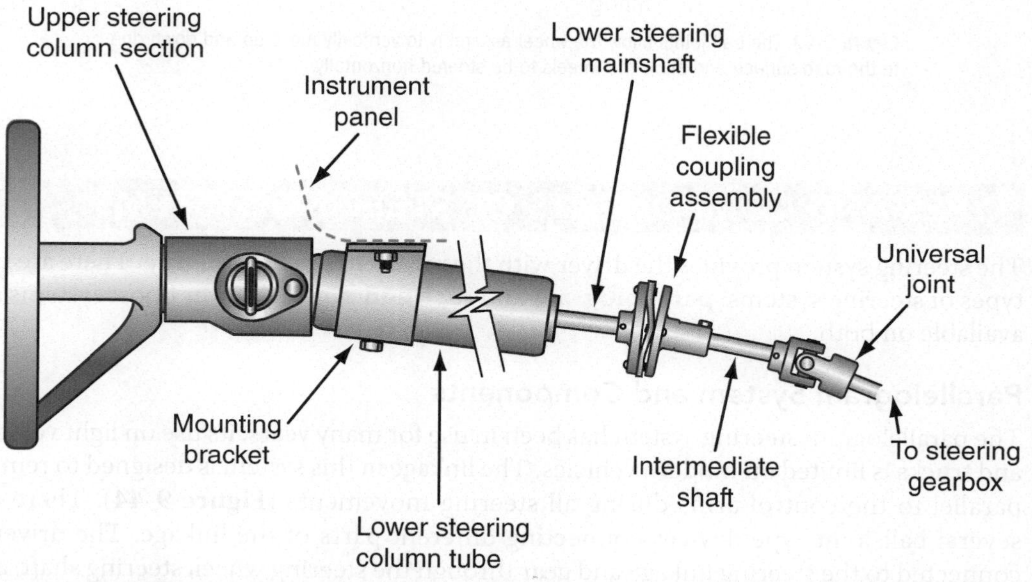

Figure 9-45 The steering shaft is inside the steering column and connects the steering wheel to the steering gear.

A **sector shaft** will only rotate within a limited arc. On steering gear boxes, this arc is approximately 60 degrees.

The **pitman arm** changes the rotational movement of the sector shaft to a linear motion for the steering linkage.

In some vehicles, the **center link** is known as the "drag link."

An **idler arm** only holds up one end of the center link during rest and steering movement.

The **tie-rod** is the last link between the steering wheel and the steering knuckle.

worm gear (**Figure 9-46**). The steering shaft rotates the worm gear. As the worm gear turns, it uses the balls as a threading device to move the nut up and down the length of the worm gear. Teeth on the nut mesh with teeth on a **sector shaft.** The lower end of the sector shaft is splined to a **pitman arm.** The gear arrangement multiplies the force delivered by the steering shaft, thereby reducing driver effort.

The pitman arm is attached to a **center link.** An **idler arm** at the opposite end of the center link from the pitman acts as a rotating anchor for the link (**Figure 9-47**). Sockets like the ball joint attach the center link to the idler arm and two tie-rods. A threaded, split, pipe-like adjusting sleeve is used to connect a tie-rod end to each of the tie-rods. The **tie-rod** ends are attached to the steering knuckle with sockets. Any rotary movement of the worm gear results in a linear movement of the pitman arm, center link, tie-rods, and ends to turn the steering knuckle and the wheel assembly from side to side. The adjusting sleeve is used in adjusting the wheel during alignment. It is clamped to the tie-rod and tie-rod end (**Figure 9-48**).

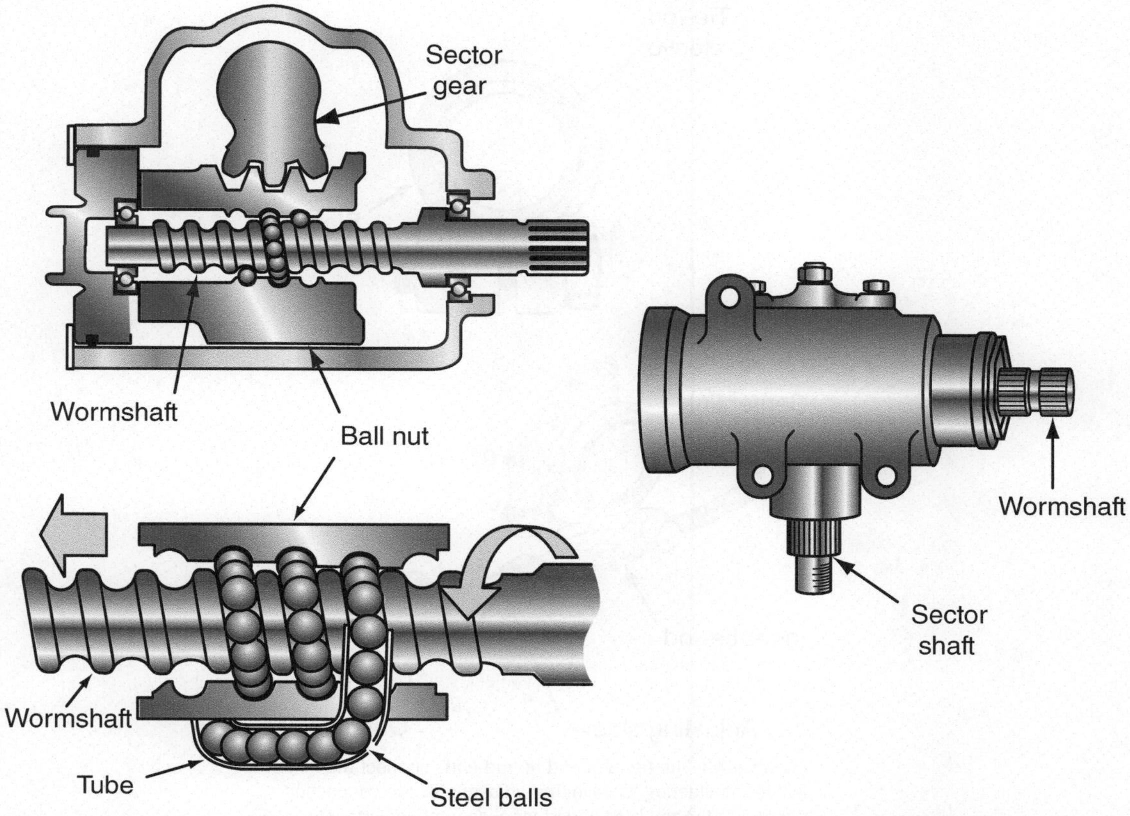

Figure 9-46 The steel balls located inside the ball nut act like the internal threads on a nut. They follow matching grooves in the worm gear and move the sector shaft in and out of the ball nut.

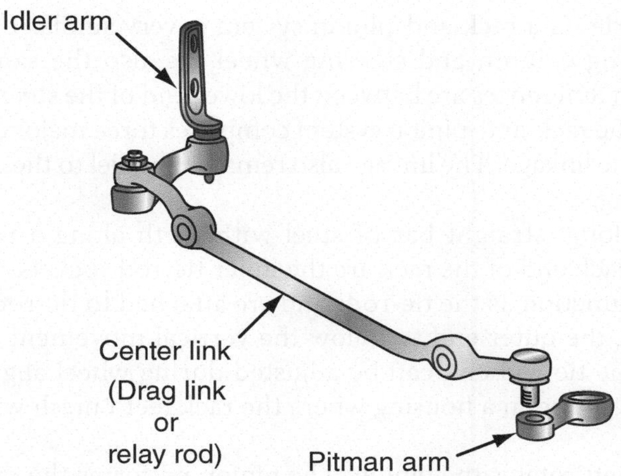

Figure 9-47 The idler arm only supports one end of the center link. Note that the center link may be referred to as a drag link or relay rod.

The steering knuckle is attached to the steering linkage and the suspension system. Attachment points at the top and bottom are used to attach the knuckle to the suspension. Another attachment point is used to connect the steering linkage. The steering knuckle may also have mounting holes or brackets for brake hardware. A spindle is a forged portion of a RWD steering knuckle. The spindle provides a machined surface (journal) for the wheel bearing, which in turn supports the wheel and tire assembly. On FWD vehicles, the steering knuckle has a machined bore to accept a bearing for the outer end of a drive axle.

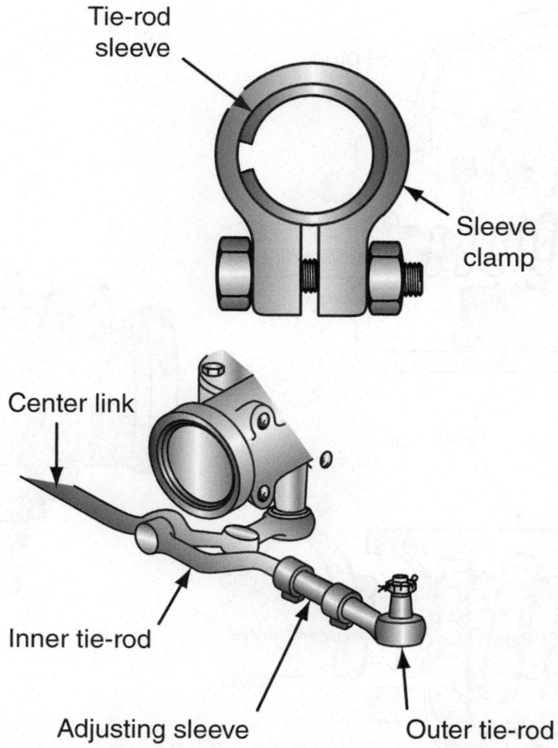

Figure 9-48 The tie-rods and tie-rod ends connect the center link to the steering knuckle. The adjusting sleeve connecting the end to the rod is threaded for alignment adjustments.

Rack-and-Pinion Systems and Components

The steering knuckle on a rack-and-pinion system is very similar to the parallelogram system. The steering column and steering wheel are also the same for all practical purposes. The main differences are between the lower end of the steering column and the steering knuckle. The rack-and-pinion system comprises three major components: a rack, a pinion gear, and the linkage. The linkage also remains parallel to the control arms during all phases of steering.

The rack is a long, straight bar of steel with teeth along a portion of one side (**Figure 9-49**). At each end of the rack are the inner tie-rod sockets. The tie-rod sockets perform the same function as the tie-rods and are attached to tie-rod ends. The sockets can pivot, allowing the outer end to follow the vertical movement of the knuckle and wheel assembly. The tie-rod end can be adjusted during wheel alignment procedures. The rack fits and is sealed in a housing where the rack teeth mesh with the pinion teeth (**Figure 9-50**).

The steering shaft rotates the pinion. The pinion performs the same function as the worm gear. The pinion moves the rack left and right as the steering wheel is turned. The rack and pinion occupies less space, is lighter, has fewer parts, and generally gives the driver better road feel. The rack-and-pinion system has very little, if any, torque multiplication. Steering even a small vehicle in traffic with manual rack and pinion can be tiring.

Power Steering

Both types of steering systems can be equipped with power boost. The boost systems for each are very similar. A pump supplies pressurized fluid to increase the driver's steering force. The pump is equipped with a pressure relief valve to prevent overpressurizing the

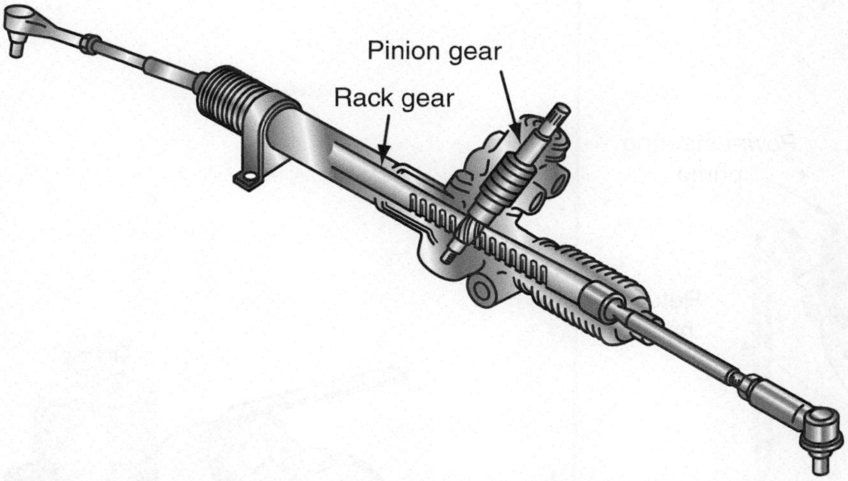

Figure 9-49 The rack-and-pinion gear are mounted and sealed inside the housing. Normally two rubber-padded clamps fasten the housing to the body or frame.

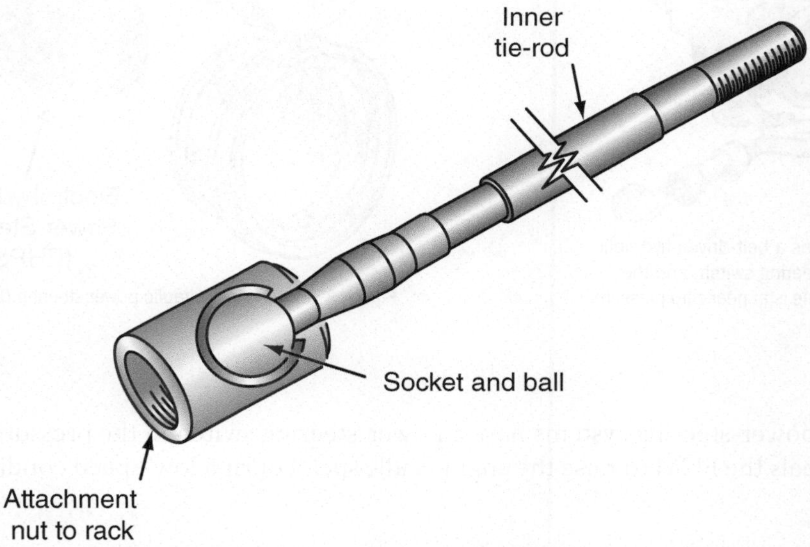

Figure 9-50 The inner tie-rod is fitted with a ball-socket section that is threaded onto one end of the rack. The outer end of the tie-rod is threaded to accept the tie-rod end.

system. Automatic transmission fluid is used many times as a power-steering fluid on older systems. Many newer models use specially blended power-steering fluid. The service manual should be consulted before topping off power-steering fluid.

A crankshaft belt-driven pump provides the pressurized fluid to the system. The fluid is delivered to the power-steering valves through steel lines and flexible hoses (**Figure 9-51**). The power-steering valve in a recirculating ball and nut gear is located at the top of the steering gear. The steering shaft rotates the valve to open a passageway to the correct side of the steering gear piston. Pressure applied to the piston assists the driver in turning the worm gear.

Pressurized fluid on a rack and pinion is delivered to the valves located at the top of the pinion gear. As the shaft and pinion are rotated, a valve opens and directs fluid to the correct side of a piston mounted to or as part of the rack (**Figure 9-52**). In this system, the power boost is applied directly to the steering linkage. In the ball-and-nut system, the boost is delivered to the steering gears.

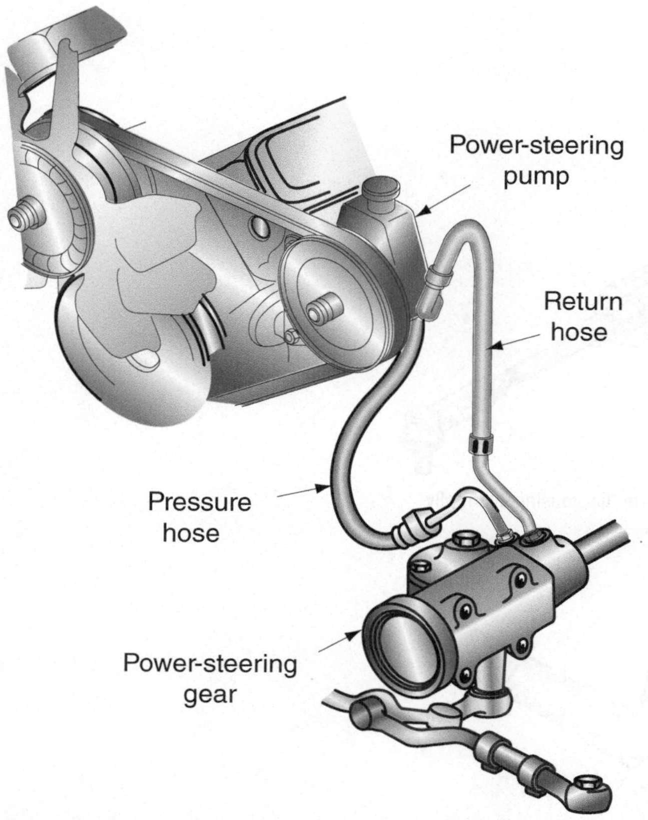

Figure 9-51 A power-steering system usually has a belt-driven hydraulic pump, a pressure line, a return line, a power-steering switch, and the necessary mounting brackets. Fluid in the return line is at near-zero pressure.

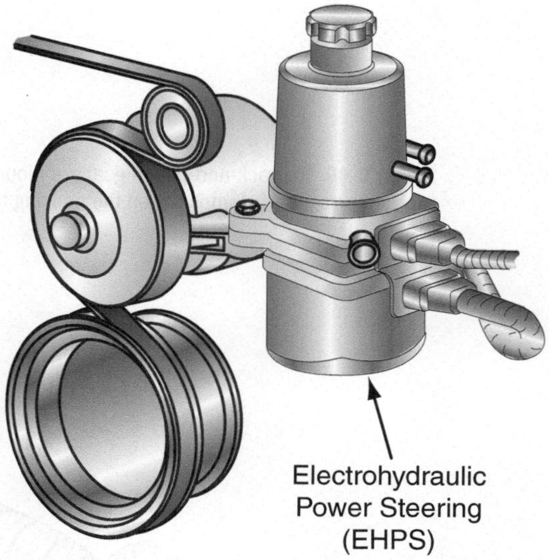

Figure 9-52 Electro-hydraulic power steering (EHPS) module.

Most power-steering systems have a power-steering switch in the pressure line. The switch signals the PCM to raise the engine's idle speed during low-speed conditions such as parking.

Electric Power Steering

There are two versions of electric power steering (EPS): electro-hydraulic and full electric systems. The electro-hydraulic system merely replaces the belt-driven pump with an electric pump (**Figure 9-53**). The pump can be switched on or off based on steering load. During parking or other slow-speed maneuvering, the electric motor will be on and full hydraulic pressure will be available. During cruise conditions, the motor may be switched off until pressure is needed. The determination of when to switch on or off is signaled by the power-steering pressure sensor and a steering angle sensor measuring steering shaft rotation speed and amount of rotation. Other factors on some vehicles may include vehicle speed, driver effort, roll angle, and throttle position. The system can be tied into the PCM to allow for speed and body angle changes.

The full electric systems like the one found on many hybrid electric vehicles and other vehicles use electronic sensors in the steering and suspension, a control module, and electric motors (**Figure 9-54**). Sensors mounted on the steering shaft measure the rotation motion and speed of the shaft plus the torque being applied to the shaft. Based on this data, the computer commands electric motors to provide steering assistance. The current to the motors is controlled, determining how much torque and speed is applied to the steering linkage.

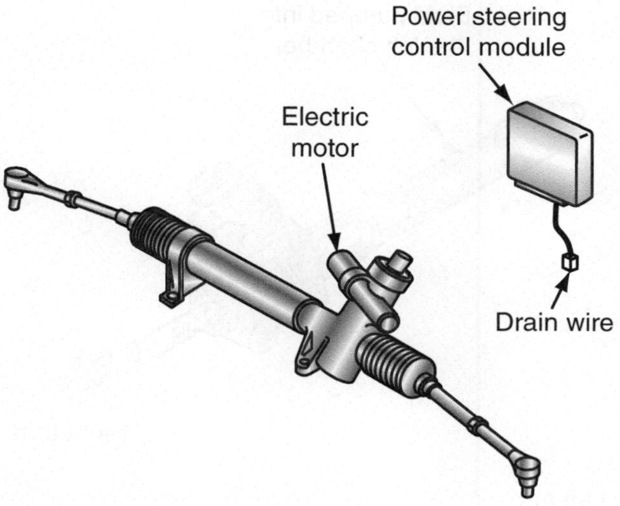

Figure 9-53 Electronic power steering.

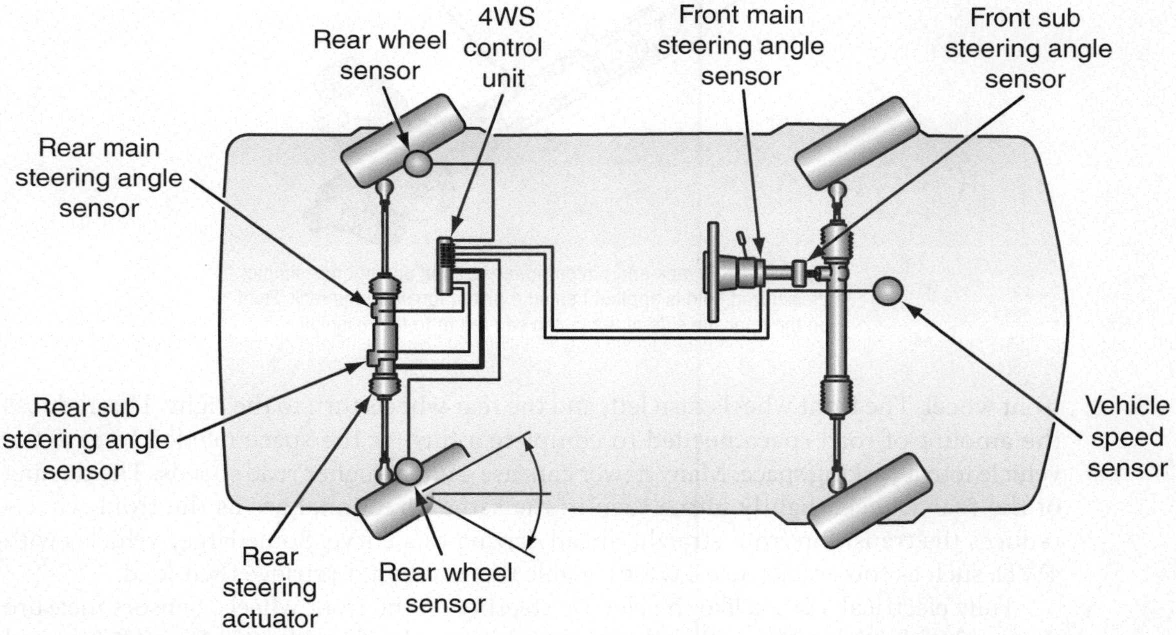

Figure 9-54 The 4WS control unit analyzes input sensor information, calculates the required rear steering angle, and operates the rear steering actuator motor to provide the proper rear steering angle.

Four-Wheel Steering (4WS)

There are several versions of four-wheel steering (4WS) systems (**Figure 9-55**). Examples of some common 4WS systems can be found on the Honda Prelude, Acura NSX, Mitsubishi 3000GT, and Chevrolet/GMC trucks and SUVs. The system allows for the normally non-steering wheels, usually the rear, to be turned so the rear of the vehicle swings in an arc similar to that of the front of the vehicle. Typical of all systems is the reduced turning arc of the rear wheels. They cannot be turned as far as the front wheels.

Earlier Honda systems used a long worm shaft extending from the front steering gear to the rear steering unit mounted online with the rear wheels. This worm shaft had a different gear ratio to reduce the degree of rear wheel turn with regard to the front wheel turning arc. At slow speeds, the rear wheels are turned in the direction opposite to the

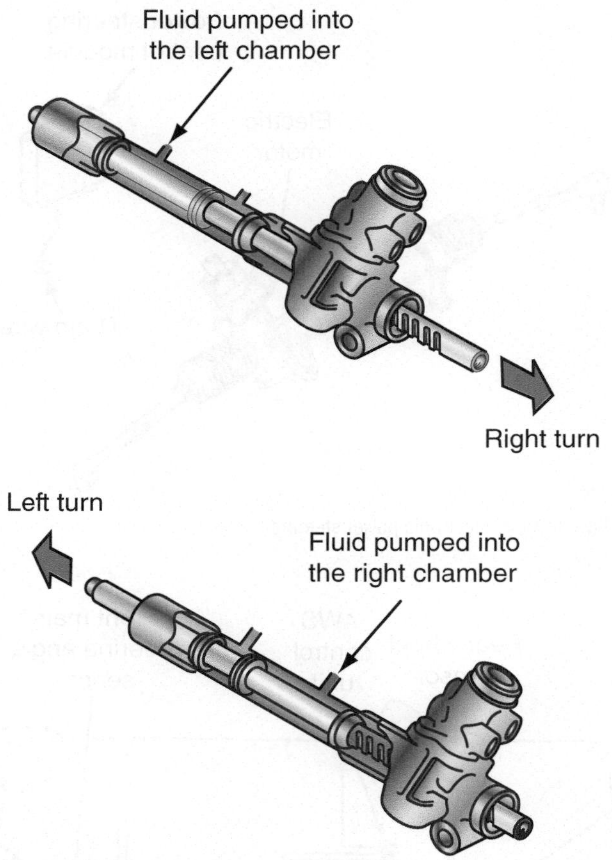

Figure 9-55 A rack-and-pinion power-steering operation is simple: Pressurized fluid is applied behind a piston forged to the rack. Fluid on the opposite side of the piston can return to the reservoir.

front wheel. The front wheels turn left, and the rear wheels turn to the right. This reduces the amount of road space needed to complete a turn or the space needed to ease the vehicle into a parking space. Many newer cars use 4WS at higher road speeds. The turning of the rear wheels slightly into a turn in the same directional arc as the front wheels reduces the transition from straight-ahead driving to a curve. Some larger vehicles with 4WD, such as tow trucks, use 4WS to enable them to better retrieve their load.

Fully electrical 4WS is like the electric steering on the front wheels. Sensors measure the steering shaft's rotation and applied torque. Information from the PCM on vehicle speed determines if 4WS is activated or not. If activated, usually at low speeds, the 4WS computer processes the sensor data and commands the electric motor(s) attached to the rear wheel steering linkage to move the linkage in the correct direction, the correct amount, and the degree of turn. Sensors in the steering motor(s) report back to the controlling module on the degree of turn achieved and where the steering linkage is in relation to the centerline of the vehicle. In this manner, the computer compares the accomplished actions with the issued commands and the signals from the front steering to determine any further actions.

Shop Manual
pages 362–368

WHEEL ALIGNMENT

Shop Manual
pages 381–384

A vehicle with perfect alignment of the four wheels will steer straight ahead and return to center after a turn. Wheels that are not in alignment with the vehicle and other tires affect the driver's control. Alignment of the wheels places all wheels parallel to each other and to the center and thrust lines of the vehicle (**Figure 9-56**).

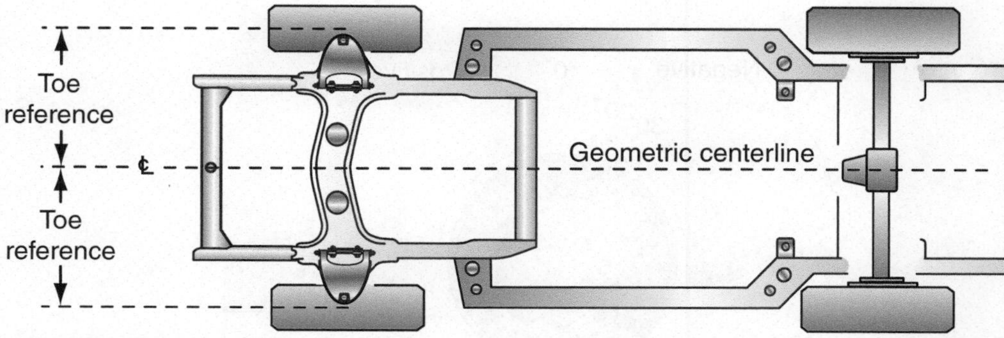

Figure 9-56 The alignment of a vehicle is the reference angles of the tires in comparison to the centerline of the vehicle.

There are usually three angles that can be adjusted by the technician, but other measurements may affect tire wear and steering. The three alignment angles that may be adjusted are **caster, camber,** and **toe.** Many FWD vehicles are not equipped for adjustment of caster. The rear wheels on many FWD vehicles can be adjusted for camber or toe.

Toe is the angle the wheels point in reference to the centerline of the vehicle (**Figure 9-57**). Toe can be measured in degrees of a circle, inches, or millimeters (from the centerline) and can be positive or negative. Toe is the difference of distances between the front edges and the rear edges of tires on the same axle (**Figure 9-58**). Older RWD vehicles usually had some toe-in, whereas most modern FWD cars have zero toe. All vehicles have some type of toe adjustment. **Toe-in** means the leading (front) edges of the tires are closer than the rear edges. **Toe-out** is the opposite. Toe sets the wheels straight ahead when the vehicle is loaded and moving. Note that the measurement is total toe, but toe adjustments are made at individual wheels, so the wheels need to be set at equal toe. For instance, if the total toe is (+) 0.12 inch, each wheel would be set at (+) 0.06 inches. Poor toe wears the tires and affects the steering.

Caster can be seen in action on the front wheels of a shopping cart.

Camber can be seen by looking at the wheel from the front. Leaning in equals negative camber; leaning out is positive camber.

When two wheels are not set equally in **toe**, the customer may complain that the steering wheel is off-centered.

Typical RWD passenger cars usually have **toe-in**. Most FWD cars have zero toe. Very few passenger cars are designed to operate with **toe-out**.

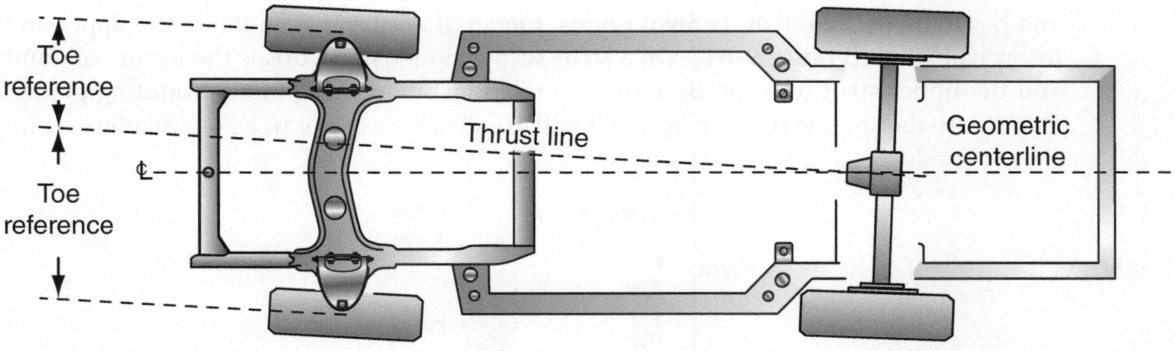

Figure 9-57 Toe can be measured as an angle or degrees of a circle, when referenced to the centerline of the vehicle.

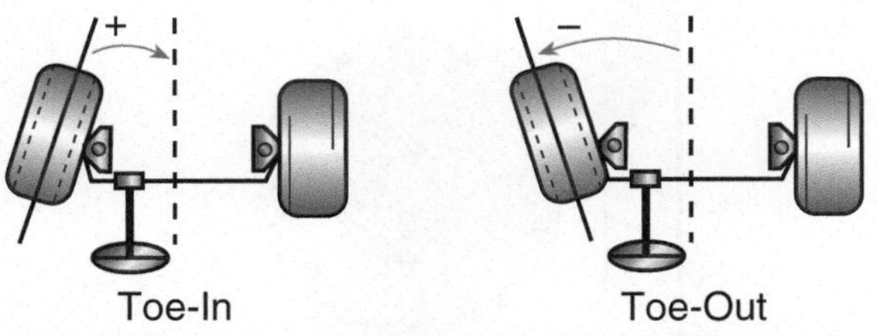

Figure 9-58 Shown is the toe on a wheel that is half of the total toe.

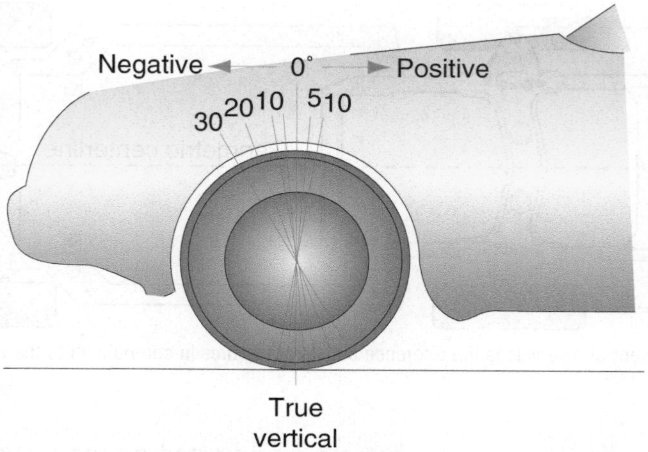

Figure 9-59 Positive and negative caster.

Caster is the forward or rearward tilt of the steering axis from vertical (**Figure 9-59**). Caster is measured as a positive or negative in degrees of a circle. Most light vehicles require a positive caster or a rearward tilt of the top of the steering axis centerline. A positive caster tends to make the wheels steer straight and assists in the return of the steering to center after making a turn. Positive caster does require more steering effort when negotiating a turn, which is why power steering is needed but mainly when turning only. Caster is not a tire wear angle.

Negative caster is used on large, heavy trucks to assist in turning the front wheels, but the wheels must be steered back to straight. Most vehicles that you will encounter will not have negative caster. Negative caster gives the vehicle a better turning ability. Think of the wheels (casters) of a shopping cart (**Figure 9-60**). The best way to remember the effects of caster is comparing caster from side to side. If caster measurements are equal on both sides, the vehicle will not pull. But if they are unequal, it will pull because of the leaning and weight shifting to one side.

Caster can be seen on the vehicle. Caster is measured as the tilting angle of the tire and has to be measured at its pivot points. On an SLA suspension, this is the upper and lower ball joints (**Figure 9-61**). On a strut suspension system, this is the lower ball joint and the upper strut bearing. Both are the relationship of the spindle mounting points. Damage to the spindle from an accident will usually cause caster to be out of adjustment.

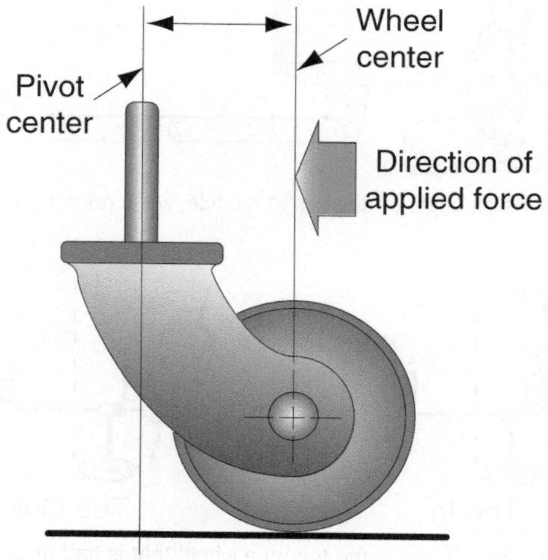

Figure 9-60 Negative caster.

The best way to identify with and understand caster is to stand in front of a vehicle while someone else is turning the wheels. Because of the tilting angle, the height of the sides of the vehicle will change (**Figure 9-62**).

Camber is the inward or outward tilt of the steering axis centerline (**Figure 9-63**). A positive camber tilts the top of the wheel away from the center of the car. Camber is used to place the load of the vehicle at the center of the tire/road contact area.

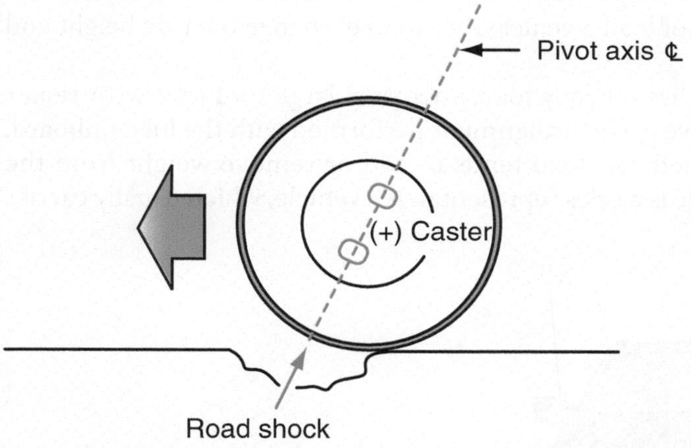

Figure 9-61 Caster angle is drawn through the centerline of the pivot points.

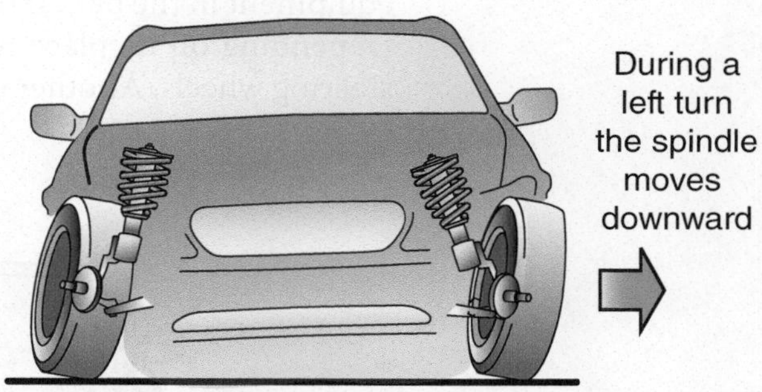

Figure 9-62 Spindle movement and position (caster angle) will cause the height change as the wheels are turned.

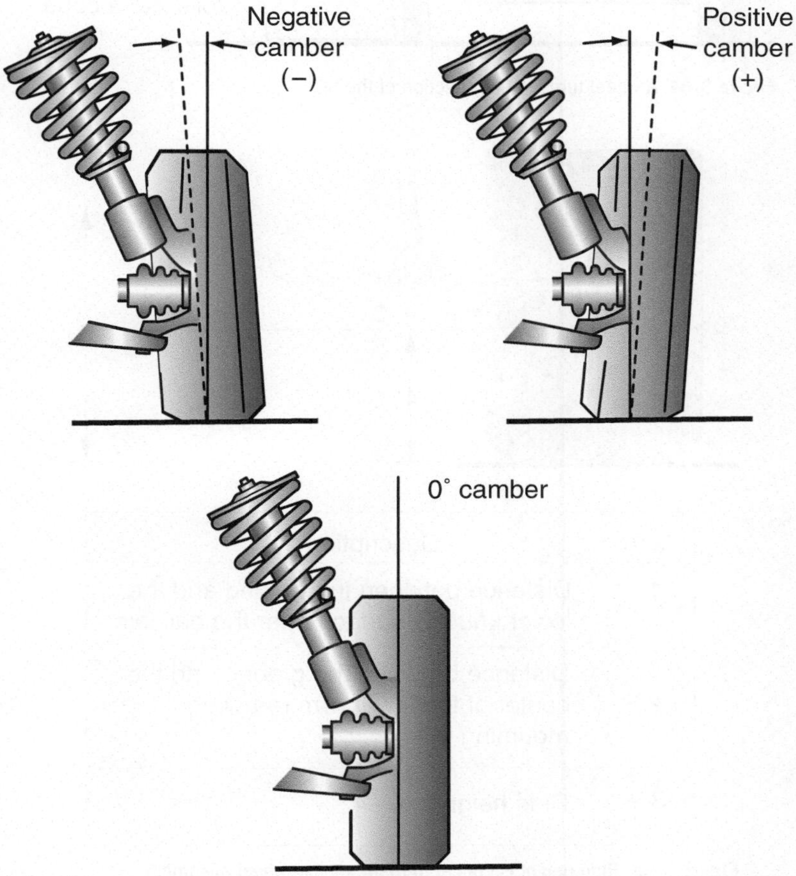

Figure 9-63 Negative and positive camber angles on a MacPherson strut front suspension system.

This prevents tire wear and assists in keeping the tires straight ahead. Incorrect camber will wear the tires, pull the vehicle, and affect steering quality (**Figure 9-64**).

A fourth measurement that may have to be adjusted before setting the caster, camber, and toe is the vehicle's ride or curb height (**Figure 9-65**). Many vehicles require the ride height to be measured and adjusted before beginning alignment procedures. The height of the vehicle affects the angles of the steering and suspension components. Earlier, in the discussion on torsion bars, it was noted that some torsion bars have an adjustment to change ride height. Other systems may require suspension components to be replaced to correct ride height. The amount of load a vehicle carries may change the ride height and suspension angles.

A vehicle that normally carries a heavy load, such as a large tool unit with repair equipment in the bed, should have a wheel alignment performed with the load onboard. Depending on its place in the bed, the load tends to add or remove weight from the steering wheels. Another example is a sales representative's vehicle, which usually carries

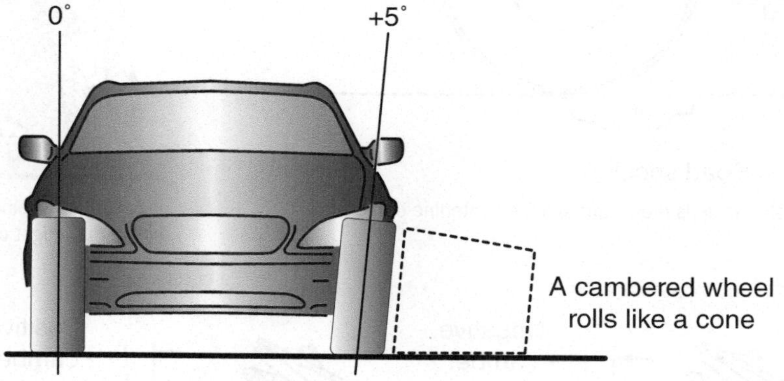

Figure 9-64 A wheel turns in the direction of the tilt.

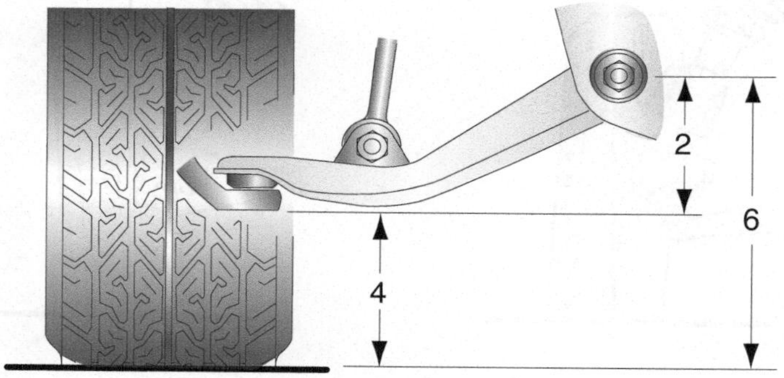

Item	Description
1	Distance between the ground and the lower knuckle surface near the ball joint
2	Distance between the ground and the center of the lower arm rearward mounting bolt
3	Ride height = 6−4

Figure 9-65 Ride height is critical to many vehicle wheel alignment procedures. It may be calculated by measuring between two different points and the ground (floor) and finding the difference between the two.

numerous product samples and sales material in the trunk or rear seat. This removes weight from the front steering wheels. In each case, the load can force the alignment out of adjustment and result in excessive tire wear and poor steering. Some factors that can cause ride height to be less than specified include worn or broken springs, broken suspension components, low tire pressure, incorrect tire size, frame damage, and more.

When performing a wheel alignment, caster should be completed first, followed by camber, then toe. Although caster and camber can be adjusted simultaneously on some vehicles, toe is always adjusted last. Toe may affect camber to some degree, but camber adjustments always affect the toe. Some vehicles require that a specified weight be placed in the passenger compartment to simulate the driver's weight. In theory, most vehicle specifications assume a 150- to 175-pound driver without any other load.

A bent or twisted frame can change the **setback, SAI**, and **toe-out-on-turn** settings without damaging suspension or steering components.

Steering axis inclination (SAI), setback angle, and **toe-out-on-turns** can only be repaired or adjusted by replacing worn or damaged parts. SAI is the angle created by the steering knuckle and control arms or struts as they are installed in the vehicle (**Figure 9-66**). If the SAI is wrong, the damaged parts, including the frame, must be replaced or repaired.

Setback is the measurement of one wheel to the other wheel on the same axle.

Setback is the measurement of one wheel to the other wheel on the same axle (**Figure 9-67**). In other words, the two wheels on one axle should be straight across from each with relation to the frame/body. Extreme setback could cause steering and possibly suspension problems. Some vehicles may have enough adjustment capability to offset some error in SAI or setback. Although this may prevent tire wear and retain good steering, the vehicle will look as though it moves sideways. This "dog tracking" is a common condition on older, half-ton pickup trucks that carried loads across rough terrain and have a bent or twisted frame.

Toe-out-on-turn is required because the turning angles between the inside wheel and the outside wheel are different. When a vehicle travels around a curve, the inside wheel

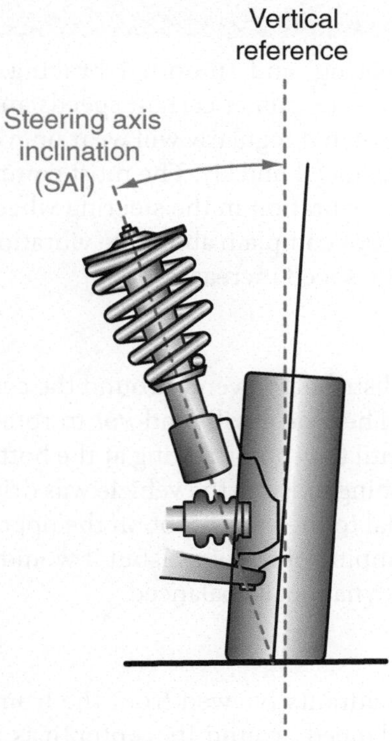

Figure 9-66 The steering axis inclination is manufactured into the suspension and steering assembly. It is not adjustable without a body shop.

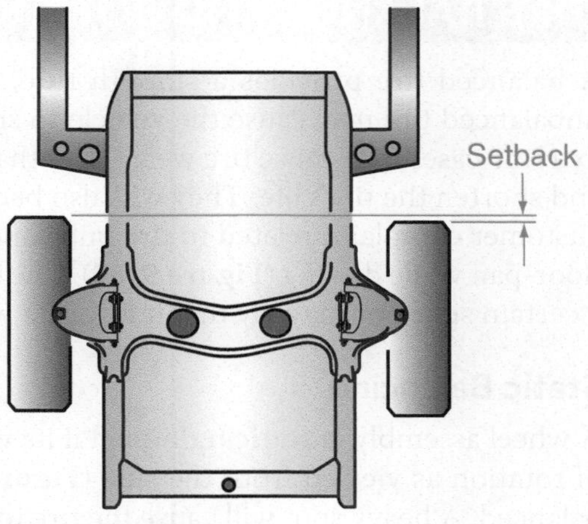

Figure 9-67 Setback.

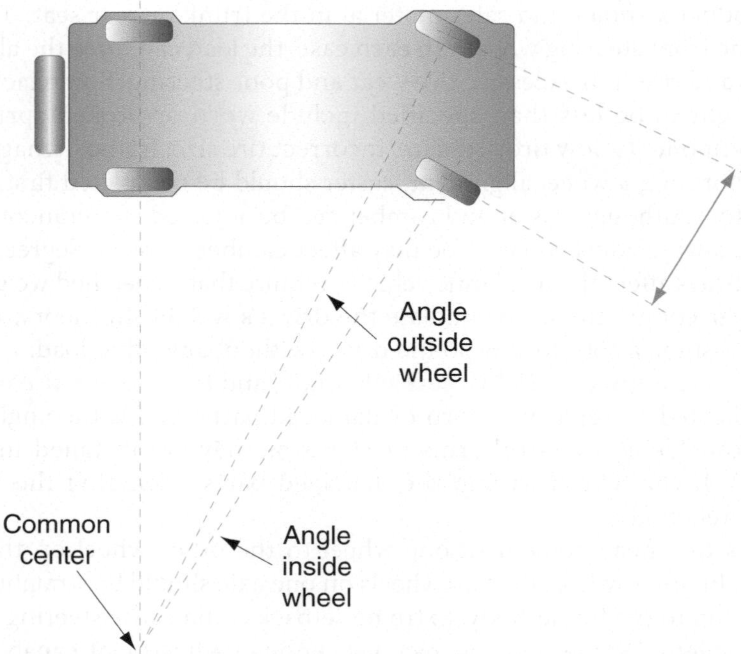

Figure 9-68 Front and rear wheels turning around a common center.

travels a shorter distance at a sharper angle than the outside wheel (**Figure 9-68**). The shaping and positioning of the steering linkage and knuckle sets the outside wheel's turning angle. If the toe-out-on-turns is wrong, some or all the components must be replaced.

TIRE BALANCE AND ROTATION

A balanced tire provides a smooth ride, better steering, and smoother braking. An unbalanced tire may cause the vehicle to shake, vibrate, or pull at certain speeds and in extreme cases may cause tire wear. Tires that are not rotated regularly will wear unevenly and shorten the tire's life. They will also become unbalanced quickly. The most common customer complaint related to tire imbalance will be a vibration in the steering wheel or floor-pan while driving (**Figure 9-69**). The customer can complain about the vibration at a certain speed or may say that it just gets worse as the speed increases.

Shop Manual
pages 376–379

Static Balancing

Static balancing is usually done on heavy truck or trailer tires using a bubble balancer.

A wheel assembly is **static balanced** if its weight is distributed evenly around the center of rotation as viewed from the side (**Figure 9-70**). The wheel will tend *not* to rotate if balanced. A heavy spot will cause the tire to rotate until the spot is resting at the bottom. The heavy spot would also cause a tramping or thumping noise as the vehicle was driven. Static balancing is performed by adding a weight equal to the heavy spot on the opposite rim of the wheel. This would stop the hopping or tramping of the wheel, but it would not eliminate a side-to-side wobble if the wheel was not dynamically balanced.

Dynamic Balancing

Dynamic balancing is almost always used on cars and light trucks.

A wheel assembly that is evenly balanced in four quadrants (viewed from the front) is **dynamic balanced.** Not only is the tire evenly balanced around its center, it is also balanced from side to side on the tire's vertical centerline. To accomplish this task, the front of the tire is divided into four equal sections and each section is balanced to the other three (**Figure 9-71**). The entire wheel assembly is now balanced.

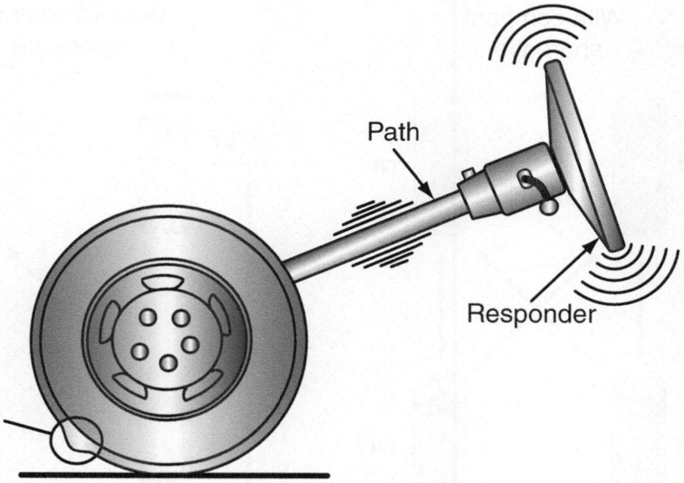

Figure 9-69 Vibration source, path, and responder.

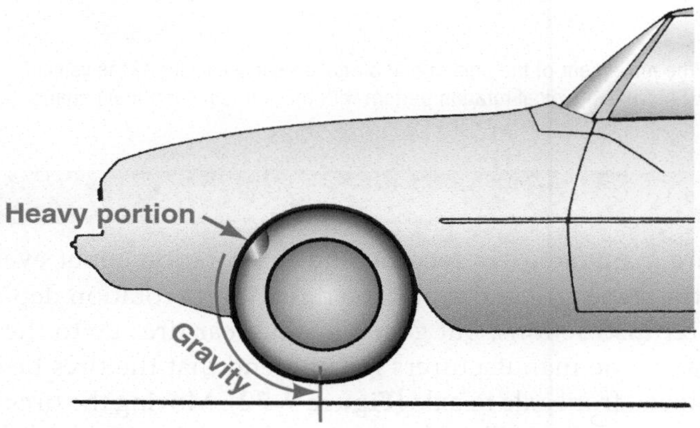

Figure 9-70 This wheel is statically unbalanced. It will rotate by itself until the heavy spot is at the lowest point.

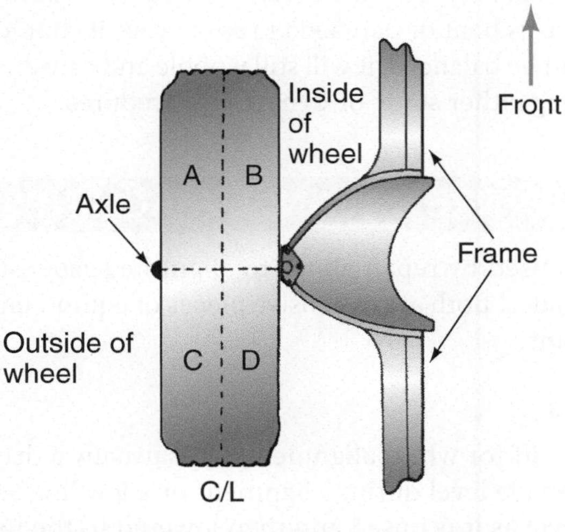

Figure 9-71 The tire is theoretically divided into four quarters and each quarter is balanced to the others in dynamic balancing. A computerized balancer is required.

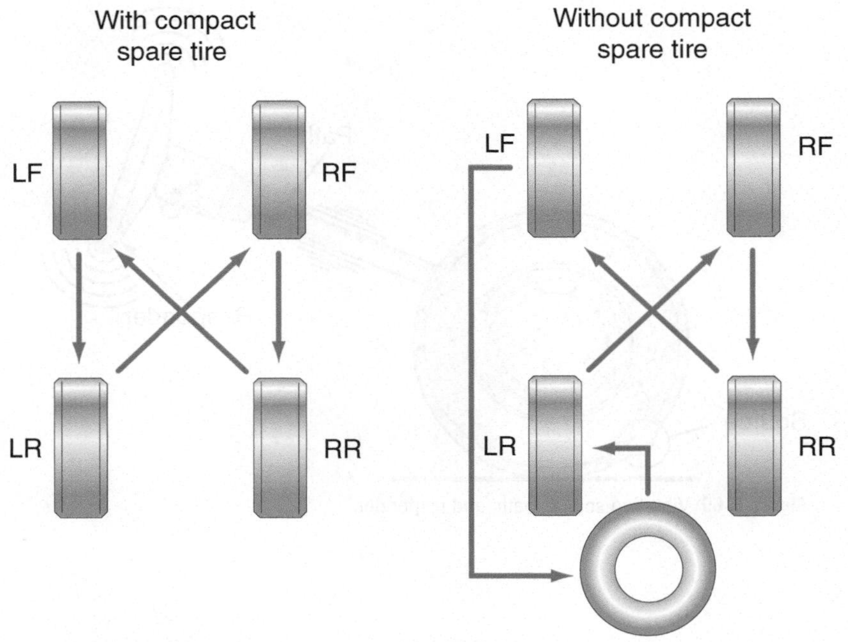

Figure 9-72 The movement of the tires during a rotation service is set by the vehicle manufacturer. Shown is a typical rotation pattern with and without a compact spare.

TIRE ROTATION

Most vehicle and tire manufacturers recommend tire rotation about every 5,000 miles. All tires should be the same size and type. The direction of rotation depends on vehicle and tire manufacturer instructions, but generally, the rear tires go to the front axle and the fronts to the rear. Some manufacturers recommend that the tires be crossed on the vehicle as they are moved from axle to axle (**Figure 9-72**). Moving the tires to the opposite sides of the vehicle reverses the rolling movement of each tire. This helps even the wear across the tire. The tires should be balanced during the rotation.

Regular tire wear will cause some unbalancing of the tire. Rough road surface, striking curbs, and other road hazards can damage a tire or wheel, thereby causing an unbalanced condition. Balancing basically means adding weights to offset a heavy spot in the wheel assembly. Most wheels are balanced by the wheel manufacturer and will remain balanced unless damaged. If a wheel is bent or damaged in some way, it should be replaced. Even if a bent wheel assembly can be balanced, it will still wobble and cause a vibration. The wheel assembly is balanced using either static or dynamic procedures.

ALIGNMENT AND TIRE BALANCE MACHINES

Shop Manual
pages 381–384

There are two machines used by repair shops to align the suspension and steering and balance the wheel assemblies. Both are expensive pieces of equipment and should be used only by trained technicians.

Alignment Machines

Most shops use a special lift for wheel alignments. It is usually a drive-on lift with drop-down legs to hold the vehicle level during alignment or a leveling sensor (**Figure 9-73**). A scissors lift will self-level as it is raised and then lowered to the locks. There are turntables for the front wheels, so they can be moved for adjustment. Most lifts of this type have long rear portions on each side that can be unlocked to allow adjustment of the rear

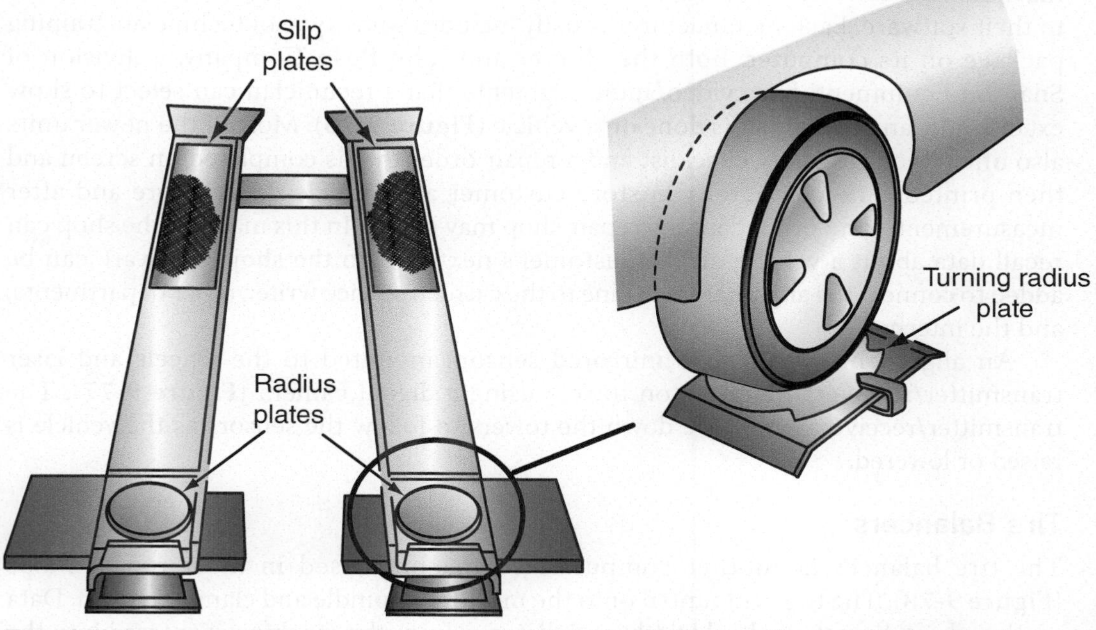

Figure 9-73 The rear skid plate can move for adjustment of the rear wheels. The turntables are for the front wheels.

wheels. The lift may have jacks at the front and rear, so the vehicle can be raised from the lift for compensating the **wheel sensors** and to make repairs or adjustments.

The **alignment machine** consists of the computer, receiver, and four-wheel sensors (**Figure 9-74**). Older versions used electrical cables to carry the sensor data to the computer. Newer machines use sensors with radio transmitters to send the data to a receiver. The receiver translates the signal so that it can be used by a personal computer. Most systems use a special Windows PC application to display the information in a readable form for the technician (**Figure 9-75**). The computer uses software designed and written for a brand of alignment machine. Hunter, Bear, John Bean, and other

Wheel sensors in this case are those devices placed on the wheel during alignment to tell the computer where the wheels are in relation to each other and to the centerline of the vehicle.

An **alignment machine** collects data from the wheel sensors and displays that data to the machine operator.

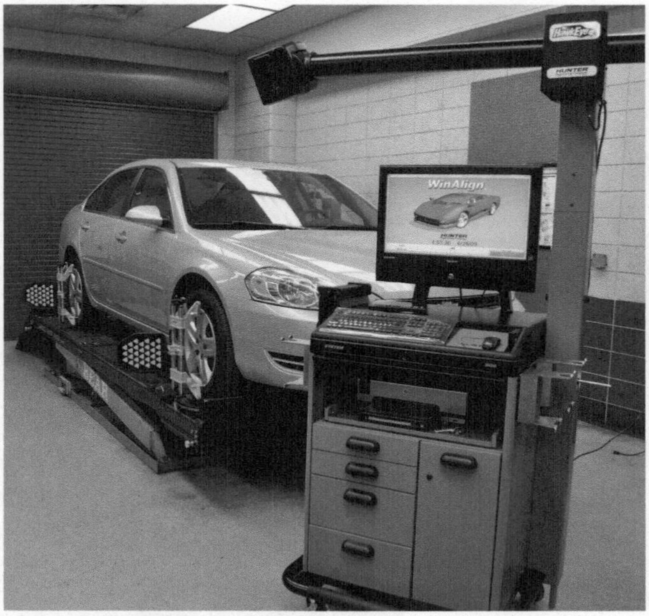

Figure 9-74 Computer wheel aligner with digital cameras and wheel targets.

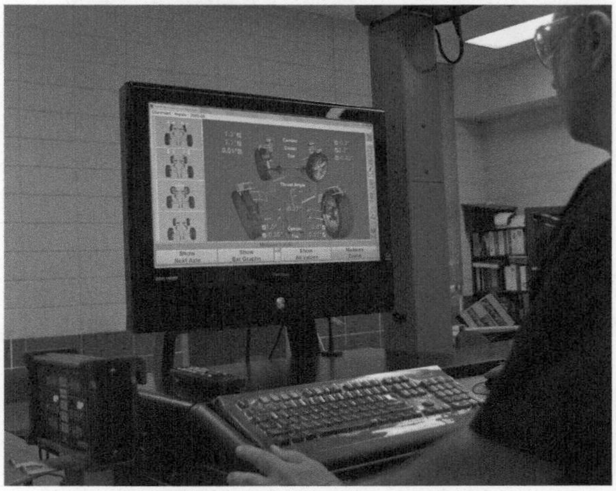

Figure 9-75 Computer wheel aligner.

machine manufacturers use the same basic data but interpret and display it according to their software. Each manufacturer usually includes some type of technician-training package on its computer. Both the Hunter and John Bean Company, a division of Snap-on Equipment, have video/audio segments that a technician can select to show exactly how an adjustment is done on a vehicle (**Figure 9-76**). Most of the newer units also offer a vehicle safety checklist and a repair order that is completed on screen and then printed. The computer can store customer and vehicle data, before and after measurements, and other data the repair shop may desire. In this manner, the shop can recall data about a vehicle on the customer's next visit to the shop. Software can be added to connect the alignment machine to the cashier, service writer, other departments, and the internet.

An alignment system uses mirrored sensors mounted to the wheels and laser transmitter/receivers mounted on towers using a digital camera (**Figure 9-77**). The transmitter/receivers can move down the towers to follow the sensors as the vehicle is raised or lowered.

Tire Balancers

The tire balancer is another computerized machine used in most repair shops (**Figure 9-78**). The tire is mounted onto the machine's spindle and clamped down. Data on the wheel diameter, wheel width, and distance from the machine is entered into the computer using buttons or dials (**Figure 9-79**). The balancer is set up to spin the tire for a set time at a set speed. In this manner, it can locate the heavy spots and select the positions for the counterweights. The location of a weight is usually displayed using a stationary marker or arrow and a light that moves with the rotation of the wheel assembly. The light is aligned with the mark and the weight is installed at the top dead center of the wheel's rim. Weights may be added on one side or both sides of the wheel assembly (**Figure 9-80**). Wheel weights are made often made of lead and should be treated and disposed of as hazardous waste when no longer usable.

The balancer can be used for static and dynamic balancing. However, most machines will only be able to balance tires up to a certain size and type. Some machines can balance a car tire up to 24 inches in diameter but may not be able to balance a medium truck tire because of the tire weight and type.

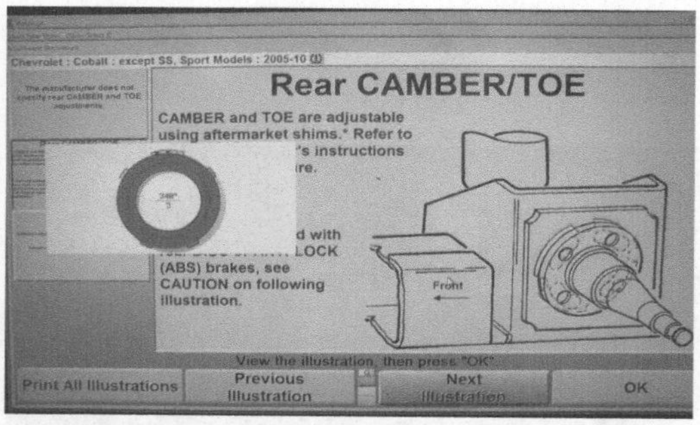

Figure 9-76 Point-and-click suspension inspection screen.

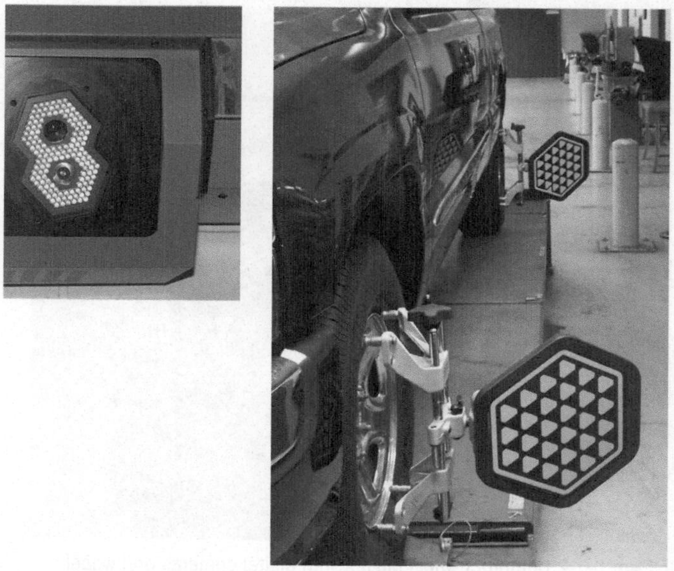

Figure 9-77 One digital camera is aimed at each wheel target.

Figure 9-78 Electronic wheel balancer.

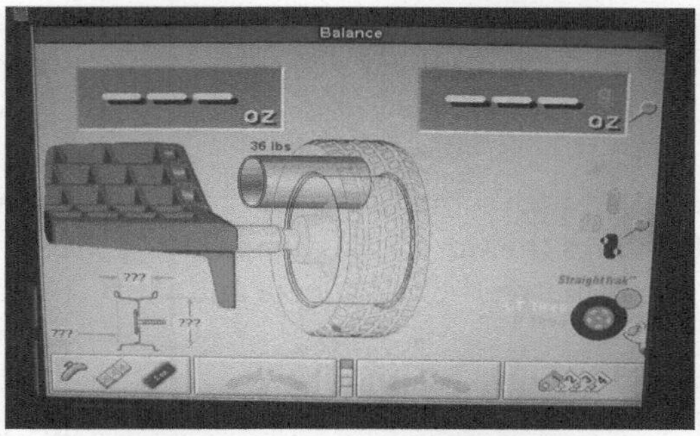

Figure 9-79 The display on a typical tire balancer.

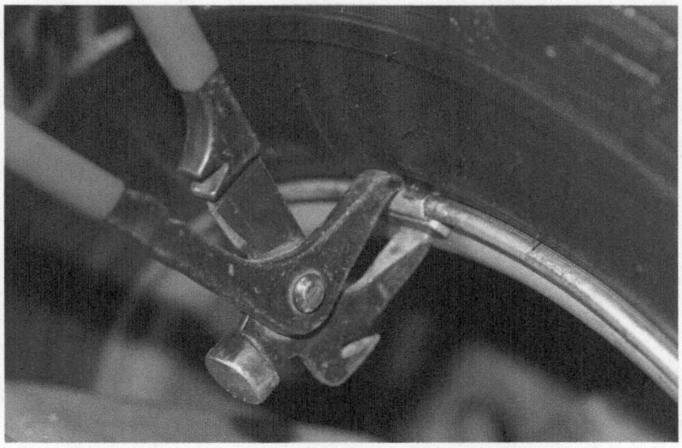

Figure 9-80 Wheel weights are removed and installed with special wheel weight pliers.

SUMMARY

- The Laws of Motion set forth the principles of the movement of objects.
- The Laws of Motion directly affect the suspension and steering systems because of their normal operations.
- Independent suspension means one wheel can react to road conditions without affecting the opposite wheel.
- Dependent suspension usually means the two opposing wheels are connected solidly to the same axle.
- Control arms are used on independent suspension systems.
- Ball joints connect the control arms to the steering knuckle.
- Springs support the vehicle's weight while the shock absorber controls spring movement.
- Jounce is the upward movement of a wheel assembly; rebound is the downward movement.
- Parallelogram steering systems require a system of linkages that remain parallel to the control arms.

- Rack-and-pinion steering uses a direct linkage between the input steering gear (pinion) and steering knuckle to turn the wheels.
- Power assist units are used to reduce driver steering efforts.
- Electric steering uses electronics to turn the steering gear.
- Four-wheel steering provides better vehicle control.
- Wheels fasten to the hub and support the tire.
- Vehicles sold after September 1, 2007, have a TPMS system.
- A TPMS system can be direct or indirect.
- Many vehicles use a tire pressure sensor installed inside the tire for the TPMS.
- The tire size includes tread width, diameter, and aspect ratio, and determines the tire's load capacity.
- Tires are graded for traction, tread wear, speed, and temperature.
- SAI, setback, and toe-out-on-turns cannot be adjusted.
- The three possible alignment adjustments are caster, camber, and toe.
- Caster should be adjusted first, followed by camber, then toe.
- Camber adjustments affect the toe.
- Checking alignment should be done using four-wheel alignment procedures.
- Tire balance can wear tires, cause the vehicle to shake or vibrate, and cause an uncomfortable ride.
- The tire balancer spins a tire to a set speed, uses a computer to calculate the heavy spots, and pinpoints the spot to add a weight.
- The wheel assembly can be balanced statically or dynamically.
- The vehicle should be aligned while carrying its normal load.
- The alignment machine uses four wheel sensors and a computer to check the alignment of each tire individually and each tire to the center or thrust line of the vehicle.

REVIEW QUESTIONS

Short Answer Essays

1. Describe the parallelogram steering system.
2. Explain how power assist is used on a rack-and-pinion steering system.
3. Describe the use of the short and long arm configuration in a suspension system.
4. Explain the purpose of caster and camber.
5. List the major components in a rack-and-pinion steering suspension.
6. Define the elements of the following data: P235/70R15.
7. Define static and dynamic balance.
8. Describe the common purpose of coil springs, leaf springs, and torsion bars.
9. Explain the use of adjusting sleeves on a parallelogram steering suspension.
10. List the wheel data entered into the tire balancer.

Fill-in-the-Blanks

1. A(n) ______________ is used to connect the pinion gear to the wheels.
2. Shock absorbers ______________ the movement of ______________.
3. Tie-rod ends are attached to the steering ______________ by a(n) ______________.
4. Control arms are mounted to the ______________ and attached to the steering knuckle with ______________ ______________.
5. Tire wear is not caused by misadjusted ______________.
6. A tire can be balanced using ______________ or ______________.
7. The parallelogram steering system normally uses a(n) ______________ gear system.
8. The control arm resting on the coil spring is the ______________ arm.
9. The ______________, ______________, and ______________ can be adjusted using an alignment machine.
10. Pressurized ______________gas is used to cause immediate action in the shock absorber.

Multiple Choice

1. The steering system is being discussed.
   Technician A says a parallelogram system uses a rack for connection to the steering knuckle.
   Technician B says the steering knuckle is used to mount the hub.
   Who is correct?
   A. A only C. Both A and B
   B. B only D. Neither A nor B
2. Each of the following may be part of a fully independent suspension EXCEPT:
   A. Coil springs. C. Struts.
   B. Shock absorbers. D. Multileaf springs.
3. Wheel alignment is being discussed.
   Technician A says ride height is a factor in alignment procedures.
   Technician B says the total toe must be divided by two during the adjustment procedures.
   Who is correct?
   A. A only C. Both A and B
   B. B only D. Neither A nor B
4. Wheel alignment is being discussed.
   Technician A says caster adjustments may affect toe.
   Technician B says the toe adjustments should be made after the caster and before the camber.
   Who is correct?
   A. A only C. Both A and B
   B. B only D. Neither A nor B
5. Steering angles are being discussed.
   Technician A says setback results from the manufacturing and assembly of the vehicle.
   Technician B says SAI angles can be adjusted on some vehicles.
   Who is correct?
   A. A only C. Both A and B
   B. B only D. Neither A nor B
6. *Technician A* says SAI angles include the three adjustable alignment angles.
   Technician B says toe-out-on-turn angles are only effective during sharp turns.
   Who is correct?
   A. A only C. Both A and B
   B. B only D. Neither A nor B
7. The suspension system is being discussed.
   Technician A says a torsion bar suspension requires a shock absorber.
   Technician B says a leaf spring system does not use a shock absorber.
   Who is correct?
   A. A only C. Both A and B
   B. B only D. Neither A nor B
8. A dependent suspension system is being discussed.
   Technician A says this type suspension is commonly used on rear-wheel-drive trucks.
   Technician B says coil springs can be used on this type of suspension.
   Who is correct?
   A. A only C. Both A and B
   B. B only D. Neither A nor B
9. Tire data is being discussed.
   Technician A says actual speed test data is molded directly onto the sidewall.
   Technician B says the required tire information is determined by the Department of Transportation.
   Who is correct?
   A. A only C. Both A and B
   B. B only D. Neither A nor B
10. Wheel alignment is being discussed.
    Technician A says a good alignment will help the driver control the vehicle with little effort.
    Technician B says adjusting the toe before adjusting the camber could cause the final toe setting to be incorrect.
    Who is correct?
    A. A only
    B. B only
    C. Both A and B
    D. Neither A nor B

CHAPTER 10
BRAKES

Upon completion and review of this chapter, you should understand and be able to:

- Understand the applications of Pascal's Law in a brake system.
- Explain the purpose and operation of the brake system.
- Explain the use of power boosters.
- Understand the purpose and operation of master cylinders.
- Explain the use of control valves.
- Explain the operation of drum brakes.
- Explain the operation of disc brakes.
- Understand the purpose of bearings.
- Understand the purpose and operation of the parking brake.
- Understand the purpose and general operation of antilock brake systems.
- Explain the basic operation of traction control systems.

Terms To Know

Amber ABS warning light
Anchor
Antilock brakes system (ABS)
Ball bearing
Bearing races
Brake pad
Brake warning light
Bushing
Caliper
Combination valve
Constant velocity (CV)
Controller
Diagonally split system
Disc pad
Double flare
Dot
Drum
Dynamic proportioning
Fixed caliper
Friction material
Hub assembly
Hygroscopic
Hydro-boost
Grease
ISO
Isolation/dump solenoid
Lining
Load-sensing proportioning valve
Master cylinder
Mechanical advantage
Metering valve
Pressure controller
Primary piston
Proportioning valve
Pushrod
Rear-wheel antilock (RWAL)
Rotor
Sealed wheel bearing
Secondary piston
Self-energizing
Shoe web
Sliding caliper
Speed sensor
Square-cut O-ring
Tone wheel
Traction control
Vacuum brake booster
Wheel cylinder

INTRODUCTION

An object in motion tends to stay in motion, which means that the brake system is the most critical system on the vehicle. Obviously, our vehicles cannot always be in motion. They must be slowed and stopped at times. The brake system supplies the necessary force for the driver to slow or stop the vehicle. The system is built on the application of hydraulics theory. This chapter will explore the theory and operation of the braking system. This chapter will also discuss the use of bearings and sealed bearing assemblies. Detailed information of all aspects of automotive brake systems can be found in Cliff Owen, *Today's Technician: Automotive Brake Systems,* Cengage Learning, Clifton Park, New York.

HISTORY OF BRAKES

The earliest automobiles used a mechanical linkage to force a block of wood, leather, or metal against the wheel for brakes. The same type was used on some horse-drawn wagons. This system was called *"scrub brakes"* and was used until rubber tires were developed.

The automobiles produced in the early 1900s were stopped by using a metal band that clamped around a drum; this system was called "*band brakes*". This system was abandoned due to only working when moving forward. The next system was *"drum brakes"*. This system was still a mechanical system using rods, links, and levers. The linkage required constant adjustment. With heavier vehicles and higher speeds, mechanical linkage was not sufficient and gave way to modern hydraulic drum brakes, which were first used in the 1920s.

Shop Manual
page 419

Hydraulic Theory and Pascal's Law Review

We will discuss further in Chapter 12, "Engine Performance Systems," liquids cannot be compressed and make ideal transmitting agents when contained in sealed systems. This is difficult to understand. A liquid cannot be compressed, but it can be pressurized. In this manner, any force applied at one end would result in an equal force being applied at the other end. The force is transmitted by pressure buildup of the fluid within the system and the fact that liquids cannot be compressed. The fluid's pressure can also be used to increase or decrease the forces at work. This is the principle of Pascal's law.

Pascal's Law states that pressure applied to fluid in a sealed system is transmitted equally in all directions and to all parts of the system. Pascal also developed a mathematical formula to calculate the use of pressure to transmit force. The formula is force *(F)* equals pressure *(P)* multiplied by area (*A*) (**Figure 10-1**). Force is measured in foot-pounds (ft.-lb.) and pressure is measured in pounds per square inch (psi). Area is in square inches (sq. in.).

AUTHOR'S NOTE Mistakes are sometimes made when recording or understanding force and pressure. Simply put, if a water hose is turned on you, pressure is pushing the stream toward you. What you feel, beside the wetness, is the actual force being applied to your body. At 60 psi, you get wet and may be colder. At 300 psi, you are probably going to get hurt because the actual delivered force is much greater.

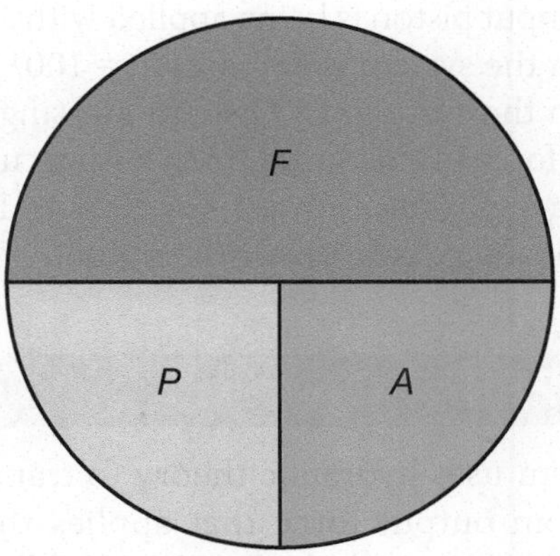

Figure 10-1 Force (ft.-lb.) is equal to pressure (psi) times area (sq. in.).

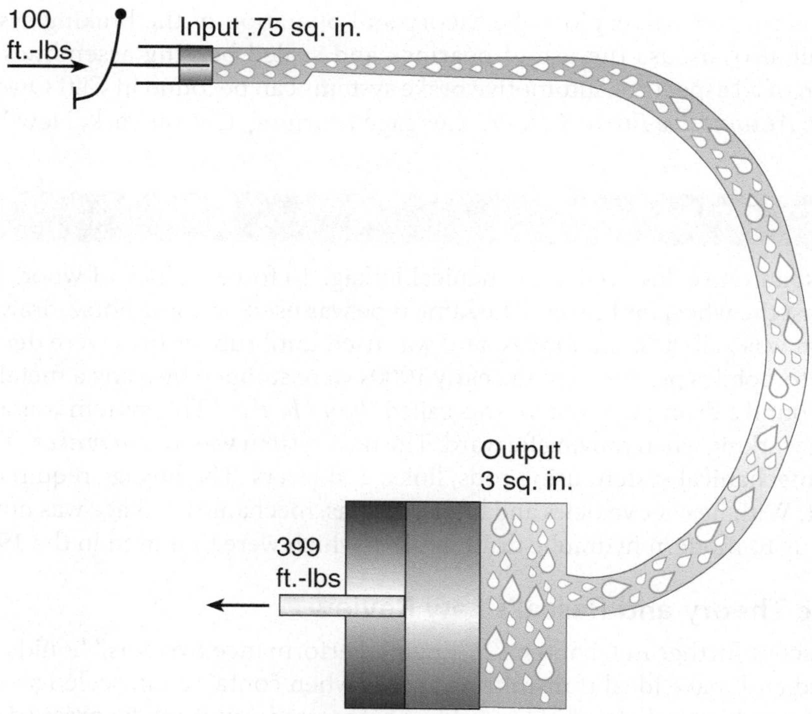

Figure 10-2 An output piston that is larger than the input piston can increase the input force.

The use of Pascal's formula is like the one used when applying Ohm's Law in Chapter 5, "Electrical Systems and Components." A major mistake during the first usage of Pascal's Law was mixing force and pressure during the calculations. An example of the correct formula is:

$F = P \times A$ where P is 10 psi and A is 2 sq. in.

$F = 10 \times 2$

$F = 20$ ft.-lbs.

In this case, a 20-ft.-lb force results when 10 psi are pressing a 2-sq.-in. piston. Another method of the same formula is $P = F/A$.

$P = F/A$ where F is 15 ft.-lb. and A is 3 sq. in.

$P = 15/3$

$P = 5$ psi

A **mechanical advantage** is the amount of force increased at the output compared to the input force.

The formula can also be used to calculate **mechanical advantage.** This is like gear ratios where a small force can be increased to a higher force within the same system. In **Figure 10-2**, the .75-sq.-in. input piston is being applied with a force of 100 ft.-lb. Using the formula, the pressure within the system is 133 psi ($P = 100/.75$). Because the pressure is the same everywhere within the system, 133 psi are pressing on each square inch of the 3-sq.-in. output piston. The force of the output piston is calculated by $F = 133$ multiplied by 3, or 399 ft.-lb. This gives a mechanical advantage of 3.99 to 1 (399/100). Another method of calculating mechanical advantage is to divide the output by the input or $3/.75 = 4:1$.

The **master cylinder** converts the driver's mechanical input force to hydraulic pressure for the brake system.

BRAKE SYSTEM OPERATION

The automotive brake system uses hydraulic theory to transfer the driver's input force throughout the system to an output force that applies the mechanical brakes. The components of a basic brake system include the pedal and linkage, the **master cylinder,** lines and hoses, controlling valves, and either drum or disc brakes at the wheels. Most systems also use a power boost to assist the driver's efforts in braking. The tires are not considered to be part of the brake system even though they provide the contact between the applied braking force and the road. If the tire is slick with little or no tread, or the road surface is slick, there may be no slowing of the vehicle even when brakes are forcibly applied. An inspection of the tires' tread should be made when the vehicle is receiving brake repair or inspection.

Shop Manual
page 420

Braking

As the driver presses downward on the brake pedal, the movement of the pedal is mechanically transmitted to the piston in the master cylinder (**Figure 10-3**). The piston pressurizes the brake fluid in the bore and attempts to force it through the brake lines. Because this is a sealed system, the pressurized fluid hydraulically transmits the applied driver's force to the output pistons at the wheels. The pistons may be in a **wheel cylinder** or disc **caliper.** The pistons act against the drum brake shoes or disc brake pads, respectively.

Wheel cylinders work with drum brakes. **Calipers** are used with disc brakes.

Releasing

When the brake pedal is released, return springs at the pedal, inside the master cylinder, and at the drum brake shoes retract each of the brake components to the rest (release) position. Within the disc brake caliper, a distorted square-cut O-ring straightens when the brake pedal is released. As the ring straightens or returns to its normal shape, the caliper piston is pulled backward slightly. This reduces the contact of the pads with the rotor.

Power Boost

Most vehicles use some type of booster on the brakes to reduce the driver's effort. The most common type used on passenger cars and light trucks is the vacuum chamber (**Figure 10-4**). The engine supplies the vacuum. A diesel engine requires the use of a vacuum pump to provide the vacuum. Movement of the brake pedal opens an air valve that allows atmospheric pressure to enter one side of the chamber. This extra force acting

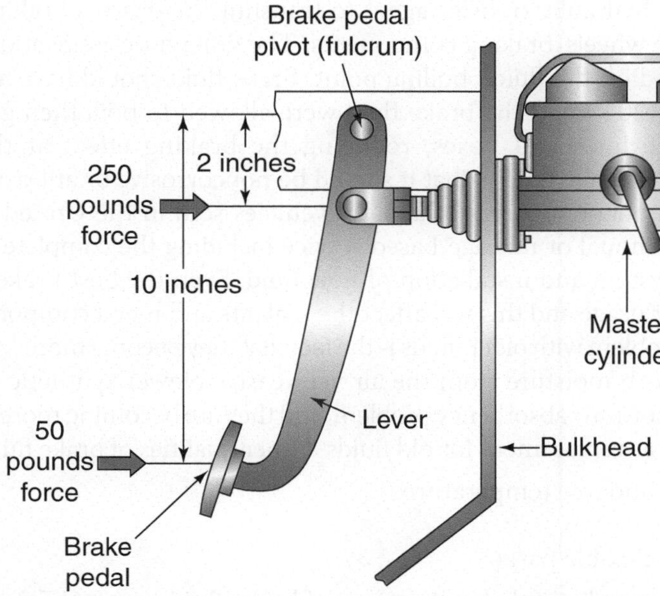

Figure 10-3 The pressure generated in the master cylinder because of the pushrod forcing against the piston will transmit the applied force to the wheel brakes (calipers or wheel cylinders).

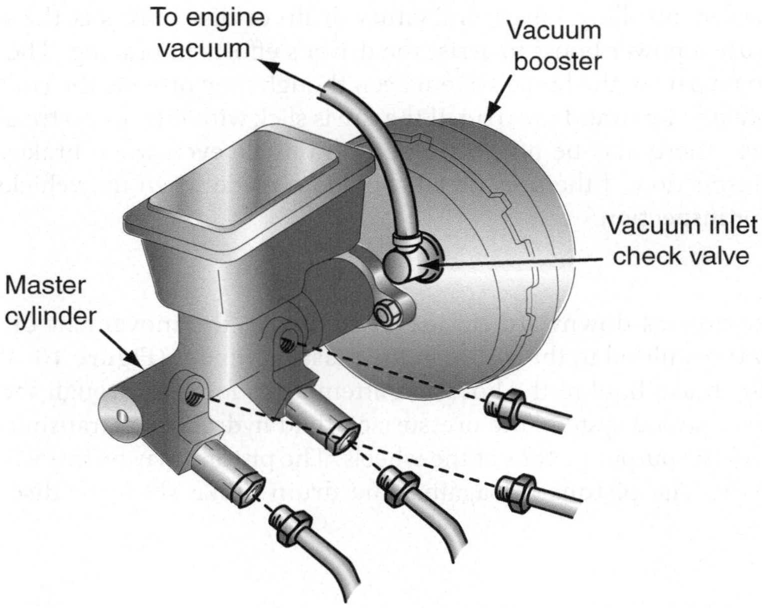

Figure 10-4 The booster check valve is mounted on the front side of the booster where the vacuum line from the engine is connected.

on the booster interior diaphragm increases the force exerted by the driver. Some vehicles use hydraulic systems to boost the driver's braking input.

The diaphragm is a large, flexible disc stretched across the interior of the vacuum brake booster to divide it into two chambers.

A dual brake system is used on every light vehicle now sold in the United States. The brake system has two hydraulic circuits including a dual chamber master cylinder. In this manner, a leak or other failure on one system will not affect the other system. It will take additional space and effort to stop the vehicle, but the driver will still be able to maintain control.

Brake Fluids

Brake fluid acts as the hydraulic transfer agent to transmit the driver's braking input to mechanical force at the wheels' braking components. The different classifications of brake fluid are generally based on the fluid's boiling point. Brake fluid should have a minimum boiling point of 400°F (204°C). If the brake fluid were allowed to boil, then gases would form within the brake lines and hoses, reducing the braking effect at the wheels. A second requirement of brake fluid is that it should be noncorrosive or at least corrosive-resistant. However, the manufacturers of all light vehicles sold in the United States and Canada recommend biannual or mileage-based service including the complete flushing of the brake's hydraulic system and installation of new fluid. Even the best brake fluids will deteriorate over time or usage and this will affect the sealants and metal components of the brake system. A third problem with older fluids is the fact that they become more ***hygroscopic***, which means they absorb moisture from the air very easily. Newer synthetic fluids have almost eliminated the moisture-absorbency problem, and they are becoming more commonly used in new vehicles or as replacement for old fluids. Other qualities of brake fluids include:

Hygroscopic A fluid that absorbs moisture from the air.

- Free flowing at high and low temperature
- Low freezing point
- Lubricate metal and flexible parts

DOT refers to the United States Department of Transportation, which issues the detailed standards for all vehicles brake fluid in the country.

The two most commonly used classifications of brake fluids are **DOT** 3 and DOT 4 (**Figure 10-5**). Both are polyglycol compounds, but DOT 4 has a higher boiling point than DOT 3, as outlined in **Table 10-1**. The two fluids can be mixed and replace each other

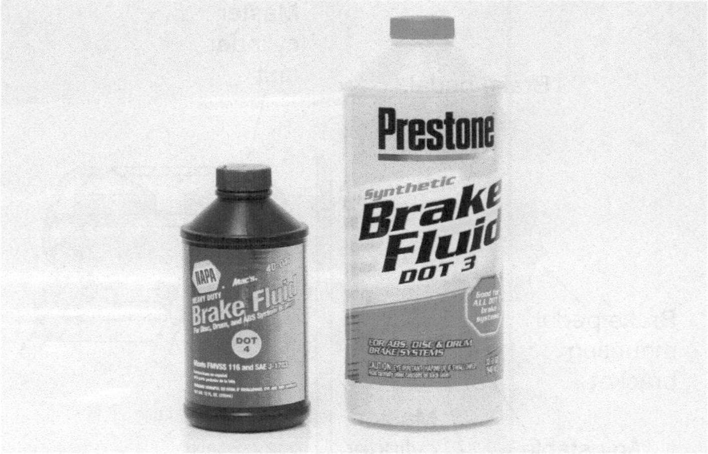

Figure 10-5 Brake fluid should be purchased only in the quantities needed for the job.

TABLE 10-1 Minimum Boiling Points for the Six Brake Fluid Grades

Boiling Point	DOT 3	DOT 4	DOT 5	DOT 3/4	DOT 5.1
Dry	401°F (205°C)	446°F (230°C)	500°F (260°C)	500°F (260°C)	585°F (307°C)
Wet	284°F (140°C)	311°F (155°C)	356°F (180°C)	347°F (175°C)	365°F (185°C)
DOT 5.1 long-life has the same dry and wet boiling point of 424°F (218°C).					

within a system. It is suggested that, if a DOT 3 brake system needs to be flushed, it is a good time to install DOT 4 fluid rather than just mixing them together. However, do not replace a higher-rated brake fluid with a lower one (e.g., DOT 3 in place of DOT 4). Three synthetic fluids are available. Each has a boiling point over 500°F (260°C) and can be mixed or can replace DOT 3 or DOT 4. They are classified as DOT 3/4, DOT 5.1, and DOT 5.1 long-life. The synthetics last longer, absorb less moisture, and are in theory better for braking. DOT 5.1 has a life of about 60 months and DOT 5.1 long-life is supposed to be good for about 10 years. DOT 3/4 has a life of about 24 months.

Shop Manual
pages 420–421

One classification of fluid, **DOT 5**, is a silicone-based fluid and is used in very few vehicles. It should **NEVER** be mixed with any other classification of brake fluids nor should it be used as a replacement for any of them. DOT 5 will damage the metal and flexible components of a typical brake system. Antilock brake systems will not work with DOT 5 fluids.

BRAKE SYSTEM COMPONENTS AND OPERATION

Each component of the brake is essential to the proper operation of the complete system. A failure of one component will affect the entire system in some manner, ranging from a substantially reduced braking action to complete system failure.

Pedals and Linkages

The pedal is connected to the master cylinder's piston with a **pushrod** (**Figure 10-6**). If the system uses a power booster, the pushrod is connected to the booster's diaphragm or valve to operate the assist. On most power boost systems, the pedal's pushrod does not act directly on the master cylinder piston.

The **pushrod** is the mechanical connection between the brake pedal and the power booster or master cylinder.

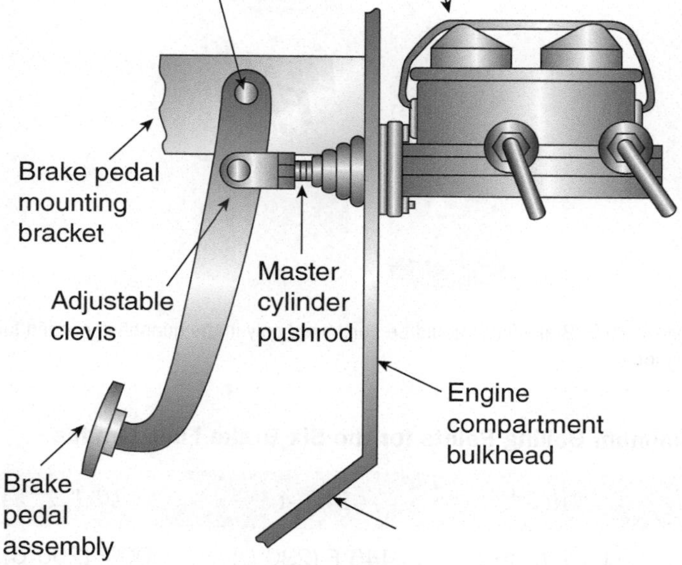

Figure 10-6 The force applied at the pedal is transmitted to the master cylinder's primary piston by the pushrod.

On some vehicles, the pushrod must be adjusted for length when the master cylinder, booster, or pedal is replaced. Better machining techniques have eliminated this adjustment requirement on many vehicles; however, most manufacturers still require an inspection of this mechanical connection when new components are installed.

The mounting for the pedal is generally a shaft from which the pedal hangs downward. A return spring is used to return the pedal to the fully up or off position after the brakes are released. Also mounted on or near the pedal shaft is the brake (stop) light switch. The switch is normally open. When the brake is applied, the pedal's shank or the pushrod presses (push) a plunger on the switch and closes the brake light circuit. At times, the position of the switch may have to be adjusted to work properly.

Antilock brake systems (ABS) help increase driver control by reducing wheel lockup during panic stops.

Modern vehicles have a multiple function brake light switch. The switch may be used to activate the **antilock brakes (ABS),** switch off the cruise control, and signal the PCM that the brakes are applied.

Power Boosters and Master Cylinders

Shop Manual
pages 429–432

The master cylinder is the point at which mechanical force is changed to hydraulic pressure. The booster is placed between the brake pedal and master cylinder. It increases the driver's ability to apply force to the master cylinder piston.

A diesel engine requires a vacuum pump for the **vacuum brake booster** and other vacuum-operated devices.

The brake's power booster may be vacuum or hydraulic operated. The **vacuum brake booster** is connected to the engine's intake manifold or vacuum pump. The operating engine draws air from the vacuum chamber and maintains a vacuum in both sides of the chamber. A check valve, normally located at the connection of the hose and chamber, maintains the vacuum during engine shutdown or hard acceleration. A large diaphragm extends across the interior of the vacuum chamber. The brake's pushrod is connected to the driver's side or rear of the diaphragm. A power piston is mounted to the front side of the diaphragm and extends forward to contact the master cylinder's primary piston (**Figure 10-7**). The power piston houses vacuum and air valves. During

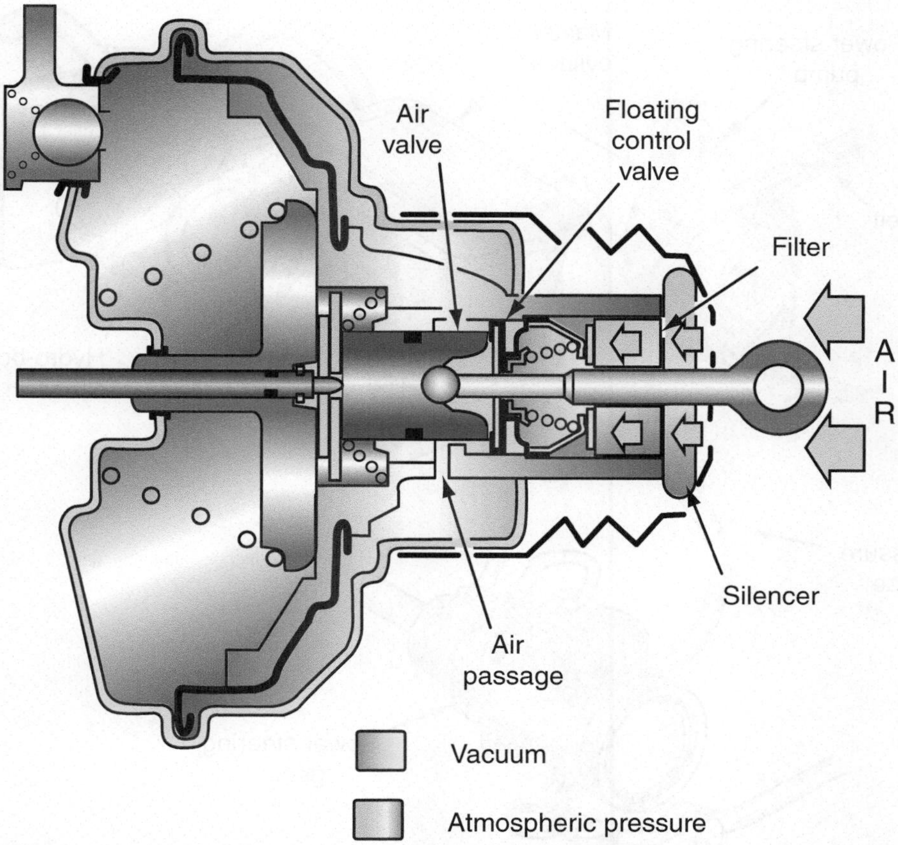

Figure 10-7 The volume of air entering the booster is controlled by the linear movement of the pedal's pushrod.

brake release, vacuum is supplied through the vacuum valve to the rear chamber of the diaphragm. This helps release the brakes quickly and helps keep the diaphragm in the neutral position.

As the brake pedal is depressed (brake apply), the air valve is opened, thereby allowing atmospheric pressure to enter the rear chamber. At the same time, the vacuum valve closes the vacuum port to the rear chamber. The controlled opening of the air valve regulates the amount of pressure that can be applied to the diaphragm. This prevents full boost and brake lockup during light braking. Because a vacuum is still applied to the front chamber, the diaphragm and power piston move forward, thereby increasing the driver's input force. The master cylinder piston is moved forward within its bore and a hydraulic pressure is created in the brake system. When the brakes are released, the rear vacuum valve is opened, the air valve is closed, and the applied vacuum and a return spring force the diaphragm to the center or off position.

Hydraulic boost is accomplished by directing a fluid under high pressure to the driver's side of a master cylinder. A **hydro-boost** system, popular now on light trucks, uses the power-steering system to supply the pressurized fluid. Fluid from the power-steering pump is routed through a valve system fitted between the brake pedal's pushrod and master cylinder (**Figure 10-8**). The valve assembly is part of the master cylinder but may be replaced individually on most units. As the pedal is depressed, the pushrod opens a valve that directs pressurized fluid to a power piston on the driver's side of the master cylinder. The valve's movement is proportional to the movement of the pedal. The further the pedal moves downward, the more the valve opens. The pressurized fluid forces a power piston and pushrod to move the master cylinder piston. This, in turn, pressurizes the brake fluid, and the brakes are applied.

Hydro-boost was originally designed, used, and owned by General Motors.

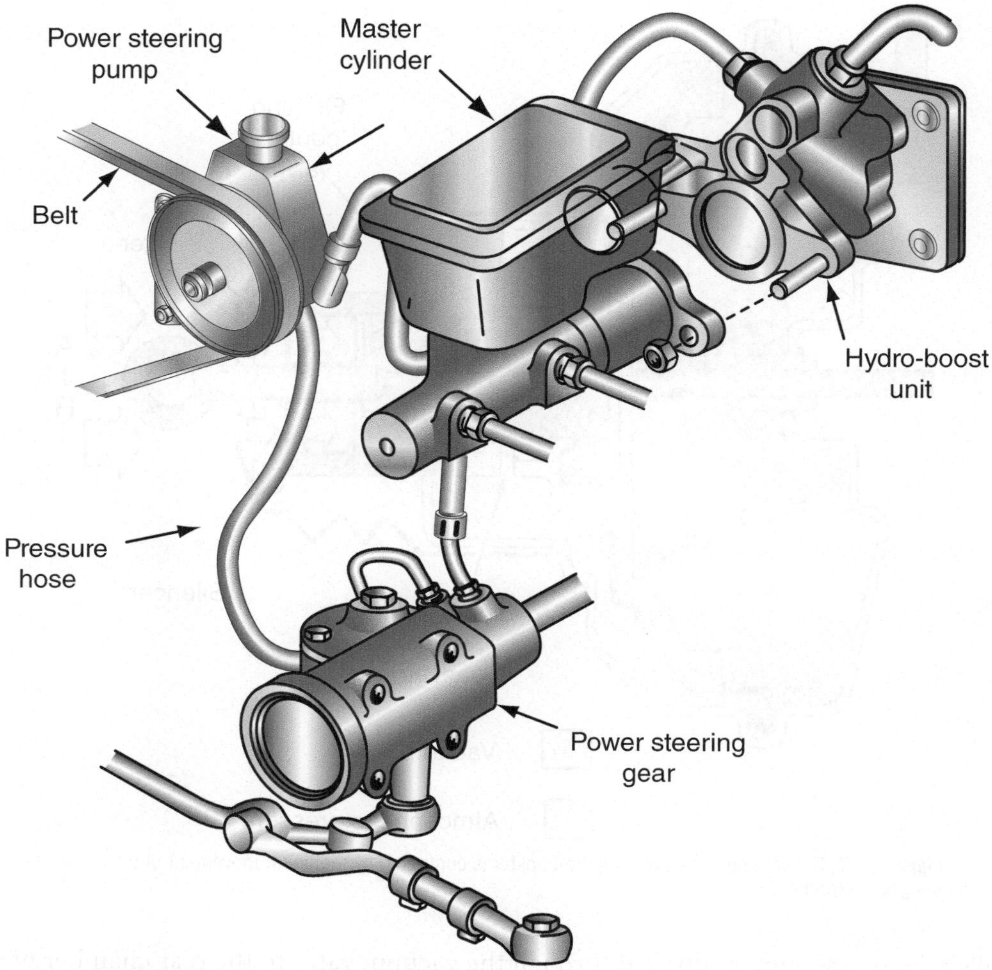

Figure 10-8 Layout of a typical hydro-boost power brake booster.

When the brakes are released, a release valve is opened to allow the pressurized fluid to flow directly back into the power steering system. A small accumulator is filled with pressurized fluid in the event of a power-steering failure. If there is not enough pressure within the boost system when the brakes are applied, the accumulator fluid is directed to the power piston and offers enough brake boost for two or three stops.

Master Cylinders

Single-piston master cylinders were installed on the original hydraulic brake systems. A single-piston master cylinder fed a single brake system. If there was a failure in any component, the complete system failed. The operation of the single-piston master cylinder works in the same manner as a dual-piston except for the extra components used in a dual system.

Dual or split brake systems have been required since 1967 (**Figure 10-9**). This setup allows half of the brake system to be available to control the vehicle if there is a leak or some other failure. The most common system is a front/rear split where one master cylinder piston feeds the brakes on one axle. A **diagonally split system** has a front and a rear wheel fed by one master cylinder piston (**Figure 10-10**).

The **diagonally split system** allows the driver to have braking power at each axle if a leak develops in the brake system.

The split system requires a dual-piston master cylinder. The master cylinder is mounted on the forward side of the brake booster or directly on the firewall. The basic components of a dual master are the primary and secondary pistons, return springs, body, and reservoir.

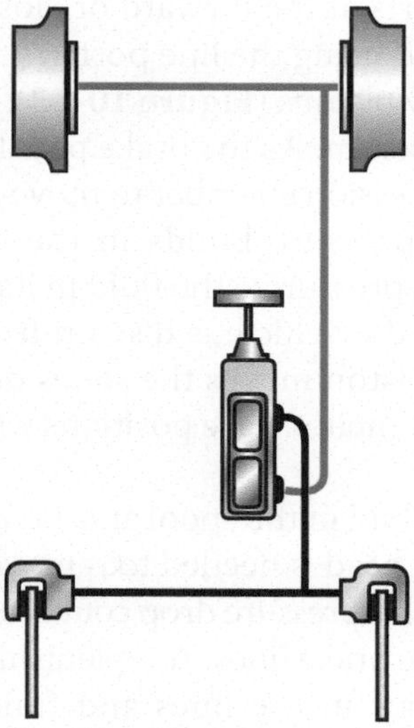

Figure 10-9 Typical front-to-rear split hydraulic system.

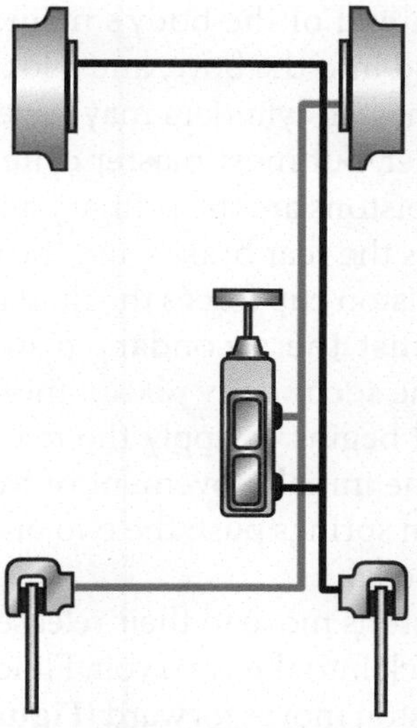

Figure 10-10 The diagonally split system requires a separate brake line from the master cylinder or combination valve to each wheel.

Some master cylinders and reservoirs are molded into one unit, but most reservoirs are plastic and can be dismounted from the master cylinder. The reservoir is mounted to the top of the master cylinder. There are two chambers in the reservoir that separately feed the two bores of the master cylinder (**Figure 10-11**). The reservoir cover can be quickly removed to add fluid. Most reservoirs are translucent, so the fluid level can be checked without opening the system. Some reservoirs have a fluid level sensor that activates a dash light to warn the driver of low fluid levels.

The master cylinder body houses the internal components and is drilled to accept the reservoir and brake lines. The threaded ports on the side of the body are connections for the brake lines and may have a metering orifice installed. There are also vents and replenishing ports drilled into the body that open into the bore and the reservoir. This allows fluid to flow to and from the bore as the pistons move.

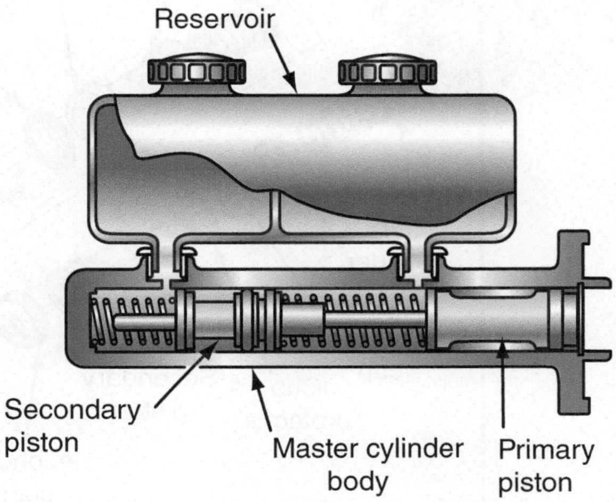

Figure 10-11 The primary piston operates the rear brakes and applies hydraulic or mechanical force against the secondary piston.

The **primary piston** is located to the rear of the master cylinder bore and the closest to the firewall. On disc/drum brake systems, it pressurizes the disc brake portion of the system.

The **secondary piston** is located to the front of the master cylinder bore. In a disc/drum brake system it pressurizes the drum portion of the system.

The open end of the body's main bore is installed toward the firewall. Pistons and springs are slid into the bore, and a locking ring is installed for retention (**Figure 10-12**). Some older master cylinders may have bleeder ports at the forward or closed end of the master cylinder, but most master cylinders are bled using the line ports.

The two pistons are the primary and secondary pistons (**Figure 10-13**). The **primary piston** applies the rear brakes and the secondary piston. As the brake pedal is depressed, the primary piston cup forces the fluid in the compression chamber to move into the brake line and against the secondary piston. As the pressure builds in the chamber and brake lines, the **secondary piston** moves, starts to pressurize the fluid in its compression chamber, and begins to apply the front brakes. If the vehicle is disc on front and drum on the rear, the initial movement of the primary piston moves the shoes out toward the drums. Return springs push the two pistons back to their release positions when the driver releases the pedal.

As the pistons move to their release positions, fluid in the spool area behind the piston head vents back into the reservoir. Fluid behind the head is needed to remove the pressure drop as the piston moves forward (**Figure 10-14**). This pressure drop could cause the piston seals to leak and no pressure would be created in the brake lines. A replenishing port keeps the compression chamber filled so there is no air in the lines and fluid is available

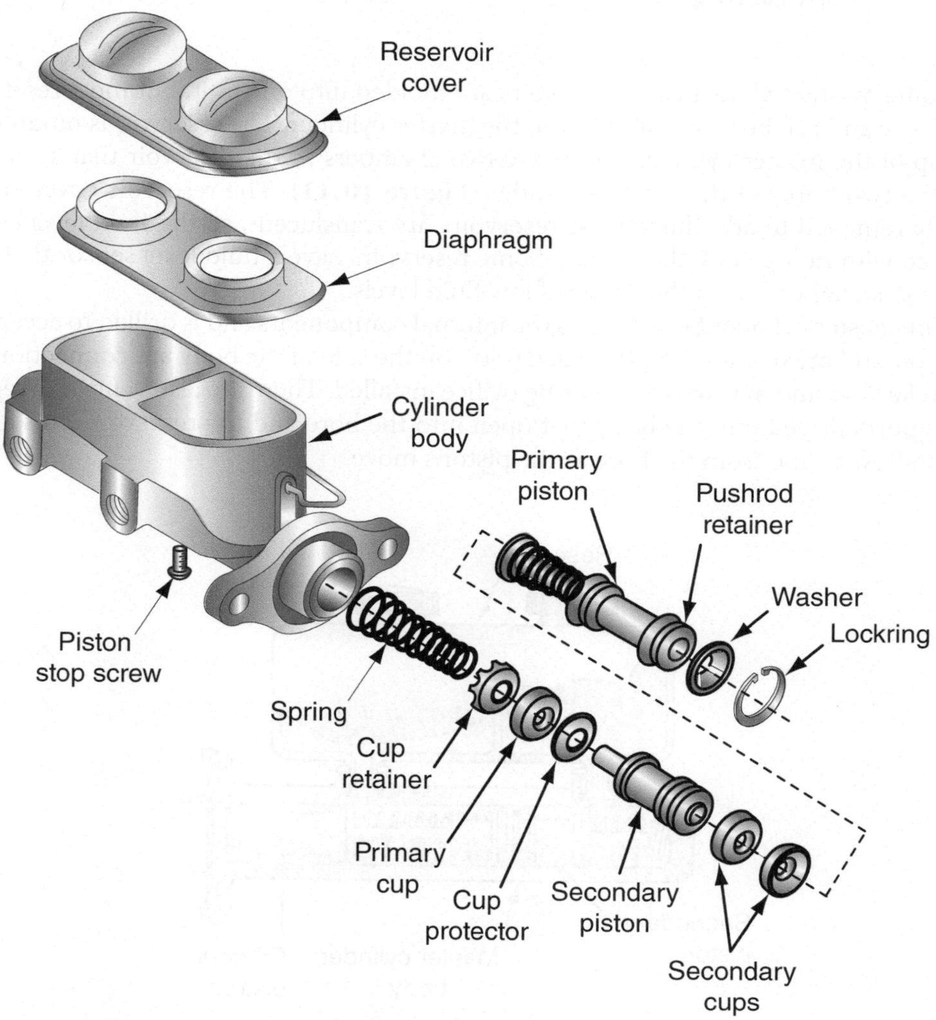

Figure 10-12 Internal components of a master cylinder.

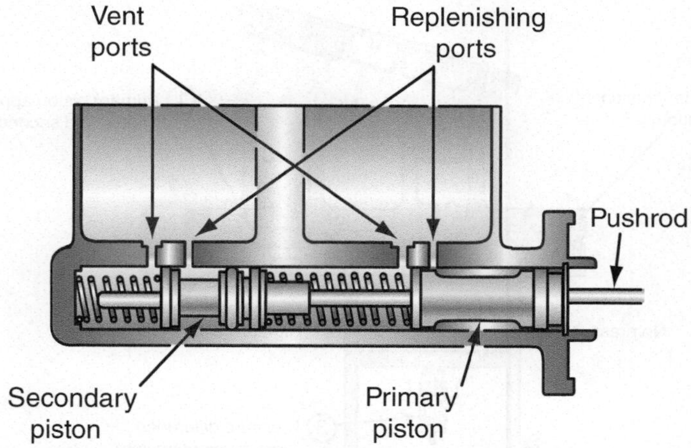

Figure 10-13 Note that each piston has its own reservoir and ports. This prevents the reservoir from being completely drained if a leak exists in one system.

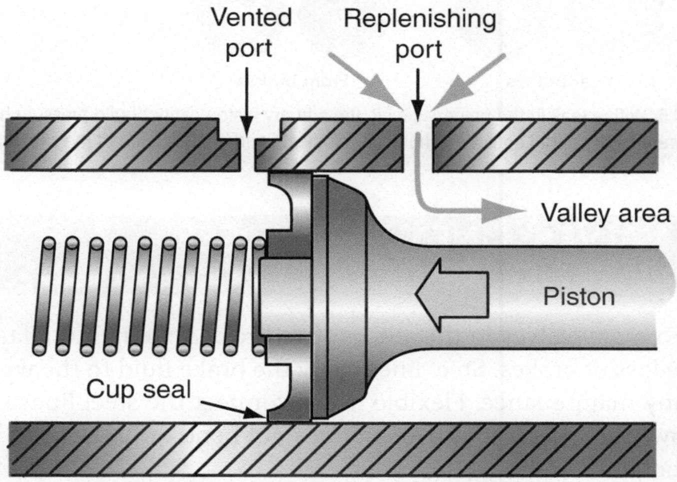

Figure 10-14 The replenishing port keeps the chamber behind the piston head full as the piston moves forward.

within the chamber for instant braking. When the brake pedal is applied, the pistons move forward, and the vent and replenishing ports are closed to the compression chamber.

The dual master cylinder was required to allow the driver to retain at least some braking if there is a leak in the system. With the single-piston system, a leak anywhere in the system may cause a complete loss of braking. The dual system reduces that possibility. The primary piston has an extension screw that extends into the compression chamber that can apply the secondary piston. If a leak occurs in the rear brake system and the brake pedal is depressed, the primary piston moves in its normal manner. However, the fluid ahead of that piston leaks out and no pressure is created. The secondary piston will not move because of the lack of pressure. The screw on the primary piston will contact and mechanically push the secondary piston forward (**Figure 10-15**). In this manner, the secondary piston can charge the front brake system and the driver will have some braking action available. If a leak occurs in the front system and the brakes are applied, the primary piston charges the rear brake system, but because the front circuit is leaking, the secondary piston must be pushed completely forward to its stop before pressure can build in the rear brakes. Both conditions will cause the brake pedal to drop drastically as the brakes are being applied.

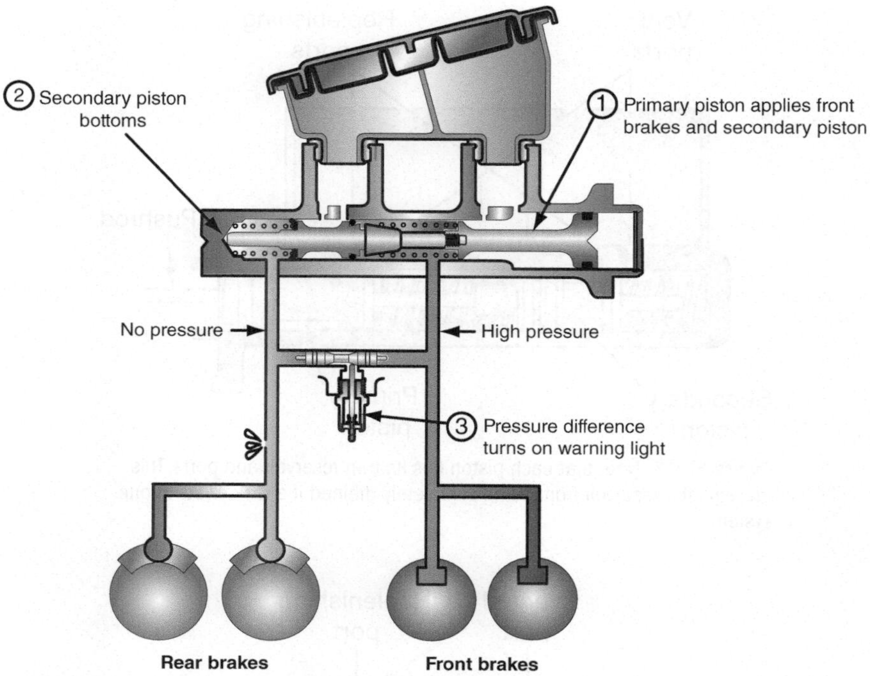

Figure 10-15 With a leak in the primary circuit, the primary piston cannot build pressure but is pushed forward by the pedal until it mechanically connects and applies force to the secondary piston.

CONTROL VALVES, LINES AND HOSES, AND WARNING LIGHTS

There are not too many valves in the automotive brake system, but a failure of one may cause a complete loss of brakes. Steel lines carry the brake fluid to the wheels and usually do not require any maintenance. Flexible hoses connect the steel lines to the individual wheels. This allows the wheels to move but does not bend or break solid lines. The hoses must be inspected and replaced at intervals.

Metering Valves

The **metering valve** controls hydraulic pressure to the front wheels during initial brake application.

Vehicles with front disc and rear drum brakes must hold the front brakes off until the rear brakes move enough to contact the drums. The **metering valve** accomplishes this task. As the brakes are applied, the disc brake reacts instantly, but the rear brake shoes must move far enough to contact the drum. In addition, enough pressure must be created to overrun the return spring on the shoes. If the front brakes were applied at the instance of brake pedal movement, they would probably lock down before the rear ones would begin to brake.

When the brakes are applied on a disc/drum system, the metering valve blocks fluid to the front until pressure builds in the rear system (**Figure 10-16**). This ensures the rear brake shoes are in contact with the drum before the front brakes begin to function. Once opened, the metering has no other function during the braking action. Operation of the metering valve could possibly be noticed during light braking or upon initial braking.

Proportioning Valves

To control the vehicle during hard braking, it is necessary to apply more brake action at the front of the vehicle than the rear of the vehicle. Many brake systems may direct up to 90 percent of the braking action to the front wheels. During heavy braking, this could still cause the rear brakes to lock up. The proportioning valve divides the braking pressure between the front and rear axle to reduce or eliminate rear-wheel lockup.

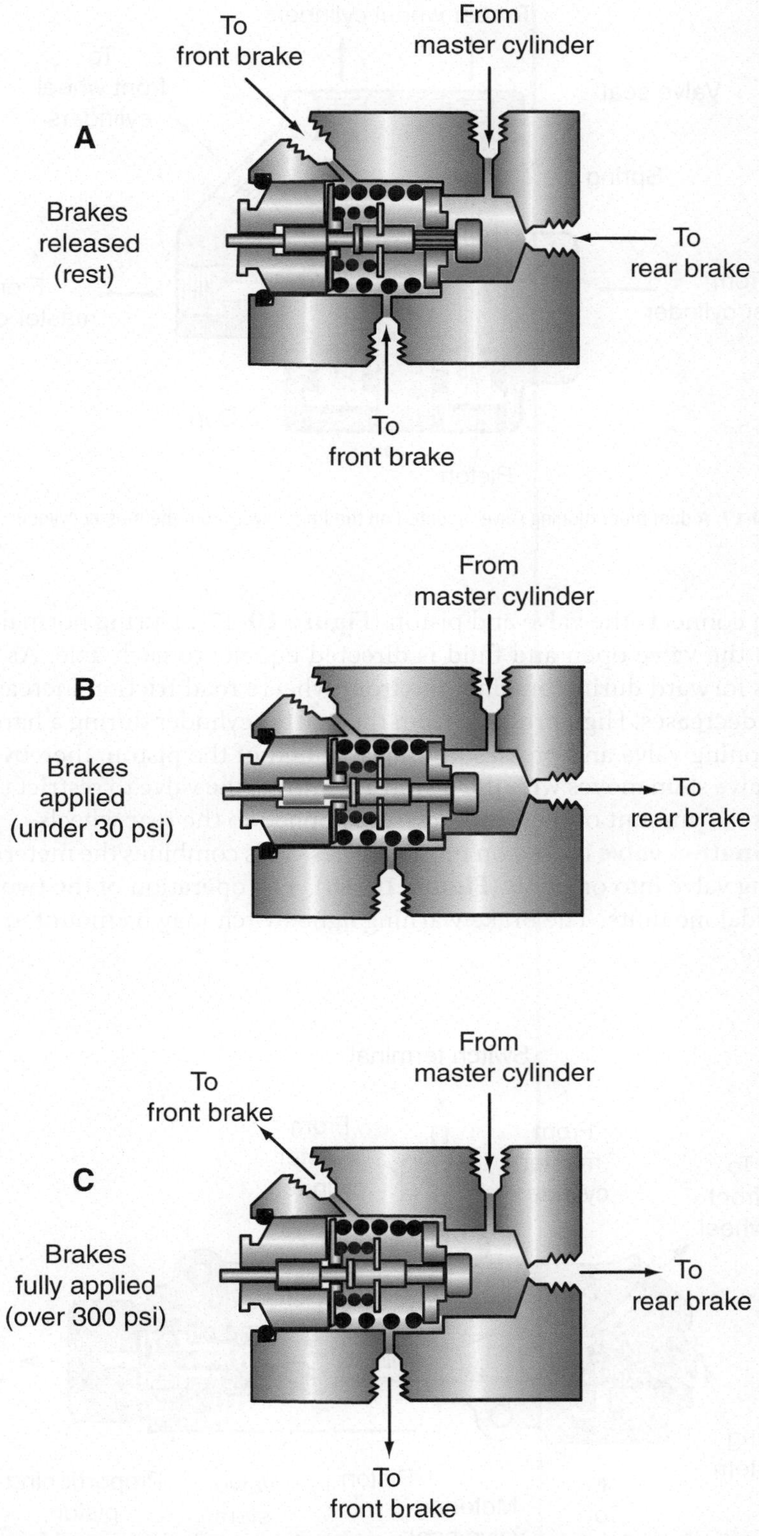

Figure 10-16 Under 30 psi (B) the valve to the front brakes is closed while allowing fluid to the rear drum brake. Over 30 psi, the larger valve is forced open, letting fluid to the front brakes.

As the pressure from the master cylinder changes, the **proportioning valve** operation also changes, thereby increasing or reducing the pressure to the rear wheels.

The **proportioning valve** is basically a block of metal drilled to accept line fittings, a valve, a piston, and a spring. The lines connected to the valve are from the master cylinder and rear brakes. The piston has a small and a large end with the small end at the valve.

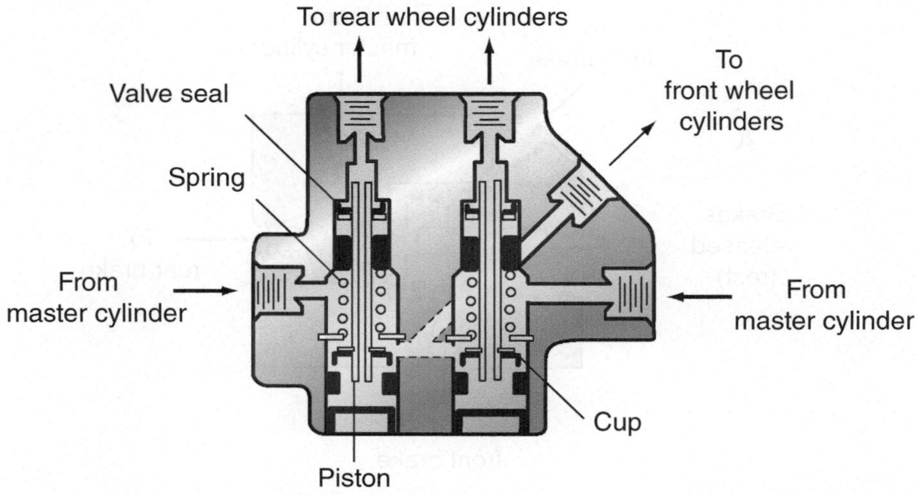

Figure 10-17 A dual proportioning valve mounted on the frame away from the master cylinder.

A valve stem connects the valve and piston (**Figure 10-17**). During normal braking, the spring holds the valve open and fluid is directed equally to each axle. As the vehicle's weight shifts forward during braking, the front wheel's road friction increases while the rear contact decreases. High pressure from the master cylinder during a hard stop enters the proportioning valve and presses on the large end of the piston, thereby forcing it to move. The valve stem moves with the piston and moves the valve to restrict the inlet port. This reduces the amount of fluid and pressure applied to the rear wheels.

A **combination valve** is a single unit that meters and proportions pressures.

A **combination valve** is used on most vehicles. This combines the metering valve and proportioning valve into one unit (**Figure 10-18**). The operation of the two valves is the same as standalone units. The brake warning light switch may be mounted on the combination valve.

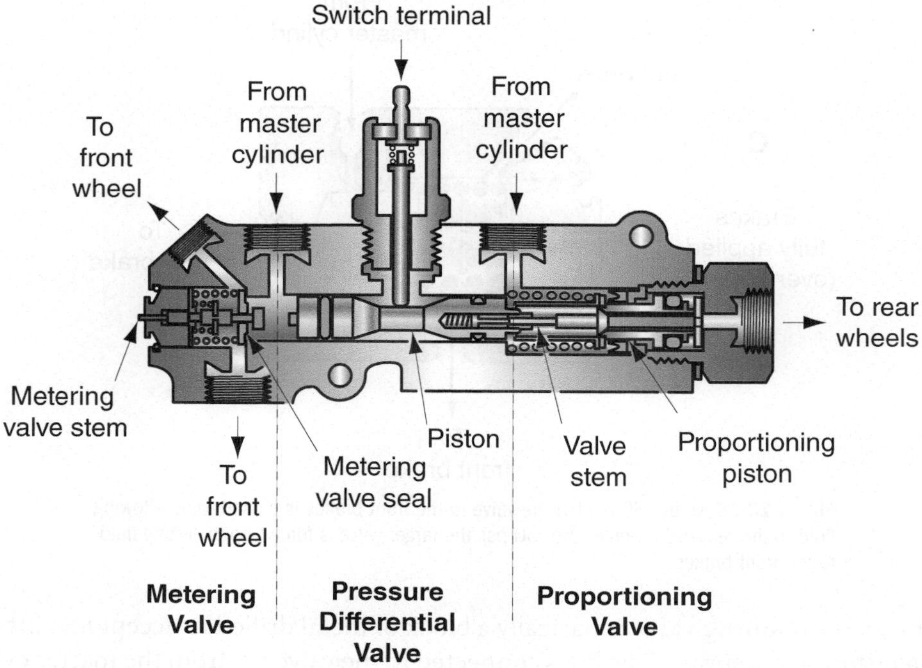

Figure 10-18 A combination or three-function valve.

Dynamic proportioning is accomplished without the mechanical and/or hydraulic valves already discussed here on vehicles equipped with antilock brake systems. Proportioning is done by electronically controlled solenoids within the ABS hydraulic modulator.

Dynamic proportioning is the use of electronically controlled valves to meter and proportion pressures.

Load-Sensing Proportioning Valves

A truck is extremely light in the rear compared to the front. This is particularly important when the truck is being heavily braked. During braking, the truck's weight shifts to the front and the lightened rear wheels lose traction with the road. If the wheels lock up and skid, the truck can go out of control. A **load-sensing proportioning valve** helps prevent the lockup by reducing the rear brake pressure during braking.

A **load-sensing proportioning valve** is a mechanically controlled pressure regulator.

One type of load-sensing valve uses the movement of the frame and rear axle to regulate pressure to the rear brakes. The valve bracket is mounted to the frame with a lever attached to the axle (**Figure 10-19**). As the rear of the frame rises during braking, the axle pulls down the lever. The valve attached to the lever restricts the amount of fluid and pressure directed to the rear wheels.

Another type of load-sensing valve is mounted on the rear of the combination valve. It has a ball that moves with the tilt of the vehicle. As the rear of the vehicle rises during braking, the ball moves forward against a ball valve. The ball valve reduces the amount of fluid and pressure to the rear wheels.

Lines and Hoses

Brake lines are made of steel with flared fittings at each end. The steel lines connect the master cylinder to the valves and the valves to the wheels. But the steel lines do not extend completely to the wheels. Flexible hoses connect the steel lines at the frame to the brake system at the wheels (**Figure 10-20**). This is required because of the almost constant movement of the wheel assembly during driving. A steel line would quickly break if it were continuously flexing. Although steel lines almost never require maintenance, the flex hoses should be inspected each time the wheels are serviced. They will deteriorate and bulge due to chemical actions and age. They can also have interval damage that is not visible.

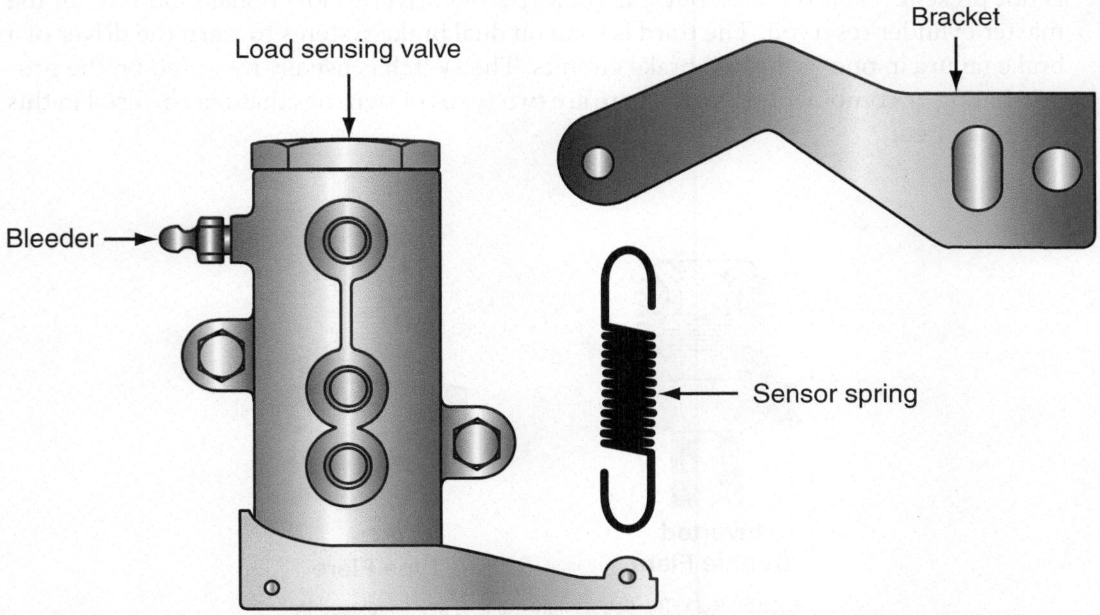

Figure 10-19 The height-sensing proportioning valve is usually mounted between the frame and either the rear axle housing or one side of the rear suspension.

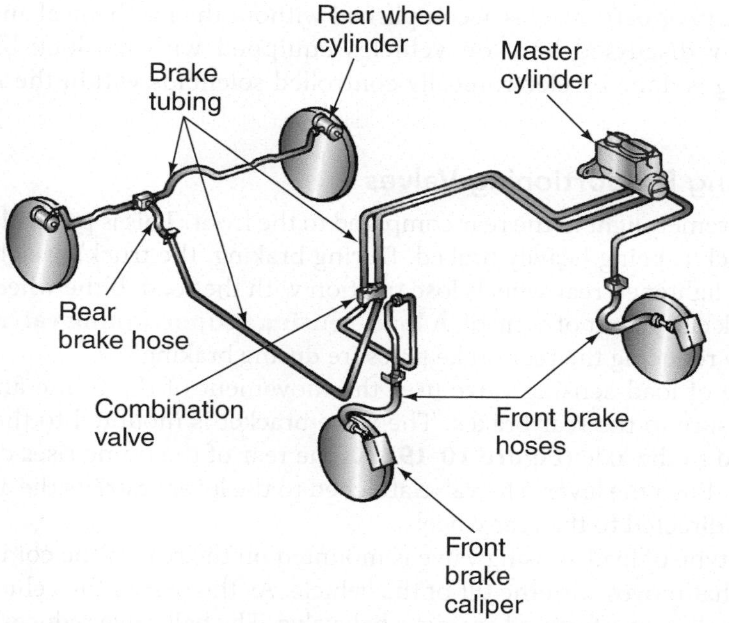

Figure 10-20 Brake fluid is contained in a sealed system consisting of metal lines and flexible hoses.

A **double flared** fitting has the end of the tube/line that is flared and flared again inwards to form a tight hydraulic seal.

ISO are the initials for the International Standards Organization.

A **brake warning light** is mounted in the dash to warn the driver of a brake system fault.

Brake lines have two types of fittings. The older type is known as a **double flare** fitting. The newer type is an **ISO**-type flare (**Figure 10-21**). Each requires a matching fitting and **CANNOT** be interchanged. The ISO-type provides a better seal.

Warning Lights

The **brake warning light** is used to alert the driver to a problem in one of the systems. There is usually one red warning light located in the instrument panel. It is turned on by any one of three possible switches or sensors. The first switch is on the parking brake and reminds the driver that the parking brake is applied. The second, a brake fluid level sensor, is not present on all vehicles, but if used, alerts the driver to low brake fluid level in the master cylinder reservoir. The third is used on dual brake systems to warn the driver of a brake failure in one of the two brake circuits. The switch is usually mounted on the proportioning or combination valve. There are two types of switches that may be used in this warning circuit.

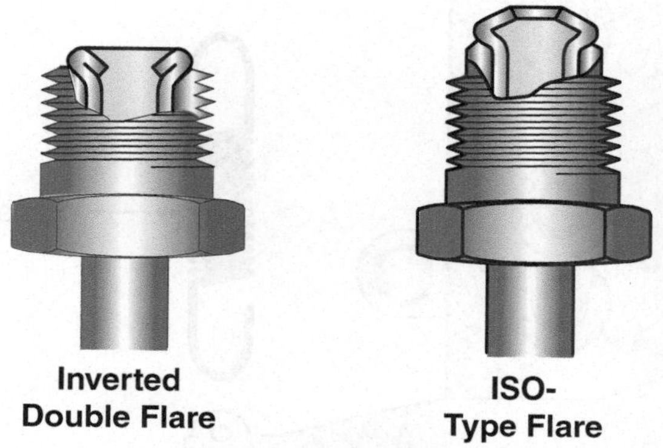

Figure 10-21 The double flare fitting is being replaced by the ISO fitting. One cannot replace the other unless both sides of the connection are changed to one type of fitting.

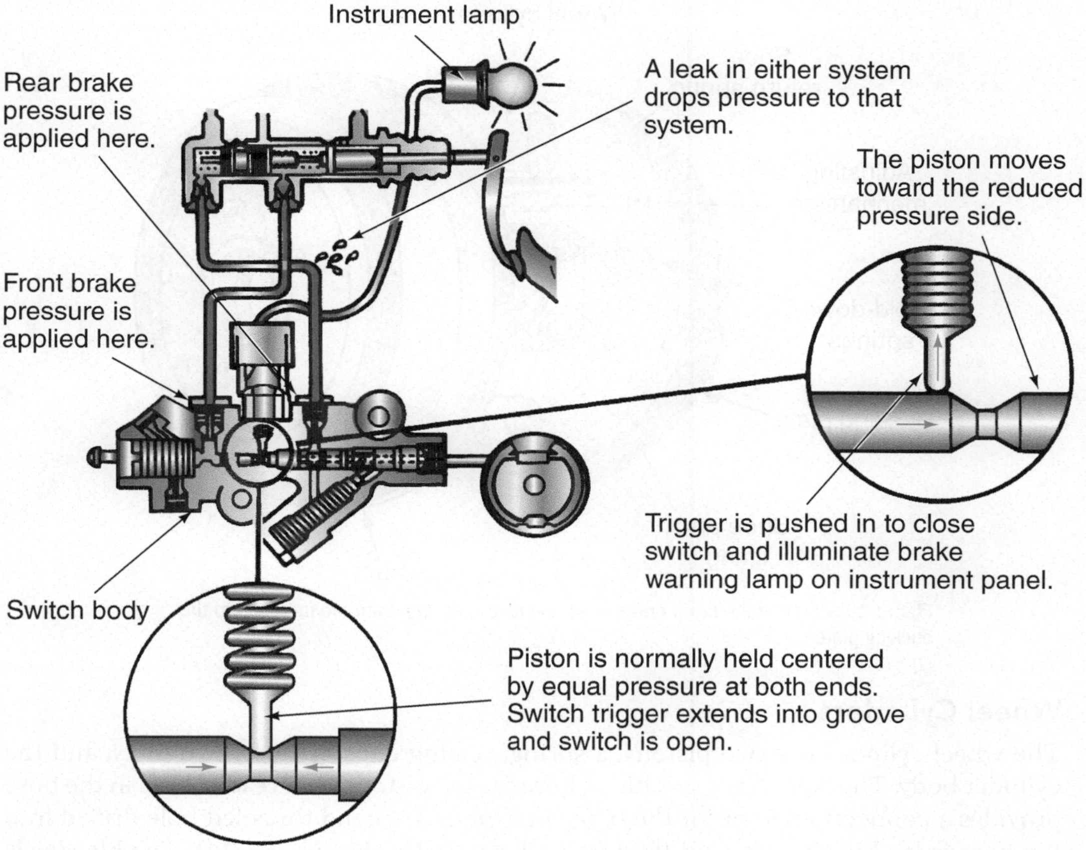

Figure 10-22 The warning light switch is closed by unequal pressures at the outer ends of the piston.

On one type, the switch plunger sets in a slot at the center of a long piston in the proportioning valve. The piston is held in the center by equal hydraulic pressures at each end of the piston. When a leak occurs, and pressure drops on one end of the piston, the pressure at the other end moves the piston out of the center. The shoulder of the piston slot pushes up on the switch plunger to close the light circuit (**Figure 10-22**). At the same time, the piston closes the outlet ports of the system that is leaking.

Another type of warning light switch is the hourglass-shaped type. The principle of operation is the same, but the switch's ground contact extends into the slot. As the piston moves off center, the shoulder of the slot touches the contact and grounds the switch, closing the circuit.

Shop Manual
pages 458–462

DRUM BRAKE COMPONENTS AND OPERATION

The brakes on all vehicles consist of a **friction material** clamped against a rotating device by hydraulic pressure. There are two basic braking systems based on how the friction material and rotating device interact with each other. The oldest system is drum brakes.

Friction material is specially formatted material used for brake pads and brake shoes. It is designed to withstand high temperature while minimizing pad and rotor wear.

Asbestos was used for many years as brake friction material. Because of the health dangers of asbestos, most brake friction material is now made of organic materials or compounds.

The wheel components in a drum brake system are the wheel cylinder, drum, shoes, adjuster, and mounting hardware (**Figure 10-23**). The hydraulic pressure is delivered from the master cylinder via steel and flexible lines. Once inside the wheel cylinder, the pressure is applied against two pistons.

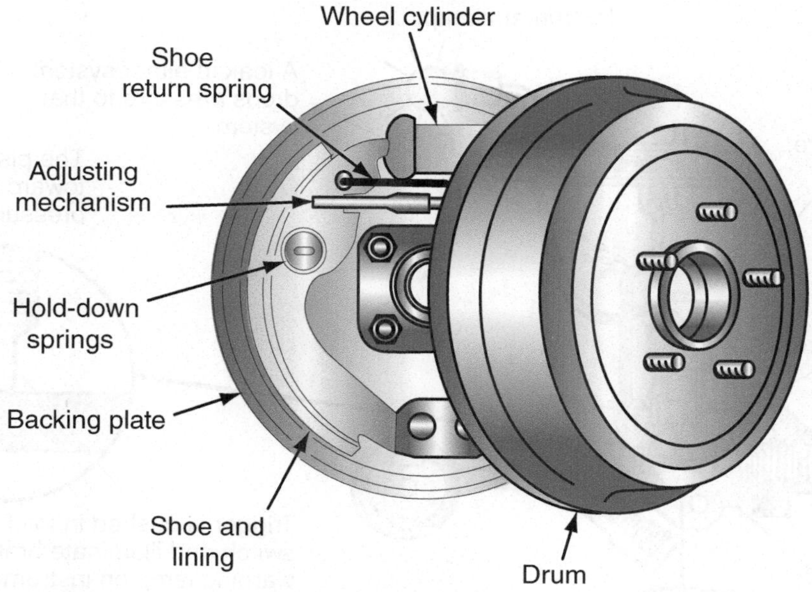

Figure 10-23 All of the drum brake components except the drum are mounted to the backing plate.

Wheel Cylinders

The wheel cylinder has two pistons, a spring, sealing cups or seals, two links, and the cylinder body. The body has a machined bore for the pistons. A threaded hole in the bore provides a connection point for the brake line while a second threaded hole drilled into the bore is for bleeding air and fluid from the cylinder (**Figure 10-24**). The bleeder is always at the highest point of the bore when the cylinder is mounted on the vehicle. Within the bore are pistons, cups, and a spring. The spring is placed between the two pistons and cups to keep all internal components in place and in contact with each other during brake release. This spring does not return the pistons or cups to their starting positions when the brakes are released.

The two cups act as seals when hydraulic pressure is applied to the bore. The interiors of the two cups face each other and the center of the bore. The outside faces of the cups are flat against the inner face of its mating piston (**Figure 10-25**). As the pressure builds

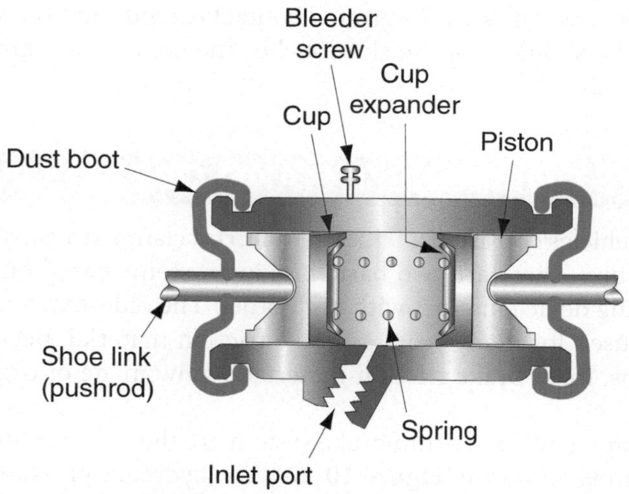

Figure 10-24 Most wheel cylinders have two operating pistons moved by hydraulic pressure trapped between them.

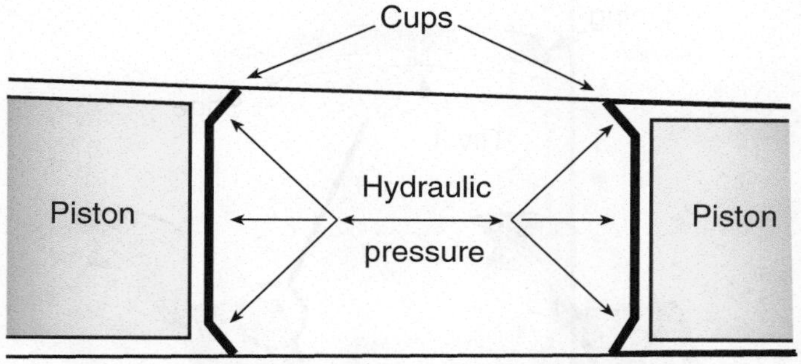

Figure 10-25 The hydraulic pressure applied against the cups seals the wheel cylinder's pressure chamber.

during brake application, the outer edges of the cup are forced outward against the bore to provide a more effective seal.

The pistons are placed at each end of the bore with the spring, cups, and hydraulic fluid trapped between them. The outward face of the piston is usually cupped to provide a mounting point for one end of a link. The links are the connecting point between the pistons and the brake shoes (**Figure 10-26**). Around each end of the cylinder is a boot, which serves as a dust cover to protect the bore. As hydraulic pressure is increased within the bore, the cups, pistons, and links move outward to apply force against the two brake shoes.

The brake **shoe web** is the metal skeleton used to mount the friction material *(lining)* and attach the shoe to its mounting.

Lining Brake lining is the part of the shoe or pad that actually contacts the metal brake disc (rotor) when engaged.

Brake Shoes

The brake shoes provide a mounting area for the friction material. The metal portion of the shoe is normally referred to as the **shoe web**. The friction material, normally referred to as **lining**, may be bonded (glued) or riveted to the web. Brake shoes are curved to match the inside wall of the drum (**Figure 10-27**). Return springs hold the upper ends of the shoes tight against the wheel cylinder links. Retainer springs attach the shoes to the backing plate, but allow the shoes to move out or in as the brakes are applied and released. The backing is bolted to the hub and acts as a rear cover and a mounting platform

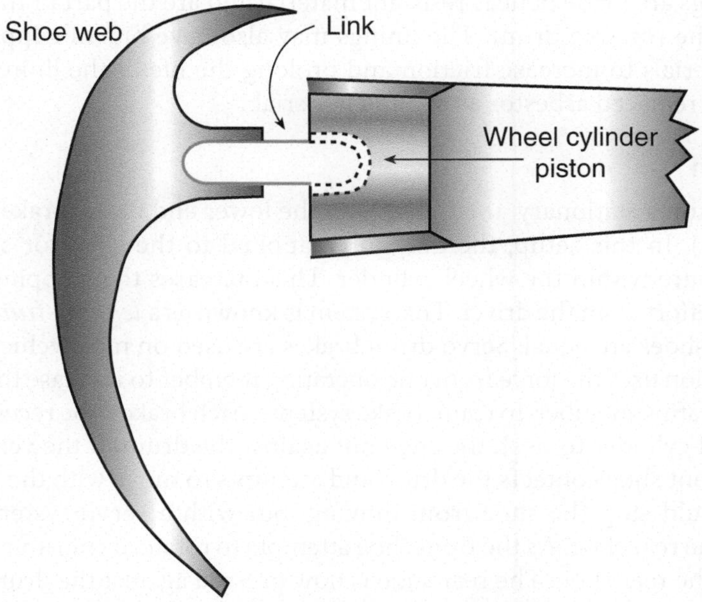

Figure 10-26 The link makes the connection between the wheel cylinder piston and the brake shoe.

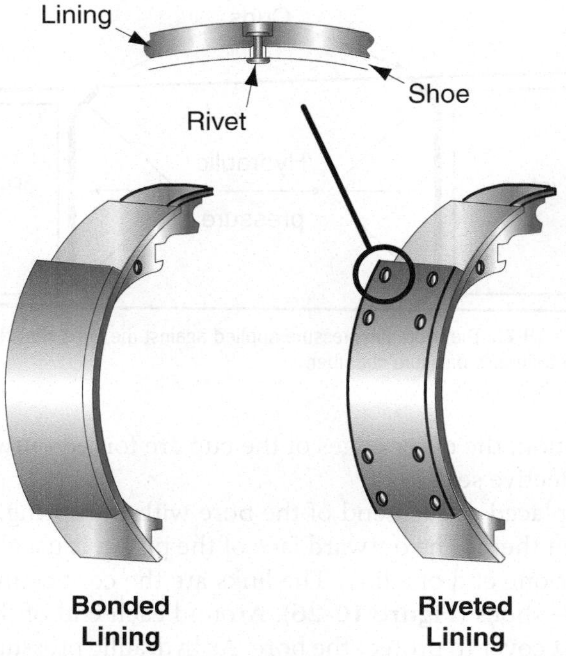

Figure 10-27 The bonded shoe is more prevalent on light vehicles, and the riveted shoe is most often, but not always, found on heavier vehicles.

for the drum brake components. There are several methods to attach the shoes at the bottom, and the type of attachment at that point can help increase braking power.

Brake Linings

The lining thickness, position on web, and shape may be different on the brake shoe set for one wheel. The secondary shoe may have a longer and thicker lining than the primary shoe. The shoes must be properly placed during installation or brake failure or excessive wear could occur.

Brake linings are made of heat-resistant material and are the part of the brake system that contacts the rotating drum. The linings may also have bits of copper, brass, zinc, and other materials to increase friction and prolong the life of the lining. This type of compound has replaced asbestos as a lining material.

Servo Action

The pressurized fluid trapped between the pistons and shoes acts as one **anchor** for the shoes.

Some systems use a stationary **anchor** to hold the lower end of the brake shoes in place (**Figure 10-28**). In this setup, the only force applied to the shoe for stopping is the hydraulic pressure within the wheel cylinder. This increases the stopping distance and requires more effort from the driver. This system is known as a *leading-trailing system* and the four brake shoes are equal. Servo drum brakes are used on most vehicles.

A servo action uses the force from one operating member to increase the force applied to another operating member. In drum brake systems, each brake shoe receives equal force from the wheel cylinder to push the shoe out against the drum. If the vehicle is moving forward, the front shoe contacts the drum and attempts to move with the rotating drum. An anchor would stop the shoe from moving, but with a servo system, this shoe is connected to the rear shoe. As the front shoe attempts to rotate, it transmits force through the linkage to the rear shoe. The rear shoe is now pressed against the drum by the wheel cylinder and the servo action. The upper end of the rear shoe cannot move inward because of the pressurized brake fluid trapped between the two-wheel cylinder pistons. In this

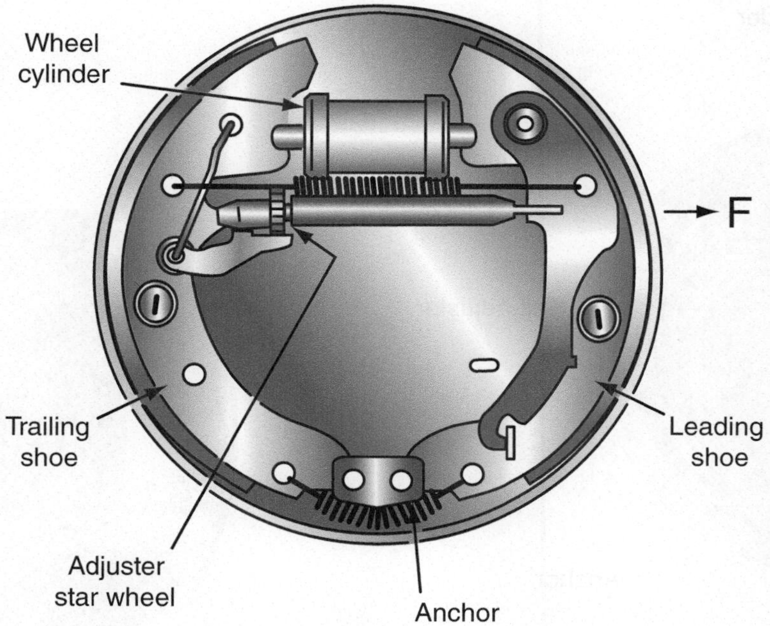

Figure 10-28 A typical setup for a leading-trailing drum brake.

manner, the driver seats the shoes against the drum, and the force of the rotating drum is also applied against the rear shoe. Stopping power is increased, and the driver's effort is decreased. Note that the secondary or rear shoe performs most of the braking effort. This effect is called **self-energizing**.

Self-energizing The actual brake lining drag force to increase the force with which the force is applied.

The majority of vehicles equipped with drum brakes usually have duo-servo action. The servo effect is possible in both forward and rearward braking with duo-servo systems, in which the secondary shoe lining is longer than the primary's. The technician who works on antique or classic vehicles may encounter a servo system where the servo action is only accomplished when the vehicle is braked during forward travel.

Brake Self-Adjusters

Drum brake systems require some type of mechanism to adjust the shoes outward as they wear. Older vehicles and some larger trucks require the technician or driver to adjust the brakes at regular intervals, normally during routine preventive maintenance. However, most light vehicles use automatic self-adjusters. The adjuster is the connecting link between the upper or lower ends of the brake shoes (**Figure 10-29**). A spring holds the two shoes and adjuster in contact with each other. The adjuster has a starred wheel, which is threaded onto an adjusting screw and is normally turned by the movement of the shoes. An adjusting lever contacts the adjusting screw and is moved by the shoe as the brakes are applied. If the shoes are worn and must travel more than a set distance for contact with the drum, the adjusting level moves further than usual. The additional motion of the lever causes a one or two tooth rotation of the adjuster. Additional adjustments are made each time the brakes are applied. If the shoe movement is short, the level cannot move far enough to turn the adjuster.

Some newer systems use the application of the parking brake to adjust drum brakes. Although this system works well, many drivers do not realize that the parking brake must be used regularly to adjust the brakes. The adjustment works the same way except for the operating member that performs the rotation of the adjusting screw. Most non-servo systems and many other vehicles have the adjuster at the top end of the shoes under the wheel cylinder. Most vehicles have a slot either in the drum or backing plate to allow a brake adjuster tool to be inserted for manual adjustment. This is primarily done to adjust the brakes after repairs have been completed.

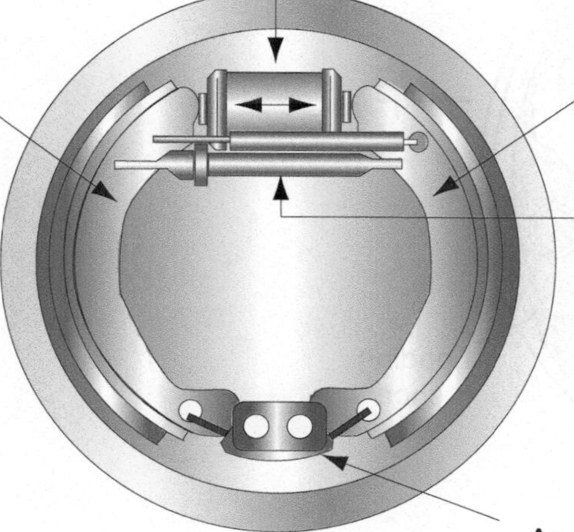

Figure 10-29 The self-adjuster for a leading-trailing drum brake is usually connected between the tops of the two shoes. On a duo-servo system, the self-adjuster replaces the anchor shown in the illustration.

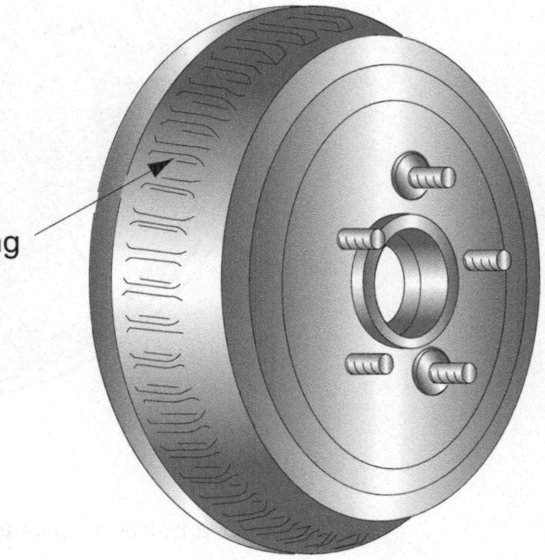

Figure 10-30 Many larger brake drums have fins molded to the outside to enhance cooling of the brakes.

Drums

The **drum** and hub may be manufactured as a one-piece component.

The **drum** covers the brake mechanism and is attached to the axle or hub (**Figure 10-30**). The drum is drilled to fit over the lug studs. The interior friction area of the drum is machined to provide a smooth braking surface. The installation of the wheel assembly and lug nuts hold the drum in place. Some manufacturers use additional fasteners that hold the drum to the axle or hub. The friction area of the drum can be machined if grooved or otherwise damaged or worn. The inside diameter must be measured and compared to the manufacturer's specifications. A drum that is too thin may break during braking.

DISC BRAKE COMPONENTS AND OPERATION

Shop Manual
pages 437–442

Disc brakes react quicker and operate differently than drum brakes. Disc brakes are pressurized and controlled the same way as drum brakes. The components of a disc brake are the pads, rotor, and caliper. Many vehicles use disc brakes at all wheels.

Pads

Brake pads and **disc pad** are names for the same component. They are friction components placed between the brake caliper and the brake rotor.

Disc **brake pads** are much smaller than drum brake shoes in area (**Figure 10-31**). The friction material is either bonded or riveted to a small metal pad. There are two pads per wheel, an inner and outer. The pads may be identical in shape and size, but most have some differences between the inner and outer pads. Between the two pads is the brake disc or rotor.

The lining on a pad is composed of materials like the ones used on drum brake linings. However, a disc pad lining is much harder than the drum linings. The pad lining must be harder because of its smaller size and the increased pressure or force applied to it during braking. Because the **disc pad** friction area is smaller than the drums' and there is no servo action, greater force is needed for braking. This is accomplished by using larger pistons than those used on drum brakes.

Figure 10-31 Typical disc brake pads.

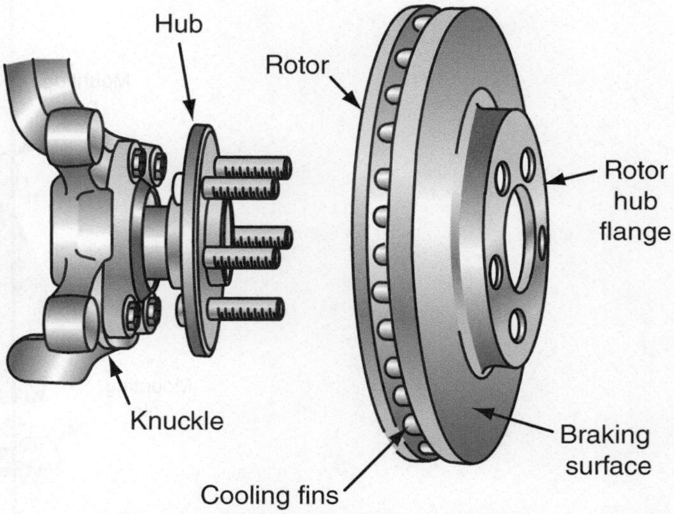

Figure 10-32 Shown is a typical floating rotor or a two-piece rotor and hub assembly. Note the cooling fins between the braking surfaces.

Rotors

The **rotor** is attached to the hub or axle and rotates with the wheel. Each side of the rotor is machined for smooth braking (**Figure 10-32**). As the brakes are applied, the two pads clamp against the rotor and slow it. When the brakes are released, the pads are moved slightly backward, but they should never lose complete contact with the rotor. The rotor is subject to great heat during braking. Excessive and hard braking will cause the rotor to warp. The rotor can be machined if the warpage is not too severe and the rotor is not too thin.

Like drum brakes, the **rotor** and hub may be manufactured as a one-piece unit.

Calipers

The caliper is mounted at the wheel and most extend over the rotor in a U-shaped mounting. Most calipers are mounted to the steering knuckle using an adapter. The caliper may have one, two, or four pistons. The pistons in most calipers are larger than those found in a wheel cylinder. The caliper may be sliding or fixed. It is drilled to accept the line fitting and a bleeder valve.

The operation of the **sliding caliper** is like the actions of a wheel cylinder except the caliper can move as a unit. As pressurized fluid enters the piston bore, the piston moves out, pushing the inner pad against the rotor (**Figure 10-33**). The outer pad is on the opposite side of the rotor and held in place by the caliper portion that extends over the rotor. Because any action creates an equal and opposite reaction, the caliper moves away from the rotor. This action draws the outer pad against the outer edge of the rotor. More force applied against the piston and inner pad results in the same amount of force being applied to the outer pad. A sliding caliper may have one or two pistons mounted in the caliper.

A **fixed caliper** usually has four pistons, two on the inner side and two on the outer side of the rotor. It requires a much larger caliper and the additional pumping to fill, equalize, and bleed the additional cavities for the two outer pistons. As the fluid enters the cavities, the pistons are forced outward, thereby clamping the rotor. This is an older system and is not seen often on newer light vehicles.

There is no manual adjustment on a disc brake system and no return springs. Each bore is fitted with a **square-cut O-ring** near the outer end of each piston. The ring acts as a seal and adjuster. As the piston moves outward on each brake application, the O-ring distorts slightly (**Figure 10-34**). When the brakes are released, the square ring returns to its original condition, pulling the piston and pad back from the rotor (**Figure 10-35**).

A **sliding caliper** is mounted so it can move on slides during brake application. It only needs one in-board piston. This type may also be known as a floating caliper.

A **fixed caliper** is connected to the steering knuckle and does not move during brake application. It requires at least two pistons, one in-board, one out-board.

A **square-cut O-ring** is used to seal the caliper piston within the caliper bore. It also acts as a means to retract the brake pads during brake release.

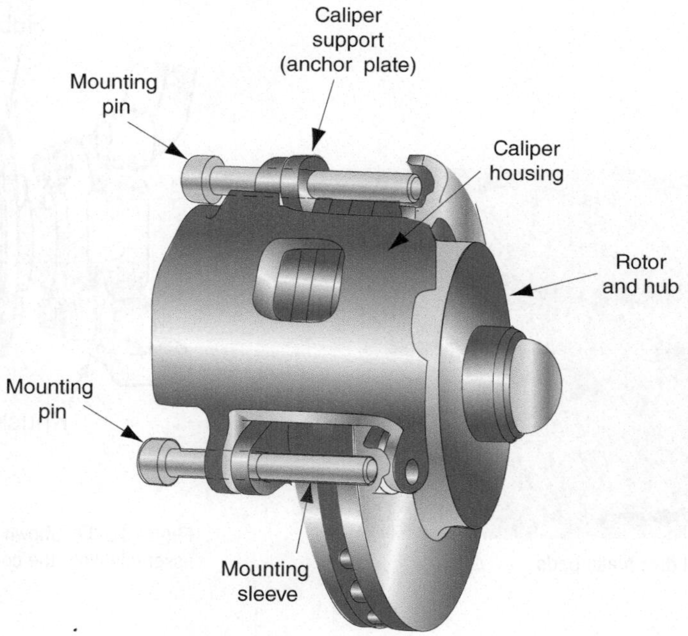

Figure 10-33 The sliding caliper moves on its mounting pins so equal force can be applied by both pads.

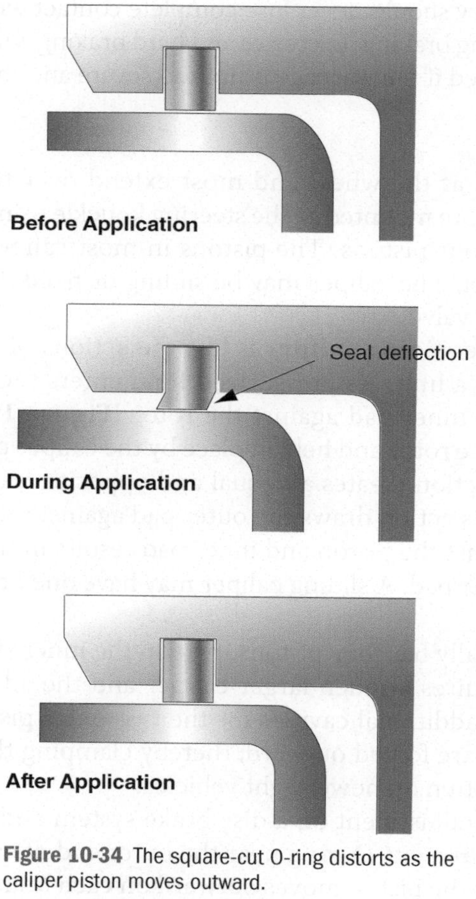

Figure 10-34 The square-cut O-ring distorts as the caliper piston moves outward.

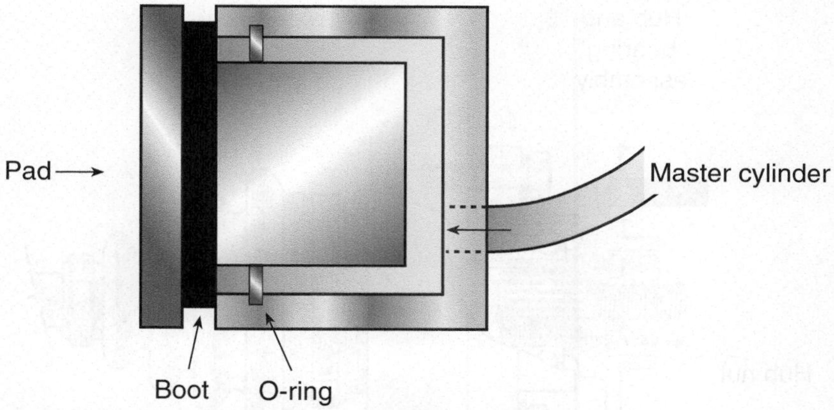

Figure 10-35 The O-ring's normal shape with brakes released.

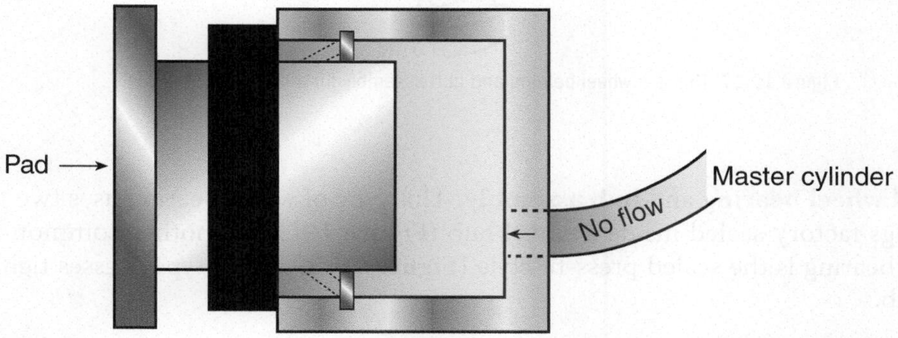

Figure 10-36 The square ring can only return the piston a short distance, which allows the piston to adjust for pad wear.

As the pads wear, the piston moves further outward and the distorted ring slides along the piston. When it recovers its shape after brake release, the piston is further down the bore (**Figure 10-36**). This action continually keeps the pads flat against the rotor and allows for the adjustment needed as the pads wear.

PARKING BRAKE OPERATION

Commonly referred to as emergency brakes, parking brakes are designed to hold the vehicle still after it has been stopped. They function only on the rear axle in passenger cars and light trucks. A foot- or hand-operated lever near the driver is used to set the brake. A foot pedal, when used, is located at the left kick panel near the driver's left foot. If a lever is used, it is usually placed between the two front seats. A typical system uses cables to connect the driver's action to a lever or cam in the rear brakes to lock the shoes to the drum. Some rear discs have small drum/shoe mechanisms on the inside of the rotor for parking. Other rear discs have screws that force the pads to clamp the rotor when the parking brake is set.

Shop Manual
pages 450–451

Wheel Bearings

A wheel bearing is usually located near or close to the brakes. It is a combination of two tapered roller bearings inside the wheel hub (**Figure 10-37**). Wheel bearings can be found in different forms. The most common type of wheel bearing on a modern vehicle is the

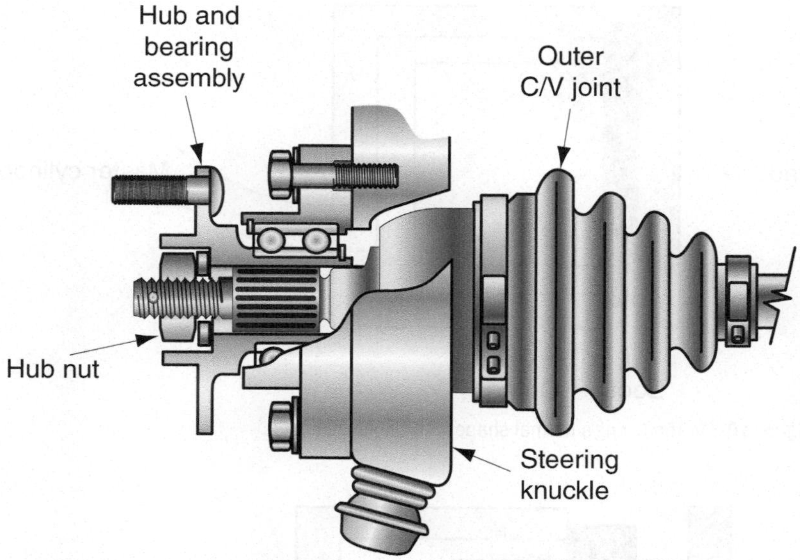

Figure 10-37 This is a wheel bearing and hub assembly for a driving axle.

Seal wheel bearing Also known as a hub assembly is a component that helps decrease friction when the wheel is turning by the use of ball bearing that are sealed from debris and moisture.

Bearing race The metal ring with grooves that hold together a set of ball bearings.

sealed wheel bearing and **hub assembly.** This type of wheel bearing uses two tapered bearings factory sealed inside a wheel hub (**Figure 10-38**). Another common type of wheel bearing is the sealed press-fit style (**Figure 10-39**). This type presses tightly into the hub.

Bearing Races

Bearing races serve the same purpose as the journal. However, races are usually purchased as part of the replacement bearing (**Figure 10-40**). A race can be assembled as part of a bearing assembly, or it can be a loose part packed and shipped as part of the assembly. Many wheel bearings have a race that must be press-fitted into the wheel hub before the bearing is installed. If either a bearing or a race fails, replace both.

Figure 10-38 This sealed hub bearing has two bearing rows inside it. This wheel bearing assembly is bolted to the knuckle.

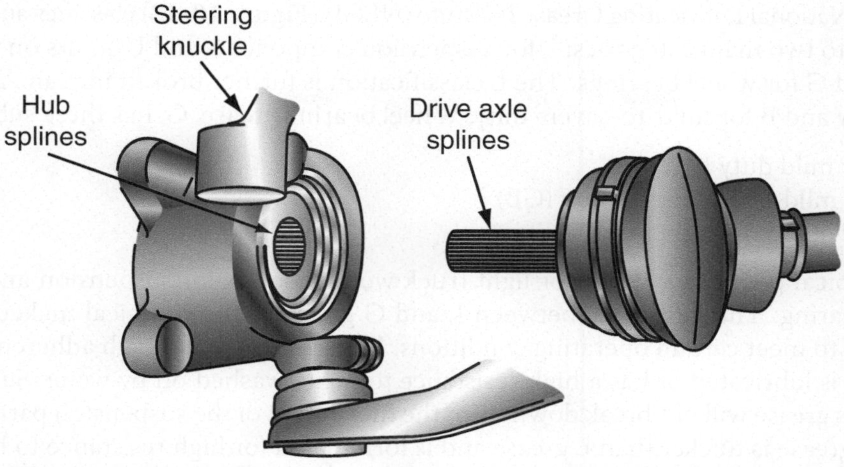

Figure 10-39 Some wheel bearings are sealed and have to be pressed into a steering knuckle.

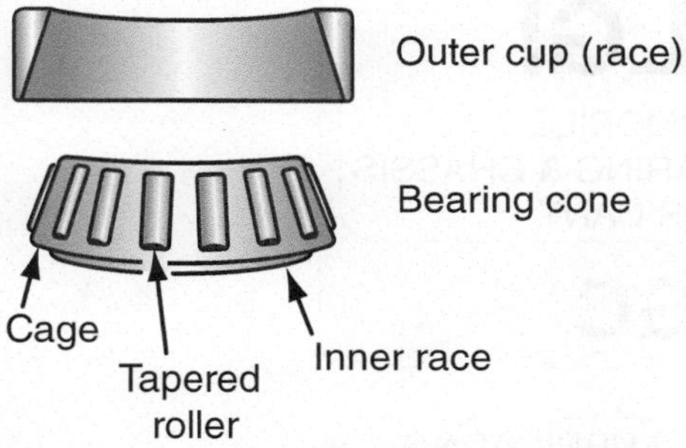

Figure 10-40 A tapered roller bearing is used almost exclusively on a vehicle's nondriving wheels. Shown is a complete tapered roller bearing assembly. The outer race is separate and must be driven into the hub with a special tool.

Figure 10-41 Bushings are normally used to support one end of a shaft that has light loads placed upon it. Some alternators have a bushing at the end of the rotor shaft.

Bushings

Bushings are made of thin steel tubing cut to specific lengths. Usually, they are copper coated, but other materials may be used for special situations. The bushing will not axially support loads very well because their purpose is to limit the radial movement of a shaft. Rotating shafts must be kept in a straight line to prevent damage to the shaft and component. Bushings form a journal on which the shaft can ride (**Figure 10-41**). However, bushings are not used in high-thrust or radial load areas like the camshaft or crankshaft. They can support a transaxle shaft or the shaft in an alternator or pump. The bushing is press-fitted into a housing and the shaft slides through it.

Bushing The inner surface between two moving parts used to reduce friction of a rotating shaft.

LUBRICANTS

Grease is a lubricant used on roller and ball bearings and suspension system components. Many bearings, like those used in the engine, transmission, and differential, are lubricated by the oil stored and used within the component. Other components, like the wheels and suspension, require the parts or bearings to be packed with grease.

Grease is a lubricant that reduces friction and smoothes movement between parts of components.

Shop Manual
page 453

The National Lubricating Grease Institute (NLGI) (**Figure 10-42**) classifies automotive grease into two main categories: L for suspension components and U-joints on the drive shaft, and G for wheel bearings. The L classification is further broken into an A class for mild duty and B for mild-to-severe duty. Wheel bearing grease, G, has three subclasses:

A for mild duty (GA)
B for mild-to-moderate duty (GB)
C for mild-to-severe duty (GC)

A typical passenger vehicle or light truck would use LA for suspension and GB for wheel bearing. The difference between L and G grease is the chemical makeup that is designed to meet certain operating conditions. L grease has a very high adherence to the part that is lubricated. It has a high resistance to being washed off by water. Suspension or chassis grease will not break down from the movement of the suspension parts. Wheel bearing grease is thicker than L grease and is formulated for high resistance to heat. The grease also resists being thrown off by the spinning bearing.

Figure 10-42 G denotes wheel bearing grease. L stands for steering and suspension components.

With the advent of front-wheel drive (FWD), specially formulated grease was needed for the **constant velocity (CV)** joints on the drive axles. If the CV protective boot or the joint itself is being replaced, new CV joint grease must be used. This grease is made to withstand the high stresses of temperature and movement inherent in a CV joint operation. Wheel bearing grease should *never* be used in CV joints.

There are special lubricants for brakes. On disc brakes, this lubricant is formulated not so much for movement, but for the high temperatures generated during braking. It is applied to the slide pins or adapters that hold the disc caliper in place and is commonly referred to as caliper lubricant. Drum brakes require a very small amount of brake lubricant to be applied at the specific places where the brake shoe web slides on the backing plate. In most cases, caliper lubricant can be used. Although there are specific greases for the suspension, bearings, and U-joints, the most commonly used grease in an automotive shop is multipurpose grease. As its name implies, multipurpose grease can be used for different purposes. On a vehicle, multipurpose grease is used for suspension, U-joints, and wheel bearings. However, it should not be used in all situations, and the service manual must be consulted for specific applications.

Constant Velocity CV) A joint that allows a drive shaft to transmit power through a variable angle at a constant speed.

ANTILOCK BRAKE SYSTEMS (ABS)

Contrary to some beliefs, antilock brake systems (ABS) were not designed to reduce the stopping distance of a vehicle. They do prevent the wheels from locking and give the driver better control over the vehicle during hard or panic braking. Stopping distance is reduced to some extent by the fact that the wheels are not sliding, and the tire maintains traction with the road. There are several different types of antilock systems. Some work only on rear wheels and are known as **rear-wheel antilock (RWAL).** RWAL is commonly used on light trucks and vans (**Figure 10-43**). The typical ABS system on a passenger vehicle is a four-wheel antilock. Four-wheel systems can be three or four channels. Three-channel systems control the braking at each front wheel and the rear axle set. Four-channel systems control the braking at each wheel (**Figure 10-44**).

Ford refers to their RWAL as RABS, or rear antilock brake systems.

Because this text covers basic theories and operation, we will only discuss the basic components and operation common to all types of antilock systems. The components are the **speed sensors, pressure controllers,** and the **controller.**

RWAL A brake system used to prevent rear wheel brake lock-up under heavy braking conditions.

Speed sensors are electrical/mechanical devices used to register the speed of a rotating component, in this case, the vehicle's wheels.

Pressure controllers are devices used to control hydraulic fluid/pressure within a sealed hydraulic system. They may be electronically or mechanically controlled or operated.

The **controller** is an electronic device that controls specific components of a system. They are basically computer units.

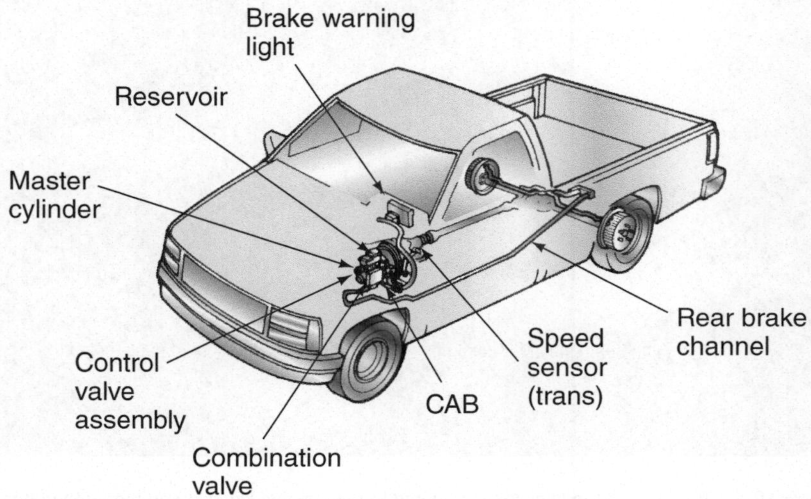

Figure 10-43 A RWAL (RABS) brake system.

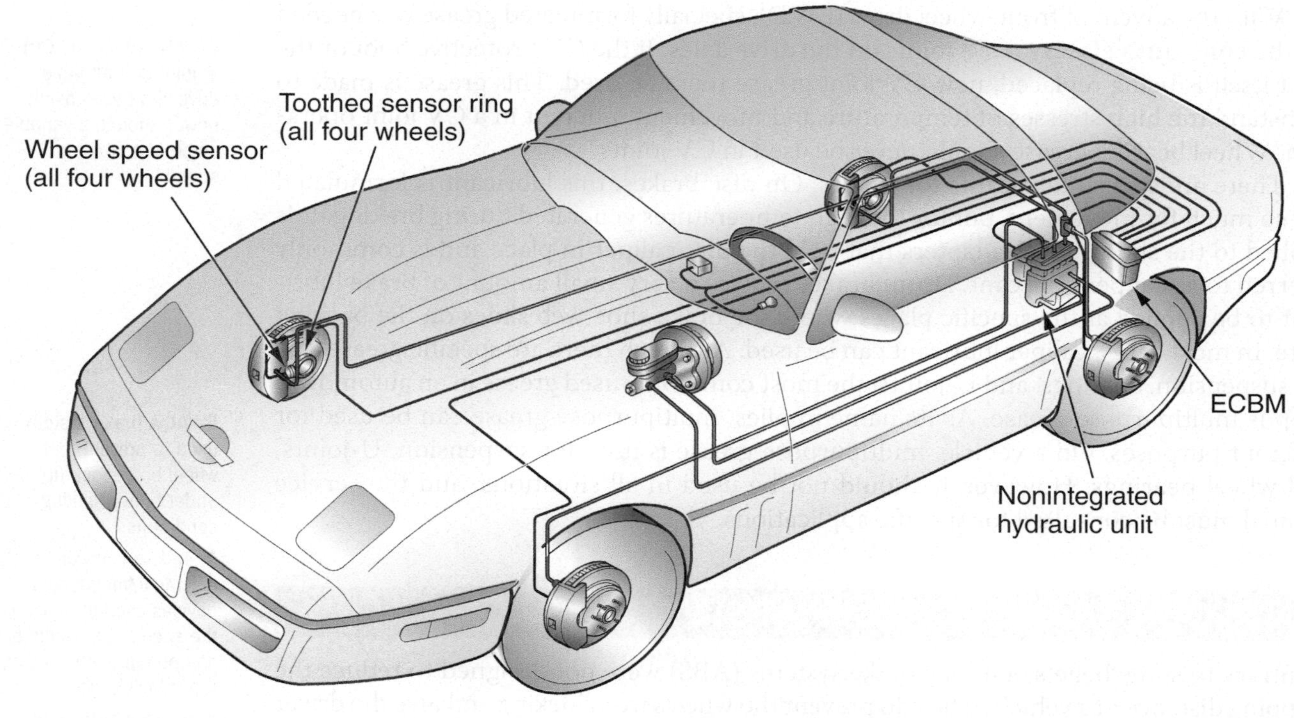

Figure 10-44 This was a typical Bosch ABS-2 used on GM vehicles up to 2000. Newer ABS components are now mounted inside the engine compartment instead of inside the trunk. However, the overall operation of ABS is virtually the same.

A **tone wheel** or ring is a toothed ring or wheel that rotates through a sensor's magnetic field generating a signal sent to a controller.

Speed Sensors

To control brake pressure, the ABS controller must know the speed of the wheels. Speed sensors generate an electrical signal each time a tooth on a toothed ring or **tone wheel** passes the sensor's magnet (**Figure 10-45**). The number of signal pulses sent to the module is used to calculate the speed of the wheel or axle.

Figure 10-45 This wheel speed sensor is mounted to the steering knuckle and the toothed ring on the outer CV joint.

On RWAL and three-channel, four-wheel ABS, the rear speed sensor is mounted to measure the speed of the rear axle. This is done by placing the sensor at the output shaft of the transmission or at the ring gear in the differential (**Figure 10-46**). The toothed ring is placed on the output shaft or ring gear. The pressure to both rear brakes is controlled as one circuit. Each brake on the rear axle will receive the same amount of fluid and pressure. A three-channel system uses sensors at each front wheel in conjunction with the rear sensor. The front tone rings mounted on the brake rotor or the outer CV joint. The three-channel system can control pressure to each of the front wheels in addition to the rear brakes. It is typed as a four-wheel ABS.

Four-wheel, four-channel ABS controls the pressure to each wheel. This system will have speed sensors and tone wheels at each wheel (**Figure 10-47**). The pressure and fluid can be regulated independently of each wheel based on that wheel's speed.

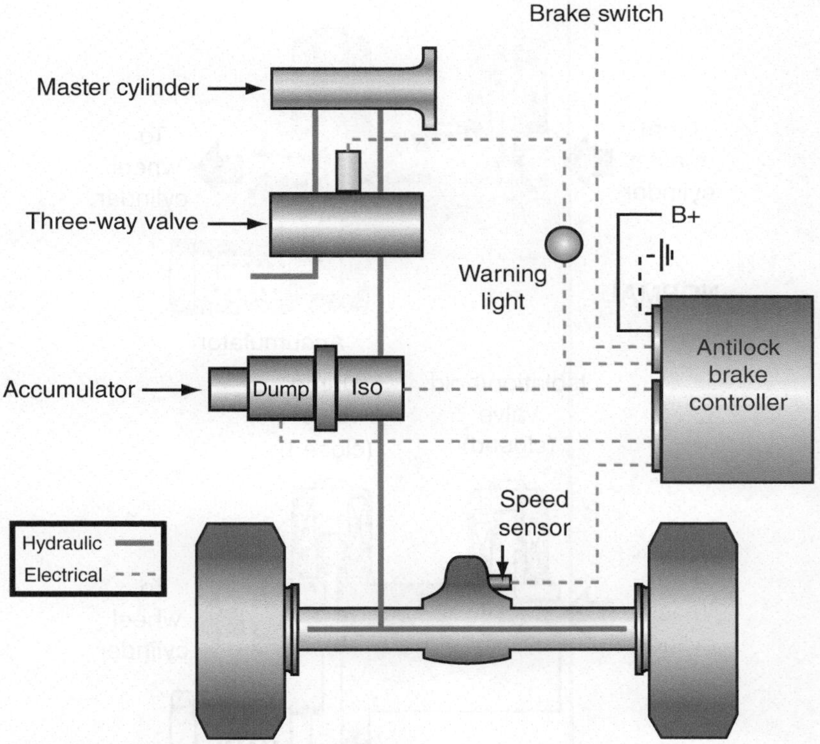

Figure 10-46 Layout of a Chrysler RWAL system.

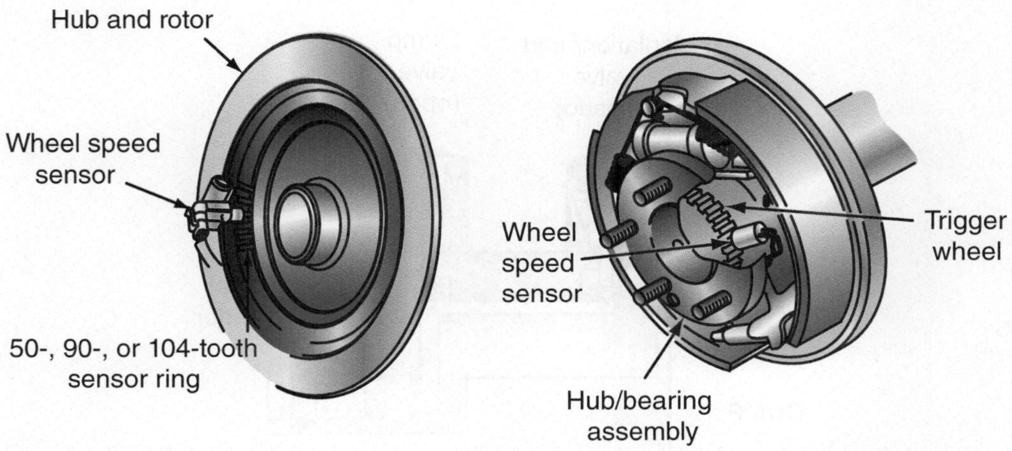

Figure 10-47 On the left the tooth ring is molded to the inside of the rotor. At the right the ring is around the outside of the axle near the backing plate.

ABS wheel speed sensors can also be used in conjunction with the Tire Pressure Monitoring System (TPMS). More information about this is covered in Chapter 9 "Suspension and Steering."

An **isolation/dump solenoid** is used to block fluid or release fluid from a portion of the brake system.

Pressure Controllers

The devices performing brake pressure regulation are solenoids commanded by the anti-lock controller. There is a solenoid for each wheel or axle with a speed sensor. A RWAL system has one solenoid commonly known as an **isolation/dump solenoid** or brake actuator (**Figure 10-48**). While different systems use different means to control the brake

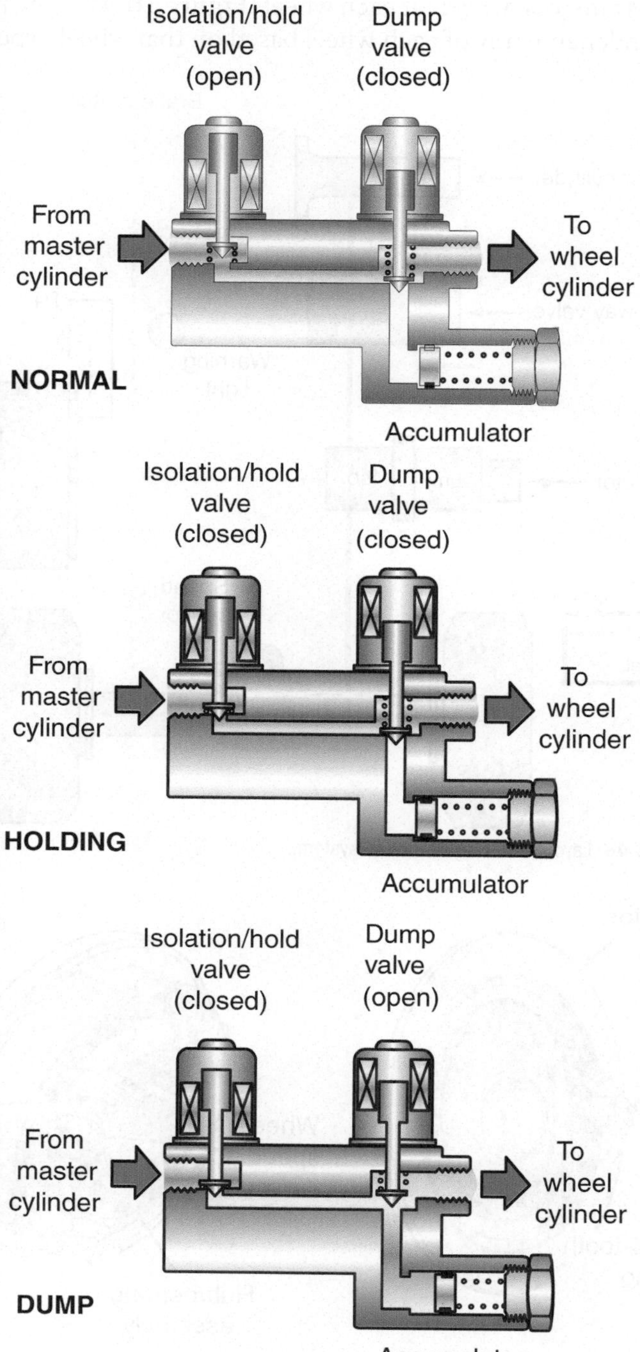

Figure 10-48 The isolation/dump valve for a RWAL system may be referred to as the RWAL actuator.

pressures, they all use the same theory of operation. Each solenoid can isolate the wheel from the master cylinder and dump the existing pressure within that brake line. In this manner, no further pressure can go to the brake from the master cylinder. At the same time, fluid can be released from the brake line and reduce the braking effect. The released fluid is held in an accumulator for future use, if necessary. The accumulator is charged with compressed gas to keep the fluid pressurized. Some systems use an electric motor and screw to draw fluid from the brake line (**Figure 10-49**). The motor is reversed to force fluid back into the line. The solenoid can also pop the master cylinder isolation valve open to allow fluid into the brake line and solenoid as required.

ABS Operation

As the brake is applied, the ABS controller senses a signal from the brake light switch. At this point, the controller begins to monitor the speed of the wheels. If a wheel locks or stops rotating, the controller commands the solenoid for that wheel/axle to isolate the master cylinder and dump or relieve brake line pressure. With the brake partially released, the wheel rotates, and the solenoid is turned off. The controller monitors all wheels equipped with a speed sensor and can control individual solenoids without affecting braking on the other wheels. The solenoid can be cycled thousands of times per minute. This can be an unnerving experience for the driver because the brake pedal and vehicle will pulsate and shudder as the solenoids are operated. The technician can help by informing the customer that this is a normal condition and the brake pedal must be kept depressed until the conditions requiring the panic braking no longer exist.

The antilock brake system has an **amber ABS warning light** that can be lit by the controller if a defect is found. The only defect that the controller can detect is an electrical

The **amber ABS warning light** is not to be confused with the red brake light. The red light means the regular brakes may fail.

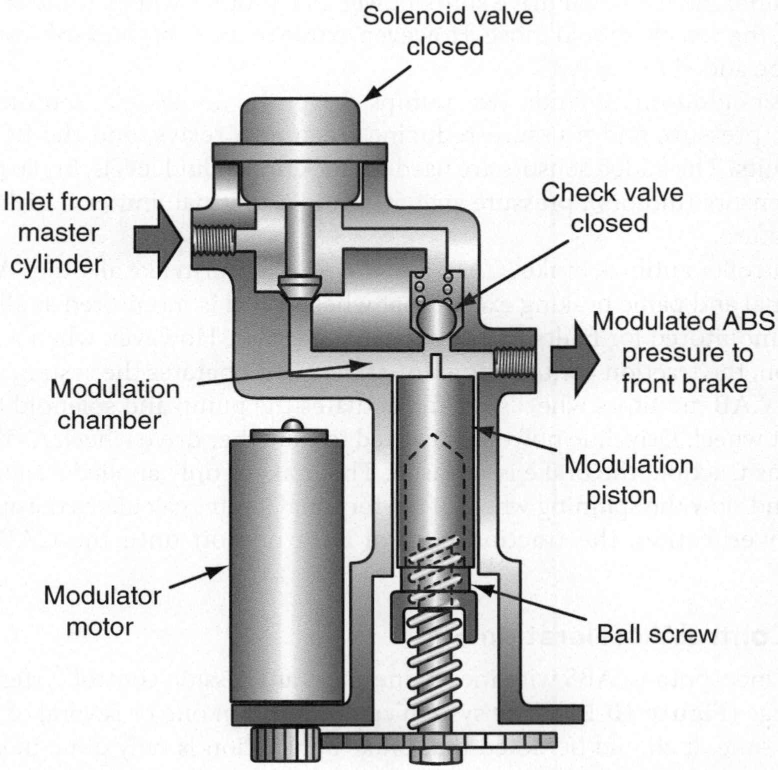

Figure 10-49 The modulator motor can move the piston up or down. Down relieves pressure to the wheel and up applies pressure to the wheel. The piston can also open the check valve so fluid can flow quickly back to the master cylinder. Note that the check valve and solenoid valve are closed, isolating the wheel or wheels from any further pressure increase from the master cylinder.

one. The controller cannot detect mechanical brake conditions such as a leaking wheel cylinder or excessively worn shoes or pads unless a mechanical condition directly affects a sensor.

The regular brake system will function normally if the ABS warning light is illuminated, but the ABS will not function. Certain brake failures may illuminate the red brake warning and the amber ABS lights at once. In this case, it is probable that the regular brakes have failed or have a fault and the vehicle must be serviced as soon as possible.

Some of the newer brake systems have a pressure sensor or pressure transducer to transmit brake system pressure to the control module. In this manner, the module can determine exactly how much force the driver is applying to the brake pedal. The control module can control lockup better by comparing vehicle speed, wheel speed, and the pressure within the brake system.

TRACTION CONTROL SYSTEMS

Traction control is a system used to electronically control wheel spin and is usually merged into the anti-lock brake system.

Traction control is a method of redirecting or transferring power from a spinning drive wheel to a wheel with more traction. There are two basic types of traction control, each relying heavily on electronics. Dependable four-wheel ABS make traction control a reality on many mid-price vehicles. However, for the driver who likes to accelerate quickly from dead stop, traction control can cause damage to the vehicle's driveline. Traction control is meant to assist on slick road surfaces, not in mud-bogging or hot-rodding.

Antilock Brakes with Traction Control

In this type of traction control, the brakes are actually applied to a wheel that has lost traction and the vehicle speed is below 30 mph. When the brake is applied to one drive wheel, the differential or final drive shifts power to the other wheel. If the second wheel has traction, the vehicle should move. However, extra components and software programming must be added to the ABS.

The basic additions include two pumps driven by an electric motor, a different component, pressure and pressure-reducing solenoids, relays, and the hydraulic and electric circuits. The added sensors are used to track brake fluid levels, brake pedal travel, rotational sensors (motors), pressure switches, and a thermal limiter to calculate brake pad temperature.

The controller antilock brake (CAB) and the ABS perform like all other ABS systems during normal and panic braking except that wheel speed is monitored at all times. The circuits are monitored for faults in the same manner also. However, when a drive wheel loses traction, the traction control program of the CAB operates the system as a braking device. The CAB monitors wheel spin and actuates the pump and solenoid to apply the brake at that wheel. Driveline power is diverted to the other drive wheel. As the spinning wheel regains traction, the brake is released. The brake is only applied enough to divert the power and slow the spinning wheel. If the terminal limiter calculates the applied brake pad to be overheating, the traction control is turned off until the CAB resets the system.

Traction Control/Acceleration Slip

This system incorporates ABS with the engine and transmission control system to control wheel slippage (**Figure 10-50**). The system can perform in one or several of four modes at the same time. It should be noted that brake application is only done in conjunction with another mode. The modes are:

- Command the PCM to retard spark timing to reduce engine output.
- Force the throttle blade closed to decrease engine speed and output.

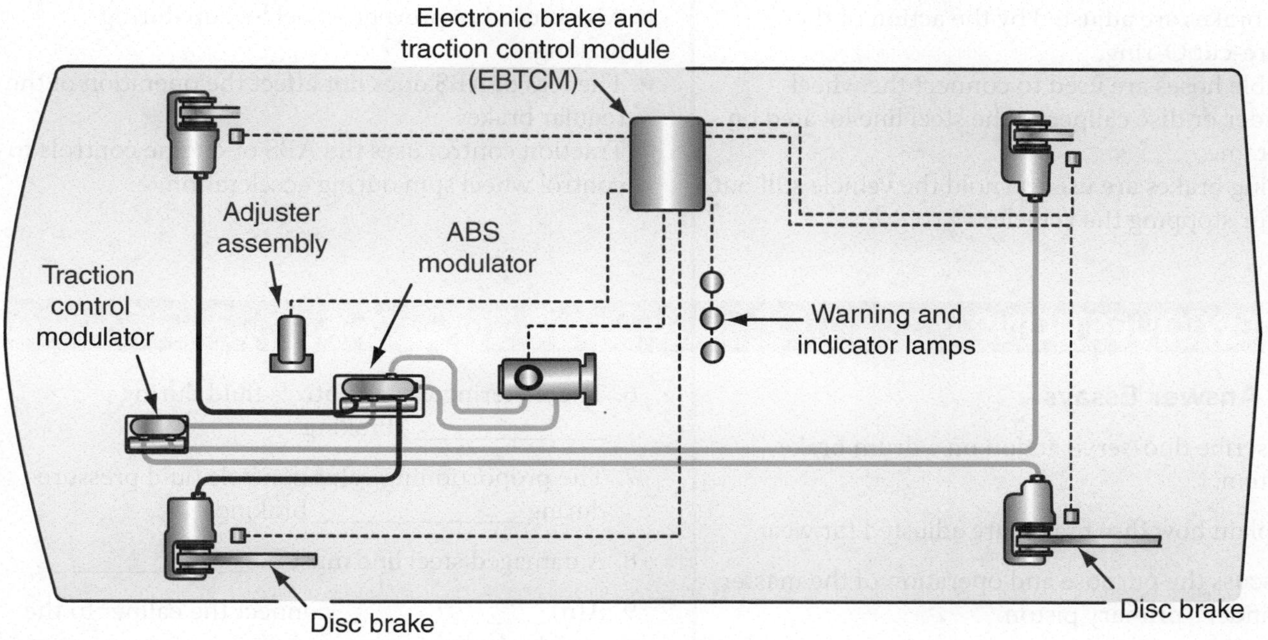

Figure 10-50 Shown is a Delphi Chassis VI ABS with traction control components installed.

- Upshift the transmission to reduce delivered torque.
- Apply the brake on a spinning wheel.

On older systems without full electronically controlled transmissions, the gear upshift is not possible. Like the previously discussed traction control and other electronic control systems, this system must be tailored, electrically and mechanically, to meet vehicle specifications. A traction control system for a Ford Crown Victoria will not work with a Lincoln Town Car, even though the vehicles are very similar in weight and overall size. Each vehicle is set up to function with electronic differences that must be considered during the design and application of all systems.

SUMMARY

- The use of hydraulics in the brake system allows force to be transferred with moving parts.
- The operator's force on the brake pedal is mechanically transmitted to the master cylinder's primary piston.
- The operator's force can be increased greatly by using a booster between the pedal and the master cylinder.
- A single-piston master cylinder may cause all of the brakes to fail if there is leak at any wheel.
- A dual-piston master cylinder allows the driver to have braking power on two wheels even if a leak is present on one wheel.
- The metering valve holds the front disc brake off until the rear drum brakes have moved to make contact with the drums.
- The proportioning valve prevents the rear brakes from locking during heavy braking by reducing the pressure.
- A combination valve houses the metering and proportioning valves.
- A brake warning light is used to alert the driver to a failure in the system.
- Wheel cylinders use two pistons and links to apply the brakes.
- Return springs on a drum brake force the shoes and pistons to their release positions.
- Most drum brakes use an automatic self-adjuster operated by the movement of the secondary shoe or parking brake.
- Disc brakes do not use a mechanical adjuster.

- Disc brakes are adjusted by the action of the square-cut O-ring.
- Flexible hoses are used to connect the wheel cylinder or disc caliper to the steel line located on the frame.
- Parking brakes are used to hold the vehicle still but not for stopping the vehicle.
- ABS is used to prevent wheel lockup during braking.
- The loss of ABS does not affect the operation of the regular brakes.
- Traction control uses the ABS or engine controls to control wheel spin during acceleration.

REVIEW QUESTIONS

Short Answer Essays

1. Describe duo-servo action on a drum brake system.
2. Explain how disc brakes are adjusted for wear.
3. Discuss the purpose and operation of the master cylinder's primary piston.
4. Explain how brake pedal force is boosted with a hydro-boost system.
5. List and explain the purpose of the wheel-mounted components of a disc brake.
6. Briefly describe the actions taken by an ABS during a panic stop.
7. Explain the purpose of the proportioning valve.
8. Describe the operation of a frame-mounted, load-sensing proportioning valve.
9. Discuss the operation of a metering valve.
10. Discuss the operation of a combination valve-mounted, load-sensing proportioning valve.

Fill-in-the-Blanks

1. The _______________ brake shoe performs most of the braking.
2. The _______________ system boosts the driver braking effort using an electric motor and pump.
3. The ABS warning light will not detect _______________ conditions.
4. The adjustment of disc brakes is done by the _______________.
5. The master cylinder secondary piston provides fluid to the _______________ brakes.
6. The metering valve controls fluid during _______________ braking.
7. The proportioning valve controls fluid pressure during _______________ braking.
8. A damaged steel line must be _______________.
9. A(n) _______________ connect the caliper to the steel brake lines.
10. If the ABS warning light is on, the regular brakes will _______________.

Multiple Choice

1. The combination valve is being discussed.
   Technician A says a leak in the front brake should turn on the warning light.
   Technician B says the brake pedal may be much lower to the floor if the brake warning light is on.
   Who is correct?
   A. A only
   B. B only
   C. Both A and B
   D. Neither A nor B
2. The diagonally split brake system is being discussed.
   Technician A says one master cylinder piston will operate the brakes on a front and a rear wheel.
   Technician B says that if a leak occurs, the operator will still have two working brakes on an axle.
   Who is correct?
   A. A only
   B. B only
   C. Both A and B
   D. Neither A nor B

3. Disc brake calipers are being discussed.
   Technician A says adjustment for pad wear cannot be performed manually.
   Technician B says the square-cut O-ring automatically adjusts for pad wear.
   Who is correct?
   A. A only
   B. B only
   C. Both A and B
   D. Neither A nor B
4. The master cylinder on a front/rear split system is being discussed.
   Technician A says a leak at the front brake will stop the primary piston from building pressure.
   Technician B says the load-sensing proportioning valve is mounted at the rear of all master cylinders.
   Who is correct?
   A. A only
   B. B only
   C. Both A and B
   D. Neither A nor B
5. Split brake systems are being discussed.
   Technician A says the diagonally split system is set so a front brake and a rear brake are available for braking.
   Technician B says the front/rear split system has the rear brakes operating from the same master cylinder piston.
   Who is correct?
   A. A only
   B. B only
   C. Both A and B
   D. Neither A nor B
6. Disc brakes are being discussed.
   Technician A says the square-cut O-ring adjusts for pad wear.
   Technician B says a fixed caliper uses only one piston.
   Who is correct?
   A. A only
   B. B only
   C. Both A and B
   D. Neither A nor B
7. A 2-square-inch input piston is being applied with 10 pounds of force. A 4-square-inch output piston is being used.
   Technician A says mechanical advantage can be calculated using the formula $F = PA$.
   Technician B says the pressure in this system is 40 psi.
   Who is correct?
   A. A only
   B. B only
   C. Both A and B
   D. Neither A nor B
8. The metering valve is being discussed.
   Technician A says the movement of the valve determines rear brake pressure.
   Technician B says the valve adjusts rear brake pressure during initial braking.
   Who is correct?
   A. A only
   B. B only
   C. Both A and B
   D. Neither A nor B
9. Drums and rotors are being discussed.
   Technician A says the drum and rotor are mounted between the wheel and axle or hub.
   Technician B says drums are machined to match the brake pads.
   Who is correct?
   A. A only
   B. B only
   C. Both A and B
   D. Neither A nor B
10. Proportioning valves are being discussed.
    Technician A says the load-sensing proportioning valve may be mounted on the frame rail.
    Technician B says some proportioning valves use a ball to control the valve.
    Who is correct?
    A. A only
    B. B only
    C. Both A and B
    D. Neither A nor B

CHAPTER 11

HEATING, VENTILATION, AND AIR-CONDITIONING SYSTEMS

Upon completion and review of this chapter, you should understand and be able to:

- Understand the general operation of heating and air-conditioning systems.
- Describe heating systems components within the heating, ventilation, and air-conditioning (HVAC) case.
- Identify the different types of antifreeze.
- Understand the difference between a radiator and a condenser.
- Understand the difference between a heater core and an evaporator.
- Describe air conditioning components within the HVAC case.
- Explain the purpose of the Clean Air Act and the control of refrigerant.
- Describe the difference between R12 and R134A refrigerant.
- Explain the requirements for working on air-conditioning systems.

Terms To Know

Accumulator/dryer
Climate control system
Compressor
Condenser
Conduction
Coolant
Desiccant
Global warming potential
Evacuating
Evaporator
Heat transfer
Heater core
HVAC
Hybrid organic acid technology
Inorganic acid technology
Organic acid technology
Ozone layer
Passenger cabin
Radiator
Receiver/dryer
Recharging
Recovery
Refrigerant
Ultraviolet

Passenger cabin The area of the vehicle designed for the seating of the driver and other passengers.

Climate control system A complex system that is able to affect and regulate the internal temperature of a vehicle.

HVAC: A complex system that it's purpose is to clean, cool. heat, regulate and dehumidify the air entering the passenger cabin.

INTRODUCTION

This chapter, and its corresponding chapter in the Shop Manual, is not designed to make the entry-level technician an electronic or climate control repair expert. Instead, the chapters will give the reader a working knowledge of the various heating and cooling systems, some basic diagnostic testing, and the ability to perform basic maintenance. Because some of the systems share common components and functions, a failure of one system or component may affect the entire heating and cooling system. Over the course of time, passenger comfort has become expected, and air quality and temperature can influence the driver's concentration and overall driving performance. This chapter will explain the process involved to regulate the temperature in the passenger compartment, also known as **passenger cabin**.

Climate control system is the inclusive term for the heating, ventilation, and air-conditioning systems (**HVAC**). The term is used to designate either a straight heating system or a heater/air-conditioning combined system. The heater system is used to warm the interior of the vehicle during cold weather. Naturally, the air-conditioning is used to do the opposite.

Within the passenger cabin of the vehicle, HVAC is designed to deliver comfortable temperature and high-quality fresh air to the passenger compartment. Depending on the outside temperature, the HVAC system can provide the desired cabin temperature by adding cooled or heated air or a combination of both with a simple command from the driver. Although these systems and components are separate, they still work together. Chapter 7 discussed how the engine can produce and control heat, and how it is transferred to the passenger cabin. This chapter will focus more on the heater controls and the components inside the vehicle, which is referred to as climate control or HVAC. It is important that a technician understand the operation of the HVAC system to properly inspect and diagnose the operation of the HVAC systems, so it can be maintained and serviced when necessary. So, we will focus more on the climate controls and understanding how the heating and cooling systems are related.

HVAC Principles

Passenger comfort devices are items such as heat, air-conditioning, heated and cooled seats, and others that make a passenger's ride in the vehicle enjoyable. They also keep the driver alert. The function of the heating system is to increase the heat and reduce the humidity inside the passenger cabin. This is achieved through the transfer of heat from the engine to the coolant. (**Figure 11-1**). The job of the air-conditioning system is to

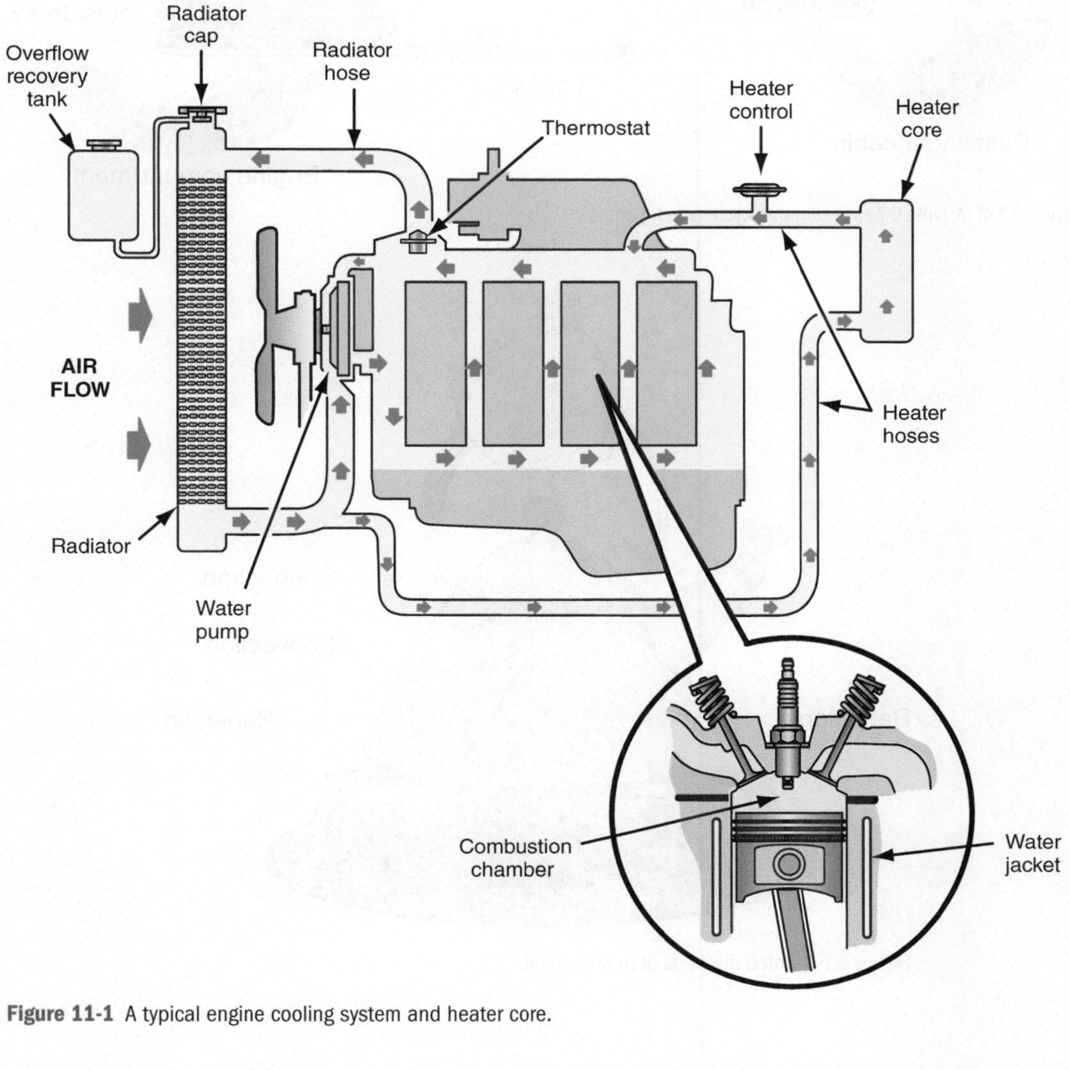

Figure 11-1 A typical engine cooling system and heater core.

Heat transfer The movement of thermal energy (heat) from one component to another component of a different temperature.

Conduction The transfer of heat directly through a substance when there is a difference in temperature.

reduce the heat and humidity inside the passenger cabin. This is achieved through a series of thermal and chemical transformations of the chemical called refrigerant (**Figure 11-2**). The transfer of heat is a principle of physics that you need to understand to properly diagnose and maintain the HVAC system. A simple way to remember is that heat always transfers from hot objects to cold objects. Another principle of physics you will need to understand is density of liquids as compared to gases. Liquids are denser than gases; liquids tend to be pulled to the bottom, and gases tend to float to the top in a sealed system. The last principle you will need to understand is the relationship between temperature and pressure. When the pressure is increased on a gas (refrigerant), the temperature of the gas will also increase. Of course, when the pressure is reduced on the gas (refrigerant), the temperature will also reduce or decrease. **Heat transfer** is the key to climate control in the passenger compartment. There are three different ways heat can be transferred: conduction, convection, and radiation (**Figure 11-3**). **Conduction** is the process of

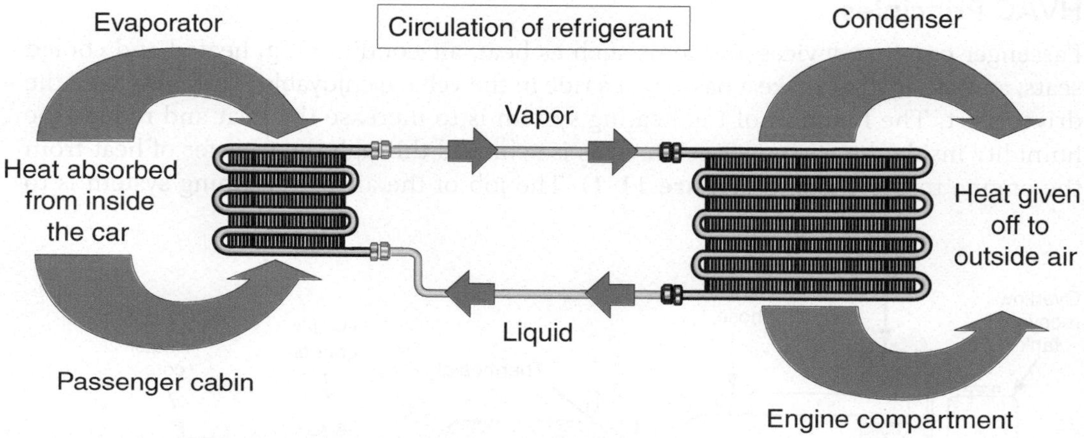

Figure 11-2 A typical heat exchange cycle of refrigerant.

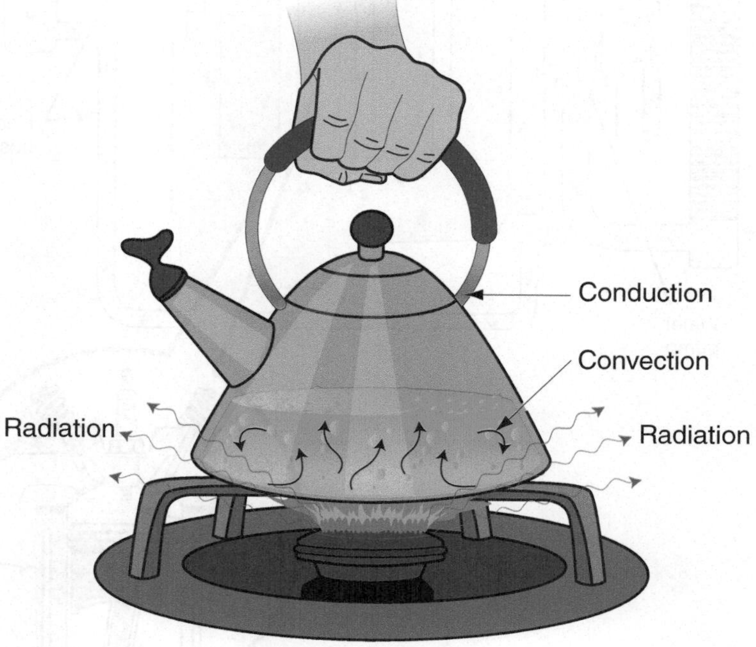

Figure 11-3 Three methods of heat transfer.

transferring heat through an object. An example of this would be if you heat one end of a metal pipe with a flame. the heat will slowly work its way across the metal pipe and the heat will transfer across the entire pipe. Convection is the process of transferring heat by the circulation of a vapor or liquid. An example of this would be if you heat a pan of water on the stove. As the water starts to warm from the bottom, the warmer water raises to the top and the colder water drops to the bottom causing the water to circulate. Now you know it's true: heat rises! Radiation is the process that moves heat from a heat source to an object by means of invisible waves. An example of this would be if you sit next to a campfire, the heat from the fire is transferred to you and any objects that may be near the flames. If your shoes and socks get wet next time you're camping, radiation is the answer to dry shoes and socks!

Shop Manual
page 503

Heater Systems

The heating system uses the engine's coolant to heat air that is being forced into the passenger compartment. Hoses run from the engine block to a small **heater core** and back to the engine block (**Figure 11-4**). The core is usually located in front of the front passenger seat, either within the dash or on the engine side of the firewall. A mechanical or vacuum-operated valve controls the flow of hot coolant into the core. A variable-speed motor drives a blower and forces air around the heater core fins and through ducts into the passenger compartment. The hot coolant in the heater core heats the passing air. The core is like a small radiator in appearance and operation.

Heater core A small radiator device used to heat the passenger cabin.

Newer heater systems, especially those installed with an air-conditioning system, have a full-flow heater core. The coolant flows through the heater core any time the engine is running. Air is directed away from or through the core by a temperature blend door. This eliminates the water valve and its controls while providing the driver more precise control over the conditioned air entering the vehicles.

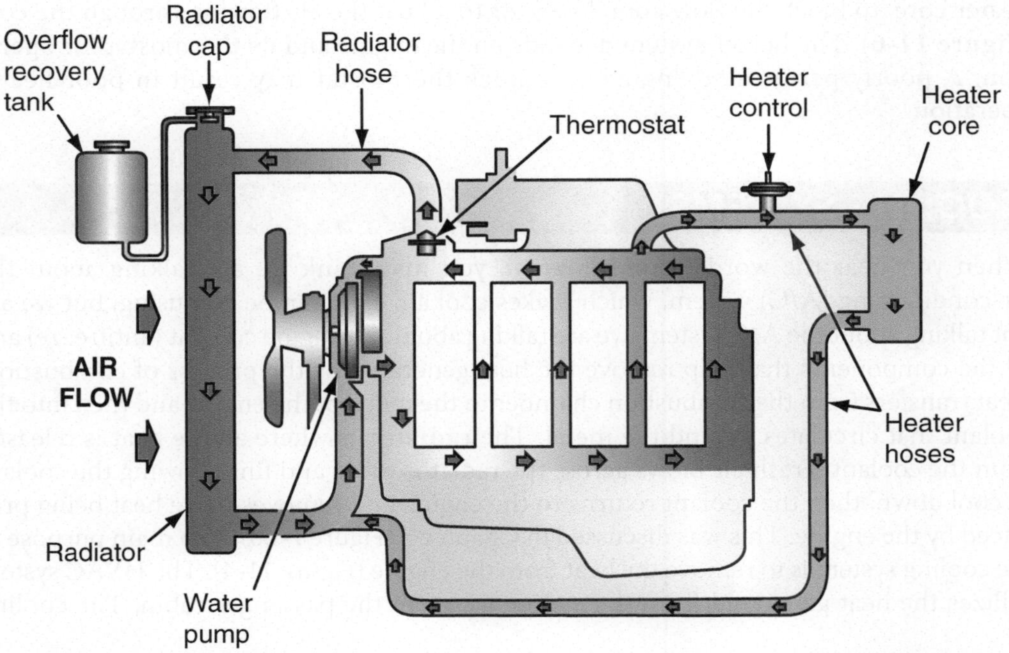

Figure 11-4 The heater hoses carry hot coolant from the engine to the heater core and back to the engine. They connect to the engine in various locations depending on engine design and placement within the engine compartment.

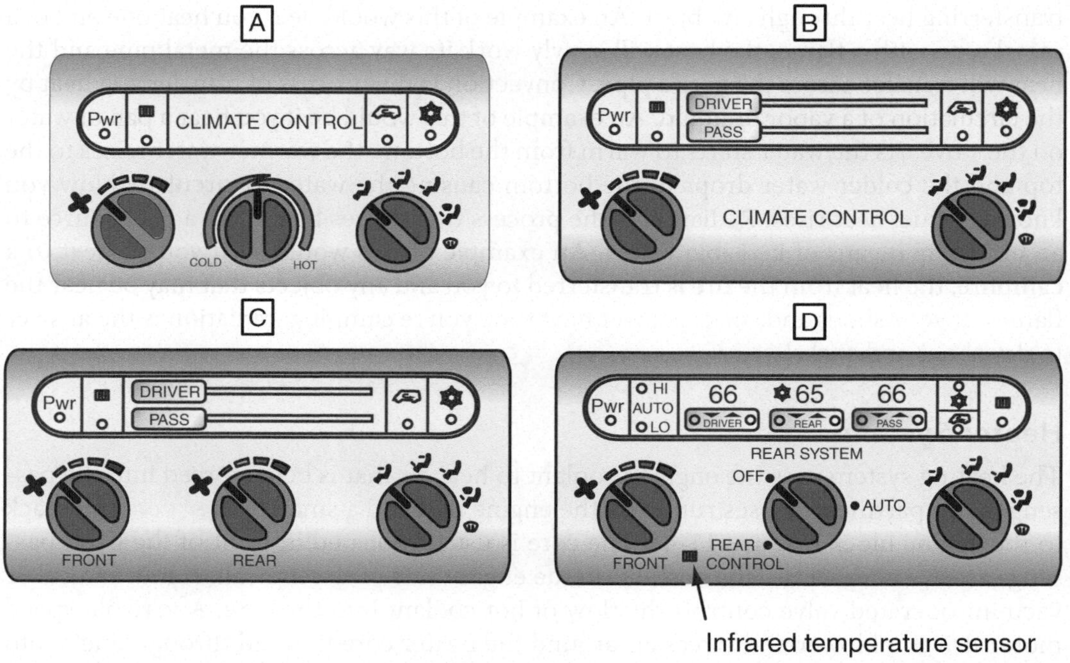

Figure 11-5 Typical control panels for HVAC systems: (A) single control system, (B) dual-zone control system, (C) three-zone system, and (D) three-zone with automatic temperature controls (ATC).

The climate control panel (**Figure 11-5**) mounted in the dash houses buttons or levers to direct the air, change blower speed, and adjust temperature. The operator can select outlets that direct warm air to the windshield, dash panel, or to the floor. The air lever moves blend doors within the ductwork mechanically with cables, small vacuum motors, or small electric motors. The blower switch has three or four speed settings. The temperature selector is used to adjust the amount of hot coolant allowed into the heater core, to block the flow completely, or to adjust the air flowing through the core (**Figure 11-6**). The heater system depends on the engine and its thermostat for operation. A poorly performing engine or a stuck thermostat may result in poor heater operation.

COOLING SYSTEMS

When you hear the words *cooling system,* you may think we are talking about the air-conditioning (A/C) system, which makes cool air. This can be confusing, but we are not talking about the A/C system, we are talking about the engine coolant (antifreeze) and all the components that help remove the heat generated by the process of combustion. Heat transfers from the combustion chamber to the metal of the engine and then into the coolant that circulates around the metal. The **radiator** is where all the heat is released from the coolant. Fresh air blows across the radiator tubes and fins allowing the coolant to cool down, then the coolant returns to the engine and removes more heat being produced by the engine. This was discussed in Chapter 7 (Figure 7-68). The main purpose of the cooling system is to remove the heat from the engine (Figure 11-4). The HVAC system utilizes the heat generated from the engine to warm the passenger cabin. The cooling

Radiator A heat exchanger designed to transfer heat from the hot coolant that flows through the radiator by blowing cooler air through the radiator with fans and the movement of the vehicle.

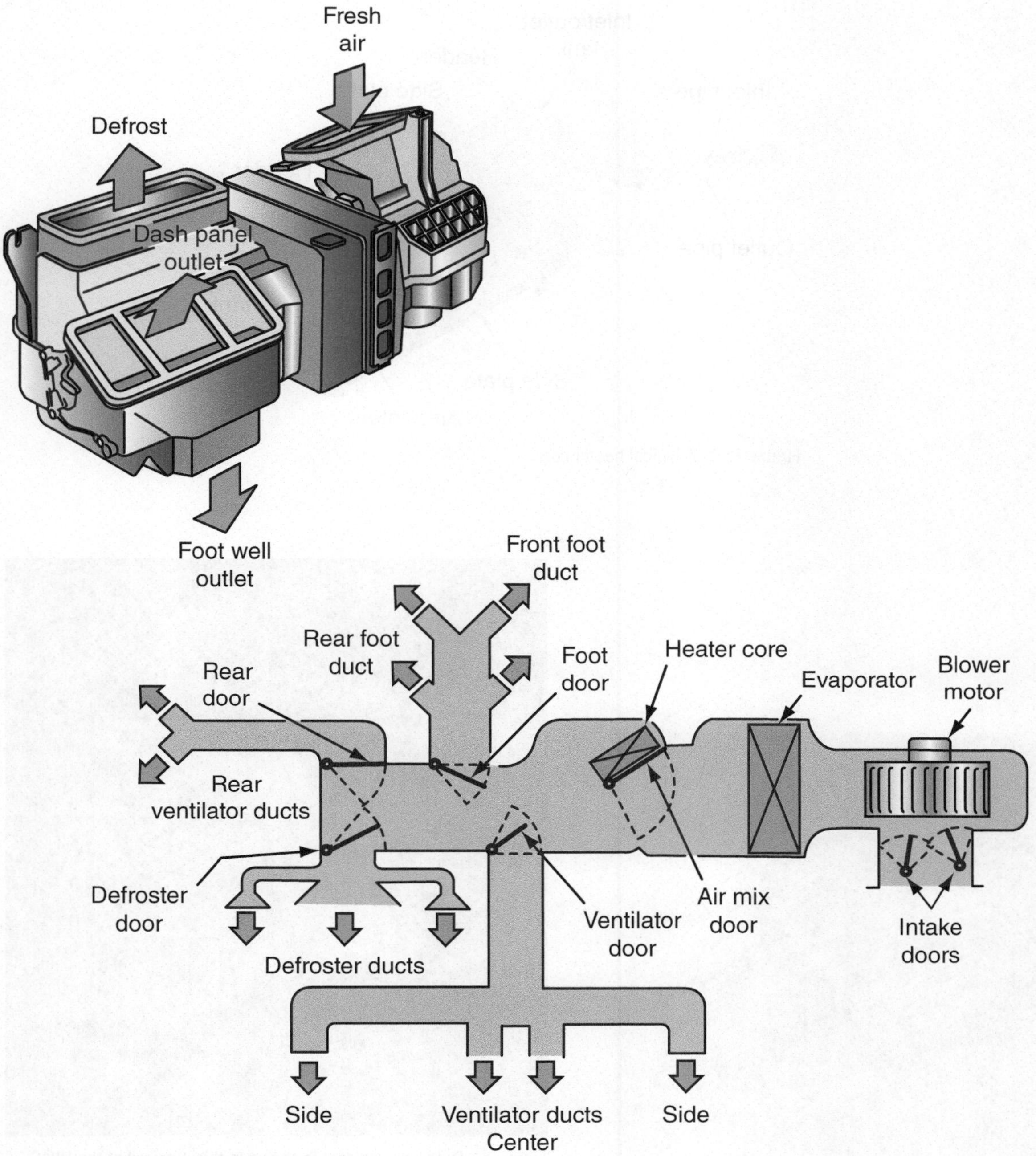

Figure 11-6 A typical HVAC case and duct system.

system uses two hoses that come from the engine through the vehicle's fire wall and into the heater core (**Figures 11-7** and **11-8**). The heater core is located inside the **HVAC** case. The HVAC case is a plastic box (**Figure 11-9**) inside the passenger compartment located under the dash. Hot coolant flows through the heater core and a blower fan blows cool cabin air across the heater core. The heat from the heater core then provides warm air to the passenger compartment.

HVAC Short for heating, ventilation and air-conditioning system, which provides heat and cooling to the passenger cabin.

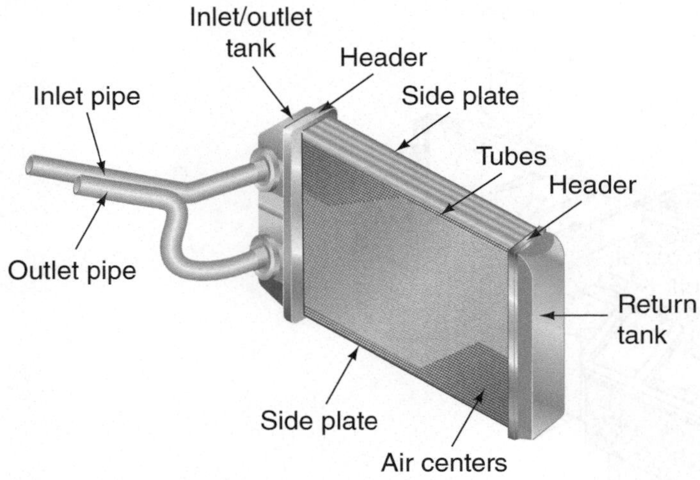

Figure 11-7 A typical heater core.

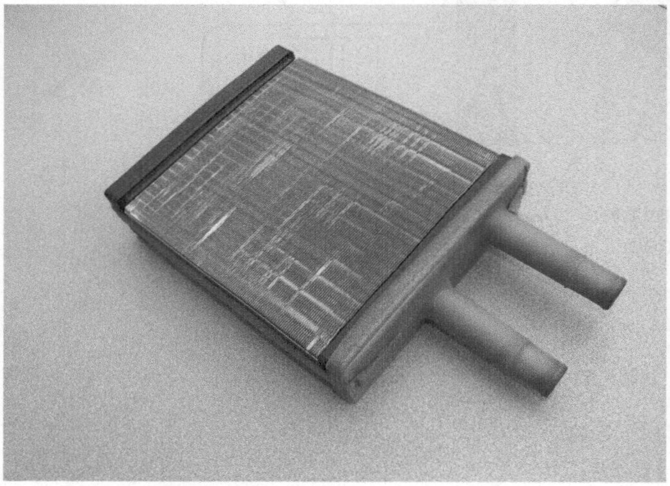

Figure 11-8 A typical plastic heater core.

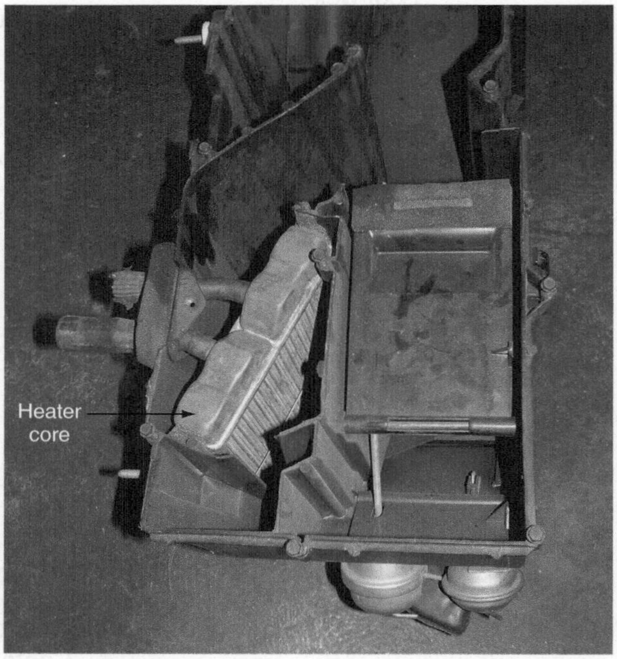

Figure 11-9 The heater core is shown in this cutaway of the HVAC case.

Coolant is a chemical mixture that reduces freezing of a liquid, usually water. It's also called antifreeze. You cannot identify coolant by the color alone!

IAT A coolant that contains ethylene glycol which is known to pollute groundwater and is harmful if swallowed by humans and animals.

Coolant

The **coolant** used in a cooling system is made up of a mixture of antifreeze and water. Antifreeze is a chemical compound that is added to water because water itself does not provide the proper characteristics needed to protect an engine. Water transfers heat quite well but freezes at 32°F and boils at 212°F. Engines must be able to operate above and below those temperatures and still transfer the heat. Today's engines use several types of coolant with many different applications. Most of the modern vehicles produced today use coolants that are incompatible with the common type of **inorganic acid technology** (IAT) or ethylene glycol-based coolant (**Figure 11-10**). For many years, we could identify coolant by the color and it was a one-size-fits-all. The most common type was IAT or ethylene glycol-based, and it was green. This coolant was not considered long-life coolant and was not compatible with the new types of materials used in modern vehicle gaskets and seals. IATs are also not considered to be environmentally friendly and are poisonous to humans and animals if swallowed. A few of the more modern types of coolant used today include **organic acid technology** (OAT) coolant (**Figure 11-11**) used by many manufacturers. This coolant is designed for long-life, and many modern engines require extended-life coolant. Another type used by many manufacturers today is **hybrid organic acid technology** (HOAT) coolant (**Figure 11-12**). It combines the best qualities of both IAT and OAT coolants. IAT is an ethylene glycol-based coolant, and OAT and HOAT are propylene glycol-based coolants. They should not be mixed because the protection levels are not the same, and you may get an incorrect reading. There are also universal types of

OAT Is a coolant that is considered a long life coolant and non-toxic, which makes them biodegradable.

HOAT This coolant combines both IAT and OAT best qualities of corrosion inhibitor of IAT and non-toxic properties of the OAT. The best of both worlds.

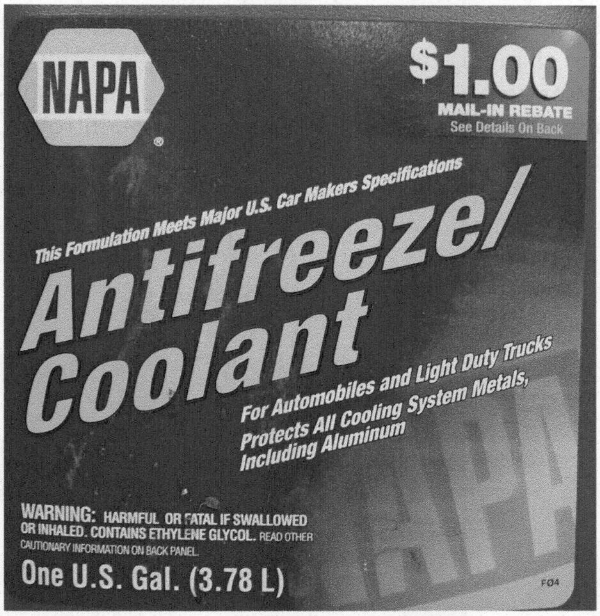

Figure 11-10 Traditional "green" ethylene glycol-based coolant.

Figure 11-11 Organic acid technology (OAT) propylene glycol.

Figure 11-12 Hybrid acid technology (HOAT) propylene glycol.

coolant that claim to work in all models. As you can see, it is very confusing to identify the different types of coolant. You should always begin by checking the owner or service manuals for the proper coolant to use when servicing the cooling system. Because of the many different types of coolant used and the different colors used, it is vitally important to use the correct coolant.

The **compressor** is a high-pressure pump built specifically for air-conditioning systems.

A **condenser** cools down a vapor and assists in changing the liquid from a vapor state to a liquid state (condensation).

The **evaporator** helps a liquid heat enough so that it will change from liquid to vapor (evaporation).

Air Conditioning

Before discussing the air-conditioning (A/C) system, a quick review of heat and pressure is needed. When a liquid is heated enough, it changes to a vapor. Adding pressure to the liquid raises the boiling point and reduces vaporization. By the same standard, removing heat from a vapor causes the element to cool and liquefy. Pressure and heat are the two principles that make air-conditioning systems work.

The A/C is usually split into two sections or sides: high and low (**Figure 11-13**). The high side extends from the **compressor,** through the **condenser,** and to the metering device. From the compressor to the condenser, the refrigerant is in the form of a high-pressure vapor before changing into a high-pressure liquid within the condenser. From the metering device, through the evaporator, and to the compressor's suction side is the low side. The amount of liquid entering the **evaporator** is controlled, which reduces the pressure. The lower pressure and heat from the passing air vaporizes the refrigerant and a low-pressure vapor exits to the compressor. The condenser and evaporator are similar in form to the radiator and heater core, respectively.

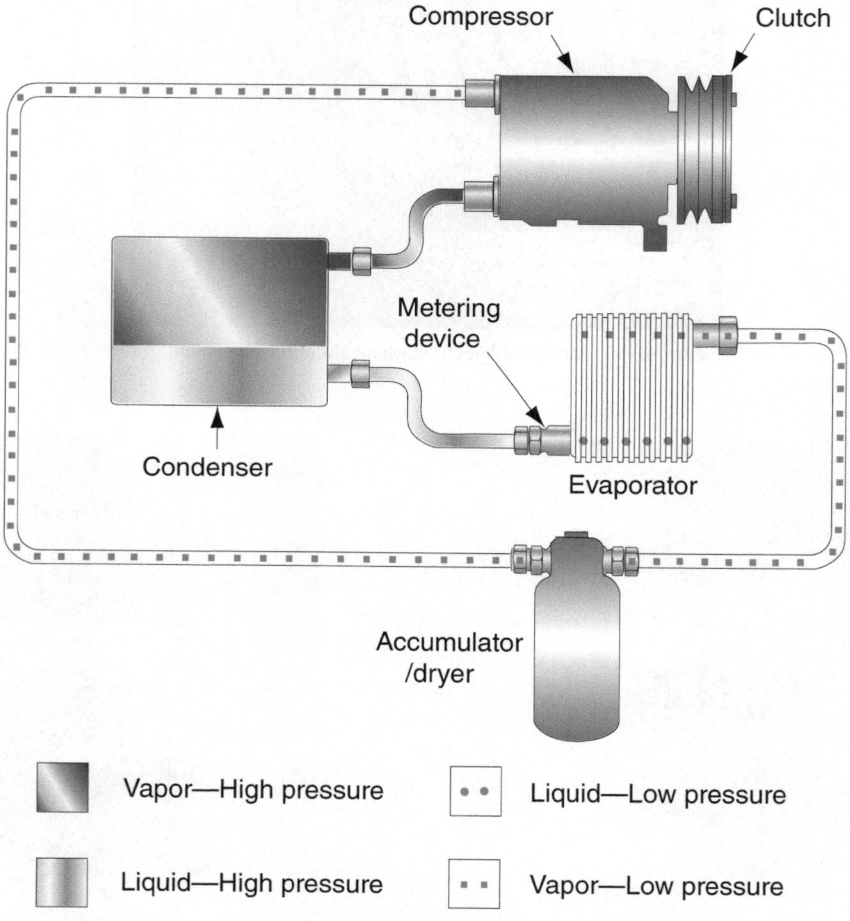

Figure 11-13 The refrigerant changes states from a liquid to a vapor (evaporator) and from a vapor to a liquid in the condenser. Differences in refrigerant pressure are a prime reason for these changes.

An A/C system consists of several more components than the heater system. The cooling agent in an A/C unit is a **refrigerant** circulating under pressure through the system. The refrigerant has properties that may affect the environment and human health. The next section covers the refrigerant and its hazards.

A **refrigerant** is a chemical that can absorb, transport, and release a great deal of heat in a small space under easy-to-create conditions.

A belt-driven compressor at the front of the engine pressurizes the refrigerant (**Figure 11-14**). Many hybrid electric vehicles do not use a belt to drive the compressor; instead the compressor is driven by an electric motor (usually AC current) and is powered by the high-voltage battery. This type of compressor is easy to identify by its orange-colored cables and connectors (**Figure 11-15**). Pressure or temperature switches protect the compressor by switching it on and off as needed. A condenser mounted in front of the radiator cools the refrigerant while a dash-mounted evaporator core allows the cool refrigerant to absorb heat from the air being directed into the passenger compartment. Depending upon the design, the A/C system may have either an expansion valve or an orifice tube to control refrigerant flow into the evaporator.

In addition, some systems will have an **accumulator/dryer** or **receiver/dryer** to filter and clean the refrigerant (**Figure 11-16**). An orifice tube is used with accumulator/drier systems, and expansion valves are used with receiver/dryers. The lines are usually made of aluminum with a high-pressure, flexible-hose section. High-side pressures may reach 300 psi or more on a hot day. Low-side pressures are proportional to the high-side pressures and usually range from 30 to 50 psi. The operating pressure rises as the outside temperature increases.

An **accumulator/dryer** separates vapor from liquid but does not capture moisture (water).

A **receiver/dryer** also separates liquid and vapor, but it will capture moisture (water) present in the refrigerant.

Figure 11-14 The air-conditioning compressor may be mounted high or low on the engine block and its positioning is based on space within the engine compartment and the design of the engine.

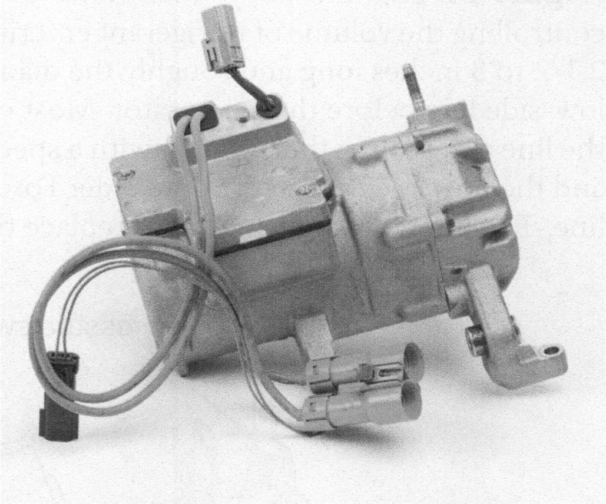

Figure 11-15 Many hybrid electric vehicle A/C compressors are driven by an electric motor instead of a belt. This allows the A/C to continue operation when the engine is not running.

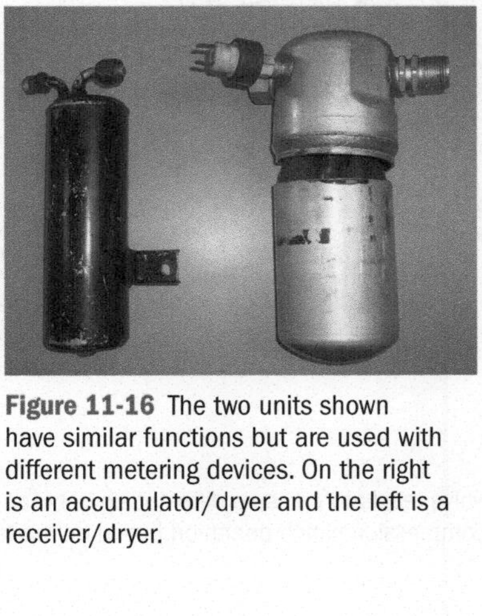

Figure 11-16 The two units shown have similar functions but are used with different metering devices. On the right is an accumulator/dryer and the left is a receiver/dryer.

When the A/C switch is closed and the engine is running, current flows through low- or high-pressure switches to the magnetic coil and clutch. The clutch locks the compressor shaft to the pulley driven by the belt. The compressor draws in low-pressure vapor refrigerant from the evaporator, pressurizes it, and pumps it to the condenser. Air flowing over the condenser fins cools and liquefies the refrigerant. The incoming high-pressure vapor at the top of the condenser forces the refrigerant down through and out of the condenser and back to the metering device. Before entering the evaporator, the refrigerant passes through an expansion valve or orifice tube. Too much refrigerant forces the evaporator pressure upward, and heat cannot be transferred. If there is too little refrigerant, ice or frost forms on the outside of the evaporator and its outlet. The warm air passing over the evaporator is cooled by transferring its heat to the liquid refrigerant. The refrigerant is heated, vaporized, and loses pressure quickly. The low-pressure vapor is forced by compressor suction through an accumulator/dryer and back into the compressor. The accumulator/dryer has a cleaning filter and a moisture-absorbing material, **desiccant,** to remove any water in the refrigerant.

Desiccant is a white, powered compound that can absorb many times its weight in water or moisture.

The expansion valve is a pressure- or temperature-sensitive metering device (**Figure 11-17**). An expansion valve is used in conjunction with a receiver/dryer. If the pressure or temperature within the evaporator or low side is too high, less refrigerant can enter the evaporator. The valve may be mounted on the evaporator's intake line or within the evaporator housing.

The orifice tube used with an accumulator/dryer is basically a series of small screens (**Figure 11-18**). The very small holes in the screens restrict the refrigerant, thereby controlling the volume of refrigerant entering the evaporator. Most orifice tubes are about 2 1/2 to 3 inches long and roughly the diameter of a pencil. The tube is inserted into the low-side line before the evaporator. Most orifice tubes can be replaced by disconnecting the line and pulling the tube out with a special wrench. The new tube is pushed into place and the lines reconnected. Some older Ford vehicles have the tube inserted deep into the line. The line must be replaced to replace the tube.

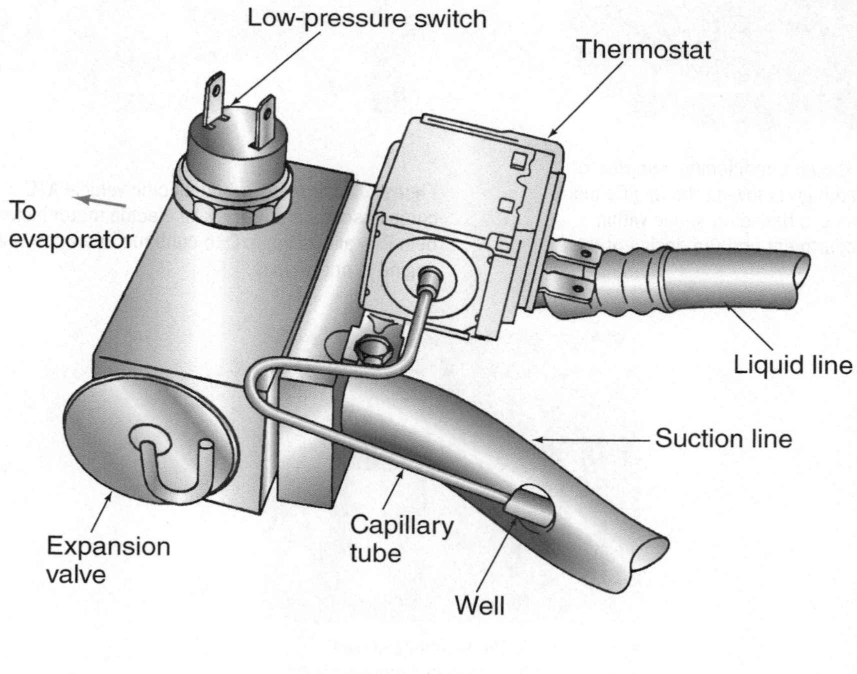

Figure 11-17 This expansion valve reacts to pressure differentials between the liquid and the suction lines. The thermostat controls compression clutch operation based on the temperature of the suction line.

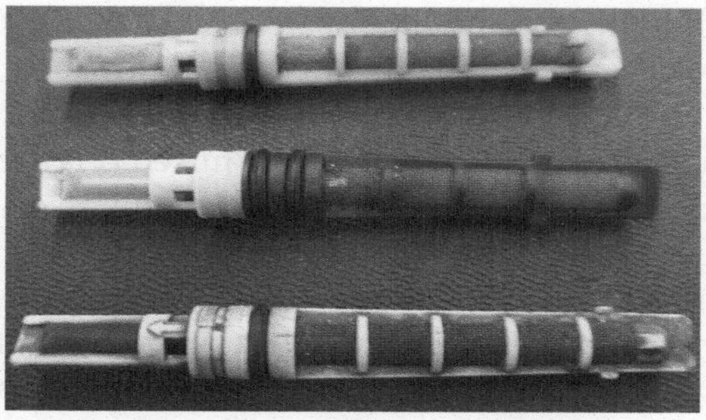

Figure 11-18 An orifice tube is a set of finely holed screens that restrict the movement of refrigerant into the evaporator. It is a fixed metering device.

The high-pressure switch is mounted in the high side, either on the line or on the compressor. The low-pressure switch is on the low side and usually mounted on or near the receiver/dryer but may be mounted on the compressor. Some units use a dual pressure switch to sense both high and low pressures (**Figure 11-19**). The switches are used to protect the system when the pressures do not meet specifications. It should be noted that many systems do not have high-pressure switches. Instead, they rely on the expansion valve to control high-side pressure. Almost all systems have a low-pressure switch to protect against low refrigerant charge and compressor damage. The compressor and other working components are lubricated in part with lubricant that is circulating through the system with the refrigerant.

AUTHOR'S NOTE Some older A/C systems without a high-pressure cutoff switch used a pressure relief valve. The valve would open suddenly to vent refrigerant if the pressure got too high. Most worked well, but some failed. In either case, the venting of refrigerant was loud, produced a cloud of steam (boiling refrigerant), and could be unnerving to the driver. Pressure relief valves are not allowed on any A/C system since the early 1990s because of ozone depletion.

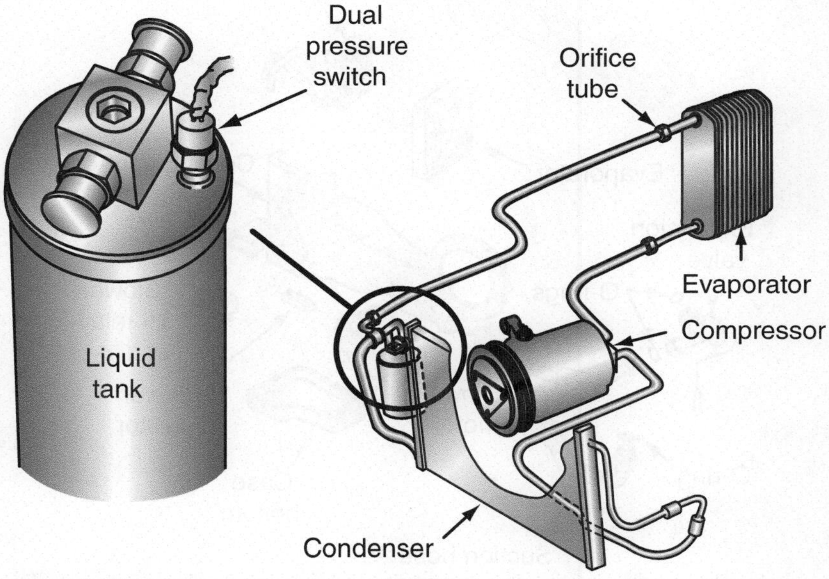

Figure 11-19 A liquid tank is common on Asian imports. The dual pressure switch is becoming more commonplace on many brands and models.

Special oil is added to the system to lubricate the compressor. Most of the lubricant is circulated with the refrigerant to lube the hoses, switches, and the valves of the compressor. The older R12 systems use different oil than the newer R134a units. The R134a units use PAG oils. The new and old oils cannot be mixed (**Figure 11-20**).

Because the air-conditioning system is installed in vehicles with a heater, the blower and ductwork are shared (**Figure 11-21**). In most cases, the A/C controls are on the same panel as the heater controls. The most common A/C controls are buttons labeled MAX and NORMAL, or something similar. The MAX switch turns on the A/C and opens the door that draws air from inside the passenger compartment. This allows cool air from the interior to be circulated back through the system. The passenger compartment is cooled quicker using this setting. The NORMAL setting turns on the A/C but draws in outside air. Normally MAX is used to quickly cool a vehicle upon startup, and NORMAL is used after the initial cooldown to keep the vehicle comfortable.

Some vehicles are equipped with automatic temperature control (ATC) systems. The operator selects the desired in-car temperature, and the ATC control module will select

Figure 11-20 The two main types of air-conditioning lubricants.

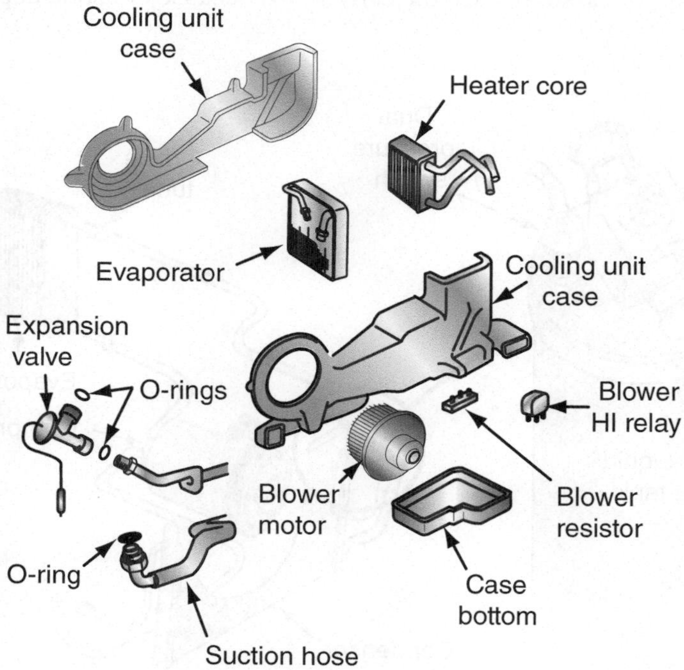

Figure 11-21 This package contains both the heater and the air-conditioning evaporator.

and operate the climate control system to fulfill the operator's request. The system may operate components of the heater and the air-conditioning system at the same time or use just one of them to keep the temperature at the desired level. The module can also control blower speed operation.

The newest ATC systems allow passengers in different seats to select a temperature for their area. This individual choice provides individual comfort to each passenger instead of just the driver.

AUTHOR'S NOTE Before we leave the air-conditioning section, let's take a look at a portion of the climate control that many technicians seem to forget. The defrost mode of the climate control uses dehumidified air to clear the windshield of moisture or fog. The air is dehumidified by the air-conditioning system. If a customer states the windshield "fogs over" in damp or cool air, turn on the defroster and check the operation of the A/C compressor. If it is not working, then the air from the defrost vents will not be dehumidified and will usually make the fogging condition worse. This is discussed more fully in the Shop Manual.

REFRIGERANT CONTROL AND USAGE

The Environmental Protection Agency (EPA) is charged with protecting the environment of the United States. One of its responsibilities is clean air. Gas- or diesel-powered vehicles are prime causes of dirty air. The Clean Air Act of 1990, as amended, is the latest law authorizing the EPA to take action concerning air quality. One of the issues addressed is A/C refrigerant.

The **ozone layer** helps protect the earth and its inhabitants from harmful **ultraviolet** rays and some of the heat from the sun. The older air-conditioning systems in homes and automobiles used carbon tetrachloride or R12 as the refrigerant (**Figure 11-22**). If

The **ozone layer** is located about 9 to 18 miles (15 to 39 km) above the earth's surface. It protects the earth from the sun's ultraviolet rays.

Ultraviolet is the term given to electromagnetic radiation and is not visible to the human eye. It is dangerous to humans and can cause skin damage.

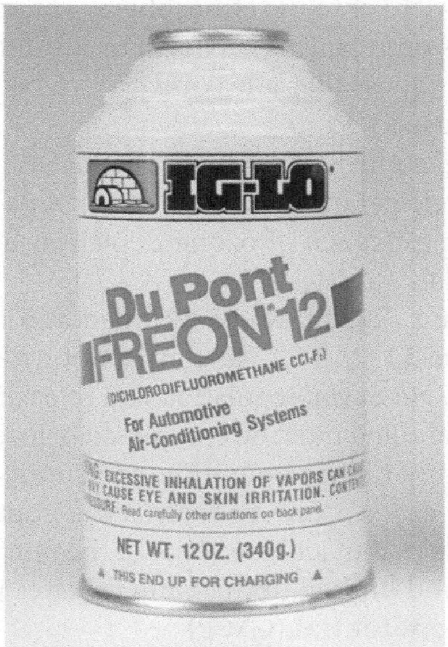

Figure 11-22 A typical container of R12 refrigerant. It can also be purchased in 30- and 50-pound containers. The containers are white or grayish-white. R12 is no longer produced in the United States and is not used on post-1996 vehicles.

refrigerant is released into the atmosphere, it breaks apart and the chloride rises to the level of the ozone layer. There, the chloride atom acts to break down the ozone molecule. The result is direct ultraviolet radiation reaching the Earth's surface.

AUTHOR'S NOTE Parts cleaners used up to the early 1990s also contained ozone-depleting chemicals. This was particularly true of hand-held sprays that were propelled by a mixture containing a chemical similar to carbon tetrachloride. Cleaning chemicals many times were based on a chloride mixture. All parts-washer chemicals containing chloride have been banned by the EPA and common sense.

The EPA and the Clean Air Act specifically prohibit the release of any refrigerant into the atmosphere, imposing very heavy fines for violators. Technicians working on A/C systems are required to pass an *EPA-approved* test before they can legally handle refrigerant. The test involves the control of refrigerant, the effect of refrigerant on the ozone layer, and some of the legal requirements of the Clean Air Act. The test does not include technical information on mechanical diagnosing and repairing an A/C system. The test is available from several approved sources including ASE and MACS. Some parts stores have the test booklet on hand for anyone interested in taking the test. The test is basically open-book, with 25 multiple-choice questions. Complete the test and mail the answer sheet with the noted fee to the address given. The results, with an identification card (if passed), are sent back. It should be noted that similar requirements are applied to fixed A/C units such as the ones used in homes and businesses and refrigerators, private and commercial. Any person desiring to purchase any refrigerant must show proof that he or she passed the EPA test by presenting the identification card.

Recovery is the process of capturing refrigerant in an A/C system.

Evacuating is removing all air and moisture from the system after recovery.

Recharging is putting refrigerant back into the system after repairs.

The EPA requires very specific actions to occur during **recovery, evacuating,** and **recharging** of an air-conditioning unit. All the requirements are based on controlling refrigerant leaks, particularly R12, into the atmosphere. The requirement allows each shop to make individual decisions on repairing a system or just adding refrigerant to a leaking system. The customer can request that a leaking system be topped off with refrigerant, but the shop may legally refuse to do so.

In addition to testing and control procedures, the EPA directed research and development of an ozone-friendly refrigerant. Research in this area has been ongoing since the mid-1970s, but the growing evidence of ozone depletion has increased the amount of funding and the number of laboratories.

A replacement refrigerant, R134a, was developed and is being used by all vehicles produced since 1995 (**Figure 11-23**). R134a consists of hydrogen-based molecules that do not affect the ozone layer. However, regulations on releasing refrigerant still apply, and the EPA test on refrigerant handling must be completed before using R134a. Systems using R134a normally run at slightly higher operating pressures, and there is a small loss in system efficiency. There are other ozone-friendly refrigerants on the market, but R134a is the one approved by vehicle manufacturers. One of the new refrigerants being produced and used widely in Europe and limited in the United States is R-1234yf. This refrigerant has a lower **global warming potential** (GWP).

Global warming potential is a relative measure of how much heat a greenhouse gas traps in the atmosphere.

To encourage the public to use the new refrigerant, federal, state, and local governments have imposed a progressively increasing tax on R12. Since 1990, the price of R12 has gone from 79 cents for a 12-ounce can to $18 or more at the parts store. Most repair

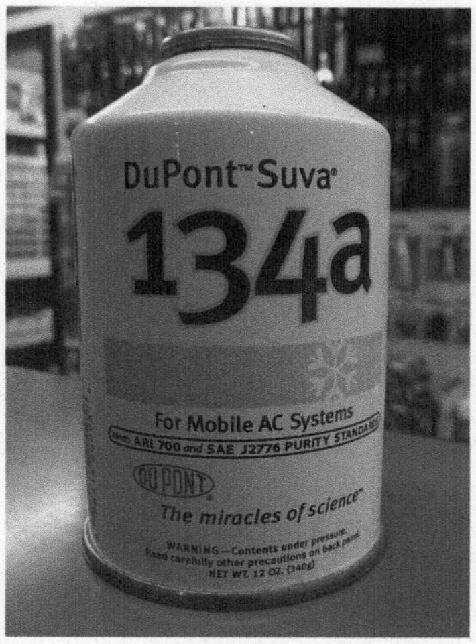

Figure 11-23 R134a containers are light blue and can be bought in 30- and 50-pound containers. The color is required by the EPA as an identification of the chemical inside.

centers charge $30 or more for a pound (12 ounces) of R12. Most R12 units can be retrofitted to R134a. During retrofit, the system is emptied of R12, and most of the old lubricant, seals, and gaskets are replaced. The R134a refrigerant and the correct oil are then charged into the systems. Many shops recommend that a damaged R12 system be retrofitted during routine repairs. The savings on the refrigerant will pay for the kit to make the change. The price of the old and new lubricants is about the same.

> **AUTHOR'S NOTE** In 2005, the price of R12 had dropped greatly and stabilized from 1999. This price drop is due to the lower number of R12-equipped vehicles now in operation. Most of the older vehicles have been retrofitted or they are not on the roads anymore. Since 1995, all models come equipped with R134. However, China and other nations in Asia and Africa were given a 10-year delay to enact the Montreal Protocol on their nonexported vehicles. They were required to ban ozone-depleting chemicals starting in the year 2000. Today, more research and testing are being done to move to a more environmentally safe refrigerant using in vehicles.

R134a is not the final refrigerant that will be used in mobile air-conditioning systems. The European Union (EU) has established a GWP requirement, which states that the refrigerant cannot exceed 150 GWP (R134a is rated at 400 GWP) while research is pending on other refrigerants. As of 2009, the EU has not implemented GWP, but it is expected to begin within the next several years. There are few "drop-in" replacements for R134a. Most of the proposed refrigerants will require massive component replacement and cost to the consumers. One refrigerant under favorable consideration is R744, commonly known as carbon dioxide (CO_2). CO_2 is environmentally friendly but requires extreme pressure to act as a good heat transfer medium.

SUMMARY

- Passenger comfort devices are designed to make the driver and passenger comfortable during vehicle operations.
- Principles of heat transfer.
- The radiator cools the antifreeze.
- The heater core warms the passenger cabin.
- Identifying the different types of antifreeze.
- Climate control includes heating and air-conditioning systems.
- Most A/C units operate at high pressure.
- The condenser cools the refrigerant.
- The evaporator cools the air by removing heat from it.
- The compressor pumps and pressurizes the refrigerant.
- Expansion valves or orifices regulate the amount of refrigerant entering the evaporator.
- The EPA is responsible for directing research and implementing procedures to protect the environment.
- The Clean Air Act of 1990, as amended, charges the EPA to clean vehicle emission discharges.
- Chloride in R12 destroys ozone molecules.
- R134a is an ozone-friendly refrigerant.
- A person must be certified by an EPA-approved test before buying refrigerant or servicing A/C systems.

REVIEW QUESTIONS

Short Answer Essays

1. Describe the three methods of transferring heat.
2. List the components that make up the cooling system.
3. Describe how warm air enters the passenger cabin.
4. List the components that make up the air conditioning system.
5. Describe how cold air enters the passenger cabin.
6. Explain what an HVAC case is and what components are contained inside the case.
7. Describe how a high-pressure liquid becomes a low-pressure vapor in an A/C system.
8. List the major components on the high side of an A/C system.
9. Explain the requirements of the Clean Air Act of 1990.
10. Describe the requirements needed to work on A/C systems.

Fill-in-the-Blanks

1. ________ ________ ________ is the inclusive term for the heating, ventilation, and air-conditioning systems (HVAC).
2. The transfer of heat is a principle of physics that you need to understand to properly ________ and ________ the HVAC system.
3. There are three different ways heat can be transferred: ________, ________, and ________.
4. The ________ ________ carry hot coolant from the engine to the heater core and back to the engine.
5. The climate control panel mounted in the dash houses buttons or levers to direct the ________, change ________, and adjust ________.
6. The ________ is where all the heat is released from the coolant.
7. Most of the modern vehicles produced today use coolants that are incompatible with the common type of ________ ________ (IAT) or ethylene glycol-based coolant.
8. The A/C is usually split into two sections or sides: ________ and ________.
9. A ________ is a chemical that can absorb, transport, and release a great deal of heat in a small space under easy-to-create conditions.
10. The EPA and the ________ ________ of ________ specifically prohibit the release of any refrigerant into the atmosphere, imposing very heavy fines for violators.

Multiple Choice

1. Climate control systems are being discussed.
   Technician A says the heater uses engine coolant to heat the passenger compartment.
   Technician B says the condenser is mounted in the dash.
   Who is correct?
   A. A only C. Both A and B
   B. B only D. Neither A nor B
2. *Technician A* says all long-life coolants are similar and can be used interchangeably.
   Technician B says that traditional green coolant can be mixed with any type of coolant.
   Who is correct?
   A. A only C. Both A and B
   B. B only D. Neither A nor B
3. *Technician A* says the climate control panel mounted in the dash and houses buttons or levers to direct the air, change blower speed, and adjust temperature.
   Technician B says the operator can select outlets that direct warm air to the windshield, dash panel, or to the floor.
   Who is correct?
   A. A only C. Both A and B
   B. B only D. Neither A nor B
4. *Technician A* says that conduction is when heat transfers through a solid.
   Technician B says that radiation is when heat transfers through space.
   Who is correct?
   A. A only C. Both A and B
   B. B only D. Neither A nor B
5. *Technician A* says R134a molecules are chloride based.
   Technician B says R134a purchase and use are regulated by the EPA.
   Who is correct?
   A. A only C. Both A and B
   B. B only D. Neither A nor B
6. Air-conditioning systems are being discussed.
   Technician A says the refrigerant changes from a liquid to a vapor in the condenser.
   Technician B says heat from the passing airflow vaporizes the refrigerant.
   Who is correct?
   A. A only C. Both A and B
   B. B only D. Neither A nor B
7. The depletion of the ozone layer is being discussed.
   Technician A says R12 released into the atmosphere destroys ozone molecules.
   Technician B says a depleted ozone layer allows more ultraviolet radiation to reach the earth.
   Who is correct?
   A. A only C. Both A and B
   B. B only D. Neither A nor B
8. *Technician A* says that R134a is the only refrigerant approved by vehicle manufacturers.
   Technician B says that R1234yf is the only refrigerant approved by vehicles manufacturers.
   Who is correct?
   A. A only C. Both A and B
   B. B only D. Neither A nor B
9. *Technician A* says the EPA and the Clean Air Act specifically prohibit the release of any refrigerant into the atmosphere.
   Technician B says some technicians working on A/C systems are required to pass an *EPA-approved* test before they can legally handle refrigerant.
   Who is correct?
   A. A only C. Both A and B
   B. B only D. Neither A nor B
10. *Technician A* says the expansion valve regulates high-side pressure on an A/C system.
    Technician B says an orifice may be used to control refrigerant flow into the evaporator.
    Who is correct?
    A. A only C. Both A and B
    B. B only D. Neither A nor B

CHAPTER 12

ENGINE PERFORMANCE THEORIES

Upon completion and review of this chapter, you should understand and be able to:

- Describe the requirement to apply theories of physics in designing and operating vehicles.
- Understand the first two laws of thermodynamics and their application in an engine.
- Explain the term *energy* and the two types of energy.
- Recognize how pressure and vacuum help operate the engine.
- Explain Newton's three Laws of Motion.
- Explain Pascal's Law and the use of hydraulics in vehicles.
- Understand performance factors in designing and operating a vehicle.
- Identify the difference between and characteristics of horsepower and torque.

Terms To Know

Aerodynamics
British thermal unit (BTU)
Combustion chamber
Crumple zones
Explosion
Foot-pounds (ft.-lb.)
Force
Fuel mileage
Horsepower
Hydraulics
Inches of mercury
Inches of water
Inertia
Kinetic energy
Mass
Perpetual energy
Potential energy
Pounds per square inch (psi)
Ratio
Thermodynamics
Torque
Work

INTRODUCTION

The automobile is made up of eight major interactive systems based on the application of certain theories of physics. The seven basic systems are the engine, body and frame, electrical, braking, steering and suspension, climate control, and driveline. The eighth is the computer system, which is not required to make a simple vehicle work but does provide the controlling element for the basic systems. The systems power the vehicle, deliver power to the wheels, and comfort and protect the passengers. Functional systems working together make it possible for the automobile to be what it is today: reliable, durable, and comfortable. In this chapter, we will discuss some of the theories of operations and examples of how they apply to the major systems. Understanding these theories will assist the technician in diagnosing problems that do not respond to the conventional practice of electronic diagnostics and direct testing of suspected failures. Understanding these theories will help in understanding engine performance and diagnosing performance related problems.

THERMODYNAMICS

Engines use the theory of **thermodynamics** to generate power. The first law of thermodynamics is a law of energy conservation: Energy cannot be created or destroyed. The amount of internal energy in a system is a direct result of the heat transferred into that system and the work done. Heat and work are means by which natural systems exchange energy with each other. Basically, it is the relationship between heat and mechanical energy (**Figure 12-1**).

The second law concerns the thermodynamic cycle. In a perfect heat engine, all of the heat produced would be completely converted to mechanical work. Nicolas Leonard Sadi Carnot, a 19th-century French scientist, proved that this ideal cycle could not exist. The best engine would lose heat and work because the engine must be exhausted. The exhaust would remove and waste a great deal of heat.

Liquids that are heated enough will change to a vapor and expand greatly in volume. A prime example of this is boiling water. As the water is heated, it expands in volume, which is the amount of space it occupies. Somewhere around 212°F, the water changes from a liquid to a gas, or vaporizes. The gas or steam greatly increases its volume. For example, place a lid on top of a container of boiling water. As the water boils more rapidly, the lid will begin to rattle and move about. We can capture the energy of the water by bolting the lid down and not allowing the steam to escape. However, if we do not allow the steam to escape in some way or use its energy, the volume of steam will continue to build until it ruptures the container and an **explosion** occurs.

Most lightweight vehicles use gasoline that is rated at about 110,000 **British thermal units (BTUs)** per gallon. BTUs are used to measure the heat-producing

Thermodynamics is the study of using radiant heat to perform mechanical work.

Shop Manual
pages 551–553

An **explosion** is a sudden, violent release of energy. An engine uses a controlled burn, not an explosion, to achieve power.

BTUs The amount of heat required to raise the temperature of one pound of water by one degree.

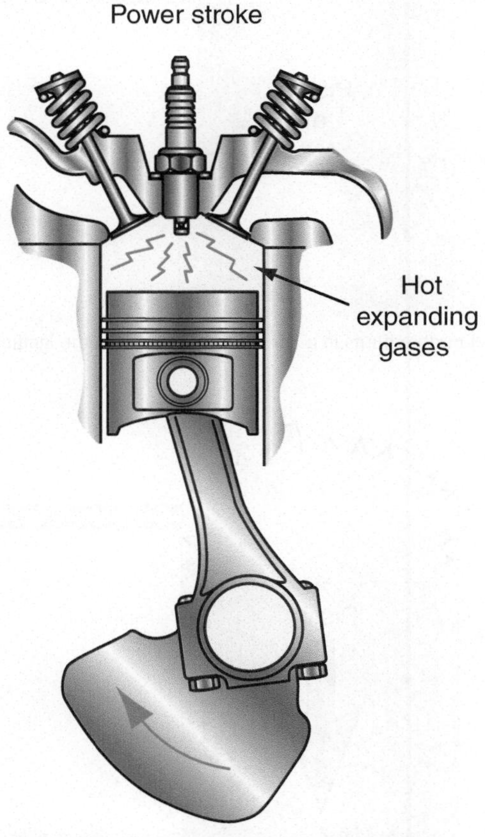

Figure 12-1 The force of hot, expanding gases is transformed into mechanical energy by the movement of the piston.

The **combustion chamber** is the open chamber at the top of an engine's cylinder. It is here that the fuel is ignited.

One **horsepower** is equal to the force needed to raise 33,000 pounds at one foot per minute.

Torque Twisting force that tends to cause rotation.

capabilities of a chemical or mechanism. Burning vaporized gasoline in a closed chamber, called a **combustion chamber**, will result in high heat and a rapid expansion of gases (**Figure 12-2**). The rapid heating and expansion of gases creates a very high force that acts against the engine's pistons and produces useable power (**Figure 12-3**). The combustion chamber is closed, but it does have one side that can move: the piston. The energy of the expanding gas is transmitted against the piston, forcing it downward. This, in turn, rotates the crankshaft, producing a twisting force or *torque*. As the torque is transmitted through the vehicle's drive system, it moves the wheels and work is performed. Work in this instance is measured either as **horsepower** or **torque**.

However, the engine loses heat and, therefore, work through the exhaust and cooling system. The exhaust is required to remove spent gases while the cooling system must remove heat from the mechanical components. Roughly two-thirds of the heat in a gallon of gas is lost through exhaust and cooling. Only about 35,000 BTUs are left to operate the engine and vehicle. After friction and overcoming the weight, the mass of components, and the vehicle, only about 19 to 20 percent of the energy in a gallon of fuel is actually applied to the drive wheels. Vehicles made in the late 1990s through 2008 do somewhat better, but the nature of the beast prevents any large increase in the efficiency of current internal combustion engines.

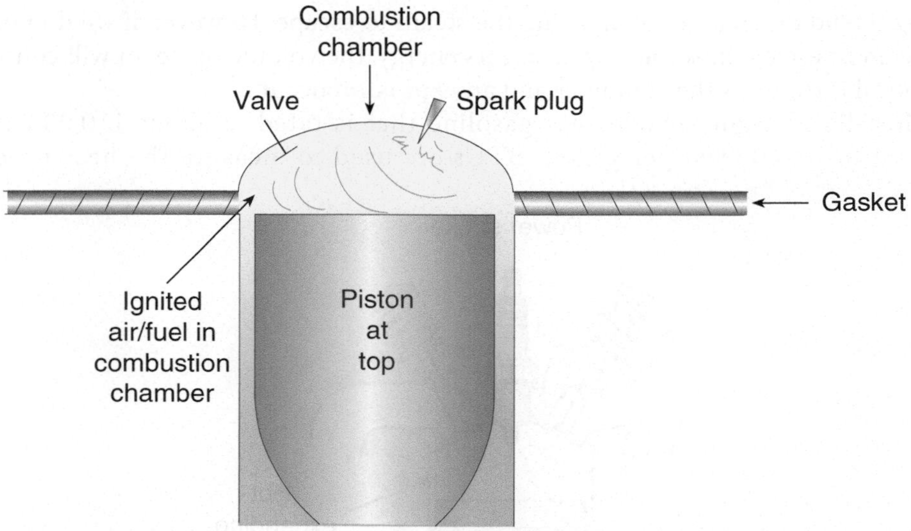

Figure 12-2 The air/fuel mixture burns in a controlled manner from the ignition point across the chamber.

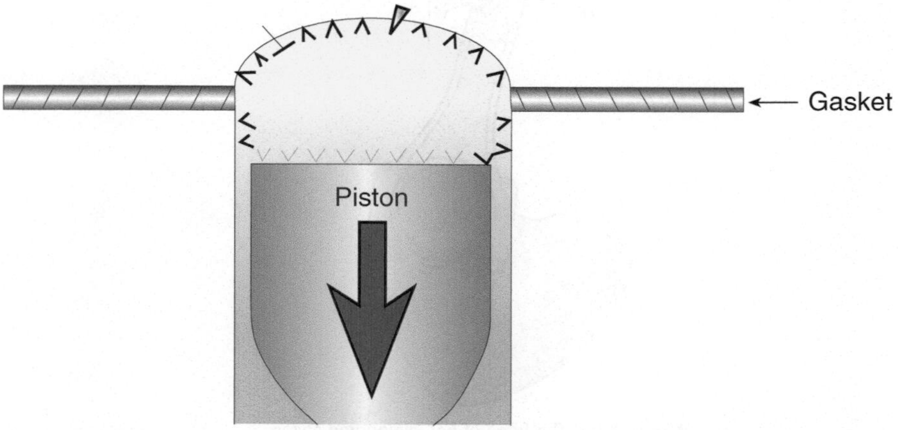

Figure 12-3 The force of the gases acts against the piston the same as it did against the projectile in Figure 12-1. Note how much pressure the gasket must endure to seal the cylinder and not leak.

A BIT OF HISTORY

A fatal example of an oxygen-rich atmosphere occurred during a ground launch test of the first Apollo spacecraft designed for travel to the moon. The atmosphere in the cockpit of the craft was almost 100 percent pure oxygen. A small electrical spark caused a fire that destroyed the cockpit and killed the three astronauts inside.

The application of thermodynamics in an engine requires fuel, oxygen, a spark, and a combustion chamber. The engine will not operate or will operate poorly if a fault occurs in any of these four elements. Almost nothing will burn without the presence of oxygen. Considering that only about 20 percent of the Earth's atmosphere is oxygen and the number of fires that occur worldwide, it is apparent that oxygen does not have to be present in large amounts. By the same token, almost everything will burn if enough oxygen is present. Steel will burn like a match if enough oxygen is available.

The control of the fuel and air (oxygen) to an engine is essential for a safe, controlled burn (**Figure 12-4**). Although an automobile does not and cannot use pure oxygen, too much or too little upsets the **ratio** of air and fuel, resulting in a poorly operating engine. An example of upsetting the air-to-fuel ratio would be a dirty air filter. The filter will not allow sufficient air to enter the combustion chamber. With insufficient oxygen, the gasoline may or may not fire. If it does fire, it will not produce the necessary energy to perform the work needed.

Ratio is the proportion or relation one element has to another. In this case, it is the relation of air to fuel.

Shop Manual
pages 551–553

AUTHOR'S NOTE The desired air/fuel mixture at idle and cruise speeds is 14.7 parts of air to 1 part of fuel (14.7:1). For cold starts and acceleration, a short-term richer mixture of about 13:1 may be best whereas a mixture of 18:1 during deceleration provides better fuel mileage. However, when mixtures are leaner or richer than 14.7:1 for long periods, fuel mileage, engine performance, and emissions output change—sometimes drastically. Note that 14.7 is generally accepted as the norm for pounds of atmospheric pressure per square inch (psi) at sea level.

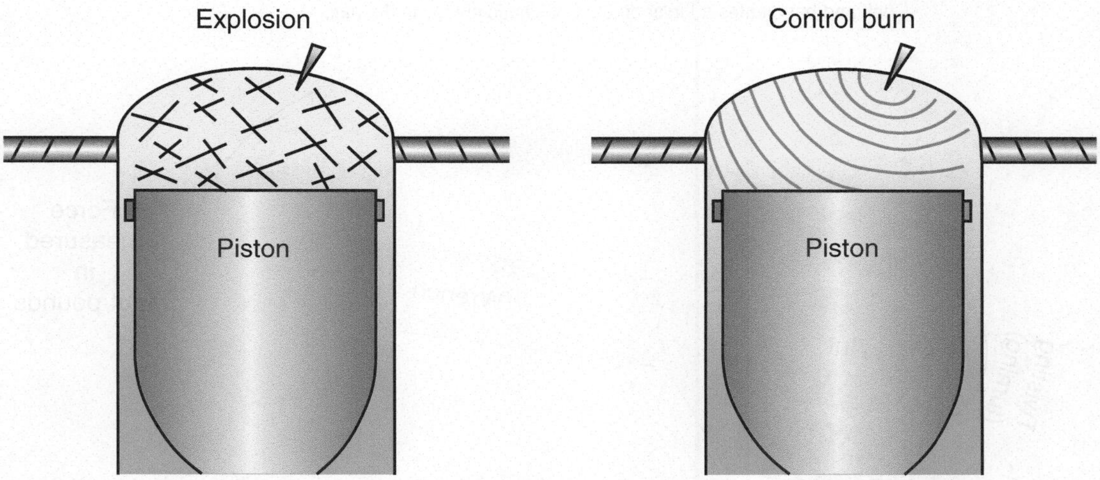

Figure 12-4 An explosion (left) creates uncontrollable energy radiating in unpredictable directions. A controlled burn radiates energy equally in all directions at once.

Potential energy is based on the mass and position of a stationary object. The speed of a moving object plus its mass creates the **kinetic energy** of the object.

Work is the expenditure of energy.

Foot-pounds Unit of energy equal to the amount required to raise one pound a distance of one foot.

Force is the pressure of one mass against another.

ENERGY AND WORK

Energy is defined as the capacity of matter to perform work through its motion or position. The motion of an object produces **kinetic energy**. **Potential energy** is available energy based on the position of an object. Consider the action of a baseball. While the pitcher is holding the ball in his or her hand, the ball represents potential energy. When the ball is released toward the hitter, it has kinetic energy based on its weight and speed. The hitter's bat has potential energy until he or she swings it. When the bat and ball connect, the kinetic energies of both result in an expenditure of energy (**Figure 12-5**). The ball is driven in a different direction with great force. An equal amount of opposing force is absorbed by the bat and the hitter (see Newton's Laws of Motion later in this chapter).

When an object moves, and energy is expended, work has been accomplished. This includes raising a hand or moving a house. **Work** is a unit of energy and may be measured in **foot-pounds** (ft.-lb.). Foot-pounds measure torque or twisting force. Work is measured by multiplying the **force** times the distance; one foot-pound will lift one pound of mass the distance of one foot (**Figure 12-6**).

Gasoline has potential energy in the fuel tank. Its rapid expansion in the combustion chamber results in great amounts of kinetic energy. This energy is used to transform a resting vehicle with potential energy into a moving vehicle with large amounts of kinetic energy. This is proven by the amounts of kinetic energy released through a traffic accident.

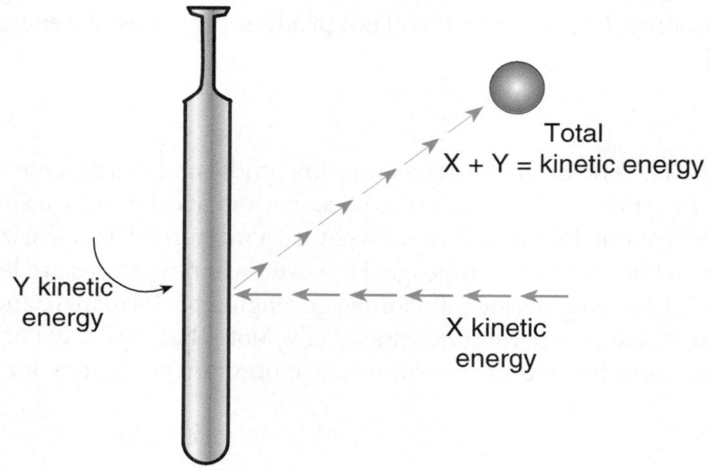

Figure 12-5 The kinetic energy created by merging or connecting a moving ball and bat creates a larger amount of kinetic energy in the ball.

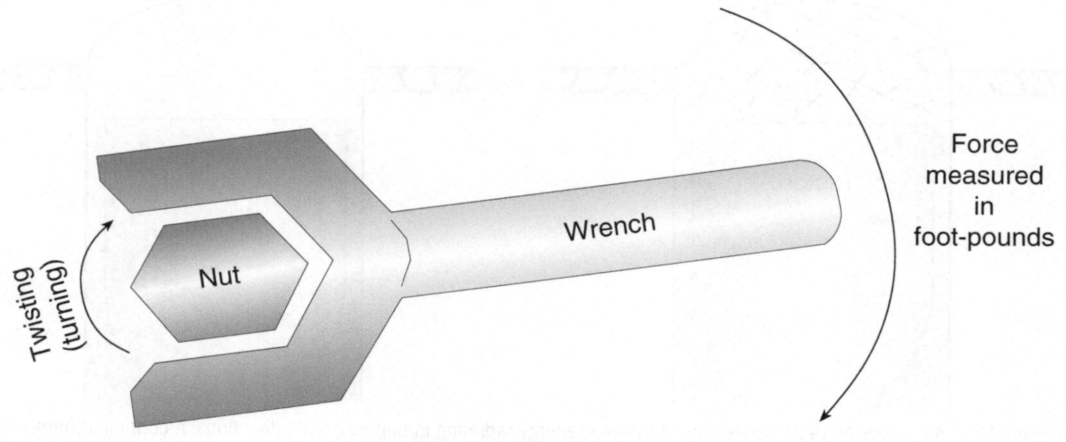

Figure 12-6 Torque is measured by determining the amount of twisting force being applied.

There is one major drawback to all energy-producing devices now used. It is a fact that a machine or device cannot produce more energy than it consumes; in other words, there is no **perpetual energy** device today. A device that seems to contradict this statement is the alternating-current generator. Without getting into depth on generator operation, the generator draws its power from the battery or its own electrical output and seems to produce more than it uses. However, when factoring the horsepower required for driving the generator's pulley, there is definitely a loss of energy. Barry Hollembeak, in *Today's Technician: Automotive Electricity and Electronics,* Cengage Learning, Clifton Park, New York, has the details for the generator's operation.

Perpetual energy is energy created using less energy than that produced.

Shop Manual
pages 52-53, 373

PRESSURE AND VACUUM

Pressure is the amount of force used to hold a gas or liquid in place. The atmosphere applies about 15 pounds of pressure per square inch to every external inch of all objects on the surface of the Earth including the human body. Atmospheric pressure can be used to achieve work in a vehicle or components and can be used to create pressure within a system to perform work.

Vacuum is defined as any pressure below atmospheric. The earth does not have sufficient natural vacuum to perform automotive work, but it can be created to some extent by certain components. One of those components is the engine.

As discussed in the section on thermodynamics, fuel and air must be present in proper amounts in the combustion chamber to create power. Because the chamber can be sealed, the downward movement of the piston increases the volume of the cylinder, creating a vacuum above it. This combination of vacuum and outside atmospheric pressure forces fresh air past a valve that is timed to piston movement (**Figure 12-7**). Fuel is pressure-injected into the airflow near the same valve or directly into the chamber during the intake of air. With the proper equipment, the amount of fuel can be regulated in a correct ratio to the amount of air entering the chamber. Older systems used the movement of air passing through a venturi tube to create a low-pressure area and draw fuel from the carburetor fuel bowl (**Figure 12-8**). Now almost all engines use electronic devices to control the fuel.

The force of the driver's foot creates pressure within a brake system (**Figure 12-9**). The force of the driver's foot is transferred by pressurized fluid to the brakes at each wheel.

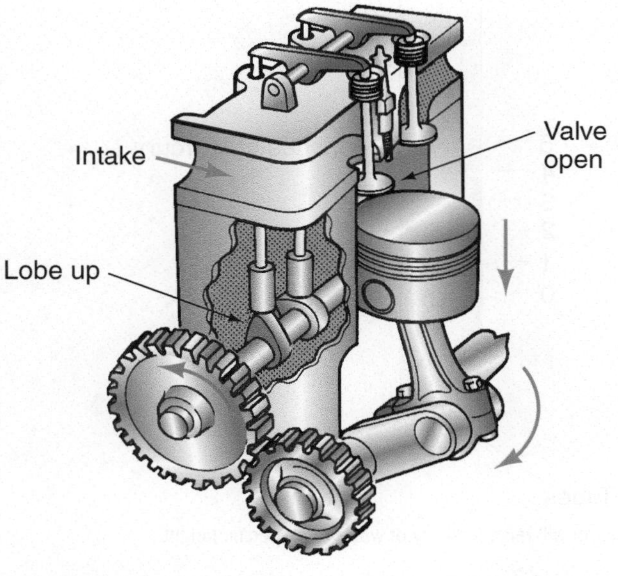

Figure 12-7 This valve opens to let the air/fuel mixture into the cylinder.

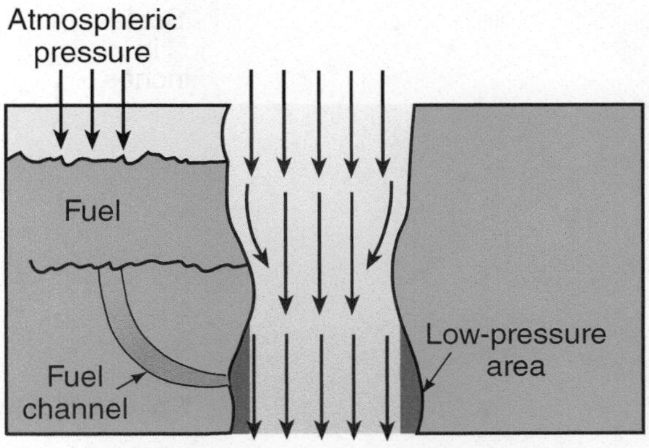

Figure 12-8 An object obstructing the airflow will create a low-pressure area downwind and behind the object.

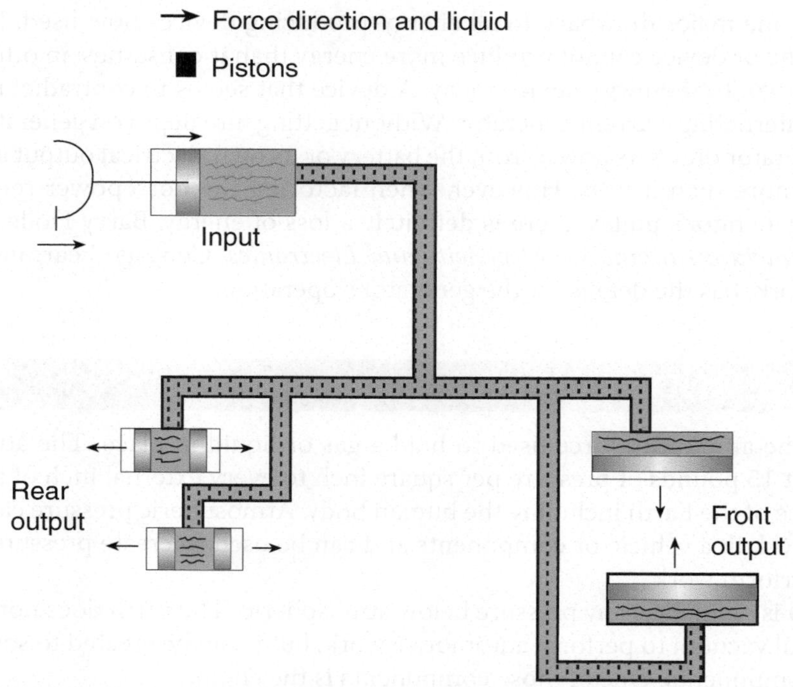

Figure 12-9 Applying force to a liquid in a sealed system will create pressure and transfer force.

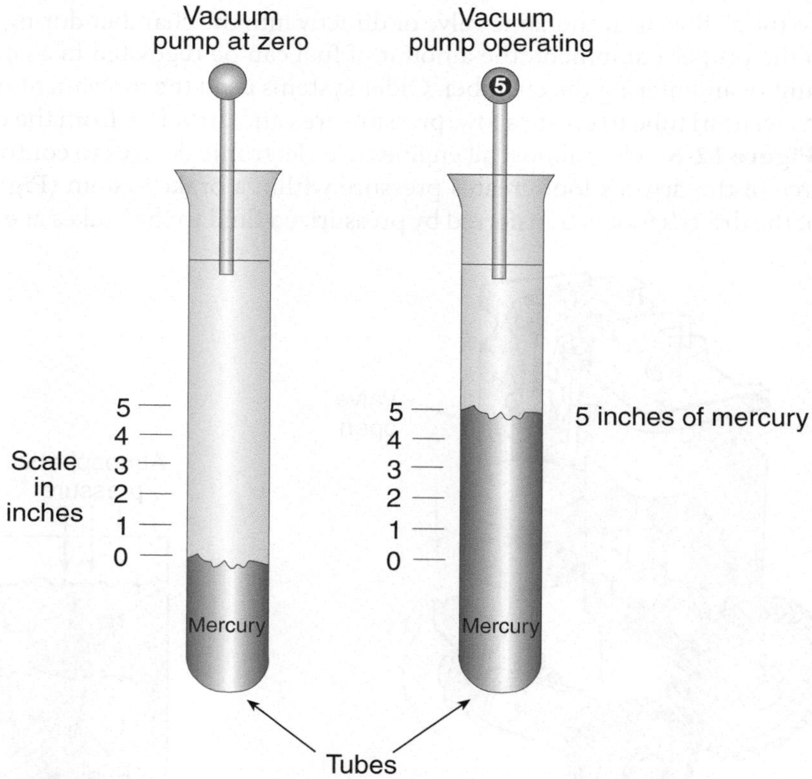

Figure 12-10 A given amount of vacuum will raise mercury or water to a specific height.

Engine vacuum and atmospheric pressure can be used to assist the driver in braking. A steering system uses a pump to supply high-pressure fluid to a power-assist unit controlled by the actions of the steering wheel. In both cases, pressure is used to assist the driver in the operation of the vehicle. Both systems are covered in earlier chapters.

Pressure is measured in **pounds per square inch (psi)**. The use of pressurized fluid is covered later in this chapter in the section on hydraulic theory. Vacuum is usually measured in **inches of mercury** or **inches of water**.

PSI A common unit of measurement for pressure. It is the amount of force that is exerted on one square inch of area surface.

Inches of mercury/water is the amount the level of mercury or water in a sealed container rises when a vacuum is applied.

The most common vacuum measurement used in automotive repair is inches of mercury. Vacuum supplied against mercury in a closed tube will cause the mercury to rise based on the amount of vacuum (**Figure 12-10**). The amount of rise is measured in inches. However, because mercury is a deadly poison and a glass tube is not practical in the shop, calibrated mechanical gauges are used to register the measurement.

NEWTON'S LAWS OF MOTION

Newton wrote three basic Laws of Motion while proving that the Earth revolved around the sun instead of the sun moving around the Earth. The laws pertain to almost every system on the vehicle.

Newton's First Law deals with **inertia**. An object at rest tends to stay at rest and an object in motion tends to stay in motion. This unchanging action is called inertia and it assumes that there is no outside force acting on or against the object. A stationary vehicle will not move unless something forces it to move. Parking a vehicle in neutral with the brakes off on level ground will result in a parked vehicle. However, parking the same vehicle on a hill, in neutral, and without brakes will allow gravity, an outside source, to act on the vehicle and cause it to roll down the hill, thereby overcoming rest inertia (potential energy) and generating motion inertia (kinetic energy).

Inertia The tendency to do nothing or to remain unchanged. "an object at rest tends to stay at rest" and "an object in motion tends to stay in motion"

Newton's Second Law: A moving object will travel in a straight line and in one direction. Try driving a car on a flat, level road. If the car is set up correctly, it will travel in a straight line. A force (steering) must act upon the car to change its direction. A second example is an inflated balloon. When an inflated balloon that has not been knotted is released, it immediately moves through the air in erratic directions. The balloon is not shaped to travel through air, and it changes direction as the air acts against it. When all the trapped air in the balloon is exhausted, the balloon falls to the Earth. The balloon's shape and the loss of pressure are two actions that work not only on the vehicle but on each moving component of the vehicle as well.

Within this law is a rule regarding acceleration of an object. The acceleration of a vehicle is directly proportional to the driving forces applied on it. This applied force must overcome the friction of the various moving components, tire against road, and the total **mass** of the vehicle and its load. Mass is not the weight of the object but the amount of matter in that object. A vehicle weighing 2 tons on Earth will weigh about 670 pounds on the moon, but its mass will be the same. When calculating the acceleration rate of a vehicle, other forces must be considered. As the drive wheels force against the road and begin to accelerate the vehicle, it is resisted by a force equal to the vehicle's mass multiplied by its acceleration rate. Then aerodynamic forces begin to act simultaneously against the vehicle in the opposite direction of the driving force, reducing the acceleration rate. In general, this reduces the amount of force available to keep the acceleration rate high. At some point, the driving forces equal the resisting forces, indicating the vehicle has reached its maximum speed. Most light vehicles will accelerate quickly to a certain speed before the rate of acceleration drops off. This can be checked easily with a

Mass is the matter or atomic structure of an item.

stopwatch and an open, clear stretch of road. Measure the time taken to accelerate from 0 to 25 miles per hour. Record the time but continue running the clock until speed reaches 50 miles per hour. On a typical passenger car or light truck, the second 25 miles per hours will be at a slower rate, even though the vehicle is already moving at the first time reading. Some individuals will tend to disagree with this scenario by quoting the 100-foot, 200-foot, and the quarter-mile acceleration rate or top-end speed of dragsters. Consider this fact: Six-thousand-plus horsepower, aerodynamic lightweight body, and a vehicle designed to deliver maximum acceleration in midrange gears is not quite your average pizza delivery vehicle.

For every action, there is an equal and opposite reaction—Newton's Third Law. Again, it is assumed that there are no outside forces in play. If a ball were thrown against a wall in outer space, it would return to the thrower with the same force and same speed. Doing the same on the Earth would result in the ball slowing on its return journey. This is due to air resistance and gravity working against the ball, so it is slowed, and its direction of travel is changed.

Earlier, we discussed kinetic energy in a vehicle accident. Consider, for example, that a 3-ton vehicle moving at 60 miles per hour with 100 tons of kinetic energy was suddenly stopped by an immovable object. The action of stopping 100 tons would immediately result in an opposite and equal reaction of 100 tons against the immovable object. That release of energy would not only affect the object but also the vehicle as well. Because energy cannot be destroyed, the 100 tons of suddenly released kinetic energy must be dissipated or changed in some manner, in this case the destruction of the vehicle and object.

The movement, weight, and operation of all vehicle components are subject to the laws of motion. To move a piston, energy must be used. In turn, the actual movement of the piston creates work. Each movement of every part requires energy and produces work. The engineering of a vehicle requires a balancing of the energy used versus the amount of work performed.

HYDRAULIC THEORY

Hydraulics is the use of a liquid to transfer force. Liquids used in hydraulic systems are selected based on the work to be done and the chemical properties of the liquid. Brake fluid will not work well in steering systems.

Blaise Pascal formulated his law on the use of **hydraulics** in 1647. Known as Pascal's Law, it states that pressure in a closed hydraulic circuit will be the same everywhere in the circuit. The law is based on the fact that liquid in a closed or sealed circuit cannot be compressed.

A hydraulic circuit is like an electrical circuit. There are hoses and lines (conductors), pumps (power source), load, and controls (**Figure 12-11**). The pump pressurizes the fluid and the control valve directs it into the hoses and lines. At the output end of the line or hose is a working device. It may be a brake component, an electronic fuel injector, or a piston that helps the operator apply force.

As with electricity, there are three measures normally used in hydraulic circuits: *force, pressure,* and *area.* The force is applied to the contained fluid and delivered by the fluid. Area is the surface on which the fluid acts. Pascal's Law uses a mathematical formula to calculate hydraulic values. It is written as $F = PA$, where F is force (ft.-lb.), P is pressure per square inch (psi), and A is area in square inches (sq. in.) (**Figure 12-12**). A hydraulic system can deliver force and increase or reduce it. An increase is called a *mechanical advantage.*

In **Figure 12-13**, the input and output pistons have the same area. Trapped between them is a liquid. Applying force on the input piston will create an equal force being delivered by the output piston because of the equal areas. However, any reaction against the

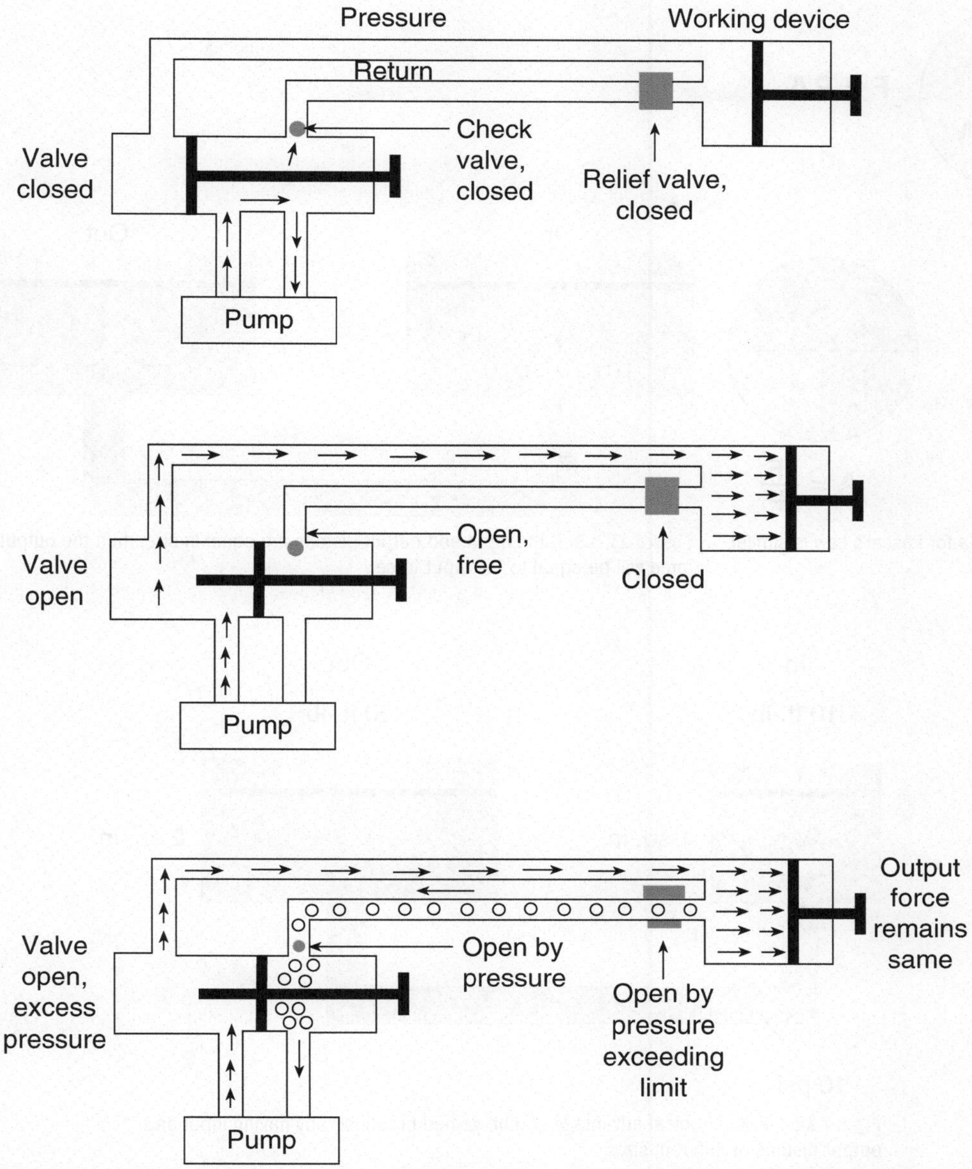

Figure 12-11 Note the actions of the check valve and the relief valve in this hydraulic circuit. Valves of this type are common in hydraulic circuits.

output piston will deliver the same force backward through the circuit to the input piston and anything that is pushing on it.

In **Figure 12-14**, the output piston has twice the area of the input piston. In theory, this means the amount of force applied at the input is doubled at the output. We can prove this by using the formula for Pascal's Law. Because the pressurized fluid between the two pistons is the *transfer median,* the amount of pressure generated must first be determined. The figure shows an area of 1 square inch for the input piston and a force of 10 pounds. Using the known values (1 and 10) and the adjusted formula $P = F/A$ ($F - PA$), we can calculate that the pressure is 10 psi ($P = 10/1$). Moving to the output piston and using $F = PA$, we can calculate the force that can be delivered by the output piston at 20 ft.-lb. ($F = 10 \times 2$). Remember that pressure is per square inch. The input is 1 square inch and the output is 2 square inches. Because there are 2 square inches on the output and 10 psi

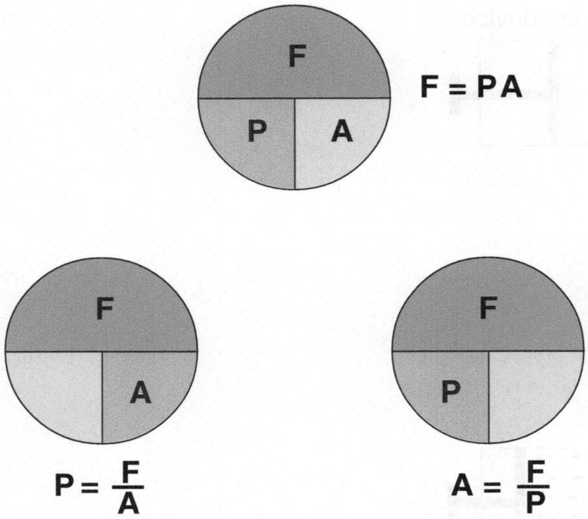

Figure 12-12 The mathematic formula for Pascal's Law is similar to the one for Ohm's Law.

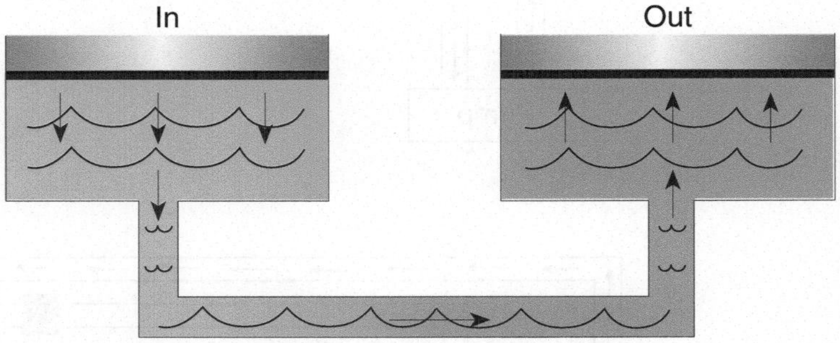

Figure 12-13 If the input and output pistons are equal in size, then the output force will be equal to the input force.

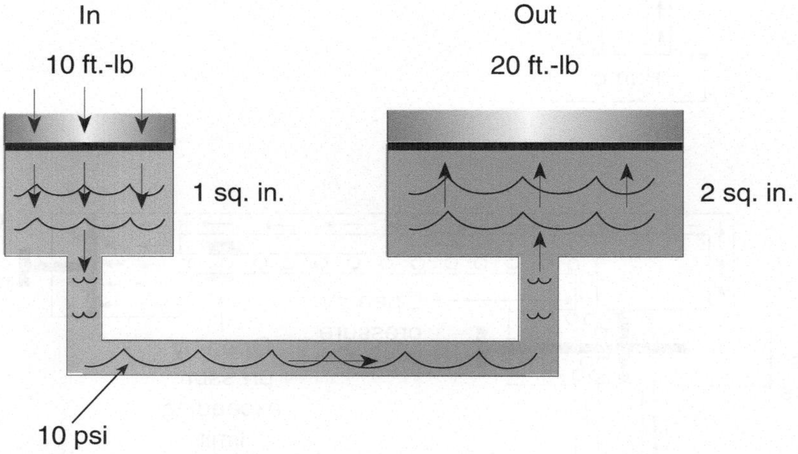

Figure 12-14 Mechanical advantage can be gained or reduced by having input and output pistons of different sizes.

on each inch, the delivered force is 20 ft.-lb. This would give a mechanical advantage of 2:1 (20/10). Calculate the total output force in the following simple braking system.

Refer to **Figure 12-15** for the known values for this problem. The figure shows a very simple hydraulic braking system. Notice that the input piston is 0.75 square inches. The rear brakes have 1-square-inch pistons at each wheel whereas the front has 2-square-inch pistons at each wheel. First, the system pressure must be calculated. The input piston is being applied with 100 ft.-lb. The known values, force and area, are used in $P = F/A$ or $P = 100/0.75$, resulting in a pressure of 133 psi rounded off. Because the pressure is the same throughout the circuit, the pressure against each square inch of the output pistons is 133 psi. Compute the rear pistons' force using $F = PA$ ($F = 1 \times 133$) or 133 ft.-lb. at each wheel or a total of 266 ft.-lb. for the rear. The front is calculated in the same manner: $F = 133 \times 2$ for 266 ft.-lb. per wheel for a total of 532 ft.-lb. in the front. The operator is applying 100 ft.-lb. on the brake pedal, and the vehicle is being stopped with a total of 798 ft.-lb. (front plus rear) of braking effort. This is a 7.98:1 mechanical advantage.

Hydraulic fluids are selected on the basis of the type of hydraulic system in use and the type of work the system must perform. In theory, you could use tap water as the fluid. It is

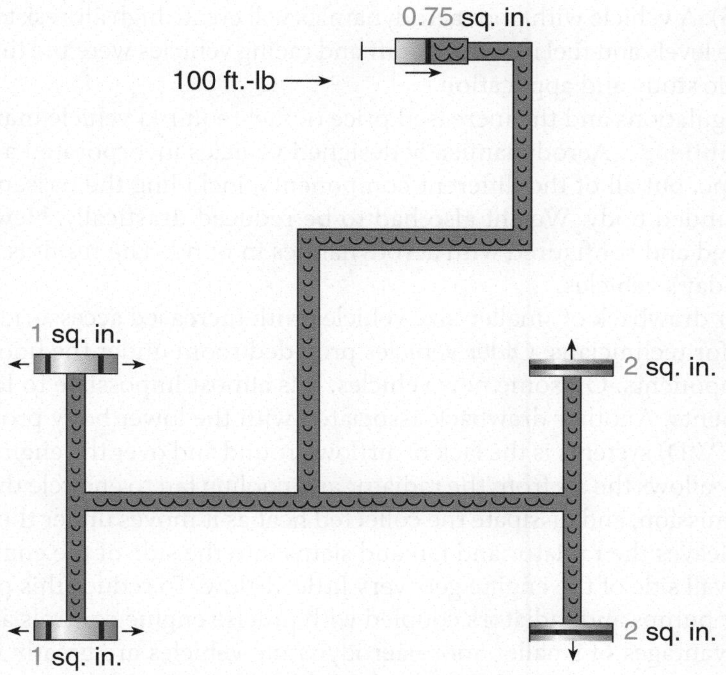

Figure 12-15 This system has 8 sq. in. of output area versus 3/4 sq. in. of input area.

non-compressible, like all liquids, and it can transmit force as long as the system is sealed. But when a liquid is compressed, its temperature goes up in direct proportion to the pressure applied. After about three or four braking efforts using water as brake fluid, the water would boil, thus creating gases (vapors) in the system and resulting in very poor braking from that point on because gases can be compressed. Brake fluid is formulated for a high boiling point and is corrosion resistant. Other hydraulic fluids are designed and selected with the same overall specifications in mind. Hydraulic fluids used in heavy equipment work under pressures as high as 2,200 psi while maintaining a relatively low outside temperature. The fluid would definitely be hot enough to cause injury if it escaped past a seal or gasket, but most of the cylinders and lines or hoses could be touched safely with regular gloves. In modern vehicles, there are different hydraulic fluids for the brakes, power steering, and automatic transmission. Each is selected based on its role in the vehicle.

Shop Manual
pages 434–437

VEHICLE DESIGN

Without going into too much detail, the engineering of an automobile covers not only the theories discussed earlier but also how different materials can be shaped and formed and their actions under different conditions. The local library has information on how the shape, composition, treatment, and the reactions of metal affect the actions of the component. For instance, aluminum warps at a lower heat than cast iron. In this book, we will not discuss vehicle design and construction at length, but be aware that the material choice, shaping, and the manufacturing treatment of each component on the vehicle were researched and designed to satisfy the laws of physics.

Aerodynamics

Aerodynamics The study of the properties of moving air between a solid body and the body moving through the air.

Until quite recently, the study and use of **aerodynamics** was not considered when building everyday vehicles. Even smaller cars were bulky, heavy, and had the aerodynamic design of a brick! Aerodynamics is the study of how an object moves through the air

(**Figure 12-16**). A vehicle with poor aerodynamics will create high air resistance, resulting in higher noise levels and fuel usage. Aircraft and racing vehicles were the first major users of aerodynamic study and application.

Fuel mileage is required to meet the standards in the federal Corporate Average Fuel Economy (CAFE) regulation. A manufacturer's vehicle line must be tested, and the average of all models offered must meet CAFE.

Federal regulations and the increased price of fuel required vehicle manufacturers to increase **fuel mileage**. Aerodynamically designed vehicles incorporated a rounded outside body shape, but all of the different components, including the passenger, had to fit inside this rounded body. Weight also had to be reduced drastically. Newer headlights were redesigned and configured with aerodynamics in mind. The result is the slick body contours of today's vehicles.

One major drawback of smaller-size vehicles with increased accessories is the lack of working area for technicians. Older vehicles provided room under the hood and dash to reach the components. On some new vehicles, it is almost impossible to locate or reach some components. Another drawback associated with the lower body profile and front-wheel-drive (FWD) systems is the lack of airflow around and over the engine. Rear-wheel drive generally allows the air from the radiator and cooling fan to encircle the engine, flow past the transmission, and dissipate the collected heat as it moves under the vehicle. With FWD, the air leaves the radiator and fan and slams into the side of the engine and transaxle. The firewall side of the engine gets very little airflow. To reduce this problem, more efficient water pumps and radiators coupled with precise engine controls are being used. The major advantages of smaller, more aerodynamic vehicles are greatly increased fuel mileage and driver protection. This, in turn, leads to lower harmful emissions emitted.

Passenger Protection

Today, almost every car on the road has seat belts (**Figure 12-17**) and most have airbags (**Figure 12-18**). Although these devices are needed to keep the passenger in his or her seat during an accident, they are minor parts of overall vehicle crash protection.

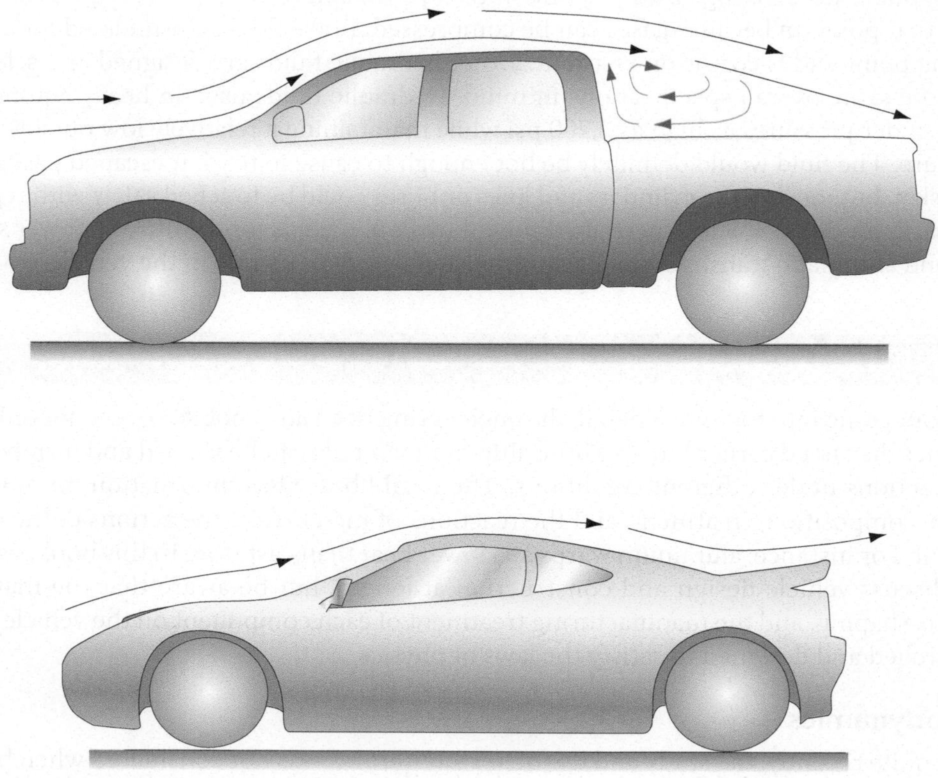

Figure 12-16 Air flows smoothly over the car. For the truck, it is forced much higher and abruptly by the truck's windshield. Note the air turbulence over the bed of the truck compared to that over the car's rear deck.

Figure 12-17 Seat belts are required to be worn in most states. Children must be in an approved seat secured in the vehicle with seat belts.

Figure 12-18 Since model year 1997, airbags have been redesigned to accommodate persons of different sizes. In most pickup trucks, the passenger side bag can be switched off, so a child seat can be safely utilized. Many newer vehicles have airbags with controllable inflation rates.

Although older vehicles had heavy bodies and frames, they usually delivered most of the shock of an accident to their passenger compartments. The automotive and insurance industries, along with various government agencies, have destroyed many vehicles to study the effect of accidents on passengers and vehicles. Newton's Laws of Motion and the dissipation of kinetic energy are key points in designing a safe vehicle.

Automotive studies have resulted in vehicles designed and built with material that will crumple on impact and absorb energy. Such areas are called **crumple zones**. More energy absorbed by the vehicle means less shock and damage delivered to the passenger

Crumple zones extend from bumper to bumper. The sections of the body and frame interact to absorb as much energy as possible before it reaches the passenger compartment.

compartment. The first point or crumple zone is energy-absorbing bumpers that incorporate devices between the bumper and frame. Under federal law, newer cars must withstand a frontal impact of 30 miles per hour before the passenger compartment begins to absorb kinetic energy from the impact. Entire vehicles have been redesigned to protect passengers and, to some extent, the vehicles themselves.

Materials

Like other things in nature, materials respond to different situations in different ways. The materials used in vehicles are selected to meet specific properties. In the engine, materials must be capable of enduring high pressure, twisting, and heavy loads and must still function properly. The most prominent materials in engines are cast iron for the block, aluminum for the cylinder heads, and treated steel for the internal components. Using different materials together presents another problem. The engine produces a great deal of heat and different materials react differently to heat. Aluminum expands much quicker than cast iron. Because they are mated to each other in the engine, engineers had to design some method to compensate for the differences. A gasket was made with materials that would react with the cast iron on one side and aluminum on the other. Gaskets are covered in Chapter 7 "Engine Systems."

Other systems of the vehicle require different materials to perform their purpose. The material may be of high strength and flexibility such as suspension components, thin sheet metal for the body, or plastics for trim and some no-load components. All of the materials are selected, shaped, treated, and used based on their physical properties and the operational theories involved.

HORSEPOWER VERSUS TORQUE

Finally, we get to the section that most drivers want to know: horsepower versus torque. One is speed; the other is pulling. Both can be obtained by designing and tuning the engine and drivetrain for a particular purpose. In construction work, a high-torque pickup truck is desirable, but fuel mileage will suffer. In a drive-to-work-and-home truck, fuel mileage is important, so great torque is not needed. Horsepower is the key in such instances.

Torque, as applied to a vehicle, is the total rotational force applied to the crankshaft by the pistons during the power stroke and is usually measured as foot-pounds at the flywheel. Horsepower is work done in a straight line. But when the work is not in a straight line, horsepower is calculated as torque times rpm divided by 5,252 (torque*rpm/5,252). Typically, maximum torque is produced at a much lower engine speed than horsepower (**Figure 12-19**).

At this point, we have only discussed how to calculate the torque and horsepower of the engine. Although this provides a good reference point to determine which vehicle we are going to purchase, it does not necessarily tell us the amount of power delivered to the drive wheels. Before determining how fast or how strong the vehicle is, the driveline components, mainly the transmission, must be calculated. The driveline changes the engine's torque or horsepower to the final actual force moving the vehicle. The selection of a particular gear may allow the vehicle to pull a heavy load at low road speed and high engine speed or cruise at high road speed and low engine speed. Generally speaking, the more torque the less speed, and vice versa.

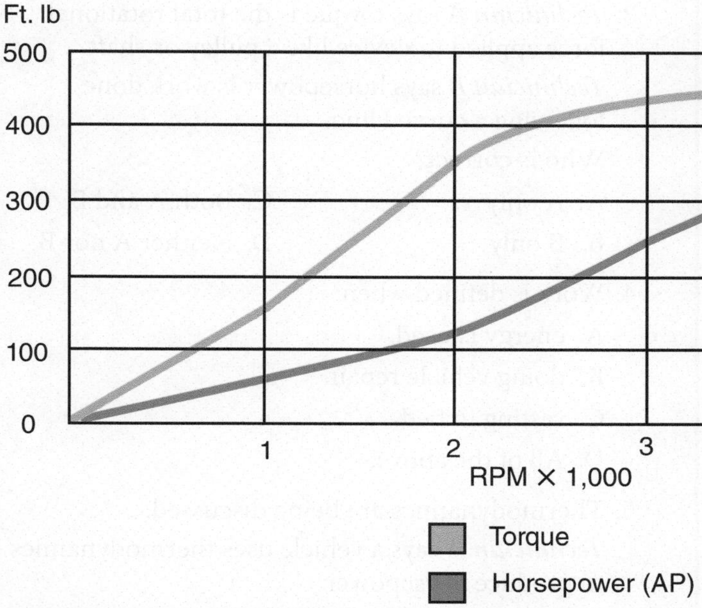

Figure 12-19 A torque/horsepower curve. Note the high torque curve with the sudden drop after 4,000 rpm. The horsepower continues to increase slightly after 4,000 rpm.

SUMMARY

- Thermodynamics is the study of using rapidly expanding gases to produce energy.
- The theory of thermodynamics is used to achieve mechanical power from a liquid fuel.
- The motion of an object is called work. Work is the use of energy.
- Pressure and vacuum are used to perform work on a vehicle.
- Vacuum is pressure below atmospheric.
- Newton's Laws of Motion apply to almost every operation on a typical vehicle.
- Hydraulic systems use liquids to transfer force.
- The measurements of hydraulics are force, area, and pressure.
- Vehicle design and manufacturing are based on the capabilities of materials used and their functions within the vehicle.
- Aerodynamics is the study of objects moving through air.
- Horsepower provides capability of high speed.
- Torque provides pulling or hauling power.

REVIEW QUESTIONS

Short Answer Essays

1. Describe how the operator's force on the brake pedal is transmitted to the wheels.
2. Briefly explain the theory of thermodynamics.
3. Define the term BTU.
4. Describe the position of the combustion chamber and its purpose in an internal combustion engine.
5. Explain why a specific amount of air (oxygen) must be mixed with a fuel to ensure ignition and power.
6. Explain the relationship between torque and horsepower.
7. Discuss some of the considerations used in designing and manufacturing a vehicle.
8. Discuss how aerodynamics may affect a vehicle.
9. Define potential energy.
10. List the discussed Laws of Motion.

Fill-in-the-Blanks

1. _______________ is a unit of energy and may be measured in foot-pounds (ft.-lb.).
2. The three hydraulic measurements are _______________, _______________, and _______________.
3. The motion of an object produces _______________ energy.
4. A machine or device cannot produce _______________ energy than it consumes.
5. Pressure is measured in _______________.
6. A hydraulic circuit consists of _______________, _______________, _______________, and _______________.
7. Vacuum is usually measured in _______________ of mercury or water.
8. The tendency of an object to stay at rest is called _______________.
9. The study of using expanding gases to perform work is called _______________.
10. Trucks will generally have less _______________ and more _______________ than a car with a comparable engine.

Multiple Choice

1. Thermodynamics is being discussed.
   Technician A says an internal combustion engine generates energy using thermodynamic theory.
   Technician B says the air/fuel mixture must be compressed before sufficient force can be created to move the piston.
   Who is correct?
   A. A only
   B. B only
   C. Both A and B
   D. Neither A nor B
2. Hydraulic systems are being discussed.
   Technician A says the force applied against a piston creates the pressure of the transferring liquid.
   Technician B says the pressure of the transferring liquid creates the force on the piston.
   Who is correct?
   A. A only
   B. B only
   C. Both A and B
   D. Neither A nor B
3. *Technician A* says torque is the total rotational force applied to device like a pulley or shaft.
   Technician B says horsepower is work done following a curved line.
   Who is correct?
   A. A only
   B. B only
   C. Both A and B
   D. Neither A nor B
4. Work is defined when:
   A. energy is used.
   B. doing vehicle repair.
   C. resting in bed.
   D. All of the above.
5. Thermodynamics are being discussed.
   Technician A says a vehicle uses thermodynamics to produce horsepower.
   Technician B says expanding gases is the study of thermodynamics.
   Who is correct?
   A. A only
   B. B only
   C. Both A and B
   D. Neither A nor B
6. All of the following conditions will affect the hydraulic pressure in a system EXCEPT:
   A. External leaks
   B. Internal leaks
   C. Weak pump operation
   D. Coiled hose or pipe
7. Fuel mileage requirements are set forth by:
   A. EPA standards
   B. DOT standards
   C. CAFE standards
   D. Vehicle manufacturer standards
8. Pascal's Law is being discussed.
   Technician A says a liquid is used to transfer force.
   Technician B says hydraulics is the study of using liquid to transfer force.
   Who is correct?
   A. A only
   B. B only
   C. Both A and B
   D. Neither A nor B

9. A brake master cylinder has a 1/2-square inch output piston. Each of the front disc brakes has a 2-square inch apply piston. What is the mechanical advantage of this portion of the brake system?

    A. 2 to 1
    B. 4 to 1
    C. 0.5 to 1
    D. 1 to 0.5

10. If the output piston generated 100-psi and the apply pistons are 2-inch square, what is the pressure at the apply piston?

    A. 100-psi
    B. 100-ft.-lb.
    C. 200-psi
    D. 200-ft.-lb.

GLOSSARY

Note: **Terms are highlighted in bold,** followed by **Spanish translation in color.**

A **bayonet** connection is so named because of its push in, twist, and lock steps to make the connection.

Una **conexión de bayoneta** se llama así debido a sus pasos de inserción, giro y bloqueo para realizar la conexión.

A **bimetallic** strip has two pieces of metal laid side by side. One piece reacts to heat faster than the other one does.

Una **tira bimetálica** tiene dos piezas de metal colocadas una al lado de la otra. Una pieza reacciona al calor más rápido que la otra.

A **brake warning light** is mounted in the dash to warn the driver of a brake system fault.

Se monta **una luz de advertencia de freno** en el tablero para advertir al conductor de una falla del sistema de frenos.

A **bypass valve** is used to redirect fluid if a passageway becomes blocked or pressure is too high.

Se usa **una válvula de derivación** para redirigir el fluido si un pasaje se bloquea o la presión es demasiado alta.

A **center punch** is made of steel and used to mark the centers of holes to be drilled.

Un **punzón central** está hecho de acero y se utiliza para marcar los centros de los agujeros que se perforarán.

A **check valve** is normally used to prevent fluid from reversing within a system.

Normalmente se usa una **válvula de retención** para evitar que el fluido se revierta dentro de un sistema.

A **circuit** provides a path for current to flow from the power source through the conductors and back to the power source.

Un **circuito** proporciona un camino para que la corriente fluya desde la fuente de energía a través de los conductores y de regreso a la fuente de energía.

A **cluster gear** cannot be repaired, and if it is damaged there is probably damage on meshing gears. Although there is no crankshaft pilot bearing or bushing used with **torque converters**, there is usually a cut out in the crankshaft, so the torque converter can expand.

Un **engranaje de racimo** no puede repararse, y si está dañado probablemente haya daños en los engranajes de malla. Aunque no se utiliza un cojinete o buje piloto del cigüeñal con los **convertidores de par**, generalmente hay un corte en el cigüeñal, por lo que el convertidor de par puede expandirse.

A **clutch pack** is the term used for hydraulic clutches in automatic transmission. The clutch pack performs the same function as the clutch on a manual transmission but is internal to the transmission and applied hydraulically.

Un **paquete de embrague** es el término utilizado para embragues hidráulicos en la transmisión automática. El paquete de embrague realiza la misma función que el embrague en una transmisión manual, pero es interno a la transmisión y se aplica hidráulicamente.

A **coil pack** consists of a primary circuit control package and an ignition coil mounted over the package. The coil is in the secondary circuit.

Un **paquete de bobinas** consta de un paquete de control de circuito primario y una bobina de encendido montada sobre el paquete. La bobina está en el circuito secundario.

A **coil spring** is a rod of steel shaped in a spiral.

Un **resorte helicoidal** es una varilla de acero con forma de espiral.

A **combination valve** is a single unit that meters and proportions pressures.

Una **válvula combinada** es una sola unidad que mide y proporciona presiones.

A **communication bus** has all sensors, computers, and modules connected to shared circuits. In this manner, each computer and module can monitor data.

Un **bus de comunicación** tiene todos los sensores, computadoras y módulos conectados a circuitos compartidos. De esta manera, cada computadora y módulo puede monitorear datos.

A **component locator** shows the location of each electrical device in a vehicle including switches, splices, and connections.

Un **localizador de componentes** muestra la ubicación de cada dispositivo eléctrico en un vehículo, incluidos interruptores, empalmes y conexiones.

A **condenser** cools down a vapor and assists in changing the liquid from a vapor state to a liquid state (condensation).

Un **condensador** enfría un vapor y ayuda a cambiar el líquido de un estado de vapor a un estado líquido (condensación).

A **conductor** is any material that will carry electrical current.

Un **conductor** es cualquier material que transportará corriente eléctrica.

A **conventional switch** has only an on/off for a single circuit.

Un **interruptor convencional** tiene solo un encendido/apagado para un solo circuito.

A **creeper** is a low, flat, wheeled platform on which the technician can lie and roll under a vehicle.

Una **camilla** es una plataforma baja, plana y con ruedas sobre la cual el técnico puede acostarse y rodar debajo de un vehículo.

A **dial indicator** uses a plunger and gear to move a needle in proportion to plunger movement.

Un **indicador de carátula** utiliza un émbolo y un engranaje para mover una aguja en proporción al movimiento del émbolo.

A **die** is used to cut external threads.

Un **troquel** se usa para cortar hilos externos.

A **distributor** directs the current from the ignition coil to the correct spark plug.

Un **distribuidor** dirige la corriente desde la bobina de encendido a la bujía correcta.

A **double flared** fitting has the end of the tube/line that is flared and flared again inwards to form a tight hydraulic seal.

Un accesorio de **doble acampanado** tiene el extremo del tubo/línea que se acampana y se vuelve a acampanar hacia adentro para formar un sello hidráulico hermético.

A **file** is a hardened, steel tool with rows of cutting edges used to remove metal for polishing, smoothing, or shaping.

Una **lima** es una herramienta de acero endurecida con hileras de bordes cortantes que se utilizan para quitar metal para pulir, alisar o dar forma.

A **fixed caliper** is connected to the steering knuckle and does not move during brake application. It requires at least two pistons, one in-board, one out-board.

Una **pinza fija** está conectada al muñón de la dirección y no se mueve durante la aplicación del freno. Requiere al menos dos pistones, uno interno y uno externo.

A **flasher** may be electrical/mechanical or fully electronic to flash the turn signals.

Una **luz intermitente** puede ser eléctrica/mecánica o completamente electrónica para hacer parpadear las señales de giro.

A **ganged switch** controls multiple circuits.

Un **interruptor agrupado** controla múltiples circuitos.

A **gear puller** is a tool used to remove gears, bearings, shafts, and other parts off shafts or out of holes.

Un **extractor de engranajes** es una herramienta utilizada para quitar engranajes, cojinetes, ejes y otras partes de los ejes o de agujeros.

A **general manager** is the person in charge of that business. In the automotive repair business, the manager supervises all aspects of the business directly or through department managers.

Un **gerente general** es la persona a cargo de ese negocio. En el negocio de reparación automotriz, el gerente supervisa todos los aspectos del negocio directamente o a través de los gerentes de departamento.

A **grounding switch** is a switch that, when closed, completes the circuit to ground.

Un **interruptor de puesta a tierra** es un interruptor que, cuando está cerrado, completa el circuito a tierra.

A **hacksaw** is made to cut metal.

Una **sierra de arco** para **metales** está hecha para cortar metal.

A Hall-effect switch measures the speed of position of a component and converts that data into an electrical signal.

Un **interruptor de efecto Hall** mide la velocidad de posición de un componente y convierte esos datos en una señal eléctrica.

A **light-emitting diode (LED)** is a diode that gives off light when current passes through it.

Un **diodo emisor de luz (LED)** es un diodo que emite luz cuando la corriente lo atraviesa.

A **load-sensing proportioning valve** is a mechanically controlled pressure regulator.

Una **válvula dosificadora con detección de carga** es un regulador de presión controlado mecánicamente.

A **lockup clutch** locks the turbine to the outside of the torque converter and is not applied until the vehicle is in high gear and engine load has decreased. The transmission should be designed to work with a lockup torque converter.

Un **embrague de bloqueo** bloquea la turbina hacia el exterior del convertidor de par y no se aplica hasta que el vehículo esté en alta velocidad y la carga del motor haya disminuido. La transmisión debe estar diseñada para funcionar con un convertidor de par de bloqueo.

A **manifold absolute pressure (MAP)** sensor converts air pressure into an electrical signal used by the computer.

Un **sensor de presión absoluta múltiple (MAP)** convierte la presión de aire en una señal eléctrica utilizada por la computadora.

A **mass airflow (MAF)** sensor converts air flow and density to an electrical signal.

Un **sensor de flujo de masa de aire (MAF)** convierte el flujo de aire y la densidad en una señal eléctrica.

A **master switch** controls multiple units from one location like the power window switch in the driver's arm rest.

Un **interruptor maestro** controla múltiples unidades desde una ubicación, como el interruptor de la ventana eléctrica en el reposabrazos del conductor.

A **meter** is the basic distance measurement of the metric system.

Un **metro** es la medida de distancia básica del sistema métrico.

A **micrometer** may be referred to as a "mike."

Un **micrómetro** puede ser referido como un "mike".

A **normally closed (NC)** is closed until it is activated.

Un interruptor **normalmente cerrado (NC)** está cerrado hasta que se activa.

A **normally open (NO)** switch is open until it is activated.

Un interruptor **normalmente abierto (NO)** está abierto hasta que se activa.

A **parallel circuit** has multiple paths for current flow.

Un **circuito paralelo** tiene múltiples caminos para el flujo de corriente.

A **pilot bearing** or **bushing** is not used with torque converters.

Un **cojinete** o **buje piloto** no se utiliza con convertidores de par.

A **potentiometer** converts mechanical movement into an electrical signal.

Un **potenciómetro** convierte el movimiento mecánico en una señal eléctrica.

A **reamer** has cutting edges used to remove a small amount of metal from a drilled hole.

Un **escariador** tiene bordes cortantes que se utilizan para eliminar una pequeña cantidad de metal de un agujero perforado.

A **receiver/dryer** also separates liquid and vapor, but it will capture moisture (water) present in the refrigerant.

Un **receptor/secador** también separa líquido y vapor, pero capturará la humedad (agua) presente en el refrigerante.

A **refrigerant** is a chemical that can absorb, transport, and release a great deal of heat in a small space under easy-to-create conditions.

Un **refrigerante** es un químico que puede absorber, transportar y liberar una gran cantidad de calor en un espacio pequeño en condiciones fáciles de crear.

A **roller** overrunning clutch is much simpler in operation than a **sprag** clutch.

Un embrague de sobrecarga de **rodillo** es mucho más simple en operación que un embrague de arrastre.

A **screw extractor** is used to remove broken screws, bolts, or studs from automotive parts.

Se utiliza un **extractor de tornillos** para quitar tornillos, pernos o espárragos rotos de las piezas de automóviles.

A **sector shaft** will only rotate within a limited arc. On steering gear boxes, this arc is approximately 60 degrees.

Un **eje de sector** solo rotará dentro de un arco limitado. En cajas de engranajes de dirección, este arco es de aproximadamente 60 grados.

A **series circuit** provides one path for current flow.

Un **circuito en serie** proporciona un camino para el flujo de corriente.

A **series-parallel circuit** typically has the series portion controlling the parallel portion.

Un **circuito en serie-paralelo** típicamente tiene la porción en serie que controla la porción en paralelo.

A **servo** is used to convert hydraulic pressure to mechanical force.

Se utiliza un **servo** para convertir la presión hidráulica en fuerza mecánica.

A **shock absorber** or **strut** is designed to absorb shock and control spring movement.

Un **amortiguador** o **puntal** está diseñado para absorber los golpes y controlar el movimiento del resorte.

A **short finder** is an electrical tool used to locate shorts in a circuit.

Un **buscador corto** es una herramienta eléctrica utilizada para localizar cortocircuitos en un circuito.

A **short-to-ground** is an unwanted connection between a positive conductor and ground.

Un **corto a tierra** es una conexión no deseada entre un conductor positivo y tierra.

A **short-to-power** is an unwanted connection between positive conductors of two different circuits.

Un **corto a la alimentación** es una conexión no deseada entre conductores positivos de dos circuitos diferentes.

A **signal** is data sent to/from a computer in an electrical form.

Una **señal** son datos enviados a/desde una computadora en forma eléctrica.

A simple **planetary gear set** has a sun gear that fits inside a planetary carrier. The carrier, in turn, fits inside a ring gear.

Un **conjunto de engranaje planetario** simple tiene un engranaje solar que cabe dentro de un portaaviones planetario. El transportador, a su vez, cabe dentro de una corona dentada.

A **single-pole, single-throw** switch only has on/off functions.

Un interruptor de un **solo polo y un solo tiro** solo tiene funciones de encendido/apagado.

A **single-throw, double throw** switch has two 'on' positions.

Un interruptor de **tiro simple y doble tiro** tiene dos posiciones de encendido.

A slant engine is the same as an in-line engine, but the cylinders are slanted and not upright.

Un motor inclinado es lo mismo que un motor en línea, pero los cilindros están inclinados y no en posición vertical.

A **sliding caliper** is mounted so it can move on slides during brake application. It only needs one in-board piston. This type may also be known as a floating caliper.

Se monta una **pinza deslizante** para que pueda moverse sobre las correderas durante la aplicación del freno. Solo necesita un pistón interno. Este tipo también puede ser conocido como un calibrador flotante.

A **snapshot** is stored data that can be retrieved and studied later. It is usually data of a particular event at a particular time, such as a misfiring cylinder under heavy load. It may be referred to as a "movie."

Una **instantánea** son datos almacenados que se pueden recuperar y estudiar más tarde. Usualmente son datos de un evento particular en un momento particular, como un cilindro que no funciona bien bajo una carga pesada. Puede ser referido como una "película".

A **square-cut O-ring** is used to seal the caliper piston within the caliper bore. It also acts as a means to retract the brake pads during brake release.

Se utiliza una junta **tórica de corte cuadrado** para sellar el pistón de la pinza dentro del orificio de la pinza. También actúa como un medio para retraer las pastillas de freno durante la liberación del freno.

A **stator** is the place where the rotating magnetic fields are converted into AC current.

Un **estator** es el lugar donde los campos magnéticos giratorios se convierten en corriente alterna.

A **tone wheel** or ring is a toothed ring or wheel that rotates through a sensor's magnetic field generating a signal sent to a controller.

Una **rueda de tonos** o anillo es un anillo dentado o rueda que gira a través del campo magnético de un sensor y genera una señal enviada a un controlador.

A **torque wrench** is designed to tighten bolts or nuts to a certain tightness or torque.

Una **llave dinamométrica** está diseñada para apretar pernos o tuercas a una cierta tensión o torque.

A **torsion bar** acts like a coil spring in that it cushions road shock for a more comfortable ride.

Una **barra de torsión** actúa como un resorte helicoidal en el sentido de que amortigua el impacto de la carretera para una conducción más cómoda.

A **transmission range** switch signals the computer which transmission gear is selected.

Un interruptor de **rango de transmisión** le indica a la computadora qué equipo de transmisión está seleccionado.

A **transponder** will transmit its code by radio or radar when actuated by a specific signal from a radio or radar transmitter (interrogator).

Un **transpondedor** transmitirá su código por radio o radar cuando sea accionado por una señal específica de un transmisor de radio o radar (interrogador).

A **trigger** is a condition that will cause an action to be taken (e.g., low oil pressure or fire a spark plug).

Un **gatillo** es una condición que hará que se tome una acción (por ejemplo, baja presión de aceite o disparar una bujía).

A **two-stroke** engine has a power stroke with each crankshaft revolution, whereas a **four-stroke** engine has one power stroke every two revolutions.

Un motor de **dos tiempos** tiene una carrera de potencia con cada revolución del cigüeñal, mientras que un motor de **cuatro tiempos** tiene una carrera de potencia cada dos revoluciones.

A **venturi** is the narrow portion of a larger bore.

Un **venturi** es la porción estrecha de un orificio más grande.

A **winding** is one wrap of wire, rope, hose, or any item that can be wrapped around itself. Coiling a hose could be considered as several windings of the hose.

Un **devanado** es una envoltura de alambre, cuerda, manguera o cualquier artículo que se pueda enrollar alrededor de sí mismo. Enrollar una manguera podría considerarse como varios devanados de la manguera.

A **wiring diagram** is a drawing of multiple electrical circuits.

Un **diagrama de cableado** es un dibujo de múltiples circuitos eléctricos.

Absorbed glass mat (AGM) batteries are becoming more common in small engine like motorcycles and garden tractors.

Las **baterías de estera de vidrio absorbido (AGM)** se están volviendo más comunes en motores pequeños como motocicletas y tractores de jardín.

Active body control (ABC) is a computer-controlled system that increases vehicle stability and reduces accidents.

El **Control activo de la carrocería (ABC)** es un sistema controlado por computadora que aumenta la estabilidad del vehículo y reduce los accidentes.

Actuators are mechanical work components controlled by computer command.

Los **actuadores** son componentes mecánicos de trabajo controlados por comando de computadora.

AERA Specialized network of professional engine builders, re-builders production re-manufacturers and installers with the expertise and connections to get you the information you need.

AERA Red especializada de constructores profesionales de motores, reconstructores de producción remanufacturadores e instaladores con la experiencia y las conexiones para obtener la información que necesitas.

Aerodynamics The study of the properties of moving air between a solid body and the body moving through the air.

Aerodinámica El estudio de las propiedades del aire en movimiento entre un cuerpo sólido y el cuerpo que se mueve a través del aire.

Air gap usually refers to the space between two magnetic components on an automobile.

El **espacio de aire** generalmente se refiere al espacio entre dos componentes magnéticos en un automóvil.

Air springs are basically tough balloons that can be used to replace metal springs. Suspension systems are called either **dependent** or **independent** based on how much of the weight on one wheel is delivered in some manner to the other wheels.

Los **resortes neumáticos** son básicamente globos resistentes que pueden usarse para reemplazar los resortes metálicos. Los sistemas de suspensión se denominan **dependientes** o **independientes en** función de la cantidad de peso en una rueda que se entrega de alguna manera a las otras ruedas.

Air-operated ratchet driver A air driven ratching wrench much like a hand ratchet.

Llave de trinquete accionada por aire Una llave de trinquete accionada por aire muy parecida a un trinquete de mano.

All **roller** and **ball bearings** provide a rolling surface for the moving components so that friction and wear are reduced.

Todos los **cojinetes de rodillos** y **bolas** proporcionan una superficie de rodadura para los componentes móviles, de modo que se reducen la fricción y el desgaste.

All-season treads are standard equipment on almost all standard cars and light trucks.

Las **bandas de rodadura para todas las estaciones** son equipos estándar en casi todos los automóviles y camionetas estándar.

An **accumulator/dryer** separates vapor from liquid but does not capture moisture (water).

Un **acumulador/secador** separa el vapor del líquido, pero no captura la humedad (agua).

An **Allen wrench** is used to tighten or loosen Allen head screws.

Se usa una **llave Allen** para apretar o aflojar los tornillos de cabeza Allen.

An **analog** signal changes from positive to negative voltage at a given rate.

Una señal **analógica** cambia de voltaje positivo a negativo a una velocidad dada.

An **automatic ride control (ARC)** system requires some events to occur in sequence, such as the ignition key turned off and the door opened, before the computer commands an action. Do not fault this system until the required events are performed in order.

Un **Sistema de Control Automático del Retardador (ARC)** requiere que ocurran algunos eventos en secuencia, como la llave de encendido apagada y la puerta abierta, antes de que la computadora ordene una acción. No critique este sistema hasta que los eventos requeridos se realicen en orden.

An **automatic transmission** changes forward gears by using valves and hydraulic pressure. A **manual transmission** requires the driver to select a gear using a shift lever.

Una **transmisión automática** cambia las marchas de avance mediante el uso de válvulas y presión hidráulica. Una **transmisión manual** requiere que el conductor seleccione una marcha con una palanca de cambios.

An electronic switch that uses low current to control a high current is a **relay**.

Un interruptor electrónico que usa corriente baja para controlar una corriente alta es un **relé**.

An **evaporative (EVAP)** system captures and recycles vapors from the fuel system.

Un **sistema evaporativo (EVAP)** captura y recicla los vapores del sistema de combustible.

An **explosion** is a sudden, violent release of energy. An engine uses a controlled burn, not an explosion, to achieve power.

Una **explosión** es una liberación repentina y violenta de energía. Un motor utiliza una combustión controlada, no una explosión, para lograr potencia.

An **idler arm** only holds up one end of the center link during rest and steering movement.

Un **brazo libre** solo sostiene un extremo del enlace central durante el movimiento de reposo y dirección.

An **integrated circuit** is a system of many transistors and other electronic devices on a special board. There may be thousands of those devices on one board.

Un **circuito integrado** es un sistema de muchos transistores y otros dispositivos electrónicos en una placa especial. Puede haber miles de esos dispositivos en una placa.

An **isolation/dump solenoid** is used to block fluid or release fluid from a portion of the brake system.

Se usa un **solenoide de aislamiento/descarga** para bloquear el fluido o liberar fluido de una parte del sistema de frenos.

An **offset engine's** cylinders are offset from the centerline of the block.

Los cilindros de un **motor compensado** están desplazados de la línea central del bloque.

An **ohm** is the standard measurement of electrical resistance.

Un **ohmio** es la medida estándar de resistencia eléctrica.

An **outside micrometer** is normally used in the automotive business to measure shaft external bearing journals or metal thicknesses.

Un **micrómetro externo** se usa normalmente en el negocio automotriz para medir los cojinetes externos del eje o el grosor del metal.

An **overhead camshaft engine (OHC)** has one camshaft per cylinder head.

Un **motor de árbol de levas en cabeza (OHC)** tiene un árbol de levas por cada cabeza de cilindro.

An **oxyacetylene welding and cutting outfit** is a system that uses oxygen and acetylene gas as a heating source for welding, cutting, and heating metal.

Un **equipo de soldadura y corte de oxiacetileno** es un sistema que utiliza oxígeno y gas acetileno como fuente de calentamiento para soldar, cortar y calentar metal.

Another name for a **stepped feeler gauge** is a go/no-go gauge.

Otro nombre para un **medidor de espesor escalonado** es un medidor de avance/no avance.

Antifreeze is a chemical mixture that reduces freezing of a liquid, usually water.

El **anticongelante** es una mezcla química que reduce la congelación de un líquido, generalmente agua.

Antilock brake systems (ABS) help increase driver control by reducing wheel lockup during panic stops. A diesel engine requires a vacuum pump for the **vacuum brake booster** and other vacuum-operated devices.

El **sistema de frenos antibloqueo (ABS)** ayudan a aumentar el control del conductor al reducir el bloqueo de las ruedas durante las paradas de pánico. Un motor diésel requiere una bomba de vacío para el **refuerzo del freno de vacío** y otros dispositivos operados por vacío.

ASE Education Foundation is a non-profit organization that evaluates and accredits entry-level automotive technology education programs.

Fundación de Educación ASE es una organización sin fines de lucro que evalúa y acredita programas de educación de tecnología automotriz de nivel básico.

Atmospheric pressure is considered to be 14.7 psi at sea level.

Se considera que la **presión atmosférica** es de 14.7 psi al nivel del mar.

Atom The smallest unit of a chemical element.

Átomo la unidad más pequeña de un elemento químico.

ATRA Worlds largest trade organization of transmission re-builders with over 2000 members throughout the United States and Canada.

ATRA La organización comercial más grande del mundo de reconstructores de transmisión con más de 2000 miembros en todo Estados Unidos y Canadá.

Axial load is produced by the component pushing directly against the bearing from inside out or vice versa.

La **carga axial** es producida por el componente que empuja directamente contra el rodamiento desde adentro hacia afuera o viceversa.

AYES Effective January 1, 2018 NATEF and AYES have merged to become ASE education Foundation.

AYES Efectivo desde Enero 1, 2018, NATEF y AYES se fusionaron para convertirse en la Fundación de Educación ASE.

Backlit means the light is coming from behind or from the side of the object being viewed.

Retroiluminado significa que la luz proviene de atrás o del lado del objeto que se está viendo.

Balanced scale indicators are usually used to measure runout.

Los **indicadores de escala equilibrada** generalmente se usan para medir la desviación.

Ball-peen hammer A hammer that has two heads, one side is flat the other is rounded (peen).

Martillo de bola Un martillo que tiene dos cabezas, un lado es plano y el otro es redondeado (peen).

BDC means bottom dead center or the piston's lowest point of travel.

BDC significa punto muerto inferior o el punto de recorrido más bajo del pistón.

BDC The point at which the piston of an engine is nearest the lowest point or bottom of the pistons travel closest to the crankshaft. This is known as bottom dead center.

BDC El punto en el que el pistón de un motor está más cerca del punto más bajo o la parte inferior de los pistones se desplaza más cerca del cigüeñal. Esto se conoce como punto muerto inferior.

Bearing race The metal ring with grooves that hold together a set of ball bearings.

Pista de rodamientos El anillo de metal con ranuras que mantienen unidos un conjunto de rodamientos de bolas.

Bearings reduce the friction, heat, and wear of moving components.

Los **cojinetes** reducen la fricción, el calor y el desgaste de los componentes móviles.

Bias ply tires on heavy load carrier have straight treads.

Los **neumáticos de cubierta diagonal** en portadores de carga pesada tienen bandas de rodadura rectas.

Biodiesel A renewable fuel manufactured domestically from vegetable oils, animals' fats, or recycled restaurant grease.

Biodiesel Un combustible renovable fabricado en el país a partir de aceites vegetales, grasas animales o grasa reciclada de restaurantes.

Block configuration is the shape of the block.

La **configuración del bloque** es la forma del mismo.

Blocker rings make the first contact with a gear and initially allow the gear to slip a little. As the blocker ring is forced against the gear, the connection becomes tighter and the gear speeds up to blocker-ring speed, allowing a smooth, quiet gear change.

Los **anillos de bloqueo** hacen el primer contacto con un engranaje e inicialmente permiten que el engranaje se deslice un poco. A medida que el anillo bloqueador se fuerza contra el engranaje, la conexión se vuelve más apretada y el engranaje se acelera hasta la velocidad del anillo bloqueador, lo que permite un cambio de engranaje suave y silencioso.

Bolts A threaded fastener with external male threads. Bolts are very closely related to screws.

Pernos Un sujetador roscado con roscas macho externas. Los pernos están muy relacionados con los tornillos.

Box-end wrenches A wrench with a closed loop that fits over a nut or bolt head.

Llaves corona Una llave con un extremo circular que se ajusta sobre una cabeza de tuerca o tornillo.

Boxer engine A horizontally opposed engine has the cylinders set horizontally at either side of the crankshaft. The cylinders are placed opposite each other, and they work lying down.

Motor bóxer Un motor opuesto horizontalmente tiene los cilindros colocados horizontalmente a cada lado del cigüeñal. Los cilindros se colocan uno frente al otro y funcionan acostados.

Brake pads and **disc pad** are names for the same component. They are friction components placed between the brake caliper and the brake rotor. Like drum brakes, the **rotor** and hub may be manufactured as a one-piece unit.

Las **pastillas de freno** y la **pastilla de disco** son nombres para el mismo componente. Son componentes de fricción colocados entre la pinza de freno y el rotor de freno.Al igual que los frenos de tambor, el **rotor** y el cubo pueden fabricarse como una unidad de una sola pieza.

Brass hammers A hammer with a brass head use to strike an object without causing a spark or damaging it.

Martillos de latón Un martillo con cabeza de latón se usa para golpear un objeto sin provocar una chispa o dañarlo.

BTUs The amount of heat required to raise the temperature of one pound of water by one degree.

BTU La cantidad de calor requerida para elevar la temperatura de una libra de agua en un grado.

Bushing The inner surface between two moving parts used to reduce friction of a rotating shaft.

Buje La superficie interna entre dos partes móviles utilizada para reducir la fricción de un eje giratorio.

Business ethics are hard to define. Because one person's idea of ethics may be different from another's, the managers set the rules. Treating customers, managers, and employees fairly is the basic rule.

Ética empresarial es difícil de definir. Debido a que la idea de ética de una persona puede ser diferente de la de otra, los gerentes establecen las reglas. Tratar a los clientes, gerentes y empleados de manera justa es la regla básica.

Butt connector A round crimp connector with a crimp on each end Crimp-on **terminal** and **butt connectors** are called "solder-less connectors."

Conector a tope Un conector de crimpado redondo con un crimpado en cada extremoLos **conectores de terminal** y los **conectores a tope** se denominan "conectores sin soldadura".

Camber can be seen by looking at the wheel from the front. Leaning in equals negative camber; leaning out is positive camber. When two wheels are not set equally in **toe**, the customer may complain that the steering wheel is off-centered. Typical RWD passenger cars usually have **toe-in**. Most FWD cars have zero toe. Very few passenger cars are designed to operate with **toe-out**. A bent or twisted frame can change the **setback**, **SAI**, and **toe-out-on-turn** settings without damaging suspension or steering components.

La inclinación se puede ver mirando la rueda desde el frente. Inclinarse hacia adentro es igual a una inclinación negativa; inclinarse hacia afuera es una inclinación positiva. Cuando dos ruedas no están igualadas en la **punta**, el cliente puede quejarse de que el volante está descentrado. Los típicos vehículos de pasajeros RWD generalmente tienen **convergencia de las ruedas**. La mayoría de los vehículos FWD no tienen convergencia. Muy pocos vehículos de pasajeros están diseñados para operar con convergencia de giro. Un chasis doblado o torcido puede cambiar la configuración del **retroceso**, la **EFS** y la **convergencia** sin dañar los componentes de la suspensión o la dirección.

Camshaft bearings for an OHV engine come with different outside diameters. The inside diameters are all the same.

Los **cojinetes del árbol de levas** para un motor OHV vienen con diferentes diámetros exteriores. Los diámetros interiores son todos iguales.

Camshaft The mechanical component of an internal combustion engine that opens and closes the intake and exhausts valves.

Árbol de levas El componente mecánico de un motor de combustión interna que abre y cierra las válvulas de admisión y escape.

Caster can be seen in action on the front wheels of a shopping cart.

La **rueda guía** se puede ver en acción en las ruedas delanteras de un carrito de compras.

Catalyst action is a chemical reaction between two or more elements. One element (catalyst) causes a chemical change in the other element(s) without changing itself.

La acción del **catalizador** es una reacción química entre dos o más elementos. Un elemento (catalizador) provoca un cambio químico en el otro elemento(s) sin cambiar a sí mismo.

Cavitation is the introduction of air into a fluid, in this case, air forced into the oil by the rotating crankshaft and connecting rod.

La **cavitación** es la introducción de aire en un fluido, en este caso, aire forzado al aceite por el cigüeñal giratorio y la biela.

Channel-lock pliers is a trade name that has been accepted as a type of pliers. These types of pliers are actually rib joint pliers.

Alicates de bloqueo de canal es un nombre comercial que ha sido aceptado como un tipo de alicate. Estos tipos de alicates son en realidad alicates de unión de costillas.

Chemical sealants and sealers should be checked for their use with electronic engine controls.

Selladores químicos y selladores deben verificarse para su uso con controles electrónicos del motor.

Chisels A long bladed hand tool with a beveled cutting edge.

Cinceles Herramienta manual de hoja larga con filo biselado.

Clearance is the space between adjacent components. The clearance between a bearing and its journal is called **oil clearance** and is usually about 0.0015 inch to 0.002 inch.

El **espacio libre** es el espacio entre componentes adyacentes. El espacio libre entre un cojinete y su gorrón se llama **espacio libre de aceite** y generalmente es de aproximadamente 0.0015 pulgadas a 0.002 pulgadas.

Climate control system A complex system that is able to affect and regulate the internal temperature of a vehicle.

Sistema de control climático Un sistema complejo que puede afectar y regular la temperatura interna de un vehículo.

Clutch position switch A switch located under the dash that prevents you from starting the vehicle while the transmission is in gear.

Interruptor de posición del embrague Un interruptor ubicado debajo del tablero que le impide arrancar el vehículo mientras la transmisión está en marcha.

Coil One or more turn of a current carrying wire designed to produce a magnetic field.

Bobina Una o más vueltas de un cable conductor de corriente diseñado para producir un campo magnético.

Coil-near-plug has a coil for each spark plug and is mounted near the spark plug with a short connecting spark plug wire.

La **bobina cercana a la bujía** tiene una bobina para cada bujía y está montada cerca de la bujía con un cable de conexión corto.

Coil-on-plug means the coil is attached directly to the spark plug terminal.

Bobina en bujía significa que la bobina está conectada directamente al terminal de la bujía.

Combination wrenches A wrench with one end open and the other end with a closed loop.

Llaves combinadas Una llave con un extremo de boca abierto y el otro extremo con una corona cerrada.

Composite refers to a mixture of materials to form a more durable gasket.

Compuesto se refiere a una mezcla de materiales para formar una junta más duradera.

Composites refer to engine components that are made of plastics and other materials resembling plastics.

Los **compuestos** se refieren a componentes del motor que están hechos de plásticos y otros materiales que se asemejan a los plásticos.

Compound gearsets A number of gear fixed together usually differ in size and number of teeth.

Engranajes compuestos Un número de engranajes unidos entre sí generalmente difieren en tamaño y número de dientes.

Compression stroke The stroke in an engine which the air/fuel mixture is compressed before ignition.

Carrera de compresión La carrera en un motor que comprime la mezcla de aire/combustible antes de la ignición.

Computerized data is information stored on a computer or accessed through the internet.

Datos computarizados son información almacenada en una computadora o a la que se accede a través de Internet.

Conduction The transfer of heat directly through a substance when there is a difference in temperature.

Conducción La transferencia de calor directamente a través de una sustancia cuando hay una diferencia de temperatura.

Constant mesh means the mating gears' teeth are running together constantly.

Malla constante significa que los dientes de los engranajes de acoplamiento se juntan constantemente.

Constant Velocity CV) A joint that allows a drive shaft to transmit power through a variable angle at a constant speed.

Velocidad constante (VC) Una junta que permite que un eje de transmisión transmita potencia a través de un ángulo variable a una velocidad constante.

Contact patch The area of the tire that makes contact with the road surface.

Parche de contacto El área del neumático que hace contacto con la superficie de la carretera.

Continuous scale indicators are normally used to measure endplay or lateral movement.

Los **indicadores de escala continua** se utilizan normalmente para medir el juego final o el movimiento lateral.

Conversion factors are used to change a measurement from one system to another measuring system.

Los **factores de conversión** se utilizan para cambiar una medición de un sistema a otro.

Coolant is a chemical mixture that reduces freezing of a liquid, usually water. It's also called antifreeze. You cannot identify coolant by the color alone!

El **refrigerante** es una mezcla química que reduce la congelación de un líquido, generalmente agua. También se llama anticongelante. ¡No puede identificar el refrigerante solo por el color!

Corporate Average Fuel Economy (CAFE) is a federal requirement concerning fuel mileage or usage. It is an average calculated from a manufacturer's vehicles using the worst (lowest mileage) versus the best (highest) mileage. A company's CAFE average also considers the total number of vehicles and vehicle type sold. The most current CAFE requirement is 27.5 miles per gallon (mpg) on cars and 22.2 mpg on light trucks. In 2016, the CAFE average for cars will go up to 38.5 mpg, and in 2025, the CAFE requirement will be 54.5 mpg. This will only be possible with advanced computers and other advances to vehicle structure.

Economía de Combustible Promedio Corporativo (CAFE) es un requisito federal en relación al millaje o el consumo de combustible. Es un promedio calculado a partir de los vehículos de un fabricante que utilizan el peor kilometraje (el más bajo) versus el mejor (el más alto) kilometraje. El promedio CAFE de una empresa también considera el número total de vehículos y el tipo de vehículo vendido. El requisito de CAFE más reciente es de 27.5 millas por galón (mpg) en automóviles y 22.2 mpg en camiones ligeros. En 2016, el promedio de CAFE para automóviles aumentará a 38.5 mpg, y en 2025, el requisito de CAFE será de 54.5 mpg. Esto solo será posible con computadoras avanzadas y otros avances en la estructura del vehículo.

Corrosion can be found most commonly on battery cables, usually resulting from a lack of maintenance.

La **corrosión** se puede encontrar más comúnmente en los cables de la batería, generalmente como resultado de la falta de mantenimiento.

Cotter pin puller A special tool that looks like a offset screwdriver but with a pointed tip, used to pull cotter pins.

Extractor de chavetas Una herramienta especial que se parece a un destornillador offset, pero con una punta puntiaguda, que se utiliza para extraer chavetas.

Crumple zones extend from bumper to bumper. The sections of the body and frame interact to absorb as much energy as possible before it reaches the passenger compartment.

Las **zonas de deformación** se extienden de parachoques a parachoques. Las secciones del cuerpo y el marco interactúan para absorber tanta energía como sea posible antes de que llegue al compartimiento de pasajeros.

Current is the number of moving electrons flowing between the ends of a circuit.

La **corriente** es el número de electrones en movimiento que fluyen entre los extremos de un circuito.

Daytime running lights were first introduced in the United States by GM as a warning signal to other drivers.

GM introdujo por primera vez las **luces de circulación diurna** en los Estados Unidos como una señal de advertencia para otros conductores.

Decimal numbers are expressions of whole numbers in hundreds, thousands, or ten-thousandths.

Los **números decimales** son expresiones de números enteros en cientos, miles o diezmilésimas.

Desiccant is a white, powered compound that can absorb many times its weight in water or moisture.

El **desecante** es un compuesto blanco y accionado que puede absorber muchas veces su peso en agua o humedad.

Diagnostic trouble codes (DTCs) are numerical data that can be retrieved from the vehicle's computer with a scan tool. Each code designates a specific fault.

Los **Códigos de diagnóstico de problemas (DTCs)** son datos numéricos que se pueden recuperar de la computadora del vehículo con una herramienta de escaneo. Cada código designa una falla específica.

Diagonal cutting pliers have cutting edges on the jaw for cutting pins or wire. These pliers are often called "dikes," which is short for "diagonals."

Los alicates de corte diagonal tienen bordes cortantes en la mandíbula para cortar pasadores o alambre. Estos alicates a menudo se llaman "diques", que es una abreviatura de "diagonales".

Differential A device that splits the engines torque two ways allowing each output to spin at a different speed.

Diferencial Un dispositivo que divide el par motor de dos maneras permitiendo que cada salida gire a una velocidad diferente.

Differential carrier A case that holds the differential gears.

Portador del diferencial Una caja que contiene los engranajes diferenciales.

Digital signals switch between on and off voltages.

Las señales **digitales** cambian entre voltajes de encendido y apagado.

Displacement is the volume of air the engine can exchange in one complete engine cycle.

El **desplazamiento** es el volumen de aire que el motor puede intercambiar en un ciclo completo del motor.

DOT refers to the United States Department of Transportation, which issues the detailed standards for all vehicles brake fluid in the country.

DOT se refiere al Departamento de Transporte de los Estados Unidos, que emite las normas detalladas para todos los líquidos de frenos de vehículos en el país.

Double-post lift A hydraulic lift that raises the vehicle from the sides for the lift giving access to the bottom of the vehicle.

Elevador de doble poste Un elevador hidráulico que eleva el vehículo desde los lados para que el elevador permita el acceso a la parte inferior del vehículo.

Dowel pin A fastener that are use to join two or more items together. Short round rods that can be tapered, slotted, or grooved.

Pasador de clavija Un sujetador que se usa para unir dos o más elementos. Varillas redondas cortas que pueden ser cónicas, ranuradas o estriado.

Drive gear Usually refers to the timing drive gear that receives forces from the power output then transfers power to the driven gear.

Engranaje de transmisión Por lo general, se refiere al engranaje de transmisión de sincronización que recibe fuerzas de la salida de potencia y luego transfiere potencia al engranaje impulsado.

Drive shaft A rotating shaft which transmits torque in an engine.

Eje de transmisión Eje giratorio que transmite el par en un motor.

Drive-by-wire is a term used in the automotive industry which means that a component is controlled by a computer-regulated system and is NOT operated by cables or mechanical linkage.

Drive-by-wire es un término utilizado en la industria automotriz que significa que un componente está controlado por un sistema regulado por computadora y NO es operado por cables o enlaces mecánicos.

Driven gear The member or a pair of gear to which motion and power are transfered by the drive gear.

Engranaje accionado El miembro o un par de engranajes a los que el engranaje de transmisión transfiere el movimiento y la potencia.

Drive-on lifts A lift that a vehicle can be driven on and raised up with the wheels on the lift.

Elevadores automáticos Un elevador sobre el cual se puede conducir un vehículo y elevarlo con las ruedas en el elevador.

Dynamic balancing is almost always used on cars and light trucks.

El **balanceo dinámico** casi siempre se usa en automóviles y camionetas.

Dynamic proportioning is the use of electronically controlled valves to meter and proportion pressures.

La **dosificación dinámica** es el uso de válvulas controladas electrónicamente para medir y proporcionar presiones.

Electrical is electricity flowing through a circuit and may be controlled by **electronic** signals from a computer.

Eléctrico es la electricidad que fluye a través de un circuito y puede ser controlada por señales **electrónicas** de una computadora.

Electricity The presence and flow of electric charge.

Electricidad presencia y flujo de carga eléctrica.

Electrolysis is the chemical process of decomposition between a liquid and metal producing a small electrical current.

La **electrólisis** es el proceso químico de descomposición entre un líquido y un metal que produce una pequeña corriente eléctrica.

Endplay is the amount of horizontal movement of a shaft within its mounts. All automotive shafts must have some endplay.

El **juego final** es la cantidad de movimiento horizontal de un eje dentro de sus monturas. Todos los ejes automotrices deben tener algo de juego final.

Engine **displacement** is the amount of air an engine can move through it. It may be measured in cubic inches displacement (cid), liters (L), or cubic centimeters (cc).

El **desplazamiento del motor** es la cantidad de aire que un motor puede mover a través de él. Se puede medir en pulgadas cúbicas de desplazamiento (cid), litros (L) o centímetros cúbicos (cc).

Ethanol Ethyl alcohol same as the alcohol found in alcoholic beverages mainly used as a biofuel additive for gasoline.

Etanol Alcohol etílico igual que el alcohol que se encuentra en las bebidas alcohólicas. Utilizado principalmente como aditivo de biocombustibles para gasolina.

Ethical work Moral principles a technician uses in their job, which encompasses traits like: Reliability, dependability, dedication, productivity, cooperation, character, integrity, responsibility, quality, teamwork and professionalism.

Trabajo ético Principios morales que un técnico utiliza en su trabajo, que abarca rasgos como: confiabilidad, dependencia, dedicación, productividad, cooperación, carácter, integridad, responsabilidad, calidad, trabajo en equipo y profesionalismo.

Evacuating is removing all air and moisture from the system after recovery.

Evacuar es eliminar todo el aire y la humedad del sistema después de la recuperación.

Express services refer to quick routine preventive maintenance, primarily oil changes, tire rotation, and vehicle safety inspections.

Servicios expresos se refieren al mantenimiento preventivo de rutina rápido, principalmente cambios de aceite, rotación de llantas e inspecciones de seguridad del vehículo.

Extension housing Also known as a tail housing it is fastened to the tail of a transmission or transfer case.

Cärcasa de extensión También conocida como carcasa de cola, está sujeta a la cola de una caja de transmisión o transferencia.

Fasteners A hardware device that mechanically joins two or more objects together.

Sujetadores Un dispositivo de equipo que une mecánicamente dos o más objetos.

Final drive The last set of gears in the drive train.

Transmisión final El último conjunto de engranajes en el tren de transmisión.

Flex fuel A flexible-fuel vehicle (FFV) is an alternative fuel for vehicles. The engine is designed to run on more than one fuel, usually gasoline blended with ethanol or methanol fuel.

Combustible flexible Un vehículo de combustible flexible (FFV) es un combustible alternativo para vehículos. El motor está diseñado para funcionar con más de un combustible, generalmente gasolina mezclada con combustible de etanol o metanol.

Floor jack is designed to roll as a vehicle is lifted, thereby keeping the lift plate in position.

El **gato de piso** está diseñado para rodar a medida que se eleva un vehículo, manteniendo así la placa de elevación en su posición.

Foot-pounds Unit of energy equal to the amount required to raise one pound a distance of one foot.

Pie-libras Unidad de energía igual a la cantidad requerida para elevar una libra a una distancia de un pie.

Force is the pressure of one mass against another.

La **fuerza** es la presión de una masa contra otra.

Franchise are businesses locally owned and operated under the policies of a larger company. For example, McDonald's is a franchise operation. A person (or group) decides to build a store and purchases the name recognition rights. They must follow rules and can buy and sell certain items directly from the manufacturer.

Franquicias son negocios de propiedad y operación local bajo las políticas de una compañía más grande. Por ejemplo, McDonald's es una operación de franquicia. Una persona (o grupo) decide construir una tienda y compra los derechos de reconocimiento de nombre. Deben seguir las reglas y pueden comprar y vender ciertos artículos directamente del fabricante.

Frequency is the number of times an action or repeated action happens within a given time period. It is also known as "duty cycle."

La **frecuencia** es el número de veces que una acción o acción repetida ocurre dentro de un período de tiempo determinado. También se conoce como "ciclo de trabajo".

Friction material is specially formatted material used for brake pads and brake shoes. It is designed to withstand high temperature while minimizing pad and rotor wear.

El **material de fricción** es un material especialmente formateado utilizado para pastillas de freno y zapatas de freno. Está diseñado para soportar altas temperaturas y minimizar el desgaste de la almohadilla y el rotor.

Fuel cell in vehicles generate electricity to power the motor, generally using oxygen and compressed hydrogen. Most fuel cell vehicles are considered zero-emission vehicles.

Pila de combustible en los vehículos genera electricidad para alimentar el motor, generalmente usando oxígeno e hidrógeno comprimido. La mayoría de los vehículos con pilas de combustible se consideran vehículos de cero emisiones.

Fuel mileage is required to meet the standards in the federal Corporate Average Fuel Economy (CAFE) regulation. A manufacturer's vehicle line must be tested, and the average of all models offered must meet CAFE.

El **millaje de combustible** se requiere para cumplir con los estándares de la regulación federal de economía de combustible promedio corporativa (CAFE). Se debe probar la línea de vehículos de un fabricante, y el promedio de todos los modelos ofrecidos debe cumplir con CAFE.

Fuses, circuit breakers, and **fusible links** are devices most commonly found in automotive circuits.

Fusibles, disyuntores y **enlaces fusibles** son dispositivos que se encuentran más comúnmente en los circuitos automotrices.

Gap usually refers to the gap between two nonelectronic components.

La **brecha** generalmente se refiere a la separación entre dos componentes no electrónicos.

Gear A toothed wheel that increase the force needed to push or pull something.

Engranaje Una rueda dentada que aumenta la fuerza necesaria para empujar o tirar de algo.

Global warming potential is a relative measure of how much heat a greenhouse gas traps in the atmosphere.

El potencial de calentamiento global es una medida relativa de cuánto calor atrapa un gas de efecto invernadero en la atmósfera.

Grade markings Lines on the head of a bolt that indicate the strength of the bolt.

Marcas de grado Líneas en la cabeza de un perno que indican la fuerza del perno.

Grease is a lubricant that reduces friction and smoothes movement between parts of components.

La **grasa** es un lubricante que reduce la fricción y suaviza el movimiento entre las partes de los componentes.

Heat transfer The movement of thermal energy (heat) from one component to another component of a different temperature.

Transferencia de calor El movimiento de energía térmica (calor) de un componente a otro componente de una temperatura diferente.

Heater core A small radiator device used to heat the passenger cabin.

Núcleo del calentador Un pequeño dispositivo radiador utilizado para calentar la cabina de pasajeros.

Helicoil A coiled wire type of threaded insert used to create inside threads.

Helicoil Es un inserto roscado de alambre enrollado que se utiliza para crear hilos internos.

Hertz is a measurement of how many times a function cycles in 1 second. In the 5-hertz instance, the suspension can change up to five times per second.

Hercio es una medida de cuántas veces una función realiza ciclos en 1 segundo. En el caso de 5 hercios, la suspensión puede cambiar hasta cinco veces por segundo.

High impedance is basically the same as high resistance. It is required to obtain an accurate measurement of electricity.

La **alta impedancia** es básicamente lo mismo que la alta resistencia. Se requiere para obtener una medición precisa de la electricidad.

HOAT This coolant combines both IAT and OAT best qualities of corrosion inhibitor of IAT and non-toxic properties of the OAT. The best of both worlds.

HOAT Este refrigerante combina las mejores cualidades de IAT y OAT como inhibidor de corrosión de IAT y propiedades no tóxicas de OAT. Lo mejor de ambos mundos.

HVAC Short for heating, ventilation and air-conditioning system, which provides heat and cooling to the passenger cabin.

HVAC Corto para sistema de calefacción, ventilación y aire acondicionado, que proporciona calor y refrigeración a la cabina de pasajeros.

Hybrid power A vehicle that uses more than one means of propulsion. Hybrid vehicles use a combination of diesel or gasoline engine with an electric motor.

Potencia híbrida Un vehículo que usa más de un medio de propulsión. Los vehículos híbridos usan una combinación de motor diésel o de gasolina con un motor eléctrico.

Hydraulics is the use of a liquid to transfer force. Liquids used in hydraulic systems are selected based on the work to be done and the chemical properties of the liquid. Brake fluid will not work well in steering systems.

La hidráulica es el uso de un líquido para transferir fuerza. Los líquidos utilizados en los sistemas hidráulicos se seleccionan en función del trabajo a realizar y las propiedades químicas del líquido. El líquido de frenos no funcionará bien en los sistemas de dirección.

Hydro-boost was originally designed, used, and owned by General Motors.

Hydro-boost fue originalmente diseñado, usado y propiedad de General Motors.

Hydrocarbons (HC), **carbon monoxide (CO)**, and **oxides of nitrogen** are harmful products of burning fossil fuels.

Los **hidrocarburos (HC)**, **el monóxido de carbono (CO)** y los **óxidos de nitrógeno** son productos nocivos de la quema de combustibles fósiles.

Hygroscopic A fluid that absorbs moisture from the air.

Higroscópico Un fluido que absorbe la humedad del aire.

IAT A coolant that contains ethylene glycol which is known to pollute groundwater and is harmful if swallowed by humans and animals.

IAT Un refrigerante que contiene etilenglicol que se sabe que contamina el agua subterránea y es dañino si es ingerido por humanos y animales.

Impact wrench An air operated impact gun that gives a rapid succession of sudden torque.

Llave de impacto Una pistola de impacto operada por aire que proporciona una rápida sucesión de torque repentino.

Impeller Vanes or blades that directs the transmission fluid against the turbine blades creates torque.

Impulsor Paletas o cuchillas que dirigen el fluido de la transmisión contra los álabes de la turbina crean torque.

In some vehicles, the **center link** is known as the "drag link."

En algunos vehículos, el **enlace central** se conoce como el "enlace de arrastre".

In theory, a **maintenance-free battery** requires no maintenance during its life.

En teoría, una **batería libre de mantenimiento** no requiere mantenimiento durante su vida útil.

Inches of mercury/water is the amount the level of mercury or water in a sealed container rises when a vacuum is applied.

Pulgadas de mercurio/agua es la cantidad que aumenta el nivel de mercurio o agua en un recipiente sellado cuando se aplica un vacío.

Index lines are where two or more markings are aligned.

Las **líneas de índice** son donde se alinean dos o más marcas.

Inertia The tendency to do nothing or to remain unchanged. "an object at rest tends to stay at rest" and "an object in motion tends to stay in motion."

Inercia La tendencia a no hacer nada o permanecer sin cambios. "Un objeto en reposo tiende a permanecer en reposo" y "un objeto en movimiento tiende a permanecer en movimiento."

In-line engine A type in internal combustion engine having its cylinders arranged in a row.

Motor en línea Un tipo de motor de combustión interna que tiene sus cilindros dispuestos en una fila.

Input devices or **sensors** sends data to a computer on circuit conditions at a given time.

Los **dispositivos de entrada** o **sensores** envían datos a una computadora en condiciones de circuito en un momento dado.

Instrument Panel The surface in front of the driver, on which are arranged an array of dials, lights and gauges.

Panel de instrumentos La superficie frente al controlador, en la que se disponen una serie de diales, luces y medidores.

Investment is the amount of time, money, and other resources used to establish a business.

Inversión es la cantidad de tiempo, dinero y otros recursos utilizados para establecer un negocio.

ISO are the initials for the International Standards Organization.

ISO son las iniciales de la Organización Internacional de Normalización.

Jam hexagon nuts A secondary thin nut used to jam or lock against a full size nut.

Tuercas Hexagonales de Mermelada Una tuerca fina secundaria utilizada para atascar o bloquear contra una tuerca de tamaño completo.

Jounce generally occurs when the wheel hits a high spot like a rock.

El **movimiento de aplastamiento** generalmente ocurre cuando la rueda golpea un punto alto como una roca.

Key A small piece of metal that fits into a slot called a keyway.

Llave Una pequeña pieza de metal que cabe en un hueco llamada ranura.

Leaf Spring Strips of heavy metal slightly curved upward and clamped together one above the other.

Ballesta Tiras de metal pesado ligeramente curvada hacia arriba y se sujeta juntos uno encima del otro.

Lifelong learning Continuing to learn and acquire the latest skills required by industry advances in technology.

Aprendizaje permanente Continuar aprendiendo y adquiriendo las últimas habilidades requeridas por los avances de la industria en tecnología.

Linear measures are measurements of distance such as a mile or a meter.

Las **medidas lineales** son medidas de distancia, como una milla o un metro.

Lining Brake lining is the part of the shoe or pad that actually contacts the metal brake disc (rotor) when engaged. The pressurized fluid trapped between the pistons and shoes acts as one **anchor** for the shoes.

Revestimiento El revestimiento del freno es la parte de la zapata o almohadilla que realmente entra en contacto con el disco de freno de metal (rotor) cuando está enganchado.El fluido presurizado atrapado entre los pistones y los zapatos actúa como un **ancla** para los zapatos.

Lip seal a circular seal encompassing a rotation shaft and preventing fluid leakage. A garter spring may be located inside the seal to help the lip seal better to the rotation shaft.

Sello de labio Un sello circular que abarca un eje de rotación y evita fugas de líquido. Se puede ubicar un resorte de liga dentro del sello para ayudar a que el labio selle mejor al eje de rotación.

Lock washers A helical spring washer designed to lock by squishing flat between a nut and the mounting surface when tightening them.

Arandelas de seguridad Una arandela de resorte helicoidal diseñada para bloquearse aplastando entre una tuerca y la superficie de montaje al apretarlas.

MacPherson struts are lighter than a coil spring and control but are almost maintenance free and can be replaced much more easily.

Los **puntales MacPherson** son más livianos que un resorte helicoidal y control, pero casi no requieren mantenimiento y se pueden reemplazar con mucha más facilidad.

MACS Empowers members to grow their businesses and delivers tangible members benefits.

MACS Empodera a los miembros a hacer crecer sus negocios y ofrece beneficios tangibles para los miembros.

Magnetic field may be natural, such as magnetic rock or something electrical. Materials that are magnetized by electricity are electromagnets.

El **campo magnético** puede ser natural, como una roca magnética o algo eléctrico. Los materiales que están magnetizados por la electricidad son electroimanes.

Main bearing The bearing that holds the crankshaft in place while still allowing rotation of the crankshaft.

Cojinete principal El cojinete que mantiene el cigüeñal en su lugar mientras permite la rotación del cigüeñal.

Main bearings hold the crankshaft in place within the engine.

Los **cojinetes principales** mantienen el cigüeñal en su lugar dentro del motor.

Many people call the **adjustable wrench** a "crescent wrench." This is due to the fact that the Crescent Tool Company made this type of wrench for many years.

Muchas personas llaman a la **llave ajustable** una "llave inglesa." Esto se debe al hecho de que la empresa Crescent Tool fabricó este tipo de llave durante muchos años.

Mass is the matter or atomic structure of an item.

La **masa** es la materia o estructura atómica de un artículo.

Measuring devices An instrument use for measuring a physical quantity, amount or degree of something. A device that requires skill for proper use.

Dispositivos de medición Un instrumento utilizado para medir una cantidad física, cantidad o grado de algo. Un dispositivo que requiere habilidad para un uso adecuado.

Mechanical linkages are the piston, connecting rod, and crankshaft.

Los **enlaces mecánicos** son el pistón, la biela y el cigüeñal.

Meshed components touch or engage each other all the time, such as two meshed gears.

Los **componentes de malla** se tocan o se engranan todo el tiempo, como dos engranajes de malla.

Metering is the controlled flow of a liquid or vapor.

La **medición** es el flujo controlado de un líquido o vapor.

Microprocessors: An electronic component that is used by a computer to do its work.

Microprocesadores: un componente electrónico que utiliza una computadora para hacer su trabajo.

Monoleaf springs are usually not found on standard passenger cars and light trucks.

Los **resortes monolíticos** generalmente no se encuentran en los automóviles de pasajeros y camiones ligeros estándar

Mud and snow tires are used occasionally in the southern United States and very common during winter in the northern United States and Canada. None of the tires mentioned here work well in icy conditions.

Los **neumáticos de barro y nieve** se usan ocasionalmente en el sur de los Estados Unidos y comúnmente durante el invierno en el norte de los Estados Unidos y Canadá. Ninguno de los neumáticos mencionados aquí funciona bien en condiciones de hielo.

Mufflers and **resonators** are exhaust system components designed to reduce exhaust noises.

Los **silenciadores** y **resonadores** son componentes del sistema de escape diseñados para reducir los ruidos de escape.

Multiplexing is a method of connecting multiple electronic devices to send/transmit data.

La **multiplexación** es un método para conectar múltiples dispositivos electrónicos para enviar/transmitir datos.

National Institute for Automotive Service Excellence (ASE) is the acronym for the National Institute for Automotive Service Excellence.

Instituto Nacional para la Excelencia en el Servicio Automotriz (ASE) es el acrónimo del Instituto Nacional para la Excelencia en el Servicio Automotriz.

Needle bearings are not used for high-load conditions.

Los **cojinetes de agujas** no se usan para condiciones de alta carga.

Needle nose pliers A pliers with a long nose end useful for reaching into small areas.

Alicates de punta fina Un alicate con un extremo de punta larga útil para alcanzar áreas pequeñas.

Nickel–cadmium (NiCad) and **nickel–metal hydrid (NiMH)** are used because of their high total output and battery life.

El **níquel-cadmio (NiCad)** y el **níquel-metal hidruro (NiMH)** se utilizan debido a su alto rendimiento total y duración de la batería.

Nonthreaded fasteners A fastener that does not have threads which includes pins, retaining rivets, push nuts.

Sujetadores sin rosca Un sujetador que no tiene roscas que incluye pasadores, remaches de retención, tuercas de empuje.

Note that, when a wheel is mounted on a vehicle, the place for the lug nuts appears to be deep inside the wheel opening. This would be a **dropped center safety rim**.

Tenga en cuenta que, cuando se monta una rueda en un vehículo, el lugar para las tuercas parece estar profundamente dentro de la abertura de la rueda. Este sería un **borde de seguridad central caído**.

Nuts CA hardware device that mechanically joins two or more objects together with internal threads.

Tuercas Dispositivo de equipo CA que une mecánicamente dos o más objetos junto con hilos internos.

OAT Is a coolant that is considered a long life coolant and non-toxic, which makes them biodegradable.

OAT Es un refrigerante que se considera un refrigerante de larga duración y no tóxico, lo que los hace biodegradables.

Offset screwdrivers A tool with a flattened, cross-shaped or star-shaped (torx) tip at a right angle that fits into a head of a screw.

Destornilladores descentrados Una herramienta con una punta aplanada, en forma de cruz o en forma de estrella (torx) en ángulo recto que encaja en la cabeza de un tornillo.

On a **spool valve**, fluid flows through the grooves when they are aligned with a hydraulic port.

En una **válvula de carrete**, el fluido fluye a través de las ranuras cuando están alineadas con un puerto hidráulico.

On the front wheels of many front-wheel-drive vehicles, tapered roller bearings have been replaced with **ball bearings**.

En las ruedas delanteras de muchos vehículos con tracción delantera, los cojinetes de rodillos cónicos han sido reemplazados por **cojinetes de bolas**.

One **horsepower** is equal to the force needed to raise 33,000 pounds at one foot per minute.

Un **caballo de fuerza** es igual a la fuerza necesaria para subir 33,000 libras a un pie por minuto.

One-way clutch A unit that rotates in one direction and locks in the opposite direction. To see **vortex flow**, place a liquid into a bowl and stir it with a mixer. The liquid will flow around the bowl (rotary flow) and is then pulled under by the mixer and inserted back into the rotary flow. Without the stator and vortex flow, there would not be enough torque to power the geartrain.

Embrague unidireccional Una unidad que gira en una dirección y se bloquea en la dirección opuesta. Para ver el **flujo de vórtice**, coloque un líquido en un tazón y revuelva con una batidora. El líquido fluirá alrededor del recipiente (flujo rotativo) y luego se mezclará con el mezclador y se insertará nuevamente en el flujo rotativo. Sin el estator y el flujo de vórtice, no habría suficiente torque para alimentar el tren de engranajes.

Open An electrical circuit that is not complete.

Abierto Un circuito eléctrico que no está completo.

Parallelogram Used on car and trucks because all the pivot points are paralleled. The steering linkage consists of pitman arm, idler arm, tie-rod ends and a center link.

Paralelogramo Se utiliza en automóviles y camiones porque todos los puntos de pivote están en paralelo. El varillaje de dirección consta de brazo pitman, brazo libre, extremos de barra de acoplamiento y un enlace central.

Passenger cabin The area of the vehicle designed for the seating of the driver and other passengers.

Cabina de pasajeros El área del vehículo diseñada para los asientos del conductor y otros pasajeros.

Payload is the amount of weight a vehicle is designed to carry including the operator. The term "gross payload" includes the total weight of the vehicle, including all passengers and the payload.

La **carga útil** es la cantidad de peso que un vehículo está diseñado para transportar, incluido el operador. El término "carga útil bruta" incluye el peso total del vehículo, incluidos todos los pasajeros y la carga útil.

Perpetual energy is energy created using less energy than that produced.

La **energía perpetua** es la energía creada usando menos energía que la producida.

Pig tail A connector with a loose wire on one end.

Cola de cerdo: Un conector con un cable suelto en un extremo.

Pin punch A punch with a very small tip used to dislodge rivets and pins.

Punzón de alfileres Un punzón con una punta muy pequeña utilizada para desalojar remaches y alfileres.

Pitch gauge A tool used to measure the distance between threads on a bolt.

Medidor de paso Una herramienta utilizada para medir la distancia entre las roscas de un perno.

Planetary carrier Connects the center of the two gears and rotates to carry one gear around the other.

Portador planetario Conecta el centro de los dos engranajes y gira para transportar un engranaje alrededor del otro.

Plastic-tip hammers A hammer with a plastic head used to strike a surface without damaging it.

Martillos con punta de plástico Un martillo con una cabeza de plástico utilizado para golpear una superficie sin dañarla.

Pneumatic The use of compressed air or gas to operate a tool or object. Filling a balloon with air from the lungs makes that balloon **pneumatic**.

Neumático El uso de aire comprimido o gas para operar una herramienta u objeto. Llenar un globo con aire de los pulmones hace que el globo sea **neumático**.

Potential energy is based on the mass and position of a stationary object. The speed of a moving object plus its mass creates the **kinetic energy** of the object.

La **energía potencial** se basa en la masa y la posición de un objeto estacionario. La velocidad de un objeto en movimiento más su masa crea la **energía cinética** del objeto.

Powertrain Control Module (PCM) A control unit used on motor vehicles. It is generally a combined control unit, consisting of the engine control unit and the transmission control unit.

Módulo de Control del Tren Motriz (PCM) Una unidad de control utilizada en vehículos automotores. Generalmente es una unidad de control combinada, que consiste en la unidad de control del motor y la unidad de control de la transmisión.

Pressure controllers are devices used to control hydraulic fluid/pressure within a sealed hydraulic system. They may be electronically or mechanically controlled or operated.

Los **controladores de presión** son dispositivos utilizados para controlar la presión / fluido hidráulico dentro de un sistema hidráulico sellado. Pueden ser controlados u operados electrónica o mecánicamente.

Pressure is one of prime requirements in the fuel injection system.

La **presión** es uno de los requisitos principales en el sistema de inyección de combustible.

Pressure lubricated is oil forced between two components by pressure.

La **presión lubricada** es el aceite forzado entre dos componentes por presión.

PSI A common unit of measurement for pressure. It is the amount of force that is exerted on one square inch of area surface.

PSI Una unidad de medida común para la presión. Es la cantidad de fuerza que se ejerce sobre una pulgada cuadrada de superficie del área.

Rack and Pinion A compact steering system that uses a gearwheel (pinion) and a flat tooth bar (rack) that is connected to the steering wheel and tires.

Cremallera y piñón Un sistema de dirección compacto que utiliza una rueda dentada (piñón) y una barra de dientes plana (cremallera) que está conectada al volante y los neumáticos.

Radial load is produced by the component pressing at 90 degrees to the sides of the bearing.

La **carga radial** es producida por el componente que presiona a 90 grados a los lados del rodamiento.

Radial ply tires are found on all modern cars and most light trucks. Some heavy load carriers are required to have radial ply tires on the steering axle.

Los **neumáticos de capa radial** se encuentran en todos los automóviles modernos y en la mayoría de los camiones ligeros. Se requiere que algunos portadores de carga pesada tengan neumáticos de capas radiales en el eje de dirección.

Radiator A heat exchanger designed to transfer heat from the hot coolant that flows through the radiator by blowing cooler air through the radiator with fans and the movement of the vehicle.

Radiador Un intercambiador de calor diseñado para transferir calor del refrigerante caliente que fluye a través del radiador soplando aire más frío a través del radiador con ventiladores y el movimiento del vehículo.

Ratio is the proportion or relation one element has to another. In this case, it is the relation of air to fuel.

Razón es la proporción o relación que un elemento tiene con otro. En este caso, es la relación del aire con el combustible.

Rebound is downward movement after a jounce.

El **rebote** es un movimiento descendente después de un **movimiento de aplastamiento**.

Recessed-head screwdrivers are made to fit recessed-head screws like Phillips, clutch, and Torx.

Los **destornilladores de cabeza empotrada** están hechos para adaptarse a tornillos de cabeza empotrada como Phillips, embrague y Torx.

Recharging is putting refrigerant back into the system after repairs.

Recargar es devolver refrigerante al sistema después de las reparaciones.

Recirculating ball and nut Also known as a worm and sector or steering gear is a steering mechanism commonly found in older vehicles.

Bola y tuerca de recirculación También conocido como gusano y sector o mecanismo de dirección es un mecanismo de dirección que se encuentra comúnmente en vehículos más antiguos.

Recovery is the process of capturing refrigerant in an A/C system.

La **recuperación** es el proceso de capturar refrigerante en un sistema de A/C.

Refrigerant is a chemical that cools hot air by changing its state from liquid to vapor to liquid.

Refrigerante es un químico que enfría el aire caliente cambiando su estado de líquido a vapor a líquido.

Resistance slows or blocks current flow. It may be an electric working load or a solid **insulator**.

La resistencia ralentiza o bloquea el flujo de corriente. Puede ser una carga de trabajo eléctrica o un **aislante sólido**.

Ring gear A large gear in the differential of an automobile that is driven by the propeller shaft pinion.

Engranaje de anillo Un engranaje grande en el diferencial de un automóvil que es impulsado por el piñón del eje de la hélice.

Rivets A permanent fastener that consists of a smooth cylindrical shaft that when installed mushrooms and holds to object together.

Remaches Un sujetador permanente que consiste en un eje cilíndrico liso que cuando se instala crea una cabeza de hongo y mantiene unido al objeto.

Rod bearings A rod bearing is also called a plain bearing, connected to the connecting rod and the crankshaft which provides a smooth snug fit.

Cojinetes de biela Un cojinete de biela también se denomina cojinete liso, conectado a la biela y al cigüeñal, lo que proporciona un ajuste suave y ceñido.

Roll pin A spring pin that secures the position of two or more parts together.

Pasador Un pasador de resorte que asegura la posición de dos o más partes juntas.

Roll pin punch A small punch that has a small point on the end to fit into the roll pin without damaging the roll pin.

Punzón de pasador de rodillo Un punzón pequeño que tiene un pequeño punto en el extremo para encajar en el pasador de rodillo sin dañar el pasador de rodillo.

Rotary engine An internal combustion engine that works completely different way than a conventional piston engine. The crankshaft does not rotate.

Motor rotativo Un motor de combustión interna que funciona de manera completamente diferente a un motor de pistón convencional. El cigüeñal no gira.

Rotary engines have no reciprocating pistons but use a rotor to move the air and fuel mixture and produce power.

Motores **rotativos** no tienen pistones alternativos, pero usan un rotor para mover la mezcla de aire y combustible y producir energía.

Round feeler gauges are made of precision manufactured wire attached to handle of some type.

Los **calibradores redondos** están hechos de alambre fabricado con precisión conectado al mango de algún tipo.

RTV A silicone that begins curing as soon as it is exposed to moisture in the air.

RTV Una silicona que comienza a curarse tan pronto como se expone a la humedad del aire.

Rubber mallets A hammer with a rubber head softer than a plastic and brass hammer.

Mazos de goma Un martillo con una cabeza de goma más suave que un martillo de plástico y latón.

Rulers may have very precise graduations but still not be considered a precision instrument.

Las **reglas** pueden tener graduaciones muy precisas, pero aún no se consideran un instrumento de precisión.

Run-Flat Tires (RFTs) have strengthened sidewalls to support a vehicle for an extended distance in the event of tire damage. Some have temporary self-sealing capabilities.

Los **neumáticos de rodamiento sin aire (RFT)** han reforzado las paredes laterales para soportar un vehículo durante una distancia prolongada en caso de daños en los neumáticos. Algunos tienen capacidades temporales de autosellado.

Runout is the amount of side-to-side movement in a rotating device. Runout may be referred to as "wobble."

La **desviación** es la cantidad de movimiento de lado a lado en un dispositivo giratorio. El agotamiento puede denominarse "bamboleo".

RWAL A brake system used to prevent rear wheel brake lock-up under heavy braking conditions.

RWAL Un sistema de frenos utilizado para evitar el bloqueo de los frenos de las ruedas traseras en condiciones de frenado intensas.

SAE provides technical support it is comprised of engineers, executives, educators and students.

SAE Proporciona soporte técnico, está compuesto por ingenieros, ejecutivos, educadores y estudiantes.

Scissor lifts A lift that vehicle can be driven on and it will lift the vehicle straight up.

Elevadores de tijera Un elevador sobre el que se puede conducir el vehículo y lo elevará directamente.

Screw A sharp-pointed metal pin with threads running around it and a slotted or hex head.

Tornillo Un pasador de metal puntiagudo con hilos que lo rodean y una cabeza ranurada o hexagonal.

Screwdriver A tool with a flattened, cross-shaped or star-shaped (torx) tip that fits into a head of a screw.

Destornillador Una herramienta con una punta aplanada, en forma de cruz o estrella (torx) que encaja en la cabeza de un tornillo.

Seal wheel bearing Also known as a hub assembly is a component that helps decrease friction when the wheel is turning by the use of ball bearing that are sealed from debris and moisture.

Sello del rodamiento de la rueda También conocido como conjunto de cubo es un componente que ayuda a disminuir la fricción cuando la rueda gira mediante el uso de un rodamiento de bolas que está sellado contra la suciedad y la humedad.

Secondary air injection forces fresh air into the catalytic converter to reduce exhaust emissions.

La **inyección de aire secundario** fuerza el aire fresco al convertidor catalítico para reducir las emisiones de escape.

Self-energizing The actual brake lining drag force to increase the force with which the force is applied.

Autoenergizante La fuerza de arrastre real del forro del freno para aumentar la fuerza con la que se aplica la fuerza.

Sensors are devices that measure mechanical or electrical actions and sends the data to a computer.

Los **sensores** son dispositivos que miden acciones mecánicas o eléctricas y envían los datos a una computadora.

Setback is the measurement of one wheel to the other wheel on the same axle.

El **retroceso** es la medida de una rueda a la otra rueda en el mismo eje.

Shank The thread-less part of a bolt between the threads and the head.

Vástago La parte sin rosca de un perno entre los hilos y la cabeza.

Shifter rails Rails that connect the shift lever to the shift fork.

Rieles de cambio Rieles que conectan la palanca de cambio a la horquilla de cambio.

Side gear Also commonly known as spider gears is a set of gears that allow the rear wheels of a vehicle to turn a different speeds.

Engranaje lateral También conocido comúnmente como engranaje de araña es un conjunto de engranajes que permiten que las ruedas traseras de un vehículo giren a una velocidad diferente.

Single-post lifts a hydraulic cylinder lift that raises the vehicle from the center of the lift limiting access to the bottom of the vehicle.

Elevadores de un solo poste un elevador de cilindro hidráulico que eleva el vehículo desde el centro del elevador limitando el acceso a la parte inferior del vehículo.

Slip joint Serves to adjust the length of a shaft when demanded by axle movement.

Junta deslizante Sirve para ajustar la longitud de un eje cuando lo exige el movimiento del eje.

Snapring A retaining ring is a fastener that holds components or assemblies onto a shaft or housing/bore.

Anillo de retención Un anillo de retención es un sujetador que sujeta componentes o ensamblajes en un eje o alojamiento/orificio.

Snapring pliers A tool designed to install and remove snap rings, which are small metal rings that hold in a bearing or other assemblies.

Alicates de presión Una herramienta diseñada para instalar y quitar anillos de presión, que son pequeños anillos metálicos que sujetan un rodamiento u otros conjuntos.

Socket wrenches and attachments are often grouped together and called sockets. They fit all the way around a bolt or nut and can be detached from a driver.

Llaves de vaso y los accesorios a menudo se agrupan y se llaman juegos de llaves de copa. Se ajustan completamente alrededor de un perno o tuerca y se pueden separar del mango.

Solder: A low melting alloy used for joining copper wires.

Soldadura: Una aleación de bajo punto de fusión utilizada para unir alambres de cobre.

Solenoids are computer-controlled mechanical devices commonly used to regulate fluid flows.

Los **solenoides** son dispositivos mecánicos controlados por computadora que se usan comúnmente para regular los flujos de fluidos.

Some **feeler gauges** are so thin, they must be used with extreme care to get an accurate reading and prevent damage.

Algunos **medidores de espesores** son tan delgados que deben usarse con extremo cuidado para obtener una lectura precisa y evitar daños.

Specific gravity is the difference between the weight of water compared to a specific fluid.

La **gravedad específica** es la diferencia entre el peso del agua en comparación con un fluido específico.

Speed sensors are electrical/mechanical devices used to register the speed of a rotating component, in this case, the vehicle's wheels.

Los **sensores de velocidad** son dispositivos eléctricos/mecánicos utilizados para registrar la velocidad de un componente giratorio, en este caso, las ruedas del vehículo.

Splash lubricated is oil randomly sprayed onto components by the actions of those components.

Salpicaduras lubricadas es aceite rociado al azar sobre los componentes por las acciones de esos componentes.

Splines Ridges or teeth on a shaft that mesh with grooves in a mating piece and transfers torque.

Estrías Crestas o dientes en un eje que se engrana con ranuras en una pieza de acoplamiento y transfiere torque.

Spring air valves are computer-controlled actuators to allow the air spring to inflate or deflate.

Las **válvulas de aire de resorte** son actuadores controlados por computadora para permitir que el resorte de aire se infle o desinfle.

Sprung The portion of the vehicle's total mass that is supported by the suspension. Load carrying.

Masa suspendida La porción de la masa total del vehículo que es soportada por la suspensión. Soporte de carga.

Sprung weight is all vehicle weight supported by the springs.

El **peso suspendido** es todo el peso del vehículo soportado por los resortes.

Standard gauge A tool used to calibrate a dial caliper to insure proper measuring.

Calibrador estándar Una herramienta utilizada para calibrar un calibrador de cuadrante para asegurar una medición adecuada.

Starter punch A short shaft punch that give you more control to break a pin loose.

Punzón de inicio Un punzón de eje corto que le brinda más control para soltar un pasador.

Static balancing is usually done on heavy truck or trailer tires using a bubble balancer.

El **balanceo estático** generalmente se realiza en neumáticos de camiones pesados o remolques con un equilibrador de burbujas.

Steel **bands** that wrap around sections of the gear train and connect to the housing.

Bandas de acero que envuelven secciones del tren de engranajes y se conectan a la carcasa.

Stud Threaded fasteners that are threaded on both ends and does not have a head.

Pernos roscados que están roscados en ambos extremos y no tienen cabeza.

Sun gear The gear in the center of a Simpson planetary gearset.

Piñón central El engranaje en el centro de un tren de engranajes planetarios Simpson.

Sway bars and **stabilizer bars** help the driver control vehicle stability during turns.

Las barras balanceo y **las barras estabilizadoras** ayudan al conductor a controlar la estabilidad del vehículo durante los giros.

Switches are mechanical or electrical sensors to signal when a device is either on or off.

Los **interruptores** son sensores mecánicos o eléctricos para señalar cuando un dispositivo está encendido o apagado.

Synchronizers are used to adjust the speeds of the mating gears, so they can be engaged without damage.

Los sincronizadores se utilizan para ajustar las velocidades de los engranajes de acoplamiento, de modo que puedan activarse sin daños.

Tap A tool designed to repair or make inside threads.

Juego de roscas con macho Una herramienta diseñada para reparar o hacer hilos internos.

TDC The position of the cylinder when it is the farthest from the crankshaft or at the top of the pistons travel.

TDC La posición del cilindro cuando está más alejado del cigüeñal o en la parte superior del recorrido de los pistones.

Tensile strength The maximum load in tension that a bolt can stretch without breaking.

Fuerza de tensión La carga máxima de tensión que un perno puede estirar sin romperse.

The aim of the headlight is set by federal regulations. The term **dimmer switch** was adopted because the headlights were much dimmer on low beam.

El objetivo del faro está establecido por las regulaciones federales. El término **interruptor de atenuación** se adoptó porque los faros eran mucho más tenues en luz de cruce.

The **amber ABS warning light** is not to be confused with the red brake light. The red light means the regular brakes may fail.

La **luz de advertencia ámbar del ABS** no debe confundirse con la luz roja de freno. La luz roja significa que los frenos normales pueden fallar.

The **ampere** is the standard measurement of moving electrons.

El **amperio** es la medida estándar de electrones en movimiento.

The area called **crush** is needed to allow for tolerances during bearing and rod manufacturing.

El área llamada **aplastamiento** es necesaria para permitir tolerancias durante la fabricación de cojinetes y bielas.

The **aspect ratio** is the ratio of the sidewall to the tread width expressed as a percentage.

La **relación de aspecto** es la relación de la pared lateral al ancho de la banda de rodadura expresada como un porcentaje.

The **bell housing** may be part of the molded transmission case or it may be bolted to the front of the transmission case.

La **carcasa de la campana** puede ser parte de la caja de transmisión moldeada o puede estar atornillada al frente de la caja de transmisión.

The **bore** of an engine is the diameter of the cylinder.

El **diámetro** de un motor es el diámetro del cilindro.

The brake **shoe web** is the metal skeleton used to mount the friction material *(lining)* and attach the shoe to its mounting.

La **banda de la zapata** de freno es el esqueleto de metal utilizado para montar el material de fricción *(revestimiento)* y unir la zapata a su montaje.

The **brushes** and **rotor** of an alternator work together to create the magnetic fields necessary to convert mechanical energy to electricity.

Los **cepillos** y el **rotor** de un alternador trabajan juntos para crear los campos magnéticos necesarios para convertir la energía mecánica en electricidad.

The camshaft is connected to and driven by the crankshaft by a **timing chain** or **timing belt**.

El árbol de levas está conectado y accionado por el cigüeñal mediante una **cadena de distribución** o **correa de distribución**.

The **carcass** is the completed tire minus the tread.

La **carcasa** es el neumático completo menos la banda de rodadura.

The **center high-mounted stop light (CHMSL)** is a federally mandated third brake light, usually located on the lower rear window or on the trunk lid of a vehicle.

La **luz de freno central montada en la parte superior (CHMSL)** es una tercera luz de freno con mandato federal, generalmente ubicada en la ventana trasera inferior o en la tapa del maletero de un vehículo.

The **center of gravity** is the balancing point of an object.

El **centro de gravedad** es el punto de equilibrio de un objeto.

The **combustion chamber** is the area of the cylinder where the air–fuel mixture is compressed and ignited.

La **cámara de combustión** es el área del cilindro donde se comprime y se enciende la mezcla de aire y combustible.

The **combustion chamber** is the open chamber at the top of an engine's cylinder. It is here that the fuel is ignited.

La **cámara de combustión** es la cámara abierta en la parte superior del cilindro de un motor. Es aquí donde se enciende el combustible.

The **compressor** is a high-pressure pump built specifically for air-conditioning systems.

El **compresor** es una bomba de alta presión construida específicamente para sistemas de aire acondicionado.

The computer uses input data to control **output devices** or **actuators**.

La computadora usa datos de entrada para controlar **dispositivos de salida** o **actuadores**.

The **controller** is an electronic device that controls specific components of a system. They are basically computer units.

El **controlador** es un dispositivo electrónico que controla componentes específicos de un sistema. Son básicamente unidades de computadora.

The **conventional theory of electricity** assumes that electrons flow from positive to negative and allows the automotive industry to use **one-wire return** circuity.

La **teoría convencional de la electricidad** supone que los electrones fluyen de positivo a negativo y permite que la industria automotriz use un **circuito de retorno de un solo cable**.

The **diagonally split system** allows the driver to have braking power at each axle if a leak develops in the brake system.

El **sistema dividido en diagonal** permite al conductor tener potencia de frenado en cada eje si se desarrolla una fuga en el sistema de frenos.

The **digital storage oscilloscope (DSO)** is a large, expensive piece of equipment. It is capable of storing and displaying enormous amounts of data in graph, wave, and numerical forms.

El **osciloscopio de almacenamiento digital (DSO)** es un equipo grande y costoso. Es capaz de almacenar y mostrar enormes cantidades de datos en gráficos, ondas y formas numéricas.

The **drum** and hub may be manufactured as a one-piece component.

El **tambor** y el cubo pueden fabricarse como un componente de una pieza.

The **electron theory of electricity** is the actual physics of electricity in that electrons flow from negative to positive.

La **teoría electrónica de la electricidad** es la física real de la electricidad en la que los electrones fluyen de negativo a positivo.

The **emergency flasher** must work with the ignition key in any position regardless of the operating conditions of the other lights.

La **luz intermitente de emergencia** debe funcionar con la llave de encendido en cualquier posición, independientemente de las condiciones de funcionamiento de las otras luces.

The **evaporator** helps a liquid heat enough so that it will change from liquid to vapor (evaporation).

El **evaporador** ayuda a que el líquido se caliente lo suficiente como para que cambie de líquido a vapor (evaporación).

The **flywheel** is driven by the starter motor during cranking to spin the crankshaft.

El **volante** es accionado por el motor de arranque durante el arranque para hacer girar el cigüeñal.

The **friction disc**, once clamped, provides the connection between the engine and the transmission.

El **disco de fricción**, una vez sujeto, proporciona la conexión entre el motor y la transmisión.

The **ignition timing** must be adjusted because the load of the engine and the speed of the pistons change as the vehicle is driven.

El **tiempo de ignición** debe ajustarse porque la carga del motor y la velocidad de los pistones cambian a medida que se conduce el vehículo.

The **input gear** is a forged part of the input shaft and is the first drive gear in a manual transmission.

El **engranaje de entrada** es una parte forjada del eje de entrada y es el primer engranaje de transmisión en una transmisión manual.

The input **pinion gear** is a long gear with spiral teeth. The pinion gears in the carrier are small, round gears with 10 to 14 teeth.

El **piñón** de entrada es un engranaje largo con dientes en espiral. Los engranajes de piñón en el transportador son pequeños y redondos con 10 a 14 dientes.

The machined portions of a basic **block** are the cylinders, mounting deck for the cylinder head, lifter bores, and the saddles or bores for the bearings.

Las partes mecanizadas de un **bloque** básico son los cilindros, la plataforma de montaje para la culata, los orificios de los elevadores y los asientos u orificios para los cojinetes.

The **master cylinder** converts the driver's mechanical input force to hydraulic pressure for the brake system.

El **cilindro maestro** convierte la fuerza de entrada mecánica del conductor en presión hidráulica para el sistema de frenos.

The **metering valve** controls hydraulic pressure to the front wheels during initial brake application. As the pressure from the master cylinder changes, the **proportioning valve** operation also changes, thereby increasing or reducing the pressure to the rear wheels.

La **válvula de medición** controla la presión hidráulica a las ruedas delanteras durante la aplicación inicial del freno. A medida que cambia la presión del cilindro maestro, la operación de la **válvula de dosificación** también cambia, lo que aumenta o reduce la presión sobre las ruedas traseras.

The **multiple function** lever is commonly referred to as the turn signal lever.

La palanca de **funciones múltiples** se conoce comúnmente como palanca de señal de giro.

The nose of the **input shaft** extends through the friction disc and into the pilot bearing.

La nariz del **eje de entrada se** extiende a través del disco de fricción y dentro del rodamiento piloto.

The **ozone layer** is located about 9 to 18 miles (15 to 39 km) above the earth's surface. It protects the earth from the sun's ultraviolet rays.

La **capa de ozono** se encuentra a unos 9 a 18 millas (15 a 39 km) sobre la superficie de la tierra. Protege la tierra de los rayos ultravioleta del sol.

The **pitman arm** changes the rotational movement of the sector shaft to a linear motion for the steering linkage.

El **brazo pitman** cambia el movimiento de rotación del eje de sector a un movimiento lineal para el enlace de dirección.

The **pressure plate** assembly is used to clamp the clutch disc to the flywheel.

El conjunto de **placa de presión** se utiliza para sujetar el disco de embrague al volante.

The **primary piston** is located to the rear of the master cylinder bore and the closest to the firewall. On disc/drum brake systems, it pressurizes the disc brake portion of the system.

El **pistón primario** está ubicado en la parte posterior del orificio del cilindro maestro y el más cercano al cortafuegos. En los sistemas de freno de disco/tambor, presuriza la porción de freno de disco del sistema.

The proper name for an **alternator** is AC generator.

El nombre propio de un **alternador** es generador de CA.

The **pushrod** is the mechanical connection between the brake pedal and the power booster or master cylinder.

La **varilla de empuje** es la conexión mecánica entre el pedal del freno y el amplificador de potencia o el cilindro maestro.

The **release bearing** provides a point of contact between a stationary component (linkage) and a moving component (pressure plate assembly).

El **cojinete de liberación** proporciona un punto de contacto entre un componente estacionario (enlace) y un componente móvil (conjunto de placa de presión).

The **reverse idler gear** changes the direction of rotation but does not change the gear ratio.

La **marcha lenta inversa** cambia la dirección de rotación, pero no cambia la relación de transmisión.

The **ride**, also known as **curb height**, is the distance an "at rest" vehicle sits above the ground. It is measured at specific points on the vehicle.

La **altura de manejo**, también conocida como **altura de la acera**, es la distancia que un vehículo "en reposo" se encuentra sobre el suelo. Se mide en puntos específicos del vehículo.

The satellite network known as the **Global Positioning Satellite (GPS)** system consists of hundreds of satellites that communicate with one another and transmitters and receivers on the Earth.

La red de satélites conocida como **Sistema de Posicionamiento Global (GPS)** consta de cientos de satélites que se comunican entre sí y con transmisores y receptores en la Tierra.

The **secondary piston** is located to the front of the master cylinder bore. In a disc/drum brake system it pressurizes the drum portion of the system.

El **pistón secundario** está ubicado en la parte delantera del orificio del cilindro maestro. En un sistema de freno de disco/tambor presuriza la parte del tambor del sistema.

The **slave cylinder** converts hydraulic pressure from the clutch master cylinder to mechanical action to operate the clutch.

El **cilindro esclavo** convierte la presión hidráulica del cilindro maestro del embrague en acción mecánica para operar el embrague.

The **stroke** is the distance that the piston travels in the cylinder.

La **carrera** es la distancia que recorre el pistón en el cilindro.

The term **dual overhead camshaft (DOHC)** does not represent a different valvetrain. It's a version of the OHC system.

El término **doble árbol de levas en cabeza (DOHC)** no representa un tren de válvulas diferente. Es una versión del sistema OHC.

The term **heads- up-display (HUD)** was adopted because the driver does not have to look down to read operational data.

El término **pantalla de visualización frontal (HUD)** se adoptó porque el conductor no tiene que mirar hacia abajo para leer los datos operativos.

The **throttle position sensor (TPS)** measures the mechanical movement of the throttle and converts the movement to an electrical signal.

El **sensor de posición del acelerador (TPS)** mide el movimiento mecánico del acelerador y convierte el movimiento en una señal eléctrica.

The **tie-rod** is the last link between the steering wheel and the steering knuckle.

El **tirante** es el último eslabón entre el volante y el muñón de la dirección.

The **Tire Performance Criteria (TPC)** is a number, molded in the sidewall, that lists important information about the tire.

Criterios de Rendimiento de los Neumáticos (TPC) es un número, moldeado en la pared lateral, que enumera información importante sobre el neumático.

The **turbine**, **stator**, and lockup clutch are not visible without cutting the two halves of the converter apart.

La **turbina**, el **estator** y el embrague de bloqueo no son visibles sin cortar las dos mitades del convertidor.

The two **planetary gearsets** in a Simpson geartrain are used together to obtain the correct gear ratio for a given situation.

Los dos **engranajes planetarios** en un tren de engranajes Simpson se usan juntos para obtener la relación de transmisión correcta para una situación dada.

The **valvetrain** consists of valves, valve springs, a camshaft, valve-retaining devices, and may have valve **lifters**, pushrods, and **rocker arms**.

El **tren de válvulas** consta de válvulas, resortes de válvula, un árbol de levas, dispositivos de retención de válvula y puede tener **elevadores** de válvula, varillas de empuje y **balancines**.

The vehicle's **secondary circuit** consists of the spark plugs, plug wires, distributor, and the secondary windings of the ignition coil.

El **circuito secundario** del vehículo consiste en las bujías, los cables de las bujías, el distribuidor y los devanados secundarios de la bobina de encendido.

The **volt** is the standard measurement of electricity.

El **voltio** es la medida estándar de electricidad.

The wire in a **variable rate** spring may be tapered, truncated cone, double cone, or barrel-shaped.

El alambre en un resorte de **velocidad variable** puede ser cónico, cono truncado, doble cono o en forma de barril.

The word **driveline** is used here in lieu of *drivetrain.* Both refer to the components that transfer engine output to the drive wheels.

La palabra **línea de transmisión** se usa aquí en lugar de *transmisión.* Ambos se refieren a los componentes que transfieren la salida del motor a las ruedas motrices.

Thermodynamics is the study of using radiant heat to perform mechanical work.

La termodinámica es el estudio del uso de calor radiante para realizar trabajos mecánicos.

Thermodynamics Physical science that deals with the relationship between heat and other forms of energy.

Termodinámica Ciencia física que se ocupa de la relación entre el calor y otras formas de energía.

Threaded fasteners A hardware device that mechanically joins two or more objects together using a bolt and a nut.

Sujetadores roscados Un dispositivo de equipo que une mecánicamente dos o más objetos mediante un perno y una tuerca.

Thread-restoring files A file used to repair thread on a damaged bolt.

Limas para restaurar la rosca Una lima utilizada para reparar la rosca en un perno dañado.

Threads per inch A measurement of the threads per inch also known as thread pitch.

Hilos por pulgada Una medida de los hilos por pulgada también conocida como paso de hilo.

Thrust bearing A bearing that can support rotational movement but also can support axial load or side to side movement.

Cojinete de empuje Un cojinete que puede soportar movimiento rotacional pero también puede soportar carga axial o movimiento de lado a lado.

Tire bead The edge of the tire that sits on the wheel or rim.

Talón del neumático El borde del neumático que se sienta en la rueda o llanta.

Tire diameter is the size of the large opening in the tire, and the wheel must match that diameter.

El **diámetro del neumático** es el tamaño de la abertura grande en el neumático y la rueda debe coincidir con ese diámetro.

To see **rotary flow**, put water into a bowl and use a spoon to make the water flow around the inside of the bowl.

Para ver el **flujo rotativo**, ponga agua en un tazón y use una cuchara para que el agua fluya por el interior del tazón.

Today **bias ply** tires are usually associated with heavy load carriers like semi-trucks.

Hoy en día, los **neumáticos de cubierta diagonal** se asocian generalmente con transportadores de carga pesada, como los semirremolques.

Tolerances set the limit that a component can be different from a perfect size.

Las **tolerancias** establecen el límite de que un componente puede ser diferente de un tamaño perfecto.

Torque Twisting force that tends to cause rotation.

Torque Fuerza de torsión que tiende a causar rotación.

TPMS An electronic system designed to monitor the air pressure inside the pneumatic tires. Using a spare tire that is oversized/undersized compared to the original tire will affect the **indirect tire pressure monitoring system (TPMS)** more than the **direct** system.

TPMS Un sistema electrónico diseñado para controlar la presión del aire dentro de los neumáticos. El uso de una llanta de refacción que sea demasiado grande/pequeña en comparación con la llanta original afectará más al sistema **indirecto de monitoreo de presión de llantas (TPMS)** que al sistema **directo**.

Traction control is a system used to electronically control wheel spin and is usually merged into the anti-lock brake system.

El **control de tracción** es un sistema utilizado para controlar electrónicamente el giro de la rueda y generalmente se combina con el sistema de frenos antibloqueo.

Transducers convert energy from one type to another.

Los **transductores** convierten la energía de un tipo a otro.

Transmission Control Module (TCM) It is an electronic module that controls the operation of the transmission.

Módulo de Control de Transmisión (TCM) Es un módulo electrónico que controla el funcionamiento de la transmisión.

Transmission housing Also known as the bellhousing is the portion of the transmission that covers the flywheel and clutch or torque converter.

Caja de transmisión También conocida como carcasa de la campana es la parte de la transmisión que cubre el volante y el convertidor de embrague o par.

Transmission jacks can be used under a lifted vehicle or with a vehicle on jack stands.

Los **gatos de transmisión** se pueden usar debajo de un vehículo elevado o con un vehículo en soportes de gato.

Tread is the molded portion of the tire that touches the road. It is supported by the **sidewall**, which extends from the bead to the tread area.

La **banda de rodamiento** es la parte moldeada del neumático que toca la carretera. Está soportado por la **pared lateral**, que se extiende desde el cordón hasta el área de la banda de rodadura.

Tread width is the measurement of the actual width of the tread, not just the part that contacts the road.

El **ancho de la banda** de rodadura es la medida del ancho real de la banda de rodadura, no solo la parte que está en contacto con la carretera.

Turbine shaft Transmits the torque to the output shaft and redirects the fluid leaving the turbine to the hydraulic system.

Eje de la turbina Transmite el par al eje de salida y redirige el fluido que sale de la turbina al sistema hidráulico.

Twist drills A drill bit with a twisted body like an auger.

Brocas helicoidales Una broca con un cuerpo retorcido como una barrena.

Twisted pair wires are used as communication wires for computer circuits.

Los **cables de par** trenzado se utilizan como cables de comunicación para circuitos de computadora.

Typically, a **dial caliper** in not used in an automotive repair shop but used extensively in automotive machining shops.

Por lo general, un **calibrador de cuadrante** no se usa en un taller de reparación de automóviles, pero se usa ampliamente en talleres de mecanizado de automóviles.

U-joint Also known as a universal joint is a device that allows one or more rotating shafts to be linked together which allows the transmission of torque.

Junta en U También conocida como junta universal, es un dispositivo que permite unir uno o más ejes rotativos, lo que permite la transmisión del par.

Ultraviolet is the term given to electromagnetic radiation and is not visible to the human eye. It is dangerous to humans and can cause skin damage.

Ultravioleta es el término dado a la radiación electromagnética y no es visible para el ojo humano. Es peligroso para los humanos y puede causar daños en la piel.

Unified System A standard thread pitch mainly used in the United States.

Sistema unificado Un paso de rosca estándar utilizado principalmente en los Estados Unidos.

Uniform Tire Quality Grading (UTQG) is a system developed to provide the customer a way to compare, select, and rate tires.

La **Clasificación Uniforme de la Calidad de Neumáticos (UTQG)** es un sistema desarrollado para proporcionar al cliente una forma de comparar, seleccionar y calificar los neumáticos.

Unsprung weight is all vehicle weight not supported by the springs including the springs themselves.

El peso no suspendido es todo el peso del vehículo no soportado por los resortes, incluidos los propios resortes.

Vacuum is essential to getting air into a gasoline engine cylinder.

El **vacío** es esencial para introducir aire en un cilindro de motor de gasolina.

Valence ring The outer path of electrons around the nucleus of the atom

Anillo de valencia El camino externo de los electrones alrededor del núcleo del átomo.

Valve overlap The period during the valve timing where both the intake and exhaust valves are open.

uperposición de válvulas El período durante la sincronización de la válvula donde las válvulas de admisión y escape están abiertas.

Valvetrain refers to the placement of the engine valves.

Tren de válvulas se refiere a la colocación de las válvulas del motor.

Variable valve timing (VVT) is a computer-controlled system that changes the timing of the valves by changing the timing and position of the camshaft.

La **sincronización variable de válvulas (VVT)** es un sistema controlado por computadora que cambia la sincronización de las válvulas cambiando la sincronización y la posición del árbol de levas.

V-block The V type engine has two rows (banks) of cylinders, usually forming an angle of 60 or 90 degrees between the two banks. see figure 7-3.

Bloque en V El motor tipo V tiene dos filas (bancos) de cilindros, que generalmente forman un ángulo de 60 o 90 grados entre los dos bancos. ver figura 7-3.

Vehicle speed sensor (VSS) converts the rotary motion of an axle or wheel to an electrical signal for the computer to determine vehicle speed.

El **sensor de velocidad del vehículo (VSS)** convierte el movimiento giratorio de un eje o rueda en una señal eléctrica para que la computadora determine la velocidad del vehículo.

Vehicle stability A system that controls and processes data in critical situations to keep the vehicle safely on course by controlling each wheel separately.

Estabilidad del vehículo Un sistema que controla y procesa datos en situaciones críticas para mantener el vehículo en curso de forma segura controlando cada rueda por separado.

Vendors are businesses that sell parts or service to other businesses and the public. For example, NAPA is an automotive vendor.

Suplidores son empresas que venden piezas o servicios a otras empresas y al público. Por ejemplo, NAPA es un suplidor de automóviles.

Vernier scale A fine measurement used to measure a ten-thousandths of an inch.

Escala de Vernier Una medida fina utilizada para medir diez milésimas de pulgada.

Viscosity is a rating of how well a fluid will flow.

La **viscosidad** es una clasificación de qué tan bien fluirá un fluido.

Vise grips A pliers that can be locked into position, using an over center toggle action.

Alicates de agarre Un alicate que se puede bloquear en su posición, utilizando una palanca central de conmutación.

Voltage is a measurement of **electromotive force (EMF)**, which is the difference between the number of free electrons at one end of the circuit compared to the other end or **potential difference**.

El **voltaje** es una medida de **la fuerza electromotriz (EMF)**, que es la diferencia entre el número de electrones libres en un extremo del circuito en comparación con el otro extremo o la **diferencia de potencial**.

Washers A thin plate with a hole that is used to distribute the load of a fastener.

Arandelas Una placa delgada con un hueco que se usa para distribuir la carga de un sujetador.

Weather-tight connector: A connector that can with stand exposure to extreme temperatures, moisture, harsh engine compartment fluids and chemicals.

Conector impermeable: Un conector que puede exponerse a temperaturas extremas, humedad, líquidos del compartimento del motor y productos químicos.

Wheel blocks are also known as wheel chocks.

Los **bloques de rueda** también se conocen como calzos de rueda.

Wheel cylinders work with drum brakes. **Calipers** are used with disc brakes.

Los **cilindros de las ruedas** funcionan con frenos de tambor. Las **pinzas** se usan con frenos de disco.

Wheel sensors in this case are those devices placed on the wheel during alignment to tell the computer where the wheels are in relation to each other and to the centerline of the vehicle. An **alignment machine** collects data from the wheel sensors and displays that data to the machine operator. A **mechanical advantage** is the amount of force increased at the output compared to the input force.

Los **sensores de las ruedas** en este caso son aquellos dispositivos colocados en la rueda durante la alineación para decirle a la computadora dónde están las ruedas entre sí y con la línea central del vehículo.Una **máquina de alineación** recopila datos de los sensores de la rueda y muestra esos datos al operador de la máquina.Una **ventaja mecánica** es la cantidad de fuerza aumentada en la salida en comparación con la fuerza de entrada.

When recording **measurements**, always include the whole number before the decimal or fraction, for example, 0.002 or 1.05.

Al registrar **mediciones**, siempre incluya el número entero antes del decimal o fracción, por ejemplo, 0.002 o 1.05.

Work is the expenditure of energy.

El **trabajo** es el gasto de energía.

Yoke A strap that holds a bearing in place and connects to the output shaft.

ugo Una correa que mantiene un rodamiento en su lugar y se conecta al eje de salida.

INDEX

Note: Page numbers followed by "f" indicate material in a figure.

D

E